QUANTUM MECHANICS IN HILBERT SPACE

SECOND EDITION

QUANTUM MECHANICS IN HILBERT SPACE

SECOND EDITION

EDUARD PRUGOVEČKI

DOVER PUBLICATIONS, INC.
MINEOLA, NEW YORK

Bibliographical Note

This Dover edition, first published in 2006, is an unabridged republication of the second edition of the work, originally published in the "Pure and Applied Mathematics" series by Academic Press, Inc., New York, in 1981.

Library of Congress Cataloging-in-Publication Data

Prugovečki, Eduard.
Quantum mechanics in Hilbert space / Eduard Prugovečki.
p. cm.
Originally published: 2nd ed. New York : Academic Press, 1981, in series: Pure and applied mathematics ; 92.
Includes index.
ISBN 0-486-45327-8 (pbk.)
1. Hilbert space. 2. Quantum theory. I. Title.

QC174.17.H55 P78 2007
530.1201'515733—dc22

2006050200

Manufactured in the United States of America
Dover Publications, Inc., 31 East 2nd Street, Mineola, N.Y. 11501

To My Parents

Contents

* One asterisk indicates sections in which all proofs can be skipped at a first reading.
** Two asterisks indicate sections that are introductory to original papers and research material.

V. Quantum Mechanical Scattering Theory

Preface to the Second Edition

Since the appearance in 1971 of the first edition, Chapters III, IV, and V of this book have served, with certain modifications, as the basis of a two-term course on functional analysis in quantum mechanics offered at the University of Toronto at the advanced undergraduate and first-year graduate level. The aforementioned modifications have now been incorporated in the present edition, together with additional material which brings the book up to date, and in some instances provides an introduction to topics of contemporary research interest.

Practically all alterations occur in the second half of the book, and especially in the part dealing with quantum scattering theory. In addition to reorganized exposition, they consist of substantial simplifications in the proofs of various theorems. Hence it is hoped that the text has now become in its entirety easy to follow for any student in mathematics and/or physics interested in the subject. Furthermore, a number of completely new sections have been included summarizing research results which have been published during the past decade and that directly pertain to the subject matter of this book. Since a detailed coverage of all the material contained in Chapters III–V would be unfeasible in a typical one-year course, asterisks have been liberally employed to indicate proofs of theorems, and even full sections, that can be skipped without making the subsequent material incomprehensible.

As was the case with the first edition, the goal of the present edition is not to supplant but rather to supplement any of the good textbooks on nonrelativistic quantum mechanics, which as a rule have a purely physics orientation. More precisely, the chief aim is to provide in a readable and self-contained form, which would be accessible even to students without an extensive mathematical background, rigorous proofs of all the main statements of a functional-analytic nature that one encounters in standard textbooks on quantum mechanics. It should be emphasized that no attempt has been made to include any material dealing with complex analysis or with group theory. Indeed, in the case of complex analysis a reasonable level of mathematical rigor is maintained in many of the better physics textbooks on quantum scattering theory, whereas the subject of group theory in quantum mechanics has already received ample attention in the form of monographs and compilations of review articles on the subject at all conceivable levels of mathematical sophistication.

Albeit a few monographs dealing in part with some of the subject matter of this book have also been published since the first edition made its initial appearance, as a rule they tend to lean heavily toward the purely mathematical aspects of the subject. It is therefore hoped that the present edition may still perform a useful function by filling the gap between purely physics and purely mathematics oriented textbooks and monographs pertaining to Hilbert space theory as applied to nonrelativistic quantum mechanics.

E. Prugovečki

Toronto, 1981

Preface to the First Edition

This book was developed from a fourth-year undergraduate course given at the University of Toronto to advanced undergraduate and first-year graduate students in physics and mathematics. It is intended to provide the inquisitive student with a critical presentation of the basic mathematics of nonrelativistic quantum mechanics at a level which meets the present standards of mathematical rigor. It should also be of interest to the mathematician working in functional analysis and related areas, who would like to see some of the applications of the basic theorems of functional analysis to quantum mechanics.

With these aims in mind, I have tried to make the book self-contained. A knowledge of advanced calculus, linear algebra, the basics of ordinary linear differential equations, and the very basic concepts of classical mechanics (such as the notions of force, momentum, angular momentum energy, etc. of an n-particle system) provide an adequate background for the reader of this work.

The material is organized in the form of definitions, theorems, and proofs of theorems. Such a format has the advantage of enabling the reader to grasp immediately the basic concepts and results, separating them from the sometimes tedious techniques. In order to facilitate the perusing of the book preceding a more thorough reading, marginal asterisks indicate those theorems whose proofs can be skipped at a first reading.

I consider the material contained in the present book to stand in the same relation to quantum physics, as the material of conventional mathematical physics stands in relation to classical physics. Therefore, the prevailing style and approach is similiar to texts on mathematical physics: the mathematical methods are in the foreground; the physical examples appear only as illustrations of these methods. Emphasis is placed on general concepts and basic techniques of proving theorems, rather than on computational methods. Consequently, the reader who is not already familiar with the material that can be found in a conventional book on quantum mechanics will find it extremely useful to take at least an occasional glance at any of the standard textbooks on quantum mechanics (some of which are listed in the References) which cover the physical aspects of quantum theory in more detail.

Since the text is oriented toward physical application, I have given the minimal amount of mathematics necessary for a good understanding of the main mathematical aspects of nonrelativistic quantum mechanics. Consequently, basic material on measure theory and functional analysis has been introduced in a selective manner, always keeping in mind its application to quantum mechanics. However, I believe that the interested reader can easily expand his knowledge of the areas of mathematics introduced in this book by using the reference books and the present text as a starting point. The list of references given at the end of each chapter is meant to help him in this task. Furthermore, since the last two chapters contain mostly material which until now was published only in original papers but not in textbooks or monographs, I have supplemented these references with a few historical notes. However, many of the mathematical tools used in the present book have already become standard. Hence, no effort has been made to provide an exhaustive bibliography or to give credit to the original source, with the exception of names which are now customarily associated with certain theorems, formulas, equations, etc., and of references related to topics of current research interest.

Most of the exercises form an integral part of the text. Consequently, though many of these exercises are relatively easy to solve, it has been deemed necessary to provide the solutions or detailed hints appearing at the end of the text. However, the reader is urged to try and solve these exercises by himself before looking up the solutions.

The book should appeal to the increasing number of students in theoretical physics who desire to understand the justification of the multitude of heuristic procedures and shortcuts ordinarily employed in the typical books and courses on nonrelativistic quantum mechanics. Such procedures, which until a few decades ago had to be taken at face value, can be now rigorously justified, due to advances in functional analysis in general, and in the Hilbert space theory in particular.

E. PRUGOVEČKI

Toronto, 1971

List of Symbols

This list contains only the symbols which consistently refer to the same concept, and which are frequently employed, and the page numbers where they first appear.

a^*	complex conjugate of the complex number a, 19
A	linear or antilinear operator, 173
A^*	adjoint of the operator A, 187
$\mathscr{A}(X)$	Boolean algebra generated by the family X of sets, 60
$\mathscr{A}_\sigma(X)$	Boolean σ algebra generated by X, 60
$\mathfrak{A}(\mathscr{H})$	the algebra of bounded operators on $\mathscr{H}$, 175, 181
B	Borel set, 62
$\mathscr{B}^n$	the family of Borel sets on $\mathbb{R}^n$, 62
$\mathscr{B}^n_0$	the family of Borel sets on $\mathbb{R}^n$ which are finite unions of intervals, 61
$\mathscr{B}^n_\Omega$	the family of Borel subsets of $\Omega \subset \mathbb{R}^n$, 111
C_Ω	the family of characteristic functions of subsets of Ω, 101
$\mathscr{C}^0(\mathbb{R}^n)$	the family of all complex-valued continuous functions defined on $\mathbb{R}^n$, 17
$\mathscr{C}^m(\mathbb{R}^n)$	the family of m-times continuously differentiable complex functions on $\mathbb{R}^n$, 45
$\mathscr{C}^m_{(2)}(\mathbb{R}^n)$	a family of square-integrable and vanishing at infinity functions in $\mathscr{C}^m(\mathbb{R}^n)$, 45
$\mathscr{C}^m_{\mathrm{b}}(\mathbb{R}^n)$	the subset of $\mathscr{C}^m(\mathbb{R}^n)$ of functions of compact support, 129
$\mathbb{C}^n$	the family of n-tuples of complex numbers, 12
$\mathscr{D}_A$	domain of definition of the operator A, 186
$d(\cdot\,,\cdot)$	metric function, 25
Δ	Laplacian, 121
$\delta_l(k)$	phase shift of lth partial wave, 497, 589, 593

δ_{ij}	Kronecker delta, 22
$E(S)$	spectral measure, 231
E_λ	spectral function, 235
E_{M}	projector onto space M, 197
$\mathscr{E}$	Euclidean space, 18
e_i	vector in orthonormal system, 22
$\mathscr{F}$	field of scalars, 12
$f_{\mathbf{k}}(\theta, \phi)$	scattering amplitude, 494, 553, 556
$\Phi_\beta^{(\pm)}(\alpha)$, $\Phi_\beta(\alpha)$	eigenfunctions, 496, 499, 500
$\Phi_{\mathbf{k}}(\mathbf{r})$	plane wave, 491
$\Phi_{\mathbf{k}}^{(\pm)}(\mathbf{r})$	distorted plane waves, 494, 503
$G_A(\alpha, \alpha'; \xi)$, $G_A^{(\pm)}(\alpha, \alpha'; \xi)$	Green functions, 501, 502
G_A	graph of an operator A, 191
H	Hamiltonian, 294
H_0	free Hamiltonian, 298
H_{S}	Schroedinger operator, 122, 368
$\mathscr{H}$	Hilbert space, 2, 31
$\underset{\sim}{h}$	Hilbert ray, 269
$h = 2\pi\hbar$	Planck's constant, 8, 46
h-lim	limit in the Hilbert–Schmidt norm on $\mathscr{L}(\mathscr{H})$, 399
$\chi_S(\cdot)$	characteristic function of S, 31
$\inf F$	infimum of F, 84
$\mathscr{I}^n$	family of intervals in $\mathbb{R}^n$, 61
$K(\cdot, \cdot)$	kernel of an integral operator, 428
$\mathscr{K}$	class of sets, 58
$L^1(\Omega, \mu)$	Banach space of μ-integrable functions on Ω, 181
$L^2(\Omega, \mu)$	Hilbert space of μ-square-integrable functions on Ω, 103
$L_{(2)}(\Omega, \mu)$	family of μ-square-integrable functions on Ω, 103
l.i.m.	limit in the mean, 105, 463
$l^2(n)$	Hilbert space of one-column complex matrices, $n = 1, \ldots, \infty$ 23
$\mathscr{L}(\mathscr{H})$	Liouville space over $\mathscr{H}$, 397
$\mathscr{M}$	metric space, 25
$\mathfrak{M}(\mathscr{F})$	monotone class on $\mathscr{F}$, 63
M	closed linear subspace of a Hilbert space $\mathscr{H}$, 197
$\mathsf{M}_\pm \mathsf{M}_0$	intial domains of $\Omega_\pm$, 438, 441
$\mu(S)$	measure of S, 67
$\mu_l^{(n)}(B)$	Lebesgue measure of B, 67, 68
$\bar{\mu}(S)$	extension of measure μ, 70
$\mathscr{N}$	normed space, 30
$\mathbb{R}^n$	the family of n-tuples of real numbers, 2, 7, 12
$R_A(\zeta)$	resolvent of the operator A at complex value ζ, 452

$\mathscr{R}_A$	range of operator A, 186
$\mathfrak{R}$	set of rational numbers, 35
$\mathsf{R}_{\pm}$	final domains $\Omega_{\pm}$, 441
ρ	density operator, 391, 404
$S, S^{(2)}$	scattering operator, 423, 438
S^A	the spectrum of the operator A, 252, 475
S_c^A	the continuous spectrum of A, 253, 475
S_p^A	the point spectrum of A, 48, 253, 475
$\mathscr{S}_V{}^1$	the exceptional set of energy values, 546
$\mathscr{S}_V{}^3$	the set exceptional set of momentum values, 546
s-lim	strong limit, 230
$\sup F$	supremum of F, 84
$\operatorname{supp} F$	support of F, 128
σ	spin variable, 308
$T, T^{(2)}$	transition operator,
$\mathscr{T}(\zeta)$	off-shell $\mathscr{T}$ operators, 511
$\operatorname{Tr} A$	trace of operator A, 380
$\theta(\lambda)$	step function, 130
U	unitary operator, 214
$U(t, t_0)$	time evolution operator, 286
U_t	one-parameter group of unitary operators, 288
u-lim	uniform limit, 230
$V(\cdot)$	potential, 121
$\mathscr{V}$	vector space, 11
$\mathscr{X}$	a set over which a measurable space is defined, 58
w-lim	weak limit, 230
$\Omega_{\pm}, \Omega_{\pm}^{(2)}$	incoming and outgoing Møller wave operators,
$\omega = (\theta, \phi)$	angles in spherical coordinates,
$\otimes_a$	algebraic tensor product, 141
$\otimes^{\mathsf{A}}$	antisymmetric tensor product, 306
$\otimes^{\mathsf{S}}$	symmetric tensor product, 306
$\|f\rangle\langle f\|$	projector onto f, 199, 376
$\langle\cdot \| \cdot\rangle$	inner product, 18
$\langle\cdot \| \cdot\rangle_2$	Hilbert–Schmidt inner product on $\mathscr{L}(\mathscr{H})$, 397
$\\|\cdot\\|$	norm, 20
$\\|\cdot\\|_2$	Hilbert–Schmidt norm, 384
$\\|V\\|_{\mathrm{R}}$	Rollnik norm of a potential V, 545

Introduction

The beginning of the era of modern quantum mechanics is marked by the year 1925 and the two almost simultaneous papers of Heisenberg [1925] and Schroedinger [1926]. The first of these papers proposes essentially the formalism of matrix mechanics, while the second one proposes the formalism of wave mechanics. It was first indicated by Schroedinger that these two formulations are physically equivalent. Both can be embraced in a more general formulation of quantum mechanics, which was first proposed in a somewhat heuristic form by Dirac [1930]. A mathematically rigorous development of this general formalism of quantum mechanics can be achieved by taking advantage of Hilbert space theory, which is done in this book.*

Though classical and quantum mechanics are very dissimilar in many respects, they do share a lot of common features when viewed from a general structural point of view. To find these common features, one must have some insight into the mechanism of a physical theory from an abstract point of view, at which one arrives by ignoring the technical details of the theory.

One can dissect any physical theory into the following main constituents: (1) formalism, (2) dynamical law, (3) correspondence rules.

Viewed from an abstract point of view, the *formalism* consists of a set of symbols and rules of deduction. With these symbols one can build statements or propositions. The rules of deduction enable us to deduce new statements from those already given.

Ideally, every scientific theory starts with a set of basic statements

* For a detailed historical survey of the origins and development of quantum theory consult any of the many standard textbooks on quantum mechanics (e.g., Messiah [1962]).

called the axioms of the theory. However, in practice, while a theory is in the growing process, it very often happens that no set of axioms is clearly stated and the axiomatic formulation of the theory is left to the future. Nonrelativistic quantum mechanics has an accepted axiomatic formulation (given in Chapter IV) which covers all the present practical needs, though proposals for new sets of axioms are still making their appearance in scientific journals. These proposals have mainly the virtue of greater generality over the conventional Hilbert space formulation adopted here (so that they contain this formulation as a special case). However, from the practical point of view they have not yet produced any new results not derivable from the conventional approach.

The two classes of basic objects in the formalism of both classical and quantum mechanics are the *states* and *observables*. We shall illustrate these concepts by a few examples.

In the Newtonian approach to classical physics, the state of a system is given by the family of trajectories of all particles constituting the system. For instance, the state of one particle moving in three dimensions is given by a vector-valued function $\mathbf{r}(t)$, $\mathbf{r}(t) \in \mathbb{R}^3$ ($\mathbb{R}^n$ denotes an n-dimensional real Euclidean space), in the time-parameter t. In the Hamiltonian (canonical) approach to classical mechanics, the state of a single particle is also a vector-valued function

$$\big(q_1(t), q_2(t), q_3(t), p_1(t), p_2(t), p_3(t)\big),$$

defined now in a six-dimensional real Euclidean space called the phase space of the system. In the Schroedinger formulation of quantum mechanics, the state is again a vector-valued function $\Psi(t)$, only this time the function assumes values from an infinite-dimensional complex Euclidean space $\mathscr{H}$, which bears the name of Hilbert space.

The observables of the formalism are symbols related to specific experimental procedures for measuring them. For instance, in Newtonian classical physics of a one-particle system, $\mathbf{r}(t)$ is the position of the particle at time t; $\mathbf{p}(t) = m\dot{\mathbf{r}}(t)$ its momentum ($\dot{\mathbf{r}}(t) = d\mathbf{r}(t)/dt$); $m\dot{\mathbf{r}}^2(t)/2$ its kinetic energy (m denotes the mass of the particle), etc. In the Hamiltonian approach $q_1(t)$, $q_2(t)$, $q_3(t)$, $p_1(t)$, $p_2(t)$, and $p_3(t)$ are, in general, related by some functional relation to the Cartesian position and momentum components of the particle. In quantum mechanics, observables are represented by so-called self-adjoint operators in the Hilbert space $\mathscr{H}$.

Every physical theory contains a particularly important ingredient called the *dynamical law* (frequently referred to as the "dynamics"). This dynamical law is, in general, a relation which some of the basic objects of the formalism must satisfy. It is the key component of the theory because it gives the theory its predictive power. From a formal logical point of view, the dynamical law could be considered a part of the formalism. However, due to its central role in predicting the future behavior of the system from a knowledge of its present or past behavior, we prefer to single it out.

From the mathematical point of view, the dynamical law is very frequently a differential equation which has to be satisfied by the state of the system, and in that case it is called the *equation of motion*. For instance, in Newtonian classical mechanics of a single particle of mass m moving under the influence of a force $\mathbf{F}$, the equation of motion is given by Newton's second postulate

$$m\ddot{\mathbf{r}}(t) = \mathbf{F}(t), \qquad \ddot{\mathbf{r}}(t) = \frac{d^2\mathbf{r}(t)}{dt^2}. \tag{0.1}$$

In the canonical formalism of a single particle we have

$$\frac{dq_k(t)}{dt} = \frac{\partial H}{\partial p_k}, \qquad \frac{dp_k(t)}{dt} = -\frac{\partial H}{\partial q_k}, \qquad k = 1, 2, 3,$$

where $H(q_k, p_k, t)$ is the Hamiltonian of the system. In the Schroedinger formulation of quantum mechanics, the equation of motion governs the time evolution of the state of the system, and is called the Schroedinger equation. However, in the Heisenberg formulation of quantum mechanics, the equation of motion, called the Heisenberg equation, does not involve the state (which in this case is time independent) but rather the observables of the system (see Chapter IV, §3).

It must be realized that the dynamical law need not always be expressed by a differential equation. Very frequently, it is expressed by an integral equation, sometimes derivable from a differential equation. In modern quantum scattering theory, the dynamical law is given by analyticity conditions that certain key functions (the so-called S-matrix elements) appearing in the theory have to satisfy.

The *correspondence rules* are the rules which assign empirical meaning to some of the symbols appearing in the formalism. As such, they provide the link between theory and experiment. The body of all correspondence rules of a theory is in physics better known under the name of the *physical interpretation* of the theory.

In classical mechanics, which has a very direct intuitive appeal, the correspondence rules are very straightforward, and they leave no room for plausible alternatives. For instance, in Newtonian one-particle mechanics, it is well known that for a particle in the state $\mathbf{r}(t)$, $t \in \mathbb{R}^1$, the components of the vectors $\mathbf{r}(t)$, $m\dot{\mathbf{r}}(t)$, etc. represent the length of the projections of the position, momentum, etc. vectors on the three axes of an inertial frame* with Cartesian coordinates.

In the case of quantum mechanics, which does not yield itself to visualization as readily as classical mechanics, the matter of physical interpretation is much more complex and therefore open to discussion. In this case the desirability and empirical consistency of the entire body of correspondence rules is still open to scrutiny and critical appraisal, being treated in its own right in the *theory of measurement* of quantum mechanics. However, the main features of the now widely accepted interpretation, sometimes called the Copenhagen school interpretation (see also Chapter IV, §1), are quite clear. The most striking of these features is that this interpretation is essentially statistical; it does not provide us with definite statements about the future behavior of a quantum mechanical system, but rather with probabilistic statements about the likelihood of different patterns of behavior. It should be mentioned that this feature has given rise to much controversy as to whether quantum mechanics provides the ultimate tool in describing the behavior of atomic and subatomic systems. The ranks of physicists who have adopted a critical attitude when considering this subject include notably the names of some of the pioneers in the field, such as Einstein, Schroedinger,† and de Broglie.

Part of the physical interpretation of the theory consists in giving correspondence rules which assign to observables experimental procedures for measuring them. From the mathematical point of view, these observables are represented (in the conventional formalism adopted here) by self-adjoint operators in a Hilbert space. These operators will be studied from a general point of view in Chapter III. At present we shall describe some experimental procedures for measuring, in principle,‡ the most common of observables: those of position, momentum, and spin of an object of atomic or subatomic size (we shall call such an object

* We remind the reader that an inertial frame is a physical body (usually the laboratory of the experimenter) in which Newton's equation of motion is valid for macroscopic bodies moving at speeds which are low by comparison with the speed of light. It is an empirical fact that a frame of reference moving during the duration of the experiment at uniform speed with respect to the sun is a very good approximation to an inertial frame.

† It is interesting to mention that one of the earlier physical interpretations of the formalism of wave mechanics (which constitutes a special case of the Hilbert space formalism of quantum mechanics) was proposed by Schroedinger, was in total disagreement with the Copenhagen school interpretation, and lacked any statistical features.

‡ The experimental procedures for measuring these observables "in principle" may be conceptually very simple, but from the purely technical point of view they may not be the most efficient or the easiest to carry out with a high degree of accuracy and at a given cost.

a microparticle, so as not to confuse it with the more special term, elementary particle).

Measurements for determining the position of a particle can be carried out by a number of detectors (Wilson chambers, bubble chambers, Geiger–Müller counters, etc.). These detectors signal the presence of a particle within the volume enclosed by the detector by a directly noticeable (macroscopic) change of their state. Conceptually the simplest of such detectors is a photographic plate on which a landing microparticle leaves a characteristic mark.

The essential parts of an apparatus for determining the momentum of a charged particle are depicted in Fig. 1. If particles of known charge e

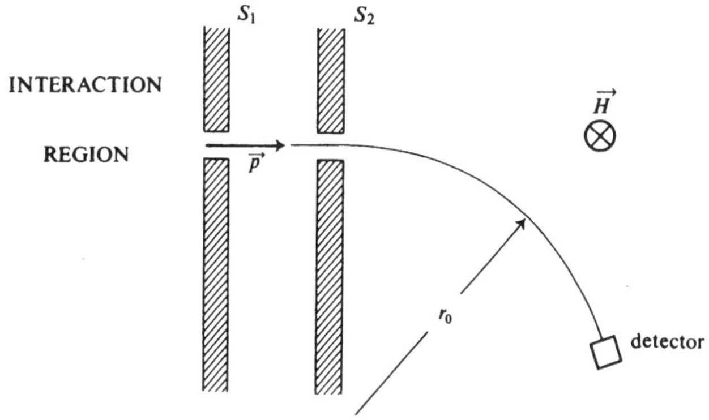

FIG. 1. Experimental arrangement for determinative measurement of the momentum of a charged microparticle.

interact to the left of the screen* S_1, and if a particle is detected by the detector depicted in Fig. 1 (which can be any of the types mentioned, or some other type best suited to the needs of a particular experiment), then the determined momentum $\mathbf{p}$ of particle is of magnitude (in suitable units)

$$|\mathbf{p}| = eHr_0, \tag{0.2}$$

and in the direction indicated in Fig. 1; in the above formula, H is the strength of a uniform magnetic field $\mathbf{H}$, which is present to the right of the screen S_2 and orthogonal to the plane of drawing.

* Both screens S_1 and S_2 should be impenetrable to all the particles used in the experiment. This impenetrability can be established by closing the apertures in the screens and checking that a detector placed to the right of each screen does not detect anything except the "background noise," which has a definite pattern and is due to the everpresent cosmic rays.

It must be stressed that the above-mentioned experimental arrangements are the ones which *give* an empirical meaning to the concepts of position and momentum of a microscopic particle. In other words, it is not meaningful to ask the question: "How do we know that the macroscopic change of state taking place in a detector (the black mark on a photoplate, or the track of condensed vapor in a Wilson chamber) is "in reality" due to a particle, and that the momentum of that particle is indeed given by Eq. (0.2)?" Such a question is empirically meaningless because, in the ultimate analysis (see Heisenberg [1925, p. 174]), the only knowledge that we possess about the microscopic world is the one we can extract from macroscopic phenomena.

The above type of measurement determines the value of observables at the instant the measurement is carried out. More precisely, from such a measurement one can compute, within bounds of accuracy inherent in the particular apparatus, what *would* be the value of a particular observable if there were no disturbances due to the interaction of the system (which in the above cases is the microparticle) and the apparatus. This type of measurement will be called a *determinative measurement.*

A determinative measurement does not tell us, in general, what the value of the measured observable, or even what the fate of the system will be after the measurement. For example, the system might be completely destroyed or cease to be an independent entity (for instance, when the detector is a photoplate and the particle is absorbed by it) after the measurement. In general, the measurement process will disturb the system or even change its nature (particle "creation and annihilation"), and such a disturbance must be taken into account if the measurement is used as a *preparation* of a system with a certain value of the measured observable. This phenomenon is characteristic of microphysics and cannot be ignored, as is done in macrophysics; in principle it exists also in macrophysics, but from the practical point of view it is completely negligible. To realize this, one should recall that measurements in classical mechanics require "seeing" the system, i.e., the reflection or emission of light from the objects constituting the system. Now, light (i.e., photons) has a certain momentum, which is imparted to the objects on which it impinges or by which it is emitted, but that momentum is completely negligible when dealing with objects of macroscopic size.

Thus, in quantum mechanics it is important to distinguish between determinative measurements of the above type and *preparatory* measurements.*

* It is more common to call a determinative measurement simply a "measurement," and a preparatory one a "preparation of state." We prefer to avoid the term "preparation of state," since it suggests that such a measurement will prepare a *quantum mechanical state,* which is by no means always true.

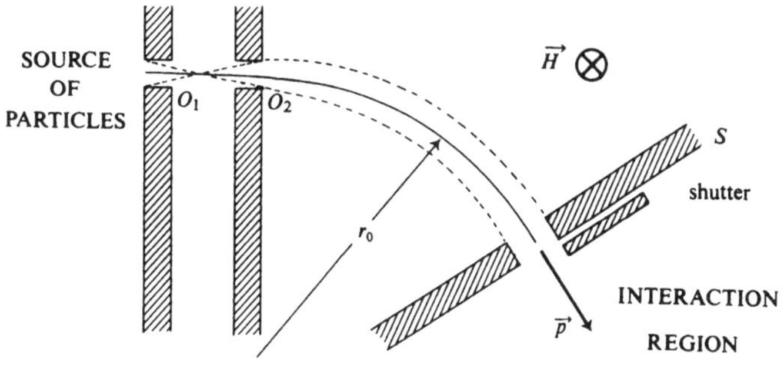

FIG. 2. Experimental arrangement for a preparatory measurement of the momentum of a charged microparticle.

For instance, in the modification in Fig. 2 of the apparatus in Fig. 1, we have a device which is meant to prepare charged particles of a given momentum. If we keep the shutter open for a very short "instant"* from t_0 to $t_0 + \Delta t$, then any particle that might have emerged during this period to the right of the screen S (i.e., in the interaction region, where it would interact with other particles adequately prepared) would have a momentum of magnitude $|\mathbf{p}|$ given by the formula (0.2), and in the direction indicated on Fig. 2.

A striking feature of the above experimental procedure is that it does not tell us, by itself, whether there was indeed a particle which had passed through the aperture while the shutter was open. This feature is common to many preparatory measurements. It is in the determinative (later) stage of the experiment that the presence or absence of a particle is established.†

We note that the above preparatory procedure cannot prepare a sharp value $\mathbf{p}$ of the momentum, but rather, due to the finite size of the apertures O_1, O_2, and O_3, it prepares a whole range $\Delta \subset \mathbb{R}^3$ of momenta. When the size of the last aperture O_3 is such that so-called "diffraction effects" can be neglected, this prepared range Δ consists of all the momenta that a particle of charge e would possess when traveling along all the imaginable paths that such a particle would have to follow according to *classical* mechanics in order to pass through the apertures O_1, O_2, and O_3.

The above-mentioned diffraction effects are again characteristic of

* By a very short "instant" is meant a time interval of duration Δt which is negligible in comparison with the other errors of measurement occurring in the particular experiment in which the described preparatory measurement is included.

† We could, of course, add a detector to the above apparatus, but such a detector would significantly disturb the prepared value of the momentum.

microphysics, and constitute the so-called "wave nature" of microparticles. They correspond to the experimental observation that if a detector (e.g., a photoplate) is placed behind two parallel impenetrable screens each having one aperture, a "particle" *might be* detected not only along any straight line passing through the two apertures, but also at other points. If we imagine that a microparticle travels along a trajectory, then it would seem that the trajectory is in such cases bent after the particle has passed through the aperture. If we have a beam of particles (i.e., many *independent* particles), then the net effect which will be observed in a photoplate placed behind the screens is a diffraction pattern, qualitatively very similar to the diffraction pattern of a beam of light.

We note that the experimental arrangement in Fig. 2 prepares not only a range Δ of momenta, but also a range Δ' of the position observables, where Δ' is the region enclosed by the aperture O_3. However, while classical mechanics assumes that the linear dimensions of Δ and Δ' could be made arbitrarily small by building sufficiently precise apparatus, it is a direct consequence of the diffraction effects that this is not the case with microparticles. This experimental finding is embodied in Heisenberg's *uncertainty principle* which states that no apparatus can be built which would prepare a particle to have the x coordinate (in some Cartesian frame of reference) within the interval Δ_x and the p_x-momentum component within the interval Δ_{p_x}, so that the inequality

$$|\Delta_x|\,|\Delta_{p_x}| \geqslant h/2\pi \tag{0.3}$$

is not satisfied, where $|\Delta_x|$ and $|\Delta_{p_x}|$ denote the lengths of the respective intervals, and h is a universal constant called the Planck constant; similar relations

$$|\Delta_y|\,|\Delta_{p_y}| \geqslant h/2\pi, \qquad |\Delta_z|\,|\Delta_{p_z}| \geqslant h/2\pi,$$

hold for the y and z Cartesian components.

It should be mentioned that the experimental arrangements in Figs. 1 and 2 determine and prepare, respectively, also the energy of the microparticle on which the measurement is carried out. By definition, if the momentum of a free particle of mass* m is $\mathbf{p}$, then its energy is

$$E = \mathbf{p}^2/2m.$$

* The mass of a particle can be measured by the experimental arrangement in Fig. 1 if we introduce an electric field **E** orthogonal to the existing magnetic field **H**. Quantities like mass, charge, and intrinsic spin are characteristic of each kind of microparticle, and, in fact, provide the only means of differentiating types of microparticles, like electrons, positrons, neutrons, protons, hydrogen atoms, water molecules, etc.

We turn now to measurements of spin. Assume that we have in classical mechanics a body which rotates, with respect to an inertial frame of reference, around a certain axis passing through it, which is stationary or moves at a uniform speed with respect to that frame of reference. Such a body has then an angular momentum $\mathbf{s}$ with respect to that axis. If the dimensions of the body are negligible with respect to other dimensions in the experiment, so that the body can be called a particle, then $\mathbf{s}$ is called the spin of the particle. If the particle carries a charge, then due to its spin it will have a magnetic (dipole) moment $\boldsymbol{\mu}$ proportional to $\mathbf{s}$. When such a particle travels through an area in which there is an inhomogeneous magnetic field $\mathbf{H}$, this field will act on it with a force

$$\mathbf{F} = \text{grad}(\boldsymbol{\mu} \cdot \mathbf{H}).$$

In the light of the above remarks, each one of the particles of equal velocity (i.e., equal momentum) and equal spin in a beam passing through the experimental arrangement depicted in Fig. 3 would be deflected

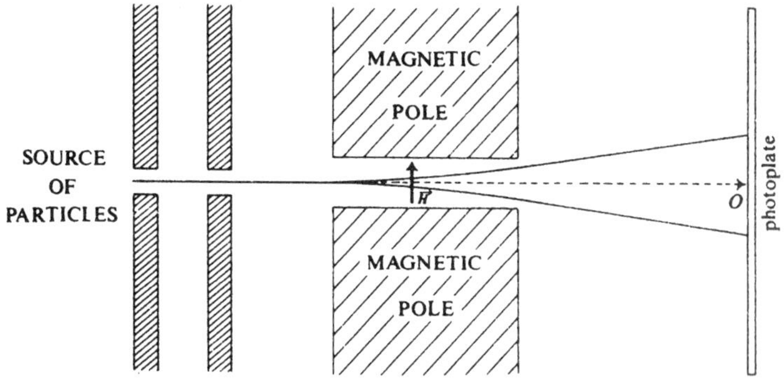

FIG. 3. The Stern–Gerlach experiment.

from a straight path by an amount proportional to the projection μ_z of the magnetic moment $\boldsymbol{\mu}$ of that particle in the direction of the gradient $\mathbf{H}'$ of the field $\mathbf{H}$. If the spins of the particles were randomly oriented in all directions (as we could expect them to be if the source were a gas of such particles), then one would predict on the basis of the above considerations a continuous distribution of marks on the photoplate. However, the *Stern–Gerlach experiment* carried out with molecular and atomic beams reveals instead, in general, n discrete lines (in Fig. 3, $n = 2$)—a phenomenon which is usually described by saying that the spin is quantized.

In quantum mechanics, by definition, the spin of the above particles is taken to be

$$s = \frac{n-1}{2}$$

(though, strictly speaking, for reasons to become clear when a systematic theoretical study of the spin is undertaken, the total spin is $[s(s+1)]^{1/2}$). Thus, s can assume only integer and half-integer values $\frac{1}{2}$, 1, $\frac{3}{2}$, 2,...; in Fig. 3 we have depicted the case of spin $\frac{1}{2}$.

The experimental arrangement of Stern and Gerlach can be used as an apparatus for a determinative measurement of the *spin component* in the direction $\mathbf{H}'$. In that case the source of particles would originate in the interaction region, where particles of known spin are interacting. If a particle of integer spin* s leaves a mark at O, then, by definition, it has spin zero in the $\mathbf{H}'$direction; the first, second,..., $(n-1)/2$ mark above O correspond to spin components in the $\mathbf{H}'$ direction equal to 1, 2,..., $(n-1)/2$, respectively, while the first, second,..., $(n-1)/2$ marks below O correspond to spin components -1, -2,..., $-(n-1)/2$, respectively. In case of a particle of half-integer spin, there will be no middle mark; the first, second,..., $(n-1)/2$ marks above or below O correspond to spin components $\frac{1}{2}$, $\frac{3}{2}$,..., $(n-1)/2$, or $-\frac{1}{2}$, $-\frac{3}{2}$,..., $-(n-1)/2$, respectively. Hence we see that according to the very definition of the spin projection onto a certain axis, that projection can assume only integer values in case of integer-spin particles, and only half-integer values in case of half-integer-spin particles.

The above experimental arrangement can be easily transformed into an apparatus for preparatory measurements of spin by replacing the photoplate with a screen which has apertures at the spots where a beam of particles from the given source had left tracks. It has to be mentioned that no simultaneous measurements of spin in two different directions can be carried out on microparticles—a feature which is in complete agreement with certain properties (noncommutativity of spin-component operators) of the formalism of quantum mechanics.

Here we end this short survey of some of the experimental procedures for measuring some of the basic observables which occur in quantum mechanics, and which will frequently appear in the pages of this book. In Chapter I we start our systematic study of the Hilbert space formalism of quantum mechanics, and related mathematics.

* This means that a beam of such particle with random-oriented spins would have $2s+1$ tracks on the photoplate.

CHAPTER I

Basic Ideas of Hilbert Space Theory

The central object of study in this chapter is the infinite-dimensional Hilbert space. The main goal is to give a rigorous analysis of the problem of expanding a vector in a Hilbert space in terms of an orthogonal basis containing a countable infinity of vectors.

We first review in §1 a few key theorems on vector spaces in general, and in §2 we investigate the basic properties of vector spaces on which an inner product is defined. In order to define convergence in an inner-product space, we introduce in §3 the concept of metric. In §4 we give the basic concepts and theorems on separable Hilbert spaces, concentrating especially on properties of orthonormal bases. We conclude the chapter by illustrating some of the physical applications of these mathematical results with the initial-value problem in wave mechanics.

1. Vector Spaces

1.1. Vector Spaces over Fields of Scalars

A mathematical space is in general a set endowed with some given structure. Such a structure can be given, for instance, by means of certain operations which are defined on the elements of that set. These operations are then required to obey certain general rules, which are called the postulates or the axioms of the mathematical space.

Definition 1.1. Any set $\mathscr{V}$ on which the operations of vector addition and multiplication by a scalar are defined is said to be a *vector* *space* (or *linear space*, or *linear manifold*). The operation of *vector addition* is a mapping,*

$$(f,g) \to f+g, \qquad (f,g) \in \mathscr{V} \times \mathscr{V}, \qquad f+g \in \mathscr{V},$$

of $\mathscr{V} \times \mathscr{V}$ into $\mathscr{V}$, while the operation of multiplication by a scalar a from a field† F is a mapping

$$(a,f) \mapsto af, \qquad (a,f) \in \mathsf{F} \times \mathscr{V}, \qquad af \in \mathscr{V},$$

of $\mathsf{F} \times \mathscr{V}$ into $\mathscr{V}$. These two vector operations are required to satisfy the following axioms for any $f, g, h \in \mathscr{V}$ and any scalars $a, b \in \mathsf{F}$:

(1) $f + g = g + f$ (commutativity of vector addition).
(2) $(f + g) + h = f + (g + h)$ (associativity of vector addition).
(3) There is a vector $\mathbf{0}$, called the *zero vector*, such that g satisfies the relation $f + g = f$ if and only if $g = \mathbf{0}$.
(4) $a(f + g) = af + ag$.
(5) $(a + b)f = af + bf$.
(6) $(ab)f = a(bf)$.
(7) $1f = f$, where 1 denotes the unit element in the field.

By following a tacit convention, we denote a mathematical space constructed from a set S by the same letter S, except where ambiguities might arise. Thus, we shall denote by $\mathscr{V}$ the vector space consisting of a set $\mathscr{V}$ together with the vector operations on $\mathscr{V}$ also by $\mathscr{V}$.

When in a vector space the multiplication by a scalar is defined for scalars which are elements of the field F, we say that we are dealing with a *vector space over the field* F. If the field F is the field of real or complex numbers the vector space is called, respectively, a *real* or a *complex vector space*.

* We remind the reader that a mapping M *of* a set S *into* a set T is any unambiguous rule assigning to each element ξ of S a single element $M(\xi)$ of T; $M(\xi)$ is called the image of ξ under the mapping M. The set S is the domain of definition of M, while the subset $T_1 \subset T$ of all image points $M(\xi)$, $T_1 = \{\eta = M(\xi)\colon \xi \in S\}$, is the range of M. If $T_1 = T$, then we say that M is a mapping of the set S *onto* the set T.

If $S_1, ..., S_n$ are sets, then $S_1 \times \cdots \times S_n$ denotes the family $(\xi_1, ..., \xi_n)$ of all n-tuples of elements $\xi_1 \in S_1, ..., \xi_n \in S_n$, and is called the Cartesian product of the sets $S_1, ..., S_n$.

† A field is a set on which field operations of summation and multiplication are defined, i.e., operations satisfying certain axioms. We do not give these axioms because in the sequel we are interested only in two special well-known fields: the field of real numbers $\mathbb{R}^1$ and the field of complex numbers $\mathbb{C}^1$ consisting, respectively, of the set of real numbers $\mathbb{R}^1$ and the set of complex numbers $\mathbb{C}^1$ on which the field operations are ordinary summation and multiplication of numbers (see Birkhoff and MacLane [1953]).

As an example (see also Exercises 1.1, 1.2, and 1.3) of a real vector space consider the family $(\mathbb{R}^n)$ of one-column real matrices and define for

$$\alpha = \begin{pmatrix} a_1 \\ \vdots \\ a_n \end{pmatrix}, \qquad \beta = \begin{pmatrix} b_1 \\ \vdots \\ b_n \end{pmatrix}$$

vector summation by the mapping

$$(1.1) \qquad (\alpha, \beta) \mapsto \alpha + \beta = \begin{pmatrix} a_1 + b_1 \\ \vdots \\ a_n + b_n \end{pmatrix},$$

and for any scalar $a \in \mathbb{R}^1$ define multiplication of α by a as the mapping

$$(1.2) \qquad (a, \alpha) \mapsto a\alpha = \begin{pmatrix} aa_1 \\ \vdots \\ aa_n \end{pmatrix}.$$

It is easy to check that Axioms 1–7 in Definition 1.1 are satisfied.

Analogously we can define the complex vector space $(\mathbb{C}^n)$ by introducing in the set $\mathbb{C}^n$ of one-column matrices vector operations defined by the mapping (1.1) and (1.2), where now $\alpha, \beta \in \mathbb{C}^n$, and therefore $a_1, \ldots, a_n$, $b_1, \ldots, b_n$, as well as the scalar a, are complex numbers.

1.2. Linear Independence of Vectors

Theorem 1.1. Each vector space $\mathscr{V}$ has only one zero vector $\mathbf{0}$, and each element f of a vector space has one and only one inverse $(-f)$. For any $f \in \mathscr{V}$,

$$0f = \mathbf{0}, \qquad (-1)f = (-f).$$

Proof. If there are two zero vectors $\mathbf{0}_1$ and $\mathbf{0}_2$, they both have to satisfy Axiom 3 in Definition 1.1,

$$f = f + \mathbf{0}_1 = f + \mathbf{0}_2$$

for all f. Hence, by taking $f = \mathbf{0}_1$ we get $\mathbf{0}_1 = \mathbf{0}_1 + \mathbf{0}_2$, and then by taking $f = \mathbf{0}_2$ we deduce that $\mathbf{0}_2 = \mathbf{0}_2 + \mathbf{0}_1 = \mathbf{0}_1 + \mathbf{0}_2 = \mathbf{0}_1$. Now

$$f = 1f = (1 + 0)f = 1f + 0f = f + 0f$$

and therefore $0f = \mathbf{0}$. We have

$$(-1)f + f = (-1)f + 1f = (-1 + 1)f = 0f = \mathbf{0},$$

which proves the existence of an inverse $(-f) = (-1)f$ for f. This inverse $(-f)$ is unique, because if there is another $f_1 \in \mathscr{V}$ such that $f + f_1 = \mathbf{0}$, we have

$$(-f) = (-f) + \mathbf{0} = (-f) + (f + f_1) = [(-f) + f] + f_1$$
$$= \mathbf{0} + f_1 = f_1 . \quad \text{Q.E.D.}$$

Definition 1.2. The vectors $f_1, \ldots, f_n$ are said to be *linearly independent* if the relation

$$c_1 f_1 + \cdots + c_n f_n = \mathbf{0}, \qquad c_1, \ldots, c_n \in \mathsf{F},$$

has $c_1 = \cdots = c_n = 0$ as the *only* solution. A subset S (finite or infinite) of a vector space $\mathscr{V}$ is called a *set of linearly independent* vectors if any finite number of *different* vectors from S are linearly independent. The *dimension* of a vector space $\mathscr{V}$ is the least upper bound (which can be finite or positive infinite) of the set of all integers ν for which there are ν linearly independent vectors in $\mathscr{V}$.

1.3. Dimension of a Vector Space

When the maximal number of linearly independent vectors in the vector space $\mathscr{V}$ is finite and equal to n, then by the above definition $\mathscr{V}$ is n dimensional; otherwise the dimension of $\mathscr{V}$ is $+\infty$, and $\mathscr{V}$ is said to be infinite dimensional.

Theorem 1.2. If the vector space $\mathscr{V}$ is n dimensional ($n < +\infty$), then there is at least one set $f_1, \ldots, f_n$ of linearly independent vectors, and each vector $f \in \mathscr{V}$ can be expanded in the form

$$f = a_1 f_1 + \cdots + a_n f_n , \tag{1.3}$$

where the coefficients $a_1, \ldots, a_n$ (which are scalars) are uniquely determined by f.

Proof. If $f = \mathbf{0}$, (1.3) is established by taking $a_1 = \cdots = a_n = 0$. For $f \neq \mathbf{0}$, the equation

$$cf + c_1 f_1 + \cdots + c_n f_n = 0 \tag{1.4}$$

should have a solution with $c \neq 0$ due to the assumption that $f_1, \ldots, f_n$ are linearly independent, while $f, f_1, \ldots, f_n$ have to be linearly dependent because $\mathscr{V}$ is n dimensional. From (1.4) we get

$$f = (-c_1/c) f_1 + \cdots + (-c_n/c) f_n ,$$

which establishes (1.3). If we also had

$$f = b_1 f_1 + \cdots + b_n f_n, \tag{1.5}$$

then by subtracting (1.5) from (1.3) we get

$$(a_1 - b_1) f_1 + \cdots + (a_n - b_n) f_n = \mathbf{0}.$$

As $f_1, \ldots, f_n$ are linearly independent we deduce that $a_1 - b_1 = 0, \ldots, a_n - b_n = 0$, thus proving that $a_1, \ldots, a_n$ are uniquely determined when f is given. Q.E.D.

Definition 1.3. We say that the (finite or infinite) set S *spans* the vector space $\mathscr{V}$ if every vector in $\mathscr{V}$ can be written as a linear combination

$$f = a_1 h_1 + \cdots + a_n h_n, \qquad h_1, \ldots, h_n \in S$$

of a finite number of vectors belonging to S; if S is in addition a set of linearly independent vectors, then S is called a *vector basis* of $\mathscr{V}$.

Theorem 1.3. If the set $\{g_1, \ldots, g_m\}$ is a vector basis of the n-dimensional ($n < +\infty$) vector space $\mathscr{V}$, then necessarily $m = n$.

Proof. As $\mathscr{V}$ is n-dimensional, there must be n linearly independent vectors $f_1, \ldots, f_n$. If the set $\{g_1, \ldots, g_m\}$ is a vector basis in $\mathscr{V}$, we can write

$$\begin{aligned} f_1 &= a_{11} g_1 + \cdots + a_{m1} g_m \\ &\vdots \\ f_n &= a_{1n} g_1 + \cdots + a_{mn} g_m. \end{aligned} \tag{1.6}$$

Thus, if we try to satisfy the equation

$$x_1 f_1 + \cdots + x_n f_n = \mathbf{0}, \tag{1.7}$$

we get by substituting $f_1, \ldots, f_n$ in (1.7) with the expressions in (1.6)

$$(a_{11} x_1 + \cdots + a_{1n} x_n) g_1 + \cdots + (a_{m1} x_1 + \cdots + a_{mn} x_n) g_m = \mathbf{0}. \tag{1.8}$$

Since $g_1, \ldots, g_m$ are assumed to be linearly independent, the above equation has a solution in $x_1, \ldots, x_n$ if and only if

$$\begin{aligned} a_{11} x_1 + \cdots + a_{1n} x_n &= 0 \\ &\vdots \\ a_{m1} x_1 + \cdots + a_{mn} x_n &= 0. \end{aligned} \tag{1.9}$$

However, as $f_1, \ldots, f_n$ are also linearly independent, (1.7) or equivalently (1.8) or (1.9) should have as the only solution the trivial one $x_1 = \cdots =$

$x_n = 0$. Now, $m \leqslant n$ because $\mathscr{V}$ is n dimensional and $g_1, \ldots, g_m$ are linearly independent (see Definition 1.2); therefore, (1.9) has only a trivial solution if and only if $m = n$. Q.E.D.

Definition 1.4. A subset $\mathscr{V}_1$ of a vector space $\mathscr{V}$ is a *vector subspace* (*linear subspace*) of $\mathscr{V}$ if it is closed under the vector operations, i.e., if $f + g \in \mathscr{V}_1$ and $af \in \mathscr{V}_1$ whenever $f, g \in \mathscr{V}_1$ and for any scalar a. A vector subspace $\mathscr{V}_1$ of $\mathscr{V}$ is said to be *nontrivial* if it is different from $\mathscr{V}$ and from the set $\{\mathbf{0}\}$.

From the very definition of the dimension of a vector space $\mathscr{V}$ we can conclude that the dimension of a vector subspace $\mathscr{V}_1$ of $\mathscr{V}$ cannot exceed the dimension of $\mathscr{V}$.

1.4. Isomorphism of Vector Spaces

Definition 1.5. Two vector spaces $\mathscr{V}_1$ and $\mathscr{V}_2$ over the same field are *isomorphic* if there is a one-to-one mapping $\mathscr{V}_1$ *onto* $\mathscr{V}_2$ which has the properties that if f_2 and g_2, $f_2, g_2 \in \mathscr{V}_2$, are the images of f_1 and g_1, $f_1, g_1 \in \mathscr{V}_1$, respectively, then for any scalar a, af_2 is the image of af_1

$$af_1 \leftrightarrow af_2,$$

and $f_2 + g_2$ is the image of $f_1 + g_1$

$$f_1 + g_1 \leftrightarrow f_2 + g_2.$$

The importance of the isomorphism of two vector spaces $\mathscr{V}_1$ and $\mathscr{V}_2$ lies in the obvious fact that two such spaces have an identical vector structure. It is easy to see that the relation of isomorphism is transitive (see Exercise 1.6), i.e., if $\mathscr{V}_1$ and $\mathscr{V}_2$ as well as $\mathscr{V}_2$ and $\mathscr{V}_3$ are isomorphic, then $\mathscr{V}_1$ and $\mathscr{V}_3$ are also isomorphic.

Theorem 1.4. All complex (real) n-dimensional $(n < +\infty)$ vector spaces are isomorphic to the vector space $(\mathbb{C}^n)$ [$(\mathbb{R}^n)$ in case of real vector spaces].

Proof. Consider the case of an n-dimensional vector space $\mathscr{V}$. According to Theorem 1.2 there is a vector basis consisting of n vectors $f_1, \ldots, f_n$, and each vector $f \in \mathscr{V}$ can be expanded in the form (1.3), where $a_1, \ldots, a_n \in \mathbb{C}^1$ are uniquely determined by f. Consequently

$$f \mapsto \alpha_f = \begin{pmatrix} a_1 \\ \vdots \\ a_n \end{pmatrix} \in (\mathbb{C}^n)$$

is a mapping of $\mathscr{V}$ into $(\mathbb{C}^n)$. Furthermore, this is a one-to-one mapping of $\mathscr{V}$ *onto* $(\mathbb{C}^n)$ because to any

$$\beta = \begin{pmatrix} b_1 \\ \vdots \\ b_n \end{pmatrix} \in (\mathbb{C}^n)$$

corresponds a unique $f = b_1 f_1 + \cdots + b_n f_n$ such that $\beta = \alpha_f$. It is also easy to see that

$$f + g \mapsto \alpha_{f+g} = \alpha_f + \alpha_g ,$$
$$af \mapsto \alpha_{af} = a\alpha_f .$$

Since isomorphism of vector spaces is a transitive relation (see Exercise 1.6) we can conclude that all n-dimensional complex vector spaces are mutually isomorphic, because each of them is isomorphic to $(\mathbb{C}^n)$. Q.E.D.

EXERCISES

1.1. Check that the set of all $m \times n$ complex matrices constitutes an $m \cdot n$ dimensional complex vector space if vector addition is defined as being addition of matrices, and multiplication by a scalar is multiplication of a matrix by a complex number.

1.2. Show that the set $\mathbb{C}^1$ of all complex numbers becomes a two-dimensional *real* vector space if vector addition is identical to addition of complex numbers, and multiplication by a scalar is multiplication of a complex number (the vector) by a real number (the scalar).

1.3. Show that the family $\mathscr{C}^0(\mathbb{R}^1)$ of all complex-valued continuous functions defined on the real line is an infinite-dimensional vector space if the vector sum $f + g$ of $f(x), g(x) \in \mathscr{C}^0(\mathbb{R}^1)$ is the function $(f + g)(x) = f(x) + g(x)$, and the product af of $f(x) \in \mathscr{C}^0(\mathbb{R}^1)$ with $a \in \mathbb{C}^1$ is the function $(af)(x) = af(x)$. The zero vector is taken to be the function $f(x) \equiv 0$.

1.4. Prove that if $\mathscr{K}$ is a family of linear subspaces L of a vector space $\mathscr{V}$, then their set intersection $\bigcap_{L \in \mathscr{K}} L$ is also a vector subspace of $\mathscr{V}$.

1.5. Show that if S is any subset of a vector space $\mathscr{V}$, then there is a unique smallest vector subspace $\mathscr{V}_S$ containing S (called the vector subspace spanned by S).

1.6. Verify that the relation of isomorphism of vector spaces is:

(a) reflexive, i.e., every vector space $\mathscr{V}$ is isomorphic to itself;

(b) symmetric, i.e., if $\mathscr{V}_1$ is isomorphic to $\mathscr{V}_2$, then $\mathscr{V}_2$ is isomorphic to $\mathscr{V}_1$;

(c) transitive, i.e., if $\mathscr{V}_1$ is isomorphic to $\mathscr{V}_2$ and $\mathscr{V}_2$ is isomorphic to $\mathscr{V}_3$, then $\mathscr{V}_1$ is isomorphic to $\mathscr{V}_3$.

1.7. Prove that the following subsets of the set $\mathscr{C}^0(\mathbb{R}^1)$ (see Exercise 1.3) are vector subspaces of the vector space $\mathscr{C}^0(\mathbb{R}^1)$:

(a) the set $\mathscr{P}_\infty$ of all polynomials with complex coefficients;
(b) the set $\mathscr{P}_n$ of all polynomials of *at most* degree n.

Show that $\mathscr{P}_\infty \supset \mathscr{P}_n$.

2. Euclidean (Pre-Hilbert) Spaces

2.1. Inner Products on Vector Spaces

A *Euclidean* (or *pre-Hilbert or inner product* or *unitary*) space $\mathscr{E}$ is a vector space on which an inner product is defined. The Euclidean space is called real or complex if the vector space on which the inner product is defined is, respectively, real or complex.

Definition 2.1. An *inner* (or *scalar*) product $\langle \cdot \mid \cdot \rangle$ on the complex vector space $\mathscr{V}$ is a mapping of the set $\mathscr{V} \times \mathscr{V}$ into the set $\mathbb{C}^1$ of complex numbers

$$(f, g) \mapsto \langle f \mid g \rangle, \qquad (f, g) \in \mathscr{V} \times \mathscr{V}, \qquad \langle f \mid g \rangle \in \mathbb{C}^1,$$

which satisfies the following requirements:

(1) $\langle f \mid f \rangle > 0, \quad$ for all $\; f \neq \mathbf{0}$,
(2) $\langle f \mid g \rangle = \langle g \mid f \rangle^*$,
(3) $\langle f \mid ag \rangle = a\langle f \mid g \rangle, \qquad a \in \mathbb{C}^1$,
(4) $\langle f \mid g + h \rangle = \langle f \mid g \rangle + \langle f \mid h \rangle$.

Note that by inserting $f = g = h = \mathbf{0}$ in Point 4 we get $\langle \mathbf{0} \mid \mathbf{0} \rangle = 0$.

Following a notation first introduced by Dirac [1930] and widely adopted by physicists, we denote the inner product of f and g by $\langle f \mid g \rangle$. Mathematicians often prefer the notation (f, g) and replace Point 3 in Definition 2.1 by

$$(af, g) = a(f, g).$$

The above definition can be easily specialized to real vector spaces, in which case the inner product $\langle f \mid g \rangle$ is a real number, and Point 2 of Definition 2.1 becomes $\langle f \mid g \rangle = \langle g \mid f \rangle$. As in quantum physics we deal almost exclusively with complex Euclidean spaces, we limit ourselves from now on to the complex case. Consequently, if not otherwise stated, whenever we talk about a Euclidean space, we shall mean a *complex* Euclidean space.

Theorem 2.1. In a Euclidean space $\mathscr{E}$, the inner product $\langle f \mid g \rangle$ satisfies the relations

(a) $\langle af \mid g \rangle = a^* \langle f \mid g \rangle$,
(b) $\langle f + g \mid h \rangle = \langle f \mid h \rangle + \langle g \mid h \rangle$.

The proof is obtained by a straightforward application of Points 1–4 in Definition 2.1:

$$\langle af \mid g \rangle = \langle g \mid af \rangle^* = [a \langle g \mid f \rangle]^* = a^* \langle g \mid f \rangle^* = a^* \langle f \mid g \rangle,$$
$$\langle f + g \mid h \rangle = \langle h \mid f + g \rangle^* = [\langle h \mid f \rangle + \langle h \mid g \rangle]^* = \langle h \mid f \rangle^* + \langle h \mid g \rangle^*$$
$$= \langle f \mid h \rangle + \langle g \mid h \rangle.$$

As an example of a finite-dimensional Euclidean space, we can take the vector space $(\mathbb{C}^n)$ defined in the preceding section, in which we introduce as the inner product of the vectors α and β with the kth components a_k and b_k,

$$\langle \alpha \mid \beta \rangle = a_1{}^* b_1 + a_2{}^* b_2 + \cdots + a_n{}^* b_n .$$

It is easy to check that the above mapping of $(\mathbb{C}^n) \times (\mathbb{C}^n)$ into $\mathbb{C}^1$ satisfies the four requirements of Definition 2.1. We shall denote the above Euclidean space with the symbol $l^2(n)$.

An example of an infinite-dimensional Euclidean space is provided by the vector space $[\mathscr{C}^0_{(2)}(\mathbb{R}^1)]$ of all continuous complex-valued functions $f(x)$ on the real line which satisfy

$$\int_{-\infty}^{+\infty} |f(x)|^2 \, dx < +\infty, \quad \lim_{x \to \pm\infty} f(x) = 0, \tag{2.1}$$

in which the inner product (see Exercise 2.1) is

$$\langle f \mid g \rangle = \int_{-\infty}^{+\infty} f^*(x) \, g(x) \, dx. \tag{2.2}$$

Theorem 2.2. Any two elements f, g of a Euclidean space $\mathscr{E}$ satisfy the Schwarz–Cauchy inequality

$$|\langle f \mid g \rangle|^2 \leqslant \langle f \mid f \rangle \langle g \mid g \rangle.$$

Proof. For any given $f, g \in \mathscr{E}$ and any complex number a we have, from property 1 in Definition 2.1 and the comment following it,

$$\langle f + ag \mid f + ag \rangle \geqslant 0.$$

In particular, if we take in the above inequality

$$a = \lambda \frac{\langle f \mid g \rangle^*}{|\langle f \mid g \rangle|}, \qquad \lambda = \lambda^*,$$

we easily show that the inequality

$$g(\lambda) = \lambda^2 \langle g \mid g \rangle + 2\lambda \, |\langle f \mid g \rangle| + \langle f \mid f \rangle \geqslant 0$$

is true for all real values of λ. A necessary and sufficient condition that $g(\lambda) \geqslant 0$ is that the discriminant of the quadratic polynomial $g(\lambda)$ is not positive

$$|\langle f \mid g \rangle|^2 - \langle f \mid f \rangle \langle g \mid g \rangle \leqslant 0,$$

from which the Schwarz–Cauchy inequality follows immediately. Q.E.D.

2.2. The Concept of Norm

The family of all Euclidean spaces is obviously contained in the family of vector spaces. There is another family of vector spaces with special properties which is of great importance in mathematics: the family of normed spaces.

Definition 2.2. A mapping

$$f \to \| f \|, \qquad f \in \mathscr{V}, \qquad \| f \| \in \mathbb{R}^1,$$

of a complex vector space $\mathscr{V}$ into the set of real numbers is called a *norm* if it satisfies the following conditions:

(1) $\| f \| > 0$ for $f \neq \mathbf{0}$,

(2) $\| \mathbf{0} \| = 0$,

(3) $\| af \| = | a | \, \| f \|$ for all $a \in \mathbb{C}^1$,

(4) $\| f + g \| \leqslant \| f \| + \| g \|$ (the triangle inequality).

We denote the above norm by $\| \cdot \|$.

For a real vector space, we require in Point 3 that $a \in \mathbb{R}^1$.

The last requirement in Definition 2.2 is known as the triangle inequality because it represents in a two- or three-dimensional real vector space a relation satisfied by the sides of a triangle formed by three vectors f, g and $f + g$.

A real (complex) vector space on which a particular norm is given is called a real (complex) normed vector space. A Euclidean space is a special case of a normed space; this can be seen from the following theorem.

Theorem 2.3. In a Euclidean space $\mathscr{E}$ with the inner product $\langle f \mid g\rangle$ the real-valued function

$$\|f\| = \sqrt{\langle f \mid f\rangle} \tag{2.3}$$

is a norm.

Proof. The only one of the four properties of a norm which is not satisfied by (2.3) in an evident way is the triangle inequality. We easily get

$$\begin{aligned}\|f+g\|^2 &= \langle f+g \mid f+g\rangle = \langle f+f\rangle + \langle f \mid g\rangle + \langle g \mid f\rangle + \langle g \mid g\rangle \\ &= \langle f \mid f\rangle + 2\mathrm{Re}\langle f \mid g\rangle + \langle g \mid g\rangle.\end{aligned} \tag{2.4}$$

From the Schwarz–Cauchy inequality we have

$$|\,\mathrm{Re}\langle f \mid g\rangle| \leqslant |\langle f \mid g\rangle| \leqslant \|f\|\,\|g\|,$$

which when inserted in (2.4) yields

$$\|f+g\|^2 \leqslant \|f\|^2 + 2\,\|f\|\,\|g\| + \|g\|^2 = (\|f\| + \|g\|)^2.$$

The above relation leads immediately to the triangle inequality. Q.E.D.

2.3. Orthogonal Vectors and Orthonormal Bases

Some elementary geometrical concepts valid for real two- or three-dimensional Euclidean spaces can be generalized in a straightforward manner to any Euclidean space.

Definition 2.3. In a Euclidean space $\mathscr{E}$ two vectors f and g are called orthogonal, symbolically $f \perp g$, if $\langle f \mid g\rangle = 0$. Two subsets R and S of $\mathscr{E}$ are said to be *orthogonal* (symbolically, $R \perp S$) if each vector in R is orthogonal to each vector in S. A set of vectors in which any two vectors are orthogonal is called an *orthogonal system* of vectors. A vector f is said to be *normalized* if $\|f\| = 1$. An orthogonal system of vectors is called an *orthonormal system* if each vector in the system is normalized.

Theorem 2.4. If S is a finite or countably infinite set of vectors in a Euclidean space $\mathscr{E}$ and (S) is the vector subspace of $\mathscr{E}$ spanned by S, then there is an orthonormal system T of vectors which spans (S), i.e., for which $(T) = (S)$; T is a finite set when S is a finite set.

Proof. As the set S is at most countable we can write it in the form

$$S = \{f_1, f_2, \ldots\}$$

by assigning each vector in S to a natural number. In general some of the vectors in S might be linearly dependent. We can build from S

another set S_0 of linearly independent vectors spanning the same subspace (S), i.e., such that $(S_0) = (S)$, by the following procedure (which should be applied consecutively on $n = 1, 2, \ldots$): if f_n is the zero vector or is linearly dependent on $f_1, \ldots, f_{n-1}$, then discard it; otherwise include it in S_0. Thus we get a set S_0 of linearly independent vectors

$$S_0 = \{g_1, g_2, \ldots\}, \qquad (S_0) = (S).$$

We can obtain from S_0 an orthonormal set T such that $(T) = (S_0)$ by the so-called *Schmidt* (or *Gram–Schmidt*) *orthonormalization procedure.*

Since $g_1 \neq \mathbf{0}$, we can introduce the vector

$$e_1 = \frac{g_1}{\| g_1 \|},$$

which is normalized. Proceeding by induction, assume that we have obtained the orthonormal system of vectors $e_1, \ldots, e_{n-1}$. Then e_n is given by

$$e_n = \frac{g_n - \langle e_{n-1} \mid g_n \rangle e_{n-1} - \cdots - \langle e_1 \mid g_n \rangle e_1}{\| g_n - \langle e_{n-1} \mid g_n \rangle e_{n-1} - \cdots - \langle e_1 \mid g_n \rangle e_1 \|}.$$

The above vector is certainly well defined, since the denominator of the above expression is different from zero; namely, if it were zero, then we would have

$$g_n - \langle e_{n-1} \mid g_n \rangle e_{n-1} - \cdots - \langle e_1 \mid g_n \rangle e_1 = 0,$$

i.e., g_n would depend on $e_1, \ldots, e_{n-1}$. However, by solving the equations for $e_1, \ldots, e_{n-1}$, it is easy to see that we have

$$\begin{aligned} g_1 &= c_{1,1} e_1 \\ &\vdots \\ g_{n-1} &= c_{n-1,1} e_1 + c_{n-1,2} e_2 + \cdots + c_{n-1,n-1} e_{n-1}, \end{aligned}$$

and therefore if g_n depended on $e_1, \ldots, e_{n-1}$, then it would also depend on $g_1, \ldots, g_{n-1}$, contrary to the fact that S_0 consists only of linearly independent vectors.

The vectors of T are obviously normalized. In order to prove that T is an orthonormal system, assume that we have proved that $\langle e_i \mid e_j \rangle = \delta_{ij}$ for $i, j = 1, \ldots, n - 1$. Then we have for $m < n$

$$\langle e_m \mid e_n \rangle = \frac{1}{\| g_n - \cdots - \langle e_1 \mid g_1 \rangle e_1 \|} \left(\langle e_m \mid g_n \rangle - \sum_{k=1}^{n-1} \langle e_k \mid g_n \rangle \cdot \delta_{km} \right) = 0,$$

which proves that $\langle e_i \mid e_j \rangle = \delta_{ij}$ for $i, j = 1, \ldots, n$. Thus, by induction T is orthonormal.

As we have for any n that $e_1, \ldots, e_n$ can be expressed in terms of $g_1, \ldots, g_n$, and vice versa, we can conclude that $(T) = (S_0)$. Q.E.D.

2.4. Isomorphism of Euclidean Spaces

We introduce now a concept of isomorphism of Euclidean spaces, which makes two isomorphic Euclidean spaces identical from the point of view of their vector structure as well as from the point of view of the structure induced by the inner product.

Definition 2.4. Two Euclidean spaces $\mathscr{E}_1$ and $\mathscr{E}_2$ with inner products $\langle \cdot \mid \cdot \rangle_1$ and $\langle \cdot \mid \cdot \rangle_2$, respectively, are *isomorphic* (or *unitarily equivalent*) if there is a mapping of $\mathscr{E}_1$ *onto* $\mathscr{E}_2$

$$f_1 \mapsto f_2, \qquad f_1 \in \mathscr{E}_1, \qquad f_2 \in \mathscr{E}_2$$

such that if for any $f_1, g_1 \in \mathscr{E}_1$ the vector $f_2 \in \mathscr{E}_2$ is the image of f_1 and the vector $g_2 \in \mathscr{E}_2$ is the image of g_1, then

$$f_1 + g_1 \mapsto f_2 + g_2,$$
$$af_1 \mapsto af_2, \qquad a \in \mathbb{C}^1,$$
$$\langle f_1 \mid g_1 \rangle_1 = \langle f_2 \mid g_2 \rangle_2 .$$

A mapping having the above properties is called a *unitary* transformation of $\mathscr{E}_1$ onto $\mathscr{E}_2$.

Theorem 2.5. All complex Euclidean n-dimensional spaces are isomorphic to $l^2(n)$, and consequently (see Exercise 2.8) mutually isomorphic.

Proof. If $\mathscr{E}$ is an n-dimensional Euclidean space, there is according to Theorem 1.2 a set of n vectors $f_1, \ldots, f_n$ spanning $\mathscr{E}$. According to Theorem 2.4, we can find an orthonormal system of n vectors $e_1, \ldots, e_n$ which also spans $\mathscr{E}$. It is easy to check (see Exercise 2.7) that the mapping

$$f \leftrightarrow \begin{pmatrix} a_1 \\ \vdots \\ a_n \end{pmatrix}, \qquad a_1 = \langle e_1 \mid f \rangle, \ldots, a_n = \langle e_n \mid f \rangle, \tag{2.5}$$

provides an isomorphism between $\mathscr{E}$ and $l^2(n)$. Q.E.D.

Obviously, a similar theorem can be proved for real Euclidean spaces.

Theorem 2.6. A unitary transformation

$$f_1 \mapsto f_2, \qquad f_1 \in \mathscr{E}_1, \qquad f_2 \in \mathscr{E}_2, \tag{2.6}$$

of the Euclidean space $\mathscr{E}_1$ onto the Euclidean space $\mathscr{E}_2$ has a unique inverse mapping which is a unitary transformation of $\mathscr{E}_2$ onto $\mathscr{E}_1$.

Proof. We note that since

$$\|f_1 - g_1\|_1 = \|f_2 - g_2\|_2,$$

the images f_2 and g_2 of f_1 and g_1, respectively, are distinct whenever $f_1 \neq g_1$. Since the unitary map of $\mathscr{E}_1$ is onto $\mathscr{E}_2$, we conclude that the inverse of the mapping (2.6) exists.

We leave to the reader the details of the remainder of the proof.

Exercises

2.1. Show that for a finite interval I

$$\langle f \mid g \rangle = \int_I f^*(x)\, g(x)\, dx$$

is an inner product on the vector space $\mathscr{C}^0(I)$.

2.2. Show that the vector space $\mathscr{C}^0_{(2)}(\mathbb{R}^1)$ introduced in Section 2.1 is a subspace of the vector space $\mathscr{C}^0(\mathbb{R}^1)$.

2.3. Prove that (2.2) is an inner product in $\mathscr{C}^0_{(2)}(\mathbb{R}^1)$.

2.4. Show that

$$|\langle f \mid g \rangle|^2 = \langle f \mid f \rangle \langle g \mid g \rangle,$$
$$\|f + g\| = \|f\| + \|g\|,$$

if and only if either f is a multiple of g, i.e., if $f = ag$, $a \in \mathbb{C}^1$, or $g = \mathbf{0}$, and if in addition $a \geqslant 0$ in case of the second relation.

2.5. Show that if T is an orthonormal system of vectors, then all the vectors in T are necessarily linearly independent.

2.6. Prove that a subspace of a Euclidean space is also a Euclidean space.

2.7. Show that the mapping (2.5) is a mapping of $\mathscr{E}$ *onto* $l^2(n)$, and that it satisfies the requirements of isomorphism given in Definition 2.4.

2.8. Show that the relation of isomorphism of inner-product spaces is an equivalence relation, i.e., it is (see Exercise 1.6) reflexive, symmetric, and transitive.

3. Metric Spaces

3.1. Convergence in Metric Spaces

In an n-dimensional Euclidean space $\mathscr{E}$ we can always find, due to Theorems 1.2 and 2.4, a basis of n vectors $e_1, \ldots, e_n$ which constitute an orthonormal system. We can then expand any vector f of $\mathscr{E}$ in that basis

$$f = \sum_{k=1}^{n} a_k e_k \,. \tag{3.1}$$

We easily see that $a_k = \langle e_k \mid f \rangle$.

In an infinite-dimensional Euclidean space not every vector can be expanded in general in terms of a finite number of vectors. We can hope, however, to replace (3.1) with the formula

$$f = \sum_{k=1}^{\infty} a_k e_k \,,$$

but then we meet with the problem of giving a precise meaning to the convergence of the above series. This problem is solved in its most general form in topology, but for our purposes it will be sufficient to solve it within the context of metric spaces.

Definition 3.1. If S is a given set, a function $d(\xi, \eta)$ on $S \times S$ is a *metric* (or *distance function*) if it fulfills the following requirements for any $\xi, \eta, \zeta \in S$:

(1) $d(\xi, \eta) > 0$ if $\xi \neq \eta$,

(2) $d(\xi, \xi) = 0$,

(3) $d(\xi, \eta) = d(\eta, \xi)$,

(4) $d(\xi, \zeta) \leqslant d(\xi, \eta) + d(\eta, \zeta)$ (triangle inequality).

A set S on which a metric is defined is called a *metric space*.

A metric space does not have to be a linear space. For instance, a bounded open domain in the plane becomes a metric space if the metric is taken to be the distance between each pair of points belonging to that domain; such a domain obviously is not closed under the operations of adding vectors in the plane, but it provides a metric space.

Generalizing from the case of one-, two-, or three-dimensional real Euclidean spaces, we introduce the following notions.

Definition 3.2. An infinite sequence $\xi_1, \xi_2, \ldots$ in a metric space $\mathscr{M}$ is said to converge to the point $\xi \in \mathscr{M}$ if for any $\epsilon > 0$ there is a positive number $N(\epsilon)$ such that $d(\xi, \xi_n) < \epsilon$ for all $n > N(\epsilon)$. An infinite sequence $\xi_1, \xi_2, \ldots$ is called a *Cauchy sequence* (or a *fundamental sequence*) if for any $\epsilon > 0$ there is a positive number $M(\epsilon)$ such that $d(\xi_m, \xi_n) < \epsilon$ for all $m, n > M(\epsilon)$.

Theorem 3.1. If a sequence $\xi_1, \xi_2, \ldots$ in a metric space $\mathcal{M}$ converges to some $\xi \in \mathcal{M}$, then its limit ξ is unique, and the sequence is a Cauchy sequence.

Proof. If $\xi_1, \xi_2, \ldots$ converges to $\xi \in \mathcal{M}$ and to $\eta \in \mathcal{M}$, then by definition, for any $\epsilon > 0$ there are $N_1(\epsilon)$ and $N_2(\epsilon)$ such that $d(\xi, \xi_n) < \epsilon$ for $n > N_1(\epsilon)$ and $d(\eta, \xi_n) < \epsilon$ for $n > N_2(\epsilon)$. Consequently, for $n > \max(N_1(\epsilon), N_2(\epsilon))$ we get by applying the triangle inequality of Definition 3.1, Point 4,

$$d(\xi, \eta) \leqslant d(\xi, \xi_n) + d(\xi_n, \eta) < 2\epsilon.$$

As $\epsilon > 0$ can be chosen arbitrarily small, we get $d(\xi, \eta) = 0$, which, according to Definition 3.1, can be true only if $\xi = \eta$.

Similarly we get

$$d(\xi_m, \xi_n) \leqslant d(\xi_m, \xi) + d(\xi, \xi_n) < \epsilon$$

if $m, n > N_1(\epsilon/2)$; i.e., the sequence $\xi_1, \xi_2, \ldots$ is also a Cauchy sequence.

Q.E.D.

3.2. Complete Metric Spaces

In case of sequences of real numbers, every Cauchy sequence is convergent, i.e., the set $\mathbb{R}^1$ of all real numbers is complete. We state this generally in Definition 3.3.

Definition 3.3. A metric space $\mathcal{M}$ is *complete* if every Cauchy sequence converges to an element of $\mathcal{M}$.

Not every metric space is complete, as exemplified by the set $\mathfrak{R}$ of all rational numbers with the metric $d(m_1/n_1, m_2/n_2) = | m_1/n_1 - m_2/n_2 |$, which is incomplete. However, we know that the set $\mathfrak{R}$ is everywhere dense in the set $\mathbb{R}^1$; we state this generally as follows:

Definition 3.4. A subset S of a metric space $\mathcal{M}$ is (*everywhere*) *dense* in $\mathcal{M}$ if for any given $\epsilon > 0$ and any $\xi \in \mathcal{M}$ there is an element η belonging to S for which $d(\xi, \eta) < \epsilon$.

We can reexpress the above definition after introducing a few topological concepts, generalized from the case of sets in one, two, or three real dimensions.

Definition 3.5. If ξ is an element of a metric space $\mathcal{M}$, then the set of all points η satisfying the inequality $d(\xi, \eta) < \epsilon$ for some $\epsilon < 0$ is called the ϵ *neighborhood* of ξ. If S is a subset of $\mathcal{M}$, a point $\zeta \in \mathcal{M}$ is called an *accumulation* (or *cluster* or *limit*) *point* of S if every ϵ neighborhood of ζ contains a point of S. The set $\bar{S}$ consisting of all the cluster points of S is called the *closure* of S. Obviously always $S \subset \bar{S}$; if $S = \bar{S}$ then S is called a *closed* set.

We say that the subset S of a metric space $\mathcal{M}$ is (*everywhere*) *dense* in $\mathcal{M}$ if and only if $\mathcal{M}$ is its closure, i.e., if and only if $\bar{S} = \mathcal{M}$.

The procedure of completing the set $\mathfrak{R}$ of rational numbers by embedding it in the set of all real numbers can be generalized.

Definition 3.6. A metric space $\mathcal{M}$ is said to be *densely embedded* in the metric space $\tilde{\mathcal{M}}$ if there is an isometric mapping of $\mathcal{M}$ into $\tilde{\mathcal{M}}$, and if the image set $\mathcal{M}'$ of $\mathcal{M}$ in $\tilde{\mathcal{M}}$ is everywhere dense in $\tilde{\mathcal{M}}$.

A one-to-one mapping $\xi \leftrightarrow \tilde{\xi}$ of a metric space $\mathcal{M}$ into another metric space $\tilde{\mathcal{M}}$ is called *isometric* if it preserves distances, i.e., if $d_1(\xi, \eta) = d_2(\tilde{\xi}, \tilde{\eta})$ for $\xi, \eta \in \mathcal{M}$ and $\tilde{\xi}, \tilde{\eta} \in \tilde{\mathcal{M}}$ whenever $\xi \leftrightarrow \tilde{\xi}$ and $\eta \leftrightarrow \tilde{\eta}$.

3.3. Completion of a Metric Space

***Theorem 3.2.** Every incomplete metric space $\mathcal{M}$ can be embedded in a complete metric space $\tilde{\mathcal{M}}$, called the *completion* of $\mathcal{M}$.

The proof of this theorem can be given by generalizing Cantor's construction, by which one builds the set of real numbers from the rational numbers.

Denote by $\tilde{\mathcal{M}}_s$ the family of all Cauchy sequences in $\mathcal{M}$. If $\tilde{\xi}' = \{\xi_1', \xi_2', \ldots\}$ and $\tilde{\xi}'' = \{\xi_1'', \xi_2'', \ldots\}$ are two such sequences, we say that they are equivalent if and only if

$$\lim_{n\to\infty} d(\xi_n', \xi_n'') = 0. \tag{3.2}$$

It is easy to see that we have thus introduced an equivalence relation in $\tilde{\mathcal{M}}_s$ (see Exercise 3.1) if we recall (see Exercises 1.6 and 2.8) the general definition of an equivalence relation.

Definition 3.7. A relation $\xi \sim \eta$ holding between any two ordered elements of a set S is called an *equivalence relation* if it is

(1) reflexive: $\xi \sim \xi$ for all $\xi \in S$;

(2) symmetric: $\xi \sim \eta$ implies that $\eta \sim \xi$;

(3) transitive: $\xi \sim \eta$ and $\eta \sim \zeta$ implies that $\xi \sim \zeta$.

A subset X of S having the property that all the elements of X are equivalent and that if $\eta \sim \xi$ and $\xi \in X$ then $\eta \in X$ is called an *equivalence class* (with respect to the equivalence relation $\sim$).

We denote the family of all equivalence classes in $\tilde{\mathscr{M}}_s$ [with respect to the equivalence relation given by (3.2)] by the symbol $\tilde{\mathscr{M}}$, and agree to denote the equivalence class containing the Cauchy sequence $\tilde{\xi}$ also by $\tilde{\xi}$. Consequently if $\tilde{\xi}', \tilde{\xi}'' \in \tilde{\mathscr{M}}$, then $\tilde{\xi}' = \tilde{\xi}''$ if and only if the Cauchy sequences $\tilde{\xi}', \tilde{\xi}'' \in \tilde{\mathscr{M}}_s$ are equivalent, i.e., satisfy (3.2).

We introduce the real function $d_s(\tilde{\xi}, \tilde{\eta})$ on $\tilde{\mathscr{M}}_s \times \tilde{\mathscr{M}}_s$ by defining for $\tilde{\xi} = \{\xi_1, \xi_2, ...\}$ and $\tilde{\eta} = \{\eta_1, \eta_2, ...\}$

$$d_s(\tilde{\xi}, \tilde{\eta}) = \lim_{n\to\infty} d(\xi_n, \eta_n). \tag{3.3}$$

In order to see that the above limit exists for any $\tilde{\xi}, \tilde{\eta} \in \tilde{\mathscr{M}}_s$ we employ the relation (see Exercise 3.2)

$$| d(\xi_m, \eta_m) - d(\xi_n, \eta_n)| \leqslant d(\xi_m, \xi_n) + d(\eta_m, \eta_n) \tag{3.4}$$

to show that $d(\xi_1, \eta_1), d(\xi_2, \eta_2), ...$ is a Cauchy sequence of numbers, and therefore has a limit; namely as $\xi_1, \xi_2, ...$ and $\eta_1, \eta_2, ...$ are Cauchy sequences, we can make $d(\xi_m, \xi_n) < \epsilon$ if $m, n > N_1(\epsilon)$, and $d(\eta_m, \eta_n) < \epsilon$ if $m, n > N_2(\epsilon)$, which, used in conjunction with (3.4), proves the statement.

We can show that $d_s(\tilde{\xi}, \tilde{\eta})$ also defines a real function on $\tilde{\mathscr{M}} \times \tilde{\mathscr{M}}$ by establishing that $d_s(\tilde{\xi}', \tilde{\eta}') = d_s(\tilde{\xi}'', \tilde{\eta}'')$ if $\tilde{\xi}' = \tilde{\xi}''$ and $\tilde{\eta}' = \tilde{\eta}''$ for $\tilde{\xi}', \tilde{\xi}'', \tilde{\eta}', \tilde{\eta}'' \in \tilde{\mathscr{M}}$. We first obtain that $d_s(\tilde{\xi}', \tilde{\eta}') = d_s(\tilde{\xi}'', \tilde{\eta}')$ from the inequality (see Exercise 3.3)

$$| d(\xi_n', \eta_n') - d(\xi_n'', \eta_n')| \leqslant d(\xi_n', \xi_n'')$$

because $d(\xi_n', \xi_n'') \to 0$ as $n \to \infty$ due to the fact that the Cauchy sequences $\tilde{\xi}'$ and $\tilde{\xi}''$ belong to the same equivalence class. Similarly we can show that $d_s(\tilde{\xi}'', \tilde{\eta}') = d_s(\tilde{\xi}'', \tilde{\eta}'')$, and thus prove that $d_s(\tilde{\xi}', \tilde{\eta}') = d_s(\tilde{\xi}'', \tilde{\eta}'')$.

It is easy to check that the function $d_s(\tilde{\xi}, \tilde{\eta})$ defines a metric on $\tilde{\mathscr{M}}$ (see Exercise 3.4). We show now that the ensuing metric space, which we denote also by $\tilde{\mathscr{M}}$, is complete.

Assume that $\tilde{\xi}^{(1)}, \tilde{\xi}^{(2)}, ...$ is a Cauchy sequence in $\tilde{\mathscr{M}}$, where $\tilde{\xi}^{(k)}$ is the equivalence class containing the Cauchy sequence $\{\xi_1^{(k)}, \xi_2^{(k)}, ...\}$ of elements of $\mathscr{M}$. Choose for each integer k an element $\eta_k = \xi_n^{(k)} \in \mathscr{M}$ such that $d(\xi_m^{(k)}, \eta_k) = d(\xi_m^{(k)}, \xi_n^{(k)}) < 1/k$ for all m greater than some N_k; this is certainly possible because $\xi_1^{(k)}, \xi_2^{(k)}, ...$ is a Cauchy sequence in $\mathscr{M}$.

Consider now the elements $\tilde{\eta}_k = \{\eta_k, \eta_k, ...\}$ and $\tilde{\xi}_m^{(k)} = \{\xi_m^{(k)}, \xi_m^{(k)}, ...\}$ of $\tilde{\mathscr{M}}_s$. We obviously have

$$d_s(\tilde{\xi}_m^{(k)}, \tilde{\eta}_k) = d(\xi_m^{(k)}, \eta_k) < 1/k.$$

If we let in the above relation $m \to \infty$, then we find that $d_s(\tilde{\xi}_m^{(k)}, \tilde{\eta}_k) \to d_s(\tilde{\xi}^{(k)}, \tilde{\eta}_k)$ since $\lim d_s(\tilde{\xi}_m^{(k)}, \tilde{\xi}^{(k)}) = 0$ as $m \to \infty$ (see Exercises 3.5 and 3.6) and consequently

$$d_s(\tilde{\xi}^{(k)}, \tilde{\eta}_k) \leqslant 1/k.$$

We can now deduce that $\tilde{\eta} = \{\eta_1, \eta_2, ...\}$ is a Cauchy sequence in $\mathscr{M}$ by writing

$$\begin{aligned} (3.5) \qquad d(\eta_m, \eta_n) &= d_s(\tilde{\eta}_m, \tilde{\eta}_n) \\ &\leqslant d_s(\tilde{\eta}_m, \tilde{\xi}^{(m)}) + d_s(\tilde{\xi}^{(m)}, \tilde{\xi}^{(n)}) + d_s(\tilde{\xi}^{(n)}, \tilde{\eta}_m) \\ &\leqslant \frac{1}{m} + d_s(\tilde{\xi}^{(m)}, \tilde{\xi}^{(n)}) + \frac{1}{n}. \end{aligned}$$

Since $\tilde{\xi}^{(1)}, \tilde{\xi}^{(2)}, ...$ is a Cauchy sequence in $\tilde{\mathscr{M}}$, we can make $d_s(\tilde{\xi}^{(m)}, \tilde{\xi}^{(n)})$, and consequently the entire right-hand side of (3.5) arbitrarily small for all sufficiently large m and n. Thus, $\tilde{\eta} \in \tilde{\mathscr{M}}_s$.

We can establish that the equivalence class $\tilde{\eta} \in \tilde{\mathscr{M}}$ containing the Cauchy sequence $\tilde{\eta} = \{\eta_1, \eta_2, ...\}$ is the limit of $\tilde{\xi}^{(1)}, \tilde{\xi}^{(2)}, ...$ if we write

$$(3.6) \qquad d_s(\tilde{\eta}, \tilde{\xi}^{(k)}) \leqslant d_s(\tilde{\eta}, \tilde{\eta}_k) + d_s(\tilde{\eta}_k, \tilde{\xi}^{(k)}).$$

The right-hand side of (3.6) can be made arbitrarily small for sufficiently large k because $d_s(\tilde{\eta}_k, \tilde{\xi}^{(k)}) \leqslant 1/k$ and $\lim_{k\to\infty} d_s(\tilde{\eta}, \tilde{\eta}_k) = 0$ (see Exercise 3.6).

In order to finish the proof of the theorem, we have to embed $\mathscr{M}$ into the complete metric space $\tilde{\mathscr{M}}$. To that purpose we map $\xi \in \mathscr{M}$ into the equivalence class $\tilde{\xi}$ containing the sequence $\{\xi, \xi, ...\}$. This mapping is obviously one-to-one and isometric, as $d(\xi, \eta) = d_s(\tilde{\xi}, \tilde{\eta})$. Furthermore, the image $\mathscr{M}'$ of $\mathscr{M}$ in $\tilde{\mathscr{M}}$ is everywhere dense in $\tilde{\mathscr{M}}$; namely if $\tilde{\eta} \in \tilde{\mathscr{M}}$ contains $\{\eta_1, \eta_2, ...\} \in \tilde{\mathscr{M}}_s$, then for arbitrary $\epsilon > 0$ we can choose an $\tilde{\eta}_k$ in $\mathscr{M}'$ containing $\{\eta_k, \eta_k, ...\}$ and such that $d_s(\tilde{\eta}, \tilde{\eta}_k) < \epsilon$.

Exercises

3.1. Show that the relation $\tilde{\xi} \sim \tilde{\eta}$ between any two Cauchy sequences $\tilde{\xi} = \{\xi_1, \xi_2, ...\}$ and $\tilde{\eta} = \{\eta_1, \eta_2, ...\}$ of a metric space $\mathscr{M}$, defined to mean that $\lim_{n\to\infty} d(\xi_n, \eta_n) = 0$, satisfies the three requirements given in Definition 3.7 for an equivalence relation.

3.2. Prove that any four elements $\xi_1, \xi_2, \eta_1, \eta_2$ of a metric space $\mathscr{M}$ satisfy the relation

$$| d(\xi_1, \xi_2) - d(\eta_1, \eta_2)| \leqslant d(\xi_1, \eta_1) + d(\xi_2, \eta_2).$$

3.3. Prove that if ξ, η, ζ are elements of a metric space $\mathscr{M}$, then

$$| d(\xi, \eta) - d(\xi, \zeta)| \leqslant d(\eta, \zeta).$$

3.4. Show that the function $d_s(\tilde{\xi}, \tilde{\eta})$ defined on $\tilde{\mathscr{M}} \times \tilde{\mathscr{M}}$ by (3.3) satisfies the four requirements for a metric (those requirements are formulated in Definition 3.1).

3.5. If in a metric space $\mathscr{M}$ the sequence $\xi_1, \xi_2, \ldots$ converges to ξ, prove that for any $\eta \in \mathscr{M}$, $\lim_{n\to\infty} d(\xi_n, \eta) = d(\xi, \eta)$.

3.6. If $\tilde{\xi}$ is the equivalence class of $\tilde{\mathscr{M}}$ (introduced in Theorem 3.2) containing the Cauchy sequence $\{\xi_1, \xi_2, \ldots\} \in \tilde{\mathscr{M}}_s$, then for any $\epsilon > 0$ there is an $N(\epsilon)$ such that $d_s(\tilde{\xi}, \tilde{\xi}_k) < \epsilon$ for all $k > N(\epsilon)$, where $\tilde{\xi}_k = \{\xi_k, \xi_k, \ldots\}$. Prove this statement!

3.7. Show that if S_1 is an everywhere dense subset of a metric space $\mathscr{M}$, and S_2 is an everywhere dense subset of S_1, then S_2 is everywhere dense in $\mathscr{M}$.

4. Hilbert Space

4.1. Completion of a Euclidean Space

It is easy to establish (see Exercise 4.1) that in normed space $\mathscr{N}$

$$d(f, g) = \|f - g\| \tag{4.1}$$

is a metric. Therefore, we can define in $\mathscr{N}$ convergence, completeness, etc. in the metric (4.1), which is then called *convergence, completeness*, etc. *in the norm.* A complete normed space bears the name of *Banach space.*

The above concepts can also be applied to Euclidean spaces, because according to Theorem 2.3 we can introduce in such spaces a norm, and therefore also a metric. A Euclidean space which is complete in the norm* is called a *Hilbert space.*

Not every Euclidean space is a Hilbert space. For instance, the

* The concept of completeness can be defined and considered for other topologies besides the norm topology.

Euclidean space $\mathscr{C}^0_{(2)}(\mathbb{R}^1)$ introduced in §2 is not complete. To see this, note that the sequence $f_1, f_2, \ldots$ of continuous functions

$$f_n(x) = \begin{cases} 1 & \text{for } |x| \leqslant a \\ \exp[-n^2(|x| - a)^2] & \text{for } |x| > a, \end{cases} \tag{4.2}$$

is a Cauchy sequence in $\mathscr{C}^0_{(2)}(\mathbb{R}^1)$ (see Exercise 4.2) but it does not converge to an element of $\mathscr{C}^0_{(2)}(\mathbb{R}^1)$. In fact, it is easy to establish that with increasing n, the functions in the above sequence approximate arbitrarily closely in norm the discontinuous step function

$$\chi(x) = \begin{cases} 1 & \text{for } |x| \leqslant a \\ 0 & \text{for } |x| > a, \end{cases}$$

which, however, does not belong to $\mathscr{C}^0_{(2)}(\mathbb{R}^1)$.

Definition 4.1. We say that the Euclidean space $\mathscr{E}$ can be *densely embedded* in the Hilbert space $\mathscr{H}$ if there is a one-to-one mapping of $\mathscr{E}$ into $\mathscr{H}$, such that the image $\mathscr{E}'$ of $\mathscr{E}$ is everywhere dense in $\mathscr{H}$, and the mapping represents an isomorphism between the Euclidean spaces $\mathscr{E}$ and $\mathscr{E}'$.

Theorem 4.1. Any incomplete Euclidean space $\mathscr{E}$ can be densely embedded in a Hilbert space.

Proof. Denote by $\tilde{\mathscr{E}}$ the complete metric space built from the set $\tilde{\mathscr{E}}_s$ of Cauchy sequences in $\mathscr{E}$ according to the procedure used in proving Theorem 3.2. Define in $\tilde{\mathscr{E}}_s$ the operations

$$\begin{aligned} \tilde{f} + \tilde{g} &= \{f_1 + g_1, f_2 + g_2, \ldots\}, \\ a\tilde{f} &= \{af_1, af_2, \ldots\} \end{aligned} \tag{4.3}$$

for any two sequences $\tilde{f} = \{f_1, f_2, \ldots\}$, $\tilde{g} = \{g_1, g_2, \ldots\}$ from $\tilde{\mathscr{E}}_s$. It is easy to check that the above operations are operations of vector addition and multiplication by scalar. Furthermore, if $\tilde{f}' = \tilde{f}''$, where $\tilde{f}' = \{f_1', f_2', \ldots\}$ and $\tilde{f}'' = \{f_1'', f_2'', \ldots\}$, i.e., if $\tilde{f}'$ and $\tilde{f}''$ belong to the same equivalence class in $\tilde{\mathscr{E}}$ and therefore

$$\lim_{n\to\infty} \| f_n' - f_n'' \| = \lim_{n\to\infty} d(f_n', f_n'') = 0,$$

then $\| af_n' - af_n'' \| = |a| \, \| f_n' - f_n'' \| \to 0$; thus, we also have that $\tilde{f}' + \tilde{g} = \tilde{f}'' + \tilde{g}$ and $a\tilde{f}' = a\tilde{f}''$. Consequently, (4.3) defines vector operations on $\tilde{\mathscr{E}}$, which thus becomes a vector space.

We now introduce the complex function on $\tilde{\mathscr{E}}_s \times \tilde{\mathscr{E}}_s$ defined by

$$\langle \tilde{f} \mid \tilde{g} \rangle_s = \lim_{n\to\infty} \langle f_n \mid g_n \rangle. \tag{4.4}$$

The limit in (4.4) exists because $\langle f_1 \mid g_1 \rangle$, $\langle f_2 \mid g_2 \rangle, \ldots$ is a Cauchy sequence of numbers, as can be seen from the following inequality:

$$\begin{aligned} |\langle f_m \mid g_m \rangle - \langle f_n \mid g_n \rangle| &= |\langle f_m - f_n \mid g_m \rangle + \langle f_n \mid g_m - g_n \rangle| \\ &\leqslant \| f_m - f_n \| \, \| g_m \| + \| f_n \| \, \| g_m - g_n \|; \end{aligned}$$

namely, $\| f_n \| \to d_s(\tilde{f}, 0) = \| \tilde{f} \|_s$ and $\| g_n \| \to \| \tilde{g} \|_s$ for $n \to \infty$, where

$$\| \tilde{f} \|_s = \sqrt{\langle \tilde{f} \mid \tilde{f} \rangle_s},$$

while $\| f_m - f_n \|$ and $\| g_m - g_n \|$ can be made arbitrarily small for sufficiently large m and n.

Furthermore, if $\tilde{f}' = \tilde{f}'' \in \tilde{\mathscr{E}}_s$, then $\langle \tilde{f}' \mid \tilde{g} \rangle_s = \langle \tilde{f}'' \mid \tilde{g} \rangle_s$, as can be concluded from the inequality

$$|\langle f_n' \mid g_n \rangle - \langle f_n'' \mid g_n \rangle| \leqslant \| f_n' - f_n'' \| \, \| g_n \|.$$

Hence, $\langle \tilde{f} \mid \tilde{g} \rangle_s$ is a uniquely defined function on $\tilde{\mathscr{E}} \times \tilde{\mathscr{E}}$. This function determines an inner product on $\tilde{\mathscr{E}}$ (see Exercise 4.4). It is obvious that the mapping $f \leftrightarrow \tilde{f} = \{f, f, \ldots\}$ of $\mathscr{E}$ into $\tilde{\mathscr{E}}$ has an image $\mathscr{E}'$ which is a linear subspace of $\tilde{\mathscr{E}}$; according to the construction, $\mathscr{E}'$ is everywhere dense in $\tilde{\mathscr{E}}$ and the above mapping provides an isomorphism between $\mathscr{E}$ and $\mathscr{E}'$. Q.E.D.

A similar theorem can be proved for normed spaces (see Exercise 4.5).

4.2. Separable Hilbert Spaces

In quantum mechanics we deal at present almost exclusively with a special class of Hilbert spaces which are called separable.

Definition 4.2. The Euclidean space $\mathscr{E}$ is called *separable* if there is a *countable* everywhere dense subset of vectors of $\mathscr{E}$.

In the early days of research on Hilbert spaces, separability was taken to be an integral part of the concept of a Hilbert space.

In quantum mechanics we are concerned primarily with separable complex Hilbert spaces. We shall agree that in the future whenever we refer to a space as a Hilbert space we mean a complex Hilbert space, except if otherwise stated.

Theorem 4.2. Every subspace of a separable Euclidean space is a separable Euclidean space.

Proof. The fact that a subspace $\mathscr{E}_1$ of a Euclidean space $\mathscr{E}$ is also a Euclidean space is easy to check (see Exercise 2.6). In order to establish the separability of $\mathscr{E}_1$, construct a countable subset $S = \{g_{11}, g_{12}, g_{21}, g_{13}, ...\}$ of $\mathscr{E}_1$ in the following way.

Let $R = \{f_1, f_2, ...\}$ be a countable everywhere dense subset of $\mathscr{E}$; there has to be such a set because of the separability of $\mathscr{E}$. Let g_{mn} denote a vector of $\mathscr{E}_1$ satisfying $\| g_{mn} - f_n \| < 1/m$, in case there is at least one such vector, or the zero vector in case there is no vector of $\mathscr{E}_1$ in the $1/m$ neighborhood of f_n.

The set S is everywhere dense in $\mathscr{E}_1$ because for any given $h \in \mathscr{E}_1$ and $m > 0$ we can find an $f_n \in R$ such that $\| h - f_n \| < 1/m$. Thus, by the above rule of constructing S we certainly have $g_{mn} \neq \mathbf{0}$ and therefore

$$\| h - g_{mn} \| \leqslant \| h - f_n \| + \| f_n - g_{nm} \| < 2/m.$$

This proves that S is everywhere dense in $\mathscr{E}_1$. Q.E.D.

4.3. l^2 Spaces as Examples of Separable Hilbert Spaces

As an important example of an infinite-dimensional separable Hilbert space we give the space $l^2(\infty)$, which is basic in matrix mechanics.

Theorem 4.3. The set $l^2(\infty)$ of all one-column complex matrices α with a countable number of elements

$$\alpha = \begin{pmatrix} a_1 \\ a_2 \\ \vdots \end{pmatrix}$$

for which

$$\sum_{k=1}^{\infty} | a_k |^2 < +\infty \tag{4.5}$$

becomes a separable Hilbert space, denoted by $l^2(\infty)$, if the vector operations are defined by

$$\alpha + \beta = \begin{pmatrix} a_1 + b_1 \\ a_2 + b_2 \\ \vdots \end{pmatrix}, \tag{4.6}$$

$$a\alpha = \begin{pmatrix} aa_1 \\ aa_2 \\ \vdots \end{pmatrix}, \qquad a \in \mathbb{C}^1, \tag{4.7}$$

and the inner product by

$$\langle \alpha \mid \beta \rangle = \sum_{k=1}^{\infty} a_k{}^* b_k . \tag{4.8}$$

Proof. The operation (4.7) maps $\mathbb{C}^1 \times l^2(\infty)$ into $l^2(\infty)$ because $\sum_{k=1}^{\infty} | aa_k |^2 = | a |^2 \sum_{k=1}^{\infty} | a_k |^2 < +\infty$ if (4.5) is satisfied.

In order to see that (4.6) maps $l^2(\infty) \times l^2(\infty)$ into $l^2(\infty)$, apply the triangle inequality on the ν-dimensional space $l^2(\nu)$, $\nu < +\infty$, in order to obtain

$$\left[\sum_{k=1}^{\nu} | a_k + b_k |^2\right]^{1/2} \leqslant \left[\sum_{k=1}^{\nu} | a_k |^2\right]^{1/2} + \left[\sum_{k=l}^{\nu} | b_k |^2\right]^{1/2}.$$

The above inequality shows that when $\nu \to \infty$, the left-hand side converges if $\alpha, \beta \in l^2(\infty)$.

Similarly, we prove that (4.8) converges absolutely by applying the Schwarz–Cauchy inequality on the ν-dimensional space $l^2(\nu)$, $\nu < +\infty$, in order to obtain

$$\sum_{k=1}^{\nu} | a_k^* b_k | = \sum_{k=1}^{\nu} | a_k | \, | b_k | \leqslant \left[\sum_{k=1}^{\nu} | a_k |^2\right]^{1/2} \left[\sum_{k=1}^{\nu} | b_k |^2\right]^{1/2}.$$

We leave as an exercise for the reader (see Exercise 4.6) to check that we deal indeed with a Euclidean space.

To prove that this Euclidean space is complete, assume that $\alpha^{(1)}, \alpha^{(2)}, \ldots$ is a Cauchy sequence, where

$$\alpha^{(n)} = \begin{pmatrix} a_1^{(n)} \\ a_2^{(n)} \\ \vdots \end{pmatrix}.$$

As we obviously have that for any $k = 1, 2, \ldots$

$$| a_k^{(m)} - a_k^{(n)} | \leqslant \| \alpha^{(m)} - \alpha^{(n)} \|,$$

we deduce that for fixed k the sequence $a_k^{(1)}, a_k^{(2)}, \ldots$ is a Cauchy sequence of complex numbers; hence, this sequence has a limit b_k. We shall prove that the one-column infinite matrix

$$\beta = \begin{pmatrix} b_1 \\ b_2 \\ \vdots \end{pmatrix}$$

is an element of $l^2(\infty)$, and that $\alpha^{(1)}, \alpha^{(2)}, \ldots$ converges in the norm to β.

By applying again the triangle inequality on the ν-dimensional space $l^2(\nu)$, $\nu < +\infty$, we obtain

(4.9)
$$\left[\sum_{k=1}^{\nu} | b_k - a_k^{(n)} |^2\right]^{1/2} \leqslant \left[\sum_{k=1}^{\nu} | b_k - a_k^{(m)} |^2\right]^{1/2} + \left[\sum_{k=1}^{\nu} | a_k^{(m)} - a_k^{(n)} |^2\right]^{1/2};$$

the above inequality is true for any $m = 1, 2, \ldots$. As $\alpha^{(1)}, \alpha^{(2)}, \ldots$ is a Cauchy sequence, there is for any given $\epsilon > 0$ an $N_0(\epsilon)$ such that for any $m, n > N_0(\epsilon)$ and any positive integer ν

$$\sum_{k=1}^{\nu} | a_k^{(m)} - a_k^{(n)} |^2 \leqslant \| \alpha^{(m)} - \alpha^{(n)} \|^2 < \epsilon^2/4.$$

On the other hand, as $b_k = \lim_{m\to\infty} a_k^{(m)}$, we can find for fixed ν an $N_\nu(\epsilon)$ such that $| b_k - a_k^{(m)} | < \epsilon/2^{(k+1)/2}$ for any $m > N_\nu(\epsilon)$ and all $k = 1, 2, \ldots, \nu$. Thus, we get from (4.9) that for all $n > N_0(\epsilon)$ and all positive integers ν

$$\left[\sum_{k=1}^{\nu} | b_k - a_k^{(n)} |^2\right]^{1/2} \leqslant \epsilon \left(\sum_{k=1}^{\nu} \frac{1}{2^{k+1}}\right) + \frac{\epsilon}{2} \leqslant \frac{\epsilon}{2}\left(\sum_{k=1}^{\infty} \frac{1}{2^k}\right) + \frac{\epsilon}{2} = \epsilon. \tag{4.10}$$

As the right-hand side of the above inequality is independent of ν and the inequality itself is true for any $n > N_0(\epsilon)$, we can let $\nu \to \infty$ in (4.10) to derive

$$\left[\sum_{k=1}^{\infty} | b_k - a_k^{(n)} |^2\right]^{1/2} \leqslant \epsilon \qquad \text{for all} \quad n > N_0(\epsilon). \tag{4.11}$$

By returning again to $l^2(\nu)$, $\nu < +\infty$, to obtain

$$\left[\sum_{k=1}^{\nu} | b_k |^2\right]^{1/2} \leqslant \left[\sum_{k=1}^{\nu} | b_k - a_k^{(n)} |^2\right]^{1/2} + \left[\sum_{k=1}^{\nu} | a_k^{(n)} |^2\right]^{1/2}$$

we establish that $\beta \in l^2(\infty)$ by letting $\nu \to \infty$. The relation (4.11) tells us now that $\alpha^{(1)}, \alpha^{(2)}, \ldots$ converges to β.

Finally, in order to prove the separability of $l^2(\infty)$, consider the set D of all the one-column matrices α from $l^2(+\infty)$ with kth components $(\alpha)_k = a_k$, where a_k has rational numbers as its real and imaginary part, i.e., $\operatorname{Re} a_k$, $\operatorname{Im} a_k \in \mathfrak{R}$, $k = 1, 2, \ldots$, and in addition

$$a_{n+1} = a_{n+2} = \cdots = 0 \tag{4.12}$$

for some integer n. The set D is countable (see Exercise 4.7). In order to prove that D is everywhere dense in $l^2(\infty)$, take any one-column matrix $\gamma \in l^2(+\infty)$ with kth component $(\gamma)_k = c_k$. As $\gamma \in l^2(\infty)$, there is for any given $\epsilon > 0$ an integer n such that

$$\sum_{k=n+1}^{\infty} | c_k |^2 < \epsilon^2/2.$$

Furthermore, as the set $\mathfrak{R}$ of rational numbers is dense in the set $\mathbb{R}^1$ of real numbers, we can choose an $\alpha \in D$ which satisfies (4.12) and is such that $| c_k - a_k | < \epsilon/\sqrt{2n}$ for all $k = 1,..., n$. Thus, we have

$$\| \gamma - \alpha \| = \left[\sum_{k=1}^{n} | c_k - a_k |^2 + \sum_{k=n+1}^{\infty} | c_k |^2 \right]^{1/2} < \epsilon,$$

which proves that $l^2(\infty)$ is separable. Q.E.D.

4.4. Orthonormal Bases in Hilbert Space

In an infinite-dimensional Euclidean space it is important to distinguish between the vector space (S) spanned by a set S, and the *closed* vector space $[S]$ spanned by S.

Definition 4.3. The *vector space* (*or linear manifold*) (S) *spanned by the subset* S of a Euclidean space $\mathscr{E}$ is the smallest* subspace of $\mathscr{E}$ containing S. The *closed vector subspace* $[S]$ *spanned by* S is the smallest closed vector subspace of $\mathscr{E}$ containing S.

In the finite-dimensional case $(S) = [S]$ because all finite-dimensional Euclidean spaces are closed (see Exercise 4.8). That this is not so in the infinite-dimensional case can be deduced from the following theorem, whose simple proof we leave to the reader (see Exercise 4.9).

Theorem 4.4. The subspace (S) of the Euclidean space $\mathscr{E}$ spanned by the set S is identical with the set of all finite linear combinations $a_1 f_1 + \cdots + a_n f_n$ of vectors from S, i.e., in customary set-theoretical notation,

$$(S) = \{a_1 f_1 + \cdots + a_n f_n \colon \; f_1, ..., f_n \in S, \quad a_1, ..., a_n \in \mathbb{C}^1, \quad n = 1, 2, ...\}.$$

The closed linear subspace $[S]$ spanned by S is identical to the closure $\overline{(S)}$ of (S).

Definition 4.4. An orthonormal system S of vectors in a Euclidean space $\mathscr{E}$ is called an *orthonormal basis* (or a *complete orthonormal system*) in the Euclidean space $\mathscr{E}$ if the closed linear space $[S]$ spanned by S is identical to the entire Euclidean space, i.e., $[S] = \mathscr{E}$.

* That is, if $\mathscr{V}$ is a subspace of $\mathscr{E}$ and $S \subset \mathscr{V}$, then necessarily $(S) \subset \mathscr{V}$.

It must be realized that an orthonormal basis T in an infinite-dimensional Euclidean space $\mathscr{E}$ is not a *vector basis* for the vector space $\mathscr{E}$, i.e., (T) is in general different from $\mathscr{E}$. For instance, this is so with the basis $e_1, e_2, \ldots$ of $l^2(\infty)$, where e_m is the vector whose nth matrix component is $(e_m)_n = \delta_{mn}$.

Theorem 4.5. A Euclidean space $\mathscr{E}$ is separable if and only if there is a *countable* orthonormal basis in $\mathscr{E}$.

Proof. (a) To prove that in a separable Euclidean space $\mathscr{E}$ there is a countable orthonormal basis, note that due to the separability of $\mathscr{E}$ there is a countable set $S = \{f_1, f_2, \ldots\}$ which is everywhere dense in $\mathscr{E}$. According to Theorem 2.4, there is a countable orthonormal system $\mathsf{T} = \{e_1, e_2, \ldots\}$ such that $(S) = (\mathsf{T})$. Due to Theorem 4.4 we have then

$$[\mathsf{T}] = \overline{(\mathsf{T})} = \overline{(S)} = [S] = \mathscr{E}.$$

(b) Conversely, to show that if there is a countable orthonormal basis $\mathsf{T} = \{e_1, e_2, \ldots\}$, then $\mathscr{E}$ is separable, consider the set

$$R = \{r_1e_1 + \cdots + r_ne_n\colon \ \operatorname{Re} r_1, \ldots, \operatorname{Re} r_n, \operatorname{Im} r_1, \ldots, \operatorname{Im} r_n \in \mathfrak{R}, \quad n = 1, 2, \ldots\}$$

which is countable, as can be established by using the technique for solving Exercise 4.7. The set R is also everywhere dense in $\mathscr{E}$; namely, if $f \in \mathscr{E}$ and $\epsilon > 0$ is given, then as $[e_1, e_2, \ldots] = \mathscr{E}$, there is a vector $g = a_1e_1 + \cdots + a_ne_n$ such that

$$\| f - a_1e_1 - \cdots - a_ne_n \| < \epsilon/2.$$

Furthermore, we can choose complex numbers $r_1, \ldots, r_n$ with rational real and imaginary parts so that

$$| r_k - a_k | < \epsilon/2\sqrt{n}, \qquad k = 1, \ldots, n.$$

Thus, we have that for $h = r_1e_1 + \cdots + r_ne_n \in R$

$$\| f - h \| \leqslant \| f - g \| + \| g - h \| < \frac{\epsilon}{2} + \frac{\epsilon}{2} = \epsilon. \quad \text{Q.E.D.}$$

There are a few very important criteria by means of which we can establish whether an orthonormal system S is a basis in a Euclidean space.

Theorem 4.6. Each of the following statements is a *sufficient and necessary* condition that the countable orthonormal system $\mathsf{T} = \{e_1, e_2, \ldots\}$ is a basis in the separable Hilbert space* $\mathscr{H}$:

* The theorem applies to the finite-dimensional as well as the infinite-dimensional case, though it is stated and proved here for infinite-dimensional $\mathscr{H}$. In the finite-dimensional case, ∞ should be replaced by the dimension of $\mathscr{H}$.

(a) The only vector f satisfying the relations

$$\langle e_k \mid f\rangle = 0, \qquad k = 1, 2, \ldots \tag{4.13}$$

is the zero vector, i.e., (4.13) implies $f = \mathbf{0}$.

(b) For any vector $f \in \mathscr{H}$,

$$\lim_{n\to+\infty} \left\| f - \sum_{k=1}^{n} \langle e_k \mid f\rangle\, e_k \right\| = 0, \tag{4.14}$$

or symbolically written

$$f = \sum_{k=1}^{\infty} \langle e_k \mid f\rangle\, e_k\,,$$

where $\langle e_k \mid f\rangle$ is sometimes called the *Fourier coefficient* of f.

(c) Any two vectors $f, g \in \mathscr{H}$ satisfy *Parseval's relation*:

$$\langle f \mid g\rangle = \sum_{k=1}^{\infty} \langle f \mid e_k\rangle\langle e_k \mid g\rangle. \tag{4.15}$$

(d) For any $f \in \mathscr{H}$

$$\| f \|^2 = \sum_{k=1}^{\infty} |\langle e_k \mid f\rangle|^2. \tag{4.16}$$

We start by proving that the criteria (a) and (b) are equivalent to the requirement that $\mathsf{T} = \{e_1, e_2, \ldots\}$ is an orthonormal basis, as that requirement was formulated in Definition 4.4. To do that we shall prove that (a) implies (b), (b) implies "T is a basis" (as formulated in Definition 4.4) and "T is a basis" implies (a).

In order to show that (a) implies (b) we need the following lemma.

Lemma 4.1. For any given vector f of a Euclidean space $\mathscr{E}$ (not necessarily separable) and any countable orthonormal system $\{e_1, e_2, \ldots\}$ in $\mathscr{E}$, the sequence $f_1, f_2, \ldots$ of vectors

$$f_n = \sum_{k=1}^{n} \langle e_k \mid f\rangle\, e_k \tag{4.17}$$

is a Cauchy sequence, and the Fourier coefficients $\langle e_k \mid f\rangle$ satisfy Bessel's inequality

$$\| f_n \|^2 = \sum_{k=1}^{n} |\langle e_k \mid f\rangle|^2 \leqslant \| f \|^2. \tag{4.18}$$

Proof. Write

$$h_n = f - f_n,$$

where f_n is given by (4.17). We have

$$\langle f_n \mid h_n \rangle = 0$$

because $\langle e_i \mid h_n \rangle = 0$ for $i = 1, 2, \ldots, n$,

$$\langle e_i \mid h_n \rangle = \Big\langle e_i \,\Big|\, f - \sum_{j=1}^{n} \langle e_j \mid f \rangle\, e_j \Big\rangle$$

$$= \langle e_i \mid f \rangle - \sum_{j=1}^{n} \langle e_j \mid f \rangle \langle e_i \mid e_j \rangle = 0,$$

as $\langle e_i \mid e_j \rangle = \delta_{ij}$. Thus

$$\langle f \mid f \rangle = \langle f_n + h_n \mid f_n + h_n \rangle = \langle f_n \mid f_n \rangle + \langle h_n \mid h_n \rangle,$$

and consequently, since $\langle h_n \mid h_n \rangle \geqslant 0$,

$$\langle f_n \mid f_n \rangle \leqslant \langle f \mid f \rangle. \tag{4.19}$$

By using (4.17) and $\langle e_i \mid e_j \rangle = \delta_{ij}$ we derive

$$\| f_n \|^2 = \langle f_n \mid f_n \rangle = \sum_{i,j=1}^{n} \langle e_i \mid f \rangle^* \langle e_i \mid e_j \rangle \langle e_j \mid f \rangle$$

$$= \sum_{i=1}^{n} |\langle e_i \mid f \rangle|^2,$$

which shows in conjunction with (4.19) that Bessel's inequality (4.18) is true. From (4.18) we can deduce that

$$\sum_{i=1}^{\infty} |\langle e_i \mid f \rangle|^2 \leqslant \| f \|^2 < +\infty, \tag{4.20}$$

i.e., the above series with nonnegative terms is bounded and therefore it converges. Since we know that for $m > n$

$$\| f_m - f_n \|^2 = \sum_{i=n+1}^{m} |\langle e_i \mid f \rangle|^2, \tag{4.21}$$

we easily see from (4.20) and (4.21) that $f_1, f_2, \ldots$ is a Cauchy sequence.

We return now to proving Theorem 4.6. If $\mathsf{T} = \{e_1, e_2, \ldots\}$ is a countable orthonormal system in the Hilbert space $\mathscr{H}$, then according to Lemma 4.1 for any given $f \in \mathscr{H}$ the sequence $f_1, f_2, \ldots$, where

$$f_n = \sum_{k=1}^{n} \langle e_k | f \rangle \, e_k$$

is a Cauchy sequence. Since $\mathscr{H}$ is complete, this sequence has a limit $g \in \mathscr{H}$.

We can now show that if the statement (a) about T in Theorem 4.6 is true, then (b) is also true, due to the fact that (a) implies $f = g$; namely, for any $k = 1, 2, \ldots$ we have

$$\langle f - g \mid e_k \rangle = \lim_{n \to \infty} \langle f - f_n \mid e_k \rangle$$

$$= \langle f \mid e_k \rangle - \lim_{n \to +\infty} \sum_{i=1}^{n} \langle e_i | f \rangle^* \langle e_i \mid e_k \rangle = 0.$$

Thus, if (a) is true, we must have $f - g = \mathbf{0}$.

It is obvious that statement (b) implies that $\mathsf{T} = \{e_1, e_2, \ldots\}$ is a basis, because according to Theorem 4.4 any $f \in \mathscr{H}$ is the limit of elements $f_1, f_2, \ldots$ from the linear space (T) spanned by T, where $f_n \in (\mathsf{T})$ is of the form (4.17).

We show now that the fact that $\mathsf{T} = \{e_1, e_2, \ldots\}$ is an orthonormal basis implies that (a) is true. Assume that some $f \in \mathscr{H}$ is orthogonal on the system $\{e_1, e_2, \ldots\}$. Since $f \in [e_1, e_2, \ldots] = \mathscr{H}$, there is a sequence $g_1, g_2, \ldots \in (e_1, e_2, \ldots)$, i.e., for some integer s_n

$$g_n = \sum_{k=1}^{s_n} a_k e_k,$$

which converges to f. Consequently, as $\langle f \mid e_k \rangle = 0$,

$$\langle f | f \rangle = \lim_{n \to \infty} \langle f | g_n \rangle = \lim_{n \to \infty} \sum_{k=1}^{s_n} a_k \langle f \mid e_k \rangle = 0,$$

and therefore $f = \mathbf{0}$.

We shall demonstrate that statement (c) is equivalent to (a) or (b) by showing that (b) implies (c), and (c) implies (a), and thus finish the proof of Theorem 4.6.

If (b) is true, then we have (see Exercise 4.10)

$$\langle f | g \rangle = \lim_{n \to \infty} \langle f_n | g_n \rangle, \tag{4.22}$$

where

$$f_n = \sum_{k=1}^{n} \langle e_k | f \rangle e_k, \qquad g_n = \sum_{k=1}^{n} \langle e_k | g \rangle e_k .$$

From the relation

$$\langle f_n | g_n \rangle = \sum_{i,j=1}^{n} \langle e_i | f \rangle^* \langle e_i | e_j \rangle \langle e_j | g \rangle = \sum_{i=1}^{n} \langle f | e_i \rangle \langle e_i | g \rangle$$

we immediately obtain Parseval's relation (4.15).

If we assume (c) to be true, then (a) is also true, because if some vector f is orthogonal on $\{e_1, e_2, \ldots\}$, i.e., $\langle f | e_k \rangle = 0$, $k = 1, 2, \ldots$, then by inserting $f = g$ in (4.15) we get

$$\langle f | f \rangle = \sum_{k=1}^{\infty} \langle f | e_k \rangle \langle e_k | f \rangle = 0,$$

which implies that $f = \mathbf{0}$.

Finally, (d) follows from (c) by taking again in (4.15) that $f = g$. Vice versa, if (d) is true then (a) has to be true, because if $\langle f | e_k \rangle = 0$ for $k = 1, 2, \ldots$, then we get from (4.16) that $\| f \|^2 = 0$, which implies that $f = \mathbf{0}$. Q.E.D.

It is easy to see that, due to the fact that every Euclidean space can be embedded in a Hilbert space (Theorem 4.1), the criteria (b) (c), and (d) are also necessary and sufficient criteria for T to be an orthonormal basis in a Euclidean space in general, while (a) is necessary but not sufficient (see Exercise 4.13).

4.5. Isomorphism of Separable Hilbert Spaces

We can now demonstrate for infinite-dimensional Hilbert spaces a theorem analogous to the Theorem 2.5 for finite-dimensional Hilbert spaces.

Theorem 4.7. All complex infinite-dimensional separable Hilbert spaces are isomorphic to $l^2(\infty)$, and consequently mutually isomorphic.

Proof. If $\mathscr{H}$ is separable, there is, according to Theorem 4.5, an orthonormal countable basis $\{e_1, e_2, \ldots\}$ in $\mathscr{H}$, which is infinite when $\mathscr{H}$ is infinite dimensional. According to Theorem 4.6 we can write for any $f \in \mathscr{H}$

$$f = \sum_{k=1}^{\infty} c_k e_k, \qquad c_k = \langle e_k | f \rangle,$$

where by (4.16)

$$\sum_{k=1}^{\infty} | c_k |^2 = \| f \|^2 < +\infty.$$

Therefore

$$\alpha_f = \begin{pmatrix} c_1 \\ c_2 \\ \vdots \end{pmatrix} \in l^2(\infty).$$

Vice versa, if

$$\beta = \begin{pmatrix} b_1 \\ b_2 \\ \vdots \end{pmatrix} \in l^2(\infty),$$

then $f_1, f_2, \ldots$

$$f_n = \sum_{k=1}^{n} b_k e_k$$

is a Cauchy sequence, because for any $m > n$

$$\| f_m - f_n \|^2 = \sum_{k=n+1}^{m} | b_k |^2$$

and $\sum_{k=1}^{\infty} | b_k |^2$ converges. Thus, due to the completeness of $\mathscr{H}$ $f_1, f_2, \ldots$ converges to a vector $f \in \mathscr{H}$ and we have

$$c_k = \langle e_k \mid f \rangle = \lim_{n \to \infty} \left\langle e_k \mid \sum_{i=1}^{n} b_i e_i \right\rangle = b_k .$$

Therefore, the inverse mapping of the mapping $f \mapsto \alpha_f$ of $\mathscr{H}$ into $l^2(\infty)$ exists, and has $l^2(\infty)$ as its domain of definition. Hence the mapping $f \mapsto \alpha_f$ is a one-to-one mapping of $\mathscr{H}$ *onto* $l^2(\infty)$. It can be easily checked (see Exercise 4.11) that this mapping supplies an isomorphism between $\mathscr{H}$ and $l^2(\infty)$. Q.E.D.

As we shall see later, the above theorem provides the basis of the equivalence of Heisenberg's matrix formulation and Schroedinger's wave formulation of quantum mechanics.

Theorem 4.8. If the mapping

$$f \mapsto f', \qquad f \in \mathscr{E}, \qquad f' \in \mathscr{E}'$$

is a unitary transformation of the separable Euclidean space into the Euclidean space $\mathscr{E}'$, and if $\{e_1, e_2, \ldots\}$ is an orthonormal basis in $\mathscr{E}$, then $\{e_1', e_2', \ldots\}$ is an orthonormal basis in $\mathscr{E}'$, where e_n' denotes the image of e_n.

Proof. Let $\mathscr{E}$ be infinite dimensional, and denote by $\langle \cdot \mid \cdot \rangle_1$ and $\langle \cdot \mid \cdot \rangle_2$ the inner products in $\mathscr{E}$ and $\mathscr{E}'$ respectively. Then

$$\langle e_i' \mid e_j' \rangle_2 = \langle e_i \mid e_j \rangle_1 = \delta_{ij},$$

i.e., $\{e_1', e_2', \ldots\}$ is an orthonormal system in $\mathscr{E}'$. Since each $f' \in \mathscr{E}'$ has a unique inverse image $f \in \mathscr{E}$, we have

$$\lim_{n\to\infty} \left\| f' - \sum_{k=1}^{n} \langle e_n' \mid f' \rangle_2 \, e_n' \right\|_2 = \lim_{n\to\infty} \left\| f - \sum_{k=1}^{n} \langle e_n \mid f \rangle_1 \, e_k \right\|_1 = 0,$$

which by Theorem 4.6(b) proves that $\{e_1', e_2', \ldots\}$ is a basis.

The case when $\mathscr{E}$ is finite dimensional can be treated in a similar manner. Q.E.D.

Exercises

4.1. Show that in a normed space $\mathscr{N}$ the real function $d(f, g) = \| f - g \|$ on $\mathscr{N} \times \mathscr{N}$ is a metric, i.e., it satisfies all the requirements of Definition 3.1.

4.2. Prove that for any $\epsilon > 0$ there is an $N(\epsilon)$ such that

$$\| f_m - f_n \| = \left(\int_{-\infty}^{+\infty} | f_m(x) - f_n(x) |^2 \, dx \right)^{1/2} < \epsilon$$

for $m, n > N(\epsilon)$, where f_n is given by (4.2).

4.3. Check that the operations (4.3) satisfy the axioms in Definition 1.1.

4.4. Check that (4.4) satisfies the requirements of Definition 2.1.

4.5. Show that if $\mathscr{N}$ is a normed space and $\tilde{\mathscr{N}}$ is the completion of $\mathscr{N}$ in the norm, then:

(a) $\tilde{\mathscr{N}}$ is a linear space with respect to the operations

$$\tilde{f} + \tilde{g} = \{f_1 + g_1, f_2 + g_2, \ldots\},$$
$$a\tilde{f} = \{af_1, af_2, \ldots\};$$

(b) the limit $\| \tilde{f} \|_s = \lim_{n\to\infty} \| f_n \|$ exists for every Cauchy sequence $\{f_1, f_2, \ldots\}$ and defines a norm in $\tilde{\mathscr{N}}$;

(c) $\tilde{\mathscr{N}}$ is a Banach space and the image $\mathscr{N}'$ of $\mathscr{N}$ in $\tilde{\mathscr{N}}$ defined by the mapping $f \leftrightarrow \{f, f, \ldots\}$ is a linear subspace of $\tilde{\mathscr{N}}$ which is everywhere dense in $\mathscr{N}$.

4.6. Show that (4.6), (4.7), and (4.8) satisfy the axioms for vector addition, multiplication by a scalar, and inner product respectively.

4.7. Show that the subset D of $l^2(\infty)$ is countable, where D consists of all vectors α which have the properties: (1) a finite number of components $a_1 ,..., a_n$ (for some integer $n = 1, 2,...$) of α are complex numbers with real and imaginary parts which are rational numbers; (2) the rest of the components vanish.

4.8. Show that every finite-dimensional Euclidean space is a separable Hilbert space.

4.9. Prove Theorem 4.4.

4.10. Show that if in a Euclidean space $f_1, f_2,...$ converges in norm to f and $g_1, g_2,...$ to g, then $\langle f \mid g\rangle = \lim_{n\to\infty} \langle f_n \mid g_n\rangle$.

4.11. Show that the mapping $f \leftrightarrow \alpha_f$ of $\mathscr{H}$ onto $l^2(\infty)$ satisfies the requirements for an isomorphism, given in Definition 2.4.

4.12. Prove that if one orthonormal system $\{e_1, e_2,...\}$ in a Euclidean space $\mathscr{E}$ satisfies either (4.14), or (4.15), or (4.16), for every vector f (or, in case of (4.15), for any two vectors f and g) from $\mathscr{E}$, then $\{e_1, e_2,...\}$ is a basis in $\mathscr{E}$.

4.13. Verify that the criterion of Theorem 4.6(a) is not sufficient to insure that an orthonormal system $\{e_1, e_2,...\}$ in a Euclidean space $\mathscr{E}$ satisfying that criterion is a basis by showing the following:

Let $\{h_1, h_2,...\}$ be an orthonormal basis in a Hilbert space $\mathscr{H}$, and let $\mathscr{E}$ be the vector subspace spanned by $(\sum_{k=1}^{\infty} (1/k)h_k)$, h_2, h_3, ..., i.e., $\mathscr{E} = (\sum_{k=1}^{\infty} (1/k)\, h_k, h_2,...)$; then $\mathscr{E}$ is a Euclidean space. Prove that:

(a) $\{e_1 = h_2, e_2 = h_3,..., e_n = h_{n+1},...\}$ is not an orthonormal basis in $\mathscr{E}$.

(b) If $f \in \mathscr{E}$ is orthogonal to $\{e_1, e_2,...\} \subset \mathscr{E}$, then $f = \mathbf{0}$.

5. Wave Mechanics of a Single Particle Moving in One Dimension

5.1. The Formalism and Its (Partial) Physical Interpretation

As an illustration of a physical application of the preceding results, we shall consider the case of a particle restricted to move in only one space dimension within a potential well. We denote the space-coordinate variable by x and the time variable by t. Assume that on our system there acts a force field $F(x)$ which can be derived from a potential $V(x)$, i.e.,

$F(x) = -(d/dx)\ V(x)$. In classical mechanics, if we denote the momentum of the particle by p, we have the following expression for the total energy E of a particle of mass m:

$$E = p^2/2m + V(x). \tag{5.1}$$

Classically the state of the particle is described by its trajectory $x(t)$, where at any moment t, $x(t) \in \mathbb{R}^1$.

As we mentioned in the Introduction, one of the postulates of quantum mechanics is that the state of a system is described by a function $\Psi(t)$, where $\Psi(t)$ is a vector in a Hilbert space. In the wave mechanics version of quantum mechanics, the state of a one-particle system is postulated to be described at time t by a "wave function" $\psi(x, t)$ which is required to satisfy the condition

$$\int_{-\infty}^{+\infty} |\psi(x, t)|^2\, dx = 1. \tag{5.2}$$

As a function of t, $\psi(x, t)$ is assumed to be once continuously differentiable in t; in addition we require for the present that $\psi(x, t)$ have a piecewise continuous second derivative in x. Thus, we can consider $\psi(x, t)$ to be at any fixed time t an element of the Euclidean space $\mathscr{C}^1_{(2)}(\mathbb{R}^1)$ (see Exercise 5.2) of all complex functions $f(x)$ which vanish at infinity and are square integrable, i.e.,

$$\int_{-\infty}^{+\infty} |f(x)|^2\, dx < +\infty,$$

as well as once continuously differentiable with $\lim_{x\to\pm\infty} f'(x) = 0$. In $\mathscr{C}^1_{(2)}(\mathbb{R}^1)$ the inner product is taken to be

$$\langle f \mid g\rangle = \int_{-\infty}^{+\infty} f^*(x)\, g(x)\, dx,$$

and consequently we recognize (5.2) to be the normalization condition

$$\|\Psi(t)\|^2 = \int_{-\infty}^{+\infty} |\psi(x, t)|^2\, dx = 1,$$

where $\Psi(t) \in \mathscr{C}^1_{(2)}(\mathbb{R}^1)$ denotes the vector represented by the function $f_t(x) = \psi(x, t)$.

As a dynamical law we have in classical mechanics an equation of motion derivable from Newton's second law, which in the present case is

$$-\frac{dV}{dx} = m\ddot{x}, \qquad \ddot{x} = \frac{d^2x(t)}{dt^2}. \tag{5.3}$$

In wave mechanics it is postulated that the wave function $\psi(x, t)$ satisfies the Schroedinger equation* [we assume throughout that $V(x)$ is piecewise continuous],

$$i\hbar \frac{\partial\psi(x, t)}{\partial t} = -\frac{\hbar^2}{2m} \frac{\partial^2\psi(x, t)}{\partial x^2} + V(x)\,\psi(x, t). \tag{5.4}$$

A heuristic recipe which leads to the above differential equation is the following: replace p in (5.1) by the differential operator $-i\hbar(\partial/\partial x)$ and E by the differential operator $i\hbar(\partial/\partial t)$. If we deal with these operators formally and write $(\partial/\partial x)^2 = \partial^2/\partial x^2$, we get the operator relation

$$i\hbar \frac{\partial}{\partial t} = -\frac{\hbar^2}{2m} \frac{\partial^2}{\partial x^2} + V(x)$$

which yields (5.4) when applied to $\psi(x, t)$.

We note the important fact that Schroedinger's equation (5.4) preserves the normalization condition (5.2):

$$\begin{aligned}\frac{d}{dt} \int_{-\infty}^{+\infty} |\psi(x, t)|^2\, dx &= \int_{-\infty}^{+\infty} \left(\frac{\partial\psi^*(x, t)}{\partial t}\, \psi(x, t) + \psi^*(x, t)\, \frac{\partial\psi(x, t)}{\partial t} \right) dx \\ &= \frac{\hbar}{2im} \int_{-\infty}^{+\infty} \left(\frac{\partial^2\psi^*(x, t)}{\partial x^2}\, \psi(x, t) - \psi^*(x, t)\, \frac{\partial^2\psi(x, t)}{\partial x^2} \right) dx \\ &= \frac{\hbar}{2im} \left(\lim_{a\to\infty} \left[\frac{\partial\psi^*}{\partial x}\, \psi - \psi^*\, \frac{\partial\psi}{\partial x} \right]_{x=-a}^{x=a} \right. \\ &\qquad \left. - \int_{-\infty}^{+\infty} \left(\frac{\partial\psi^*}{\partial x} \frac{\partial\psi}{\partial x} - \frac{\partial\psi^*}{\partial x} \frac{\partial\psi}{\partial x} \right) dx \right) = 0.\end{aligned}$$

This is important in view of the correspondence rule proposed by Born [1926a, b], which constitutes the generally accepted physical interpretation of quantum mechanics. A very special case of this interpretation, which will be stated in its most general form in Chapter IV, §1, is in the present case given by the rule that for any interval I in $\mathbb{R}^1$

$$P_t(I) = \int_I |\psi(x, t)|^2\, dx$$

represents the *probability* of detecting the system (i.e., the one particle) within the interval I, if a measurement of position is performed at time t. Thus

$$P_t(\mathbb{R}^1) = \int_{-\infty}^{+\infty} |\psi(x, t)|^2\, dx$$

represents the probability of finding the system anywhere and, by

* $\hbar = h/2\pi$, h is the Planck constant; numerically $\hbar = 1.054 \times 10^{-27}$ erg sec.

definition, this has to be equal to one.* In view of this interpretation the necessity of (5.2) being fulfilled at all times is obvious.

As we shall see in Chapter IV, another consequence of Born's interpretation is that whenever $\Psi(t)$ represents a physical state and c is a constant for which $|c| = 1$, the vector-valued function $(c\Psi(t))$ represents the *same* physical state.

5.2. The Wave Mechanical Initial-Value Problem

In the case of classical mechanics the basic problem which is encountered is the initial-condition problem: given the position x_0 and velocity v_0 of the system at some initial time t_0, find the state of the system at all times; i.e., find the unique trajectory $x(t)$ which satisfies the initial conditions

$$x(t_0) = x_0, \qquad \dot{x}(t_0) = v_0 \qquad (\dot{x} = dx/dt).$$

In practice this problem reduces to solving the initial-condition problem of the differential equation (5.3), which is of the second order in the variable t.

The analogous initial-value problem in the quantum mechanical case consists in giving the state $\psi_0(x)$ at some time t_0, and requiring to find a state $\psi(x, t)$ at all times t, which is such that

$$\psi(x, t_0) = \psi_0(x). \tag{5.5}$$

Since the dynamics is now given by the Schroedinger equation (5.4), which is of the first order in t, the above initial-condition problem has a unique solution if $\psi_0(x) \in \mathscr{C}^1_{(2)}(\mathbb{R}^1)$.

The general procedure of solving the above problem consists in reducing it to an "eigenvalue" problem. One tries first to find a solution $\psi(x, t)$ of the form

$$\psi(x, t) = \psi(x) \exp[-(i/\hbar) Et], \tag{5.6}$$

where E is a yet undetermined constant. When (5.6) is inserted in (5.4), we get a so-called *eigenvalue equation* for $\psi(x)$ and E

$$-\frac{\hbar^2}{2m}\frac{d^2\psi(x)}{dx^2} + V(x)\psi(x) = E\psi(x), \tag{5.7}$$

* A somewhat naive interpretation of the statement that an event E (in this case the event is the detection of the system within the interval I at time t) has the probability P, is the following "frequency interpretation": if the same experiment (in our case, the measurement of the position of the particle) is repeated under identical conditions a large number N of times, then the number of times the event E occurs should be $\nu \approx PN$, i.e., $P \approx \nu/N$, where the approximation $\approx$ should be increasingly better with increasing N (on different concepts of probability see, e.g., Pap [1962], Fine [1973]).

which in physics is called the *time-independent Schroedinger equation.* This equation has in general a family $\mathscr{E}_b$ of functions $\psi(x) \in \mathscr{C}^1_{(2)}(\mathbb{R}^1)$ as solutions for a set S_p of values of the parameter E. The numbers belonging to the set S_p are called, from the general mathematical point of view, the *eigenvalues* of the differential equation (5.7).

Theorem 5.1. If the functions $\psi_1(x)$, $\psi_2(x)$ and their first derivatives $d\psi_1(x)/dx$, $d\psi_2(x)/dx$, as well as $V(x)\,\psi_1(x)$ and $V(x)\,\psi_2(x)$, are from $\mathscr{C}^1_{(2)}(\mathbb{R}^1)$, then

$$\text{(5.8)} \qquad \left\langle \psi_1(x) \,\middle|\, \left(-\frac{\hbar^2}{2m}\frac{d^2\psi_2(x)}{dx^2} + V(x)\,\psi_2(x)\right)\right\rangle = \left\langle \left(-\frac{\hbar^2}{2m}\frac{d^2\psi_1(x)}{dx^2} + V(x)\,\psi_1(x)\right) \,\middle|\, \psi_2(x)\right\rangle.$$

If each solution $\psi(x)$ of (5.7) has the property that $\psi(x)$, $d\psi(x)/dx$, $V(x)\,\psi(x) \in \mathscr{C}^1_{(2)}(\mathbb{R}^1)$, then each eigenvalue E of the time-independent Schroedinger equation (5.7) is a real number, and if $\psi_1(x)$ and $\psi_2(x)$ are two eigenfunctions of (5.7) corresponding to two different eigenvalues $E_1 \neq E_2$, then $\psi_1(x)$ and $\psi_2(x)$ are orthogonal.

Proof. Integrating by parts twice we get

(5.9)

$$\int_{-a}^{+a} \psi_1{}^* \left(-\frac{\hbar^2}{2m}\frac{d^2\psi_2}{dx^2} + V\psi_2\right) dx = -\frac{\hbar^2}{2m}\left(\psi_1{}^*\frac{d\psi_2}{dx}\Big|_{-a}^{+a} - \frac{d\psi_1{}^*}{dx}\psi_2\Big|_{-a}^{+a}\right) + \int_{-a}^{+a}\left(-\frac{\hbar^2}{2m}\frac{d^2\psi_1{}^*}{dx^2} + V\psi_1{}^*\right)\psi_2\, dx.$$

When we let above $a \to +\infty$, we obtain (5.8) because both integrals converge due to the fact that $\psi_1(x)$, $\psi_2(x)$, $d\psi_1(x)/dx$, $d\psi_2(x)/dx$, $V(x)\,\psi_1(x)$, $V(x)\,\psi_2(x) \in \mathscr{C}^1_{(2)}(\mathbb{R}^1)$, while the first two terms on the right-hand side of (5.9) vanish due to the fact that, by definition, any function $f(x)$ in $\mathscr{C}^1_{(2)}(\mathbb{R}^1)$ has to vanish at infinity together with its first derivative.

If $\psi_1(x)$ and $\psi_2(x)$ satisfy the conditions of the theorem so that (5.8) is true, and if in addition they are solutions of (5.7) corresponding to eigenvalues E_1 and E_2, respectively, then we can write (5.8) in the form

$$\text{(5.10)} \qquad \langle \psi_1(x) \mid E_2\psi_2(x)\rangle = \langle E_1\psi_1(x) \mid \psi_2(x)\rangle.$$

By taking in (5.10) $\psi_1(x) = \psi_2(x) = \psi(x)$ and consequently $E_1 = E_2 = E$, we obtain

$$E\langle \psi(x) \mid \psi(x)\rangle = E^*\langle \psi(x) \mid \psi(x)\rangle.$$

If $\psi(x)$ is a nontrivial solution of (5.10) [i.e., if $\psi(x) \neq 0$], we have $\langle \psi \mid \psi \rangle > 0$ and therefore we must have $E = E^*$; i.e., we have proved that the eigenvalues of (5.7) are real under the specified conditions.

In order to prove the last point of the theorem, namely that $\psi_1 \perp \psi_2$ when $E_1 \neq E_2$, it is sufficient to note that due to the fact that E_1 and E_2 must be real numbers, we can write (5.10) in the form

$$(E_2 - E_1)\langle \psi_1(x) \mid \psi_2(x) \rangle = 0.$$

Since $E_1 \neq E_2$, the above implies that $\langle \psi_1(x) \mid \psi_2(x) \rangle = 0$. Q.E.D.

We shall establish in Chapter IV that the above theorem is true under more general circumstances than the ones mentioned so far.

5.3. Bound States of the System

From the physical point of view the wave functions in $\mathscr{E}_b$ are considered to be *bound states** and the numbers in S_p are taken to be (as a special case of Born's correspondence rule given in Chapter IV) the only possible energy values that a system in a bound state can assume; they are therefore said to be the *energy eigenvalues of the bound states*, and the set S_p is called the *point energy spectrum.*

We note (see Exercise 5.3) that for a given eigenvalue $E \in \mathsf{S}_p$ the corresponding set of eigenfunctions is a linear subspace M_E of $\mathscr{C}^1_{(2)}(\mathbb{R}^1)$. If the subspace M_E is one dimensional then the energy eigenvalue E is called *nondegenerate*; otherwise E is said to be degenerate.

It is easy to see that the Euclidean space $\mathscr{C}^1_{(2)}(\mathbb{R}^1)$ is not complete (see Exercise 4.2). According to Theorem 4.1 there is a completion of $\mathscr{C}^1_{(2)}(\mathbb{R}^1)$ which we denote by $\mathscr{H}^{(1)}$. We shall establish in Chapter II that $\mathscr{H}^{(1)}$ is a *separable* Hilbert space. It is assumed that each vector of $\mathscr{H}^{(1)}$ can represent a physical state at a certain time.

We shall limit ourselves in the rest of this section to the closed subspace $\mathscr{H}^{(1)}_b$ spanned by† $\mathscr{E}_b$. By definition, each one of the vectors in $\mathscr{E}_b$ represents a bound state at a given time. As $\mathscr{H}^{(1)}_b$ is a subspace of $\mathscr{H}^{(1)}$ which is separable, it has to be also separable according to Theorem 4.2.

Each one of the closed subspaces M_E of $\mathscr{H}^{(1)}_b$ corresponding to the eigenvalue E must also be separable. Consequently, according to

* Classically, a particle moving in a central force field is said to be in a bound state if its trajectory is a closed curve.

† To simplify the notation we denote by the same symbols the incomplete Euclidean spaces and their isomorphic images in their completion; we do the same with the elements of these spaces and the image points of these elements in the completion.

Theorem 4.5 there is in each M_E an at most countably infinite orthonormal system T_E spanning M_E. The closed subspace spanned by

$$\mathsf{T} = \bigcup_{E \in \mathrm{S_p}} \mathsf{T}_E$$

is obviously identical to $\mathscr{H}_{\mathrm{b}}^{(1)}$. The set T is an orthonormal system because each T_E, $E \in \mathrm{S_p}$, is an orthonormal system, while for $E_1 \neq E_2$ the sets M_{E_1} and M_{E_2}, and therefore also T_{E_1} and T_{E_2}, are orthogonal. Since all the vectors in T are linearly independent (see Exercise 4.12) while $\mathscr{H}_{\mathrm{b}}^{(1)}$ is separable, it follows (see Exercise 5.5) that T is a countable set.

In practice the elements of each T_E can be chosen to belong to $\mathscr{C}^1_{(2)}(\mathbb{R}^1)$ and therefore satisfy (5.7). Consequently, if we write T in the form

$$\mathsf{T} = \{\Psi_1, \Psi_2, ...\},$$

the vector $\Psi_k \in \mathscr{H}_{\mathrm{b}}^{(1)}$ will be related by the construction outlined in proving Theorem 4.1 to a wave function $\psi_k(x) \in \mathscr{C}^1_{(2)}(\mathbb{R}^1)$ satisfying the time-independent Schroedinger equation

$$-\frac{\hbar^2}{2m}\frac{d^2\psi_k(x)}{dx^2} + V(x)\,\psi_k(x) = E_k\psi_k(x).$$

As T is an orthonormal basis in $\mathscr{H}_{\mathrm{b}}^{(1)}$, we can expand according to Theorem 4.6 every vector in $\mathscr{H}_{\mathrm{b}}^{(1)}$ in the form

$$\Psi = \sum_{k=1}^{\infty} \langle \Psi_k \mid \Psi \rangle \Psi_k .$$

It must be remembered that the precise meaning of the above expansion is

$$\lim_{n\to\infty} \left\| \Psi - \sum_{k=1}^{n} \langle \Psi_k \mid \Psi \rangle \Psi_k \right\| = 0.$$

We claim that the general solution of the initial-value problem for bound states, when the intial state at some time t_0 is represented by a vector $\Psi_0 \in \mathscr{H}_{\mathrm{b}}^{(1)}$, is given by

$$\Psi(t) = \sum_{k=1}^{\infty} \exp\left[-\frac{i}{\hbar} E_k(t - t_0)\right] \langle \Psi_k \mid \Psi_0 \rangle \Psi_k . \tag{5.11}$$

The above series is again required to be a limit in the norm of its partial sums; E_k is the eigenvalue corresponding to the eigenfunction $\psi_k(x) \in \mathscr{C}^1_{(2)}(\mathbb{R}^1)$ which represents the vector Ψ_k.

First we have to establish that the series in (5.11) really converges for every t to an element $\Psi(t)$ of $\mathscr{H}_b^{(1)}$.

Theorem 5.2. For any fixed t the sequence $\Phi_1(t)$, $\Phi_2(t)$,...

$$\Phi_n(t) = \sum_{k=1}^{n} c_k(t)\, \Psi_k\,,$$

$$c_k(t) = \exp\left[-\frac{i}{\hbar} E_k(t - t_0)\right] \langle \Psi_k \mid \Psi_0 \rangle,$$

is convergent in the norm to some vector $\Psi(t) \in \mathscr{H}_b^{(1)}$. For $t = t_0$ its limit $\Psi(t_0)$ satisfies the initial condition $\Psi(t_0) = \Psi_0$.

Proof. As Ψ_1, Ψ_2,... is an orthonormal basis in $\mathscr{H}_b^{(1)}$, we have according to Theorem 4.6

$$\sum_{k=1}^{\infty} | c_k(t)|^2 = \sum_{k=1}^{\infty} |\langle \Psi_k \mid \Psi_0\rangle|^2 = \| \Psi_0 \|^2 < +\infty.$$

Thus, the sequence $\Phi_1(t)$, $\Phi_2(t)$,... is a Cauchy sequence, and consequently it has a limit $\Psi(t)$, since $\mathscr{H}_b^{(1)}$ is complete. As $c_k(t_0) = \langle \Psi_k \mid \Psi_0 \rangle$, $k = 1, 2,...$, are the Fourier coefficients of Ψ_0, we get from Theorem 4.6 $\Psi(t_0) = \Psi_0$. Q.E.D.

In order to justify* the statement that (5.11) represents the solution of the initial value problem, *assume* that the series

$$\sum_{k=1}^{\infty} \exp\left[-\frac{i}{\hbar} E_k(t - t_0)\right] c_k\, \psi_k(x), \qquad c_k = \langle \Psi_k \mid \Psi_0 \rangle, \tag{5.12}$$

converges in the norm as well as point by point for every value of x and t to a limit function $\psi(x, t)$, and that $\partial^2\psi(x, t)/\partial x^2$ and $\partial\psi(x, t)/\partial t$ can be obtained by differentiating the series (5.12) term by term twice in x and once in t, respectively. If we insert this $\psi(x, t)$ into the Schroedinger equation (5.4) and interchange the order of summation and differentiation, then we easily find that $\psi(x, t)$ satisfies (5.4) by taking into consideration that each $\psi_k(x, t)$ satisfies (5.7) for $E = E_k$.

* A complete and general proof will be given in Chapter IV.

5.4. A Particle Moving in a Square-Well Potential

In order to provide a simple example of the procedure outlined above, let us consider the case of a particle moving in a so-called square-well potential $V(x)$, given by

$$V(x) = \begin{cases} 0 & \text{for } 0 \leqslant x \leqslant L \\ V_0 > 0 & \text{for } x < 0 \text{ and } x > L. \end{cases}$$

This potential represents the idealized case of a force which is everywhere zero except at the "walls" $x = 0$ and $x = +L$ of the "potential well," where it is infinite. The time-independent Schroedinger equation (5.7) becomes in this case

$$\begin{aligned} &\frac{d^2\psi(x)}{dx^2} + \frac{2m}{\hbar^2} E\psi(x) = 0, && 0 \leqslant x \leqslant L \\ &\frac{d^2\psi(x)}{dx^2} + \frac{2m}{\hbar^2}(E - V_0)\,\psi(x) = 0, && x < 0, \quad x > L. \end{aligned} \tag{5.13}$$

The general solution of the system (5.13) is

$$\psi(x) = \begin{cases} ce^{ikx} + de^{-ikx}, & k = \left(\dfrac{2mE}{\hbar^2}\right)^{1/2}, \quad 0 \leqslant x \leqslant L \\ a_1 e^{-ik'x} + b_1 e^{ik'x}, & k' = \left[\dfrac{2m(E - V_0)}{\hbar^2}\right]^{1/2}, \quad x < 0 \\ a_2 e^{ik''x} + b_2 e^{-ik''x}, & k'' = \left[\dfrac{2m(E - V_0)}{\hbar^2}\right]^{1/2}, \quad x > L. \end{cases}$$

The above expressions should represent the same wave function $\psi(x)$ in the three different regions. Due to the normalization condition (5.2) a necessary condition for this is that $\psi(x) \to 0$ as $|x| \to +\infty$. This is possible only if $k' = k''$ is a pure imaginary number:

$$k' = k'' = i\kappa, \qquad \kappa = (1/\hbar)[2m(V_0 - E)]^{1/2} > 0$$

and $b_1 = b_2 = 0$.

Since $\psi(x) \in \mathscr{C}^1_{(2)}(\mathbb{R}^1)$, $\psi(x)$ also has to be continuous and have a continuous first derivative at $x = 0$ and $x = L$; i.e., we have to impose the following conditions:

$$\begin{aligned} &\psi(x)\,|_{x\to-0} = a_1 = \psi(x)\,|_{x\to+0} = c + d, \\ &\left.\frac{d\psi}{dx}\right|_{x\to-0} = \kappa a_1 = \left.\frac{d\psi}{dx}\right|_{x\to+0} = ik(c - d), \\ &\psi(x)\,|_{x\to L-0} = ce^{ikL} + de^{-ikL} = \psi(x)\,|_{x\to L+0} = a_2 e^{-\kappa L}, \\ &\left.\frac{d\psi}{dx}\right|_{x\to L-0} = ik(ce^{ikL} - de^{-ikL}) = \left.\frac{d\psi}{dx}\right|_{x\to L+0} = -a_2\kappa e^{-\kappa L}. \end{aligned} \tag{5.14}$$

By eliminating a_1 from the first pair and a_2 from the second pair of equations (5.14) we are led to the following linear system of equations for c and d:

$$
\begin{aligned}
(\kappa - ik)c + (\kappa + ik)d &= 0, \\
(\kappa + ik)\, e^{2ikL}c + (\kappa - ik)d &= 0.
\end{aligned}
\tag{5.15}
$$

In order to have a nontrivial solution ($|\, c\,| + |\, d\,| > 0$) of the system (5.15), the determinant

$$
\begin{vmatrix} \kappa - ik & \kappa + ik \\ (\kappa + ik)\, e^{2ikL} & \kappa - ik \end{vmatrix}
\tag{5.16}
$$

must vanish. Since

$$
k = (1/\hbar)\sqrt{2mE}, \qquad \kappa = (1/\hbar)\sqrt{2m(V_0 - E)},
\tag{5.17}
$$

and $\hbar$, m, and V_0 are fixed constants of the problem, the condition that (5.16) should be zero is a transcendental equation

$$
e^{2ikL} = \left(\frac{\kappa - ik}{\kappa + ik}\right)^2
\tag{5.18}
$$

to be satisfied by E. As we have

$$
\left|\frac{\kappa - ik}{\kappa + ik}\right| = 1,
$$

we can write

$$
\frac{\kappa - ik}{\kappa + ik} = e^{i\phi},
$$

where

$$
\begin{aligned}
\cos\phi &= \frac{\kappa^2 - k^2}{\kappa^2 + k^2} = 1 - 2\frac{E}{V_0}, \\
\sin\phi &= -\frac{2\kappa k}{\kappa^2 + k^2} = -2\left[\frac{E}{V_0}\left(1 - \frac{E}{V_0}\right)\right]^{1/2}.
\end{aligned}
\tag{5.19}
$$

The relation (5.18) is equivalent to

$$
\phi = kL = (1/\hbar)\sqrt{2mE},
$$

where (5.19) is determined by (5.18) up to a multiple of 2π, i.e.,

$$
\phi = \arccos\left(1 - 2\frac{E}{V_0}\right) + 2\pi n, \qquad n = 0, \pm 1, \pm 2, \ldots .
$$

Thus, we have an entire family of eigenvalues E_n, constituting the point spectrum of energy levels of a single particle moving inside the square-well potential. The energy values E_n can be obtained by solving numerically the equation

$$\frac{1}{\hbar}\sqrt{2mE} = \arccos\left(1 - 2\frac{E}{V_0}\right) + 2\pi n$$

for each n. Each eigenvalue E_n is nondegenerate, and the corresponding eigenfunction $\psi_n(x)$ is

$$\psi_n(x) = \begin{cases} c_n \dfrac{2ik_n}{\kappa_n + ik_n} e^{\kappa_n x}, & x \leqslant 0 \\ c_n \left(e^{ik_n x} - \dfrac{\kappa_n - ik_n}{\kappa_n + ik_n} e^{-ik_n x}\right), & 0 \leqslant x \leqslant L \\ c_n \left(e^{(\kappa_n + ik_n)L} - \dfrac{\kappa_n - ik_n}{\kappa_n + ik_n} e^{(\kappa_n - ik_n)L}\right) e^{-\kappa_n x}, & x \geqslant L, \end{cases} \tag{5.20}$$

where the expressions for k_n and κ_n are obtained by taking in (5.17) $E = E_n$. The absolute value $|c_n|$ of the constant c_n has to be determined from the normalization condition

$$\int_{-\infty}^{+\infty} |\psi_n(x)|^2 \, dx = 1.$$

The remarkable feature of the eigenfunction (5.20) is that $\psi_n(x)$ does not vanish outside the region $0 \leqslant x \leqslant L$. This means, in view of the physical interpretation of $|\psi_n(x)|^2$, that there is a nonvanishing probability of finding the particle outside the potential well—a typical quantum mechanical phenomenon which is not met in the classical treatment of the same problem, where the particle in a bound state is trapped inside the potential well.

Exercises

5.1. Check that the set $\mathscr{C}^n([-a, +a])$ of n-times continuously differentiable functions $f(x)$ defined on the finite interval $[-a, +a]$ becomes a Euclidean space if the vector operations are defined by

$$(af + bg)(x) = af(x) + bg(x)$$

and the inner product by

$$\langle f \mid g \rangle = \int_{-a}^{+a} f^*(x)\, g(x)\, dx.$$

5.2. Show that the set $\mathscr{C}^1_{(2)}(\mathbb{R}^1)$ of once continuously differentiable functions $f(x)$ that vanish at infinity together with their first derivatives and are square-integrable is a complex Euclidean space if the vector operations are defined for $f, g \in \mathscr{C}^1_{(2)}(\mathbb{R}^1)$ by

$$(f+g)(x) = f(x) + g(x),$$
$$(af)(x) = af(x), \qquad a \in \mathbb{C}^1,$$

and the inner product by

$$\langle f \mid g \rangle = \int_{-\infty}^{+\infty} f^*(x)\, g(x)\, dx.$$

5.3. Show that if $\psi_1(x), \psi_2(x) \in \mathscr{C}^1_{(2)}(\mathbb{R}^1)$ are eigenfunctions of (5.7) corresponding to the eigenvalue E, then for any given complex numbers a and b the function $a\psi_1(x) + b\psi_2(x)$ is also an eigenfunction with the eigenvalue E.

5.4. A linear harmonic oscillator in one dimension is a particle of mass m whose potential energy is

$$V(x) = \tfrac{1}{2} m\omega^2 x^2,$$

where ω denotes in classical mechanics the characteristic frequency of oscillations which the particle performs. Show that in the quantum mechanical case the energy spectrum S_{p} of bound states is nondegenerate and equal to

$$\mathsf{S}_{\mathrm{p}} = \{\tfrac{1}{2}\hbar\omega, \tfrac{3}{2}\hbar\omega, \ldots, (n+\tfrac{1}{2})\,\hbar\omega, \ldots\}.$$

Verify that the energy eigenvalue $E_n = (n+\frac{1}{2})\,\hbar\omega$ corresponds to the normalized eigenfunction

$$\psi_n(x) = (m\omega/\pi\hbar)^{1/4}(2^n n!)^{-1/2} \exp[-(m\omega/2\hbar)x^2]\, H_n\left[x\left(\frac{m\omega}{\hbar}\right)^{1/2}\right],$$

where $H_n(u)$, $n = 0, 1, 2, \ldots$, are the Hermite polynomials

$$H_n(u) = (-1)^n e^{u^2} \frac{d^n}{du^n}(e^{-u^2})$$
$$= (2u)^n - \frac{n(n-1)}{1}(2u)^{n-2} + \frac{n(n-1)(n-2)(n-3)}{1 \cdot 2}(2u)^{n-4} - \cdots.$$

5.5. If T is an orthonormal system in a separable Hilbert space, prove that T is a countable set.

References for Further Study

On finite-dimensional vector spaces: Birkhoff and MacLane [1953], and Nering [1963]. For §§2–4 consult Dennery and Krzywicki [1967], von Neumann [1955, Chapter II], and for a more advanced approach, Naimark [1959, Chapter I]. For §5 see Landau and Lifshitz [1958], Messiah [1962], or any other of the many standard textbooks on quantum mechanics.

CHAPTER II

Measure Theory and Hilbert Spaces of Functions

In the last chapter we met examples of Euclidean spaces whose elements were square-integrable functions. These spaces were not complete, and in order to find their completion we had to employ the constructions outlined in §3 and Theorem 4.1 of Chapter I. We can embed, however, Euclidean spaces of functions of a very general type in Hilbert spaces whose elements are also functions. In order to study these spaces it is necessary to familiarize ourselves with the basics of the modern theory of integration in general, and the theory of the Lebesgue integral in particular.

In §§1, 2, and 3 we define and study the basic constituents of measure spaces in order to be able to define in §3 integration on such spaces. The knowledge of the methods used in proving the theorems formulated in these three sections is not absolutely essential for an understanding of the rest of the material in this book. Consequently, at the first reading, the reader may skip the proofs and limit himself to reading the various definitions and theorems in order to familiarize himself with the basic concepts and results.

In §4 we introduce Hilbert spaces of square-integrable functions and in §5 we study the application of the theory of these spaces to the wave mechanics of n particles of different kinds. The introduction of the concepts of direct sums and tensor products of Hilbert spaces in §6 proves desirable before a more detailed study of the two-body bound-state problem is carried out in §7.

*1. Measurable Spaces

1.1. Boolean Algebras and σ Algebras of Sets

While discussing the physical interpretation of wave mechanics, we have seen that for each interval I

$$P_t(I) = \int_I |\psi(x, t)|^2 \, dx \tag{1.1}$$

is the probability of finding the system [which is in the state $\psi(x, t)$] within I at time t. The formula (1.1) provides an example of a real function whose domain of definition is a family of sets. Such a function is called a *set function*. In §2 we shall be concerned with a special class of set functions, which are called *measures*, and which are basic in the modern theory of integration. In order to define the concept of measure we need first the concept of measurable space.

In case of measurable spaces we deal again with a set $\mathscr{X}$ in which a certain type of structure is specified. The structure is now determined by giving a family of subsets of $\mathscr{X}$ which satisfies certain axioms, and which is then called a Boolean σ algebra.

Definition 1.1. A nonempty class $\mathscr{K}$ of subsets of a set $\mathscr{X}$ is called a *Boolean algebra* (or *field* or *additive class*) if we have

$$R \cup S \in \mathscr{K}, \qquad \text{whenever} \quad R, S \in \mathscr{K}, \tag{1.2}$$

$$R' \in \mathscr{K}, \qquad \text{whenever} \quad R \in \mathscr{K}, \tag{1.3}$$

where $R' = \mathscr{X} - R$ denotes the complement of R with respect to $\mathscr{X}$ (i.e., the set of all elements of $\mathscr{X}$ which do not belong to R).* The class $\mathscr{K}$ of subsets of $\mathscr{X}$ is called a *Boolean σ algebra* (or *σ field*, or *completely additive class*) if in addition to being a Boolean algebra it has the property that the union $\bigcup_{k=1}^{\infty} S_k$ of any *countable* number of sets $S_1, S_2, \ldots$ from $\mathscr{K}$ also belongs to $\mathscr{K}$.

Theorem 1.1. If the class $\mathscr{K}$ of subsets of a set $\mathscr{X}$ is a Boolean algebra, then

(a) the entire set $\mathscr{X}$ and the empty set $\varnothing$ belong to $\mathscr{K}$,

(b) the intersection $R \cap S$ belongs to $\mathscr{K}$ whenever $R, S \in \mathscr{K}$,

(c) the difference $R - S$ and symmetric difference, $R \triangle S = (R - S) \cup (S - R)$ belongs to $\mathscr{K}$ whenever $R, S \in \mathscr{K}$.

* We prefer the name Boolean algebra to that of field because thus no confusion with the concept of field of scalars can arise.

The fact that $\mathscr{X} \in \mathscr{K}$ if $\mathscr{K}$ is a Boolean algebra follows from the requirement that there is a subset R of $\mathscr{X}$ which belongs to $\mathscr{K}$. Consequently, $R' \in \mathscr{K}$ and therefore $\mathscr{X} = R \cup R' \in \mathscr{K}$, as well as $\varnothing = \mathscr{X}' \in \mathscr{K}$

In order to prove the other two statements of Theorem 1.1 we need to establish some of the general rules of dealing with sets.

Lemma 1.1. If $\mathscr{F}$ is a family of sets, and R is any given set, then

$$R \cap \left(\bigcup_{S \in \mathscr{F}} S\right) = \bigcup_{S \in \mathscr{F}} \left(R \cap S\right).$$

Proof. If $\xi \in R \cap (\bigcup_{S\in\mathscr{F}} S)$ then by definition $\xi \in R$ and $\xi \in \bigcup_{S\in\mathscr{F}} S$, i.e., ξ belongs to at least one set S_1 from $\mathscr{F}$. Thus, $\xi \in R \cap S_1$, $S_1 \in \mathscr{F}$, and consequently $\xi \in \bigcup_{S\in\mathscr{F}}(R \cap S)$.

Conversely, if $\eta \in \bigcup_{S\in\mathscr{F}}(R \cap S)$, then η belongs to at least one of the sets $(R \cap S)$, i.e., $\eta \in R$ and $\eta \in \bigcup_{S\in\mathscr{F}} S$. Thus, $\eta \in R \cap (\bigcup_{S\in\mathscr{F}} S)$.

Q.E.D.

It is interesting to note that if we look upon the operation $\cup$ as set addition and $\cap$ as set multiplication, then we obviously have that these operations are commutative

$$R \cup S = S \cup R, \qquad R \cap S = S \cap R,$$

and associative

$$R \cup (S \cup T) = (R \cup S) \cup T, \qquad R \cap (S \cap T) = (R \cap S) \cap T,$$

like the ordinary addition and multiplication of numbers. Furthermore, as follows from the above lemma, they are also distributive:

$$R \cap (S \cup T) = (R \cap S) \cup (R \cap T).$$

Lemma 1.2. If $\mathscr{F}$ is any family of subsets of a set $\mathscr{X}$, and if for any given set S we denote by S' the *complement* of S with respect to $\mathscr{X}$, then

$$\left(\bigcup_{S\in\mathscr{F}} S\right)' = \bigcap_{S\in\mathscr{F}} S', \qquad \left(\bigcap_{S\in\mathscr{F}} S\right)' = \bigcup_{S\in\mathscr{F}} S'. \tag{1.4}$$

Proof. To establish the first relation, assume that ξ is an element of $(\bigcup_{S\in\mathscr{F}} S)'$. Then $\xi \notin \bigcup_{S\in\mathscr{F}} S$, which means that ξ does not belong to any of the sets $S \in \mathscr{F}$, i.e., $\xi \in S'$ for all $S \in \mathscr{F}$; thus $\xi \in \bigcap_{S\in\mathscr{F}} S'$.

Conversely, if $\eta \in \bigcap_{S\in\mathscr{F}} S'$, then $\eta \in S'$ for each $S \in \mathscr{F}$, i.e., $\eta \notin S$ for all $S \in \mathscr{F}$. Therefore, $\eta \notin \bigcup_{S\in\mathscr{F}} S$, which implies that η is an element of $(\bigcup_{S\in\mathscr{F}} S)'$, thus proving completely the first of the relations (1.4).

We obtain the second of the relations (1.4) by taking in the first of the relations (1.4) the complement of the left- and right-hand side, and noting that for any subset R of $\mathscr{X}$, $(R')' = R'' = R$. Q.E.D.

It is now easy to finish the proof of Theorem 1.1. In order to prove Theorem 1.1(b), write

$$R \cap S = (R' \cup S')'.$$

Since we have $R, S \in \mathscr{K}$, we also have by (1.3) $R', S' \in \mathscr{K}$, and consequently from (1.2) we get $R' \cup S' \in \mathscr{K}$.

Similarly we get

$$R - S = R \cap S' = (R' \cup S'')' = (R' \cup S)' \in \mathscr{K},$$

by noting that $R - S = R \cap S'$. The result that $R - S$, $S - R \in \mathscr{K}$ whenever $R, S \in \mathscr{K}$ yields

$$R \triangle S = (R - S) \cup (S - R) \in \mathscr{K}.$$

Theorem 1.2. For any given nonempty family $\mathscr{F}$ of subsets of a set $\mathscr{X}$ there is a unique smallest* Boolean algebra $\mathscr{A}(\mathscr{F})$ and a unique smallest Boolean σ algebra $\mathscr{A}_\sigma(\mathscr{F})$ containing $\mathscr{F}$; $\mathscr{A}(\mathscr{F})$ and $\mathscr{A}_\sigma(\mathscr{F})$ are called, respectively, the *Boolean algebra* and the *Boolean σ algebra* generated by the family $\mathscr{F}$.

Proof. We give the proof for $\mathscr{A}(\mathscr{F})$; the proof for the case of $\mathscr{A}_\sigma(\mathscr{F})$ is completely analogous.

Denote by $\mathfrak{F}$ the family of all Boolean algebras $\mathscr{A}$ containing $\mathscr{F}$, i.e., if $\mathscr{A} \in \mathfrak{F}$ then $\mathscr{F} \subset \mathscr{A}$; $\mathfrak{F}$ is not empty because it certainly contains the family $\mathfrak{S}_{\mathscr{X}}$ of all subsets of $\mathscr{X}$, which is obviously a Boolean algebra. The family

$$\mathscr{A}(\mathscr{F}) = \bigcap_{\mathfrak{S} \in \mathfrak{F}} \mathfrak{S}$$

is a Boolean algebra; namely, $\mathscr{A}(\mathscr{F})$ is not empty since it contains $\mathscr{F}$, and if $R, S \in \mathscr{A}(\mathscr{F})$, then for each $\mathfrak{S} \in \mathfrak{F}$ we have $R \cup S$, $R' \in \mathfrak{S}$, and consequently $R \cup S$, $R' \in \mathscr{A}(\mathscr{F})$. In addition $\mathscr{A}(\mathscr{F})$ is the smallest Boolean algebra containing $\mathscr{F}$, since according to its construction, it lies within any other Boolean algebra containing $\mathscr{F}$.

There is only one smallest Boolean algebra $\mathscr{A}(\mathscr{F})$ containing

* The set S belonging to a family $\mathscr{F}$ of sets, is said to be the smallest set in that family $\mathscr{F}$ if it is contained in any other set belonging to $\mathscr{F}$.

$\mathscr{F}$; namely, if $\mathscr{A}_1(\mathscr{F})$ and $\mathscr{A}_2(\mathscr{F})$ were two such algebras, then $\mathscr{A}_1(\mathscr{F}) \cap \mathscr{A}_2(\mathscr{F})$ would also be a Boolean algebra containing $\mathscr{F}$. That is possible only if $\mathscr{A}_1(\mathscr{F}) = \mathscr{A}_2(\mathscr{F})$. Q.E.D.

1.2. Boolean Algebras of Intervals

For our purposes the most important Boolean algebra is the one generated by the family $\mathscr{I}^n$ of all intervals—including the degenerate intervals, which contain only one point, and the empty set $\varnothing$—in the n-dimensional vector space ($\mathbb{R}^n$).

Theorem 1.3. The family $\mathscr{B}_0^n$ of all unions

$$I_1 \cup \cdots \cup I_k, \qquad I_1, \ldots, I_k \in \mathscr{I}^n, \qquad k = 1, 2, \ldots$$

of intervals in $\mathscr{I}^n$ is identical to the Boolean algebra $\mathscr{A}(\mathscr{I}^n)$ generated by $\mathscr{I}^n$.

Proof. We obviously have $\mathscr{B}_0^n \subset \mathscr{A}(\mathbb{R}^n)$. Therefore, in order to show that $\mathscr{B}_0^n \equiv \mathscr{A}(\mathbb{R}^n)$ it is sufficient to prove that $\mathscr{B}_0^n$ is a Boolean algebra; namely, if we had that result established, then $\mathscr{B}_0^n$ would be a Boolean algebra containing $\mathscr{I}^n$, and since $\mathscr{A}(\mathbb{R}^n)$ is the smallest Boolean algebra containing $\mathscr{I}^n$, we would also have $\mathscr{B}_0^n \supset \mathscr{A}(\mathbb{R}^n)$, i.e., $\mathscr{B}_0^n = \mathscr{A}(\mathbb{R}^n)$.

If $R, S \in \mathscr{B}_0^n$, we can write

$$R = I_1 \cup \cdots \cup I_k, \qquad S = J_1 \cup \cdots \cup J_l,$$

where $I_1, \ldots, J_l \in \mathscr{I}^n$. Hence

$$R \cup S = I_1 \cup \cdots \cup I_k \cup J_1 \cup \cdots \cup J_l \in \mathscr{B}_0^n.$$

To prove that $R' \in \mathscr{B}_0^n$, we proceed by induction in k. In the case $k = 1$ we have $R = I_1$ is an interval, and it is easy to check that I_1' can be written in the form

$$I_1' = I^{(1)} \cup \cdots \cup I^{(\nu)}, \qquad I^{(1)}, \ldots, I^{(\nu)} \in \mathscr{I}^n, \qquad \nu \leqslant 2n3^{n-1},$$

i.e., $R' \in \mathscr{B}_0^n$. Assume now that the statement $R' \in \mathscr{B}_0^n$ is true for all R of the form $I_1 \cup \cdots \cup I_k$, i.e., that we can write

$$(I_1 \cup \cdots \cup I_k)' = \bigcup_{m=1}^{p} J_m, \qquad J_1, \ldots, J_p \in \mathscr{I}^n. \tag{1.5}$$

Then in the case $k + 1$, by writing

$$I_{k+1}' = J^{(1)} \cup \cdots \cup J^{(\nu)}, \qquad J^{(1)}, \ldots, J^{(\nu)} \in \mathscr{I}^n, \tag{1.6}$$

we get from (1.5) and (1.6), with the help of the Lemmas 1.1 and 1.2,

$$\begin{aligned}(I_1 \cup \cdots \cup I_{k+1})' &= (I_1 \cup \cdots \cup I_k)' \cap I'_{k+1} \\ &= \left[\bigcup_{m=1}^{p} (J^{(1)} \cap J_m)\right] \cup \cdots \cup \left[\bigcup_{m=1}^{p} (J^{(\nu)} \cap J_m)\right].\end{aligned}$$

Since $J^{(1)} \cap J_m, \ldots, J^{(\nu)} \cap J_m \in \mathscr{I}^n$, we conclude that $(I_1 \cup \cdots \cup I_{k+1})' \in \mathscr{B}^n$.

Q.E.D.

Notice that in constructing $\mathscr{B}_0^n$ it would have been sufficient to consider only *disjoint* finite unions of intervals (see Exercise 1.5).

1.3. Borel Sets in $\mathbb{R}^n$

A set belonging to the Boolean σ algebra $\mathscr{A}_\sigma(\mathscr{I}^n)$ generated by $\mathscr{I}^n$ bears the name of a *Borel set* in *n* dimensions. We cannot generalize Theorem 1.3 to Borel sets by saying that every Borel set is a countable union of intervals, because there are Borel sets which are not that. An example of such a Borel set in $\mathscr{A}_\sigma(\mathbb{R}^1)$ is the set $\mathfrak{J}$ of all irrational numbers; $\mathfrak{J}$ is a Borel set because it is the complement of the set $\mathfrak{R}$ of all rational numbers, and $\mathfrak{R}$ is the countable union of all degenerate intervals containing only one rational number, and therefore is a Borel set.

The fact that the family of $\mathscr{B}^n$ of all Borel sets in $\mathbb{R}^n$ is very large, though by no means contains every subset of $\mathbb{R}^n$, is confirmed by the following statement.

Theorem 1.4. Every open and every closed set in the Euclidean space $\mathbb{R}^n$ is a Borel set.

Proof. Assume that O is an open set* in $\mathbb{R}^n$. For every natural number m we consider the following open intervals in $\mathbb{R}^n$

$$I_{k_1,\ldots,k_n}^{(m)} = \left\{x = (x_1, \ldots, x_n) \colon \frac{k_1 - 1}{m} < x_1 < \frac{k_1 + 1}{m}, \ldots, \frac{k_n - 1}{m} < x_n < \frac{k_n + 1}{m}\right\}, \quad k_1, \ldots, k_n = 1, 2, \ldots,$$

which obviously cover $\mathbb{R}^n$. The union

$$R^{(m)} = \{I_{k_1,\ldots,k_n}^{(m)} \colon \; I_{k_1,\ldots,k_n}^{(m)} \subset O, \quad k_1, \ldots, k_n = 1, 2, \ldots\}$$

* O is an open set in a metric space if for every point $x \in O$ there is an ϵ neighborhood of x which lies within O.

of all intervals $I^{(m)}_{k_1,\ldots,k_n}$ lying within O is evidently countable. Furthermore

$$O = \bigcup_{m=1}^{\infty} R^{(m)};$$

namely, if $x \in O$ is any given point of O, then there is an ϵ neighborhood of x contained in O, and consequently for a sufficiently large m, $m > 2n/\epsilon$, there will be an interval $I^{(m)}_{k_1,\ldots,k_n}$ containing x and lying within that ϵ neighborhood.

Since we have written O as a countable union of Borel sets $R^{(m)}$, it follows that O is a Borel set.

Any closed set C in $\mathbb{R}^n$ is the complement (with respect to $\mathbb{R}^n$) of the open set $C' = O$. Since $O \in \mathscr{B}^n$, it follows that $C = O' \in \mathscr{B}^n$. Q.E.D.

Given the Boolean algebra $\mathscr{A}(\mathscr{F})$ generated by some family $\mathscr{F}$, we can construct the Boolean σ algebra $\mathscr{A}_\sigma(\mathscr{F})$ generated by $\mathscr{F}$, by means of monotone sequences of sets from $\mathscr{A}(\mathscr{F})$.

1.4. Monotone Classes of Sets

Definition 1.2. An infinite sequence $S_1, S_2, \ldots$ of sets is called *monotonically increasing* (or *expanding*) if $S_1 \subset S_2 \subset \cdots$, and *monotonically decreasing* (or *contracting*) if $S_1 \supset S_2 \supset \cdots$; in the first case we write

$$\lim_{k\to\infty} S_k = \bigcup_{k=1}^{\infty} S_k,$$

while in the second case we write

$$\lim_{k\to\infty} S_k = \bigcap_{k=1}^{\infty} S_k.$$

A nonempty class $\mathscr{K}$ of subsets of a set $\mathscr{X}$ is called a *monotone class* if every monotone sequence $S_1, S_2, \ldots \in \mathscr{K}$ has its limit in $\mathscr{K}$, i.e., $\lim_{k\to\infty} S_k \in \mathscr{K}$.

Theorem 1.5. If $\mathscr{F}$ is a nonempty family of subsets of a set $\mathscr{X}$, there is a unique smallest monotone class $\mathfrak{M}(\mathscr{F})$ containing $\mathscr{F}$, which is called the monotone class *generated* by $\mathscr{F}$.

The proof of the above theorem can be carried out in the same fashion as the similar proof for $\mathscr{A}(\mathscr{F})$ in Theorem 1.2, by noting that the family $\mathfrak{S}_{\mathscr{X}}$ of all subsets of $\mathscr{X}$ is a monotone class.

***Theorem 1.6.** Every Boolean σ algebra is a monotone class, and every Boolean algebra which is a monotone class is a Boolean σ algebra.

Proof. (a) If $\mathscr{A}_\sigma$ is a Boolean σ algebra and $R_1, R_2, \ldots \in \mathscr{A}_\sigma$ is monotonically increasing, then by Definition 1.1 of a Boolean σ algebra,

$$\lim_{k\to\infty} R_k = \bigcup_{k=1}^{\infty} R_k \in \mathscr{A}_\sigma . \tag{1.7}$$

In the case that $S_1, S_2, \ldots$ is a monotonically decreasing sequence in $\mathscr{A}_\sigma$, then $S_1 - S_1, S_1 - S_2, \ldots$ is a monotonically increasing sequence in $\mathscr{A}_\sigma$, and we have according to (1.7) (see also Exercise 1.3)

$$S_1 - \lim_{k\to\infty} S_k = \lim_{k\to\infty}(S_1 - S_k) \in \mathscr{A}_\sigma .$$

Therefore

$$\lim_{k\to\infty} S_k = S_1 - (S_1 - \lim_{k\to\infty} S_k) \in \mathscr{A}_\sigma .$$

(b) If $\mathscr{A}$ is a Boolean algebra and a monotone class, and $S_1, S_2, \ldots$ is an infinite sequence of sets from $\mathscr{A}$, then $R_1, R_2, \ldots$,

$$R_n = \bigcup_{k=1}^{n} S_k , \qquad n = 1, 2, \ldots,$$

is a monotonically increasing sequence in $\mathscr{A}$. Since $\mathscr{A}$ is a monotone class, we have

$$\bigcup_{k=1}^{\infty} S_k = \lim_{n\to\infty} R_n \in \mathscr{A},$$

and therefore $\mathscr{A}$ is a Boolean σ algebra. Q.E.D.

***Theorem 1.7.** If $\mathscr{A}$ is a Boolean algebra and $\mathfrak{M}(\mathscr{A})$ is the monotone class generated by $\mathscr{A}$, then $\mathfrak{M}(\mathscr{A})$ is identical with the Boolean σ algebra $\mathscr{A}_\sigma(\mathscr{A})$ generated by the family $\mathscr{A}$ of sets.

Proof. Since, according to Theorem 1.6, $\mathscr{A}_\sigma(\mathscr{A})$ is a monotone class, while by definition $\mathfrak{M}(\mathscr{A})$ is the smallest monotone class containing $\mathscr{A}$, we must have

$$\mathfrak{M}(\mathscr{A}) \subset \mathscr{A}_\sigma(\mathscr{A}). \tag{1.8}$$

Since $\mathscr{A}_\sigma(\mathscr{A})$ is the smallest Boolean σ algebra containing $\mathscr{A}$, by proving that $\mathfrak{M}(\mathscr{A})$ is a Boolean σ algebra, we could also deduce that

$$\mathscr{A}_\sigma(\mathscr{A}) \subset \mathfrak{M}(\mathscr{A}),$$

which would imply in conjunction with (1.8) that $\mathscr{A}_\sigma(\mathscr{A}) = \mathfrak{M}(\mathscr{A})$. Now, in order to prove that $\mathfrak{M}(\mathscr{A})$ is a Boolean σ algebra, it is sufficient, due to the second part of Theorem 1.6, to establish that $\mathfrak{M}(\mathscr{A})$ is a Boolean algebra.

For a given $R \in \mathfrak{M}(\mathscr{A})$ denote by $\mathfrak{N}(R)$ the family of all sets S in $\mathfrak{M}(\mathscr{A})$ which are such that $S', R \cup S \in \mathfrak{M}(\mathscr{A})$. The family $\mathfrak{N}(R)$ is a monotone class. As a matter of fact, if $S_1, S_2, \ldots \in \mathfrak{N}(R)$ is a monotone sequence, then since $\mathfrak{M}(\mathscr{A})$ is a monotone class (see Exercise 1.6),

$$(\lim_{n\to\infty} S_n)' = \lim_{n\to\infty} S_n' \in \mathfrak{M}(\mathscr{A}),$$

while the sequence of the sets $R \cup S_1, R \cup S_2, \ldots \in \mathfrak{M}(\mathscr{A})$ is also a monotone sequence and we have (see Exercise 1.6)

$$(\lim_{n\to\infty} S_n) \cup R = \lim_{n\to\infty}(S_n \cup R) \in \mathfrak{M}(\mathscr{A}).$$

In addition, if $R \in \mathscr{A}$, then $\mathfrak{N}(R)$ is certainly not empty because it contains $\mathscr{A}$. Since $\mathfrak{M}(\mathscr{A})$ is the smallest monotone class containing $\mathscr{A}$, we must have $\mathfrak{N}(R) \equiv \mathfrak{M}(\mathscr{A})$. Thus, $\mathfrak{N}(\mathscr{X}) \equiv \mathfrak{M}(\mathscr{A})$, and therefore $S' \in \mathfrak{M}(\mathscr{A})$ for every $S \in \mathfrak{N}(\mathscr{X}) \equiv \mathfrak{M}(\mathscr{A})$. Furthermore, if $R \in \mathscr{A}$, then for an arbitrary $S \in \mathfrak{M}(\mathscr{A})$ we have $S \in \mathfrak{N}(R) \equiv \mathfrak{M}(\mathscr{A})$, i.e., $R \cup S = S \cup R \in \mathfrak{M}(\mathscr{A})$; consequently, we also have $R \in \mathfrak{N}(S)$ for any $R \in \mathscr{A}$, i.e., $\mathfrak{N}(S) \supset \mathscr{A}$. Since $\mathfrak{N}(S)$ is a monotone class, while $\mathfrak{M}(\mathscr{A})$ is the smallest monotone class containing $\mathscr{A}$, it follows that $\mathfrak{N}(S) \equiv \mathfrak{M}(\mathscr{A})$ for arbitrary $S \in \mathfrak{M}(\mathscr{A})$. This means that $R \cup S$, $R' \in \mathfrak{M}(\mathscr{A})$ whenever $R, S \in \mathfrak{N}(S) \equiv \mathfrak{M}(\mathscr{A})$, i.e., $\mathfrak{M}(\mathscr{A})$ is indeed a Boolean algebra. Q.E.D.

Definition 1.3. The ordered pair $(\mathscr{X}, \mathscr{A})$ consisting of a set $\mathscr{X}$ and a Boolean σ algebra $\mathscr{A}$ of subsets of $\mathscr{X}$ is called a *measurable space.* Every set S belonging to $\mathscr{A}$ is said to be a *measurable set* in the measurable space $(\mathscr{X}, \mathscr{A})$.

By following the tacit convention adopted from the beginning with regard to mathematical spaces, we shall denote the measurable space $(\mathscr{X}, \mathscr{A})$, consisting of the set $\mathscr{X}$ and a structure on $\mathscr{X}$ given by the Boolean σ algebra $\mathscr{A}$, also by the letter $\mathscr{X}$—except where ambiguities might arise. Thus, due to the properties of Boolean σ algebras, we can say that if R is a measurable subset of $\mathscr{X}$, then R' is also measurable, and if $S_1, S_2, \ldots$ is a countable family of measurable subsets of $\mathscr{X}$, then their union $\bigcup_{k=1}^{\infty} S_k$ is measurable in $\mathscr{X}$.

Exercises

1.1. Show that if R and S are any two sets, then

$$R \cap S = (R \cup S) - (R \triangle S).$$

1.2. A nonempty class $\mathscr{K}$ of subsets of a set $\mathscr{X}$ is called a *Boolean ring* (or *additive class* of sets) if $R \cup S$, $R - S \in \mathscr{K}$ whenever $R, S \in \mathscr{K}$; a Boolean ring $\mathscr{R}$ is a Boolean σ ring if $S_1 \cup S_2 \cup \cdots \in \mathscr{R}$ whenever $S_1, S_2, \ldots \in \mathscr{R}$. Explain why every Boolean algebra (σ algebra) is a Boolean ring (σ ring).

1.3. Show that if $\mathscr{R}$ is a Boolean ring (see Exercise 1.2) of subsets of the set $\mathscr{X}$, then $R - S$, $R \triangle S$, $R \cap S \in \mathscr{R}$ whenever $R, S \in \mathscr{R}$.

1.4. Prove that if $\mathscr{F}$ is a family of subsets of a set $\mathscr{X}$, there is a smallest Boolean ring $\mathscr{R}(\mathscr{F})$ and a smallest Boolean σ ring $\mathscr{R}_\sigma(\mathscr{F})$ containing $\mathscr{F}$; these rings are said to be generated by $\mathscr{F}$.

1.5. Show that the family $\mathscr{B}_0^n$ of all unions of intervals is identical with the family of all *disjoint* unions of intervals.

1.6. Show that if $S_1, S_2, \ldots$ is a monotone sequence of subsets of a set $\mathscr{X}$, then for any given set R, the sequences $R \cup S_1, R \cup S_2, \ldots$, $R \cap S_1, R \cap S_2, \ldots$ and $S_1', S_2', \ldots$ are monotonic and $R \cup \lim_{k\to+\infty} S_k = \lim_{k\to+\infty}(R \cup S_k)$, $R \cap \lim_{k\to+\infty} S_k = \lim_{k\to+\infty} R \cap S_k$, $(\lim_{k\to+\infty} S_k)' = \lim_{k\to+\infty} S_k'$.

1.7. Show that Theorem 1.5 is true not only for Boolean algebras, but also, more generally, for Boolean rings.

1.8. Prove that if $\mathscr{A}$ is a Boolean algebra of subsets of $\mathscr{X}$, and R is any given subset of $\mathscr{X}$, then $\mathscr{A}_R = \{R \cap S: S \in \mathscr{A}\}$ is a Boolean algebra of subsets of R. Formulate and demonstrate the same result for Boolean σ algebras. Show that if $\mathscr{A}$ generates the Boolean σ algebra $\mathscr{A}_\sigma$ of subsets of $\mathscr{X}$, then $\mathscr{A}_R$ generates the Boolean σ algebra $\mathscr{A}_{\sigma,R} = \{R \cap S: S \in \mathscr{A}_\sigma\}$ of subsets of R.

1.9. Let $\mathscr{A}$ be a Boolean σ algebra and $S_1, S_2, \ldots$ a countable family of sets from $\mathscr{A}$. Show that $\bigcap_{k=1}^{\infty} S_k$ belongs to $\mathscr{A}$.

*2. Measures and Measure Spaces

2.1. The Concept of Measure

In the last section we mentioned the concept of a *set function* which is, in general, a mapping of a family $\mathscr{K}$ of sets into $\mathbb{R}^1$ or $\mathbb{C}^1$. A set function $F(S)$, $S \in \mathscr{K}$, is said to be *additive* if for any *disjoint* $S_1, S_2 \in \mathscr{K}$

$$F(S_1 \cup S_2) = F(S_1) + F(S_2);$$

it is said to be *σ additive* or *countably additive* if for any countable family $S_1, S_2, \ldots$ of disjoint sets in $\mathscr{K}$ we have

$$F(S_1 \cup S_2 \cup \cdots) = F(S_1) + F(S_2) + \cdots .$$

Definition 2.1. A *measure* $\mu(S)$ is an extended real-valued set function whose domain of definition is a Boolean algebra $\mathscr{A}$, and which satisfies the following requirements:

(1) $\mu(\varnothing) = 0$,

(2) $\mu(S) \geqslant 0$ for all $S \in \mathscr{A}$,

(3) $\mu(\bigcup_{k=1}^{\infty} S_k) = \sum_{k=1}^{\infty} \mu(S_k)$ if the sets $S_1, S_2, \ldots$ are disjoint, i.e., if $S_i \cap S_j = \varnothing$ when $i \neq j$.

We see that a measure is an extended real-valued nonnegative σ-additive set function. The adjective "extended" indicates that for some sets $S \in \mathscr{A}$ we might have $\mu(S) = +\infty$. The symbol $+\infty$ is thus formally adjoined to the set $\mathbb{R}^1$, and in the extended set $\mathbb{R}^1 \cup \{+\infty\}$ we define the formal operations $a + \infty = \infty$, $a \cdot \infty = \infty$ if $a > 0$, and $0 \cdot \infty = 0$, $a/\infty = 0$, for all $a \in \mathbb{R}^1 \cup \{+\infty\}$.

Definition 2.2. A *measure space* $(\mathscr{X}, \mathscr{A}, \mu)$ is a measurable space $(\mathscr{X}, \mathscr{A})$ for which a measure $\mu(S)$ is defined on the Boolean σ algebra $\mathscr{A}$.

For any $S \in \mathscr{A}$, the number $\mu(S)$ is said to be the measure of the set S. In the case that $\mu(S) < +\infty$, S is said to have a *finite measure.* If the set S is a countable union $\bigcup_{n=1}^{\infty} S_n$ of sets S_n of finite measure, then the measure $\mu(S)$ of S is said to be *σ finite*. A measure $\mu(S)$ on $\mathscr{A}$ for which the set $\mathscr{X}$ has a finite (σ-finite) measure is called a *finite* (*σ finite*) measure; the measure space $(\mathscr{X}, \mathscr{A}, \mu)$ is then also called finite (σ finite).

As a simple example of a measure whose domain of definition is $\mathscr{B}_0^1$ (see §1), consider the set function $\mu_l^{(1)}(B)$ which for an interval I (open or closed) from a to b on the real line is equal to the length $b - a$ of the interval. Since any element $B \in \mathscr{B}_0^1$ can be written as a finite union of disjoint intervals (see Exercise 1.5)

$$B = I_1 \cup \cdots \cup I_k, \qquad I_i \cap I_j = \varnothing \quad \text{for} \quad i \neq j, \tag{2.1}$$

we can define $\mu_l^{(1)}(B)$ by

$$\mu_l^{(1)}(B) = \mu_l^{(1)}(I_1) + \cdots + \mu_l^{(1)}(I_k) \geqslant 0. \tag{2.2}$$

The number $\mu_l^{(1)}(B)$ is called the *Lebesgue measure* of $B \in \mathscr{B}_0^1$. It is obvious that the value $\mu_l^{(1)}(B)$ is uniquely determined regardless of the way we split B into disjoint intervals. It is easy to see that the set function $\mu_l^{(1)}(B)$ is σ additive. Hence, if we define $\mu_l^{(1)}(\varnothing) = 0$, we have a measure on $\mathscr{B}_0^1$; $\mu_l^{(1)}(B)$ is a σ-finite measure because we can write $\mathbb{R}^1 = \bigcup_{n=-\infty}^{+\infty}[n, n+1]$ and the Lebesgue measure of any bounded interval is finite.

The type of measure with which we shall frequently have to deal is the probability measure. A probability measure $\mu_p(S)$ is a measure on a measurable space $(\mathscr{X}, \mathscr{A})$, which assigns to each $S \in \mathscr{A}$ a number which represents the probability that an event, whose possible outcome could be any point in $\mathscr{X}$ and *only* a point of $\mathscr{X}$, will happen to be in S. Due to this interpretation it is required that a probability measure satisfy the "normalization" condition $\mu_p(\mathscr{X}) = 1$; thus, every probability measure is necessarily finite.

An example of a probability measure—of the type we shall encounter very frequently in the future—is the measure $P_t(B)$, $B \in \mathscr{B}_0^1$, defined on $\mathscr{B}_0^1$, whose value for $B = I \in \mathscr{I}^1$ is given by (1.1); for a general $B \in \mathscr{B}_0^1$, which can always be written in the form (2.1), we define (see Exercise 2.3)

$$P_t(B) = P_t(I_1) + \cdots + P_t(I_k).$$

2.2. Basic Properties of Measures

Definition 2.3. An extended real-valued set function $F(S)$ is said to be *continuous from above* (*below*) if for every monotonically decreasing (increasing) sequence $S_1, S_2, \ldots$ of sets from the domain of definition of $F(S)$ we have

$$F(\lim_{k\to\infty} S_k) = \lim_{k\to\infty} F(S_k)$$

whenever $F(\lim_{k\to\infty} S_k)$ is defined and, in the case that $S_1, S_2, \ldots$ is decreasing, whenever $|F(S_n)| < +\infty$ for at least one value of n (see Exercise 2.4).

Theorem 2.1. Every measure is continuous from above and below.

Proof. If $R_1, R_2, \ldots$ is a monotonically increasing sequence from the measure space $(\mathscr{X}, \mathscr{A}, \mu)$ and $\lim_{k\to\infty} R_k \in \mathscr{A}$, then we can write

$$\lim_{k\to\infty} R_k = \bigcup_{k=1}^{\infty} (R_k - R_{k-1}), \qquad R_0 = \varnothing,$$

where the sets $R_1 - R_0, R_2 - R_1, \ldots \in \mathscr{A}$ ($R_k - R_{k-1} \in \mathscr{A}$ by Theorem

1.1) are disjoint. Hence, if $\lim_{k\to\infty} R_k \in \mathscr{A}$,

$$\mu(\lim_{k\to\infty} R_k) = \mu\left(\bigcup_{k=1}^{\infty} (R_k - R_{k-1})\right) = \sum_{k=1}^{\infty} \mu(R_k - R_{k-1})$$

(2.3)

$$= \lim_{k\to+\infty} \mu\left(\bigcup_{n=1}^{k} (R_n - R_{n-1})\right) = \lim_{k\to\infty} \mu(R_k),$$

proving that μ is continuous from below.

If $S_1, S_2, \ldots \in \mathscr{A}$ is monotonically decreasing and $\mu(S_{n_0}) < +\infty$ for some n_0, we have that $S_{n_0} - S_1$, $S_{n_0} - S_2, \ldots$ is monotonically increasing. According to (2.3), if $\lim_{k\to\infty} S_k \in \mathscr{A}$ and therefore

$$\lim_{k\to\infty}(S_{n_0} - S_k) = S_{n_0} - \lim_{k\to\infty} S_k \in \mathscr{A},$$

we have

$$\mu(\lim_{k\to\infty}(S_{n_0} - S_k)) = \lim_{k\to\infty} \mu(S_{n_0} - S_k).$$

Since $\mu(S_{n_0} - S_k) = \mu(S_{n_0}) - \mu(S_k)$ for $k \geqslant n_0$ (see Exercise 2.1), we get from the above relations

$$\mu(S_{n_0} - \lim_{k\to\infty} S_k) = \mu(S_{n_0}) - \lim_{k\to\infty} \mu(S_k).$$

Therefore, $\mu(S)$ is continuous from above, as can be seen from the following:

$$\mu(\lim_{k\to\infty} S_k) = \mu(S_{n_0} - (S_{n_0} - \lim_{k\to\infty} S_k))$$

$$= \mu(S_{n_0}) - \mu(S_{n_0} - \lim_{k\to\infty} S_k) = \lim_{k\to\infty} \mu(S_k). \quad \text{Q.E.D.}$$

Theorem 2.2. Every finite, nonnegative and additive set function $F(S)$ defined on a Boolean algebra $\mathscr{A}$ and satisfying $F(\varnothing) = 0$, which is either continuous from below at every $R \in \mathscr{A}$, or continuous from above at $\varnothing \in \mathscr{A}$, is necessarily also σ additive, i.e., it is a measure.

Proof. Let $S_1, S_2, \ldots$ be any infinite sequence of disjoint sets from $\mathscr{A}$, which is such that $R = \bigcup_{k=1}^{\infty} S_k \in \mathscr{A}$. Then the sequence $R_1, R_2, \ldots$ with

$$R_n = \bigcup_{k=1}^{n} S_k$$

is monotonically increasing. Since $F(S)$ is additive, we have

$$F(R_n) = \sum_{k=1}^{n} F(S_k).$$

If $F(S)$ is continuous from below at $R = \bigcup_{k=1}^{\infty} S_k$, then

$$F\left(\bigcup_{k=1}^{\infty} S_k\right) = F(\lim_{n\to\infty} R_n) = \lim_{n\to\infty} F(R_n)$$

$$= \lim_{n\to\infty} \sum_{k=1}^{n} F(S_k) = \sum_{k=1}^{\infty} F(S_k),$$

which proves that $F(S)$ is σ additive.

If $F(S)$ is continuous from below at $\varnothing$, then since $R - R_1, R - R_2, \ldots$ is a monotonically decreasing sequence with $\lim_{n\to\infty}(R - R_n) = \varnothing$, we have

$$0 = \lim_{n\to\infty} F(R - R_n) = F(R) - \lim_{n\to\infty} F(R_n),$$

i.e.,

$$F(R) = \lim_{n\to\infty} F(R_n) = \sum_{k=1}^{\infty} F(S_k). \quad \text{Q.E.D.}$$

2.3. Extensions of Measures and Outer Measures

We saw in §1, Theorem 1.6, that a Boolean algebra is a monotone class if and only if it is a Boolean σ algebra. It would obviously be convenient if we could extend a measure μ defined on a Boolean algebra $\mathscr{A}$ of subsets of $\mathscr{X}$ to the Boolean σ algebra $\mathscr{A}_\sigma(\mathscr{A})$ generated by $\mathscr{A}$, and thus obtain a measure space $(\mathscr{X}, \mathscr{A}_\sigma(\mathscr{A}), \mu)$ in the sense of Definition 2.2.

***Theorem 2.3.** Let $\mu(S)$ be a measure defined on the Boolean algebra $\mathscr{A}$ of subsets of a given set $\mathscr{X}$. The set function

$$\text{(2.4)} \quad \bar{\mu}(R) = \inf\left\{\sum_{k=1}^{\infty} \mu(S_k)\colon \ \bigcup_{k=1}^{\infty} S_k \supset R, \quad S_1, S_2, \ldots \in \mathscr{A}\right\}, \qquad R \in \mathscr{A}_\sigma,$$

defined on the Boolean σ algebra $\mathscr{A}_\sigma = \mathscr{A}_\sigma(\mathscr{A})$ generated by $\mathscr{A}$, i.e., for all $R \in \mathscr{A}_\sigma$, is a measure on $\mathscr{A}_\sigma$ which coincides with $\mu(S)$ on $\mathscr{A}$:

$$\bar{\mu}(S) = \mu(S), \qquad S \in \mathscr{A}.$$

If $\mu(S)$ is a σ-finite measure, then $\bar{\mu}(S)$ is also σ finite, and $\bar{\mu}(S)$ is the only measure on $\mathscr{A}_\sigma$ which coincides with $\mu(S)$ on $\mathscr{A}$. The measure $\bar{\mu}(S)$ is called the *extension* of $\mu(S)$.

Consider the extended real-valued set function

$$\text{(2.5)} \qquad \mu^+(R) = \inf\left\{\sum_{k=1}^{\infty} \mu(S_k)\colon \ R \subset \bigcup_{k=1}^{\infty} S_k, \quad S_1, S_2, \ldots \in \mathscr{A}\right\}$$

for any given subset R of $\mathscr{X}$. Since for any such R there are countable families $S_1, S_2, \ldots \in \mathscr{A}$ which *cover* R, i.e., whose unions contain R, we can always take the infimum (greatest lower bound) of $\sum_{k=1}^{\infty} \mu(S_k)$ for all such families, i.e., $\mu^+(R)$ is defined on the family $\mathfrak{S}_{\mathscr{X}}$ of all subsets of $\mathscr{X}$, it is obviously nonnegative, and it satisfies the relation

$$\mu^+\left(\bigcup_{n=1}^{\infty} R_n\right) \leqslant \sum_{n=1}^{\infty} \mu^+(R_n) \tag{2.6}$$

for any $R_1, R_2, \ldots \in \mathfrak{S}_{\mathscr{X}}$.

In order to see that (2.6) is true, note that for each given $\epsilon > 0$ and each R_n there is, according to the definition in (2.5), a sequence $S_{n1}, S_{n2}, \ldots \in \mathscr{A}$ such that $R_n \subset \bigcup_{k=1}^{\infty} S_{nk}$, for which

$$\mu^+(R_n) \leqslant \sum_{k=1}^{\infty} \mu(S_{nk}) \leqslant \mu^+(R_n) + \frac{\epsilon}{2^n}. \tag{2.7}$$

Since $\{S_{nk}, n, k = 1, 2, \ldots\}$ is a countable family of sets from $\mathscr{A}$ and $\bigcup_{n=1}^{\infty} R_n \subset \bigcup_{n,k=1}^{\infty} S_{nk}$, we have by (2.7)

$$\begin{aligned} \mu^+\left(\bigcup_{n=1}^{\infty} R_n\right) &\leqslant \sum_{n=1}^{\infty} \sum_{k=1}^{\infty} \mu(S_{nk}) \\ &\leqslant \sum_{n=1}^{\infty} \left(\mu^+(R_n) + \frac{\epsilon}{2^n}\right) = \epsilon + \sum_{n=1}^{\infty} \mu^+(R_n). \end{aligned} \tag{2.8}$$

The relation (2.8) is true for an arbitrarily small $\epsilon > 0$ and consequently (2.6) follows.

Definition 2.4. A nonnegative set function $M(R)$, defined on each subset $R \in \mathfrak{S}_{\mathscr{X}}$ of a set $\mathscr{X}$, for which $M(\varnothing) = 0$ and which is *countably subadditive*, i.e., for which

$$M\left(\bigcup_{n=1}^{\infty} S_n\right) \leqslant \sum_{n=1}^{\infty} M(S_n) \tag{2.9}$$

for any sequence $S_1, S_2, \ldots \in \mathfrak{S}_{\mathscr{X}}$, is called an *outer measure in* $\mathscr{X}$. A subset S of $\mathscr{X}$ is said to be *measurable with respect to the outer measure* $M(R)$, or simply M *measurable*, if it satisfies the relation

$$M(R) = M(R \cap S) + M(R \cap S') \tag{2.10}$$

for any subset R of $\mathscr{X}$.

We have proved thus far that $\mu^+(R)$ is an outer measure in $\mathscr{X}$. In order to demonstrate that $\mu^+(R)$ is a measure on $\mathscr{A}_\sigma = \mathscr{A}_\sigma(\mathscr{A})$ we have to prove the next two lemmas.

***Lemma 2.1.** If $M(R)$, $R \in \mathfrak{S}_\mathscr{X}$, is an outer measure in $\mathscr{X}$, then the class $\mathscr{A}_M$ of all M-measurable sets $S \in \mathfrak{S}_\mathscr{X}$ is a Boolean σ algebra, and the outer measure $M(S)$, $S \in \mathscr{A}_M$, restricted to $\mathscr{A}_M$ is a measure.

Proof. We start by proving first that $\mathscr{A}_M$ is a Boolean algebra.
Since for any $R \in \mathfrak{S}_\mathscr{X}$,

$$M(R) = M(R \cap \mathscr{X}) = M(R \cap \varnothing) + M(R \cap \mathscr{X}),$$

we have $\varnothing \in \mathscr{A}_M$.

From the relation (2.10) it is evident that if S is M measurable, then S' is also M measurable. Therefore, we only have to prove that $S_1 \cup S_2$ is M measurable whenever S_1 and S_2 are M measurable.

If $S_1, S_2 \in \mathscr{A}_M$, then for any $R \in \mathfrak{S}_\mathscr{X}$

$$M(R) = M(R \cap S_1) + M(R \cap S_1'); \tag{2.11}$$

since also $R \cap S_1$, $R \cap S_1' \in \mathfrak{S}_\mathscr{X}$ we have

$$M(R \cap S_1) = M((R \cap S_1) \cap S_2) + M((R \cap S_1) \cap S_2'), \tag{2.12}$$

$$M(R \cap S_1') = M((R \cap S_1') \cap S_2) + M((R \cap S_1') \cap S_2'). \tag{2.13}$$

By substituting (2.12) and (2.13) into (2.11), and noting that

$$S_1' \cap S_2' = (S_1 \cup S_2)',$$

we arrive at

$$\begin{aligned} M(R) = {} & [M(R \cap S_1 \cap S_2) + M(R \cap S_1 \cap S_2') + M(R \cap S_1' \cap S_2)] \\ & + M(R \cap (S_1 \cup S_2)'). \end{aligned} \tag{2.14}$$

Since the above relation is true for any subset R of $\mathscr{X}$, we can replace R in (2.14) by $R \cap (S_1 \cup S_2)$. Then the last term on the right-hand side of (2.14) becomes $M(\varnothing) = 0$, while the terms in the brackets remain unchanged, because we easily get by means of Lemmas 1.1 and 1.2

$$[R \cap (S_1 \cup S_2)] \cap S_1 \cap S_2 = R \cap S_1 \cap S_2,$$

$$[R \cap (S_1 \cup S_2)] \cap S_1 \cap S_2' = R \cap S_1 \cap S_2',$$

$$[R \cap (S_1 \cup S_2)] \cap S_1' \cap S_2 = R \cap S_1' \cap S_2.$$

Therefore, we have

(2.15) $$M(R \cap (S_1 \cup S_2)) = M(R \cap S_1 \cap S_2) + M(R \cap S_1 \cap S_2') + M(R \cap S_1' \cap S_2),$$

i.e., (2.14) reads

(2.16) $$M(R) = M(R \cap (S_1 \cup S_2)) + M(R \cap (S_1 \cup S_2)'),$$

proving that $S_1 \cup S_2$ is M measurable.

In order to prove that $\mathscr{A}_M$ is also a Boolean σ algebra it is sufficient to consider sequences of *disjoint* sets $S_1, S_2, \ldots \in \mathscr{A}_M$; namely if $R_1, R_2, \ldots \in \mathscr{A}_M$ is any sequence, then it is easy to show that

$$\bigcup_{n=1}^{\infty} R_n = \bigcup_{n=1}^{\infty} \left(R_n - \bigcup_{k=1}^{n-1} R_k\right),$$

where the sets $S_n = R_n - \bigcup_{k=1}^{n-1} R_k$, $n = 1, 2, \ldots$, also belong to $\mathscr{A}_M$, which was shown to be a Boolean algebra, and are obviously disjoint.

If S_1 and S_2 are disjoint, we have $S_1 \cap S_2 = \varnothing$, $S_1 \cap S_2' = S_1$, $S_1' \cap S_2 = S_2$, and (2.15) becomes

$$M(R \cap (S_1 \cup S_2)) = M(R \cap S_1) + M(R \cap S_2).$$

By induction we get

(2.17) $$M\left(R \cap \left(\bigcup_{k=1}^{n} S_k\right)\right) = \sum_{k=1}^{n} M(R \cap S_k).$$

From the subadditivity of $M(R)$ it easily follows that

(2.18) $$M\left(R \cap \left(\bigcup_{k=1}^{n} S_k\right)'\right) \geqslant M\left(R \cap \left(\bigcup_{k=1}^{\infty} S_k\right)'\right)$$

because $(\bigcup_{k=1}^{n} S_k)' \supset (\bigcup_{k=1}^{\infty} S_k)'$. Since $\bigcup_{k=1}^{n} S_k$ is M measurable due to the established fact that $\mathscr{A}_M$ is a Boolean algebra, we get by using (2.17) and (2.18)

(2.19) $$M(R) = M\left(R \cap \left(\bigcup_{k=1}^{n} S_k\right)\right) + M\left(R \cap \left(\bigcup_{k=1}^{n} S_k\right)'\right)$$
$$\geqslant \sum_{k=1}^{n} M(R \cap S_k) + M\left(R \cap \left(\bigcup_{k=1}^{\infty} S_k\right)'\right).$$

Due to the countable subadditivity of $M(R)$, we have

$$M\left(R \cap \left(\bigcup_{k=1}^{\infty} S_k\right)\right) = M\left(\bigcup_{k=1}^{\infty} (R \cap S_k)\right) \leqslant \sum_{k=1}^{\infty} M(R \cap S_k). \tag{2.20}$$

Since (2.19) is true for any n, we obtain by letting $n \to \infty$ in (2.19) and using (2.20),

$$\begin{aligned} M(R) &\geqslant \sum_{k=1}^{\infty} M(R \cap S_k) + M\left(R \cap \left(\bigcup_{k=1}^{\infty} S_k\right)'\right) \\ &\geqslant M\left(R \cap \left(\bigcup_{k=1}^{\infty} S_k\right)\right) + M\left(R \cap \left(\bigcup_{k=1}^{\infty} S_k\right)'\right). \end{aligned} \tag{2.21}$$

Due to the subadditivity of $M(R)$, we also have

$$M(R) \leqslant M\left(R \cap \left(\bigcup_{k=1}^{\infty} S_k\right)\right) + M\left(R \cap \left(\bigcup_{k=1}^{\infty} S_k\right)'\right). \tag{2.22}$$

The relation (2.21) in conjunction with (2.22) shows that $\bigcup_{k=1}^{\infty} S_k$ is M measurable, i.e., $\bigcup_{k=1}^{\infty} S_k \in \mathscr{A}_M$, and hence $\mathscr{A}_M$ is a Boolean σ algebra!

In order to prove that $M(R)$ is σ additive on $\mathscr{A}_M$, we insert in (2.21) $R = \bigcup_{k=1}^{\infty} S_k$ to derive

$$M\left(\bigcup_{k=1}^{\infty} S_k\right) \geqslant \sum_{k=1}^{\infty} M(S_k),$$

which in conjunction with (2.9) yields the desired conclusion. Q.E.D.

***Lemma 2.2.** Every set R in the Boolean σ algebra $\mathscr{A}_\sigma$ generated by $\mathscr{A}$ is μ^+ measurable, i.e., $\mathscr{A}_\sigma$ is contained in $\mathscr{A}_{\mu^+}$.

Proof. Take any set $R \in \mathfrak{S}_{\mathscr{X}}$. By the definition (2.5) of $\mu^+(R)$, for any $\epsilon > 0$ there is a sequence $S_1, S_2, \ldots \in \mathscr{A}$ such that $R \subset \bigcup_{k=1}^{\infty} S_k$, for which

$$\mu^+(R) \leqslant \sum_{k=1}^{\infty} \mu(S_k) \leqslant \mu^+(R) + \epsilon. \tag{2.23}$$

For any set $S \in \mathscr{A}$ we have [since $S_k = (S_k \cap S) \cup (S_k \cap S')$, where $S_k \cap S$ and $S_k \cap S'$ are disjoint]

$$\begin{aligned} \sum_{k=1}^{\infty} \mu(S_k) &= \sum_{k=1}^{\infty} \mu(S_k \cap S) + \sum_{k=1}^{\infty} \mu(S_k \cap S') \\ &\geqslant \mu^+(R \cap S) + \mu^+(R \cap S'), \end{aligned} \tag{2.24}$$

where the last inequality is a consequence of the fact that $R \cap S \subset \bigcup_{k=1}^{\infty} (S_k \cap S)$ and $R \cap S' \subset \bigcup_{k=1}^{\infty} (S_k \cap S')$. By combining (2.23) and (2.24) we obtain

$$\mu^+(R) + \epsilon \geqslant \mu^+(R \cap S) + \mu^+(R \cap S'), \tag{2.25}$$

which holds for any $\epsilon > 0$. Since $\mu^+(R)$ is an outer measure and therefore is subadditive, we get

$$\mu^+(R) \leqslant \mu^+(R \cap S) + \mu^+(R \cap S'),$$

which in conjunction with (2.25) shows that

$$\mu^+(R) = \mu^+(R \cap S) + \mu^+(R \cap S'),$$

i.e., that $S \in \mathscr{A}$ is measurable.

Thus we have arrived at the result that the Boolean algebra $\mathscr{A}$ is contained in the Boolean σ algebra $\mathscr{A}_{\mu^+}$ of all μ^+-measurable subsets of $\mathscr{X}$. Since $\mathscr{A}_\sigma$ is the smallest Boolean σ algebra containing $\mathscr{A}$, we must have that $\mathscr{A}_\sigma \subset \mathscr{A}_{\mu^+}$. Q.E.D.

Lemma 2.2 tells us, *de facto*, that the set function $\bar{\mu}(R)$ defined by (2.4) is indeed an extension of $\mu(S)$, $S \in \mathscr{A}$, to $\mathscr{A}_\sigma$. From the very definition (2.4) of $\bar{\mu}(R)$ it is obvious that $\bar{\mu}(R)$ is σ finite in case that $\mu(S)$ is σ finite.

In order to see that $\bar{\mu}(S)$ is the only extension of $\mu(S)$ when $\mu(S)$ is σ finite, assume that $\bar{\mu}_1(R)$, $R \in \mathscr{A}_\sigma$, is also an extension of $\mu(S)$, i.e.,

$$\bar{\mu}_1(S) = \mu(S) = \bar{\mu}(S), \qquad S \in \mathscr{A}.$$

Let $R \in \mathscr{A}_\sigma$ be a set for which at least one of the two extensions is finite, say $\bar{\mu}(R) < +\infty$. Denote by $\mathfrak{M}(R)$ the family of all sets in $\mathscr{A}_\sigma$ which are subsets of R, i.e., they are of the form $R \cap S$, $S \in \mathscr{A}_\sigma$, and for which $\bar{\mu}_1$ and $\bar{\mu}$ are equal,

$$\bar{\mu}_1(R \cap S) = \bar{\mu}(R \cap S), \qquad R \cap S \in \mathfrak{M}(R).$$

$\mathfrak{M}(R)$ is a monotone class, because if $S_1, S_2, \ldots \in \mathfrak{M}(R)$ is a monotone sequence then, since $\bar{\mu}(S_n)$ is finite, we have

$$\bar{\mu}(\lim_{n\to\infty} S_n) = \lim_{n\to\infty} \bar{\mu}(S_n) = \lim_{n\to\infty} \bar{\mu}_1(S_n) = \bar{\mu}_1(\lim_{n\to\infty} S_n).$$

$\mathfrak{M}(R)$ obviously contains the Boolean algebra (see Exercise 1.8) $\mathscr{A}_R = \{R \cap S: S \in \mathscr{A}\}$. Since $\mathfrak{M}(R)$ is a monotone class, it must coincide with the Boolean σ algebra $\mathscr{A}_{\sigma,R} = \{R \cap S: S \in \mathscr{A}_\sigma\}$ generated (see

Exercise 1.8) by $\mathscr{A}$. Obviously $R \in \mathscr{A}_{\sigma,R}$ and therefore $R \in \mathfrak{M}(R)$, which means that $\bar{\mu}(R) = \bar{\mu}_1(R)$.

If $R_0 \in \mathscr{A}_\sigma$ is a set of infinite $\bar{\mu}$ measure, it can be covered by an infinite sequence $R_1, R_2, \ldots \in \mathscr{A}_\sigma$,

$$R_0 \subset \bigcup_{n=1}^{\infty} R_k,$$

of sets R_k of finite $\bar{\mu}$ measure because $\bar{\mu}(R_0)$ is assumed σ finite. It is easy to check that

$$R_0 = \bigcup_{n=1}^{\infty} \left[R_0 \cap \left(R_n - \bigcup_{k=1}^{n-1} R_k\right)\right],$$

and since $S^{(n)} = R_0 \cap (R_n - \bigcup_{k=1}^{n-1} R_k)$ is obviously a set from $\mathscr{A}_\sigma$ and subset of R_n, its $\bar{\mu}$ measure is finite. According to the above we have then

$$\bar{\mu}_1(S^{(n)}) = \bar{\mu}(S^{(n)}),$$

and therefore

$$\bar{\mu}_1(R_0) = \sum_{n=1}^{\infty} \bar{\mu}_1(S^{(n)}) = \sum_{n=1}^{\infty} \bar{\mu}(S^{(n)}) = \bar{\mu}(R_0),$$

which proves the uniqueness of $\bar{\mu}(R)$.

Theorem 2.3 is of great importance to us because it tells us that all the σ-finite measures defined on $\mathscr{B}_0^n$ have a unique extension to the family $\mathscr{B}^n$ of all Borel sets. This conclusion applies in particular to the Lebesgue measure introduced at the beginning of this section for the sets in $\mathscr{B}_0^1$, as well as to any probability measure (which is always finite) that has been defined on $\mathscr{B}_0^n$.

Due to Theorem 2.3, when studying measures we can restrict ourselves from now on, without loss of generality, to measures defined on Boolean σ algebras, i.e., to the study of *measure* spaces (see Definition 2.2).

2.4. Cartesian Products of Measure Spaces

An important way of building a new measure space from two or more measure spaces $\mathscr{X}_1, \ldots, \mathscr{X}_n$ is by introducing the product measure on the direct product $\mathscr{X}_1 \times \cdots \times \mathscr{X}_n$ of the sets $\mathscr{X}_1, \ldots, \mathscr{X}_n$.

Theorem 2.4. If $(\mathscr{X}_1, \mathscr{A}_1, \mu_1), \ldots, (\mathscr{X}_n, \mathscr{A}_n, \mu_n)$ are n measure spaces, there is a measure $\mu(S)$ defined on the Boolean σ algebra $\mathscr{A}$ of subsets of the set $\mathscr{X} = \mathscr{X}_1 \times \cdots \times \mathscr{X}_n$, such that $\mathscr{A}$ is generated by the family $\mathscr{A}_1 \times \cdots \times \mathscr{A}_n$ of subsets of $\mathscr{X}$, and has the property that for any $S_1 \in \mathscr{A}_1, \ldots, S_n \in \mathscr{A}_n$

$$\mu(S_1 \times \cdots \times S_n) = \mu(S_1) \cdots \mu(S_n); \tag{2.26}$$

this measure is unique if the measure spaces $(\mathscr{X}_1, \mathscr{A}_1, \mu_1), \ldots, (\mathscr{X}_n, \mathscr{A}_n, \mu_n)$ are σ finite.

The above measure μ is called the *product* of the measures $\mu_1, \ldots, \mu_n$, and the measure space $(\mathscr{X}, \mathscr{A}, \mu)$ is called the *Cartesian product* of the measure spaces $(\mathscr{X}_1, \mathscr{A}_1, \mu_1), \ldots, (\mathscr{X}_n, \mathscr{A}_n, \mu_n)$.

In order to prove the above theorem, we establish first the following lemma.

***Lemma 2.3.** If $\mathscr{A}_1, \ldots, \mathscr{A}_n$ are Boolean algebras of subsets of the sets $\mathscr{X}_1, \ldots, \mathscr{X}_n$, respectively, then the family $\mathscr{P}(\mathscr{A}_1, \ldots, \mathscr{A}_n)$ of finite unions of sets belonging to $\mathscr{A}_1 \times \cdots \times \mathscr{A}_n$ is a Boolean algebra of subsets of $\mathscr{X} = \mathscr{X}_1 \times \cdots \times \mathscr{X}_n$, and $\mathscr{P}(\mathscr{A}_1, \ldots, \mathscr{A}_n)$ is identical with the family of *disjoint* unions of sets from $\mathscr{A}_1 \times \cdots \times \mathscr{A}_n$.

Proof. It is obvious that $\mathscr{P}(\mathscr{A}_1, \ldots, \mathscr{A}_n)$ is closed under the operations of taking unions of sets from $\mathscr{P}(\mathscr{A}_1, \ldots, \mathscr{A}_n)$. Furthermore, if $S \in \mathscr{P}(\mathscr{A}_1, \ldots, \mathscr{A}_n)$, then S can be written in the form

$$S = \bigcup_{k=1}^{m} (S_1^k \times \cdots \times S_n^k), \qquad S_1^k \in \mathscr{A}_1, \ldots, S_n^k \in \mathscr{A}_n, \qquad k = 1, \ldots, m.$$

Since for any $R = R_1 \times \cdots \times R_n \in \mathscr{A}_1 \times \cdots \times \mathscr{A}_n$, we have

$$\begin{aligned} R' = \bigcup_{j=1}^{n} \bigcup_{Z_1 \in \{R_1, R_1'\}} \cdots \bigcup_{Z_n \in \{R_n, R_n'\}} (Z_1 \times \cdots \times R_j' \times \cdots \times Z_n) \\ \in \mathscr{P}(\mathscr{A}_1, \ldots, \mathscr{A}_n), \end{aligned} \tag{2.27}$$

we get by application of Lemmas 1.1 and 1.2

$$S' = \bigcap_{k=1}^{m} (S_1^k \times \cdots \times S_n^k)' = \bigcup_{i=1}^{r} T^i,$$

where r is some positive integer, and each T^i is an intersection of a finite number of sets from $\mathscr{A}_1 \times \cdots \times \mathscr{A}_n$. But, since for an intersection T of sets $R_1^j \times \cdots \times R_n^j$, $j = 1, \ldots, s$, from $\mathscr{A}_1 \times \cdots \times \mathscr{A}_n$ we have

$$T = \bigcap_{j=1}^{s} (R_1^j \times \cdots \times R_n^j) = \left(\bigcap_{j_1=1}^{s} R_1^{j_1}\right) \times \cdots \times \left(\bigcap_{j_n=1}^{s} R_n^{j_n}\right) \tag{2.28}$$

[i.e., T is also an element of $\mathscr{P}(\mathscr{A}_1, \ldots, \mathscr{A}_n)$], we conclude that $S' \in \mathscr{P}(\mathscr{A}_1, \ldots, \mathscr{A}_n)$. Thus, $\mathscr{P}(\mathscr{A}_1, \ldots, \mathscr{A}_n)$ is a Boolean algebra.

Any element

$$S = \bigcup_{k=1}^{m} S^k, \qquad S^1, \ldots, S^m \in \mathscr{A}_1 \times \cdots \times \mathscr{A}_n$$

of $\mathscr{P}(\mathscr{A}_1, \ldots, \mathscr{A}_n)$ can be written as the disjoint union of sets $R^j = S^j - \bigcup_{k=1}^{j-1} S^k$,

$$S = S^1 \cup (S^1 - S^2) \cup \cdots \cup \left(S^m - \bigcup_{k=1}^{m-1} S^k\right)$$

$$= S^1 \cup (S^1 \cap (S^2)') \cup \cdots \cup \left[S^m \cap \left(\bigcap_{k=1}^{m-1} (S^k)'\right)\right].$$

By applying (2.27) and (2.28) on each of the sets $R^1, \ldots, R^m$ in the above union, we arrive by induction at the conclusion that $R^1, \ldots, R^m$ are disjoint unions of sets from $\mathscr{A}_1 \times \cdots \times \mathscr{A}_n$, i.e., that S can be written as a disjoint union of sets from $\mathscr{A}_1 \times \cdots \times \mathscr{A}_n$. Q.E.D.

The Boolean algebra $\mathscr{P}(\mathscr{A}_1, \ldots, \mathscr{A}_n)$ is obviously the Boolean algebra generated by $\mathscr{A}_1 \times \cdots \times \mathscr{A}_n$. It is easy to check that the set function defined on a set $R \in \mathscr{P}(\mathscr{A}_1, \ldots, \mathscr{A}_n)$

$$R = \bigcup_{k=1}^{m} R^k, \qquad R^k = R_1^k \times \cdots \times R_n^k \in \mathscr{A}_1 \times \cdots \times \mathscr{A}_n,$$

$$R^i \cap R^k = \varnothing \qquad \text{for} \quad i \neq k,$$

by the expression

$$\mu(R) = \sum_{k=1}^{m} \mu_1(R_1^k) \cdots \mu_n(R_n^k) \tag{2.29}$$

is a measure. Due to Lemma 2.3, $\mu(R)$ is defined by (2.29) on the entire Boolean algebra $\mathscr{P}(\mathscr{A}_1, \ldots, \mathscr{A}_n)$. Evidently, $\mu(R)$ is σ additive if $\mu_1(R_1), \ldots, \mu_n(R_n)$ are σ additive.

The Boolean σ algebra $\mathscr{A}$ generated by $\mathscr{A}_1 \times \cdots \times \mathscr{A}_n$ is obviously identical with the Boolean σ algebra generated by $\mathscr{P}(\mathscr{A}_1, \ldots, \mathscr{A}_n)$. Consequently, by Theorem 2.3, there is an extension of $\mu(R)$, $R \in \mathscr{P}(\mathscr{A}_1, \ldots, \mathscr{A}_n)$, to $\mathscr{A}$, which is unique when $\mu(R)$ is σ finite. Thus, the proof of Theorem 2.4 is completed.

Note that we can apply Theorem 2.4 to $\mu_1(B) = \cdots = \mu_n(B) = \mu_l^{(1)}(B)$, $B \in \mathscr{B}^1$, in order to arrive at a measure $\mu_l^{(n)}(B)$, $B \in \mathscr{B}^n$, called the Lebesgue measure in $\mathscr{B}^n$ (see also Exercise 2.3). For an interval

$I = (a_1, b_1) \times \cdots \times (a_n, b_n) \in \mathscr{B}^n$, we have according to (2.29) and the definition of $\mu_l^{(1)}(B)$

$$\mu_l^{(n)}(I) = (b_1 - a_1) \cdots (b_n - a_n).$$

EXERCISES

2.1. Show that every measure $\mu(S)$ on a Boolean algebra $\mathscr{A}$ is a *monotone set function*, i.e., if $S \subset R$, $R, S \in \mathscr{A}$, then $\mu(R) \geqslant \mu(S)$, and a *subtractive set function*, i.e., $\mu(R - S) = \mu(R) - \mu(S)$ whenever $S \subset R$, $R, S \in \mathscr{A}$.

2.2. Define a set function $\mu_\delta^{(n)}(B)$ on the Boolean σ algebra $\mathscr{B}^n$ of all Borel sets in $\mathscr{B}^n$ by writing $\mu_\delta^{(n)}(B) = 1$ if $B \in \mathscr{B}^n$ contains the origin of $\mathbb{R}^n$, and $\mu_\delta^{(n)}(B) = 0$ if B does not contain the origin. Show that this set function is a finite measure; we shall call this measure the δ *measure*.

2.3. Let $\rho(x_1, ..., x_n)$ be any nonnegative Riemann integrable function on $\mathbb{R}^n$. Show that the set function $\mu_\rho(B)$ on the Boolean algebra $\mathscr{B}_0^n$, which is defined for intervals $I \in \mathscr{I}^n$ by the Riemann integral

$$\mu_\rho(I) = \int_I \rho(x_1, ..., x_n)\, dx_1 \cdots dx_n ,$$

and for a general element $B = I_1 \cup \cdots \cup I_k$, $(I_1, ..., I_k \in \mathscr{I}^n, I_i \cap I_j = \varnothing$ if $i \neq j)$ of $\mathscr{B}_0^n$ by

$$\mu_\rho(B) = \mu_\rho(I_1) + \cdots + \mu_\rho(I_k)$$

is a measure (called a *Lebesgue–Stieltjes measure* induced by ρ) if we take $\mu_l(\varnothing) = 0$. Why is $\mu_l(B)$ necessarily σ finite?

Note. If $\rho(x_1, ..., x_n) \equiv 1$, the corresponding measure $\mu_l(B) = \mu_l^{(n)}(B)$ is called the *Lebesgue measure* on $\mathscr{B}^n$.

2.4. Is it true for $\mu_l^{(1)}(B)$ defined by (2.2) that $\mu^{(1)}(\lim_{k\to\infty} B_k) = \lim_{k\to\infty} \mu_l^{(1)}(B_k)$ for *every* monotonically decreasing sequence of Borel sets in $\mathscr{B}^1$?

2.5. If $(\mathscr{X}, \mathscr{A}, \mu)$ is a measure space, the set function

$$\mu_+(R) = \sup\{\mu(S): S \subset R, S \in \mathscr{A}\}$$

is obviously an extended real-valued function defined for every subset $R \in \mathfrak{S}_\mathscr{X}$ of $\mathscr{X}$; this set function is called the *inner measure* induced by μ. Show that if $R_1, R_2, ..., \in \mathfrak{S}_\mathscr{X}$ is any disjoint sequence of sets, then

$$\mu_+\left(\bigcup_{n=1}^{\infty} R_n\right) \geqslant \sum_{n=1}^{\infty} \mu_+(R_n).$$

*3. Measurable and Integrable Functions

3.1. The Concept of a Measurable Function

In the present section our main purpose is to define and study the integral

$$\int f(\xi)\, d\mu(\xi)$$

of a function $f(\xi)$, $\xi \in \mathscr{X}$, defined on a measure space* $\mathscr{X} = (\mathscr{X}, \mathscr{A}, \mu)$. To do this we need the concept of a measurable function.

Definition 3.1. A real-valued function $f(\xi)$, $\xi \in \mathscr{X}$, defined on a measurable subset R of the measurable space $\mathscr{X} = (\mathscr{X}, \mathscr{A})$ is said to be *measurable on R* if for every open interval $I \subset \mathbb{R}^1$ the subset

$$f^{-1}(I) = \{\xi\colon \xi \in R, f(\xi) \in I\} \tag{3.1}$$

of $\mathscr{X}$ is measurable, i.e., it belongs to $\mathscr{A}$. If $f(\xi)$ is an extended real-valued function, i.e., if it can also assume the values $\pm\infty$, then $f(\xi)$ is said to be measurable on R if in addition to the sets (3.1) the sets

$$f^{-1}(+\infty) = \{\xi\colon \xi \in R, f(\xi) = +\infty\},$$

$$f^{-1}(-\infty) = \{\xi\colon \xi \in R, f(\xi) = -\infty\}$$

are also measurable. An extended complex-valued function $h(\xi)$ is measurable if its real part $\operatorname{Re} h(\xi)$ and imaginary part $\operatorname{Im} h(\xi)$ are measurable.

In case that the function $f(\xi)$, $\xi \in \mathbb{R}^n$, is measurable in $(\mathbb{R}^n, \mathscr{B}^n)$, i.e., with respect to the family of all Borel sets, then $f(\xi)$ is said to be *Borel measurable*; if $f(\xi)$ is measurable with respect to the larger class of *Lebesgue measurable* sets—i.e., with respect to the sets $S \subset \mathbb{R}^n$ which are measurable (see Definition 2.4) with respect to the Lebesgue outer measure $\mu_l^{+(n)}(R)$, $R \in \mathfrak{S}_{\mathbb{R}^n}$, defined by (2.5) from the measure $\mu_l^{(n)}(B)$, $B \in \mathscr{B}_0^n$,

$$\mu_l^{(n)}(B) = \int_B dx_1 \cdots dx_n \qquad \text{(Riemann integration on } B\text{)}$$

—then $f(\xi)$ is said to be Lebesgue measurable. Note that the class of all Lebesgue measurable sets is, according to Theorem 2.3, a Boolean

* According to the agreement we have followed until now, we denote the measure space $(\mathscr{X}, \mathscr{A}, \mu)$—consisting of the set $\mathscr{X}$ on which a Boolean σ algebra $\mathscr{A}$ of subsets of $\mathscr{X}$ and a measure $\mu(R)$, $R \in \mathscr{A}$, are defined—also by the letter $\mathscr{X}$.

σ algebra which contains* the family $\mathscr{B}^n$ of all Borel sets in $\mathbb{R}^n$; therefore, every Borel measurable function is also Lebesgue measurable.

The *characteristic function* $\chi_S(\xi)$ of any subset S of a set $\mathscr{X}$ is defined by

$$\chi_S(\xi) = \begin{cases} 1 & \text{for} \quad \xi \in S \\ 0 & \text{for} \quad \xi \notin S. \end{cases} \tag{3.2}$$

As a simple example of a measurable function we can take the characteristic function $\chi_R(\xi)$ of a measurable set R.

A slightly more complicated example is provided by a simple function on a measurable space $\mathscr{X}$.

Definition 3.2. A function $f(\xi)$ on a measurable space is called *simple* if it is of the form

$$f(\xi) = \sum_{k=1}^{n} a_k \chi_{S_k}(\xi), \qquad S_i \cap S_j = \varnothing \quad \text{for} \quad i \neq j,$$

where $a_1, \ldots, a_n$ are, in general, complex numbers and $\chi_{S_k}(\xi)$ is the characteristic function (3.2) of the set S_k.

As an example of Borel measurable functions on $\mathbb{R}^n$ we have the continuous functions. Namely, if $f(x)$ is a real continuous function defined on an open subset of $\mathbb{R}^n$ and I is an open interval in $\mathbb{R}^1$, then $f^{-1}(I)$ is an open set in $\mathbb{R}^n$; therefore $f^{-1}(I)$ is a Borel set in $\mathbb{R}^n$ by Theorem 1.4. Since the real and imaginary parts of a continuous complex function defined on an open subset of $\mathbb{R}^n$ are also continuous, we can state the following.

Theorem 3.1. Every continuous complex function defined on an open subset R of $\mathbb{R}^n$ is Borel measurable on R.

3.2. Properties of Measurable Functions

The measurable functions are important to us only to the extent that an integrable function has to be first of all measurable. In the theory of integration we shall need certain general properties of measurable functions, which we state in Theorems 3.2–3.5.

Theorem 3.2. If the extended real-valued function $f(\xi)$ defined on the measurable subset R of the measurable space $\mathscr{X} = (\mathscr{X}, \mathscr{A})$ is measurable, and if B is a Borel set on the real line, then $f^{-1}(B)$ is a measurable set in $\mathscr{X}$.

* It can be proved [Halmos, 1950, §13] that every Lebesgue measurable set is of the form $B \triangle S$, where $B \in \mathscr{B}^n$ and S is a subset of a Borel set B_1, with B_1 of zero Lebesgue measure, i.e., $\mu_l^{(n)}(B_1) = 0$.

Proof. Denote by $\mathscr{F}$ the family of all subsets S of $\mathbb{R}^1$ whose inverse images $f^{-1}(S)$ in $\mathscr{X}$ are measurable sets. We will show that $\mathscr{F}$ is a Boolean σ algebra.

Since we can write for any $S_1, S_2, \ldots \in \mathscr{F}$

$$f^{-1}\left(\bigcup_{k=1}^{\infty} S_k\right) = \bigcup_{k=1}^{\infty} f^{-1}(S_k),$$

we conclude that $\bigcup_{k=1}^{\infty} S_k \in \mathscr{F}$. Furthermore, for any subset S of $\mathbb{R}^1$ we have

$$f^{-1}(S') = [f^{-1}(S)]', \qquad S' = \mathbb{R}^1 - S, \qquad [f^{-1}(S)]' = \mathscr{X} - f^{-1}(S).$$

Consequently, if $S \in \mathscr{F}$, then we see that $f^{-1}(S') \in \mathscr{A}$, i.e., $f^{-1}(S')$ is measurable, and therefore $S' \in \mathscr{F}$.

Thus, $\mathscr{F}$ is a Boolean σ algebra which contains the family of all open intervals in $\mathbb{R}^1$. Every closed interval I_c in $\mathbb{R}^1$ has as a complement either an open interval (if I_c is infinite) or the union of two open intervals (if I_c is finite) and therefore it has to belong to the Boolean σ algebra $\mathscr{F}$. Since any other element of $\mathscr{I}^1$ can be written as the difference of two closed intervals (e.g., $[a, b) = [a, b] - \{b\}$), we conclude that $\mathscr{I}^1 \subset \mathscr{F}$. Consequently, $\mathscr{F}$ must contain the Boolean σ algebra $\mathscr{B}^1$ generated by $\mathscr{I}^1$.

Q.E.D.

Theorem 3.3. Let $f(\xi)$, $\xi \in R$, be a real measurable function on the measurable set $R \subset \mathscr{X}$, and let $g(x)$, $x \in \mathbb{R}^1$, be a Borel measurable real function defined on a subset of the real line, which contains the range of $f(\xi)$. Then the composite function $h(\xi) = g[f(\xi)]$ is a measurable function.

Proof. Since $g(x)$ is Borel measurable, for every open interval $I \in \mathscr{I}^1$ the set $g^{-1}(I)$ is a Borel set. Consequently $f^{-1}[g^{-1}(I)]$ is measurable in $\mathscr{X}$ by Theorem 3.2, because $f(\xi)$ is measurable, i.e.,

$$h^{-1}(I) = f^{-1}[g^{-1}(I)]$$

is measurable. Q.E.D.

Note that it follows immediately from the above theorem that for any real number a the functions $af(\xi)$, $a + f(\xi)$ and $|f(\xi)|^a$ are measurable whenever $f(\xi)$ is measurable; we get this result by taking $g(x)$, $x \in \mathbb{R}^1$, to be equal to, respectively, the continuous (except at $x = 0$ in the case of $|x|^a$, $a < 0$) and therefore Borel measurable functions ax, $a + x$, and $|x|^a$.

In order to prove Theorem 3.4 we need Lemmas 3.1 and 3.2 (see also Exercise 3.1).

Lemma 3.1. If $f(\xi)$ and $g(\xi)$, $\xi \in R$, are real-valued measurable functions on the measurable set R, then the set

$$\{\xi: \ f(\xi) > g(\xi)\}$$

is a measurable set.

Proof. Since the set $\mathfrak{R}$ of all rational numbers is countable, we can write it in the form of a sequence

$$\mathfrak{R} = \{r_1, r_2, ...\}.$$

For each $r_k \in \mathfrak{R}$ the sets

$$f^{-1}((r_k, +\infty)) = \{\xi: f(\xi) > r_k\},$$
$$g^{-1}((-\infty, r_k)) = \{\xi: g(\xi) < r_k\}$$

are measurable on $\mathscr{X}$, because they are inverse images of open intervals. It is easy to see that

$$\{\xi: f(\xi) > g(\xi)\} = \bigcup_{k=1}^{\infty} (\{\xi: f(\xi) > r_k\} \cap \{\xi: g(\xi) < r_k\}) \tag{3.3}$$

by noting that whenever for some $\xi \in R$ we have $f(\xi) > g(\xi)$, there must be a rational number r for which $f(\xi) > r > g(\xi)$. Since $\{\xi: f(\xi) > r_k\} \cap \{\xi: g(\xi) < r_k\}$ as the intersection of two measurable sets is also measurable, we conclude from (3.3) that $\{\xi: f(\xi) > g(\xi)\}$ is measurable. Q.E.D.

Lemma 3.2. A real-valued function $h(\xi)$, $\xi \in \mathscr{X}$, on the measurable set R is measurable if for any real number c

$$h^{-1}((-\infty, c)) = \{\xi: h(\xi) < c\}$$

is a measurable subset of $\mathscr{X}$.

Proof. For any finite open interval (a, b) we have

$$\begin{aligned} h^{-1}((a, b)) &= \{\xi: a < h(\xi) < b\} \\ &= \{\xi: h(\xi) < b\} - \{\xi: h(\xi) \leqslant a\}. \end{aligned} \tag{3.4}$$

Since we can write $\{\xi: h(\xi) \leqslant a\}$ as an intersection of a countable number of sets which are measurable according to the assumption

$$\{\xi: h(\xi) \leqslant a\} = \bigcap_{n=1}^{\infty} \{\xi: h(\xi) < a + 1/n\},$$

we conclude that $\{\xi: h(\xi) \leqslant a\}$ is measurable (see Exercise 1.9), i.e., both sets appearing on the right-hand side of (3.4) are measurable, and therefore $h^{-1}((a, b))$ is measurable. Q.E.D.

Theorem 3.4. If $g_1(x), g_2(x),...$ is a sequence of extended complex-valued functions which are measurable on the measurable set $R \subset \mathscr{X}$, and if

$$g(x) = \lim_{n\to\infty} g_n(x)$$

exists, then $g(x)$ is an extended complex-valued function which is measurable on R.

Proof. It is easy to check with the help of Lemma 3.1 (see Exercise 3.3) that the functions

$$h_n(x) = \sup_{k=n,n+1,\ldots} \operatorname{Re} g_k(x)$$

are measurable on R. [$\operatorname{Im} g_k(x)$ are treated similarly]. Since we have

$$\operatorname{Re} g(x) = \lim_{n\to\infty} \operatorname{Re} g_n(x) = \lim_{n\to\infty} h_n(x) = \inf_{n=1,2,\ldots} h_n(x)$$

because $h_1(x) \geqslant h_2(x) \geqslant \cdots$, and as $\inf_{n=1,2,\ldots} h_n(x)$ is measurable if $h_1(x), h_2(x),...$ are measurable functions (see Exercise 3.3), we conclude that $g(x)$ is measurable on R. Q.E.D.

Theorem 3.5. If $f(\xi)$ and $g(\xi)$, $\xi \in R$, are extended real-valued measurable functions defined on the measurable set R, and if

$$\{f(\xi) = \pm\infty\} \cap \{g(\xi) = \mp\infty\} = \varnothing, \tag{3.5}$$

then the functions $f(\xi) + g(\xi)$ and $f(\xi)\, g(\xi)$ are also measurable.

Proof. Due to the requirement (3.5), the functions $f(\xi) + g(\xi)$ and $f(\xi)\, g(\xi)$ are defined everywhere. The sets

$$\{f(\xi) + g(\xi) = \pm\infty\} = \{f(\xi) = \pm\infty\} \cup \{g(\xi) = \pm\infty\},$$

$$\{f(\xi)\, g(\xi) = \pm\infty\}$$
$$= [\{f(\xi) = \pm\infty\} \cap \{g(\xi) > 0\}] \cup [\{f(\xi) > 0\} \cap \{g(\xi) = \pm\infty\}]$$

obviously are measurable. Therefore, we can concentrate on the set of points at which $f(\xi)$ and $g(\xi)$ are finite.

For any real number c we can write

$$\begin{aligned} &\{\xi:\ -\infty < f(\xi) + g(\xi) < c\} \\ &= \{\xi:\ -\infty < f(\xi) < c - g(\xi)\} \cap \{\xi:\ g(\xi) > -\infty\} \end{aligned} \tag{3.6}$$

and since $c - g(\xi)$ is measurable, we conclude from Lemma 3.1 that the set on the left-hand side of (3.6) is measurable. Therefore, $f(\xi) + g(\xi)$ is measurable by Lemma 3.2.

The measurability of $f(\xi)\,g(\xi)$ follows immediately from the relation

$$f(\xi)\,g(\xi) = \tfrac{1}{4}\{|f(\xi) + g(\xi)|^2 - |f(\xi) - g(\xi)|^2\},$$

which is true whenever $f(\xi) \neq \pm\infty$, $g(\xi) \neq \pm\infty$. Q.E.D.

3.3. Positive-Definite Integrable Functions

In discussing the concept of measurable functions it was sufficient to consider only measurable spaces $\mathscr{X} = (\mathscr{X}, \mathscr{A})$. For the study of the general theory of integration we need functions $f(\xi)$, $\xi \in \mathscr{X}$, defined on *measure* spaces $\mathscr{X} = (\mathscr{X}, \mathscr{A}, \mu)$. In order to introduce the general concept of integral of such functions we have to start gradually by defining first integration for the simple functions introduced in Definition 3.2.

Definition 3.3. We say that the simple function (see Definition 3.2)

$$s(\xi) = \sum_{k=1}^{n} a_k \chi_{S_k}(\xi),$$

defined on a measurable set R of the measure space $\mathscr{X} = (\mathscr{X}, \mathscr{A}, \mu)$ is *integrable* on R if

$$\mu(R \cap \{\xi: s(\xi) \neq 0\}) < \infty.$$

Its integral with respect to μ is then defined* to be

$$\int_R s(\xi)\, d\mu(\xi) = \sum_{k=1}^{n} a_k \mu(S_k \cap R). \tag{3.7}$$

Next we define the concept of an integral on $\mathscr{X}$ for nonnegative functions:

Definition 3.4. An extended real-valued *nonnegative* function $f(\xi)$ defined on a measurable subset R of a measure space $\mathscr{X} = (\mathscr{X}, \mathscr{A}, \mu)$ is said to be *integrable on* R if $f(\xi)$ is measurable on R, and if there is a nondecreasing sequence

$$s_1(\xi) \leqslant s_2(\xi) \leqslant \cdots \tag{3.8}$$

* In the case that $\mu(R \cap \{\xi : s(\xi) = 0\}) = +\infty$, the sum appearing in (3.7) is still well defined, because we have agreed in §2 to write $0 \cdot (+\infty) = 0$.

of simple functions, each integrable on R, which converges at each point $\xi \in R$ to $f(\xi)$,

$$f(\xi) = \lim_{n\to\infty} s_n(\xi), \tag{3.9}$$

and is such that $\lim_{n\to\infty} \int s_n(\xi)\, d\mu(\xi)$ exists and is finite. The integral of $f(\xi)$ on R is defined as

$$\int_R f(\xi)\, d\mu(\xi) = \lim_{n\to\infty} \int_R s_n(\xi)\, d\mu(\xi). \tag{3.10}$$

Note that the above definition by which $f(\xi)$ is the limit of simple integrable functions implies that the measure of $R \cap \{\xi : f(\xi) > 0\}$ is σ finite, because

$$R \cap \{\xi : f(\xi) > 0\} = \bigcup_{k=1}^{\infty} [R \cap \{\xi : s_k(\xi) > 0\}],$$

and since $s_k(\xi)$ is integrable, $\mu(\{\xi : s_k(\xi) > 0\}) < +\infty$.

Conversely, if $S_0 = R \cap \{\xi : f(\xi) > 0\}$ is σ finite, we can write $S_0 = \bigcup_{n=1}^{\infty} R_n$, where $R_1, R_2, \ldots$ is a monotonically increasing sequence of sets of finite measure. The sequence $s_n(\xi)$, $n = 1, 2, \ldots$, of simple functions defined by

$$s_n(\xi) = \sum_{k=1}^{1+n2^n} \frac{k-1}{2^n} \chi_{S_k^{(n)}}(\xi),$$

$$\begin{aligned} S_1^{(n)} &= S_0 - R_n, \\ S_k^{(n)} &= R_n \cap \left\{\xi : \frac{k-1}{2^n} \leqslant f(\xi) < \frac{k}{2^n}\right\}, \qquad k = 2, \ldots, n2^n, \\ S_{1+n2^n}^{(n)} &= R_n \cap \{\xi : f(\xi) \geqslant n\} \end{aligned} \tag{3.11}$$

is a sequence of *integrable* simple functions which satisfies the requirements (3.8) and (3.9). Thus, if $\mu(R)$ is σ finite, we can always define the limit appearing in (3.10).

Now we have to establish that if the limit (3.10) exists, then it is independent of the choice of the sequence of simple functions. For that we need the following lemma.

Lemma 3.3. If $0 \leqslant s_1(\xi) \leqslant s_2(\xi) \leqslant \cdots$, $\xi \in R$, is a nondecreasing sequence of simple functions integrable on the measurable set R in the measure space $\mathscr{X} = (\mathscr{X}, \mathscr{A}, \mu)$, and if $s(\xi)$, $\xi \in R$, is also a simple function integrable on R, and such that

$$0 \leqslant s(\xi) \leqslant \lim_{n\to\infty} s_n(\xi), \tag{3.12}$$

then

$$\int_R s(\xi)\, d\mu(\xi) \leqslant \lim_{n\to\infty} \int_R s_n(\xi)\, d\mu(\xi). \tag{3.13}$$

Proof. Since $s(\xi)$ is integrable on R, the set

$$R_0 = \{\xi : s(\xi) > 0\}$$

is a measurable subset of R, and we obviously have by (3.7)

$$\int_R s(\xi)\, d\xi = \int_{R_0} s(\xi)\, d\xi.$$

Assume that $\mu(R_0) > 0$—because if $\mu(R_0) = 0$, (3.13) would already be true.

For each $n = 1, 2, \ldots$ the set

$$S_n = \left\{\xi : s_n(\xi) \geqslant s(\xi) - \frac{\epsilon}{\mu(R_0)}\right\} \cap R_0$$

is μ measurable (see Lemma 3.1 and Exercise 3.1). Since $s_1(\xi) \leqslant s_2(\xi) \leqslant \cdots$, we have $S_1 \subset S_2 \subset \cdots$; due to (3.12) we can deduce that $\lim_{n\to\infty} S_n = R_0$. By the continuity from below of $\mu(S)$ (Theorem 2.1), we get

$$\lim_{n\to\infty} \mu(S_n) = \mu(R_0). \tag{3.14}$$

The simple function $s(\xi)$ is necessarily bounded: $s(\xi) \leqslant M$. Thus, by taking into consideration the rules for dealing with integrals of simple functions (see Exercises 3.5 and 3.6), and since $\int_{R-S_n} s_n(\xi)\, d\mu(\xi) \geqslant 0$ because $s_n(\xi) \geqslant 0$, we get

$$\begin{aligned} \int_R s_n(\xi)\, d\mu(\xi) &\geqslant \int_{S_n} s_n(\xi)\, d\mu(\xi) \\ &\geqslant \int_{S_n} \left(s(\xi) - \frac{\epsilon}{\mu(R_0)}\right) d\mu(\xi) \\ &= \int_{R_0} s(\xi)\, d\mu(\xi) - \int_{R_0-S_n} s(\xi)\, d\mu(\xi) - \frac{\epsilon}{\mu(R_0)}\, \mu(S_n). \end{aligned} \tag{3.15}$$

When $n \to \infty$, we find from (3.14) that the second term on the right-hand side of (3.15) vanishes because

$$\int_{R_0-S_n} s(\xi)\, d\mu(\xi) \leqslant M\mu(R_0 - S_n),$$

while the third term converges to ϵ. Thus we have arrived at the conclusion that

$$\lim_{n\to\infty} \int_R s_n(\xi)\, d\mu(\xi) \geqslant \int_R s(\xi)\, d\mu(\xi) - \epsilon$$

is satisfied for every $\epsilon > 0$; therefore, (3.13) holds. Q.E.D.

Theorem 3.6. If $s_1(\xi), s_2(\xi),\ldots$ and $s_1'(\xi), s_2'(\xi),\ldots$ are two nondecreasing sequences of nonnegative simple functions converging at each $\xi \in R$ to $f(\xi) \geqslant 0$, then the limits $\lim_{n\to\infty} \sigma_n$ and $\lim_{n\to\infty} \sigma_n'$ of the integrals

$$\sigma_n = \int_R s_n(\xi)\, d\mu(\xi),$$

$$\sigma_n' = \int_R s_n'(\xi)\, d\mu(\xi)$$

are equal in case that at least one of the sequences $\sigma_1, \sigma_2,\ldots$ and $\sigma_1', \sigma_2',\ldots$ converges.

Proof. Assume that $\sigma_1, \sigma_2,\ldots$ converges. For any n we have $s_n'(\xi) \leqslant f(\xi) = \lim_{k\to\infty} s_k(\xi)$, and therefore by Lemma 3.3

$$\sigma_n' \leqslant \lim_{k\to\infty} \sigma_k .$$

It is easy to establish from (3.7) that $\sigma_1', \sigma_2',\ldots$ is a monotonic nondecreasing sequence, which consequently converges because it is bounded. We have

$$\lim_{n\to\infty} \sigma_n' \leqslant \lim_{k\to\infty} \sigma_k .$$

By noting that again according to Lemma 3.3

$$\sigma_k \leqslant \lim_{n\to\infty} \sigma_n'$$

for any $k = 1, 2,\ldots$, we conclude that

$$\lim_{k\to\infty} \sigma_k \leqslant \lim_{n\to\infty} \sigma_n'.$$

Hence $\lim_{k\to\infty} \sigma_k = \lim_{n\to\infty} \sigma_n'$. Q.E.D.

Lemma 3.4. If $f(\xi)$, $\xi \in R$, is an extended real-valued measurable function, then the functions

$$(3.16)\qquad \begin{aligned} f^+(\xi) &= \max\{f(\xi), 0\}, \\ f^-(\xi) &= f^+(\xi) - f(\xi) = \max\{-f(\xi), 0\} \end{aligned}$$

are also measurable.

The above lemma follows immediately from Theorem 3.3 if we note that the function $g(x) = \max\{x, 0\}$ is a continuous function on $\mathbb{R}^1$.

3.4. Real and Complex Integrable Functions

Definition 3.5. The extended *real-valued* measurable function $f(\xi)$ defined on a measurable subset R of a measure space $\mathscr{X} = (\mathscr{X}, \mathscr{A}, \mu)$ is *μ integrable on R* if its positive part $f^+(\xi)$ and its negative part $f^-(\xi)$, defined by (3.16), are μ integrable on R. Then its integral is defined to be

$$\int_R f(\xi)\, d\mu(\xi) = \int_R f^+(\xi)\, d\mu(\xi) - \int_R f^-(\xi)\, d\mu(\xi). \tag{3.17}$$

An extended *complex-valued* measurable function $h(\xi)$, $\xi \in R$, is *integrable on R* if $\operatorname{Re} h(\xi)$ and $\operatorname{Im} h(\xi)$ are integrable on R, and its integral is taken to be

$$\int_R h(\xi)\, d\mu(\xi) = \int_R \operatorname{Re} h(\xi)\, d\mu(\xi) + i \int_R \operatorname{Im} h(\xi)\, d\mu(\xi). \tag{3.18}$$

We have thus arrived at a very general concept of an integral. When we consider integration on the real line, i.e., $\mathscr{X} = \mathbb{R}^1$, and we take $R = [a, b]$, we obtain a generalization of the Riemann integral

$$\int_a^b f(x)\, dx$$

of the function $f(x)$ if we take the measure space $(\mathbb{R}^1, \mathscr{B}^1, \mu_l^{(1)})$, with the Lebesgue measure $\mu_l^{(1)}(B)$, $B \in \mathscr{B}^1$. For instance, assume that $f(x) \geqslant 0$ is continuous in $[a, b]$. Then by Theorem 3.1 $f(x)$ is measurable [note that $\mu_l^{(1)}(\{a\}) = \mu_l^{(1)}(\{b\}) = 0$]. If we build for a partition $a = x_0 < x_1 < \cdots < x_n = b$ of $[a, b]$ the Riemann sum

$$\sigma_n = \sum_{k=1}^{n} a_k(x_k - x_{k-1}), \qquad a_k = \inf\{f(x): x_{k-1} \leqslant x < x_k\},$$

we note that σ_n corresponds to the integral (3.7),

$$\sigma_n = \int_{[a,b]} s_n(x)\, d\mu_l^{(1)}(x),$$

of the simple function

$$s_n(x) = \sum_{k=1}^{n} a_k \chi_{S_k}(x), \qquad S_k = \{x: x_{k-1} \leqslant x < x_k\}.$$

If we take a sequence of finer and finer partitions of $[a, b]$ in which the $(n+1)$st partition is a subpartition of the nth one, so that $\max_{k=1,\ldots,n} | x_k - x_{k+1} | \to 0$ when $n \to \infty$, then obviously the non-decreasing sequence $s_1(x) \leqslant s_2(x) \leqslant \cdots$ converges to $f(x)$, and by Definition 3.4

$$\int_{[a,b]} f(x)\, d\mu_l^{(1)}(x) = \lim_{n\to\infty} \sigma_n . \tag{3.19}$$

On the other hand, from the definition of the Riemann integral we also have

$$\int_a^b f(x)\, dx = \lim_{n\to\infty} \sigma_n ,$$

proving the equality of these two integrals.

The integral

$$\int_R f(x)\, d\mu_l^{(1)}(x)$$

of a function on $\mathbb{R}^1$ is called the Lebesgue integral. It is ordinarily denoted by

$$\int_R f(x)\, dx,$$

like the Riemann integral, to which it is equal in the case that the Riemann integral exists. However, as we can see in simple examples (see Exercise 3.4), the Lebesgue integral exists also in cases in which the Riemann integral does not exist. In fact, the following theorem (first proved by Lebesgue, whose proof we do not reproduce*) is true.

Theorem 3.7. A bounded function defined on $[a, b] \subset \mathbb{R}^1$ is Riemann integrable on $[a, b]$ if and only if it is continuous almost everywhere on $[a, b]$. In that case the Lebesgue integral of $f(x)$ on $[a, b]$ exists and is equal to the Riemann integral of $f(x)$ on $[a, b]$.

In Theorem 3.7 the expression "almost everywhere" appears. This expression is very frequently used in the theory of measure.

Definition 3.6. A statement concerning every element $\xi \in S$ of a measurable set S in measure space $\mathscr{X}$ is said to be true *almost everywhere* in S, if it is true on a set $R \subset S$ whose complement $S - R$ with respect to S is of measure zero, i.e., $\mu(R) = \mu(S)$.

The fact that sets of measure zero are not essential in computing integrals can be observed in the following theorem.

* See Munroe [1953, Theorems 24.4, 24.5].

Theorem 3.8. If $h(\xi)$, $\xi \in R$, is any measurable* function on the set R of measure zero, then $h(\xi)$ is integrable on R and

$$\int_R h(\xi)\, d\mu(\xi) = 0. \tag{3.20}$$

Consequently, if $f(\xi) = g(\xi)$ almost everywhere on a measurable set S, and $f(\xi)$ is integrable on S, then $g(\xi)$ is integrable on S and

$$\int_S f(\xi)\, d\mu(\xi) = \int_S g(\xi)\, d\mu(\xi). \tag{3.21}$$

Proof. In case that $h(\xi) \geqslant 0$, (3.20) can be very easily proved by resorting to sequences (3.8) of simple functions, and noting that in our case (3.7) is always zero. By means of Definition 3.5 this result can be immediately generalized to any real, and consequently any complex measurable function. The second part of the theorem and (3.21) follow by taking $h(\xi) = f(\xi) - g(\xi)$. Q.E.D.

Theorem 3.9. If $f(\xi)$ and $g(\xi)$ are measurable functions defined on the measurable set R, if $|f(\xi)| \leqslant g(\xi)$ almost everywhere on R, and if $g(\xi)$ is integrable on R, then $f(\xi)$ is also integrable on R, and

$$\left| \int_R f(\xi)\, d\mu(\xi) \right| \leqslant \int_R |f(\xi)|\, d\mu(\xi) \leqslant \int_R g(\xi)\, d\mu(\xi).$$

Proof. From Definition 3.5 of the integral for real and complex functions it is obvious that the theorem is true in general if it is true for non-negative functions $f(\xi)$, since $(\mathrm{Re}\, f(\xi))^{\pm} \leqslant |f(\xi)|$, $(\mathrm{Im}\, f(\xi))^{\pm} \leqslant |f(\xi)|$.

Consider therefore the case when $f(\xi) \geqslant 0$. Since $|f(\xi)| \leqslant g(\xi)$ on a set $R_0 \subset R$, where $\mu(R - R_0) = 0$, we have

$$[R_0 \cap \{\xi: f(\xi) > 0\}] \subset [R_0 \cap \{\xi: g(\xi) > 0\}].$$

Therefore, as $R \cap \{\xi: g(\xi) > 0\}$ has a σ-finite measure because $g(\xi)$ is integrable on R, the set $R \cap \{\xi: f(\xi) > 0\}$ has to be also of σ-finite measure. Consequently, the sequence $s_1(\xi) \leqslant s_2(\xi) \leqslant \cdots$ of simple functions (3.11) which are integrable on R and for which $\lim_{n\to\infty} s_n(\xi) = f(\xi)$, can be built.

* In measure theory the measure spaces are usually extended to so-called complete measure spaces, having the property that *any* subset of a set of measure zero is also measurable and of measure zero. This device makes any function $h(\xi)$, $\xi \in R$, automatically measurable on the set R when $\mu(R) = 0$.

By Lemma 3.3, given any nondecreasing sequence $s_1'(\xi), s_2'(\xi),\ldots$ of simple functions converging to $g(\xi)$ for every $\xi \in R$, we have

$$\sigma_n = \int_R s_n(\xi)\, d\mu(\xi) \leqslant \lim_{k\to+\infty} \int_R s_k'(\xi)\, d\mu(\xi) = \int_R g(\xi)\, d\mu(\xi)$$

for every $n = 1, 2,\ldots$. Consequently, the sequence $\sigma_1 \leqslant \sigma_2 \leqslant \cdots$ is bounded from above and therefore it has a finite limit, which by definition is

$$\lim_{n\to\infty} \sigma_n = \int_R f(\xi)\, d\mu(\xi). \quad \text{Q.E.D.}$$

3.5. Infinite Sequences and Sums of Integrals

***Theorem 3.10** (*Lebesgue monotone convergence theorem*). Let $f_1(\xi), f_2(\xi),\ldots$ be a monotone sequence of nonnegative functions, each integrable on the measurable set R. The extended real-valued function

$$f(\xi) = \lim_{n\to\infty} f_n(\xi), \qquad \xi \in R, \tag{3.22}$$

is integrable on R and

$$\int_R f(\xi)\, d\mu(\xi) = \lim_{n\to\infty} \int_R f_n(\xi)\, d\mu(\xi), \tag{3.23}$$

provided that in case of a monotonically increasing sequence the limit in (3.23) is finite.

Proof. Consider first the case when $f_1(\xi) \leqslant f_2(\xi) \leqslant \cdots$. For each $n = 1, 2,\ldots$, let $s_{n1}(\xi), s_{n2}(\xi),\ldots$ be a nondecreasing sequence of simple functions for which

$$\lim_{k\to\infty} s_{nk}(\xi) = f_n(\xi),$$

and consequently

$$\lim_{k\to\infty} \int_R s_{nk}(\xi)\, d\mu(\xi) = \int_R f_n(\xi)\, d\mu(\xi).$$

For each positive integer k the function

$$s'_{nk}(\xi) = \max_{m=1,\ldots,n} s_{mk}(\xi)$$

is evidently a simple integrable function. From the above definition and the fact that $s_{nk}(\xi) \leqslant s_{nk+1}(\xi)$ we get

$$s'_{kk}(\xi) \leqslant s'_{kk+1}(\xi) \leqslant s'_{k+1k+1}(\xi),$$

i.e., $s'_{11}(\xi) \leqslant s'_{22}(\xi) \leqslant \cdots$. Since $s'_{kk}(\xi) \leqslant f_k(\xi)$, we have for each n

$$f_n(\xi) \leqslant \lim_{k\to\infty} s'_{kk}(\xi) \leqslant f(\xi), \qquad n = 1, 2, \ldots .$$

By using (3.22) we deduce from the above inequality that

$$\lim_{k\to\infty} s'_{kk}(\xi) = f(\xi). \tag{3.24}$$

Since $s'_{kk}(\xi) \leqslant f_k(\xi)$, we can infer that

$$\lim_{k\to\infty} s'_{kk}(\xi) \leqslant \lim_{k\to\infty} f_k(\xi) < +\infty. \tag{3.25}$$

The function $f(\xi)$ is measurable over R because $f(\xi) = \sup_{n\to\infty} f_n(\xi)$ (see Exercise 3.3). Furthermore, (3.24) and (3.25) show that $f(\xi)$ is integrable over R and that

$$\int_R f(\xi)\, d\mu(\xi) = \lim_{k\to\infty} \int_R s'_{kk}(\xi)\, d\mu(\xi) \leqslant \lim_{n\to\infty} \int_R f_n(\xi)\, d\mu(\xi). \tag{3.26}$$

On the other hand, since $f_n(\xi) \leqslant f(\xi)$, we have according to Theorem 3.9

$$\int_R f_n(\xi)\, d\mu(\xi) \leqslant \int_R f(\xi)\, d\mu(\xi), \qquad n = 1, 2, \ldots . \tag{3.27}$$

The inequalities (3.26) and (3.27) yield (3.23).

In case that $f_1(\xi) \geqslant f_2(\xi) \geqslant \cdots$, the sequence $g_1(\xi), g_2(\xi), \ldots$,

$$g_n(\xi) = f_1(\xi) - f_n(\xi), \qquad n = 1, 2, \ldots,$$

is nondecreasing, and according to the above, its limit

$$g(\xi) = \lim_{n\to\infty} g_n(\xi) = f_1(\xi) - \lim_{n\to\infty} f_n(\xi)$$

is integrable over R:

$$\int_R g(\xi)\, d\mu(\xi) = \lim_{n\to\infty} \int_R g_n(\xi)\, d\mu(\xi) \leqslant \int_R f_1(\xi)\, d\mu(\xi).$$

Consequently, the function

$$\lim_{n\to\infty} f_n(\xi) = f_1(\xi) - g(\xi)$$

is integrable over R, and since

$$\int_R \lim_{n\to\infty} f_n(\xi)\, d\mu(\xi) = \int_R f_1(\xi)\, d\mu(\xi) - \int_R g(\xi)\, d\mu(\xi),$$

we see that (3.23) is again true. Q.E.D.

***Theorem 3.11.** Let $R_1, R_2, \ldots$ be an infinite sequence of disjoint measurable sets, and let

$$R = \bigcup_{n=1}^{\infty} R_n$$

be their union. If the extended real-valued function $f(\xi)$ is integrable on each set R_n, then $f(\xi)$ is integrable on R if and only if

$$\sum_{n=1}^{\infty} \int_{R_n} |f(\xi)|\, d\mu(\xi) < +\infty,$$

and in that case

$$\int_R f(\xi)\, d\mu(\xi) = \sum_{n=1}^{\infty} \int_{R_n} f(\xi)\, d\mu(\xi).$$

Proof. In case that $f(\xi) \geqslant 0$, we can introduce the functions

$$f_n(\xi) = \begin{cases} f(\xi) & \text{for } \xi \in R_1 \cup \cdots \cup R_n \\ 0 & \text{for } \xi \in R - \bigcup_{k=1}^{n} R_k, \end{cases}$$

which are obviously integrable on R. The sequence $f_1(\xi) \leqslant f_2(\xi) \leqslant \cdots$ has $f(\xi)$ as a limit, and satisfies the conditions of Theorem 3.10. Thus we have shown that $f(\xi)$ is integrable on R if and only if

$$\lim_{n\to\infty} \sum_{k=1}^{n} \int_{R_k} f(\xi)\, d\mu(\xi) = \lim_{n\to\infty} \int_R f_n(\xi)\, d\mu(\xi) = \int_R f(\xi)\, d\mu(\xi),$$

which proves the theorem for the case when $f(\xi) \geqslant 0$.

The more general case when $f(\xi)$ is not nonnegative everywhere in R can be reduced to the already considered case by treating $f^+(\xi)$ and $f^-(\xi)$ separately. Q.E.D.

3.6. Integration on Cartesian Products of Measure Spaces

We shall study now the problem of integration on Cartesian products of measure spaces in relation to integration on each of the spaces entering in such a Cartesian product. In order to simplify the notation, we shall limit ourselves to the statement and proof of results for products of two measure spaces. These results can very easily be generalized to the case of products of any finite number of measure spaces.

***Theorem 3.12.** If the extended real-valued function $f(\xi, \eta)$, $\xi \in \mathscr{X}$, $\eta \in \mathscr{Y}$, is measurable on the Cartesian product $(\mathscr{X} \times \mathscr{Y}, \mathscr{A})$ of the measurable spaces $\mathscr{X} = (\mathscr{X}, \mathscr{A}_1)$ and $\mathscr{Y} = (\mathscr{Y}, \mathscr{A}_2)$, then the functions

$$f_\eta(\xi) = f(\xi, \eta), \qquad \xi \in \mathscr{X},$$
$$f_\xi(\eta) = f(\xi, \eta), \qquad \eta \in \mathscr{Y},$$

are measurable functions on $\mathscr{X}$ and $\mathscr{Y}$, respectively.

Proof. We shall give only the proof for $f_\eta(\xi)$ since the proof for $f_\xi(\eta)$ proceeds along identical lines.

Let $I \in \mathscr{I}^1$ be a one-dimensional interval. We have for fixed $\eta \in \mathscr{Y}$

$$\begin{aligned} f_\eta^{-1}(I) &= \{\xi : f_\eta(\xi) \in I\} = \{\xi : f(\xi, \eta) \in I\} \\ &= \{\xi : (\xi, \eta) \in f^{-1}(I)\} = [f^{-1}(I)]_\eta; \end{aligned} \tag{3.28}$$

we have introduced above the notation in which, for any subset S of $\mathscr{X} \times \mathscr{Y}$, we denote by S_η the subset of $\mathscr{X}$ containing the points ξ for which (ξ, η) belongs to S, i.e.,

$$S_\eta = \{\xi : (\xi, \eta) \in S\}, \qquad S \subset \mathscr{X} \times \mathscr{Y}.$$

We shall prove now that if $S \subset \mathscr{X} \times \mathscr{Y}$ is measurable in $\mathscr{X} \times \mathscr{Y}$, then for any given $\eta \in \mathscr{Y}$ the set $S_\eta \subset \mathscr{X}$ is measurable in $\mathscr{X}$.

If $S \subset \mathscr{X} \times \mathscr{Y}$ is of the form $S^1 \times S^2$ and if S is measurable, i.e., if $S \in \mathscr{A}$, then by definition $S^1 \in \mathscr{A}_1$ and $S^2 \in \mathscr{A}_2$. Since in this case

$$S_\eta = \begin{cases} S^1, & \text{if } \eta \in S^2 \\ \varnothing, & \text{if } \eta \notin S^2, \end{cases}$$

and $\varnothing \in \mathscr{A}_1$, it follows that S_η is measurable in $\mathscr{X}$.

Denote by $\mathscr{F}$ the family of all subsets S of $\mathscr{X} \times \mathscr{Y}$ which are such that $S_\eta \in \mathscr{A}^1$ for all $\eta \in \mathscr{Y}$. It is easy to verify that $\mathscr{F}$ is a Boolean σ algebra. Since, according to the above, $\mathscr{F}$ contains $\mathscr{A}_1 \times \mathscr{A}_2$, while by definition $\mathscr{A}$ is the smallest Boolean σ algebra containing $\mathscr{A}_1 \times \mathscr{A}_2$, it follows that $\mathscr{A} \subset \mathscr{F}$.

Thus, we have proved that $S_\eta \in \mathscr{A}_1$ for all $\eta \in \mathscr{Y}$ whenever $S \in \mathscr{A}$. In particular we can deduce from (3.28) that

$$f_\eta^{-1}(I) = [f^{-1}(I)]_\eta \in \mathscr{A}_1$$

for all $I \in \mathscr{I}^1$, since $f(\xi, \eta)$ is measurable, and consequently $f^{-1}(I) \in \mathscr{A}$.

Q.E.D.

***Theorem 3.13** (*Fubini's theorem*). Let $(\mathscr{X}, \mathscr{A}_1, \mu_1)$ and $(\mathscr{Y}, \mathscr{A}_2, \mu_2)$ be σ-finite measure spaces and $(\mathscr{X} \times \mathscr{Y}, \mathscr{A}, \mu)$ the Cartesian product of these spaces. If the function $f(\xi, \eta)$ is integrable on the measurable set $S \in \mathscr{A}$, then the functions

$$g(\xi) = \int_{S_\xi} f(\xi, \eta)\, d\mu_2(\eta), \qquad \xi \in S^1 = \{\xi' : (\xi', \eta') \in S\}$$

(3.29)

$$h(\eta) = \int_{S_\eta} f(\xi, \eta)\, d\mu_1(\xi), \qquad \eta \in S^2 = \{\eta' : (\xi', \eta') \in S\}$$

are almost everywhere defined and integrable on $S^1 \subset \mathscr{X}$ and $S^2 \subset \mathscr{Y}$ respectively, and

$$(3.30)\quad \int_S f(\xi, \eta)\, d\mu(\xi, \eta) = \int_{S^1} g(\xi)\, d\mu_1(\xi) = \int_{S^1} d\mu_1(\xi) \int_{S_\xi} f(\xi, \eta)\, d\mu_2(\eta)$$

$$= \int_{S^2} h(\eta)\, d\mu_2(\eta) = \int_{S^2} d\mu_2(\eta) \int_{S_\eta} f(\xi, \eta)\, d\mu_1(\xi).$$

Proof. We shall first prove the theorem for the case when $f(\xi, \eta)$ is the characteristic function $\chi_S(\xi, \eta)$ of a measurable set $S \in \mathscr{A}$.

Consider the case when $S = S^1 \times S^2$, $S^1 \in \mathscr{A}_1$, $S^2 \in \mathscr{A}_2$. We have

$$g(\xi) = \chi_{S^1}(\xi)\, \mu_2(S^2), \qquad h(\eta) = \chi_{S^2}(\eta)\, \mu_1(S^1),$$

and consequently

$$\int_{S^1} g(\xi)\, d\mu_1(\xi) = \mu_1(S^1)\, \mu_2(S^2) = \int_{S^2} h(\eta)\, d\mu_2(\eta).$$

Since

$$\int_S \chi_S(\xi, \eta)\, d\xi\, d\eta = \mu(S) = \mu(S^1 \times S^2) = \mu_1(S^1)\, \mu_2(S^2),$$

the relation (3.30) is verified for the case where $S \in \mathscr{A}_1 \times \mathscr{A}_2$.

Denote by $\mathscr{K}$ the class of sets $S \in \mathscr{A}$ which are such that for their characteristic functions $\chi_S(\xi, \eta)$ the theorem is true. It can be inferred directly from Theorem 3.10 that $\mathscr{K}$ is a monotone class. We have seen that each $S \in \mathscr{A}_1 \times \mathscr{A}_2$ belongs to $\mathscr{K}$. Furthermore, we can extend this result to any finite (or even countably infinite) disjoint union of sets from $\mathscr{A}_1 \times \mathscr{A}_2$ by using Theorem 3.10. Since the family of all finite disjoint unions of sets from $\mathscr{A}_1 \times \mathscr{A}_2$ is identical to the Boolean algebra generated by $\mathscr{A}_1 \times \mathscr{A}_2$ (see Lemma 2.3), which is therefore contained in $\mathscr{K}$, we arrive at the desired result that $\mathscr{A} \subset \mathscr{K}$ by employing Theorem 1.7.

Thus, we have proved that the theorem is true whenever $f(\xi, \eta)$ is the characteristic function of a measurable set; this result is correct even where the integration is carried out over a set S which is different from the set R whose characteristic function is being integrated, as can be easily derived from the relation

$$\int_S \chi_R(\xi)\, d\mu(\xi) = \int_{R \cap S} \chi_{R \cap S}(\xi)\, d\mu(\xi).$$

From this statement we can infer immediately that the theorem is true for any simple function which is integrable on $S \in \mathscr{A}$.

Now let $f(\xi, \eta)$ be a nonnegative function integrable on $S \in \mathscr{A}$. Choose a sequence $f_1(\xi, \eta) \leqslant f_2(\xi, \eta) \leqslant \cdots$ of simple functions integrable on R, for which almost everywhere

$$\lim_{n\to\infty} f_n(\xi, \eta) = f(\xi, \eta),$$

and consequently

$$\lim_{n\to\infty} \int_S f_n(\xi, \eta)\, d\mu(\xi, \eta) = \int_S f(\xi, \eta)\, d\mu(\xi, \eta). \tag{3.31}$$

The sequence of functions $g_1(\xi), g_2(\xi),\ldots,$

$$g_n(\xi) = \int_{S_\xi} f_n(\xi, \eta)\, d\mu_2(\eta),$$

is nondecreasing, and by Theorem 3.10 we have, almost everywhere,

$$\lim_{n\to\infty} g_n(\xi) = \int_S f(\xi, \eta)\, d\mu_2(\eta) = g(\xi).$$

A second application of Theorem 3.10 yields the facts that $g(\xi)$ is integrable on S^1 and that

$$\int_{S^1} g(\xi)\, d\mu_1(\xi) = \lim_{n\to\infty} \int_{S^1} g_n(\xi)\, d\mu_1(\xi). \tag{3.32}$$

Since $f_n(\xi, \eta)$, $n = 1, 2,\ldots,$ are simple functions for which the theorem has been already established, we have

$$\int_{S^1} g_n(\xi)\, d\mu_1(\xi) = \int_S f_n(\xi, \eta)\, d\mu(\xi, \eta); \tag{3.33}$$

hence, by combining (3.31)–(3.33) we get

$$\int_{S^1} g(\xi)\, d\mu_1(\xi) = \lim_{n\to\infty} \int_S f(\xi, \eta)\, d\mu(\xi, \eta).$$

The equivalent result for $h(\xi)$ can be derived by the same procedure.

With the theorem established for nonnegative functions, it is trivial to extend it to real and then complex functions by using Definition 3.5.

Q.E.D.

It is essential to realize that when we are presented with an iterated integral

$$\int_{S_1} d\mu_1(\xi) \int_{S_2} d\mu_2(\eta) f(\xi, \eta), \tag{3.34}$$

we cannot apply immediately Fubini's theorem to invert the order of integration in ξ and η. We could justify such a procedure only if we knew not only that (3.34) exists, but also that $f(\xi, \eta)$ is integrable on $S_1 \times S_2$ in the measure $\mu_1 \times \mu_2$. The following theorem frequently provides help in such situations.

***Theorem 3.14** (*Tonelli's theorem*). Let $(\mathscr{X}, \mathscr{A}_1, \mu_1)$ and $(\mathscr{Y}, \mathscr{A}_2, \mu_2)$ be σ-finite measure spaces and $(\mathscr{X} \times \mathscr{Y}, \mathscr{A}, \mu)$ their Cartesian product. Suppose the function $f(\xi, \eta)$ on $S_1 \times S_2 \subset \mathscr{X} \times \mathscr{Y}$ is measurable on $S_1 \times S_2$, that $|f(\xi, \eta)|$ is integrable on S_2 for all $\xi \in S_1$, with the possible exception of a set $R_0 \subset S_1$ with $\mu_1(R_0) = 0$, and that the iterated integral

$$\int_{S_1 - R_0} d\mu_1(\xi) \int_{S_2} d\mu_2(\eta)\, |f(\xi, \eta)| \tag{3.35}$$

exists. Then the integral

$$\int_{S_1 \times S_2} f(\xi, \eta)\, d\mu(\xi, \eta) = \int_{S_1 \times S_2} f(\xi, \eta)\, d\mu_1(\xi)\, d\mu_2(\eta) \tag{3.36}$$

also exists.

Proof. Let $g(\xi, \eta)$ be a function which is positive and integrable on $S_1 \times S_2$; for example, if $S^{(1)}, S^{(2)},\ldots$ are sets of finite μ measure which cover* $S_1 \times S_2$, i.e., $S^{(1)} \cup S^{(2)} \cup \cdots \equiv S_1 \times S_2$, then the function

$$g(\xi, \eta) = \sum_n (1/n^2)\, [\mu(S^{(n)})]^{-1} \chi_{S^{(n)}}(\xi, \eta)$$

provides an adequate choice if $\mu(S^{(n)}) > 0$ for all n.

* Recall that $\mathscr{X} \times \mathscr{Y}$, and therefore also $S_1 \times S_2$, is of σ-finite μ measure, and therefore such a covering exists.

Let us introduce the functions

$$f_n(\xi, \eta) = \inf\{|f(\xi, \eta)|, ng(\xi, \eta)\}, \qquad n = 1, 2, \ldots .$$

The function $f_n(\xi, \eta)$ is measurable, since it is equal to the measurable function $ng(\xi, \eta)$ on the measurable set (see Lemma 3.1)

$$\{(\xi, \eta): ng(\xi, \eta) < f(\xi, \eta)\}$$

and equal to $f(\xi, \eta)$ on the complement of that set. Moreover, by Theorem 3.9 $f_n(\xi, \eta)$ is integrable in $S_1 \times S_2$, since $|f_n(\xi, \eta)| \leqslant ng(\xi, \eta)$ and $g(\xi, \eta)$ is integrable on $S_1 \times S_2$. Hence, by Fubini's theorem, the function $f_n(\xi, \eta)$ is integrable in η over S_2, with the possible exception of a set $R_n \subset S_1$ with $\mu_1(R_n) = 0$.

Let us introduce the set $R_0 = R_1 \cup R_2 \cup \cdots$ of μ_1 measure zero. All functions $f_n(\xi, \eta)$ are integrable on $(S_1 - R_0) \times S_2$. Since $f_n(\xi, \eta) \leqslant |f(\xi, \eta)|$, we obtain by applying Fubini's theorem and Theorem 3.9

$$\begin{aligned} \int_{S_1 \times S_2} f_n(\xi, \eta)\, d\mu(\xi, \eta) &= \int_{S_1 - R_0} d\mu_1(\xi) \int_{S_2} f_n(\xi, \eta)\, d\mu_2(\eta) \\ &\leqslant \int_{S_1 - R_0} d\mu_1(\xi) \int_{S_2} |f(\xi, \eta)|\, d\mu_2(\eta), \end{aligned}$$

which shows that the integrals of $f_n(\xi, \eta)$ on $S_1 \times S_2$ are bounded. Hence, Theorem 3.10 can be applied to infer that since $|f(\xi, \eta)| = \lim f_n(\xi, \eta)$, the function $|f(\xi, \eta)|$ is integrable on $S_1 \times S_2$. Q.E.D.

If the conditions of the above theorem are fulfilled, then the order of integration in (3.34) can be inverted. This conclusion can be reached by using Tonelli's theorem to infer that $f(\xi, \eta)$ is integrable on $S_1 \times S_2$, and then applying Fubini's theorem to justify the reversal of the order of integration.

Exercises

3.1. By employing Lemma 3.1, show that if the functions $f(\xi)$ and $g(\xi)$ are measurable, then the sets

$$\{\xi: f(\xi) \geqslant g(\xi)\}, \qquad \{\xi: f(\xi) = g(\xi)\}$$

are measurable.

3.2. By employing the same procedure as in proving Lemma 3.2, show that an extended real-valued function is measurable if and only if for every real number c the set $\{\xi: -\infty \leqslant f(\xi) \leqslant c\}$ is measurable. Prove that the same statement is true if we consider instead of the sets $\{\xi: -\infty \leqslant f(\xi) \leqslant c\}$, the sets $\{\xi: c \leqslant f(\xi) \leqslant +\infty\}$, or the sets $\{\xi: c < f(\xi) \leqslant +\infty\}$.

3.3. Prove that if $f_1(\xi), f_2(\xi), \ldots$ is a sequence of extended real-valued measurable functions on the measurable space $\mathscr{X}$, then the functions $\sup_{n=1,2,\ldots} f_n(\xi)$ and $\inf_{n=1,2,\ldots} f_n(\xi)$ are also measurable.

3.4. Show that for the simple function defined on $R = [0, 1]$ by

$$f(x) = \begin{cases} 4 & \text{for } x \text{ irrational} \\ 2 & \text{for } x \text{ rational} \end{cases}$$

the Lebesgue integral over R exists and

$$\int_R f(x)\, d\mu_l^{(1)}(x) = 4.$$

3.5. Derive directly from the Definition 3.3 that if $s_1(\xi)$ and $s_2(\xi)$ are two simple functions integrable on R, then

$$\int_R (s_1(\xi) + s_2(\xi))\, d\mu(\xi) = \int_R s_1(\xi)\, d\mu(\xi) + \int_R s_2(\xi)\, d\mu(\xi),$$

and that in case $s_1(\xi) \geqslant s_2(\xi)$, $\xi \in R$, we have

$$\int_R s_1(\xi)\, d\mu(\xi) \geqslant \int_R s_2(\xi)\, d\mu(\xi).$$

3.6. Show, by using only the Definition 3.3, that if R_1 and R_2 are two disjoint measurable sets and the simple function $s(\xi)$ is μ integrable on R_1 and R_2, then $s(\xi)$ is μ integrable on $R_1 \cup R_2$ and

$$\int_{R_1 \cup R_2} s(\xi)\, d\mu(\xi) = \int_{R_1} s(\xi)\, d\mu(\xi) + \int_{R_2} s(\xi)\, d\mu(\xi).$$

3.7. Explain why a real function $f(x)$ is μ integrable on a set R if and only if it is absolutely μ integrable on R, i.e., if $\int_R |f(x)|\, d\mu(x)$ exists.

3.8. Show that if $f(\xi)$ and $g(\xi)$ are μ integrable functions on R, then for any $a, b \in \mathbb{C}^1$, the function $af(\xi) + bg(\xi)$ is μ integrable in R and

$$\int_R (af(\xi) + bg(\xi))\, d\mu(\xi) = a \int_R f(\xi)\, d\mu(\xi) + b \int_R g(\xi)\, d\mu(\xi).$$

3.9. Show that if $f(\xi) \geqslant g(\xi)$ on the measurable set R, and if $f(\xi)$ and $g(\xi)$ are integrable on R, then

$$\int_R f(\xi)\, d\mu(\xi) \geqslant \int_R g(\xi)\, d\mu(\xi).$$

4. Spaces of Square-Integrable Functions

4.1. Square-Integrable Functions

Let Ω be a Borel subset of $\mathbb{R}^n$. Denote by $\mathscr{B}^n_\Omega$ the family of all Borel sets in $\mathbb{R}^n$ which are subsets of Ω. Clearly, $\mathscr{B}^n_\Omega$ is a Boolean σ algebra (see Exercise 1.8). Let $\mu(B)$, $B \in \mathscr{B}^n_\Omega$, be a measure on $\mathscr{B}^n_\Omega$, and denote by Ω the measure space $(\Omega, \mathscr{B}^n_\Omega, \mu)$.

Definition 4.1. We say that an extended complex-valued function $f(x)$, $x \in \Omega$, defined almost everywhere on Ω is *square integrable* whenever $f(x)$ is measurable [i.e., $|f(x)|^2$ is also measurable by Theorem 3.3] and $|f(x)|^2$ is integrable on Ω, i.e.,

$$\int_\Omega |f(x)|^2\, d\mu(x)$$

exists and is finite.

If a square-integrable function $f(x)$ is not defined everywhere on Ω, we can extend it to Ω by writing $f(x) = 0$ at the points x where it was not previously defined. Hence, we shall assume in the sequel that each square integrable function on Ω is thus suitably defined everywhere in Ω.

A function that is square integrable is not necessarily integrable! Witness the case of the function $f(x) = x^{-3/4}$, $x \in \Omega = [1, +\infty)$, for integrals in the Lebesgue measure on $\mathscr{B}^1_\Omega$:

$$\int_{[1,\infty)} |f(x)|^2\, d\mu_l^{(1)}(x) = \int_1^\infty x^{-3/2}\, dx = 2,$$

$$\int_{[1,\infty)} f(x)\, d\mu_l^{(1)}(x) = \int_1^\infty x^{-3/4}\, dx = +\infty.$$

However, in the case $\mu(\Omega) < +\infty$, a square-integrable function will also be integrable.

Theorem 4.1. If the measure $\mu(R)$ of a measurable set R is finite, then every function $f(\xi)$ which is square integrable on R is also integrable on R.

Theorem 4.1 can be deduced immediately from Theorem 3.9 and the inequality

$$|f(\xi)| \leqslant \tfrac{1}{2}(1 + |f(\xi)|^2).$$

Theorem 4.2. The set $L_{(2)}(\Omega, \mu)$ of all complex-valued functions which are square integrable on Ω is a complex vector space with respect to the vector operations

$$(af)(x) = af(x), \qquad a \in \mathbb{C}^1, \qquad f(x) \in L_{(2)}(\Omega, \mu), \tag{4.1}$$

$$(f + g)(x) = f(x) + g(x), \qquad f(x), g(x) \in L_{(2)}(\Omega, \mu). \tag{4.2}$$

Proof. If $f(x)$ is square integrable on Ω then $af(x)$ is obviously also square integrable.

In order to establish that $f(x) + g(x) \in L_{(2)}(\Omega, \mu)$ whenever $f(x)$, $g(x) \in L_{(2)}(\Omega, \mu)$ it is sufficient to use the easily verifiable inequality

$$|f(x) + g(x)|^2 \leqslant 2(|f(x)|^2 + |g(x)|^2)$$

in conjunction with Theorems 3.5 and 3.9.

Once it is established that the operations (4.1) and (4.2) leave $L_{(2)}(\Omega, \mu)$ invariant, it is a trivial task to check that all the axioms for a vector space, given in Chapter I, Definition 1.1, are satisfied, if in addition we take as the zero vector the function $f(x) \equiv 0$, $x \in \Omega$. Q.E.D.

The family of all spaces $L_{(2)}(\Omega, \mu)$ includes as a special case the vector space $l^2(\infty)$ introduced in Chapter I, §4; $l^2(\infty)$ is obtained when we take $\Omega = \{1, 2, \ldots\}$ and when the measure μ is defined (uniquely) as the σ-additive set function for which

$$\mu(\{n\}) = 1, \qquad n = 1, 2, \ldots .$$

We get by applying Theorem 3.11

$$\int_\Omega |f(x)|^2 \, d\mu(x) = \sum_{n=1}^{\infty} \int_{\{n\}} |f(x)|^2 \, d\mu(x) = \sum_{n=1}^{\infty} |f(n)|^2 \mu(\{n\}) = \sum_{n=1}^{\infty} |f(n)|^2.$$

Lemma 4.1. Two extended complex-valued measurable functions $f(\xi)$ and $g(\xi)$, defined on the measurable set R in a measure space, are equal almost everywhere in R if and only if

$$\int_R |f(\xi) - g(\xi)|^2 \, d\mu(\xi) = 0. \tag{4.3}$$

Proof. If the above two functions are equal almost everywhere, then $|f(\xi) - g(\xi)| = 0$ almost everywhere and (4.3) follows from Theorem 3.8.

Conversely, if (4.3) is true, then for any integer $n > 0$ the set $S_n = \{\xi: |f(\xi) - g(\xi)| > 1/n\} \cap R$ has to be of measure zero, because since $|f(\xi) - g(\xi)| > (1/n)\, \chi_{S_n}(\xi)$, we have as a special case of Theorem 3.9,

$$0 \leqslant \mu(S_n) = \int_R \chi_{S_n}(\xi)\, d\mu(\xi) \leqslant n^2 \int_R |f(\xi) - g(\xi)|^2\, d\mu(\xi) = 0.$$

Consequently

$$\mu(\{\xi: \ |f(\xi) - g(\xi)| > 0\}) = \mu\left(\bigcup_{n=1}^{\infty} S_n\right) \leqslant \sum_{n=1}^{\infty} \mu(S_n) = 0. \quad \text{Q.E.D.}$$

4.2. Hilbert Spaces of Square-Integrable Functions

It is easy to check (see Exercise 4.1) that the relation of "being almost everywhere equal" that can hold between pairs of functions from $L_{(2)}(\Omega, \mu)$ is an equivalence relation. It is customary to denote the family of all equivalence classes of functions which are almost everywhere equal and belong to $L_{(2)}(\Omega, \mu)$ also by the symbol $L_{(2)}(\Omega, \mu)$, but in order to avoid confusion, we shall denote that family by $L^2(\Omega, \mu)$.

We denote the equivalence class which contains the square integrable function $f(x)$ by f. Thus, if $f(x)$ and $g(x)$ are almost everywhere equal, then f and g are identical equivalence classes, and we can write $f = g$; conversely, $f = g$ implies that $f(x) = g(x)$ almost everywhere.

Theorem 4.3. If $f(x), g(x) \in L_{(2)}(\Omega, \mu)$ belong to the equivalence classes $f, g \in L^2(\Omega, \mu)$, and a is a complex number, denote by af and $f + g$ the equivalence classes containing the function $af(x)$ and $f(x) + g(x)$, respectively. The operations

$$\text{(4.4)} \quad \begin{aligned} &(a, f) \to af, \qquad (a, f) \in \mathbb{C}^1 \times L^2(\Omega, \mu), \qquad af \in L^2(\Omega, \mu) \\ &(f, g) \to f + g, \qquad (f, g) \in L^2(\Omega, \mu) \times L^2(\Omega, \mu), \qquad f + g \in L^2(\Omega, \mu) \end{aligned}$$

are vector operations, and the set $L^2(\Omega, \mu)$ with these vector operations becomes a Hilbert space when the inner product of $f, g \in L^2(\Omega, \mu)$ is defined by

$$\text{(4.5)} \qquad \langle f \mid g \rangle = \int_\Omega f^*(x)\, g(x)\, d\mu(x).$$

The proof of this theorem, which is the central theorem of this section, will be given in a few stages.

First we have to note that (4.4) are indeed mappings, i.e., that af and $f + g$ are uniquely determined by f and g. That is quite obvious, because if $f_1 = f_2$ and $g_1 = g_2$, then $af_1(x) = af_2(x)$ almost everywhere, i.e., $af_1 = af_2$; and since

$$\begin{aligned}(4.6)\qquad &\{x: f_1(x) + g_1(x) \neq f_2(x) + g_2(x)\}\\ &\subset (\{x: f_1(x) \neq f_2(x)\} \cup \{x: g_1(x) \neq g_2(x)\}),\end{aligned}$$

the measure of the set on the left-hand side of (4.6) is zero because the measures of both sets on the right-hand side of (4.6) are zero; therefore, $f_1(x) + g_1(x) = f_2(x) + g_2(x)$ almost everywhere, i.e., $f_1 + g_1 = f_2 + g_2$.

From Theorem 4.2 we know that $af(x), f(x) + g(x) \in L_{(2)}(\Omega, \mu)$ whenever $f(x), g(x) \in L_{(2)}(\Omega, \mu)$, so that both mappings (4.4) are defined for all $(a, f) \in \mathbb{C}^1 \times L^2(\Omega, \mu)$ and $(f, g) \in L^2(\Omega, \mu) \times L^2(\Omega, \mu)$. Theorem 4.2 implies in an obvious way that we deal indeed with vector operations, if we adopt as the zero vector in $L^2(\Omega, \mu)$ the equivalence class of square integrable functions that vanish almost everywhere on Ω.

From the inequality

$$(|f(x)| - |g(x)|)^2 \geqslant 0$$

we can immediately derive

$$(4.7)\qquad |f^*(x)\, g(x)| = |f(x)|\, |g(x)| \leqslant \tfrac{1}{2}(|f(x)|^2 + |g(x)|^2).$$

Therefore, the function $|f^*(x)\, g(x)|$ is integrable on Ω if $f(x)$ and $g(x)$ are square integrable on Ω, as a consequence of Theorems 3.3 and 3.9. Thus, $f^*(x)\, g(x)$ is integrable on Ω (see Exercise 3.7), and (4.5) is defined for any $f(x), g(x) \in L_{(2)}(\Omega, \mu)$. Furthermore, $\langle f \mid g\rangle$ is uniquely defined for any $f, g \in L^2(\Omega, \mu)$ because if $f_1(x) = f_2(x)$ almost everywhere, then

$$\int_\Omega (f_1(x) - f_2(x))^*\, g(x)\, d\mu(x)) = \int_R (f_1(x) - f_2(x))^*\, g(x)\, d\mu(x) = 0,$$
$$R = \{x: f_1(x) \neq f_2(x)\},$$

by Theorem 3.8. It is easy to check (see Exercise 4.2) that (4.5) satisfies the axioms for an inner product.

Thus far we have proved that $L^2(\Omega, \mu)$ is a Euclidean space. We turn now to proving the completeness of $L^2(\Omega, \mu)$.

Definition 4.2. A sequence of functions $f_1(x), f_2(x),... \in L_{(2)}(\Omega, \mu)$ is said to be *fundamental in the mean* if for any $\epsilon > 0$ there is a positive integer $N(\epsilon)$ such that

$$(4.8)\qquad \int_\Omega |f_m(x) - f_n(x)|^2\, d\mu(x) < \epsilon$$

for all $m, n > N(\epsilon)$. We say that such a sequence *converges in the mean* to $f(x) \in L_{(2)}(\Omega, \mu)$ if

$$\lim_{n\to+\infty} \int_\Omega |f(x) - f_n(x)|^2 \, d\mu(x) = 0,$$

and in that case we write

$$f(x) = \underset{n\to+\infty}{\text{l.i.m.}} f_n(x),$$

where "l.i.m." stands for "limit in the mean."

According to (4.5) for any $f(x) \in L_{(2)}(\Omega, \mu)$

$$\| f \|^2 = \langle f \mid f \rangle = \int_\Omega |f(x)|^2 \, d\mu(x).$$

Hence, if $f_1(x), f_2(x),\ldots$ is fundamental in the mean, then $f_1, f_2, \ldots \in L^2(\Omega, \mu)$ is a Cauchy (or fundamental) sequence in the norm in $L^2(\Omega, \mu)$, and convergence in the mean in $L_{(2)}(\Omega, \mu)$ corresponds to convergence in the norm in $L^2(\Omega, \mu)$.

***Theorem 4.4** (*Riesz–Fischer theorem*). If a sequence $f_1(x)$, $f_2(x),\ldots \in L_{(2)}(\Omega, \mu)$ is fundamental in the mean, then it converges in the mean to a function $f(x) \in L_{(2)}(\Omega, \mu)$.

Proof. If $f_1(x), f_2(x),\ldots$ is fundamental in the mean, by taking in (4.8)

$$n_k = 1 + N(1/8^k), \qquad k = 1, 2,\ldots,$$

we get a nondecreasing sequence $n_1 \leqslant n_2 \leqslant \cdots$ for which

$$\int_\Omega |f_{n_{k+1}}(x) - f_{n_k}(x)|^2 \, d\mu(x) < 1/8^k. \tag{4.9}$$

Since $f_1(x), f_2(x),\ldots$ are μ-measurable functions, the sets

$$S_k = \{x: |f_{n_{k+1}}(x) - f_{n_k}(x)| \geqslant 1/2^k\} \tag{4.10}$$

are μ measurable. Furthermore, by (4.9)

$$\int_{S_k} |f_{n_{k+1}}(x) - f_{n_k}(x)|^2 \, d\mu(x) \leqslant \int_\Omega |f_{n_{k+1}}(x) - f_{n_k}(x)|^2 \, d\mu(x) < 1/8^k$$

and since

$$\int_{S_k} |f_{n_{k+1}}(x) - f_{n_k}(x)|^2 \, d\mu(x) \geqslant (1/4^k)\, \mu(S_k),$$

we have

$$\mu(S_k) < 1/2^k. \tag{4.11}$$

Therefore, the sets

$$R_k = S_k \cup S_{k+1} \cup \cdots, \qquad k = 1, 2, \ldots \tag{4.12}$$

have measures

$$\mu(R_k) \leqslant 1/2^k + 1/2^{k+1} + \cdots = 1/2^{k-1}. \tag{4.13}$$

A point $x_1 \in \Omega$ belongs to R_k if and only if for some $m \geqslant k$

$$|f_{n_{m+1}}(x_1) - f_{n_m}(x_1)| \geqslant 1/2^m.$$

Therefore, a point $x \in \Omega$ does not belong to R_k if and only if

$$|f_{n_{m+1}}(x) - f_{n_m}(x)| < 1/2^m, \qquad m = k, k+1, \ldots .$$

Thus, if $x \in R_k' = \Omega - R_k$, then the series

$$f_{n_1}(x) + (f_{n_2}(x) - f_{n_1}(x)) + (f_{n_2}(x) - f_{n_3}(x)) + \cdots \tag{4.14}$$

converges absolutely and uniformly on R_k'. Since this is true for every $k = 1, 2, \ldots$, we conclude that

$$f(x) = \lim_{m\to\infty} f_{n_m}(x) \tag{4.15}$$

everywhere on

$$\bigcup_{k=1}^{\infty} R_k' = \left(\bigcap_{k=1}^{\infty} R_k\right)' = R', \qquad R = \bigcap_{k=1}^{\infty} R_k .$$

It is obvious from (4.12) that $R_1 \supset R_2 \supset \cdots$. Consequently, due to the continuity from above of every measure (Theorem 2.1), we get by using (4.13)

$$\mu(R) = \lim_{k\to\infty} \mu(R_k) \leqslant \lim_{k\to\infty} 1/2^k = 0,$$

i.e., R is of measure zero.

The function $f(x)$ is defined everywhere on R' by (4.15), and being the limit of functions measurable on R' it is also measurable on R' by Theorem 3.4. We can extend the domain of definition of $f(x)$ to Ω by writing $f(x) = 0$ for $x \in R$. The function so derived is still measurable because R is measurable.

Since $f_{n_1}(x), f_{n_2}(x), \ldots$ converges uniformly to $f(x)$ on R_k', the convergence

$$(4.16) \qquad g_{n1}(x) = |f_n(x) - f_{n_1}(x)|^2, \qquad g_{n2}(x) = |f_n(x) - f_{n_2}(x)|^2, \ldots$$
$$\to g_n(x) = |f_n(x) - f(x)|^2$$

is also uniform on R_k' for any fixed $n = 1, 2, \ldots$. By assumption $g_{nm}(x) = |f_n(x) - f_{n_m}(x)|^2$ are integrable on Ω, and therefore also on R_k', so that the set

$$T_k^{(n)} = \bigcup_{m=1}^{\infty} \{x: g_{nm}(x) > 0\} - R_k$$

is of σ-finite measure, and can be therefore written as the disjoint union of sets $T_{k1}^{(n)}, T_{k2}^{(n)}, \ldots$ of finite measure. By Theorem 3.11 we can write

$$(4.17) \quad \int_{R_k'} g_{nm}(x)\, d\mu(x) = \int_{T_k^{(n)}} g_{nm}(x)\, d\mu(x) = \sum_{i=1}^{\infty} \int_{T_{ki}^{(n)}} g_{nm}(x)\, d\mu(x).$$

Due to the uniform convergence on R_k' of the sequence (4.16), for each $\epsilon > 0$ there is an $N_n(\epsilon)$ such that

$$|g_{nm}(x) - g_n(x)| < \epsilon, \qquad m > N_n(\epsilon)$$

for all $x \in R_k'$. Thus, for every $i = 1, 2, \ldots$,

$$\int_{T_{ki}^{(n)}} |g_{nm}(x) - g_n(x)|\, d\mu(x) < \epsilon\mu(T_{ki}^{(n)}), \qquad m > N_n(\epsilon),$$

and consequently

$$(4.18) \qquad \int_{T_{ki}^{(n)}} g_n(x)\, d\mu(x) = \lim_{m\to\infty} \int_{T_{ki}^{(n)}} g_{nm}(x)\, d\mu(x).$$

Therefore, by using (4.17), (4.18), and Theorem 3.11 we get

$$(4.19) \quad \lim_{m\to\infty} \int_{R_k'} g_{nm}(x)\, d\mu(x) = \lim_{m\to\infty} \int_{T_k^{(n)}} g_{nm}(x)\, d\mu(x)$$
$$= \sum_{i=1}^{\infty} \lim_{m\to\infty} \int_{T_{ki}^{(n)}} g_{nm}(x)\, d\mu(x)$$
$$= \sum_{i=1}^{\infty} \int_{T_{ki}^{(n)}} g_n(x)\, d\mu(x)$$
$$= \int_{T_k^{(n)}} g_n(x)\, d\mu(x) = \int_{R_k'} g_n(x)\, d\mu(x).$$

On the other hand, since the sequence $f_1(x), f_2(x),\ldots$ is fundamental in the mean, for any $\epsilon > 0$ there is an $N(\epsilon)$ such that when n, $n_m > N(\epsilon)$, we have

$$\int_\Omega g_{nm}(x)\, d\mu(x) = \int_\Omega |f_n(x) - f_{n_m}(x)|^2\, d\mu(x) < \epsilon.$$

Since $g_{nm}(x) \geqslant 0$, we obtain that when n, $n_m > N(\epsilon)$

$$\int_{R_k'} g_{nm}(x)\, d\mu(x) = \int_\Omega g_{nm}(x)\, d\mu(x) < \epsilon.$$

Consequently, by using (4.19) we conclude that for $n > N(\epsilon)$

$$\int_{R_k'} |f_n(x) - f(x)|^2\, d\mu(x) = \int_{R_k'} g_n(x)\, d\mu(x) = \lim_{m\to\infty} \int_{R_k'} g_{nm}(x)\, d\mu(x) \leqslant \epsilon$$

for every $k = 1, 2,\ldots$. As we have

$$\Omega - R = \bigcup_{k=1}^{\infty} R_k' = R_1' \cup (R_2' - R_1') \cup (R_3' - R_2') \cup \cdots,$$

we obtain by employing Theorem 3.8 (note that $\mu(R) = 0$) and Theorem 3.11

$$\begin{aligned}\int_\Omega g_n(x)\, d\mu(x) &= \int_{\Omega - R} g_n(x)\, d\mu(x)\\ &= \int_{R_1'} g_n(x)\, d\mu(x) + \sum_{k=1}^{\infty} \int_{R_{k+1}' - R_k'} g_n(x)\, d\mu(x)\\ &= \lim_{m\to\infty} \int_{R_m'} g_n(x)\, d\mu(x),\end{aligned}$$

i.e., for $n > N(\epsilon)$,

$$\int_\Omega |f_n(x) - f(x)|^2\, d\mu(x) \leqslant \epsilon. \tag{4.20}$$

Since we can write

$$|f_n(x)|^2 \leqslant |f(x)|^2 + |f_n(x) - f(x)|^2,$$

(4.20) implies that $f(x)$ is square integrable and that $f_1(x), f_2(x),\ldots$ converges to $f(x)$ in the mean. Q.E.D.

From Theorem 4.4 we can immediately conclude that $L^2(\Omega, \mu)$ is complete, because if $f_1, f_2,\ldots \in L^2(\Omega, \mu)$ is a Cauchy sequence, then

$f_1(x), f_2(x),... \in L_{(2)}(\Omega, \mu)$ is a sequence which is fundamental in the mean; therefore, by the Riesz–Fisher theorem, $f_1(x), f_2(x),...$ converges in the mean to a function $f(x) \in L_{(2)}(\Omega, \mu)$, and consequently $f_1, f_2,...$ converges in the norm to $f \in L^2(\Omega, \mu)$.

We note that in the course of proving Theorem 4.4 we have also obtained the following results.

Theorem 4.5. If $f_1(x), f_2(x),... \in L_{(2)}(\Omega, \mu)$ is a sequence which converges in the mean to $f(x) \in L_{(2)}(\Omega, \mu)$ then there is a subsequence $f_{n_1}(x), f_{n_2}(x),...$ of the original sequence which converges almost everywhere point by point to $f(x)$, i.e., there is a set R of measure zero such that

$$\lim_{m\to\infty} f_{n_m}(x) = f(x), \qquad x \in R' = \Omega - R.$$

Furthermore, there is a monotonically decreasing sequence of sets

$$R_1 \supset R_2 \supset \cdots \supset R, \qquad R = \bigcap_{k=1}^{\infty} R_k,$$

which is such that $f_{n_1}(x), f_{n_2}(x),...$ converges *uniformly* to $f(x)$ on each $R_k' = \Omega - R_k$.

*4.3. The Separability of L^2 Spaces

Thus far we have proved that $L^2(\Omega, \mu)$ is a Hilbert space, i.e., we have proved the key Theorem 4.3. There are, however, spaces $L^2(\Omega, \mu)$ which are not separable (see Exercise 4.6). We would like to investigate under what conditions $L^2(\Omega, \mu)$ is separable. In order to do that, we need the following theorem.

Theorem 4.6. The vector subspace (C_Ω) of $L_{(2)}(\Omega, \mu)$, spanned* by the family $C_\Omega = \{\chi_B(x), B \in \mathscr{B}_\Omega^{(n)}, \mu(B) < +\infty\}$ of all characteristic functions $\chi_B(x)$, $x \in \Omega$, of Borel sets $B \subset \Omega$ of finite measure, is everywhere dense in the mean in $L_{(2)}(\Omega, \mu)$, i.e., for a given $f(x) \in L_{(2)}(\Omega, \mu)$ and any $\epsilon > 0$ there is a function $s(x) \in C_\Omega$ such that

$$\int_\Omega |f(x) - s(x)|^2 \, d\mu(x) < \epsilon. \tag{4.21}$$

Proof. Due to Definition 3.5, it is obviously sufficient if we prove

* Note that C_Ω is identical to the family of all complex simple functions which are integrable on Ω.

the theorem for nonnegative functions $f(x) \geqslant 0$ from $L_{(2)}(\Omega, \mu)$. Given such a function $f(x)$, the set

$$R_0 = \{x: f(x) > 0\}$$

has to be of σ-finite measure because $|f(x)|^2$ is integrable on Ω, and therefore on R_0. Thus, we can write R_0 as a disjoint union

$$R_0 = \bigcup_{k=1}^{\infty} R_k, \qquad \mu(R_k) < +\infty, \qquad k = 1, 2,...,$$

of sets of finite measure. By Theorem 3.11

$$\int_{\Omega} |f(x)|^2 \, d\mu(x) = \int_{R_0} |f(x)|^2 \, d\mu(x) = \sum_{k=1}^{\infty} \int_{R_k} |f(x)|^2 \, d\mu(x).$$

Consequently, given an $\epsilon > 0$, there is an integer $N(\epsilon) \geqslant 0$ such that

$$\sum_{k=N(\epsilon)+1}^{\infty} \int_{R_k} |f(x)|^2 \, d\mu(x) < \epsilon/2. \tag{4.22}$$

Since $f(x)$ is square integrable on R_k and $\mu(R_k) < +\infty$, by Theorem 4.1 $f(x)$ is also integrable on R_k. Therefore, according to Definition 3.4, there is a nondecreasing sequence $s_1^{(k)}(x), s_2^{(k)}(x),...$ of integrable simple functions vanishing outside R_k such that

$$\lim_{n\to\infty} s_n^{(k)}(x) = \begin{cases} f(x) & \text{for } x \in R_k \\ 0 & \text{for } x \in R_k'. \end{cases}$$

Consequently, $|f(x) - s_1^{(k)}(x)|^2$, $|f(x) - s_2^{(k)}(x)|^2$,... is a nonincreasing sequence on R_k for which

$$\lim_{n\to\infty} |f(x) - s_n^{(k)}(x)|^2 = 0, \qquad x \in R_k.$$

Therefore, by Theorem 3.10

$$\lim_{n\to\infty} \int_{R_k} |f(x) - s_n^{(k)}(x)|^2 \, d\mu(x) = 0$$

so that there is an $n = n_0$ for which

$$\int_{R_k} |f(x) - s_{n_0}^{(k)}(x)|^2 \, d\mu(x) < \epsilon/2N(\epsilon). \tag{4.23}$$

The function

$$s(x) = \sum_{k=1}^{N(\epsilon)} s_{n_0}^{(k)}(x)$$

is obviously an element of C_Ω, for which we have, because of (4.22) and (4.23),

$$\int_\Omega |f(x) - s(x)|^2 \, d\mu(x) = \sum_{k=1}^{N(\epsilon)} \int_{R_k} |f(x) - s(x)|^2 \, d\mu(x)$$

$$+ \sum_{k=N(\epsilon)+1}^{\infty} \int_{R_k} |f(x)|^2 \, d\mu(x) < \epsilon. \quad \text{Q.E.D.}$$

Consider now the family $\mathscr{I}_r^n$ of all closed intervals $I_{a_1,\ldots,a_n}^{b_1,\ldots,b_n}$,

(4.24) $$I_{a_1,\ldots,a_n}^{b_1,\ldots,b_n} = [a_1, b_1] \times \cdots \times [a_n, b_n]$$

with rational end points, i.e., $a_1, \ldots, a_n$, $b_1, \ldots, b_n \in \mathfrak{R}$. Denote by C_Ω^I the subset of C_Ω consisting of all characteristic functions $\chi_R(x)$, $x \in \Omega$, corresponding to all sets $R = (I_1 \cup \cdots \cup I_m) \cap \Omega$, where $I_1, \ldots, I_m \in \mathscr{I}_r^n$, i.e.,

(4.25) $$C_\Omega^r = \{\chi_R(x) \colon R = (I_1 \cup \cdots \cup I_m) \cap \Omega, I_1, \ldots, I_m \in \mathscr{I}_r^n, \; m = 1, 2, \ldots\}.$$

We claim the following.

***Lemma 4.2.** If the measure $\mu(B)$, $B \in \mathscr{B}_\Omega^n$, is σ finite in Ω, the family C_Ω^r defined in (4.25) is everywhere dense in the set $C_\Omega \subset L_{(2)}(\Omega, \mu)$, i.e., for every $B \in \mathscr{B}_\Omega^n$ and any $\epsilon > 0$, there can be found a $\chi_R(x) \in C_\Omega^r$ such that

(4.26) $$\int_\Omega |\chi_B(x) - \chi_R(x)|^2 \, d\mu(x) < \epsilon.$$

Proof. Note that $|\chi_B(x) - \chi_R(x)|$ can assume only the values 0 and 1 and consequently

(4.27) $$\int_\Omega |\chi_B(x) - \chi_R(x)|^2 \, d\mu(x) = \int_\Omega |\chi_B(x) - \chi_R(x)| \, d\mu(x)$$

$$= \mu(B - R) + \mu(R - B) = \mu(B \triangle R).$$

Consider now the Boolean σ algebra $\mathscr{A}_\sigma(\mathscr{B}_r^n)$ generated by the family $\mathscr{B}_r^n$ of all finite unions of intervals with rational ends. Since any interval $I \in \mathscr{I}^n$ can be written as a countable union of disjoint intervals with rational end points, we have $\mathscr{A}_\sigma(\mathscr{B}_r^n) \supset \mathscr{I}^n$. Consequently

$$\mathscr{A}_\sigma(\mathscr{B}_r^n) = \mathscr{A}_\sigma(\mathscr{I}^n) = \mathscr{B}^n$$

because $\mathscr{A}(\mathscr{I}^n) \supset \mathscr{B}_r^n$. Thus, we also have

$$\mathscr{A}_\sigma(C_\Omega^r) = \{B \cap \Omega: B \in \mathscr{A}_\sigma(\mathscr{B}_r^n)\} = \{B \cap \Omega: B \in \mathscr{B}^n\} = \mathscr{B}_\Omega^n,$$

i.e., $\mathscr{B}_\Omega^n$ is identical to the Boolean σ algebra generated by C_Ω^r.

We can easily establish (see Exercise 4.7) that C_Ω^r is a Boolean algebra. According to Theorem 2.3, the set function

$$\bar{\mu}(B) = \inf \left\{ \sum_{k=1}^{\infty} \mu(R_k), \bigcup_{k=1}^{\infty} R_k \supset B, R_1, R_2, \ldots \in C_\Omega^r \right\}, \qquad B \in \mathscr{B}_\Omega^n,$$

is a measure on the Boolean σ algebra $\mathscr{B}_\Omega^n$ generated by the Boolean algebra C_Ω^r. Furthermore, if μ is a σ-finite measure, then $\mu(B)$, $B \in C_\Omega^r$, has a *unique* extension to $\mathscr{B}_\Omega^n$ so that μ and $\bar{\mu}$ are identical:

$$\bar{\mu}(B) = \mu(B), \qquad B \in \mathscr{B}_\Omega^n.$$

From the definition of $\bar{\mu}(B)$ it follows that for any $\epsilon > 0$ we can choose $R_1, R_2, \ldots \in C_\Omega^r$ so that

$$\mu(B) \leqslant \sum_{k=1}^{\infty} \mu(R_k) < \mu(B) + \epsilon/2.$$

If we write

$$S_n = \bigcup_{k=1}^{n} R_k, \qquad S = \bigcup_{k=1}^{\infty} R_k,$$

we have $S_1 \subset S_2 \subset \cdots$, and since $B \subset S$,

$$\lim_{n\to\infty} \mu(B \triangle S_n) = \lim_{n\to\infty} \mu(B - S_n) + \lim_{n\to\infty} \mu(S_n - B) = \mu(S - B)$$

$$= \mu(S) - \mu(B) \leqslant \sum_{k=1}^{\infty} \mu(R_k) - \mu(B) < \epsilon/2.$$

Consequently, there is an n_0 for which

$$\mu(B \triangle S_{n_0}) \leqslant \epsilon/2,$$

and therefore, for $R = S_{n_0}$ we have

$$\mu(B \triangle R) < \epsilon.$$

The above result in conjunction with (4.27) proves that $\chi_R(x)$ satisfies (4.26). Q.E.D.

***Lemma 4.3.** If μ is a σ-finite measure, the family $\mathfrak{D}_S$ of all simple μ-integrable functions

$$s(x) = \sum_{k=1}^{m} a_k \chi_{R_k}(x), \qquad R_1, ..., R_m \in C_\Omega^r, \qquad m = 1, 2, ...,$$

with $a_1, ..., a_m \in \mathbb{C}^1$ having rational numbers for their real and imaginary parts (i.e. Re a_1, Im a_1,..., Re a_m, Im $a_m \in \mathfrak{R}$), is a countable subset of $L_{(2)}(\Omega, \mu)$ which is everywhere dense in the mean in $L_{(2)}(\Omega, \mu)$.

Proof. The result that $\mathfrak{D}_S$ is countable can be deduced easily (see Exercise 4.4) from the statement that the set $\mathfrak{R}$ of all rational numbers is countable.

If $f(x) \in L_{(2)}(\Omega, \mu)$, according to Theorem 4.6 we can find an integrable simple function $s(x) \in (C_\Omega)$

$$s(x) = \sum_{k=1}^{m} c_k \chi_{B_k}(x), \qquad B_i \cap B_j = \varnothing \quad \text{for} \quad i \neq j,$$

$$\chi_{B_1}(x), ..., \chi_{B_m}(x) \in C_\Omega$$

such that

$$\| f - s \|^2 = \int_\Omega | f(x) - s(x)|^2 \, d\mu(x) < \epsilon/9. \tag{4.28}$$

We can now choose complex numbers $a_1, ..., a_m$ with real and imaginary parts which are rational numbers so that

$$| a_k - c_k |^2 < \epsilon/9m\mu(B_k), \qquad | a_k | > 0.$$

The simple function

$$s_1(x) = \sum_{k=1}^{m} a_k \chi_{B_k}(x)$$

is integrable because $s(x)$ is integrable, and

$$\| s - s_1 \|^2 = \int_\Omega | s(x) - s_1(x)|^2 \, d\mu(x) = \sum_{k=1}^{m} | a_k - c_k |^2 < \epsilon/9. \tag{4.29}$$

Finally, according to Lemma 4.2, for each $\chi_{B_k}(x)$ we can find a $\chi_{R_k}(x)$ such that

$$\| \chi_{B_k} - \chi_{R_k} \|^2 = \int_\Omega | \chi_{B_k}(x) - \chi_{R_k}(x)|^2 \, d\mu(x) < \frac{\epsilon}{9m^2 | a_k |^2}, \tag{4.30}$$

$$k = 1, ..., m.$$

We see that

$$s_r(x) = \sum_{k=1}^{m} a_k \chi_{R_k}(x)$$

belongs to $\mathfrak{D}_S$, and since $\| h \|$, $h(x) \in L_{(2)}(\Omega, \mu)$, is a norm in $L^2(\Omega, \mu)$, by applying the triangle inequality and using (4.30), we get

$$\| s_1 - s_r \| = \left| \sum_{k=1}^{m} a_k(\chi_{B_k} - \chi_{R_k}) \right| \leqslant \sum_{k=1}^{m} | a_k | \| \chi_{B_k} - \chi_{R_k} \| < \sqrt{\epsilon}/3.$$

Thus, by combining this relation with (4.28) and (4.29), we arrive at

$$\| f - s_r \| \leqslant \| f - s \| + \| s - s_1 \| + \| s_1 - s_r \| < \sqrt{\epsilon},$$

i.e., we have an $s_r(x) \in \mathfrak{D}_S$ for which

$$\| f - s_r \|^2 = \int_\Omega | f(x) - s_r(x) |^2 \, d\mu(x) < \epsilon. \quad \text{Q.E.D.}$$

Lemma 4.3 establishes the separability of $L^2(\Omega, \mu)$ when μ is σ finite because the subset $\tilde{\mathfrak{D}}_S$ of all equivalence classes $s \in L^2(\Omega, \mu)$ containing a function $s(x) \in \mathfrak{D}_S$ is obviously a countable everywhere dense set in $L^2(\Omega, \mu)$. Thus, we have proved the following theorem.

Theorem 4.7. The Hilbert space $L^2(\Omega, \mu)$ is separable when the measure $\mu(B)$, $B \in \mathscr{B}^n_\Omega$, is σ finite.

The following statement can be immediately deduced from Lemma 4.3 by noting that every finite union R_k of finite intervals can be written as a finite union of disjoint finite intervals.

Theorem 4.8. If the set Ω is of σ-finite μ measure, the vector subspace spanned by all the characteristic functions $\chi_R(x)$, $x \in \Omega$, corresponding to the sets $R = I \cap \Omega$, where $I \in \mathscr{I}^n$ is a finite interval, is everywhere dense in $L^2(\Omega, \mu)$.

A special class of Hilbert spaces $L^2(\Omega, \mu)$ which are of great importance in quantum mechanics corresponds to the choice $\mu(B) = \mu_l^{(n)}(B)$, $B \in \mathscr{B}^n_\Omega$, where $\mu_l^{(n)}$ is the Lebesgue measure on $\mathscr{B}^n_\Omega$. The Hilbert space $L^2(\Omega, \mu_l^{(n)})$ is separable because $\mu_l^{(n)}$ is a σ-finite measure. Such a Hilbert space $L^2(\Omega, \mu_l^{(n)})$ is simply denoted by $L^2(\Omega)$, and correspondingly $L_{(2)}(\Omega, \mu_l^{(n)})$ is denoted by $L_{(2)}(\Omega)$.

In the case $\Omega = \mathbb{R}^n$, the set $L_{(2)}(\mathbb{R}^n)$ is the space of all Borel measurable functions $f(x)$, for which the Lebesgue integral

$$\int_{\mathbb{R}^n} | f(x) |^2 \, d^n x = \int_{\mathbb{R}^n} | f(x) |^2 \, d\mu_l^{(n)}(x)$$

exists. The subset $L_R^2(\mathbb{R}^n)$ of all $g \in L_R^2(\mathbb{R}^n)$ corresponding to functions $g(x)$ whose Riemann integral $\int_{-\infty}^{+\infty} \cdots \int_{-\infty}^{+\infty} |g(x)|^2 \, d^n x$ exists and is finite is obviously a linear subspace of $L_R^2(\mathbb{R}^n)$. Due to Theorem 3.7 the Euclidean space $L_R^2(\mathbb{R}^n)$ is isomorphic to the space $\mathscr{C}_{(2)}^0(\mathbb{R}^n)$ of all continuous functions which are square integrable on $\mathbb{R}^n$. We know that $\mathscr{C}_{(2)}^0(\mathbb{R}^1)$ is incomplete (see Chapter I, §4), and the same can be proved about $\mathscr{C}_{(2)}^0(\mathbb{R}^n)$. Thus, the space $L_R^2(\mathbb{R}^n)$ is incomplete, which shows that the introduction of the Lebesgue integral is essential for the validity of the Riesz–Fischer theorem, and thus for the construction of complete Euclidean spaces of square integrable functions.

4.4. Change of Variables of Integration

In order to prove the next theorem we need the following lemma.

***Lemma 4.4.** If the functions $\phi_1(x),\ldots, \phi_n(x)$, $x = (x_1, \ldots, x_n) \in B_0$, mapping the Borel set $B_0 \in \mathscr{B}^n$ into $\mathbb{R}^n$, are Borel measurable, then the mapping

$$x \to x^1 = \phi(x) = (\phi_1(x),\ldots, \phi_n(x))$$

is a *Borel measurable mapping*, i.e., the inverse image $\phi^{-1}(B)$ of any Borel set $B \in \mathscr{B}^n$ is again a Borel set.

Proof. For an n-dimensional interval $I = I_1 \times \cdots \times I_n$, $I_1, \ldots, I_n \in \mathscr{I}^1$, we can write

$$\phi^{-1}(I) = \bigcap_{k=1}^{n} \phi_k^{-1}(I_k).$$

Since $\phi_k(x)$ is a Borel measurable function, the set $\phi_k^{-1}(I_k)$ is a Borel set; hence, $\phi^{-1}(I)$ is a Borel set for any interval $I \in \mathscr{I}^n$.

Denote by $\mathscr{F}_\phi$ the family of all sets S for which $\phi^{-1}(S) \in \mathscr{B}^n$. $\mathscr{F}_\phi$ is a Boolean σ algebra: if $S_1, S_2, \ldots \in \mathscr{F}_\phi$, then

$$\phi^{-1}(S_1 \cup S_2 \cup \cdots) = \phi^{-1}(S_1) \cup \phi^{-1}(S_2) \cup \cdots \in \mathscr{B}^n,$$

i.e., $S_1 \cup S_2 \cup \cdots \in \mathscr{F}_\phi$, and for any $S \in \mathscr{F}_\phi$

$$\phi^{-1}(S') = (\phi^{-1}(S))' \in \mathscr{B}^n,$$

i.e., $S' \in \mathscr{F}_\phi$. Since we have already established that $\mathscr{F}_\phi \supset \mathscr{I}^n$, and since $\mathscr{B}^n$ is the smallest Boolean σ algebra containing $\mathscr{I}^n$, it follows that $\mathscr{B}^n \subset \mathscr{F}_\phi$, i.e., $\phi^{-1}(B) \in \mathscr{B}^n$ for any $B \in \mathscr{B}^n$. Q.E.D.

Theorem 4.9. (a) Let Ω be a Borel subset of $\mathbb{R}^n$ whose boundary* is of Lebesgue measure zero and on which a Lebesgue–Stieltjes measure

$$\mu(B) = \int_B \rho(x)\, d^n x, \qquad \rho(x) \geqslant 0, \qquad B \in \mathscr{B}_\Omega^n,$$

is defined, and let $x \to x^1 = \phi(x)$ be a mapping of Ω onto the set $\Omega^1 \subset \mathbb{R}^n$, given by

$$\begin{aligned} x_1^1 &= \phi_1(x_1, \ldots, x_n), \\ &\vdots \\ x_n^1 &= \phi_n(x_1, \ldots, x_n), \end{aligned} \tag{4.31}$$

$$x = (x_1, \ldots, x_n) \in \Omega, \qquad x^1 = (x_1^1, \ldots, x_n^1) \in \Omega^1,$$

where $\phi_1(x), \ldots, \phi_n(x)$ are almost everywhere in Ω continuously differentiable. If the Jacobian of the transformation

$$J(x) = \frac{\partial(\phi_1, \ldots, \phi_n)}{\partial(x_1, \ldots, x_n)} = \begin{vmatrix} \dfrac{\partial\phi_1}{\partial x_1} & \cdots & \dfrac{\partial\phi_1}{\partial x_n} \\ \vdots & & \vdots \\ \dfrac{\partial\phi_n}{\partial x_1} & \cdots & \dfrac{\partial\phi_n}{\partial x_n} \end{vmatrix} = \frac{1}{J^1(x^1)}$$

is almost everywhere positive, then

$$\mu^1(B^1) = \int_{\phi^{-1}(B^1)} d\mu(x) = \int_{B^1} J^1(x^1)\rho(\phi^{-1}(x^1))\, d^n x^1, \qquad B^1 \in \mathscr{B}_{\Omega^1}^n, \tag{4.32}$$

determines a Lebesgue–Stieltjes measure on $\mathscr{B}_{\Omega^1}^n$.

(b) If the mapping $x \to x^1 = \phi(x)$ has almost everywhere in Ω an inverse $x^1 \to x = \phi^{-1}(x^1)$, then for each function $f(x)$, $x \in \Omega$, the function

$$f^1(x^1) = f(\phi^{-1}(x^1)) \tag{4.33}$$

is defined almost everywhere on Ω^1, and is μ^1 integrable on B^1 if $f(x)$ is μ integrable in $B = \phi^{-1}(B^1)$; we have

$$\int_{B^1} f^1(x^1)\, d\mu^1(x^1) = \int_{B=\phi^{-1}(B^1)} f(x)\, d\mu(x), \qquad B^1 \in \mathscr{B}_{\Omega^1}^n. \tag{4.34}$$

* Recall that a boundary point of a subset of a metric space is any point which is not an interior point but it is an accumulation point of that subset; an interior point of a set Ω is a point which has an ϵ neighborhood which lies in Ω.

The mapping

$$f \to f^1, \qquad f \in L^2(\Omega, \mu), \qquad f^1 \in L^2(\Omega^1, \mu^1), \tag{4.35}$$

is a unitary transformation (see Definition 2.4 in Chapter I) of the Hilbert space $L^2(\Omega, \mu)$ onto the Hilbert space $L^2(\Omega^1, \mu^1)$.

Proof. (a) Denote by Ω_0 the set of all interior points $x \in \Omega$ at which $\phi_1(x),..., \phi_n(x)$ in (4.31) are continuously differentiable and at which $J(x) > 0$. If $x \in \Omega_0$, then $\phi_1(x),..., \phi_n(x)$ are continuously differentiable and $J(x) > 0$ also in some neighborhood of x; thus, the set Ω_0 is open. According to the assumptions, $\mu(\Omega - \Omega_0) = 0$.

The set Ω_0 is mapped by (4.31) onto a set $\Omega_0^1 \subset \Omega^1$. Since $J^1(x^1)$ is continuous in Ω_0, it is also bounded and therefore integrable on every compact set $B \subset \Omega_0$ of finite measure. According to Lemma 4.4 the inverse image B, under the mapping (4.31), of every Borel set $B^1 \subset \Omega_0^1$ is also a Borel set, due to the fact that $\phi_1(x),..., \phi_n(x)$ are continuous and therefore measurable by Theorem 3.1.

Thus, (4.32) really determines a measure on Ω_0^1, which can be extended to Ω^1 if we write

$$\mu^1(\Omega^1 - \Omega_0^1) = 0.$$

(b) It is easy to see that the function $f^1(x^1) = f(\phi^{-1}(x^1))$ is μ^1 measurable on* Ω^1 if $f(x)$ is μ measurable on Ω. In order to establish (4.34), note that for

$$f^1(x^1) = \chi_{B^1}(x^1), \qquad B^1 \in \mathscr{B}^n_{\Omega^1},$$

we get from (4.32) that (4.34) is equivalent to the relation

$$\mu^1(B^1) = \mu(\phi^{-1}(B^1)). \tag{4.36}$$

If we choose B^1 to be an open subset of Ω^1, then the set B is also open if B denotes the inverse image $\phi^{-1}(B^1)$ of B^1 after a possible subtraction from $\phi^{-1}(B^1)$ of the set $\phi^{-1}(B^1) \cap (\Omega - \Omega_0)$ of measure zero. When B^1 is an interval in Ω, we can deduce from well-known theorems of the calculus that (4.36) is true by noting (first for continuous ρ, then generalizing) that (4.36) can be written in the form

$$\int_{B^1} \rho(\phi^{-1}(x^1)) J^1(x^1)\, d^n x^1 = \int_B \rho(x)\, d^n x. \tag{4.37}$$

* $f^1(x^1)$ is defined almost everywhere by the relation $f^1(x^1) = f(\phi^{-1}(x^1))$. On a remaining set of μ^1 measure zero we can define $f^1(x^1)$ arbitrarily.

Thus, due to the additivity of $\mu^1(B^1)$ and $\mu(B)$, (4.36) is also established for any set B^1 which is a finite union of disjoint intervals lying in Ω^1, i.e., on the Boolean algebra $\mathscr{B}^n_{0,\Omega}$ generated by the family $\mathscr{I}^n_\Omega$ of all intervals lying in Ω (see Exercise 4.8). Since $\mu^1(B^1)$ is a σ-finite measure and the Boolean σ algebra generated by $\mathscr{B}^n_{0,\Omega^1}$ is $\mathscr{B}^n_{\Omega^1}$ (see Exercise 4.8), we conclude from Theorem 2.3 that (4.36) is true for any $B^1 \in \mathscr{B}^n_{\Omega^1}$. Thus, as the next obvious result, we have that (4.34) is true when $f^1(x^1)$—and consequently also $f(x)$—are simple functions. From Definitions 3.4 and 3.5 we can infer in a straightforward manner that (4.34) is true whenever $f(x)$ is μ integrable on B.

If $f_1(x) = f_2(x)$ almost everywhere in the μ measure, then $f_1^1(x^1) = f_2^1(x^1)$ almost everywhere in the μ^1 measure, where $f_i^1(x^1)$, $i = 1, 2$, are defined almost everywhere by (4.33). We conclude that (4.35) defines a mapping of $L^2(\Omega, \mu)$ onto $L^2(\Omega^1, \mu^1)$. It is quite evident that this mapping is linear, has an inverse and preserves the inner products of the spaces $L^2(\Omega, \mu)$ and $L^2(\Omega^1, \mu^1)$, i.e., that it is a unitary transformation. Q.E.D.

Exercises

4.1. Show that the relation, "$f(x) = g(x)$ almost everywhere," is an equivalence relation in $L_{(2)}(\Omega, \mu)$.

4.2. Check that

$$\langle f \mid g \rangle = \int f^*(x)\, g(x)\, d\mu(x), \qquad f, g \in L^2(\Omega, \mu),$$

satisfies the axioms for an inner product.

4.3. How can you deduce directly from Theorem 4.3 that $l^2(+\infty)$ is a separable Hilbert space?

4.4. Show that the set C^r_Ω appearing in Lemma 4.2, and the set $\mathfrak{D}_S$ introduced in Lemma 4.3 are both countable sets.

4.5. Show that the set $\mathscr{B}^n_\Omega$, $\mu(\Omega) < +\infty$, in which we identify any two sets whose symmetric difference if of measure zero, is a metric space if the distance of $R, S \in \mathscr{B}^n_\Omega$ is taken to be $d(R, S) = \mu(R \triangle S)$.

4.6. Check that the set function $\mu(B)$, $B \in \mathscr{B}^1$, defined as being equal to the number of elements in B, in case B is finite, and $+\infty$ in case that B is an infinite set, is a measure on $\mathscr{B}^1$. Show that $L^2(\mathbb{R}^1, \mu)$ is not separable.

4.7. Prove that the set C^r_Ω introduced in Lemma 4.2 is a Boolean algebra of subsets of Ω.

4.8. Show that the family $\mathscr{B}^n_{0,\Omega}$ of all unions of disjoint intervals $I \in \mathscr{I}^n$ lying within an open set Ω, $\Omega \subset \mathbb{R}^n$, is a Boolean algebra. Prove that the Boolean σ algebra generated by $\mathscr{B}^n_{0,\Omega}$ is identical to $\mathscr{B}^n_{\Omega}$.

5. The Hilbert Space of Systems of n Different Particles in Wave Mechanics

5.1. The Schroedinger Equation of n-Particle Systems

We shall study now the general framework used in wave mechanics to describe a system of n particles in which each one of the particles is of a different kind from the rest. The requirement that the particles are of different kinds is essential because systems containing two or more particles of the same kind of atomic or subatomic size have peculiarities of behavior reflected in certain statistical features, which require that they conform to either Bose–Einstein or Fermi–Dirac statistics—subjects that we will discuss in Chapter IV. For instance, a two-particle system of the kind we are considering now would be an electron–proton system which, when in a bound state, constitutes a hydrogen atom. An example of a three-body system of different particles is an electron–proton–neutron system, which is called a deuterium atom when it is in a bound state. However, tritium, consisting of one electron, one proton, and *two* neutrons, already is not a system of the considered type, except if we treat its nucleus as *one* particle and the electron as another particle.

We must mention that at present we ignore the possible internal degrees of freedom that the particles might have. These internal degrees of freedom are reflected in the existence of spin, a concept we shall introduce in Chapter IV.

We assume that our n particles move in three dimensions, and denote by $\mathbf{r}_k$ the position vector of the kth particle. If we choose in the three-dimensional real Euclidean space $\mathbb{R}^3$ a fixed inertial reference system of coordinates* characterized by three orthonormal vectors $\mathbf{e}_x$, $\mathbf{e}_y$, and $\mathbf{e}_z$, we can expand $\mathbf{r}_k$ in terms of $\mathbf{e}_x$, $\mathbf{e}_y$, and $\mathbf{e}_z$,

$$\mathbf{r}_k = x_k \mathbf{e}_x + y_k \mathbf{e}_y + z_k \mathbf{e}_z, \qquad k = 1, 2, \ldots, n.$$

In wave mechanics it is postulated that the state of a system of n particles is described at any given time t by a function $\psi(\mathbf{r}_1, \ldots, \mathbf{r}_n; t)$

* In practice the reference frame is usually tied to the laboratory in which the experiments are performed. For the duration of the typical microscopic experiment such a frame moves uniformly with respect to the sun, i.e., it is approximately inertial. Such a system is called the *laboratory frame of reference.*

defined on the configuration space $\mathbb{R}^{3n}$ of the coordinate vectors $\mathbf{r}_1, \ldots, \mathbf{r}_n$. It is required in addition that $\psi(\mathbf{r}_1, \ldots, \mathbf{r}_n; t)$ is once continuously differentiable in t and that for fixed t it is square integrable,* i.e., $\psi(\mathbf{r}_1, \ldots, \mathbf{r}_n; t) \in L_{(2)}(\mathbb{R}^{3n})$, and normalized

$$\int_{\mathbb{R}^{3n}} |\psi(\mathbf{r}_1, \ldots, \mathbf{r}_n; t)|^2 \, d\mathbf{r}_1 \cdots d\mathbf{r}_n = 1. \tag{5.1}$$

In general, if a function $\psi(\mathbf{r}_1, \ldots, \mathbf{r}_n; t)$ can represent a state of the given system, then $\psi(\mathbf{r}_1, \ldots, \mathbf{r}_n; t)$ is called a *wave function* of that system.

As in the one-particle, one-dimensional case treated in Chapter I, §5, the wave function $\psi(\mathbf{r}_1, \ldots, \mathbf{r}_n; t)$ does not have any direct physical meaning. Instead, we have again (as a special case of the Born correspondence rule, which will be given in Chapter IV) that the Lebesgue–Stieltjes measure

$$P_t(B) = \int_B |\psi(\mathbf{r}_1, \ldots, \mathbf{r}_n; t)|^2 \, d\mathbf{r}_1 \cdots d\mathbf{r}_n, \qquad B \in \mathscr{B}^{3n}, \tag{5.2}$$

is interpreted as a probability measure: $P_t(B)$ gives, for every Borel set B, the *probability* of having within B the outcome of a measurement at time t of the positions $\mathbf{r}_1, \ldots, \mathbf{r}_n$ of the n particles of a system prepared in a state represented by $\psi(\mathbf{r}_1, \ldots, \mathbf{r}_n; t)$.

In view of Born's interpretation, it turns out, as we shall prove in general in Chapter IV, that if a state is described at time t by the wave function $\psi(\mathbf{r}_1, \ldots, \mathbf{r}_n; t)$, then the function $c\psi_1(\mathbf{r}_1, \ldots, \mathbf{r}_n; t)$, where $|c| = 1$ and $\psi(\mathbf{r}_1, \ldots, \mathbf{r}_n; t) = \psi_1(\mathbf{r}_1, \ldots, \mathbf{r}_n; t)$ almost everywhere (with respect to the Lebesgue measure on $\mathbb{R}^{3n}$), describes the same state. This implies that each function in the equivalence class $\psi(t) \in L^2(\mathbb{R}^{3n})$, containing $\psi(\mathbf{r}_1, \ldots, \mathbf{r}_n; t)$ describes the same state as $\psi(\mathbf{r}_1, \ldots, \mathbf{r}_n; t)$. Thus, we can state as the first assumption† of wave mechanics:

Postulate W1. The state of a system of n different particles is described at any time t by a normalized vector $\psi(t)$ from the Hilbert space $L^2(\mathbb{R}^{3n})$. The time-dependent vector function $c\psi(t)$, $|c| = 1$, represents the same state as $\psi(t)$.

* Throughout this section integrability and measurability are only with respect to the measure space $(\mathbb{R}^{3n}, \mathscr{B}^{3n}, \mu_l^{(3n)})$.

† The postulates we shall state in this section represent a convenient way of systematically formulating *some* of the basic assumptions of wave mechanics. These postulates are by no means exhaustive. A complete study of the postulates of quantum mechanics will be given in Chapter IV. The reader should find it easy to adapt those general axioms to the special wave mechanical case.

As an equation of motion we have now a Schroedinger equation of a more general form than (5.4) of Chapter I. If the considered system is such that in the classical mechanics case the potential energy of the system* is $V(\mathbf{r}_1, \ldots, \mathbf{r}_n)$, then we arrive at the Schroedinger equation by a generalization of the heuristic procedure given in Chapter I, §5.

Take the energy conservation equation

$$E = \sum_{k=1}^{n} \frac{\mathbf{p}_k^2}{2m_k} + V, \tag{5.3}$$

where $\mathbf{p}_k$ and m_k are, respectively, the momentum and the mass of the kth particle, and make in (5.3) the following formal substitution:

$$E \to i\hbar \frac{\partial}{\partial t}, \qquad \mathbf{p}_k \to -i\hbar \nabla_k,$$

$$\nabla_k = \mathbf{e}_x \frac{\partial}{\partial x_k} + \mathbf{e}_y \frac{\partial}{\partial y_k} + \mathbf{e}_z \frac{\partial}{\partial z_k}.$$

We arrive at the formal operator relation

$$\begin{aligned} i\hbar \frac{\partial}{\partial t} &= - \sum_{k=1}^{n} \frac{\hbar^2}{2m_k} \Delta_k + V, \\ \Delta_k &= \frac{\partial^2}{\partial x_k^2} + \frac{\partial^2}{\partial y_k^2} + \frac{\partial^2}{\partial z_k^2}. \end{aligned} \tag{5.4}$$

Now we postulate the following.

Postulate W2. Each wave function $\psi(\mathbf{r}_1, \ldots, \mathbf{r}_n; t)$, $\psi(t) \in L^2(\mathbb{R}^{3n})$, $\|\psi(t)\| = 1$, is once differentiable in t. If $\psi(\mathbf{r}_1, \ldots, \mathbf{r}_n; t)$ is twice differentiable in the configuration space $\mathbb{R}^{3n}$ of the variables $\mathbf{r}_1, \ldots, \mathbf{r}_n$, then it satisfies the time-dependent Schroedinger equation,

$$\begin{aligned} i\hbar \frac{\partial}{\partial t} \psi(\mathbf{r}_1, \ldots, \mathbf{r}_n; t) \\ = \Big(- \sum_{k=1}^{n} \frac{\hbar^2}{2m_k} \Delta_k + V(\mathbf{r}_1, \ldots, \mathbf{r}_n)\Big)\psi(\mathbf{r}_1, \ldots, \mathbf{r}_n; t), \end{aligned} \tag{5.5}$$

whenever the function on the right-hand side of (5.5) belongs to $L_{(2)}(\mathbb{R}^{3n})$.

In view of the interpretation of $P_t(B)$ in (5.2), it is essential that a

* We are assuming no external force, so that the potential energy $V(\mathbf{r}_1, \ldots, \mathbf{r}_n)$ is time independent. Classically, $V(\mathbf{r}_1, \ldots, \mathbf{r}_n)$ completely describes the interaction. The force $\mathbf{F}_k$ with which the rest of the particles are acting on the kth particle is $\mathbf{F}_k = -[\mathbf{e}_x(\partial V/\partial x_k) + \mathbf{e}_y(\partial V/\partial y_k) + \mathbf{e}_z(\partial V/\partial z_k)]$.

wave function satisfy (5.1) at all times. We leave to the reader (see Exercise 5.2) to prove the following theorem.

Theorem 5.1. If the function $\psi(\mathbf{r}_1, \ldots, \mathbf{r}_n\,;\,t)$, $\psi(t) \in L^2(\mathbb{R}^{3n})$, is once continuously differentiable in t and twice continuously differentiable in the configuration space variables, and such that $(\partial/\partial t)\,\psi(\mathbf{r}_1, \ldots, \mathbf{r}_n\,;\,t)$ and $\sum_{k=1}^{n} \Delta_k \psi(\mathbf{r}_1, \ldots, \mathbf{r}_n\,;\,t)$ are square integrable, and if $\psi(\mathbf{r}_1, \ldots, \mathbf{r}_n\,;\,t)$ satisfies the Schroedinger equation (5.5), then

$$\| \psi(t)\|^2 = \int_{\mathbb{R}^{3n}} | \psi(\mathbf{r}_1, \ldots, \mathbf{r}_n; t)|^2 \, d\mathbf{r}_1 \cdots d\mathbf{r}_n = 1$$

provided $\| \psi(0)\| = 1$ and $| \psi \nabla_k \psi | = O(| \mathbf{r}_k |^{-\alpha_k})$, $\alpha_k > 2$, as $| \mathbf{r}_k | \to \infty$.

Using the same procedure as in §5 of Chapter I, we shall try to find solutions of (5.5) of the form

$$\psi(\mathbf{r}_1, \ldots, \mathbf{r}_n) \exp[-(i/\hbar)\, Et] \tag{5.6}$$

where $\psi \in L^2(\mathbb{R}^{3n})$. By inserting (5.6) in (5.5) we arrive at the *time-independent Schroedinger equation* for our problem, which can be written in symbolic form:

$$H_S \psi(\mathbf{r}_1, \ldots, \mathbf{r}_n) = E\psi(\mathbf{r}_1, \ldots, \mathbf{r}_n), \tag{5.7}$$

where H_S is a differential operator form*

$$H_S = - \sum_{k=1}^{n} \frac{\hbar^2}{2m_k} \Delta_k + V(\mathbf{r}_1, \ldots, \mathbf{r}_n) \tag{5.8}$$

called a *Schroedinger operator.*

Thus we have arrived again at an eigenvalue problem (5.7). The same terminology applies now as in the one-dimensional, single-particle case treated in Chapter I.

5.2. The Center-of-Mass Frame of Reference

There is, however, an essential difference (besides dimension and number of particles) between the present case and the one already

* It must be realized that a differential operator form such as (5.8) is essentially only a recipe for mapping a certain class $\mathscr{D}$ of functions into another class $\mathscr{K}$ of functions by applying to each function from $\mathscr{D}$ the procedure symbolically embodied in the operator form. Once the class of functions $\mathscr{D}$ is specified, we deal with a well-defined mapping, which is customarily called an operator. However, the same differential operator form can be applied to different classes of functions, thus defining *different* operators. The importance, from the mathematical as well as physical point of view, of specifying in a given problem the class of functions on which the differential operator form acts will be explained in Chapter III, §2, and (5.8) will be studied in Chapter IV, §7.

treated in Chapter I: the one particle was moving inside an external field of force described by the potential $V(x)$, while at present the interaction originates inside the system by having the particles interact with one another with forces described by the potential $V(\mathbf{r}_1, \dots, \mathbf{r}_n)$, and we assume *no external forces*. Thus, classically speaking, the total force on the center of mass

$$\text{(5.9)} \qquad \mathbf{R} = \frac{m_1\mathbf{r}_1 + \cdots + m_n\mathbf{r}_n}{M}, \qquad M = m_1 + \cdots + m_n ,$$

of the system will be zero (see Exercise 5.1), and the center of mass will be in uniform motion in any inertial reference frame. Quantum mechanically, this means that (5.7) does not have any solutions for which $\psi \in L^2(\mathbb{R}^{3n})$. To see this, let us replace the variables $\mathbf{r}_1, \dots, \mathbf{r}_n$ with new variables $\mathbf{r}_1', \dots, \mathbf{r}_{n-1}', \mathbf{r}_n' = \mathbf{R}$, which include the center-of-mass position vector $\mathbf{R}$. To be specific, let us choose the following variables:

$$\begin{aligned}
\mathbf{r}_1' &= \mathbf{r}_2 - \mathbf{r}_1 ,\\
\mathbf{r}_2' &= \mathbf{r}_3 - \mathbf{R}_1 ,\\
&\vdots\\
\text{(5.10)} \quad \mathbf{r}_{n-1}' &= \mathbf{r}_n - \mathbf{R}_{n-2} ,\\
\mathbf{R} &= \frac{1}{M}(m_1\mathbf{r}_1 + \cdots + m_n\mathbf{r}_n),\\
\mathbf{R}_k &= \frac{m_{k+1}\mathbf{r}_{k+1} + (m_1 + \cdots + m_k)\,\mathbf{R}_{k-1}}{m_1 + \cdots + m_{k+1}}, \qquad k = 1, 2, \dots, n-2.
\end{aligned}$$

In order to establish how the Schroedinger operator form looks in these new variables, we shall proceed in a few stages. First note that if we make in $\psi(\mathbf{r}_1, \dots, \mathbf{r}_n)$ the substitution

$$\text{(5.11)} \qquad \begin{aligned}
\mathbf{r}_1' &= \mathbf{r}_2 - \mathbf{r}_1 ,\\
\mathbf{R}_1 &= \frac{m_1\mathbf{r}_1 + m_2\mathbf{r}_2}{m_1 + m_2},
\end{aligned}$$

thus obtaining the function $\psi_{(1)}(\mathbf{r}_1', \mathbf{R}_1, \mathbf{r}_3, \dots, \mathbf{r}_n)$, we have

$$\begin{aligned}
\text{(5.12)} \quad &\left(\frac{1}{2m_1}\Delta_1 + \frac{1}{2m_2}\Delta_2\right)\psi(\mathbf{r}_1, \dots, \mathbf{r}_n)\\
&= \left(\frac{1}{2M_1}\Delta_1' + \frac{1}{2(m_1 + m_2)}\Delta_{R_1}\right)\psi_{(1)}(\mathbf{r}_1', \mathbf{R}_1, \mathbf{r}_3, \dots, \mathbf{r}_n),
\end{aligned}$$

where M_1 is the reduced mass of the first two particles:

$$M_1 = \frac{m_1 m_2}{m_1 + m_2}$$

$$(5.13)\quad \Delta_1' = \frac{\partial^2}{\partial x_1'^2} + \frac{\partial^2}{\partial y_1'^2} + \frac{\partial^2}{\partial z_1'^2}, \qquad \mathbf{r}_1' = x_1'\mathbf{e}_x + y_1'\mathbf{e}_y + z_1'\mathbf{e}_z,$$

$$\Delta_{\mathbf{R}_1} = \frac{\partial^2}{\partial X_1^2} + \frac{\partial^2}{\partial Y_1^2} + \frac{\partial^2}{\partial Z_1^2}, \qquad \mathbf{R}_1 = X_1\mathbf{e}_x + Y_1\mathbf{e}_y + Z_1\mathbf{e}_z .$$

We shall proceed by induction in k. By performing in the function $\psi_{(k-1)}(\mathbf{r}_1',\ldots, \mathbf{r}_{k-1}', \mathbf{R}_{k-1}, \mathbf{r}_{k+1},\ldots, \mathbf{r}_n)$ the substitution

$$(5.14)\quad \begin{aligned} \mathbf{r}_k' &= \mathbf{r}_{k+1} - \mathbf{R}_{k-1}, \\ \mathbf{R}_k &= \frac{m_{k+1}\mathbf{r}_{k+1} + (m_1 + \cdots + m_k)\,\mathbf{R}_{k-1}}{m_1 + \cdots + m_{k+1}}, \end{aligned}$$

we get the function $\psi_{(k)}(\mathbf{r}_1',\ldots, \mathbf{r}_k', \mathbf{R}_k, \mathbf{r}_{k+2},\ldots, \mathbf{r}_n)$, and we have

$$(5.15)\quad \begin{aligned} &\left(\frac{1}{2m_{k+1}}\,\Delta_{k+1} + \frac{1}{2(m_1 + \cdots + m_k)}\,\Delta_{\mathbf{R}_{k-1}}\right) \\ &\quad\times\ \psi_{(k-1)}(\mathbf{r}_1',\ldots, \mathbf{r}_{k-1}', \mathbf{R}_{k-1}, \mathbf{r}_{k+1},\ldots, \mathbf{r}_n) \\ &= \left(\frac{1}{2M_k}\,\Delta_k' + \frac{1}{2(m_1 + \cdots + m_{k+1})}\,\Delta_{\mathbf{R}_k}\right) \\ &\quad\times\ \psi_{(k)}(\mathbf{r}_1',\ldots, \mathbf{r}_k', \mathbf{R}_k, \mathbf{r}_{k+2},\ldots, \mathbf{r}_n), \end{aligned}$$

where, for $k = 1, 2,\ldots, n-1$,

$$M_k = \frac{(m_1 + \cdots + m_k)m_{k+1}}{m_1 + \cdots + m_{k+1}}$$

$$(5.16)\quad \Delta_k' = \frac{\partial^2}{\partial x_k'^2} + \frac{\partial^2}{\partial y_k'^2} + \frac{\partial^2}{\partial z_k'^2}, \qquad \mathbf{r}_k' = x_k'\mathbf{e}_x + y_k'\mathbf{e}_y + z_k'\mathbf{e}_z,$$

$$\Delta_{\mathbf{R}_k} = \frac{\partial^2}{\partial X_k^2} + \frac{\partial^2}{\partial Y_k^2} + \frac{\partial^2}{\partial Z_k^2}, \qquad \mathbf{R}_k = X_k\mathbf{e}_x + Y_k\mathbf{e}_y + Z_k\mathbf{e}_z .$$

Thus, by induction in k we easily get from (5.12) and (5.15)

$$(5.17)\quad \begin{aligned} &\left(\sum_{k=1}^{n} \frac{1}{2m_k}\,\Delta_k\right)\psi(\mathbf{r}_1,\ldots, \mathbf{r}_n) \\ &= \left(\frac{1}{2M_1}\,\Delta_1' + \cdots + \frac{1}{2M_{n-1}}\,\Delta_{n-1}' + \frac{1}{2M}\,\Delta_{\mathbf{R}}\right) \\ &\quad\times\ \psi_{(n-1)}(\mathbf{r}_1',\ldots, \mathbf{r}_{n-1}', \mathbf{R}), \end{aligned}$$

since from (5.11) and (5.14) it follows that

$$\mathbf{R}_{n-1} = \frac{m_n\mathbf{r}_n + m_{n-1}\mathbf{r}_{n-1} + \cdots + m_1\mathbf{r}_1}{m_1 + \cdots + m_n} = \mathbf{R}.$$

Now, in physics the only potentials used are those which, from the classical point of view, correspond to forces which satisfy the action–reaction principle. This implies (see Exercise 5.3) that when $V(\mathbf{r}_1, \ldots, \mathbf{r}_n)$ is expressed in terms of the variables (5.10), it does not depend on $\mathbf{R}$. Thus, in the new variables the Schroedinger operator form is

$$H_S = -\frac{\hbar^2}{2M}\Delta_{\mathbf{R}} - \sum_{j=1}^{n-1}\frac{\hbar^2}{2M_j}\Delta_j' + V(\mathbf{r}_1', \ldots, \mathbf{r}_{n-1}'). \tag{5.18}$$

If we seek now a solution of (5.7) of the form

$$\psi(\mathbf{r}_1, \ldots, \mathbf{r}_n) = \psi_c(\mathbf{R})\,\Psi(\mathbf{r}_1', \ldots, \mathbf{r}_{n-1}') \tag{5.19}$$

we arrive at the following system of differential equations:

$$-\frac{\hbar^2}{2M}\Delta_R\psi_c(\mathbf{R}) = E_c\psi_c(\mathbf{R}), \tag{5.20}$$

$$\left[-\sum_{j=1}^{n-1}\frac{\hbar^2}{2M_j}\Delta_j' + V(\mathbf{r}_1', \ldots, \mathbf{r}_{n-1}')\right]\Psi(\mathbf{r}_1', \ldots, \mathbf{r}_{n-1}') = E_b\Psi(\mathbf{r}_1', \ldots, \mathbf{r}_{n-1}'), \tag{5.21}$$

$$E = E_c + E_b\,.$$

Following the classical tradition, E_c is called the energy of the center of mass of the system, E_b is the internal energy of the system, and E is the total energy.

The transformations (5.10) obviously map $\mathbb{R}^{3n}$ into itself. Because of the linearity in $\mathbf{r}_1, \ldots, \mathbf{r}_n$ of those transformations, the Jacobian

$$J = \frac{\partial(\mathbf{r}_1, \ldots, \mathbf{r}_n)}{\partial(\mathbf{r}_1', \ldots, \mathbf{r}_{n-1}', \mathbf{R})}$$

of the transformation is obviously a constant in our case. Thus, we can write (Theorem 4.9)

$$\int_{\mathbb{R}^{3n}} |\psi(\mathbf{r}_1, \ldots, \mathbf{r}_n)|^2\, d\mathbf{r}_1 \cdots d\mathbf{r}_n = J\int_{\mathbb{R}^{3n}} |\psi_1(\mathbf{r}_1', \ldots, \mathbf{r}_{n-1}', \mathbf{R})|^2\, d\mathbf{r}_1' \cdots d\mathbf{r}_{n-1}'\, d\mathbf{R},$$

where $\psi_1(\mathbf{r}_1',..., \mathbf{R})$ is the function which we obtain when we express in $\psi(\mathbf{r}_1,..., \mathbf{r}_n)$ all $\mathbf{r}_1,..., \mathbf{r}_n$ in terms of $\mathbf{r}_1',..., \mathbf{R}$. If ψ_1 is of the form (5.19), then we have by Fubini's theorem (Theorem 3.13)

$$\int_{\mathbb{R}^{3n}} |\psi(\mathbf{r}_1,..., \mathbf{r}_n)|^2 \, d\mathbf{r}_1 \cdots d\mathbf{r}_n$$
$$= J \int_{\mathbb{R}^3} |\psi_c(\mathbf{R})|^2 \, d\mathbf{R} \int_{\mathbb{R}^{3(n-1)}} \Psi(\mathbf{r}_1',..., \mathbf{r}_{n-1}') \, d\mathbf{r}_1' \cdots d\mathbf{r}_{n-1}',$$

i.e., $\psi \in L^2(\mathbb{R}^{3n})$ if and only if $\psi_c \in L^2(\mathbb{R}^3)$ and $\Psi \in L^2(\mathbb{R}^{3(n-1)})$. However, the equation (5.20) for the energy of the center of mass of the system does not have any eigenfunctions $\psi_c \in L^2(\mathbb{R}^3)$. Its eigenfunctions

$$-\frac{\hbar^2}{2M} \Delta_{\mathbf{R}} \exp\left(\frac{i}{\hbar} \mathbf{p}_c\mathbf{R}\right) = \frac{\mathbf{p}_c^2}{2M} \exp\left(\frac{i}{\hbar} \mathbf{p}_c\mathbf{R}\right), \qquad \mathbf{p}_c \in \mathbb{R}^3,$$

are not square integrable, corresponding to the fact that, as we shall see in Chapter IV, §7, the operator $\Delta_{\mathbf{R}}$ does not have a point spectrum, but only a continuous one. Physically, this is due to the fact that, because of the lack of external forces, the center of mass of the system is in free motion.

5.3. The Bound States of n-Particle Systems

Postulate W3. The set S_p of all eigenvalues of the equation (5.21) which correspond to square-integrable eigenfunctions are the only internal energy values which the n-particle system in a bound state can assume. The closed linear subspace $\mathscr{H}_b^{(n)}$ of $L^2(\mathbb{R}^{3n})$, spanned by all $\psi \in L^2(\mathbb{R}^{3n})$ for which

$$\Psi_{\mathbf{R}}(\mathbf{r}_1',..., \mathbf{r}_{n-1}') = \psi(\mathbf{r}_1,..., \mathbf{r}_n), \qquad \Psi_{\mathbf{R}} \in L^2(\mathbb{R}^{3(n-1)}),$$

is an eigenfunction of (5.21) for every $\mathbf{R} \in \mathbb{R}^3$, contains all the Hilbert vectors which can represent, at a given time t, a bound state of the n-particle system* interacting via the potential

$$V(\mathbf{r}_1,..., \mathbf{r}_n) = V(\mathbf{r}_1',..., \mathbf{r}_{n-1}').$$

The interpretation of an eigenvalue E_b of (5.21), with eigenfunction $\Psi \in L^2(\mathbb{R}^{3(n-1)})$, as internal bound energy is possible only if E_b is a real number. This will be so if we can generalize Theorem 5.1 in Chapter I.

* We talk about bound states of the *entire* n-particle system, which leaves out, e.g., when $n = 3$, the case when *two* particles are bound, while the third one is free! We shall study this more general problem in scattering theory.

Theorem 5.2. The Schroedinger operator form H_S, the center-of-mass energy operator form

(5.22) $$H_c = -\frac{\hbar^2}{2M}\Delta_{\mathbf{R}}$$

and the internal energy operator form

(5.23) $$H_i = -\sum_{j=1}^{n-1}\frac{\hbar^2}{2M_j}\Delta_j' + V(\mathbf{r}_1',\ldots,\mathbf{r}_{n-1}')$$

define Hermitian operators when applied to twice continuously differentiable functions, which together with their first derivatives vanish faster than $|\mathbf{r}_k|^{-1}$, $k = 1,\ldots, n$, as $|\mathbf{r}_k| \to \infty$.

A differential operator A defined on $L^2(\mathbb{R}^m)$ is said to be *Hermitian* if, whenever Af and Ag are defined for $f, g \in L^2(\mathbb{R}^m)$ and belong to $L^2(\mathbb{R}^m)$, then

(5.24) $$\langle Af \mid g\rangle = \langle f \mid Ag\rangle.$$

Proof. Denote by S_r the sphere in $\mathbb{R}^3$ of radius r and with center at the origin, and by B_r its surface. By applying Green's theorem twice on $S_r^n = S_r \times \cdots \times S_r \in \mathbb{R}^{3n}$, we get

(5.25) $$\begin{aligned}\int_{S_r^n} (H_S f)^* g\, d\mathbf{r}_1 \cdots d\mathbf{r}_n &= -\sum_{k=1}^{n}\frac{\hbar^2}{2m_k}\int_{B_r}\left[\left(\frac{\partial}{\partial n_k} f^*\right) g - f^* \frac{\partial}{\partial n_k} g\right] dS \\ &\quad + \int_{S_r^n} f^*(H_S g)\, d\mathbf{r}_1 \cdots d\mathbf{r}_n,\end{aligned}$$

where $\partial/\partial n_k$ is the derivative in the coordinates $\mathbf{r}_k$ in the direction of the outer normal to B_r. When we let $r \to +\infty$ in (5.25), we get

$$\int_{\mathbb{R}^{3n}} (H_S f)^* g\, d\mathbf{r}_1 \cdots d\mathbf{r}_n = \int_{\mathbb{R}^{3n}} f^*(H_S g)\, d\mathbf{r}_1 \cdots d\mathbf{r}_n$$

because the surface integrals entering in the sum on the right-hand side of (5.25) have to vanish, since the functions $|f|$, $|g|$, $|\partial f/\partial n_k|$, and $|\partial g/\partial n_k|$ are assumed to vanish faster than $|\mathbf{r}_k|^{-1}$ when $|\mathbf{r}_k| \to +\infty$.

The proof for H_c and H_i is completely analogous. Q.E.D.

If $\Psi_1, \Psi_2 \in L^2(\mathbb{R}^{3(n-1)})$ are eigenfunctions of (5.23) corresponding to the eigenvalues $E_b^{(1)}$ and $E_b^{(2)}$, respectively, then we also have, by (5.23), that $H_i\Psi_1$, $H_i\Psi_2 \in L^2(\mathbb{R}^{3(n-1)})$. Consequently, by Theorem 5.2 (in Chapter IV we shall see that the condition at infinity can be dropped):

$$(5.26)\quad E_b^{(1)*}\langle \Psi_1 \mid \Psi_2 \rangle = \langle H_i\Psi_1 \mid \Psi_2 \rangle = \langle \Psi_1 \mid H_i\Psi_2 \rangle = E_b^{(2)}\langle \Psi_1 \mid \Psi_2 \rangle.$$

In case that $\Psi_1 = \Psi_2$, and therefore $E_b^{(1)} = E_b^{(2)} = E_b$, the above relation implies that $E_b = E_b{}^*$. If $E_b^{(1)} \neq E_b^{(2)}$ we get from (5.26) that $\langle \Psi_1 \mid \Psi_2 \rangle = 0$. Thus, we have arrived at the following conclusion.

Theorem 5.3. Each eigenvalue E of the internal energy operator H_i in (5.23), belonging to an eigenfunction $\Psi \in L^2(\mathbb{R}^{3(n-1)})$, is a real number. Eigenfunctions corresponding to different eigenvalues are mutually orthogonal.

Thus, to each eigenvalue belongs at least one nonzero eigenvector, and the eigenvectors belonging to different eigenvalues are orthogonal and therefore linearly independent. Since $L^2(\mathbb{R}^{3(n-1)})$ is separable, any orthogonal system of vectors contains at most a countable number of elements. Hence, we can state Theorem 5.4.

Theorem 5.4. The number of bound-state energy eigenvalues is at most countably infinite, i.e., the point energy spectrum S_p contains a countable number of elements.

5.4. Properties of the n-Particle Schroedinger Operator

An immediately noticeable feature of the Schroedinger operator form in (5.8) is that it is not defined on all the elements of $L^2(\mathbb{R}^{3n})$. The class of functions from $L^2(\mathbb{R}^{3n})$ on which H_S is defined depends on the particular form of the potential $V(\mathbf{r}_1, \ldots, \mathbf{r}_n)$. Broadly speaking, the more singularities $V(\mathbf{r}_1, \ldots, \mathbf{r}_n)$ has the narrower the class on which H_S can be defined. In the next chapter we shall find Theorem 5.5 very useful.

Theorem 5.5. If $V(\mathbf{r}_1, \ldots, \mathbf{r}_n)$ is square integrable on every bounded measurable set in $\mathbb{R}^{3n}$, then the differential form H_S when applied on the functions of the family $\mathscr{C}_b^\infty(\mathbb{R}^{3n})$ of infinitely many-times differentiable functions with compact support* in $\mathbb{R}^{3n}$, maps linearly the set $\mathscr{C}_b^\infty(\mathbb{R}^{3n})$ (where obviously $\mathscr{C}_b^\infty(\mathbb{R}^{3n}) \subset L_{(2)}(\mathbb{R}^{3n})$) into $L_{(2)}(\mathbb{R}^{3n})$.

Proof. If $f \in \mathscr{C}_b^\infty(\mathbb{R}^{3n})$, then f can be represented by a function $f(\mathbf{r}_1, \ldots, \mathbf{r}_n)$ which is certainly continuous and thus measurable, and vanishes outside a bounded open set S, i.e., it is integrable and square

* The support supp f of a function f is the closure of the set of all points at which the function is not zero. In $\mathbb{R}^n$ a set is compact if and only if it is closed and bounded.

integrable on $\mathbb{R}^{3n}$. By the hypothesis of the theorem, $V(\mathbf{r}_1, ..., \mathbf{r}_n)$ is square integrable on S. Thus, $V(\mathbf{r}_1, ..., \mathbf{r}_n) f(\mathbf{r}_1, ..., \mathbf{r}_n)$ is integrable and therefore also square integrable, as we can convince ourselves by using an inequality like (4.7) and applying Theorem 3.9. Furthermore, we evidently have that for every $k = 1, ..., n$

$$\Delta_k f \in \mathscr{C}_{\mathrm{b}}^{\infty}(\mathbb{R}^{3n}) \subset L_{(2)}(\mathbb{R}^{3n}),$$

and therefore we can conclude that $H_S f \in L^2(\mathbb{R}^{3n})$. Q.E.D.

The importance of the preceding theorem is related to Theorem 5.6.

Theorem 5.6. The set $\mathscr{C}_{\mathrm{b}}^{\infty}(\mathbb{R}^m)$ of all elements of $L^2(\mathbb{R}^m)$ which can be represented by infinitely many times differentiable functions of compact support is a vector subspace of $L^2(\mathbb{R}^m)$, which is dense in $L^2(\mathbb{R}^m)$.

Proof. The set $\mathscr{C}_{\mathrm{b}}^{\infty}(\mathbb{R}^m)$ is obviously linear. That $\mathscr{C}_{\mathrm{b}}^{\infty}(\mathbb{R}^m)$ is dense in $L^2(\mathbb{R}^m)$ can be demonstrated in the following way:

According to Theorem 4.8, for any given $f \in L^2(\mathbb{R}^m)$ we can find a simple function

$$s(x) = \sum_{k=1}^{\nu} c_k \chi_{I_k}(x)$$

(where each I_k is a finite interval) such that

$$\| f - s \| < \epsilon/2.$$

If we could prove that the characteristic function of any finite interval can be approximated in the mean arbitrarily well by an infinitely many-times differentiable function of compact support, then for each I_k we could find a $g_k \in \mathscr{C}_{\mathrm{b}}^{\infty}(\mathbb{R}^m)$ such that

$$\| \chi_{I_k} - g_k \| < \frac{\epsilon}{2\nu \mid c_k \mid}. \tag{5.27}$$

Therefore, the fact that $\overline{\mathscr{C}_{\mathrm{b}}^{\infty}(\mathbb{R}^m)} = L^2(\mathbb{R}^m)$ would be established because we have

$$g = \sum_{k=1}^{\nu} c_k g_k \in \mathscr{C}_{\mathrm{b}}^{\infty}(\mathbb{R}^m),$$

and for the arbitrarily chosen $f \in L^2(\mathbb{R}^m)$

$$\begin{aligned} \| f - g \| &\leqslant \| f - s \| + \| s - g \| \\ &\leqslant \| f - s \| + \sum_{k=1}^{\nu} \mid c_k \mid \| \chi_{I_k} - g_k \| < \epsilon. \end{aligned}$$

Any finite interval I in $\mathbb{R}^m$ is a direct product of m finite intervals in $\mathbb{R}^1$. Consider a semiclosed finite interval

$$I = [a_1, b_1) \times \cdots \times [a_m, b_m) \tag{5.28}$$

whose characteristic function can be written as

$$\chi_I(x_1, \ldots, x_m) = \prod_{k=1}^{m} [\theta(x_k - a_k) - \theta(x_k - b_k)], \tag{5.29}$$

where $\theta(x)$ is the step function

$$\theta(x) = \begin{cases} 0 & \text{for } x < 0 \\ 1 & \text{for } x \geqslant 0. \end{cases}$$

Assume that we have at our disposal a sequence $h_1(x), h_2(x), \ldots, x \in \mathbb{R}^1$, of functions which are monotonically nondecreasing

$$h_n(x') \leqslant h_n(x''), \qquad x' < x'', \qquad n = 1, 2, \ldots,$$

infinitely many times differentiable in x, and such that

$$h_n(x) = \begin{cases} 0 & \text{for } x < -\epsilon_n \\ 1 & \text{for } x \geqslant 0, \end{cases}$$

where $\epsilon_n > 0$ and

$$\lim_{n \to \infty} \epsilon_n = 0.$$

An example of such a sequence is provided by

$$h_n(x) = \begin{cases} 0 & \text{for } x \leqslant -1/n \\ \exp\left(-\dfrac{n^2x^2}{1 - n^2x^2}\right) & \text{for } -1/n < x \leqslant 0 \\ 1 & \text{for } x > 0. \end{cases}$$

We have

$$\int_{-\infty}^{+\infty} | h_n(x) - \theta(x)|^2 \, dx \leqslant \epsilon_n \to 0, \qquad n \to +\infty. \tag{5.30}$$

The function

$$h_n(x_1, \ldots, x_m) = \prod_{k=1}^{m} [h_n(x_k - a_k) - h_n(x_k - b_k)]$$

obviously is infinitely many times differentiable and of compact support. It is easy to derive from (5.29) and (5.30) that

$$\lim_{n\to\infty} \int_{\mathbb{R}^m} | \chi_I(x_1 ,..., x_m) - h_n(x_1 ,..., x_m)|^2 \, dx_1 \cdots dx_m = 0.$$

Thus, we have proved that a $g_k \in \mathscr{C}_b^\infty(\mathbb{R}^m)$ satisfying (5.27) can always be found when I_k is a semiclosed interval like (5.28). Since the difference between I in (5.28) and any other interval with the *same* end points as I is of Lebesgue measure zero, our result applies to any finite interval in $\mathbb{R}^m$. Q.E.D.

5.5. The Initial-Value Problem

The time-dependent Schroedinger equation (5.5) is of the first order in t. Hence given the initial condition at some time $t = t_0$ in the form of a twice differentiable function $\psi^{(0)}(\mathbf{r}_1 ,..., \mathbf{r}_n)$, there will be at most one solution $\psi(\mathbf{r}_1 ,..., \mathbf{r}_n ; t)$ of (5.5) which satisfies the initial condition

(5.31) $$\psi(\mathbf{r}_1 ,..., \mathbf{r}_n; t_0) = \psi^{(0)}(\mathbf{r}_1 ,..., \mathbf{r}_n).$$

In the case where $\Psi_{\mathbf{R}}^{(0)} \in L^2(\mathbb{R}^{3(n-1)})$,

$$\Psi_{\mathbf{R}}^{(0)}(\mathbf{r}_1{}' ,..., \mathbf{r}_{n-1}') = \psi^{(0)}(\mathbf{r}_1 ,..., \mathbf{r}_n),$$

is a bound state, i.e., $\psi^{(0)} \in \mathscr{H}_b^{(n)}$, we have the result

(5.32) $$\Psi_{\mathbf{R}}(t) = \sum_\nu \int \frac{d\mathbf{K}}{(2\pi\hbar)^{3/2}} \times \exp\left[\frac{i}{\hbar}\mathbf{RK} - \frac{i}{\hbar}\left(E_b^{(\nu)} + \frac{\mathbf{K}^2}{2M}\right)(t - t_0)\right] \langle \Psi_\nu \mid \tilde{\Psi}_{\mathbf{K}}^{(0)} \rangle \Psi_\nu ,$$

where (see Chapter III, §4.3 on Fourier transforms)

$$\psi(\mathbf{r}_1 ,..., \mathbf{r}_n; t) = \Psi_{\mathbf{R}}(\mathbf{r}_1{}' ,..., \mathbf{r}_{n-1}'; t),$$

$$\langle \Psi_\nu \mid \tilde{\Psi}_{\mathbf{K}}^{(0)} \rangle = \int_{\mathbb{R}^{3(n-1)}} \Psi_\nu{}^*(\mathbf{r}_1{}' ,..., \mathbf{r}_{n-1}') \, \tilde{\Psi}_{\mathbf{K}}^{(0)}(\mathbf{r}_1{}' ,..., \mathbf{r}_{n-1}') \, d\mathbf{r}_1{}' \cdots d\mathbf{r}_{n-1}' ,$$

$$\tilde{\Psi}_{\mathbf{K}}^{(0)}(\mathbf{r}_1{}' ,..., \mathbf{r}_{n-1}') = \frac{1}{(2\pi\hbar)^{3/2}} \int_{\mathbb{R}^3} e^{-i\mathbf{KR}} \Psi_{\mathbf{R}}^{(0)}(\mathbf{r}_1{}' ,..., \mathbf{r}_{n-1}') \, d\mathbf{R},$$

is the solution of the initial value problem (5.31). In (5.32), $\{\Psi_1 , \Psi_2 ,...\}$ is an orthonormal basis (in general, countably infinite) of the space $\mathscr{H}_b^{(n-1)}$ (spanned by all the eigenvectors $\Psi \in L^2(\mathbb{R}^{3(n-1)})$ of H_i) which consists of eigenvectors of H_i :

$$H_i \Psi_\nu = E_b^{(\nu)} \Psi_\nu .$$

One can arrive at the conclusion that such a basis exists by a straightforward generalization of the reasoning employed in §5 of Chapter I.

By duplicating the proof of Theorem 5.2 in Chapter I, it can be established that if the series (5.32) is infinite, it has to converge in the norm, and that its limit $\psi(t)$ has the property that $\psi(t_0) = \psi_0$. The fact that (5.32) really satisfies the equation (5.21) can be checked by inserting (5.32) in (5.5) and differentiating term by term. Naturally, in this procedure we have to assume that the order of summation and differentiation can be interchanged—a very arbitrary, and sometimes incorrect assumption, when the series is infinite. However, it follows from the general theory of linear operators expounded in Chapter III that (5.32) really represents the general solution of the initial-value problem for bound states.

Exercises

5.1. Show that if there are no external forces acting on an n-particle system, then classically, the total force $\mathbf{F} = M\ddot{\mathbf{R}}$ ($\mathbf{R}$ and M are given in (5.9)) acting on the center of mass $\mathbf{R}$ of the system is a zero vector.

5.2. Prove Theorem 5.1, and justify each step in the proof by applying well-known theorems of the calculus.

5.3. Show that as a consequence of the action–reaction principle $\nabla_R V = 0$ in the absence of external forces.

5.4. Prove that the series in (5.32) converges in the norm, and that its limit $\Psi_{\mathbf{R}}(t_0)$ for $t = t_0$ is equal to $\Psi_{\mathbf{R}}^{(0)}$.

6. Direct Sums and Tensor Products of Hilbert Spaces

6.1. Direct Sums of Euclidean Spaces

In this section we shall study two important procedures of constructing new Hilbert spaces from given Hilbert spaces. These constructions—the direct sum and the tensor product of Hilbert spaces—have a great variety of applications to problems in quantum mechanics. With these applications in mind, we shall not treat the most general concept of such spaces that appears in the mathematical literature, but rather restrict ourselves to the level of generality which is sufficient for treating the mathematical problems to which nonrelativistic quantum mechanics gives rise. Such applications will occur frequently in §7 and in Chapter IV.

Theorem 6.1. Let $\mathscr{E}_1, \mathscr{E}_2, \ldots$ be a finite or countably infinite sequence of Euclidean spaces, with inner products $\langle\cdot|\cdot\rangle_1, \langle\cdot|\cdot\rangle_2, \ldots$, respectively. The family of all sequences

$$f = (f_1, f_2, \ldots) \in \mathscr{E}_1 \times \mathscr{E}_2 \times \cdots$$

for which

$$\sum_k \langle f_k | f_k \rangle_k < +\infty \tag{6.1}$$

is a Euclidean space—customarily denoted by

$$\bigoplus_k \mathscr{E}_k = \mathscr{E}_1 \oplus \mathscr{E}_2 \oplus \cdots$$

and called the direct sum of $\mathscr{E}_1, \mathscr{E}_2, \ldots$—in which the vector operations and inner product of $f, g \in \bigoplus_k \mathscr{E}_k$, $f = (f_1, f_2, \ldots)$, $g = (g_1, g_2, \ldots)$, are defined in the following way:

$$\begin{aligned} f + g &= (f_1 + g_1, f_2 + g_2, \ldots), \\ af &= (af_1, af_2, \ldots), \\ \langle f | g \rangle &= \sum_k \langle f_k | g_k \rangle_k . \end{aligned} \tag{6.2}$$

Sequences $(f_1, f_2, \ldots)$ satisfying (6.1) are denoted by $f_1 \oplus f_2 \oplus \cdots$.

Proof. If we have a finite number of Euclidean spaces $\mathscr{E}_1, \ldots, \mathscr{E}_n$, then each element f of $\mathscr{E}_1 \times \cdots \times \mathscr{E}_n$ obviously satisfies (6.1) and therefore f is also an element of $\mathscr{E}_1 \oplus \cdots \oplus \mathscr{E}_n$. It is a straightforward task to check that $f + g$, af and $\langle f | g \rangle$ are defined for all $f, g \in \mathscr{E}_1 \oplus \cdots \oplus \mathscr{E}_n$ and scalars a, and satisfy all the axioms for vector operations and inner product, respectively.

If we have a countably infinite number of Euclidean spaces $\mathscr{E}_1, \mathscr{E}_2, \ldots$, we have to check first whether the expressions in (6.2) are really defined for all $f, g \in \bigoplus_{k=1}^{\infty} \mathscr{E}_k$. Since we have

$$\sum_{k=1}^{\infty} \| af_k \|_k^2 = | a |^2 \sum_{k=1}^{\infty} \| f_k \|_k^2, \qquad \| f_k \|_k = (\langle f_k | f_k \rangle)^{1/2},$$

it follows that $af \in \bigoplus_{k=1}^{\infty} \mathscr{E}_k$ whenever $f \in \bigoplus_{k=1}^{\infty} \mathscr{E}_k$.

In order to establish that $f + g \in \bigoplus_{k=1}^{\infty} \mathscr{E}_k$ when $f, g \in \bigoplus_{k=1}^{\infty} \mathscr{E}_k$, we apply the triangle inequality to the space $\bigoplus_{k=1}^{\nu} \mathscr{E}_k$ to obtain

$$\left(\sum_{k=1}^{\nu} \| f_k + g_k \|_k^2 \right)^{1/2} \leqslant \left(\sum_{k=1}^{\nu} \| f_k \|_k^2 \right)^{1/2} + \left(\sum_{k=1}^{\nu} \| g_k \|_k^2 \right)^{1/2}.$$

From the above inequality we can immediately conclude that $f + g \in \bigoplus_{k=1}^{\infty} \mathscr{E}_k$ by letting $\nu \to +\infty$.

In order to show that

$$\sum_{k=1}^{\infty} \langle f_k \mid g_k \rangle_k \tag{6.3}$$

is absolutely convergent (and therefore also convergent) whenever $(f_1, f_2, \ldots)$ and $(g_1, g_2, \ldots)$ belong to $\bigoplus_{k=1}^{\infty} \mathscr{E}_k$, we use the Schwarz–Cauchy inequality in each of the spaces $\mathscr{E}_k$ to obtain

$$\sum_{k=1}^{\nu} |\langle f_k \mid g_k \rangle_k| \leqslant \sum_{k=1}^{\nu} \| f_k \|_k \| g_k \|_k . \tag{6.4}$$

Since $\| f_1 \|_1, \ldots, \| f_\nu \|_\nu$ and $\| g_1 \|_1, \ldots, \| g_\nu \|_\nu$ are numbers, by applying the Schwarz–Cauchy inequality to $l^2(\nu)$ we get

$$\sum_{k=1}^{\nu} \| f_k \|_k \| g_k \|_k \leqslant \left[\sum_{k=1}^{\nu} \| f_k \|_k^2 \right]^{1/2} \left[\sum_{k=1}^{\nu} \| g_k \|_k^2 \right]^{1/2} . \tag{6.5}$$

From (6.4) and (6.5) we can obviously deduce the absolute convergence of (6.3).

Thus, we have established that the expressions (6.2) are really defined for every $f, g \in \bigoplus_{k=1}^{\infty} \mathscr{E}_k$. We leave to the reader the straightforward task of checking that all the eaxioms of a Euclidean space are fulfilled when in addition we take the sequence $(\mathbf{0}_1, \mathbf{0}_2, \ldots)$ of the zero vectors $\mathbf{0}_1, \mathbf{0}_2, \ldots$ of $\mathscr{E}_1, \mathscr{E}_2, \ldots$ respectively, to be the zero vector of $\bigoplus_{n=1}^{\infty} \mathscr{E}_n$. Q.E.D.

As examples of direct sums of Hilbert spaces, note that $l^2(m + n)$ is isomorphic to $l^2(m) \oplus l^2(n)$, and that

$$l^2(+\infty) = l^2(1) \oplus l^2(1) \oplus \cdots .$$

6.2. Separability and Completeness of Direct Sums of Hilbert Spaces

In proving the next two theorems, we shall use the same method as in proving the separability and completeness of $l^2(+\infty)$ in Theorem 4.3 of Chapter I.

***Theorem 6.2.** If $\mathscr{E}_1, \mathscr{E}_2, \ldots$ is a finite or infinite sequence of Euclidean spaces, their direct sum $\bigoplus_k \mathscr{E}_k$ is a separable Euclidean space if and only if each of the spaces $\mathscr{E}_1, \mathscr{E}_2, \ldots$ is separable.

Proof. We shall consider only the case where we deal with an infinite sequence $\mathscr{E}_1, \mathscr{E}_2, \ldots$ of Euclidean spaces.

Let $S_1, S_2, \ldots$ be countable everywhere dense subsets of $\mathscr{E}_1, \mathscr{E}_2, \ldots$, respectively. Now, all sets of the form

$$R_\nu = S_1 \times \cdots \times S_\nu \times \{\mathbf{0}_{\nu+1}\} \times \{\mathbf{0}_{\nu+2}\} \times \cdots$$

are countable subsets of $\bigoplus_{k=1}^{\infty} \mathscr{E}_k$ ($\mathbf{0}_k$ is the zero vector of $\mathscr{E}_k$). Hence

$$R = \bigcup_{\nu=1}^{\infty} R_\nu \subset \bigoplus_{k=1}^{\infty} \mathscr{E}_k$$

is a countable set, which is also dense in $\bigoplus_{k=1}^{\infty} \mathscr{E}_k$. Namely, for any given $f = (f_1, f_2, \ldots) \in \bigoplus_{k=1}^{\infty} \mathscr{E}_k$ and $\epsilon > 0$, there is a positive integer n for which

$$\sum_{k=n+1}^{\infty} \| f_k \|_k^2 < \epsilon/2.$$

Furthermore, there are $g_1 \in S_1, \ldots, g_n \in S_n$ such that

$$\| f_k - g_k \|_k^2 < \epsilon/2n, \qquad k = 1, \ldots, n,$$

because S_k is dense in $\mathscr{E}_k$. Therefore for

$$g = (g_1, \ldots, g_n, \mathbf{0}_{n+1}, \mathbf{0}_{n+2}, \ldots) \in R_n$$

we have

$$\| f - g \|^2 = \langle f - g \mid f - g \rangle = \sum_{k=1}^{n} \| f_k - g_k \|^2 + \sum_{k=n+1}^{\infty} \| f_k \|^2 < \epsilon. \qquad \text{Q.E.D.}$$

***Theorem 6.3.** If $\mathscr{H}_1, \mathscr{H}_2, \ldots$ is a finite or countably infinite sequence of Hilbert spaces, their direct sum $\bigoplus_k \mathscr{H}_k$ is also a Hilbert space.

Proof. We shall consider explicitly only the more difficult case of an infinite sequence of Hilbert spaces $\mathscr{H}_1, \mathscr{H}_2, \ldots$.

Let $f^{(1)}, f^{(2)}, \ldots \in \bigoplus_{k=1}^{\infty} \mathscr{H}_k$,

$$f^{(n)} = (f_1^{(n)}, f_2^{(n)}, \ldots),$$

be a Cauchy sequence in $\bigoplus_{k=1}^{\infty} \mathscr{H}_k$. Since

$$\| f_k^{(m)} - f_k^{(n)} \|_k \leqslant \left[\sum_{i=1}^{\infty} \| f_i^{(m)} - f_i^{(n)} \|_i^2 \right]^{1/2} = \| f^{(m)} - f^{(n)} \|,$$

the sequence $f_k^{(1)}, f_k^{(2)},\ldots$ is a Cauchy sequence in $\mathscr{H}_k$, and has a limit $f_k \in \mathscr{H}_k$, because $\mathscr{H}_k$ is complete.

By using the triangle inequality in $l^2(\nu)$ we can derive that

$$\left[\sum_{k=1}^{\nu} \|f_k - f_k^{(n)}\|_k^2\right]^{1/2} \leqslant \left[\sum_{k=1}^{\nu} \|f_k - f_k^{(m)}\|_k^2\right]^{1/2} + \left[\sum_{k=1}^{\nu} \|f_k^{(m)} - f_k^{(n)}\|_k^2\right]^{1/2} \tag{6.6}$$

for each $m = 1, 2,\ldots$. Since $f^{(1)}, f^{(2)},\ldots \in \bigoplus_{k=1}^{\infty} \mathscr{H}_k$ is a Cauchy sequence, for any $\epsilon > 0$ there is an integer $N(\epsilon)$ such that

$$\sum_{k=1}^{\nu} \|f_k^{(m)} - f_k^{(n)}\|_k^2 \leqslant \|f^{(m)} - f^{(n)}\|^2 < \epsilon^2/4$$

for all $m, n > N(\epsilon)$. Furthermore, since $f_k = \lim_{m\to\infty} f_k^{(m)}$ for each fixed positive integer, there is an integer $N_\nu(\epsilon)$ such that

$$\|f_k - f_k^{(m)}\|_k < 2^{-(k+2)/2}\epsilon, \qquad k = 1, 2,\ldots, \nu,$$

for $m > N(\epsilon)$. Thus, from (6.6) we find that

$$\left(\sum_{k=1}^{\nu} \|f_k - f_k^{(n)}\|_k^2\right)^{1/2} \leqslant \epsilon\left(\sum_{k=1}^{\nu} \frac{1}{2^{k+2}}\right)^{1/2} + \frac{\epsilon}{2} \leqslant \frac{\epsilon}{2}\left(\sum_{k=1}^{\infty} \frac{1}{2^k}\right)^{1/2} + \frac{\epsilon}{2} = \epsilon$$

for all $n > N_\nu(\epsilon)$. By letting $\nu \to \infty$ in the above relation we derive for any $n > N(\epsilon)$

$$\sum_{k=1}^{\infty} \|f_k - f_k^{(n)}\|_k^2 < \epsilon^2. \tag{6.7}$$

The triangle inequality in $\mathscr{H}_1 \oplus \cdots \oplus \mathscr{H}_\nu$ yields

$$\left[\sum_{k=1}^{\nu} \|f_k\|_k^2\right]^{1/2} \leqslant \left[\sum_{k=1}^{\nu} \|f_k - f_k^{(n)}\|_k^2\right]^{1/2} + \left[\sum_{k=1}^{\nu} \|f_k^{(n)}\|_k^2\right]^{1/2}.$$

From the above we can immediately infer that

$$\sum_{k=1}^{\infty} \|f_k\|_k^2 < +\infty$$

by choosing a fixed $n = n_0 > N(\epsilon)$ and employing (6.7). Thus, $f = (f_1, f_2,\ldots) \in \bigoplus_{k=1}^{\infty} \mathscr{H}_k$, and by (6.7), $f = \lim_{n\to\infty} f^{(n)}$. Q.E.D.

It is easy to see that the operation of constructing direct sums of Euclidean spaces is commutative and associative (see Exercises 6.2–6.4).

6.3. Bilinear Forms on Vector Spaces

In order to be able to study *tensor products* of Euclidean spaces, we need to familiarize ourselves with some of the properties of bilinear forms.

Definition 6.1. A *bilinear form* $(\cdot \mid \cdot)$ on a vector space* is a complex-valued function defined on $\mathscr{V} \times \mathscr{V}$, having the property that for any $f, g, h \in \mathscr{V}$ and any complex number a

$$(f \mid ag) = a(f \mid g) = (a^*f \mid g),$$
$$(f \mid g + h) = (f \mid g) + (f \mid h),$$
$$(f + g \mid h) = (f \mid h) + (g \mid h).$$

A bilinear form $(f \mid g)$ is called *Hermitian* if for all $f, g \in \mathscr{V}$

$$(f \mid g) = (g \mid f)^*.$$

A Hermitian bilinear form is said to be *positive definite* if

$$(f \mid f) \geqslant 0$$

for all $f \in \mathscr{V}$.

If $(\cdot \mid \cdot)$ is a bilinear form, the function $(f \mid f)$ on $\mathscr{V}$ is called the *quadratic form associated with* $(\cdot \mid \cdot)$ (see also Exercise 6.6).

We note that an inner product on $\mathscr{V}$ is a positive-definite bilinear form for which $(f \mid f) = 0$ only when $f = \mathbf{0}$.

As an example of a bilinear form on $l^2(n)$ we can take

$$(6.8) \qquad (\alpha \mid \beta)_A = \sum_{i,k=1}^{n} a_i^* A_{ik} b_k, \qquad \alpha = \begin{pmatrix} a_1 \\ \vdots \\ a_n \end{pmatrix}, \quad \beta = \begin{pmatrix} b_1 \\ \vdots \\ b_n \end{pmatrix},$$

where A is an $n \times n$ matrix

$$A = \begin{pmatrix} A_{11} & \cdots & A_{1n} \\ \vdots & & \vdots \\ A_{n1} & \cdots & A_{nn} \end{pmatrix}.$$

Anticipating the results we shall obtain in the next chapter (see Chapter III, Exercise 2.5), we can state that, vice versa, to each bilinear form $(\cdot \mid \cdot)$ on $l^2(n)$ corresponds a unique matrix A in terms of which

* Throughout this section we consider only complex vector spaces and on them $(\cdot \mid \cdot)$ is sometimes called a sesquilinear form. All the definitions and theorems can be easily specialized to real vector spaces by limiting the field of scalars to $\mathbb{R}^1$.

$(\alpha \mid \beta)$ can be written in the form (6.8). Furthermore, $(\cdot \mid \cdot)$ is Hermitian if and only if the matrix A is Hermitian.

By replacing in the proof of Theorem 2.2 in Chapter I the inner product $\langle f \mid g \rangle$ by a positive-definite bilinear form $(f \mid g)$, we arrive at the following conclusion.

Theorem 6.4. If $(\cdot \mid \cdot)$ is a positive-definite bilinear form on the vector space $\mathscr{V}$, then $(f \mid g)$ satisfies the Schwarz-Cauchy inequality

$$|(f \mid g)|^2 \leqslant (f \mid f)(g \mid g)$$

for any two vectors $f, g \in \mathscr{V}$.

Theorem 6.5. If $(f \mid g)$ is a positive-definite bilinear form on the vector space $\mathscr{V}$, the set

$$\mathscr{V}_0 = \{f: (f \mid f) = 0\}$$

is a vector subspace of $\mathscr{V}$.

Proof. If $f \in \mathscr{V}_0$ and $a \in \mathbb{C}^1$, then $(af \mid af) = |a|^2(f \mid f) = 0$, i.e., $af \in \mathscr{V}_0$.

Take any two vectors $f, g \in \mathscr{V}_0$. From Theorem 6.4 we get

$$|(f \mid g)|^2 \leqslant (f \mid f)(g \mid g) = 0$$

and therefore $(f \mid g) = 0$. Consequently

$$(f + g \mid f + g) = (f \mid f) + (g \mid f) + (f \mid g) + (g \mid g) = 0,$$

i.e., $f + g \in \mathscr{V}_0$. Q.E.D.

Lemma 6.1. Let $\mathscr{V}$ be a vector space and $\mathscr{V}_1$ a vector subspace of $\mathscr{V}$; then the family $\mathscr{V}/\mathscr{V}_1$ of all sets* $\tilde{f}$ of the form

$$\tilde{f} = \{f + h: h \in \mathscr{V}_1\}, \qquad f \in \mathscr{V},$$

is also a vector space—called the *quotient space* of $\mathscr{V}$ and $\mathscr{V}_1$—if the vector operations are defined in the following way:

$$a\tilde{f} = \{a(f + h): h \in \mathscr{V}_1\},$$
$$\tilde{f} + \tilde{g} = \{f + g + h: h \in \mathscr{V}_1\} \tag{6.9}$$

and the zero vector $\mathbf{0}$ of $\mathscr{V}/\mathscr{V}_1$ is chosen to be

$$\tilde{0} = \{0 + h: h \in \mathscr{V}_1\} = \mathscr{V}_1. \tag{6.10}$$

* Here, $\tilde{f}$ is called the equivalence class of f modulo $\mathscr{V}_1$.

Proof. It is easy to see that the above operations (6.9) are not dependent on the chosen representative from $\mathscr{V}$ of an equivalence class belonging to $\mathscr{V}/\mathscr{V}_1$. For instance, if $f_1 = f_2$ modulo $\mathscr{V}_1$, i.e.,

$$\tilde{f}_1 = \{f_1 + h: h \in \mathscr{V}_1\} \equiv \{f_2 + h: h \in \mathscr{V}_1\} = \tilde{f}_2,$$

then $f_2 \in \tilde{f}_1$ so that there is an $h_1 \in \mathscr{V}_1$ for which $f_1 + h_1 = f_2$. Therefore

$$f_1 + g = f_2 + g \ (\text{mod } \mathscr{V}_1)$$

because for any $h \in \mathscr{V}_1$

$$(f_1 + g) + h = (f_2 + g) + h', \qquad h' = h - h_1,$$

where $h' \in \mathscr{V}_1$, since $h, h_1 \in \mathscr{V}_1$ and $\mathscr{V}_1$ is a linear space.

It is a straightforward task to check that the operations (6.9) satisfy all the axioms for vector operations if (6.10) is taken to be the zero vector in $\mathscr{V}_1$. Q.E.D.

Theorem 6.6. If $(\cdot \mid \cdot)$ is a positive-definite bilinear form on the vector space $\mathscr{V}$, and $\mathscr{V}_0$ is the vector subspace on which the quadratic form $(f \mid f)$ vanishes (see Theorem 6.5), then

$$\langle \tilde{f} \mid \tilde{g} \rangle = (f \mid g), \qquad \tilde{f}, \tilde{g} \in \mathscr{V}/\mathscr{V}_0, \qquad f, g \in \mathscr{V}, \tag{6.11}$$

is an inner product on the quotient space $\mathscr{V}/\mathscr{V}_0$.

Proof. First, we have to establish that $\langle \tilde{f} \mid \tilde{g} \rangle$ is uniquely defined for every $\tilde{f}, \tilde{g} \in \mathscr{V}/\mathscr{V}_0$, i.e., that when $f_1 = f_2 \ (\text{mod } \mathscr{V}_0)$ and $g_1 = g_2 \ (\text{mod } \mathscr{V}_0)$, then $(f_1 \mid g_1) = (f_2 \mid g_2)$. To that purpose write

$$\begin{aligned} (f_1 \mid g_1) &= (f_1 - f_2 \mid g_1) + (f_2 \mid g_1) \\ &= (f_1 - f_2 \mid g_1) + (f_2 \mid g_1 - g_2) + (f_2 \mid g_2). \end{aligned} \tag{6.12}$$

Since $(f \mid g)$ is positive definite, we can use the Schwarz–Cauchy inequality (Theorem 6.4) to write

$$|(f_1 - f_2 \mid g_1)|^2 \leqslant (f_1 - f_2 \mid f_1 - f_2)(g_1 \mid g_1),$$
$$|(f_2 \mid g_1 - g_2)|^2 \leqslant (f_2 \mid f_2)(g_1 - g_2 \mid g_1 - g_2),$$

and then note that the right-hand sides of the above inequalities vanish because $(f_1 - f_2 \mid f_1 - f_2) = (g_1 - g_2 \mid g_1 - g_2) = 0$; thus, we get from (6.12) that $(f_1 \mid g_1) = (f_2 \mid g_2)$.

It is easy to check that $\langle \tilde{f} \mid \tilde{g} \rangle$ is a positive-definite bilinear form on $\mathscr{V}/\mathscr{V}_0$. Furthermore, if $\langle \tilde{f} \mid \tilde{f} \rangle = (f \mid f) = 0$ it follows that $f \in \mathscr{V}_0$, i.e., $\tilde{f} = \mathscr{V}_0$, where $\mathscr{V}_0$ is the zero vector of $\mathscr{V}/\mathscr{V}_0$. This establishes the theorem, since a positive-definite bilinear form whose quadratic form vanishes only on the zero vector is an inner product. Q.E.D.

It is worth noting that in §4 we implicitly used Theorem 6.6 when we constructed the Hilbert space $L^2(\Omega, \mu)$ from the vector space $L_{(2)}(\Omega, \mu)$.

*6.4. Algebraic Tensor Products of Vector Spaces

Consider now the Cartesian product $\mathscr{V}_1 \times \cdots \times \mathscr{V}_n$ of n sets $\mathscr{V}_1, ..., \mathscr{V}_n$, which are vector spaces. Introduce in $\mathscr{V}_1 \times \cdots \times \mathscr{V}_n$ equivalence classes (see Exercise 6.7) by the identification

$$(6.13) \quad (ah^{(1)}) \times h^{(2)} \times \cdots \times h^{(n)} = h^{(1)} \times (ah^{(2)}) \times \cdots \times h^{(n)} = \cdots = h^{(1)} \times h^{(2)} \times \cdots \times (ah^{(n)})$$

for all $a \in \mathbb{C}^1$ and all $h^{(1)} \times \cdots \times h^{(n)} \in \mathscr{V}_1 \times \cdots \times \mathscr{V}_n$, and denote by $(\mathscr{V}_1 \times \cdots \times \mathscr{V}_n)^\sim$ the family of all such equivalence classes. We shall use the symbol $h^{(1)} \otimes \cdots \otimes h^{(n)}$ to denote the equivalence class containing $h^{(1)} \times \cdots \times h^{(n)}$, i.e.,

$$h^{(1)} \otimes \cdots \otimes h^{(n)} = \{(c_1 h^{(1)}) \times \cdots \times (c_n h^{(n)}) : c_1 \cdots c_n = 1,\ c_1, ..., c_n \in \mathbb{C}^1\}.$$

Define in $(\mathscr{V}_1 \times \cdots \times \mathscr{V}_n)^\sim$ an operation of multiplication by a scalar $a \in \mathbb{C}^1$,

$$(6.14) \qquad a(h^{(1)} \otimes \cdots \otimes h^{(n)}) = (ah^{(1)}) \otimes \cdots \otimes h^{(n)}.$$

Note that if $\mathbf{0}_k$ is the zero vector of $\mathscr{V}_k$, then

$$ah^{(1)} \otimes \cdots \otimes \mathbf{0}_k \otimes \cdots \otimes h^{(n)} = h^{(1)} \otimes \cdots \otimes \mathbf{0}_k \otimes \cdots \otimes h^{(n)}$$

for all $a \in \mathbb{C}^1$, i.e.,

$$(6.15) \qquad \mathbf{0}_1 \otimes h^{(2)} \otimes \cdots \otimes h^{(n)} = \cdots = h^{(1)} \otimes h^{(2)} \otimes \cdots \otimes \mathbf{0}_n$$

for all $h^{(1)} \in \mathscr{V}_1, ..., h^{(n)} \in \mathscr{V}_n$. We shall call (6.15) the zero element of $(\mathscr{V}_1 \times \cdots \times \mathscr{V}_n)^\sim$, and denote it by $\mathbf{0}$.

Consider now the family of all ν-tuples $(f_1, ..., f_\nu)$ of elements $f_1, ..., f_\nu$ from $(\mathscr{V}_1 \times \cdots \times \mathscr{V}_n)^\sim$, and introduce in this family equivalence classes (see Exercise 6.8) by the following identifications:

$$(6.16) \qquad (f_1, ..., f_\nu) = (f_{k_1}, ..., f_{k_\nu})$$

for any permutation $f_{k_1},\ldots,f_{k_\nu}$ of $f_1,\ldots,f_\nu \in (\mathscr{V}_1 \times \cdots \times \mathscr{V}_n)^\sim$,

$$(f_1,\ldots,f_\nu,0) = (f_1,\ldots,f_\nu), \tag{6.17}$$

for any $f_1,\ldots,f_\nu \in (\mathscr{V}_1 \times \cdots \times \mathscr{V}_n)^\sim$, and

$$\begin{aligned}(h^{(1)} \otimes \cdots \otimes h_1^{(k)} \otimes \cdots \otimes h^{(n)}, h^{(1)} \otimes \cdots \otimes h_2^{(k)} \otimes \cdots \otimes h^{(n)}, f_1,\ldots,f_\nu)\\ = (h^{(1)} \otimes \cdots \otimes (h_1^{(k)} + h_2^{(k)}) \otimes \cdots \otimes h^{(n)}, f_1,\ldots,f_\nu)\end{aligned} \tag{6.18}$$

for any $f_1,\ldots,f_\nu$, $h^{(1)} \otimes \cdots \otimes h^{(n)} \in (\mathscr{V}_1 \times \cdots \times \mathscr{V}_n)^\sim$. Denote the family of all such equivalence classes by $\mathscr{V}_1 \otimes_a \cdots \otimes_a \mathscr{V}_n$. Introduce in $(\mathscr{V}_1 \times \cdots \times \mathscr{V}_n)^\sim$ operations defined in the following way:

$$(f_1,\ldots,f_\mu) + (g_1,\ldots,g_\nu) = (f_1,\ldots,f_\mu, g_1,\ldots,g_\nu), \tag{6.19}$$

$$a(f_1,\ldots,f_\mu) = (af_1,\ldots,af_\mu), \qquad a \in \mathbb{C}^1. \tag{6.20}$$

It is a straightforward though tedious procedure to check that (6.19) and (6.20) leave the equivalence classes of $\mathscr{V}_1 \otimes_a \cdots \otimes_a \mathscr{V}_n$ invariant, and that they satisfy all the axioms imposed on vector operations.

Theorem 6.7. The set $\mathscr{V}_1 \otimes_a \cdots \otimes_a \mathscr{V}_n$ of equivalence classes constructed from elements of the Cartesian product $\mathscr{V}_1 \times \cdots \times \mathscr{V}_n$ of vector spaces $\mathscr{V}_1,\ldots,\mathscr{V}_n$ according to the equivalence relations (6.13)–(6.14) and (6.16)–(6.18) is a vector space (called the *algebraic tensor product* of $\mathscr{V}_1,\ldots,\mathscr{V}_n$) under the operations (6.19) and (6.20)—having as a zero vector the equivalence class containing $(\mathbf{0})$, where $\mathbf{0}$ is the zero element of $(\mathscr{V}_1 \times \cdots \times \mathscr{V}_n)^\sim$.

Since we can obtain for any $f = (f_1,\ldots,f_\nu) \in \mathscr{V}_1 \otimes_a \cdots \otimes_a \mathscr{V}_n$, by repeated application of (6.19), that

$$(f_1,\ldots,f_\nu) = (f_1) + \cdots + (f_\nu),$$

we can state another lemma.

Lemma 6.2. The subset $(\mathscr{V}_1 \times \cdots \times \mathscr{V}_n)^\sim$ of $\mathscr{V}_1 \otimes_a \cdots \otimes_a \mathscr{V}_n$ spans the vector space $\mathscr{V}_1 \otimes_a \cdots \otimes_a \mathscr{V}_n$; every $f \in \mathscr{V}_1 \otimes_a \cdots \otimes_a \mathscr{V}_n$ can be expanded in the form

$$\begin{aligned} f = h_1^{(1)} \otimes \cdots \otimes h_1^{(n)} + \cdots + h_\nu^{(1)} \otimes \cdots \otimes h_\nu^{(n)},\\ h_1^{(1)},\ldots,h_\nu^{(1)} \in \mathscr{V}_1,\ldots, h_1^{(n)},\ldots,h_\nu^{(n)} \in \mathscr{V}_n.\end{aligned} \tag{6.21}$$

In the above relation we have simplified the notation by agreeing to write $h = (h)$ for any $h \in \mathscr{V}_1 \times \cdots \times \mathscr{V}_n$.

Assume now that the bilinear forms $(\cdot \mid \cdot)_1, \ldots, (\cdot \mid \cdot)_n$ are defined on the vector spaces $\mathscr{V}_1, \ldots, \mathscr{V}_n$, respectively. We can define a complex function $(f \mid g)$ for all $f, g \in \mathscr{V}_1 \times \cdots \times \mathscr{V}_n$ in the following way:

$$(f^{(1)} \otimes \cdots \otimes f^{(n)} \mid g^{(1)} \otimes \cdots \otimes g^{(n)}) = \prod_{k=1}^{n} (f^{(k)} \mid g^{(k)})_k . \tag{6.22}$$

Note that (6.22) really determines a unique function on $(\mathscr{V}_1 \times \cdots \times \mathscr{V}_n)^\sim \times (\mathscr{V}_1 \times \cdots \times \mathscr{V}_n)^\sim$:

$$\begin{aligned} &((af^{(1)}) \otimes f^{(2)} \otimes \cdots \mid g^{(1)} \otimes g^{(2)} \otimes \cdots) \\ &\quad = (af^{(1)} \mid g^{(1)})_1 (f^{(2)} \mid g^{(2)})_2 \cdots = (f^{(1)} \mid g^{(1)})_1 (af^{(2)} \mid g^{(2)})_2 \cdots \\ &\quad = (f^{(1)} \otimes (af^{(2)}) \otimes \cdots \mid g^{(1)} \otimes g^{(2)} \otimes \cdots). \end{aligned}$$

By Lemma 6.2 we can write any $f, g \in \mathscr{V}_1 \otimes_a \cdots \otimes_a \mathscr{V}_n$ in the form

$$\begin{aligned} f &= \sum_{i=1}^{\nu_1} f_i^{(1)} \otimes \cdots \otimes f_i^{(n)}, \\ g &= \sum_{j=1}^{\nu_2} g_j^{(1)} \otimes \cdots \otimes g_j^{(n)}. \end{aligned} \tag{6.23}$$

If we define

$$(f \mid g) = \sum_{i=1}^{\nu_1} \sum_{j=1}^{\nu_2} (f_i^{(1)} \otimes \cdots \otimes f_i^{(n)} \mid g_j^{(1)} \otimes \cdots \otimes g_j^{(n)}), \tag{6.24}$$

it is easy to check that the above expression does not change when any of the substitutions suggested by (6.16)–(6.18) are performed, i.e., that (6.24) extends the domain of definition of $(\cdot \mid \cdot)$ to $(\mathscr{V}_1 \otimes_a \cdots \otimes_a \mathscr{V}_n) \times (\mathscr{V}_1 \otimes_a \cdots \otimes_a \mathscr{V}_n)$. Furthermore, it is very easy to verify that $(\cdot \mid \cdot)$ is a bilinear form on $\mathscr{V}_1 \otimes_a \cdots \otimes_a \mathscr{V}_n$.

If the bilinear forms $(\cdot \mid \cdot)_1, \ldots, (\cdot \mid \cdot)_n$ are Hermitian, it is evident from (6.22)–(6.24) that $(\cdot \mid \cdot)$ will also be Hermitian.

We shall prove now that $(\cdot \mid \cdot)$ is a positive-definite bilinear form whenever $(\cdot \mid \cdot)_1, \ldots, (\cdot \mid \cdot)_n$ are positive definite. We need Lemma 6.3.

Lemma 6.3. If $(f_1, f_2, \ldots)$ is a finite or countably infinite system of vectors in a vector space $\mathscr{V}$, then there is a countable system $\{e_1, e_2, \ldots\}$, such that $(e_i \mid e_j) = \delta_{ij}$ and each vector f_i can be expanded in the following way:

$$f_i = \sum_{j=1}^{n_i} a_{ij} e_j + f_i', \qquad (f_i' \mid f_i') = 0. \tag{6.25}$$

Proof. In the quotient space $\mathscr{V}/\mathscr{V}_0$,

$$\mathscr{V}_0 = \{f: (f \mid f) = 0,\ f \in \mathscr{V}\}$$

the bilinear form $(f \mid g)$ induces, according to Theorem 6.6, the inner product $\langle \tilde{f} \mid \tilde{g} \rangle$ defined in (6.11). We can apply on $\{\tilde{f}_1, \tilde{f}_2, \ldots\}$ Theorem 2.4 in Chapter I to deduce the existence of a system $\{\tilde{e}_1, \tilde{e}_2, \ldots\}$ for which $\langle \tilde{e}_i \mid \tilde{e}_j \rangle = \delta_{ij}$ and

$$\tilde{f}_i = \sum_{i=1}^{n_i} a_{ij}\tilde{e}_j . \tag{6.26}$$

If e_i is a chosen vector of $\mathscr{V}$ belonging to the equivalence class $\tilde{e}_i$, (6.26) implies, according to (6.9),

$$f_i + g_i = \sum_{j=1}^{n_i} a_{ij}e_j + h_i, \qquad h_i, g_i \in \mathscr{V}_0 .$$

We get (6.25) from the above by noting that $f_i' = h_i - g_i \in \mathscr{V}_0$. Q.E.D.

In order to show that $(f \mid f) \geqslant 0$ for any given $f \in \mathscr{V}_1 \otimes_a \cdots \otimes_a \mathscr{V}_n$ when $(f \mid g)_1, \ldots, (f \mid g)_n$ are positive definite, expand f in the form (6.21). According to Lemma 6.3 we can find for each $k = 1, \ldots, n$ vectors $e_1^{(k)}, \ldots, e_{p_n}^{(k)}$ in terms of which

$$h_i^{(k)} = \sum_{l=1}^{p_k} a_{il}^{(k)} e_l^{(k)} + g_i^{(k)}, \tag{6.27}$$

where $(g_i^{(k)} \mid g_i^{(k)})_k = 0$, and therefore $(f^{(k)} \mid g_i^{(k)}) = 0$ for all $f^{(k)} \in \mathscr{V}_k$. Thus, we have

$$\begin{aligned} (f \mid f) &= \sum_{i,j=1}^{\nu} (h_i^{(1)} \mid h_j^{(1)})_1 \cdots (h_i^{(n)} \mid h_j^{(n)})_n \\ &= \sum_{i,j=1}^{\nu} \sum_{l_1=1}^{p_1} \cdots \sum_{l_n=1}^{p_n} a_{il_1}^{(1)*} a_{jl_1}^{(1)} \cdots a_{il_n}^{(n)*} a_{jl_n}^{(n)} \\ &= \sum_{l_1=1}^{p_1} \cdots \sum_{l_n=1}^{p_n} \left(\sum_{i=1}^{\nu} a_{il_1}^{(1)} \cdots a_{il_n}^{(n)} \right)^* \left(\sum_{j=1}^{\nu} a_{jl_1}^{(1)} \cdots a_{jl_n}^{(n)} \right) \geqslant 0. \end{aligned} \tag{6.28}$$

In particular, if $(\cdot \mid \cdot)_1, \ldots, (\cdot \mid \cdot)_n$ are inner products in $\mathscr{V}_1, \ldots, \mathscr{V}_n$ respectively, then we have in (6.27) $g_i^{(k)} = \mathbf{0}_k$. Consequently, from (6.28) we can deduce that $(f \mid f) = 0$ if and only if $f = \mathbf{0}$.

In summary, we can state the following.

Theorem 6.8. If $(\cdot \mid \cdot)_1, \ldots, (\cdot \mid \cdot)_n$ are bilinear forms on $\mathscr{V}_1, \ldots, \mathscr{V}_n$, respectively, then the complex function $(f \mid g)$ defined by (6.22)–(6.24) is a bilinear form on $\mathscr{V}_1 \otimes_a \cdots \otimes_a \mathscr{V}_n$; $(\cdot \mid \cdot)$ is Hermitian, positive definite, or an inner product if *all* $(\cdot \mid \cdot)_1, \ldots, (\cdot \mid \cdot)_n$ are, respectively, Hermitian, positive definite, or inner products.

*6.5. Hilbert Tensor Products of Hilbert Spaces

Definition 6.2. Let $\mathscr{H}_1, \ldots, \mathscr{H}_n$ be Hilbert spaces with inner products $\langle \cdot \mid \cdot \rangle_1, \ldots, \langle \cdot \mid \cdot \rangle_n$ respectively. In the algebraic tensor product $\mathscr{H}_1 \otimes_a \cdots \otimes_a \mathscr{H}_n$ of $\mathscr{H}_1, \ldots, \mathscr{H}_n$ denote by $\langle \cdot \mid \cdot \rangle$ the inner product which has the value

$$\langle f \mid g \rangle = \prod_{k=1}^{n} \langle f^{(k)} \mid g^{(k)} \rangle_k \tag{6.29}$$

for f and g of the form

$$f = f^{(1)} \otimes \cdots \otimes f^{(n)}, \qquad g = g^{(1)} \otimes \cdots \otimes g^{(n)};$$

(6.29) defines a unique inner product by Theorem 6.8. The *Hilbert tensor product* $\mathscr{H}_1 \otimes \cdots \otimes \mathscr{H}_n$ of the Hilbert spaces $\mathscr{H}_1, \ldots, \mathscr{H}_n$ is the Hilbert space which is the completion of the Euclidean space $\mathscr{E}$ with the inner product (6.29).

Consider now two Hilbert spaces $L^2(\Omega_1, \mu_1)$ and $L^2(\Omega_2, \mu_2)$. Denote by $\mu_1 \times \mu_2$ the product of the measures μ_1 and μ_2 (as defined in Theorem 2.4) on the Borel subsets of $\Omega_1 \times \Omega_2$. If $f \in L^2(\Omega_1, \mu_1)$ and $g \in L^2(\Omega_2, \mu_2)$, then $f(x_1)\, g(x_2)$ represents an element of $L^2(\Omega_1 \times \Omega_2, \mu_1 \times \mu_2)$, which we denote by $f \cdot g$:

$$(f \cdot g)(x_1, x_2) = f(x_1)\, g(x_2). \tag{6.30}$$

Theorem 6.9. The linear mapping

$$\begin{gathered} h \mapsto \hat{h}, \qquad h \in \mathscr{H}_1 \otimes_a \mathscr{H}_2, \qquad \hat{h} \in \mathscr{H}_3, \\ h = \sum_{k=1}^{n} a_k f_k \otimes g_k, \qquad \hat{h} = \sum_{k=1}^{n} a_k f_k \cdot g_k, \end{gathered} \tag{6.31}$$

of the algebraic tensor product of $\mathscr{H}_1 = L^2(\Omega_1, \mu_1)$ and $\mathscr{H}_2 = L^2(\Omega_2, \mu_2)$ into $\mathscr{H}_3 = L^2(\Omega_1 \times \Omega_2, \mu_1 \times \mu_2)$ can be extended uniquely to a unitary transformation of $\mathscr{H}_1 \otimes \mathscr{H}_2$ onto $\mathscr{H}_3$.

Proof. It is easy to see that (6.31) is a linear mapping of $\mathscr{H}_1 \otimes_a \mathscr{H}_2$ into $\mathscr{H}_3$.

According to Theorem 4.8, the linear set

$$\mathscr{V}_i = \left\{ \sum_{k=1}^{n} a_k \chi_{S_k}(x) \colon S_k = I_k \cap \Omega_i;\ a_k \in \mathbb{C}^1,\ k = 1,\dots, n,\ n = 1, 2,\dots \right\},$$

(where $\Omega_0 = \Omega_1 \times \Omega_2$, and I_n are intervals in the Euclidean space containing the set Ω_i, $i = 1, 2$) is dense in $\mathscr{H}_i$. Therefore (see Exercise 6.9), $\mathscr{V}_1 \otimes_a \mathscr{V}_2$ is dense in $\mathscr{H}_1 \otimes_a \mathscr{H}_2$, and consequently also in $\mathscr{H}_1 \otimes \mathscr{H}_2$.

The mapping (6.31) maps $\mathscr{V}_1 \otimes_a \mathscr{V}_2$ into $\mathscr{V}_3$. Furthermore, for any $\chi_{R_i}, \chi_{S_i} \in \mathscr{V}_i$ we can write, if we denote by $\langle \cdot \mid \cdot \rangle$ the inner product in $\mathscr{H}_1 \otimes \mathscr{H}_2$,

$$\begin{aligned}
&\langle \chi_{R_1} \otimes \chi_{R_2} \mid \chi_{S_1} \otimes \chi_{S_2} \rangle \\
&\quad = \int_{\Omega_1} \chi_{R_1}(x_1)\, \chi_{S_1}(x_1)\, d\mu_1(x_1) \int_{\Omega_2} \chi_{R_2}(x_2)\, \chi_{S_2}(x_2)\, d\mu_2(x_2) \\
&\quad = \int_{\Omega_3} \chi_{R_1}(x_1)\, \chi_{R_2}(x_2)\, \chi_{S_1}(x_1)\, \chi_{S_2}(x_2)\, d(\mu_1 \times \mu_2)(x_1, x_2).
\end{aligned}$$

From the above we can deduce that if $f \mapsto \hat{f}$ and $g \mapsto \hat{g}$, then

$$\langle f \mid g \rangle = \int_{\Omega_3} \hat{f}(x)\, \hat{g}(x)\, d(\mu_1 \times \mu_2)(x), \qquad f, g \in \mathscr{V}_1 \otimes_a \mathscr{V}_2,$$

i.e., the mapping (6.30) is an isomorphism of $\mathscr{V}_1 \otimes \mathscr{V}_2$ onto $\mathscr{V}_3$. Since $\mathscr{V}_1 \otimes \mathscr{V}_2$ is dense in $\mathscr{H}_1 \otimes \mathscr{H}_2$ and $\mathscr{V}_3$ is dense in $\mathscr{H}_3$, this isomorphism can be extended uniquely (cf. Exercise 6.10) to an isomorphism between $\mathscr{H}_1 \otimes \mathscr{H}_2$ and $\mathscr{H}_3$. Q.E.D.

Theorem 6.10. The Hilbert tensor product $\mathscr{H}_1 \otimes \cdots \otimes \mathscr{H}_n$ of separable Hilbert spaces $\mathscr{H}_1, \dots, \mathscr{H}_n$ is separable; if $\{e_i^{(k)} \colon i \in \mathfrak{U}_k\}$ is an orthonormal basis in the $\mathscr{H}_k$, then $\{e_{i_1}^{(1)} \otimes \cdots \otimes e_{i_n}^{(n)} \colon i_1 \in \mathfrak{U}_1, \dots, i_n \in \mathfrak{U}_n\}$ is an orthonormal basis in $\mathscr{H}_1 \otimes \cdots \otimes \mathscr{H}_n$.

Proof. If $\mathscr{H}_k$ is separable, then there is a countable orthonormal basis $\{e_i^{(k)} \colon i \in \mathfrak{U}_k\}$ in $\mathscr{H}_k$. The set

$$\mathsf{T} = \{e_{i_1}^{(1)} \otimes \cdots \otimes e_{i_n}^{(n)} \colon i_1 \in \mathfrak{U}_1, \dots, i_n \in \mathfrak{U}_n\}$$

is also countable. In addition, T is an orthonormal system:

$$\begin{aligned}
\langle e_{i_1}^{(1)} \otimes \cdots \otimes e_{i_n}^{(n)} \mid e_{j_1}^{(1)} \otimes \cdots \otimes e_{j_n}^{(n)} \rangle &= \langle e_{i_1}^{(1)} \mid e_{j_1}^{(1)} \rangle_1 \cdots \langle e_{i_n}^{(n)} \mid e_{j_n}^{(n)} \rangle_n \\
&= \delta_{i_1 j_1} \cdots \delta_{i_n j_n}.
\end{aligned}$$

In order to see that T is a basis in $\mathscr{H}_1 \otimes \cdots \otimes \mathscr{H}_n$, note that the linear manifold $\mathscr{E}_k = (e_1^{(k)}, e_2^{(k)},\ldots)$ spanned by $\{e_1^{(k)}, e_2^{(k)},\ldots\}$ is dense in $\mathscr{H}_k$, and therefore $\mathscr{E}_1 \otimes_a \cdots \otimes_a \mathscr{E}_n$, which is contained in the linear space (T) spanned by T, is dense (see Exercise 6.9) in $\mathscr{H}_1 \otimes \cdots \otimes \mathscr{H}_n$.

Q.E.D.

In the next section we shall find useful the following generalization of the preceding theorem.

Theorem 6.11. If $\{e_i: i \in \mathfrak{U}\}$ is an orthonormal basis in the separable Hilbert space $\mathscr{H}_1$ and $\{e'_{ij} : j \in \mathfrak{B}\}$ is for each value of the index $i \in \mathfrak{U}$ an orthonormal basis in the separable Hilbert space $\mathscr{H}_2$, then $\mathsf{T} = \{e_i \otimes e'_{ij} : j \in \mathfrak{B}, i \in \mathfrak{U}\}$ is an orthonormal basis in $\mathscr{H}_1 \times \mathscr{H}_2$.

Proof. T is an orthonormal system because

$$\begin{aligned}\langle e_{i_1} \otimes e_{i_1 j_1} \mid e_{i_2} \otimes e_{i_2 j_2}\rangle &= \langle e_{i_1} \mid e_{i_2}\rangle_1 \langle e_{i_1 j_1} \mid e_{i_2 j_2}\rangle_2 = \delta_{i_1 i_2}\langle e_{i_1 j_1} \mid e_{i_1 j_2}\rangle_2 \\ &= \delta_{i_1 i_2}\delta_{j_1 j_2}.\end{aligned}$$

Let f be any given element of $\mathscr{H}_1 \otimes \mathscr{H}_2$. Since $\{e_i \otimes e'_{1k} : i \in \mathfrak{U}, k \in \mathfrak{B}\}$ is by Theorem 6.10 a basis in $\mathscr{H}_1$, for any given $\epsilon > 0$ there is a

$$g = \sum_{i,k=1}^{m} a_{ik} e_i \otimes e'_{1k}$$

for which

$$\| f - g \| < \epsilon/2.$$

As each $\mathsf{T}_i = \{e_{ij} : j \in \mathfrak{B}\}$ is a basis in $\mathscr{H}_2$, we can find finite linear combinations $\sum_{j=1}^{n_i} b_{ikj} e'_{ij}$ approximating e'_{1k}:

$$\| e'_{1k} - \sum_{j=1}^{n_i} b_{ikj} e'_{ij} \|_2 < \frac{\epsilon}{2}\left(\sum_{i,k=1}^{m} | a_{ik} |\right)^{-1}.$$

Consequently

$$\begin{aligned}&\left\| f - \sum_{i=1}^{m}\sum_{j=1}^{n_i}\left(\sum_{k=1}^{m} a_{ik} b_{ikj}\right) e_i \otimes e'_{ij} \right\| \\ &\quad\leqslant \| f - g \| + \sum_{i,k=1}^{m} | a_{ik} | \left\| e_i \otimes \left(e'_{1k} - \sum_{j=1}^{n_i} b_{ikj} e'_{ij}\right)\right\| < \epsilon.\end{aligned}$$

Thus, (T) is also dense in $\mathscr{H}_1 \otimes \mathscr{H}_2$, i.e., $[\mathsf{T}] = \mathscr{H}_1 \otimes \mathscr{H}_2$. Q.E.D.

EXERCISES

6.1. Prove that $L^2((-\infty, 0]) \oplus L^2([0, +\infty))$ is isomorphic to $L^2(\mathbb{R}^1)$.

6.2. Show that if $\mathscr{E}_1, ..., \mathscr{E}_n$ are Euclidean spaces and $k_1, ..., k_n$ is any permutation of the indices $1, ..., n$, then $\mathscr{E}_1 \oplus \cdots \oplus \mathscr{E}_n$ is isomorphic to $\mathscr{E}_{k_1} \oplus \cdots \oplus \mathscr{E}_{k_n}$. Prove the same result for the countably infinite case, when instead of $k_1, ..., k_n$ one takes an infinite sequence $k_1, k_2, ...$ in which each positive integer appears once and only once.

6.3. Show that the Euclidean spaces $\mathscr{E}_1 \oplus (\mathscr{E}_2 \oplus \mathscr{E}_3)$, $\mathscr{E}_1 \oplus \mathscr{E}_2 \oplus \mathscr{E}_3$ and $(\mathscr{E}_1 \oplus \mathscr{E}_2) \oplus \mathscr{E}_3$ are isomorphic to one another.

6.4. Show that $(\bigoplus_{k=1}^{\infty} \mathscr{E}_k') \oplus (\bigoplus_{n=1}^{\infty} \mathscr{E}_n'')$ and $\bigoplus_{k=1}^{\infty} (\mathscr{E}_k' \oplus \mathscr{E}_k'')$ are isomorphic Euclidean spaces.

6.5. Show that if the complex function $(f \mid g)$ satisfies $(f \mid g + h) = (f \mid g) + (f \mid h)$ and $(f \mid ag) = a(f \mid g)$, $(f \mid g) = (g \mid f)^*$, then $(f \mid g)$ is a Hermitian bilinear form.

6.6. Prove that a bilinear form $(\cdot \mid \cdot)$ is Hermitian if and only if its quadratic form $Q(f)$ is real for all $f \in \mathscr{V}$.

6.7. Show that if we state that "$h^{(1)} \times \cdots \times h^{(n)} \sim g^{(1)} \times \cdots \times g^{(n)}$ if and only if $h^{(i)} = cg^{(i)}$, $h^{(j)} = (1/c)\, g^{(j)}$, for some $c \in \mathbb{C}^1$ and for a certain value of i and a certain value of j $(i, j = 1, ..., n)$," then the relation $\sim$ so defined is an equivalence relation.

6.8. Prove that the "equality" relations in (6.16), (6.17), and (6.18) are equivalence relations in $\mathscr{E}_1 \times \cdots \times \mathscr{E}_n$.

6.9. Prove that if S_1 and S_2 are dense vector subspaces of $\mathscr{H}_1$ and $\mathscr{H}_2$, respectively, then $S_1 \otimes_a S_2$ is dense in $\mathscr{H}_1 \otimes_a \mathscr{H}_2$, and therefore also in $\mathscr{H}_1 \otimes \mathscr{H}_2$.

6.10. Suppose that $\mathscr{V}_1$ and $\mathscr{V}_2$ are dense vector subspaces of the Hilbert spaces $\mathscr{H}_1$ and $\mathscr{H}_2$, respectively. Prove that a linear mapping $f \mapsto \tilde{f}$ of $\mathscr{V}_1$ onto $\mathscr{V}_2$, which preserves the inner product, can be extended uniquely to a unitary transformation of $\mathscr{H}_1$ onto $\mathscr{H}_2$.

7. The Two-Body Bound-State Problem with a Spherically Symmetric Potential

7.1. TWO PARTICLES INTERACTING VIA A SPHERICALLY SYMMETRIC POTENTIAL

When a system consists of only two particles of different kinds, the states are represented by wave functions $\psi(\mathbf{r}_1, \mathbf{r}_2, t)$, $\psi(t) \in L^2(\mathbb{R}^6)$, in the six configuration space variables $\mathbf{r}_1$ and $\mathbf{r}_2$. The time-independent Schroedinger equation is in this case

$$\left(-\frac{\hbar^2}{2m_1}\Delta_1 - \frac{\hbar^2}{2m_2}\Delta_2 + V(\mathbf{r}_1 - \mathbf{r}_2)\right)\psi(\mathbf{r}_1, \mathbf{r}_2) = E\psi(\mathbf{r}_1, \mathbf{r}_2), \tag{7.1}$$

where, on physical grounds, only potentials depending on $\mathbf{r}_1 - \mathbf{r}_2$ exclusively are considered* (see §5).

After going over to new coordinates, specified by the relative position vector $\mathbf{r}$ and the center of mass position vector $\mathbf{R}$,

$$\begin{aligned} \mathbf{r} &= \mathbf{r}_2 - \mathbf{r}_1, \\ \mathbf{R} &= \frac{m_1\mathbf{r}_1 + m_2\mathbf{r}_2}{m_1 + m_2}, \end{aligned} \tag{7.2}$$

we get as a special case of (5.18)

$$\begin{aligned} H_S &= -\frac{\hbar^2}{2M_0}\Delta_{\mathbf{R}} - \frac{\hbar^2}{2m_0}\Delta + V(\mathbf{r}), \\ \Delta_{\mathbf{R}} &= \frac{\partial^2}{\partial X^2} + \frac{\partial^2}{\partial Y^2} + \frac{\partial^2}{\partial Z^2}, \qquad \mathbf{R} = X\mathbf{e}_x + Y\mathbf{e}_y + Z\mathbf{e}_z, \\ \Delta &= \frac{\partial^2}{\partial x^2} + \frac{\partial^2}{\partial y^2} + \frac{\partial^2}{\partial z^2}, \qquad \mathbf{r} = x\mathbf{e}_x + y\mathbf{e}_y + z\mathbf{e}_z, \end{aligned} \tag{7.3}$$

where M_0 and m_0 are, respectively, the *total mass* and the *reduced mass* of the system:

$$\begin{aligned} M_0 &= m_1 + m_2, \\ m_0 &= \frac{m_1 m_2}{m_1 + m_2}. \end{aligned} \tag{7.4}$$

If we seek solutions of (7.1) of the form

$$\psi(\mathbf{r}_1, \mathbf{r}_2) = \psi_{\mathrm{c}}(\mathbf{R})\,\Psi(\mathbf{r}),$$

we get as a special case of (5.20) and (5.21)

$$-\frac{\hbar^2}{2M_0}\Delta_{\mathbf{R}}\psi_{\mathrm{c}}(\mathbf{R}) = E_{\mathrm{c}}\psi_{\mathrm{c}}(\mathbf{R}), \tag{7.5}$$

$$\left(-\frac{\hbar^2}{2m_0}\Delta + V(\mathbf{r})\right)\Psi(\mathbf{r}) = E_{\mathrm{b}}\Psi(\mathbf{r}). \tag{7.6}$$

* In phenomenological treatments of nuclear forces, velocity-dependent potentials are also considered. In that case the more general requirement of translational invariance is always imposed.

The equation (7.5) is uninteresting to us, because it is related, as we already know from §5, only to the motion of the center of mass of the two particles.

Concerning (7.6), in practice we are most often faced with a *spherically symmetric* or *central* potential, i.e., a potential

$$V(\mathbf{r}) = V_0(r), \qquad r = |\,\mathbf{r}\,| = \sqrt{x^2 + y^2 + z^2},$$

depending only on the distance r of the two particles. This considerably simplifies the problem.

We shall treat this problem in the rest of this section in a detailed manner, because it provides a fair sample of the main methods used in solving the eigenvalue problem of the time-independent Schroedinger equation. It should be noted that in attacking this problem we will find that it is necessary to employ most of the main theorems obtained thus far.

7.2. The Equation of Motion in Spherical Coordinates

Let us write (7.6) in the form

$$\Delta\Psi(\mathbf{r}) + (2m_0/\hbar^2)(E - V_0(r))\,\Psi(\mathbf{r}) = 0, \tag{7.7}$$

where we have simplified the notation by leaving out the subscript b in the internal bound-energy eigenvalue E_b.

It is natural to try to solve (7.7) by going over to the spherical coordinates r, θ, ϕ,

$$x = r\cos\phi\sin\theta, \qquad y = r\sin\phi\sin\theta, \qquad z = r\cos\theta, \tag{7.8}$$

in terms of which the Laplacian Δ becomes (see Exercise 7.1) equal to

$$\frac{1}{r^2}\left[\frac{\partial}{\partial r}\left(r^2\frac{\partial}{\partial r}\right) + \frac{1}{\sin\theta}\frac{\partial}{\partial\theta}\left(\sin\theta\frac{\partial}{\partial\theta}\right) + \frac{1}{\sin^2\theta}\frac{\partial^2}{\partial\phi^2}\right]$$

almost everywhere, i.e., with the exception of the z axis, which is a set of zero Lebesgue measure.

If we express the function $f(x, y, z)$, $f \in L^2(\mathbb{R}^3)$, in terms of the spherical coordinates r, θ, ϕ,

$$f_\mathrm{p}(r, \theta, \phi) = f(r\sin\theta\cos\phi,\, r\sin\theta\sin\phi,\, r\cos\theta), \tag{7.9}$$

we note that $f_\mathrm{p}(r, \theta, \phi)$ is defined on the closed subset Ω_p of $\mathbb{R}^3$,

$$\Omega_\mathrm{p} = \{(r, \theta, \phi)\colon\ 0 \leqslant r < +\infty,\ 0 \leqslant \theta \leqslant \pi,\ 0 \leqslant \phi \leqslant 2\pi\}, \tag{7.10}$$

and satisfies the periodicity condition

$$f_{\mathrm{P}}(r, \theta, 0) = f_{\mathrm{P}}(r, \theta, 2\pi). \tag{7.11}$$

It is well known that the Jacobian of the transformation $(x, y, z) \to (r, \theta, \phi)$ is

$$J(r, \theta, \phi) = r^2 \sin\theta.$$

Thus, if we introduce the Lebesgue–Stieltjes measure

$$\mu_{\mathrm{P}}(B) = \int_B r^2 \sin\theta \, dr \, d\theta \, d\phi, \qquad B \in \mathscr{B}^3_{\Omega_{\mathrm{P}}}, \tag{7.12}$$

on the Borel subsets of Ω_{P}, then by Theorem 4.9 we have

$$\int_{\mathbb{R}^3} |f(x, y, z)|^2 \, dx \, dy \, dz = \int_{\Omega_{\mathrm{P}}} |f_{\mathrm{P}}(r, \theta, \phi)|^2 \, d\mu_{\mathrm{P}}(r, \theta, \phi),$$

i.e., $f_{\mathrm{P}} \in L^2(\Omega_{\mathrm{P}}, \mu_{\mathrm{P}})$. It is easy to see that, in fact, $f \to f_{\mathrm{P}}$ is an isometric linear mapping of $L^2(\Omega)$ into $L^2(\Omega_{\mathrm{P}}, \mu_{\mathrm{P}})$. Furthermore, the range of this mapping is $L^2(\Omega_{\mathrm{P}}, \mu_{\mathrm{P}})$, because if $f_{\mathrm{P}}(r, \theta, \phi) \in L_{(2)}(\Omega_{\mathrm{P}}, \mu_{\mathrm{P}})$ satisfies the periodicity conditions (7.11), then there is an $f \in L^2(\Omega)$ which is mapped into f_{P}; but since the sets $\{(r, \theta, \phi): 0 \leqslant r < +\infty, \theta = \pi, 0 \leqslant \phi \leqslant 2\pi\}$ and $\{(r, \theta, \phi): 0 \leqslant r < +\infty, 0 \leqslant \theta \leqslant \pi, \phi = 2\pi\}$ obviously are of μ_{P} measure zero, every equivalence class $f_{\mathrm{P}} \in L^2(\Omega_{\mathrm{P}}, \mu_{\mathrm{P}})$ contains a function which satisfies (7.10). Thus, we can state Lemma 7.1.

Lemma 7.1. The Hilbert spaces $L^2(\mathbb{R}^3)$ and $L^2(\Omega_{\mathrm{P}}, \mu_{\mathrm{P}})$ are isomorphic, and the mapping $f \to f_{\mathrm{P}}$ defined by (7.9) is a unitary transformation of $L^2(\mathbb{R}^3)$ onto $L^2(\Omega_{\mathrm{P}}, \mu_{\mathrm{P}})$.

We can now try to solve (7.7) by the method of separation of variables. We look for solutions of (7.7) of the form

$$\Psi(\mathbf{r}) = R(r) \, Y(\theta, \phi). \tag{7.13}$$

Thus, we arrive at the system of equations

$$\frac{1}{\sin\theta} \frac{\partial}{\partial\theta} \left(\sin\theta \, \frac{\partial Y(\theta, \phi)}{\partial\theta} \right) + \frac{1}{\sin^2\theta} \frac{\partial^2 Y(\theta, \phi)}{\partial\phi^2} = \lambda Y(\theta, \phi), \tag{7.14}$$

$$\frac{1}{r^2} \frac{d}{dr} \left(r^2 \frac{dR(r)}{dr} \right) + \frac{\lambda}{r^2} R(r) + \frac{2m_0}{\hbar^2} (E - V_0(r)) \, R(r) = 0. \tag{7.15}$$

We note that (7.14) is an eigenvalue equation for λ, in which the potential does not enter. Therefore, we are able to investigate (7.14) without knowing anything more specific about our two-body problem, except for the assumption that the potential is spherically symmetric, i.e., it depends only on r.

On the other hand, the equation (7.15) for the *radial part* $R(r)$ of the wave function contains $V_0(r)$, and its solution can be approached only when the potential is specified. We shall investigate at the end of this section the solution of (7.15) for the Coulomb potential $V_0(r) = c/r$, where c is a constant.

Consider the sets

$$\Omega_r = [0, +\infty) \subset \mathbb{R}^1,$$
$$\Omega_s = [0, \pi] \times [0, 2\pi] \subset \mathbb{R}^2,$$

and the Lebesgue–Stieltjes measures μ_r on $\mathscr{B}^1_{\Omega_r}$ and μ_s on $\mathscr{B}^1_{\Omega_s}$ defined by

$$\mu_r(B) = \int_B r^2 dr, \qquad B \in \mathscr{B}^1_{\Omega_r}, \tag{7.16}$$
$$\mu_s(B) = \int_B \sin\theta \, d\theta \, d\phi, \qquad B \in \mathscr{B}^2_{\Omega_s}.$$

If $\Psi(\mathbf{r})$, $\Psi \in L^2(\mathbb{R}^3)$, is square integrable in the Riemann sense and if it can be written in the form (7.13), then we have

$$\begin{aligned}\int_{\mathbb{R}^3} |\Psi(\mathbf{r})|^2 \, d\mathbf{r} &= \int_{\Omega_p} |R(r)\, Y(\theta, \phi)|^2 \, r^2 \sin\theta \, dr \, d\theta \, d\phi \\ &= \int_0^\infty |R(r)|^2 r^2 \, dr \int_0^\pi \int_0^{2\pi} |Y(\theta, \phi)|^2 \sin\theta \, d\theta \, d\phi \\ &= \int_{\Omega_r} |R(r)|^2 \, d\mu_r(r) \int_{\Omega_s} |Y(\theta, \phi)|^2 \, d\mu_s(\theta, \phi),\end{aligned}$$

which implies that $R \in L^2(\Omega_r, \mu_r)$ and $Y \in L^2(\Omega_s, \mu_s)$. More generally, by using Theorems 2.4 and 6.9 we can infer the following result.

Lemma 7.2. The measure space $(\Omega_p, \mathscr{B}^3_{\Omega_p}, \mu_p)$ is the Cartesian product of the measure spaces $(\Omega_r, \mathscr{B}^1_{\Omega_r}, \mu_r)$ and $(\Omega_s, \mathscr{B}^2_{\Omega_s}, \mu_s)$, and therefore the Hilbert space $L^2(\Omega_p, \mu_p)$ is isomorphic to the tensor product $L^2(\Omega_r, \mu_r) \otimes L^2(\Omega_s, \mu_s)$ of the Hilbert spaces $L^2(\Omega_r, \mu_r)$ and $L^2(\Omega_s, \mu_s)$.

7.3. Spherical Harmonics on the Unit Sphere

We note that since θ and ϕ, $(\theta, \phi) \in \Omega_s$, can be considered to provide a system of orthogonal coordinates on the unit sphere in $\mathbb{R}^3$, we can always visualize a function $Y(\theta, \phi)$ for which

$$Y(\theta, 0) = Y(\theta, 2\pi) \tag{7.17}$$

as being defined on the unit sphere. A special class of function of the above type are the spherical harmonics

$$Y_l^m(\theta, \phi) = (-1)^m \left[\frac{2l+1}{4\pi} \frac{(l-|m|)!}{(l+|m|)!}\right]^{1/2} P_l^{|m|}(\cos\theta)\, e^{im\phi},$$

(7.18)

$$l = 0, 1, 2, \ldots, \qquad m = -l, -l+1, \ldots, +l,$$

where $P_l^m(u)$ are the *associated Legendre functions* of the mth order,* which can be given by *Rodrigues' formula*:

$$P_l^m(u) = \frac{(1-u^2)^{m/2}}{2^l l!} \frac{d^{l+m}}{du^{l+m}} (u^2-1)^l, \qquad m = 0, 1, 2, \ldots . \tag{7.19}$$

The complete solution of the eigenvalue problem for (7.14) results in the following important theorem.

Theorem 7.1. The spherical harmonics (7.18) are eigenfunctions of (7.14)

$$\frac{1}{\sin\theta} \frac{\partial}{\partial\theta}\left(\sin\theta \frac{\partial Y_l^m(\theta, \phi)}{\partial\theta}\right) + \frac{1}{\sin^2\theta} \frac{\partial^2 Y_l^m(\theta, \phi)}{\partial\phi^2} = -l(l+1)\, Y_l^m(\theta, \phi),$$

and they determine an orthonormal basis

$$\mathfrak{T} = \{Y_l^m,\ l = 0, 1, 2, \ldots,\ m = -l, -l+1, \ldots, +l\},$$

(7.20)

$$\int_0^\pi \int_0^{2\pi} (Y_l^m(\theta, \phi))^*\, Y_{l'}^{m'}(\theta, \phi) \sin\theta\, d\theta\, d\phi = \delta_{ll'}\delta_{mm'},$$

in the Hilbert space $L^2(\Omega_S, \mu_S)$.

We shall carry out the proof of Theorem 7.1 in a few stages, thus being able to state and prove a few results which are of independent interest.

In trying to solve (7.14) by the method of separation of variables, we look for solutions of the form

$$Y(\theta, \phi) = \Theta(\theta)\, \Phi(\phi). \tag{7.21}$$

Thus, we arrive at the following system of independent differential equations:

$$\frac{d^2\Phi(\phi)}{d\phi^2} + \mu\Phi(\phi) = 0, \tag{7.22}$$

$$\sin\theta \frac{d}{d\theta}\left(\sin\theta \frac{d\Theta(\theta)}{d\theta}\right) - (\lambda \sin^2\theta + \mu)\, \Theta(\theta) = 0. \tag{7.23}$$

* The associated Legendre functions $P_l^0(u)$ of the zero order are called the *Legendre polynomials*.

By applying again Theorems 2.4 and 6.9 we get Lemma 7.3.

Lemma 7.3. The Hilbert space $L^2(\Omega_s, \mu_s)$ is isomorphic to $L^2([0, \pi], \mu_\theta) \otimes L^2([0, 2\pi])$, where

$$\mu_\theta(B) = \int_B \sin\theta \, d\theta, \qquad B \in \mathscr{B}^1_{(0,2\pi)} .$$

7.4. The Completeness of Trigonometric Functions

When the periodicity condition (7.17) is imposed on (7.21), it becomes

$$\Phi(0) = \Phi(2\pi). \tag{7.24}$$

The only linearly independent solutions of (7.22) satisfying (7.24) are

$$e^{im\phi}, \qquad m = 0, +1, +2, \ldots,$$

and they do belong to $L_{(2)}([0, 2\pi])$.

Theorem 7.2. The family $\{\Phi_m : m = 0, \pm 1, \pm 2, \ldots\}$ of elements of $L^2([0, 2\pi])$ corresponding to the functions

$$\Phi_m(\phi) = (1/\sqrt{2\pi})\, e^{im\phi}$$

is an orthonormal basis in $L^2([0, 2\pi])$.

The fact that $\{\Phi_0, \Phi_{\pm 1}, \ldots\}$ is an orthonormal system is easy to establish:

$$\langle \Phi_m \mid \Phi_{m'} \rangle = \frac{1}{2\pi} \int_0^{2\pi} \exp[i(m' - m)\phi] \, d\phi = \delta_{mm'} .$$

In order to prove its completeness we need Theorem 7.3.

Theorem 7.3. If $f(\phi)$, $\phi \in [0, 2\pi]$, is a continuous periodic function

$$f(0) = f(2\pi), \tag{7.25}$$

with a piecewise continuous first derivative, then

$$f(\phi) = \lim_{n\to\infty} \frac{1}{\sqrt{2\pi}} \sum_{m=-n}^{n} \langle \Phi_m \mid f \rangle \, e^{im\phi}, \tag{7.26}$$

and the convergence of $\sigma_1(\phi), \sigma_2(\phi), \ldots,$

$$\sigma_n(\phi) = \frac{1}{\sqrt{2\pi}} \sum_{m=-n}^{n} \langle \Phi_m \mid f \rangle \, e^{im\phi},$$

to $f(\phi)$ is uniform in $[0, 2\pi]$.

Proof. Since $f' \in L^2([0, 2\pi])$, $f'(\phi) = df(\phi)/d\phi$, and $\{\Phi_0, \Phi_{\pm 1}, ...\}$ is an orthonormal system, we have by Bessel's inequality (see Chapter I, Lemma 4.1),

$$\sum_{m=-\infty}^{+\infty} |\langle \Phi_m \mid f' \rangle|^2 \leqslant \|f'\|^2. \tag{7.27}$$

On the other hand, integrating by parts we get

$$\begin{aligned}\langle \Phi_m \mid f' \rangle &= \frac{1}{\sqrt{2\pi}} \int_0^{2\pi} f'(\phi)\, e^{-im\phi}\, d\phi \\ &= \frac{im}{\sqrt{2\pi}} \int_0^{2\pi} f(\phi)\, e^{-im\phi}\, d\phi = im \langle \Phi_m \mid f \rangle.\end{aligned}$$

Therefore, by (7.27)

$$\sum_{m=-\infty}^{+\infty} m^2 |\langle \Phi_m \mid f \rangle|^2 \leqslant \|f'\|^2. \tag{7.28}$$

To prove the uniform convergence of

$$\sigma_n(\phi) = \sum_{m=-n}^{+n} c_m \Phi_m(\phi), \qquad c_m = \langle \Phi_m \mid f \rangle,$$

use the Schwarz–Cauchy inequality in $l^2(2\nu)$ to obtain

$$\begin{aligned}| \sigma_{n+\nu}(\phi) - \sigma_n(\phi)|^2 &= \left| \sum_{|m|=n+1}^{n+\nu} c_m \Phi_m(\phi) \right|^2 \\ &= \left| \sum_{|m|=n+1}^{n+\nu} (mc_m) \frac{\Phi_m(\phi)}{m} \right|^2 \\ &\leqslant \left(\sum_{|m|=n+1}^{n+\nu} m^2 \mid c_m \mid^2 \right) \left(\sum_{|m|=n+1}^{n+\nu} \frac{\mid \Phi_m(\phi) \mid^2}{m^2} \right) \\ &\leqslant \frac{1}{2\pi} \|f'\|^2 \sum_{|m|=n+1}^{n+\nu} \frac{1}{m^2},\end{aligned}$$

where the last inequality follows from (7.28). Since the series $\sum_{m=1}^{\infty} 1/m^2$ converges, for any $\epsilon > 0$ we can find an $N(\epsilon)$ such that

$$\sum_{|m|=n+1}^{n+\nu} \frac{1}{m^2} \leqslant 2 \sum_{m=n+1}^{\infty} \frac{1}{m^2} < 2\pi\epsilon \|f'\|^{-2}$$

for all $n > N(\epsilon)$, and therefore the uniform convergence of $\sigma_n(x)$ follows.

In order to show that $\sigma_1(\phi), \sigma_2(\phi),\ldots$ converges to $f(\phi)$, note that

$$\begin{aligned}
\sigma_n(\phi) &= \frac{1}{2\pi}\sum_{m=-n}^{n}\exp(im\phi)\int_0^{2\pi} f(\tau)\exp(-im\tau)\,d\tau \\
&= \frac{1}{2\pi}\int_0^{2\pi} f(\tau)\left(\sum_{m=-n}^{n}\exp[im(\phi-\tau)]\right)d\tau \\
&= \frac{1}{\pi}\int_0^{2\pi} f(\tau)\operatorname{Re}(\tfrac{1}{2}+\exp[i(\phi-\tau)]+\cdots+\exp[in(\phi-\tau)])\,d\tau \\
&= \frac{1}{\pi}\int_0^{2\pi} f(\tau)\operatorname{Re}\left(\frac{\exp[i(n+1)(\phi-\tau)]-1}{\exp[i(\phi-\tau)]-1}-\frac{1}{2}\right)d\tau \\
&= \frac{1}{2\pi}\int_0^{2\pi} f(\tau)\operatorname{Re}\left(\frac{1}{i}\,\frac{\exp[i(n+\frac{1}{2})(\phi-\tau)]-\exp[-\frac{1}{2}i(\phi-\tau)]}{\sin\frac{1}{2}(\phi-\tau)}-1\right)d\tau \\
&= \frac{1}{2\pi}\int_0^{2\pi} f(\tau)\,\frac{\sin(n+\frac{1}{2})(\phi-\tau)}{\sin\frac{1}{2}(\phi-\tau)}\,d\tau \\
&= \frac{1}{2\pi}\int_{-\phi}^{2\pi-\phi} f(\omega+\phi)\,\frac{\sin(n+\frac{1}{2})\omega}{\sin(\omega/2)}\,d\omega,
\end{aligned}$$

where in the last integral we have introduced the new variable $\omega = \tau - \phi$.

We extend the domain of definition of $f(\phi)$ to the entire real line:

$$f(\phi + 2\pi n) = f(\phi), \qquad n = \pm 1, \pm 2, \ldots .$$

The function so extended is periodic (with a period 2π), everywhere continuous, due to (7.25), and has a piecewise continuous derivative. We can now write

$$\sigma_n(\phi) = \frac{1}{2\pi}\int_0^{2\pi} f(\phi+\omega)\,\frac{\sin(n+\frac{1}{2})\omega}{\sin(\omega/2)}\,d\omega.$$

Since we have

$$\frac{1}{2\pi}\int_0^{2\pi}\frac{\sin(n+\frac{1}{2})\omega}{\sin(\omega/2)}\,d\omega = \frac{1}{\pi}\int_0^{2\pi}(\tfrac{1}{2}+\cos\omega+\cdots+\cos n\omega)\,d\omega = 1,$$

we deduce from the above two formulas that

$$\begin{aligned}
(7.29)\qquad \sigma_n(\phi) - f(\phi) &= \frac{1}{2\pi}\int_0^{2\pi}\frac{f(\phi+\omega)-f(\phi)}{\sin(\omega/2)}\sin(n+\tfrac{1}{2})\omega\,d\omega \\
&= \frac{1}{2\pi}\int_0^{2\pi}\left(\frac{f(\phi+\omega)-f(\phi)}{\sin(\omega/2)}\cos(\omega/2)\right)\sin n\omega\,d\omega \\
&\quad + \frac{1}{2\pi}\int_0^{2\pi}(f(\phi+\omega)-f(\phi)]\cos n\omega\,d\omega,
\end{aligned}$$

where the function

$$g_1(\omega) = \frac{f(\phi + \omega) - f(\phi)}{\sin(\omega/2)} \cos(\omega/2)$$

is piecewise continuous, since

$$\lim_{\omega \to \pm 0} g_1(\omega) = \lim_{\omega \to \pm 0} \frac{f(\phi + \omega) - f(\phi)}{\omega} \frac{\omega}{\sin(\omega/2)} \cos(\omega/2) = 2f'(\phi \pm),$$

while

$$g_2(\omega) = f(\phi + \omega) - f(\omega)$$

is obviously continuous.

Thus, both functions $g_1(\omega)$ and $g_2(\omega)$ are integrable and we have

$$\text{(7.30)} \quad \sigma_n(\phi) - f(\phi) = \frac{1}{4\pi} \int_0^{2\pi} [e^{in\omega}(g_2(\omega) - ig_1(\omega)) + e^{-in\omega}(g_2(\omega) + ig_1(\omega))] d\omega$$
$$= \frac{1}{2(2\pi)^{1/2}} \langle \Phi_{-n} \mid g_2 - ig_1 \rangle + \frac{1}{2(2\pi)^{1/2}} \langle \Phi_n \mid g_2 + ig_1 \rangle.$$

Since on the right-hand side of (7.30) are the Fourier coefficients of functions belonging to $L^2([0, 2\pi])$, from Bessel's inequality we can infer that they converge to zero when $n \to +\infty$. Q.E.D.

It is easy to finish the proof of Theorem 7.2 with the help of Theorem 7.3.

If f is an arbitrarily chosen element of $L^2([0, 2\pi])$ and $\epsilon > 0$ is any given number, then by Theorem 5.6 there is a $g_1 \in \mathscr{C}_{\mathrm{b}}^{\infty}(\mathbb{R}^1)$, which can be represented by an infinitely many times differentiable function $g_1(x)$, $x \in \mathbb{R}^1$, such that

$$\| f_1 - g_1 \|_1^2 = \int_{-\infty}^{+\infty} | f_1(x) - g_1(x)|^2 \, dx < \epsilon^2/9,$$

where

$$f_1(x) = \begin{cases} f(x), & 0 \leqslant x \leqslant 2\pi \\ 0, & x \notin [0, 2\pi]. \end{cases}$$

If we introduce the function

$$g(x) = \begin{cases} g_1(x), & 0 \leqslant x \leqslant 2\pi \\ 0, & x \notin [0, 2\pi], \end{cases}$$

which is infinitely many times differentiable in $[0, 2\pi]$, we have

$$\text{(7.31)} \quad \| f - g \|^2 = \int_0^{2\pi} | f(\phi) - g(\phi)|^2 \, d\phi \leqslant \int_{-\infty}^{+\infty} | f_1(x) - g_1(x)|^2 \, dx < \epsilon^2/9.$$

We can approximate $g(\phi)$ in the mean as closely as desired by a continuous and piecewise differentiable function $h(\phi)$, $\phi \in [0, 2\pi]$, which is periodic, i.e., $h(0) = h(2\pi)$. For instance

$$h(\phi) = \begin{cases} g(2\pi) + \dfrac{g(\phi_0) - g(2\pi)}{\phi_0}\phi, & 0 \leqslant \phi \leqslant \phi_0 \\ g(\phi), & \phi_0 < \phi \leqslant 2\pi, \end{cases}$$

is such a function, for which

$$\| g - h \|^2 = \int_0^{\phi_0} | g(2\pi) + \frac{g(\phi_0) - g(2\pi)}{\phi_0}\phi - g(\phi)|^2 \, d\phi < \epsilon^2/9 \tag{7.32}$$

if ϕ_0 is chosen sufficiently small. Since $h(\phi)$ satisfies the conditions of Theorem 7.3, for sufficiently large n we shall have

$$| h(\phi) - \sum_{m=-n}^{n} c_m \Phi_m(\phi)| < \epsilon/\sqrt{18\pi}, \qquad c_m = \langle \Phi_m \mid h \rangle,$$

and consequently

$$\left| h - \sum_{m=-n}^{n} c_m \Phi_m \right|^2 = \int_0^{2\pi} \left| h(\phi) - \sum_{m=-n}^{n} c_m \Phi_m(\phi) \right|^2 d\phi < \epsilon^2/9. \tag{7.33}$$

By combining (7.31), (7.32), and (7.33) we deduce that

$$\left\| f - \sum_{m=-n}^{n} c_m \Phi_m \right\| \leqslant \| f - g \| + \| g - h \| + \left\| h - \sum_{m=-n}^{n} c_m \Phi_m \right\| < \epsilon,$$

which establishes Theorem 7.2.

7.5. The Completeness of Legendre Polynomials

We can use Theorem 7.2 to prove the *Weierstrass approximation theorem*, which will be used later.

Theorem 7.4 (*Weierstrass*). Any complex function $f(x)$, $x \in I$, which is continuous in a closed finite interval $I = [a, b]$ can be uniformly approximated in I by a polynomial $P(x)$ in x; i.e., for any given $\epsilon > 0$ there is a polynomial $P(x)$ with complex coefficients, such that

$$| f(x) - P(x)| < \epsilon, \qquad a \leqslant x \leqslant b. \tag{7.34}$$

Proof. Consider the case when $0 < a < b < 2\pi$. Choose a partition

$$0 = x_0 < x_1 = a < x_2 < \cdots < x_{n-1} = b < x_n = 2\pi$$

of $[0, 2\pi]$. The function

$$
g(x) = \begin{cases} \dfrac{x}{a} f(a), & x_0 = 0 \leqslant x < x_1 = a \\ \dfrac{x - x_{k-1}}{x_k - x_{k-1}} [f(x_k) - f(x_{k-1})] + f(x_{k-1}), & \\ \qquad x_{k-1} \leqslant x < x_k, \; k = 2, \ldots, n-1 & \\ -\dfrac{x - 2\pi}{2\pi - b} f(b), & x_{n-1} = b \leqslant x \leqslant 2\pi, \end{cases} \tag{7.35}
$$

represents a polygon inscribed in the graph of $f(x)$ by joining the points $(x_1, f(x_1)), \ldots, (x_{n-1}, f(x_{n-1}))$ with straight lines, and then joining the ends to the real axis at $x_0 = 0$ and $x_n = 2\pi$. Such functions $g(x)$ can approximate uniformly and arbitrarily well the continuous function $f(x)$ (recall that a continuous function is uniformly continuous in a closed finite interval). Since the function $g(x)$ in (7.35) satisfies the conditions of Theorem 7.3, it can be approximated uniformly by a trigonometric polynomial

$$
h(x) = \sum_{m=-n}^{n} c_m e^{imx}.
$$

The above trigonometric polynomial can be, in its turn, approximated uniformly and arbitrarily well by polynomials, since due to Taylor's formula

$$
e^{imx} = 1 + \frac{imx}{1!} + \cdots + \frac{(imx)^k}{k!} + \frac{(imx)^{k+1}}{(k+1)!} (\cos m\theta_1 x + i \sin m\theta_2 x), \qquad 0 < \theta_{1,2} < 1,
$$

each e^{imx} can be approximated uniformly in $[0, 2\pi]$ by ordinary polynomials. Thus, the theorem is proved for the case when $0 < a < b < 2\pi$.

The general case of $[a, b]$ can be reduced to the above case by mapping linearly $[a, b]$ into $[0, 2\pi]$, e.g., by using the mapping

$$
x \mapsto \pi \frac{x - a}{b - a} + \frac{\pi}{2},
$$

and by noting that a linear mapping maps polynomials into polynomials. Q.E.D.

We note that if (7.34) is true, then

$$
\| f - P \| = \left[\int_a^b | f(x) - P(x) |^2 \, dx \right]^{1/2} < \epsilon \sqrt{b - a}.
$$

Consequently, from Theorems 5.6 and 7.4 we can easily infer the following lemma.

Lemma 7.4. The vector subspace $(p_0, p_1, \ldots)$ of $L^2([a, b])$, spanned by the elements of $L^2([a, b])$ represented by the polynomials

$$p_0(x) = 1, \qquad p_1(x) = x, \ldots, \qquad p_n(x) = x^n, \ldots,$$

is everywhere dense in $L^2([a, b])$.

We can turn now to solving the differential equation (7.23), which in the new variable

$$u = -\cos\theta, \qquad -1 \leqslant u \leqslant 1, \tag{7.36}$$

becomes (with $\mu = -m^2$) the associated Legendre equation

$$(1 - u^2)\frac{d}{du}\left[(1 - u^2)\frac{dw(u)}{du}\right] - [\lambda(1 - u^2) - m^2]\,w(u) = 0$$

for the new function

$$w(u) = \Theta(\operatorname{arc}\cos(-u)). \tag{7.37}$$

The above associated Legendre equation (called simply the Legendre equation when $m = 0$) can be written in the more customary form

$$(1 - u^2)\,w''(u) - 2u\,w'(u) - \left[\lambda - \frac{m^2}{1 - u^2}\right] w(u) = 0. \tag{7.38}$$

We can state the following direct consequence of Theorem 4.9.

Lemma 7.5. The mapping $\Theta \to w$, defined by (7.37), is a unitary transformation of $L^2([0, \pi], \mu_\theta)$ onto $L^2([-1, +1])$.

Theorem 7.5. The Legendre polynomials

$$P_l(u) = \frac{1}{2^l l!}\frac{d^l(u^2 - 1)^l}{du^l}, \qquad l = 0, 1, 2, \ldots, \tag{7.39}$$

satisfy the Legendre equation with $\lambda = -l(l + 1)$:

$$(1 - u^2)\,P_l''(u) - 2uP_l'(u) + l(l + 1)\,P_l(u) = 0. \tag{7.40}$$

The set $\{P_0, P_1, \ldots\} \subset L^2([-1, 1])$ is an orthogonal basis in $L^2([-1, +1])$, with

$$\langle P_l \mid P_{l'}\rangle = \int_{-1}^{+1} P_l(u)\,P_{l'}(u)\,du = \frac{2}{2l + 1}\,\delta_{ll'}\,. \tag{7.41}$$

Proof. To derive the result that $P_l(u)$ satisfies (7.40), we start with the identity

$$(1 - u^2)(d/du)(u^2 - 1)^l - 2lu(u^2 - 1)^l = 0.$$

By taking, with the help of the Leibniz rule, the $l+1$ derivative of the above equation we arrive at (7.40).

By employing the binomial formula we get

$$P_l(u) = \frac{1}{2^l l!} \frac{d^l}{du^l} \left(\sum_{k=0}^{l} \binom{l}{k} (-1)^k (u^2)^{l-k} \right)$$

$$= \frac{1}{2^l l!} \sum_{k=0}^{l} \frac{(-1)^k\, l!}{k!(l-k)!} \frac{d^l}{du^l} u^{2(l-k)}$$

from which we derive, after carrying out all the differentiations,

(7.42)
$$P_l(u) = \sum_{k=0}^{(1/2)(l-\delta_l)} \frac{(-1)^k (2l-2k)!}{2^l k!(l-k)!(l-2k)!} u^{l-2k},$$

$$\delta_l = \begin{cases} 0, & l \text{ even} \\ 1, & l \text{ odd.} \end{cases}$$

The above formula shows that P_l is a polynomial of degree l. Integrating by parts l times consecutively and using the fact that

$$\frac{d^k}{du^k} (u^2 - 1)^l \bigg|_{u=\pm 1} = 0, \qquad k = 0, 1, \ldots, l-1,$$

we arrive at the relation

$$\langle P_l \mid p_m \rangle = \frac{1}{2^l l!} \int_{-1}^{+1} \left[\frac{d^l}{du^l} (u^2 - 1)^l \right] u^m \, du$$

$$= \frac{(-1)^l}{2^l l!} \int_{-1}^{+1} (u^2 - 1)^l \frac{d^l u^m}{du^l} \, du.$$

From the above we get $\langle P_l \mid p_m \rangle = 0$ for $m = 1, \ldots, l-1$. Consequently, $\langle P_l \mid P_{l'} \rangle = 0$ for $l' < l$, i.e.,

$$\langle P_l \mid P_{l'} \rangle = \int_{-1}^{+1} P_l(u)\, P_{l'}(u)\, du = 0, \qquad l \neq l',$$

which proves that the system $P_0, P_1, \ldots$ consists of mutually orthogonal elements of $L^2([-1, +1])$. It is easy to check (see Exercise 7.2) that (7.41) is true for $l = l'$, i.e., the Legendre polynomials are mutually orthogonal but they are not normalized.

From (7.39) it is obvious that $p_0(u) = 1$, $p_1(u) = u, \ldots, p_n(u) = u^n$ can be expressed in terms of $P_0(u), \ldots, P_n(u)$. Consequently the vector subspace $(p_0, p_1, \ldots)$ of $L^2([-1, +1])$ spanned by $p_0, p_1, \ldots$ is identical to the vector subspace $(P_0, P_1, \ldots)$ spanned by the elements of $L^2([-1, +1])$ represented by Legendre polynomials. Since according to Lemma 7.4 $(p_0, p_1, \ldots)$ is dense in $L^2([-1, +1])$, we have

$$[P_0, P_1, \ldots] = [p_0, p_1, \ldots] = L^2([-1, +1]). \quad \text{Q.E.D.}$$

Theorem 7.6. The associated Legendre functions

$$P_l^m(u) = \frac{(1 - u^2)^{m/2}}{2^l l!} \frac{d^{l+m}}{du^{l+m}} (u^2 - 1)^l, \qquad l = m, m + 1, \ldots, \tag{7.43}$$

defined for all positive integral values of m, are solutions of the associated Legendre equation,

$$(1 - u^2) \frac{d^2 P_l^m(u)}{du^2} - 2u \frac{dP_l^m(u)}{du} + \left(l(l + 1) - \frac{m^2}{1 - u^2} \right) P_l^m(u) = 0, \tag{7.44}$$
$$m = 1, 2, 3, \ldots .$$

For a fixed integer value of m the set $\{P_m^m, P_{m+1}^m, P_{m+2}^m, \ldots\}$ is an orthogonal basis in $L^2([-1, +1])$, with

$$\langle P_l^m \mid P_{l'}^m \rangle = \int_{-1}^{+1} P_l^m(u) \, P_{l'}^m(u) \, du = \frac{2}{2l + 1} \frac{(l + m)!}{(l - m)!} \delta_{ll'} . \tag{7.45}$$

Proof. By substituting in the left-hand side of (7.44)

$$P_l^m(u) = (1 - u^2)^{m/2} \frac{d^m}{du^m} P_l(u)$$

we get by repeated use of the Leibniz rule

$$\frac{d^m}{du^m} \left[(1 - u^2) \frac{d^2 P_l(u)}{du^2} - 2u \frac{dP_l(u)}{du} + l(l + 1) P_l(u) \right] = 0.$$

Thus, we deduce that (7.44) is true from the fact that (7.40) is true.

It can be shown by direct integration that (7.45) is true, i.e., $\{P_m^m, P_{m+1}^m, \ldots\}$ constitutes an orthogonal system in $L^2([-1, +1])$. In order to prove that this system constitutes a basis in $L^2([-1, +1])$, note that for any given function $f(x)$, $f \in L^2([-1, +1])$, we can write

$$\langle P_l^m \mid f \rangle = \int_{-1}^{+1} (1 - u^2)^{m/2} f(u) \frac{d^m}{du^m} P_l(u) \, du. \tag{7.46}$$

From (7.42) we can easily see that $p_0(u) = 1$, $p_1(u) = u, \ldots, p_n(u) = u^n$ can be expressed in terms of

$$\frac{d^m P_m(u)}{du^m}, \frac{d^m P_{m+1}(u)}{du^m}, \ldots, \frac{d^m P_{m+n}(u)}{du^m},$$

i.e., the vector spaces spanned by these two systems are identical:

$$(p_0, p_1, \dots) = \left(\frac{d^m P_m}{du^m}, \frac{d^m P_{m+1}}{du^m}, \dots\right).$$

Thus, we deduce from (7.46) that the relations

$$\langle P_l^m \mid f \rangle = 0, \qquad l = m, m+1, \dots \tag{7.47}$$

imply that for $h(u) = (1 - u^2)^{m/2} f(u)$ we have $\langle g \mid h \rangle = 0$ for all $g \in (p_0, p_1, \dots)$. Since according to Lemma 7.4, $(p_0, p_1, \dots)$ is dense in $L^2([-1, +1])$ and since obviously $h \in L^2([-1, +1])$ when $f \in L^2([-1, +1])$, there is a sequence $g_1, g_2, \dots \in (p_0, p_1, \dots)$ which converges to h. Hence, if (7.47) is true, we shall have

$$\langle h \mid h \rangle = \lim_{k\to\infty} \langle g_k \mid h \rangle = 0.$$

The above relation implies that $h = 0$, i.e., $h(u) = 0$ almost everywhere. Consequently, $f(u) = (1 - u^2)^{-m/2} h(u) = 0$ almost everywhere, because $(1 - u^2)^{-m/2}$ is finite and different from zero everywhere in $(-1, +1)$, while the Lebesgue measure of $\{-1, +1\}$ is zero.

Thus, we conclude that (7.47) implies that $f = 0$; consequently, $\{P_m^m, P_{m+1}^m, \dots\}$ is according to Theorem 4.6(a) in Chapter I an orthogonal basis in $L^2([-1, +1])$. Q.E.D.

7.6. Completeness of the Spherical Harmonics

Now we are in the position to carry through to the end the proof of Theorem 7.1. Due to Lemma 7.5 we can deduce from Theorems 7.5 and 7.6 that for each fixed nonnegative integer m the functions

$$\left\{\Theta_l^m(\theta) = \left[\frac{2}{2l+1} \frac{(l+m)!}{(l-m)!}\right]^{-1/2} P_l^m(-\cos\theta),\ l = m, m+1, \dots\right\}$$

determine an orthonormal basis in $L^2([0, \pi], \mu_\theta)$. By applying Theorem 6.11 to the orthonormal bases $\{\Phi_m : m = 0, \pm 1, \dots\}$ (see Theorem 7.2) and $\{\Theta_l^{|m|}, l = |m|, |m| + 1, \dots\}$, $m = 0, \pm 1, \dots$, we deduce that the system

$$\{\Theta_l^m \otimes \Phi_m, l = 0, 1, 2, \dots, m = -l. -l+1, \dots, +l\}$$

is an orthonormal basis in $L^2([0, \pi], \mu_\theta) \otimes L^2([0, 2\pi])$. From this result and Lemma 7.3 (see also Theorem 4.8 in Chapter I) we can finally conclude that the spherical harmonics determine an orthonormal system in $L^2(\Omega_s, \mu_s)$.

We can easily derive from Theorem 7.1 the following result.

Theorem 7.7. The set

$$\{0, -2, -6, \ldots, -l(l+1), \ldots\}$$

contains all the eigenvalues λ of the equation (7.14) belonging to eigenfunctions $Y(\theta, \phi)$, $(\theta, \phi) \in \Omega_s$, having continuous second derivatives in Ω_s, and satisfying the following periodicity conditions:

$$Y(\theta, 0) = Y(\theta, 2\pi), \qquad \left.\frac{\partial Y(\theta, \phi)}{\partial \phi}\right|_{\phi=0+} = \left.\frac{\partial Y(\theta, \phi)}{\partial \phi}\right|_{\phi=2\pi-0}.$$

Denote by $\mathscr{C}_p^2(\Omega_s)$ the set of all elements of $L^2(\Omega_s, \mu_s)$ which are represented by such functions $Y(\theta, \phi)$. Any eigenfunction $Y(\theta, \phi)$, $Y \in \mathscr{C}_p^2(\Omega_s)$, corresponding to the eigenvalue $\lambda = -l(l+1)$ is equal, almost everywhere, to a linear combination of the spherical harmonics

$$Y_l^{-l}(\theta, \phi), \ldots, Y_l^{+l}(\theta, \phi).$$

Proof. It is easy to check explicitly by integrating by parts that the differential operator* form

$$-\hbar^{-2}\mathbf{L}^2 = \frac{1}{\sin\theta}\frac{\partial}{\partial\theta}\left(\sin\theta\frac{\partial}{\partial\theta}\right) + \frac{1}{\sin^2\theta}\frac{\partial^2}{\partial\phi^2} \tag{7.48}$$

defines a Hermitian operator when applied to twice continuously differentiable functions $Y(\theta, \phi)$, where $Y \in \mathscr{C}_p^2(\Omega_s)$,

$$\int_{\Omega_s} Y_1^*(\theta, \phi)(\mathbf{L}^2 Y_2)(\theta, \phi)\, d\mu_s(\theta, \phi)$$
$$= \int_{\Omega_s} (\mathbf{L}^2 Y_1)^*(\theta, \phi)\, Y_2(\theta, \phi)\, d\mu_s(\theta, \phi), \qquad Y_1, Y_2 \in C_p^2(\Omega_s).$$

Consequently, we can infer, by the procedure used in proving Theorem 5.3, that any two eigenfunctions of $-\mathbf{L}^2$ corresponding to different eigenvalues are orthogonal.

If $Y_{(\lambda)}(\theta, \phi)$ is an eigenfunction of (7.14)

$$(-\mathbf{L}^2 Y_{(\lambda)})(\theta, \phi) = \lambda\hbar^2 Y_{(\lambda)}(\theta, \phi), \qquad Y_{(\lambda)} \in L^2(\Omega_s, \mu_s),$$

we can expand it, due to Theorem 7.1, in terms of the spherical harmonics

$$Y_\lambda = \sum_{l=0}^{\infty} \sum_{m=-l}^{l} \langle Y_l^m \mid Y_{(\lambda)} \rangle\, Y_l^m. \tag{7.49}$$

* As we shall see in Section 7.8 of Chapter IV, the operator $\mathbf{L}^2$ corresponds to the square of the angular momentum $\mathbf{L}$ of the system.

However, according to the above-mentioned property of eigenfunctions corresponding to different eigenvalues

$$\langle Y_l^m \mid Y_{(\lambda)} \rangle = 0, \qquad \lambda \neq -l(l+1).$$

Consequently, $Y_{(\lambda)}$ can be a nontrivial eigenfunction only if λ is equal to $-l(l+1)$ for some $l = 0, 1, \ldots$. In that case we are left in (7.49) with the finite vector sum

$$Y_{[-l(l+1)]} = \sum_{m=-l}^{+l} c_m Y_l^m, \qquad c_m = \langle Y_l^m \mid Y_{[-l(l+1)]} \rangle.$$

From the definition of equality of functions in L^2 spaces we get $Y_{[-l(l+1)]}(\theta, \phi) = \sum_{m=-l}^{l} c_m Y_l^m(\theta, \phi)$ almost everywhere in Ω_s. Q.E.D.

This completes the systematic study of the angular dependence of internal energy eigenfunctions $\Psi(\mathbf{r})$ of two-body quantum mechanical systems interacting via central potentials. We turn now to the study of the complete solution of the eigenvalue problem of Equation (7.7) for the special case of Coulomb (attractive) potentials

$$V_0(r) = -c/r, \qquad c > 0.$$

7.7. The Two-Body Problem with a Coulomb Potential

When we are dealing with the hydrogen atom, our two-body system consists of an electron carrying one elementary unit of negative charge $-e$, and a nucleus carrying the positive charge $+e$. The constant c is in this case equal to e^2. We shall denote the positive constant c also in the general case by e^2.

Since we know that the only possible values of λ in (7.14) are $\lambda = -l(l+1)$, $l = 0, 1, 2, \ldots$, the equation (7.15) for the radial part of the wave function assumes the form

$$\text{(7.50)} \qquad \frac{d^2R_l(r)}{dr^2} + \frac{2}{r}\frac{dR_l(r)}{dr} + \left[\frac{2m_0E}{\hbar^2} - \frac{l(l+1)}{r^2} + \frac{2m_0}{\hbar^2}\frac{e^2}{r}\right] R_l(r) = 0,$$

where we introduce the subscript l in $R_l(r)$ to indicate that the solutions of (7.50) depend implicitly on the parameter l.

If we look for negative energy solutions, $E < 0$, we can introduce the new variable

$$\text{(7.51)} \qquad \rho = \frac{\sqrt{-8m_0E}}{\hbar}\, r$$

and in (7.50) make the substitution

$$\text{(7.52)} \qquad R_l(r) = \rho^l \exp(-\rho/2)\, v_l(\rho).$$

After some simple algebra we find that $v_l(\rho)$ has to satisfy the *confluent hypergeometric equation*

$$\rho \frac{d^2 v_l(\rho)}{d\rho^2} + (2l + 2 - \rho) \frac{dv_l(\rho)}{d\rho} - (l + 1 - n)\, v_l(\rho) = 0, \tag{7.53}$$

$$n = \frac{e^2}{\hbar}\left(-\frac{m_0}{2E}\right)^{1/2}.$$

We shall attempt to find solutions of the above equation which can be written as a series of the following form:

$$v_l(\rho) = \rho^\nu \sum_{k=0}^{\infty} a_k \rho^k = \sum_{k=0}^{\infty} a_k \rho^{k+\nu}. \tag{7.54}$$

If we insert the above series in (7.53) and assume that we can differentiate term by term, we get an expansion which has to vanish identically, and therefore the coefficients of this expansion have to be zero. By equating these coefficients to zero we get the relations

$$\nu(\nu + 2l + 1)\, a_0 = 0, \tag{7.55}$$

$$(k + \nu)(k + \nu + 2l + 1)\, a_k = (k + \nu + l - n)\, a_{k-1}, \qquad k = 1, 2,..., \tag{7.56}$$

which can be looked upon as recursive relations for determining $a_1, a_2, \ldots$. We can satisfy the *indicial equation* (7.55) by choosing $\nu = 0$. By solving (7.56), starting with $a_0 = 1$, we get as a solution of (7.53) a *confluent hypergeometric function* given by the *Kummer's series*

$$F(l + 1 - n \mid 2l + 2 \mid \rho) = 1 + \sum_{k=1}^{\infty} \frac{(l - n + 1) \cdots (l - n + k)}{(2l + 2) \cdots (2l + k + 1)} \frac{\rho^k}{k!}. \tag{7.57}$$

It can be verified (see Exercise 7.3) that the above power series has an infinite radius of convergence, and since the nth derivative of a function represented by a Taylor series can be obtained within the radius of convergence by differentiating the series term by term n times, (7.57) is indeed one of the solutions of (7.53).

The equation (7.53) is an ordinary linear differential equation of the second order; consequently it has only two linearly independent solutions. A second solution of (7.53), which is linearly independent from (7.56), can be obtained by taking the solution $\nu = -2l - 1$ of the indicial equation (7.55). If $l \neq 0, \frac{1}{2}, 1, \frac{3}{2},...$, we can derive from (7.56) the coefficients of the series (7.54) for this second solution. We note that this second solution behaves like $\approx \rho^{-2l-1}$ when $\rho \to 0$.

Unfortunately, in our case $l = 0, 1, 2, \ldots$, and (7.56) does not have any solutions because it cannot be satisfied for $k = -\nu = 2l + 1$. However, a second linearly independent solution of (7.53) can still be obtained (see Bateman [1953, Vol. I, Section 6.7.1, p. 261, Eq. (13)]), and its behavior for small ρ is again $\approx \rho^{-2l-1}$. Consequently, $R_l(r)$ for this second solution [see (7.52)] behaves like ρ^{-l-1} for small ρ. Thus, for $l = 1, 2, \ldots$ the corresponding function (7.13) is not integrable in the μ_{p} measure (7.12) because of its behavior at $\rho = 0$, and therefore these solutions are ruled out. For $l = 0$ this solution provides a square-integrable function $\Psi(\mathbf{r})$, but it is still ruled out on the following grounds: if a wave function $\psi_{\mathrm{c}}(\mathbf{R})\, \Psi(\mathbf{r})$, with $\Psi(\mathbf{r})$ behaving like $1/r$ for $r \to 0$, would be included in the family of functions on which we apply the Schroedinger operator form H_{S} [see (7.3)–(7.6)], then H_{S} would not define a Hermitian operator (see Exercise 7.4)—a requirement which constitutes one of the postulates* of quantum mechanics.

We return to the solution (7.57) of (7.53). If $l + 1 - n \notin \{0, -1, -2, \ldots\}$, then all the coefficients in the power series (7.57) are different from zero. Therefore, when $r \to +\infty$, i.e., for $\rho \to +\infty$, the function $F(l + 1 - n) \mid 2l + 2 \mid \rho)$ in (7.57) obviously diverges faster than any finite power of ρ, and from the similarity of the Kummer's series to the exponential series for e^ρ, one can suspect that $F(l + 1 - n \mid 2l + 2 \mid \rho)$ diverges about as fast as e^ρ. In fact it can be shown (see Bateman [1953, Vol. 1, Section 6.13.1, p. 278]) that in this case the function $F(l + 1 - n \mid 2l + 2 \mid \rho)$ behaves like $\rho^{-l-1-n} e^\rho$ for large ρ. Consequently

$$R_l(r) = \rho^l \exp(-\rho/2) F(l + 1 - n \mid 2l + 2 \mid \rho)$$

has the asymptotic behaviour

$$R_l(r) \sim \rho^{-n-1} \exp(\rho/2), \qquad \rho \to +\infty,$$

which indicates that $R_l \notin L^2(\Omega_{\mathrm{r}}, \mu_{\mathrm{r}})$, i.e., the corresponding function (7.13) is not square integrable in that case.

Thus, we are left only with the possibility that $l + 1 - n \in \{0, -1, \ldots, -n', \ldots\}$. When

$$l + 1 - n = -n', \qquad n' = 0, 1, 2, \ldots, \tag{7.58}$$

all the coefficients in the series (7.57) from the n'th on vanish, and the

* See Chapter IV, §1. We shall analyze this problem more completely in §7 of that chapter.

series reduces to a polynomial* of the n'th degree. In this case the radial function

$$R_{nl}(r) = \rho^l \exp(-\rho/2) F(l + 1 - n \mid 2l + 2 \mid \rho),$$
(7.59)
$$\rho = \frac{2me^2}{n\hbar^2} r, \qquad n = l + 1, l + 2, \ldots,$$

is square integrable in the measure μ_r. Furthermore, for each fixed l, $l = 0, 1, 2, \ldots$, the set $\{R_{nl}, n = l + 1, l + 2, \ldots\}$ is an orthogonal system in $L^2(\Omega_r, \mu_r)$,

$$\int_0^\infty R_{nl}(r)\, R_{n_1 l}(r)\, r^2\, dr = \frac{1}{4} \frac{n!(n - l - 1)!(2l + 1)!}{(n + l)!} \left(\frac{\hbar^2}{m_0 e^2}\right)^3 \delta_{nn_1}.$$

Consequently, the corresponding wave functions

(7.60) $$\Psi_{nlm}(\mathbf{r}) = 2 \left(\frac{\hbar^2}{m_0 e^2}\right)^{-3/2} \left[\frac{(n + l)!}{n!(n - l - 1)!(2l + 1)!}\right]^{1/2} R_{nl}(r)\, Y_l^m(\theta, \phi),$$
$$n = 1, 2, \ldots, \qquad l = 0, 1, 2, \ldots, n - 1, \qquad m = -l, -l + 1, \ldots, +l,$$

constitute an orthonormal system in $L^2(\mathbb{R}^3)$; $\Psi_{nlm}(\mathbf{r})$ is an eigenfunction of the internal energy Schroedinger operator

(7.61) $$H_{\mathrm{i}} = -\frac{\hbar^2}{2m_0} \Delta - \frac{e^2}{r}$$

and has as an eigenvalue

(7.62) $$E_n = -\frac{m_0 e^4}{2\hbar^2} \frac{1}{n^2}, \qquad n = 1, 2, \ldots,$$

as can be seen from the definition of n in (7.53). It is intuitively obvious from the above procedure that the above values are the only eigenvalues of H_{i} corresponding to square-integrable functions—as long as we restrict the domain of definition of H_{i} by requiring that (7.61) defines a Hermitian linear operator on $L^2(\mathbb{R}^3)$. A rigorous justification of this statement requires some familiarity with the theory of linear operators on Hilbert spaces, and will be discussed in Section 7.7 of Chapter IV.

* Equal to $n'!\, L_{n'}^{2l+1}(\rho)$, where $L_{n'}^{2l+1}(\rho)$ is an *associated* Laguerre polynomial,

$$L_{n'}^{k}(\rho) = (-1)^k \frac{d^k}{d\rho^k} L_{n'+k}(\rho),$$

and $L_\nu(\rho)$ is a Laguerre polynomial (see Exercise 7.6).

Thus, we have arrived at the conclusion that the internal energy discrete spectrum of the hydrogen atom is

$$\mathsf{S}_{\mathrm{p}} = \left\{ -\frac{m_0 e^4}{2\hbar^2}, -\frac{m_0 e^4}{8\hbar^2}, \dots, -\frac{m_0 e^4}{2n^2\hbar^2}, \dots \right\}.$$

In spectroscopy it is customary to denote the eigenfunctions corresponding to the above energy levels, in the above order, by the letters s, p, d, f, g,... . The numbers n, l, and m, are called, respectively, the *principal*, the *azimuthal*, and the *magnetic quantum numbers*. When an eigenstate of the hydrogen atom is given by writing one of the above letters s, p, d,... preceded by a natural number n, then n denotes the principal quantum number. Thus, $3p$ refers to any of the eigenstates of (7.61) with energy E_3, [obtained by taking $n = 3$ in (7.62)], which is also an eigenfunction of the angular momentum operator $\mathbf{L}^2$ defined essentially by (7.48), with an eigenvalue $l(l + 1)$, where $l = 1$.

It is worth mentioning that the above terminology has become standard in atomic physics, and is used, with certain modifications, even in nuclear physics. It is applied not only to the hydrogen atom, but also to other atoms. This application to more complicated atoms with more than one electron in their electronic "shell" is made possible by an approximate model of such an atom. According to this model the complicated many-body interactions between each particular electron and the rest of the electrons are neglected, and each electron is envisaged to move in a Coulomb potential generated by the nucleus and all the rest of the electrons which are closer to the nucleus than the considered electron. This model, though too rough for quantitative analysis, provides an adequate qualitative picture allowing a simple classification of the electron "orbits" (i.e., eigenstates of the Schroedinger operator).

According to *Pauli's exclusion principle*,* there can be in this model at most $2 \sum_{l=0}^{n-1} (2l + 1)$ electrons in states Ψ_{nlm} with a principal number n. All the electrons in such states are said to belong to the same "electronic shell," and when there are precisely $2 \sum_{l=0}^{n-1} (2l + 1)$ electrons in a shell, it is said that the shell has been "filled out." In such a "shell model" of an atom, the ground state† of the atom has the property that no electrons can be found in the nth shell if the $(n - 1)$st shell is not filled out.

* Pauli's exclusion principle states that in an atom there are at most two electrons in a state specified by given n, l, and m.

† The *ground state* of a quantum mechanical system is the state in which the internal energy of the system is at its lowest possible value; e.g., the ground state of the hydrogen atom corresponds to the principal quantum number $n = 1$.

The above model leads in a natural way to the periodic table of chemical elements if it is postulated that the number of electrons in the electrically neutral atoms of a chemical element is equal to the atomic number (in the table) of that element. Many of the regularities in the chemical properties of the elements, which were first observed by Mendeleev in the 19th century, can be deduced from the above picture of the ground state of an atom.

Exercises

7.1. Check that for any twice-differentiable function $f(x, y, z)$ the identity

$$\left(\frac{\partial^2}{\partial x^2} + \frac{\partial^2}{\partial y^2} + \frac{\partial^2}{\partial z^2}\right) f(x, y, z)$$

$$= \frac{1}{r^2}\left[\frac{\partial}{\partial r}\left(r^2 \frac{\partial}{\partial r}\right) + \frac{1}{\sin\theta}\frac{\partial}{\partial\theta}\left(\sin\theta \frac{\partial}{\partial\theta}\right)\right.$$

$$\left. + \frac{1}{\sin^2\theta}\frac{\partial^2}{\partial\phi^2}\right] f(r\sin\theta\cos\phi, r\sin\theta\sin\phi, r\cos\theta)$$

is satisfied, except for the points where $\theta = 0$ and $\theta = \pi$, i.e., for $\mathbf{r} = z\mathbf{e}_z$, $-\infty < z < +\infty$.

7.2. Show that

$$\int_{-1}^{+1} (P_l(u))^2\, du = \frac{1}{4^l (l!)^2} \int_{-1}^{+1} \left[\frac{d^l (u^2 - 1)^l}{du^l}\right]^2 du = \frac{2}{2l + 1}$$

by integrating by parts l times.

7.3. Prove that Kummer's series (7.57) converges for every $\rho \in \mathbb{C}^1$.

7.4. Denote by $\mathscr{D}$ the subset of all elements $f \in L^2(\Omega_{\mathrm{p}}, \mu_{\mathrm{p}})$ [see (7.10) and (7.12)] which can be represented by functions $f(r, \theta, \phi)$ which are twice continuously differentiable everywhere in Ω_{p}, except maybe for $r = 0$, and such that $Af \in L^2(\Omega_{\mathrm{p}}, \mu_{\mathrm{p}})$, where

$$(Af)(r, \theta, \phi) = \begin{cases} 0 & \text{for } r = 0 \\ \dfrac{1}{r^2}\dfrac{\partial}{\partial r}\left(r^2 \dfrac{f(r, \theta, \phi)}{\partial r}\right) & \text{for } r \neq 0. \end{cases}$$

Show that $\mathscr{D}$ is a linear subspace of $L^2(\Omega_{\mathrm{p}}, \mu_{\mathrm{p}})$ and that A is a linear operator (see Definition 1.1 in Chapter III) mapping $\mathscr{D}$ into $L^2(\Omega_{\mathrm{p}}, \mu_{\mathrm{p}})$.

Prove that this linear operator is not Hermitian, i.e.,

$$\int_{\Omega_S} f^*(r,\theta,\phi)(Ag)(r,\theta,\phi)\,d\mu_P(r,\theta,\phi) \neq \int_{\Omega_S} (Af)^*(r,\theta,\phi)\,g(r,\theta,\phi)\,d\mu_P(r,\theta,\phi)$$

for some $f, g \in \mathscr{D}$.

7.5. Find the internal energy spectrum and the corresponding eigenfunctions of a two-particle system, moving in three dimensions and interacting by means of a central square-well potential

$$V(r) = \begin{cases} 0 & \text{for } 0 \leqslant r \leqslant L \\ V_0 > 0 & \text{for } L < r < +\infty \end{cases}$$

by solving the radial equation (7.15) with the above potential. Compare the present solutions with the one-dimensional case treated in §5 of Chapter I.

Note. This problem is discussed by Messiah [1962, Chapter IX, Section II, pp. 359–361].

7.6. Prove by using the Leibniz rule for the nth derivative of $\rho^n e^{-\rho}$ that the *Laguerre polynomials*

$$\begin{aligned} L_n(\rho) &= n!F(-n \mid 1 \mid \rho) \\ &= n!\left(1 + \sum_{k=1}^{n} (-1)^k \frac{n(n-1)\cdots(n-k+1)}{k!}\right)\frac{\rho^k}{k!} \\ &= \sum_{k=0}^{n} (-1)^k \frac{(n!)^2}{(n-k)!k!}\,\frac{\rho^k}{k!} \end{aligned}$$

can be written as

$$L_n(\rho) = e^{\rho}\,\frac{d^n}{d\rho^n}\,(\rho^n e^{-\rho}).$$

Use the above formula to show that the set $\{l_0, l_1, \ldots\}$, where

$$l_n(\rho) = (1/n!)\,e^{-\rho/2}L_n(\rho)$$

are the *Laguerre functions*, is an orthonormal system in $L^2([0, +\infty))$.

7.7. Derive from the results of Exercise 7.6 that

$$\sum_{n=0}^{\infty} \frac{L_n(\rho)}{n!}\,s^n = \frac{\exp[-\rho s/(1-s)]}{1-s}$$

(the function on the right-hand side is called the *generating function* of Laguerre polynomials). Use this formula and the orthonormality of $\{l_0, l_1, \ldots\}$ to show that

$$\lim_{N\to\infty} \int_0^\infty \left(\frac{1}{1-s} \exp\left[-\frac{1}{2}\frac{1+s}{1-s}\rho\right] - \sum_{n=0}^{N} s^n l_n(\rho)\right)^2 d\rho$$

$$= \lim_{N\to\infty} \left(\frac{1}{1-s^2} - \sum_{n=0}^{N} s^{2n}\right) = 0.$$

7.8. Prove that the orthonormal system $\{l_0, l_1, \ldots\}$ (see Exercise 7.6) is a basis in $L^2([0, +\infty))$.

7.9. Prove that the system $\{h_0, h_1, \ldots\}$ of Hermite functions defined by

$$h_n(u) = e^{-u^2/2} H_n(u),$$

where $H_n(u)$ are the Hermitian polynomials (see Chapter I, Exercise 5.4), is an orthonormal basis in $L^2(\mathbb{R}^1)$.

References for Further Study

On measure theory (§§1–3) consult Halmos [1950], and Munroe [1953]. A less general but very readable book is the work of Kolmogorov and Fomin [1961], which is also useful for §4.

On spaces of square integrable functions (§4) see Riesz and Nagy [1955], and von Neumann [1955].

For §5 and §7 consult Landau and Lifshitz [1958], Messiah [1962], and Gottfried [1966].

A general treatment of tensor products of Hilbert spaces can be found in the work of Dixmier [1952].

CHAPTER III

Theory of Linear Operators in Hilbert Spaces

In Chapters I and II we met a large number of linear differential operators and other linear transformations. Linearity and a few other properties (such as boundedness, self-adjointness, etc.) of operators, which can be formulated and studied in a general and abstract manner without having to specify the operators, imply that the operators possessing these properties have many other additional properties. These properties are derived in the theory of linear operators. The object of the next six sections of this chapter is the study of that part of the general theory of linear operators in Hilbert spaces which is of immediate importance in quantum mechanics. The spectral theorem, formulated and proved in §6, is the main result of this theory.

A knowledge of the theory of linear operators in Hilbert spaces will enable us to study the structure of quantum mechanics from an abstract and very general point of view. In such a context the formalism of wave mechanics becomes a particular realization of the general formalism.

1. Linear and Antilinear Operators on Euclidean Spaces

1.1. Linear and Antilinear Transformations

A mapping of a vector space $\mathscr{V}_1$ into another vector space $\mathscr{V}_2$ is usually called a *transformation* of $\mathscr{V}_1$ into $\mathscr{V}_2$ or an *operator* from $\mathscr{V}_1$ into $\mathscr{V}_2$.

It is evident from the preceding two chapters that a special class of operators, called linear operators, are of great significance in quantum mechanics. Besides linear operators there is a second class of operators which play a role in quantum mechanics, though a very minor one by comparison with the linear operators. These are the antilinear operators.

Definition 1.1. A mapping

$$f \mapsto f' = A(f), \qquad f \in \mathscr{V}_1, \qquad f' \in \mathscr{V}_2,$$

of the vector space $\mathscr{V}_1$ over the field F into the vector space $\mathscr{V}_2$ over the same field F is called a *linear transformation* (*linear operator* in the case that* $\mathscr{V}_1 \equiv \mathscr{V}_2$) if

$$A(af + bg) = aA(f) + bA(g)$$

for all $a, b \in \mathsf{F}$ and all $f, g \in \mathscr{V}_1$. For vector spaces $\mathscr{V}_1$ and $\mathscr{V}_2$ over the field $\mathbb{C}^1$ of complex numbers, the mapping

$$f \mapsto f'' = B(f), \qquad f \in \mathscr{V}_1, \qquad f'' \in \mathscr{V}_2,$$

is called an *antilinear transformation*† (*semilinear transformation*) if

$$B(af + bg) = a^*B(f) + b^*B(g)$$

for all $a, b \in \mathbb{C}^1$ and all $f, g \in \mathscr{V}_1$.

It is customary to denote the image $A(f)$ of f under a linear or antilinear transformation simply by Af.

We have met many linear transformations in the preceding chapters. For instance, the unitary transformations used in defining the concept of isomorphism of Euclidean spaces (see Chapter I, Definition 2.4) are linear.

It is important to realize that the simplest type of linear operator is multiplication by a scalar a from the field F:

$$f \mapsto af, \qquad f \in \mathscr{V}.$$

We shall denote the above *operator* simply by a. However, the reader should realize that there is a conceptual distinction between a as an element of the field, and a as an operator!

As an example of an antilinear operator on the vector space $\mathscr{C}^0(\mathbb{R}^1)$ of continuous complex-valued functions on $\mathbb{R}^1$ consider the operator C of complex conjugation

$$(Cf)(x) = f^*(x), \qquad f \in \mathscr{C}^0(\mathbb{R}^1).$$

* This particular distinction between transformation and operator is adopted by us as very convenient, but it is by no means standard. It seems to be the case that, depending on the author, either the term "transformation" or the term "operator" is used exclusively.

† Note that an antilinear transformation is distinct from a linear transformation only for complex vector spaces, or, more generally, for vector spaces over fields F in which a * *operation* (or *involution operation*) is defined by a mapping $a \to a^*$, $a, a^* \in \mathsf{F}$, obeying the conditions that $(a^*)^* = a$, $a + b \to a^* + b^*$ and $ab \to b^*a^*$. Since the case of complex vector spaces is of sufficient generality for quantum mechanics, we ignore the more general possibilities.

We have

$$C(af(x) + bg(x)) = a^*f^*(x) + b^*g^*(x) = a^*(Cf)(x) + b^*(Cg)(x).$$

1.2. Algebraic Operations with Linear Transformations

Let A_1 and A_2 be linear transformations of $\mathscr{V}_1$ into $\mathscr{V}_2$. Then, as can be easily checked, the transformation A defined by

$$Af = A_1f + A_2f, \qquad f \in \mathscr{V}_1,$$

is also a linear transformation, called the *sum of the transformations* A_1 and A_2 and denoted by $A_1 + A_2$.

Let now A be a linear transformation of $\mathscr{V}_1$ into $\mathscr{V}_2$, and B a linear transformation of $\mathscr{V}_2$ into $\mathscr{V}_3$. Then it is easily seen that the transformation C of $\mathscr{V}_1$ into $\mathscr{V}_3$, defined by

$$Cf = B(Af), \qquad f \in \mathscr{V}_1,$$

is also linear; C is called the *product* of B and A, and is denoted by BA (which is, in general, different from AB).

Finally, if A is a linear transformation of the vector space $\mathscr{V}_1$ into the vector space $\mathscr{V}_2$ over the same field F, and $a \in \mathsf{F}$, the transformation aA defined by

$$(aA)f = a(Af), \qquad f \in \mathscr{V}_1,$$

is again a linear transformation.

When we are dealing with the set $\mathfrak{A}$ of all linear operators on a vector space over a field F, we see from the above definitions that for any $A, B \in \mathfrak{A}$ and $a \in \mathsf{F}$ the operators aA, $A + B$ and AB are always defined. These operations in $\mathfrak{A}$ endow $\mathfrak{A}$ with a certain algebraic structure. Such structures have been systematically studied—very much like the vector structure of vector spaces. Sets with such structures belong to mathematical spaces called *associative algebras* (to be distinguished from Boolean algebras), or simply algebras.

Definition 1.2. An *algebra* over a field F is a linear space $\mathfrak{A}$ over that field on which, in addition to the vector operations, an operation of *algebraic multiplication* is defined. An *algebraic multiplication* operation is a mapping

$$(A, B) \mapsto AB, \qquad (A, B) \in \mathfrak{A} \times \mathfrak{A}, \qquad AB \in \mathfrak{A},$$

of $\mathfrak{A} \times \mathfrak{A}$ into $\mathfrak{A}$ having the properties

(1) $(AB)\,C = A(BC)$, associativity,

(2) $(A + B)C = AC + BC$, $C(A + B) = CA + CB$, distributivity,

(3) $a(AB) = (aA)B = A(aB)$, associativity of scalar multiplication,

for all $A, B, C \in \mathfrak{A}$, and for each element a from the field F.

We say that the algebra $\mathfrak{A}$ has an identity if there is an *identity element* 1 satisfying

$$1A = A1 = A$$

for all $A \in \mathfrak{A}$.

The algebra $\mathfrak{A}$ is said to be *commutative* if $AB = BA$ for all $A, B \in \mathfrak{A}$. It is called *real* or *complex* if the field on which it is defined is the field of real or complex numbers, respectively.

It is easy to see that all the complex $n \times n$ matrices with matrix multiplication as the operation of algebraic multiplication form a complex algebra.

Definition 1.3. Two algebras $\mathfrak{A}_1$ and $\mathfrak{A}_2$ over the same field F are *isomorphic* if there is a one-to-one mapping

$$A_1 \leftrightarrow A_2, \qquad A_1 \in \mathfrak{A}_1, \qquad A_2 \in \mathfrak{A}_2,$$

which has the property that for any $A_1, B_1 \in \mathfrak{A}_1$ and $a \in \mathsf{F}$.

$$\begin{aligned} A_1 + B_1 &\leftrightarrow A_2 + B_2 \\ aA_1 &\leftrightarrow aA_2, \qquad a \in \mathsf{F}, \\ A_1B_1 &\leftrightarrow A_2B_2, \end{aligned}$$

where $A_2, B_2 \in \mathfrak{A}_2$ are the images of $A_1, B_1 \in \mathfrak{A}_1$, respectively. Such a mapping is then called an *isomorphism* between $\mathfrak{A}_1$ and $\mathfrak{A}_2$. If $\mathfrak{A}_2$ is an algebra of matrices (with the algebraic operations determined by matrix summation and multiplication) then the isomorphism is called a *faithful matrix representation* of $\mathfrak{A}_1$.

Theorem 1.1. The set $\mathfrak{A}(\mathscr{H})$ of all linear operators defined on a real (complex) finite-dimensional Hilbert space $\mathscr{H}$ is a real (complex) algebra with identity under the operations of operator summation, multiplication and multiplication of operators by scalars, with these operations defined, respectively, by the relations

$$\begin{aligned} &(A + B)f = Af + Bf, \qquad A, B \in \mathfrak{A}(\mathscr{H}) \\ &(AB)f = A(Bf), \qquad A, B \in \mathfrak{A}(\mathscr{H}) \\ &(aA)f = a(Af), \qquad A \in \mathfrak{A}(\mathscr{H}), \qquad a \in \mathbb{R}^1(\mathbb{C}^1), \end{aligned} \tag{1.1}$$

valid for all $f \in \mathscr{H}$. If the dimension of $\mathscr{H}$ is n, and $\{e_1, \ldots, e_n\}$ is any given orthonormal basis in $\mathscr{H}$, the mapping

$$A \to \begin{pmatrix} a_{11} & \cdots & a_{1n} \\ \vdots & & \vdots \\ a_{n1} & \cdots & a_{nn} \end{pmatrix}, \qquad A \in \mathfrak{A}(\mathscr{H}),$$

$$a_{ij} = \langle e_i \mid Ae_j \rangle, \qquad i, j = 1, \ldots, n \tag{1.2}$$

is a faithful representation of $\mathfrak{A}(\mathscr{H})$ by the real (complex) algebra of all $n \times n$ matrices; this representation is called the *matrix representation* of $\mathfrak{A}(\mathscr{H})$ in the basis $\{e_1, \ldots, e_n\}$.

Proof. It is a straightforward task to check that $\mathfrak{A}(\mathscr{H})$ is an algebra with identity, if the identity is taken to be the linear operator 1 defined by the identity mapping

$$f \mapsto 1f = f, \qquad f \in \mathscr{H}.$$

The matrix in (1.2) is obviously real if $\mathscr{H}$ is a real Hilbert space; for a complex Hilbert space $\mathscr{H}$, (1.2) is in general complex. If $\{e_1, \ldots, e_n\}$ is an orthonormal basis, by taking advantage of the properties of an orthonormal basis (see Chapter I, Theorem 4.6), we can write

$$Ae_j = \sum_{i=1}^{n} \langle e_i \mid Ae_j \rangle \, e_i = \sum_{i=1}^{n} a_{ij} e_i \,. \tag{1.3}$$

The above relation shows that the mapping (1.2) is one-to-one, since to every matrix corresponds only one linear operator A satisfying (1.3),

$$\begin{aligned} Af &= A \sum_{j=1}^{n} \langle e_j \mid f \rangle \, e_j = \sum_{j=1}^{n} \langle e_j \mid f \rangle \, Ae_j \\ &= \sum_{j=1}^{n} \langle e_j \mid f \rangle \left(\sum_{i=1}^{n} \langle e_i \mid Ae_j \rangle \, e_i \right) \\ &= \sum_{i=1}^{n} \left(\sum_{j=1}^{n} \langle e_i \mid Ae_j \rangle \langle e_j \mid f \rangle \right) e_i \,, \end{aligned} \tag{1.4}$$

i.e., the image Af under A of any vector f is completely determined by giving the matrix $\| a_{ij} \|$ in (1.2).

If $B \in \mathfrak{A}(\mathscr{H})$ and $\| b_{ij} \|$, $b_{ij} = \langle e_i \mid Be_j \rangle$, is the image of B under the mapping (1.2), we get by using the linearity of A and B and the definitions (1.1)

$$A + B \mapsto \| \langle e_i \mid (A + B)\, e_j \rangle \| = \| \langle e_i \mid Ae_j + Be_j \rangle \|$$

$$= \| \langle e_i \mid Ae_j \rangle + \langle e_i \mid Be_j \rangle \| = \| a_{ij} \| + \| b_{ij} \|,$$

$$AB \mapsto \| \langle e_i \mid ABe_j \rangle \| = \left\| \left\langle e_i \,\middle|\, A \sum_{k=1}^{n} \langle e_k \mid Be_j \rangle\, e_k \right\rangle \right\|$$

$$= \left\| \left\langle e_i \,\middle|\, \sum_{k=1}^{n} \langle e_k \mid Be_j \rangle\, Ae_k \right\rangle \right\|$$

$$= \left\| \left(\sum_{k=1}^{n} \langle e_k \mid Be_j \rangle \langle e_i \mid Ae_k \rangle \right) \right\|$$

$$= \left\| \left(\sum_{k=1}^{n} a_{ik} b_{kj} \right) \right\| = \| a_{ik} \| \cdot \| b_{kj} \|,$$

i.e., $A + B$ and AB are mapped by (1.2) into, respectively, the sum and product of the matrices $\| a_{ij} \|$ and $\| b_{ij} \|$. Since

$$aA \mapsto \| \langle e_i \mid (aA)\, e_j \rangle \| = \| a \langle e_i \mid Ae_j \rangle \| = a \, \| a_{ij} \|,$$

we have established that (1.2) is an isomorphism. Q.E.D.

We recall that every finite-dimensional Euclidean space is a Hilbert space. In general, in any given n-dimensional vector space $\mathscr{V}$ we can easily introduce an inner product, e.g., by choosing a *vector* basis $\{f_1, \ldots, f_n\} \subset \mathscr{V}$ and writing for every $f, g \in \mathscr{V}$,

$$\langle f \mid g \rangle = \sum_{k=1}^{n} a_k^* b_k ,$$

where $a_1, \ldots, a_n$ and $b_1, \ldots, b_n$ are the coefficients of f and g, respectively, in the basis $\{f_1, \ldots, f_n\}$,

$$f = \sum_{k=1}^{n} a_k f_k , \qquad g = \sum_{k=1}^{n} b_k f_k .$$

Consequently, the above theorem can be applied, *de facto*, to any finite-dimensional vector space. In fact, Theorem 1.1 can be proved directly for finite-dimensional vector spaces, without the introduction of an inner product (see Exercise 1.4).

1.3. Continuous and Bounded Transformations

As the concepts of continuous functions and bounded functions are of great significance in the theory of functions, so are the analogous concepts which can be defined for operators acting on Euclidean spaces, or more generally, on normed spaces.*

Definition 1.4. A transformation

$$f \mapsto T(f), \qquad f \in \mathscr{N}_1, \qquad Tf \in \mathscr{N}_2,$$

of the normed space $\mathscr{N}_1$ into the normed space $\mathscr{N}_2$ is said to be *continuous at* f_0, $f_0 \in \mathscr{N}_1$, if for any $\epsilon > 0$ there is an $\delta(\epsilon)$ such that

$$\| T(f) - T(f_0)\|_2 < \epsilon \qquad \text{for} \quad \| f - f_0 \|_1 < \delta(\epsilon),$$

where $\| h \|_i$ denotes the norm of $h \in \mathscr{N}_i$, $i = 1, 2$. The transformation T is said to be *continuous* if it is continuous at all points $f_0 \in \mathscr{N}_1$.

Definition 1.5. A set S in a normed space $\mathscr{N}$ is *bounded* if there is a constant C such that $\| f \| \leqslant C$ for all $f \in S$. A transformation T of $\mathscr{N}_1$ into $\mathscr{N}_2$ is called a *bounded transformation* if it maps *each* bounded set in $\mathscr{N}_1$ into a bounded set in $\mathscr{N}_2$.

The requirement that a linear transformation T be bounded is obviously equivalent to the requirement that there is a constant C such that

$$\| Tf \|_2 \leqslant C \| f \|_1$$

for all $f \in \mathscr{N}_1$ since $\| T(af)\|_2 = | a | \, \| Tf \|_2$ for any $a \in \mathbb{C}^1$.

If A is a linear operator on a normed space $\mathscr{N}$, then the supremum

$$\| A \| = \sup_{f \neq 0} \frac{\| Af \|}{\| f \|}, \qquad f \in \mathscr{N}_1, \tag{1.5}$$

is called the *bound of the operator* A; obviously $\| A \| \geqslant 0$.

Note that if A is bounded, then its bound $\| A \|$ is finite. Vice versa, if $\| A \| < +\infty$ then $\| Af \| \leqslant c \| A \|$ for all $\| f \| \leqslant c$, i.e., A is bounded.

For linear transformations we have an intimate connection between the concepts of boundedness and continuity. This connection is stated and proved in the second of the following two theorems.

Theorem 1.2. If the linear or antilinear transformation T of the normed space $\mathscr{N}_1$ into the normed space $\mathscr{N}_2$ is continuous at one point f_0 of $\mathscr{N}_1$, then it is continuous everywhere in $\mathscr{N}_1$.

* These concepts are defined in their most general form for topological spaces. The present definition is, however, sufficiently general for our purposes.

Proof. If T is continuous at $f_0 \in \mathcal{N}_1$, then for any $\epsilon > 0$ there is a $\delta(\epsilon)$ such that

$$\| T(f - f_0)\|_2 = \| Tf - Tf_0 \|_2 < \epsilon \tag{1.6}$$

for all $\| f - f_0 \|_1 < \delta(\epsilon)$. For arbitrary $g_0 \in \mathcal{N}_1$ we obtain

$$\| Tg - Tg_0 \|_2 = \| T(g - g_0)\|_2 < \epsilon$$

for all $\| g - g_0 \|_1 < \delta(\epsilon)$ by taking $g - g_0 = f - f_0$ in (1.6). Q.E.D.

Theorem 1.3. A linear or antilinear transformation T of the normed space $\mathcal{N}_1$ into the normed space $\mathcal{N}_2$ is continuous if and only if it is bounded.

Proof. If $\| T \| < +\infty$, and $\mathbf{0}_1$ denotes the zero vector in $\mathcal{N}_1$, then

$$\| Tf - \mathbf{0}_1 \|_2 = \| Tf \|_2 < \epsilon$$

for all $\| f \|_1 < \delta(\epsilon) = \epsilon \| T \|^{-1}$. Consequently, T is continuous at the origin $\mathbf{0}_1 \in \mathcal{N}_1$, and by the preceding theorem, it is continuous everywhere in $\mathcal{N}_1$.

Conversely, if T is assumed to be continuous at $\mathbf{0}_1$, there is such a $\delta_0 > 0$ that $\| Tg \|_2 \leqslant 1$ for all $\| g \|_1 \leqslant \delta_0$. For any $f \in \mathcal{N}_1$ we have that $g = \delta_0 \| f \|_1^{-1} f$ satisfies $\| g \|_1 = \delta_0$ and consequently

$$\delta_0 \| f \|_1^{-1} \| Tf \|_2 = \| T(\delta_0 \| f \|_1^{-1} f)\|_2 \leqslant 1.$$

Thus, if $\| f \|_1 \leqslant c$ for any given $c > 0$, then

$$\| Tf \|_2 \leqslant c\delta_0^{-1},$$

i.e., T is bounded. Q.E.D.

1.4. Examples of Bounded and Unbounded Operators

The study of operators on finite-dimensional Euclidean spaces is considerably simplified owing to the fact stated in Theorem 1.4.

Theorem 1.4. Any linear transformations of a finite-dimensional Euclidean space $\mathscr{E}_1$ into another Euclidean space $\mathscr{E}_2$ is bounded.

Proof. Let the subindices $i = 1, 2$ in $\langle \cdot \mid \cdot \rangle_i$ refer to the inner products in $\mathscr{E}_1$ and $\mathscr{E}_2$, respectively. If $\dim \mathscr{E}_1 = n$ and $e_1, \ldots, e_n$ is an orthonormal basis in $\mathscr{E}_1$, we have for any $f \in \mathscr{E}_1$

$$\text{(1.7)} \qquad \langle Tf \mid Tf \rangle_2 = \left\langle \sum_{i=1}^{n} \langle e_i \mid f \rangle_1 \, Te_i \,\middle|\, \sum_{j=1}^{n} \langle e_j \mid f \rangle_1 \, Te_j \right\rangle_2$$

$$= \sum_{i,j=1}^{n} \langle f \mid e_i \rangle_1 \langle Te_i \mid Te_j \rangle_2 \langle e_j \mid f \rangle_1 .$$

Since according to the Schwarz–Cauchy inequality

$$|\langle e_i \mid f \rangle_1 | \leqslant \| e_i \|_1 \, \| f \|_1 = \| f \|_1 ,$$

we get from (1.7)

$$\| Tf \|_2^2 \leqslant \left(\sum_{i,j=1}^{n} |\langle Te_i \mid Te_j \rangle_2 | \right) \| f \|_1^2 .$$

The above relation shows that $\| Tf \|_2 \leqslant C \| f \|_1$ for all $f \in \mathscr{N}_1$, which implies that T is bounded. Q.E.D.

The above theorem does not apply to infinite-dimensional spaces. In fact, most of the operators of interest in quantum physics are unbounded. For instance, consider the space $\mathscr{C}_b^0(\mathbb{R}^1)$ of all continuous functions $f(x)$, $x \in \mathbb{R}^1$, of compact support. The operator A

$$(Af)(x) = xf(x)$$

is obviously linear, but it is not bounded. To see that, take a function $f_0(x) \geqslant 0$ with support inside $[0, 1]$ which does not vanish identically. For each $a \in \mathbb{C}^1$ the function $f_a(x) = f(x - a)$, has its support inside $[a, a + 1]$, and $f_a(x) \in \mathscr{C}_b^0(\mathbb{R}^1)$. Since $\| f_a \| = \| f_0 \| = \text{const}$, while

$$\| Af_a \|^2 = \int_{-\infty}^{+\infty} x^2 f_0^2(x - a) \, dx = \int_{a}^{a+1} x^2 f_0^2(x - a) \, dx$$

$$\geqslant a^2 \int_{-\infty}^{+\infty} f_a^2(x) \, dx = a^2 \| f_0 \|^2 \xrightarrow[a \to +\infty]{} +\infty,$$

we see that $\| A \| = +\infty$.

Exercises

1.1. Let $a_0(x), \ldots, a_n(x)$, $x \in \mathbb{C}^1$, $n \geqslant 1$, be continuous functions. Show that the mapping

$$f(x) \mapsto g(x) = a_n(x) \frac{d^n f(x)}{dx^n} + \cdots + a_1(x) \frac{df(x)}{dx} + a_0(x),$$

$$f \in \mathscr{C}^n(\mathbb{R}^1), \qquad g \in \mathscr{C}^0(\mathbb{R}^1),$$

is a linear transformation of the linear space $\mathscr{C}^n(\mathbb{R}^1)$ (of n-times continuously differentiable functions) into $\mathscr{C}^0(\mathbb{R}^1)$.

Remark. The above linear transformation can be characterized by the differential operator form

$$a_n(x)\frac{d^n}{dx^n} + \cdots + a_1(x)\frac{d}{dx} + a_0(x)$$

and the specification of its domain of definition $\mathscr{C}^n(\mathbb{R}^1)$.

1.2. Let $I \in \mathscr{I}^1$ be a finite interval, and let $K(x, y)$ be a continuous function on $I \times I$. Prove that the mapping (Lebesgue integration)

$$f(x) \mapsto (Kf)(x) = \int_I K(x, y) f(y)\, dy, \qquad f \in L^2(I),$$

defines a linear operator on $L^2(I)$.

1.3. Prove that the linear operator defined in Exercise 1.2 is bounded.

1.4. Let $\mathscr{V}$ be a complex n-dimensional vector space, and $f_1, \ldots, f_n$ a vector basis in $\mathscr{V}$. Assign to each linear operator on $\mathscr{V}$ the complex matrix $\| a_{ik} \|$, defined by expanding Af_k, $k = 1, \ldots, n$, in the vector basis $f_1, \ldots, f_n$

$$Af_k = \sum_{i=1}^{n} a_{ik} f_i \,.$$

Show that the mapping $A \mapsto \| a_{ik} \|$ is an isomorphism between the algebra of all linear operators on the vector space $\mathscr{V}$ and the algebra of all $n \times n$ complex matrices.

1.5. Prove that the linear operator

$$(Bf)(x) = \frac{df(x)}{dx}, \qquad f(x) \in \mathscr{C}_{\mathrm{b}}^{\infty}(\mathbb{R}^1),$$

defined on the space $\mathscr{C}_{\mathrm{b}}^{\infty}(\mathbb{R}^1)$ of infinitely many times continuously differentiable functions of compact support, is unbounded.

1.6. Denote by $\mathfrak{A}(\mathscr{N})$ the algebra of all bounded linear operators acting on a normed space $\mathscr{N}$ (such as the Banach space $L^2(\Omega, \mu)$ of all μ-integrable functions on Ω). Show that the operator bound $\| A \|$, $A \in \mathfrak{A}(\mathscr{N})$, defined by (1.5), has all the properties of a norm (see Chapter I, Definition 2.2) on $\mathfrak{A}(\mathscr{N})$.

2. Linear Operators in Hilbert Spaces

2.1. Linear Functionals on Normed Spaces

A *functional* on a vector space $\mathscr{V}$ is a real- or complex-valued function defined on that space. As such, a real or complex functional can be thought of as a transformation of the vector space $\mathscr{V}$ into the vector space $(\mathbb{R}^1)$ or $(\mathbb{C}^1)$ of real or complex numbers, respectively. Hence, we can introduce the fundamental concepts of continuity, linearity, etc. by specializing the corresponding concepts, defined for transformations in general in §1, to the present special case of transformations.

Definition 2.1. A functional ϕ,

$$\phi: f \mapsto \phi(f), \qquad f \in \mathscr{N}, \qquad \phi(f) \in \mathbb{C}^1,$$

defined on the normed space $\mathscr{N}$ is *continuous at the point* $f_0 \in \mathscr{N}$ if for any $\epsilon > 0$ there is a $\delta(\epsilon)$ such that $|\phi(f) - \phi(f_0)| < \epsilon$ for all $\|f - f_0\| < \delta(\epsilon)$. The functional ϕ is said to be *continuous* if it is continuous at all points of $\mathscr{N}$; ϕ is said to be *bounded* if it maps every bounded set of $\mathscr{N}$ into a bounded set in $\mathbb{R}^1$ or $\mathbb{C}^1$, respectively, i.e., if for every $N > 0$ there is a constant C_N such that $|\phi(f)| \leqslant C_N$ for all $f \in \mathscr{N}$ such that $\|f\| \leqslant N$.

An important special class of functionals on vector spaces are the ones which possess linearity properties, e.g., the functionals

$$\phi_0(f) = f(0), \qquad f \in \mathscr{C}^0(\mathbb{R}^1),$$

$$\phi(f) = \int_a^b f(x)\, dx, \qquad f \in \mathscr{C}^0([a, b]).$$

Definition 2.2. The functional ϕ defined on the real or complex vector space $\mathscr{V}$ is *linear* if

$$\phi(f + g) = \phi(f) + \phi(g), \qquad \phi(af) = a\phi(f),$$

for all $f, g \in \mathscr{V}$ and all scalars a from $\mathbb{R}^1$ or $\mathbb{C}^1$, respectively.

We can think of the vector space $(\mathbb{R}^1)$ and $(\mathbb{C}^1)$ as normed spaces by taking the absolute value of a number to be its norm. By specializing Theorems 1.2 and 1.3 to the case of functionals, we can state the following theorem.

Theorem 2.1. (a) If the linear functional ϕ, $f \in \mathscr{N}$, defined in the normed space $\mathscr{N}$ is continuous at one point, then it is continuous everywhere in $\mathscr{N}$; (b) The linear functional ϕ is continuous if and only if it is bounded.

2.2. The Dual of a Hilbert Space

The family of all *continuous linear* functionals defined on a topological vector space* $\mathscr{V}$ is called the *dual* (or *conjugate*) space of $\mathscr{V}$, and is usually denoted by $\mathscr{V}'$. The dual space $\mathscr{V}'$ is also a linear space, and plays an important role in the theory of topological vector spaces. By specializing the above statements to normed spaces we get Theorem 2.2.

Theorem 2.2. Let $\mathscr{N}$ be a real (complex) normed space, and denote by $\mathscr{N}'$ the dual of $\mathscr{N}$, i.e., the set of all continuous real (complex) linear functionals on $\mathscr{N}$. If the sum $\phi_1 + \phi_2$ of two functionals ϕ_1 and ϕ_2 on $\mathscr{N}$ is defined by

$$(\phi_1 + \phi_2)(f) = \phi_1(f) + \phi_2(f),$$

and the product $a\phi$, of a functional ϕ on $\mathscr{N}$ with a scalar $a \in \mathbb{R}^1(\mathbb{C}^1)$ is defined by

$$(a\phi)(f) = a\phi(f),$$

then $\mathscr{N}'$ is a real (complex) vector space with the above operations as vector operations. Moreover, $\mathscr{N}'$ becomes a normed space if $\| \cdot \|$ is taken to be $\| \phi \| = \sup | \phi (f)|$ for $\| (f) \| \leqslant 1$.

The proof of this theorem is trivial and it consists in noting that $\phi_1 + \phi_2$ and $a\phi$ are continuous linear functionals if ϕ, ϕ_1, ϕ_2 are continuous linear functionals.

For the special case of normed spaces which are Hilbert spaces we have a very close relation between the space and its dual, as reflected in Theorems 2.3 and 2.4. To prove the first of these theorems we need the following lemma.

***Lemma 2.1.** If M is a closed linear subspace of a Hilbert space $\mathscr{H}$ and f is a vector in $\mathscr{H}$, there is a vector $f' \in \mathsf{M}$ such that $f'' = f - f'$ is orthogonal to M.

Proof. Denote by d the distance between f and M, i.e.,

$$d = \inf_{g \in \mathsf{M}} \| f - g \|.$$

If $d = 0$, there must be a sequence $g_1, g_2, \ldots \in \mathsf{M}$ such that $\| f - g_n \| \to 0$ when $n \to \infty$; since M is closed, we have in this case $f \in \mathsf{M}$ and the lemma is proved by taking $f' = f$ and $f'' = \mathbf{0}$.

* The concept of topology provides the most general means of introducing neighborhoods, and therefore the notion of convergence in an arbitrary set. See, e.g., Treves [1967] on the theory of topological vector spaces.

Assume that $d > 0$, and let $g_1, g_2, \ldots \in \mathsf{M}$ be the sequence for which

$$d = \lim_{n\to\infty} \| f - g_n \|.$$

It is easy to check by using the algebraic properties of the inner product that

$$\tfrac{1}{2} \| f - g_m \|^2 + \tfrac{1}{2} \| f - g_n \|^2 = \| f - \tfrac{1}{2}(g_m + g_n) \|^2 + \| \tfrac{1}{2}(g_m - g_n) \|^2.$$

If we let above $m, n \to \infty$ and note that

$$\| f - \tfrac{1}{2}(g_m + g_n) \|^2 \geqslant d^2$$

because $\tfrac{1}{2}(g_m + g_n) \in \mathsf{M}$, we get $\| g_m - g_n \| \to 0$. Thus, $g_1, g_2, \ldots$ is a Cauchy sequence, which has a limit $f' \in \mathsf{M}$, since M is closed. Moreover, we have

$$\| f - f' \| = \lim_{n\to\infty} \| f - g_n \| = d.$$

In order to show that $f'' \perp \mathsf{M}$, it is sufficient to prove that $\langle f'' \mid h \rangle = 0$ for all normalized $h \in \mathsf{M}$. By means of the identity

$$\langle h \mid f'' - \langle h \mid f'' \rangle \, h \rangle = 0$$

we get

$$\| f'' \|^2 = \| f'' - \langle h \mid f'' \rangle \, h \|^2 + \| \langle h \mid f'' \rangle \, h \|^2,$$

and since $\| f'' - \langle h \mid f'' \rangle h \| \geqslant d$ because $\langle h \mid f'' \rangle h \in \mathsf{M}$, we have

$$d^2 = \| f'' \|^2 = \| f'' - \langle h \mid f'' \rangle \, h \|^2 + |\langle h \mid f'' \rangle|^2 \geqslant d^2 + |\langle h \mid f'' \rangle|^2,$$

i.e., $|\langle h \mid f'' \rangle| = 0$. Q.E.D.

Theorem 2.3. (*Riesz' theorem*). To every continuous linear functional ϕ on a Hilbert space $\mathscr{H}$ corresponds a unique vector $g \in \mathscr{H}$, such that

$$\phi(f) = \langle g \mid f \rangle \tag{2.1}$$

for all $f \in \mathscr{H}$.

Proof. Denote by M_ϕ the null space of ϕ:

$$\mathsf{M}_\phi = \{f: \phi(f) = 0, f \in \mathscr{H}\}.$$

M_ϕ is a closed linear subspace of $\mathscr{H}$; its linearity is evident, while the fact that M_ϕ is closed follows from the continuity of ϕ: if $f_1, f_2, \ldots \in \mathsf{M}_\phi$ and if $f = \lim_{n\to\infty} f_n$, then

$$\phi(f) = \phi(f_n) + \phi(f - f_n) = \phi(f - f_n),$$

and since $\lim_{n\to\infty} \phi(f - f_n) = 0$, we have $\phi(f) = 0$, i.e., $f \in \mathsf{M}_\phi$.

If $M_\phi \equiv \mathscr{H}$, the theorem is established by taking $g = 0$. Therefore, assume that $M_\phi \neq \mathscr{H}$. Then there is at least one nonzero vector $h \notin M_\phi$. Since M_ϕ is closed, we can use Lemma 2.1 to state that h can be written as the sum of two vectors g' and g'', where $g'' \perp M_\phi$ while $g' \in M_\phi$. The vector g'' must be different from the zero vector because otherwise we would have $h \equiv g' \in M_\phi$. Thus, $\phi(g'') \neq 0$, and the vector $f - (\phi(f)/\phi(g''))g''$ exists. This vector belongs to M_ϕ because

$$\phi\left(f - \frac{\phi(f)}{\phi(g'')} g''\right) = 0.$$

Since $g'' \perp M_\phi$, we have

$$\left\langle g'' \middle| f - \frac{\phi(f)}{\phi(g'')} g'' \right\rangle = 0$$

for any $f \in \mathscr{H}$. From the above relation we get

$$\phi(f) = \left\langle \frac{\phi^*(g'')}{\langle g'' \mid g'' \rangle} g'' \middle| f \right\rangle;$$

we obtain (2.1) by setting

$$g = \frac{\phi^*(g'')}{\langle g'' \mid g'' \rangle} g''.$$

In order to establish that g is unique, assume that $\phi(f) = \langle g_1 \mid f \rangle$ for all f. Then we have

$$\langle g - g_1 \mid f \rangle = \langle g \mid f \rangle - \langle g_1 \mid f \rangle = 0;$$

by choosing $f = g - g_1$ we get $\langle g - g_1 \mid g - g_1 \rangle = 0$, i.e., $g = g_1$.

Q.E.D.

Theorem 2.4. Let $\mathscr{H}$ be a Hilbert space and $\mathscr{H}'$ its dual. There is a one-to-one antilinear* transformation

$$\phi \to g_\phi, \qquad \phi \in \mathscr{H}', \qquad g_\phi \in \mathscr{H}, \tag{2.2}$$

of $\mathscr{H}'$ *onto* $\mathscr{H}$ such that for each $\phi \in \mathscr{H}'$

$$\phi(f) = \langle g_\phi \mid f \rangle$$

for all $f \in \mathscr{H}$.

* In the case that $\mathscr{H}$ is real, $\mathscr{H}'$ is real too, and there is no distinction between linear and antilinear transformations of $\mathscr{H}'$ into $\mathscr{H}$.

Proof. According to Riesz' theorem, the above mapping is indeed a one-to-one mapping of $\mathscr{H}'$ into $\mathscr{H}$. Since for any $g \in \mathscr{H}$ the functional ϕ defined by

$$\phi(f) = \langle g \mid f \rangle$$

is continuous, this mapping is *onto* $\mathscr{H}$.

The antilinearity of this mapping is evident:

$$\begin{aligned} a\phi_1(f) + b\phi_2(f) &= a\langle g_{\phi_1} \mid f\rangle + b\langle g_{\phi_2} \mid f\rangle \\ &= \langle a^* g_{\phi_1} + b^* g_{\phi_2} \mid f\rangle. \quad \text{Q.E.D.} \end{aligned}$$

The above theorem justifies the notation introduced by Dirac, which has become very popular in physical literature. In this notation a vector f of a Hilbert space $\mathscr{H}$ is denoted by $\mid f\rangle$, and called a *ket* vector. The functional $\phi \in \mathscr{H}'$, which under the mapping (2.2) has f as an image is denoted by $\langle f \mid$ and called a *bra* vector. The inner product $\langle f \mid g\rangle$ of two vectors $f, g \in \mathscr{H}$ can be looked upon as a product of the bra $\langle f \mid$ and the ket $\mid g\rangle$.

The above notation neither presents any real advantages, nor does it enrich the techniques when dealing with Hilbert spaces.* Consequently, it will not be adopted in this book.

2.3. Adjoints of Linear Operators in Hilbert Spaces

Riesz' theorem enables us to prove the existence of the adjoint of any bounded operator acting on a Hilbert space. In order to give a sufficiently general definition of the adjoint of an operator, we enlarge our definition of operators *on* Hilbert spaces to operators *in* Hilbert spaces, by dropping the requirement that an operator be defined on the *entire* Hilbert space.

Definition 2.3. A *linear operator* A *in* a Hilbert space $\mathscr{H}$ is a linear transformation of a linear subspace $\mathscr{D}_A$ of $\mathscr{H}$ into $\mathscr{H}$; $\mathscr{D}_A$ is called the *domain of definition* of A and its image under A

$$\mathscr{R}_A = \{Af : f \in \mathscr{D}_A\}$$

is the *range* of A. If another linear operator B defined on $\mathscr{D}_B \supset \mathscr{D}_A$ coincides with A on $\mathscr{D}_A$, i.e., $Bf = Af$ for all $f \in \mathscr{D}_A$, then B is called an *extension* of A (or A a *restriction* of B), and we write $A \subseteq B$.

* There have been, however, proposals to incorporate this notation in the theory of *rigged* and *extended* Hilbert spaces. Then a ket is an element of $\mathscr{H}$ and a bra is an element of a larger scale containing $\mathscr{H}$ (see, e.g., the articles by Antoine [1969], Prugovečki [1973b]).

Theorem 2.5. Let A be an operator on $\mathscr{H}$, with domain $\mathscr{D}_A$. Denote by $\mathscr{D}_{A^*}$ the set of all vectors $g \in \mathscr{H}$ which are such that for each g there is one and only one vector g^* which satisfies the equation

$$\langle g^* \mid f \rangle = \langle g \mid Af \rangle, \qquad f \in \mathscr{D}_A, \tag{2.3}$$

for all $f \in \mathscr{D}_A$. The mapping

$$g \to g^* = A^*g, \qquad g \in \mathscr{D}_{A^*},$$

is a linear operator, called the adjoint of A, and it exists, i.e., $\mathscr{D}_{A^*} \neq \varnothing$ if and only if $\mathscr{D}_A$ is dense in $\mathscr{H}$.

Proof. In the case that $\mathscr{D}_A$ is dense in $\mathscr{H}$ and g, g^* are two vectors satisfying (2.3), g^* is uniquely determined by g; namely, if $g_1{}^*$ also satisfies

$$\langle g \mid Af \rangle = \langle g_1{}^* \mid f \rangle$$

for all $f \in \mathscr{D}_A$, then

$$\langle g^* - g_1{}^* \mid f \rangle = \langle g^* \mid f \rangle - \langle g_1{}^* \mid f \rangle = 0$$

for all $f \in \mathscr{D}_A$, and since $\mathscr{D}_A$ is dense in $\mathscr{H}$, we must have $g^* - g_1{}^* = \mathbf{0}$. Thus, if (2.3) has any solutions, we can define an operator A^*, with a domain $\mathscr{D}_{A^*}$, consisting of all g for which there is a g^* such that g and g^* satisfy (2.3), and mapping g into g^*.

The operator A^* is defined at least for $g = \mathbf{0}$, since then we can take $g^* = \mathbf{0}$. Moreover, if $g_1, g_2 \in \mathscr{D}_{A^*}$, then

$$\begin{aligned}\langle A^*(ag_1 + bg_2) \mid f \rangle &= \langle ag_1 + bg_2 \mid Af \rangle \\ &= a^*\langle g_1 \mid Af \rangle + b^*\langle g_2 \mid Af \rangle \\ &= a^*\langle A^*g_1 \mid f \rangle + b^*\langle A^*g_2 \mid f \rangle \\ &= \langle aA^*g_1 + bA^*g_2 \mid f \rangle,\end{aligned}$$

which proves that $ag_1 + bg_2 \in \mathscr{D}_{A^*}$ and A^* is linear.

Assume now that $\mathscr{D}_A$ is not dense in $\mathscr{H}$. Then its closure $\bar{\mathscr{D}}_A$ is different from $\mathscr{H}$, and there is in $\mathscr{H}$ a vector $h \notin \bar{\mathscr{D}}_A$. According to Lemma 2.1 we can write $h = h' + h''$, $h' \perp \bar{\mathscr{D}}_A$ where $h' \neq \mathbf{0}$. We have then

$$\langle h' \mid f \rangle = 0 = \langle \mathbf{0} \mid Af \rangle$$

for all $f \in \mathscr{D}_A$. This shows that when $\bar{\mathscr{D}}_A \neq \mathscr{H}$, we can take in (2.3) $g^* = h'$. Hence, we have at least two distinct solutions, $\mathbf{0}$ and h', belonging to $\mathbf{0}$. Thus, A^* does not exist, because (2.3) can never have a unique g^*

for any g, since if for some g there is a g^* satisfying (2.3), then $g^* + h'$ will also satisfy (2.3). Q.E.D.

An operator A might not have an adjoint A^*, but there might still be many operators B satisfying the relations

$$\langle Bg \mid f\rangle = \langle g \mid Af\rangle$$

for all $f \in \mathscr{D}_A$ and $g \in \mathscr{D}_B$. Two such operators A and B are said to be *adjoint to one another*. We note that if *the* adjoint A^* of A exists, and A and B are adjoint to one another, then necessarily $B \subseteq A^*$, i.e., A^* is a maximal extension of all the operators adjoint to A.

2.4. Bounded Linear Operators in Hilbert Spaces

If A is bounded linear operator which has an adjoint A^*, then from Theorems 2.6 and 2.7 it is easy to see that A^* is also bounded, and that $\mathscr{D}_{A^*} = \mathscr{H}$.

Theorem 2.6 (*The extension principle*). A bounded linear transformation T from the Hilbert space $\mathscr{H}_1$ into the Hilbert space $\mathscr{H}_2$, defined on a dense domain $\mathscr{D}_T(\bar{\mathscr{D}}_T = \mathscr{H}_1)$, has a unique continuous linear extension to the entire Hilbert space $\mathscr{H}_1$.

Proof. Let f be any given vector in $\mathscr{H}_1$. Since $\bar{\mathscr{D}}_T = \mathscr{H}_1$, there is a sequence $f_1, f_2, \ldots \in \mathscr{D}_T$ converging to f. The sequence $Tf_1, Tf_2, \ldots$ is a Cauchy sequence, since according to Definition 1.4 there is a constant C such that

$$\| Tf_m - Tf_n \|_2 \leqslant C \| f_m - f_n \|_1 .$$

Hence, the limit of $Tf_1, Tf_2, \ldots$ exists. Moreover, this limit is independent of the chosen sequence converging to f because if $f_1', f_2', \ldots \in \mathscr{D}_T$ also converges to f, then $\lim_{n\to\infty} \| Tf_n - Tf_n' \|_2 = 0$, since

$$\| Tf_n - Tf_n' \|_2 \leqslant C \| f_n - f_n' \|_1 .$$

Thus, we can define a transformation T' for all $f \in \mathscr{H}_1$ by choosing for each vector f a sequence $f_1, f_2, \ldots \in \mathscr{D}_T$ converging to f and setting $T'f$ to be the limit in the Hilbert space norm of $Tf_1, Tf_2, \ldots$,

$$\lim_{n\to\infty} \| T'f - Tf_n \|_2 = 0.$$

This transformation T', with $\mathscr{D}_{T'} = \mathscr{H}_1$, is evidently linear:

$$T'(af + bg) = \operatorname*{s-lim}_{n\to\infty} T(af_n + bg_n)$$
$$= \operatorname*{s-lim}_{n\to\infty}(aTf_n + bTg_n) = aT'f + bT'g,$$

where $g_1, g_2, \ldots \in \mathscr{D}_T$ is any sequence converging to g, and "s-lim" denotes the limit in the norm of $\mathscr{H}_2$.

The fact that T is bounded follows from the inequality

$$\| T'f \|_2 \leqslant \| T'f - Tf_n \|_2 + \| Tf_n \|_2$$
$$\leqslant \| T'f - Tf_n \|_2 + C \| f_n \|_1,$$

since when we let $n \to \infty$, $\| T'f - Tf_n \| \to 0$, while $\| f_n \|_1 \to \| f \|_1$.

It is evident that T' is the only continuous extension of T to the entire space $\mathscr{H}_1$. Q.E.D.

Theorem 2.7. If A is a bounded linear operator defined on the entire Hilbert space $\mathscr{H}$, its adjoint A^* is also a bounded linear operator defined everywhere in $\mathscr{H}$, and $\| A \| = \| A^* \|$.

Proof. For a fixed $g \in \mathscr{H}$ the functional ϕ defined by $\phi(f) = \langle g \mid Af\rangle$, $f \in \mathscr{H}$, is obviously linear. Moreover, ϕ is also continuous, as can be seen by using the continuity of A (see Theorem 1.3)

$$\lim_{\|f-f_0\|\to 0} \phi(f) = \langle g \mid \operatorname*{s-lim}_{\|f-f_0\|\to 0} Af\rangle = \langle g \mid Af_0\rangle = \phi(f_0).$$

Consequently, we can apply to $\phi(f)$ Riesz' theorem (Theorem 2.3) to deduce the existence of a unique g^*, satisfying

$$\langle g^* \mid f\rangle = \phi(f) = \langle g \mid Af\rangle.$$

By Theorem 2.5, the mapping

$$g \mapsto g^* = A^*g, \qquad g \in \mathscr{H},$$

defines the adjoint of A as a linear operator, for which we see that $\mathscr{D}_{A^*} = \mathscr{H}$.

From the inequality

$$|\langle A^*g \mid f\rangle| = |\langle g \mid Af\rangle| \leqslant \| A \| \| f \| \| g \|$$

we get by setting $f = A^*g$

$$\| A^*g \| \leqslant \| A \| \| g \|.$$

The above shows that A^* is bounded and $\| A^* \| \leqslant \| A \|$. By reversing the roles of A and A^* we arrive at $\| A \| \leqslant \| A^* \|$, thus proving that $\| A \| = \| A^* \|$. Q.E.D.

2.5. Dirac Notation for Linear Operators

It is interesting to take a look at the above theorem from the point of view of Dirac's terminology. To realize the nontrivial implications of this terminology, consider the more general case of a bounded linear operator A defined everywhere on a normed space $\mathscr{N}$. If ϕ is an element of the dual $\mathscr{N}'$, then the functional ϕ^1 defined by

$$\phi^1(f) = \phi(Af), \qquad f \in \mathscr{N},$$

is again a continuous linear functional, i.e., $\phi^1 \in \mathscr{N}'$, as can be easily proved by using the linearity and continuity of ϕ and A (A is bounded and therefore continuous, by Theorem 1.3),

$$\phi^1(af + bg) = \phi(aAf + bAg) = a\phi(Af) + b\phi(Ag) = a\phi^1(f) + b\phi^1(g),$$

$$\lim_{n\to+\infty} \phi^1(f_n) = \lim_{n\to+\infty} \phi(Af_n) = \phi(\operatorname*{s-lim}_{n\to+\infty} Af_n) = \phi(A(\operatorname*{s-lim}_{n\to+\infty} f_n)) = \phi^1(\operatorname*{s-lim}_{n\to+\infty} f_n).$$

Consequently, the mapping

$$\phi \mapsto \phi^1 = A^1\phi, \qquad \phi, \phi^1 \in \mathscr{N}',$$

is a transformation of $\mathscr{N}'$ into itself. Furthermore, it is easily seen that A^1 is a linear operator on $\mathscr{N}'$:

$$\begin{aligned}(A^1[a\phi_1 + b\phi_2])(f) &= (a\phi_1 + b\phi_2)(Af)\\ &= a\phi_1(Af) + b\phi_2(Af) = aA^1\phi_1(f) + bA^1\phi_2(f).\end{aligned}$$

Thus, every bounded linear operator A on $\mathscr{N}$ induces a linear operator A^1 on $\mathscr{N}'$. In case of a Hilbert space $\mathscr{H}$, to every element $\phi \in \mathscr{H}'$ corresponds, according to Theorem 2.4, a unique vector $g_\phi \in \mathscr{H}$. We have seen that in Dirac notation one denotes ϕ by the bra symbol $\langle g_\phi |$, so that $\phi(f) = \langle g_\phi | f \rangle$. Moreover, in this notation one writes

$$A^1\phi = \langle g_\phi | A,$$

agreeing to denote the above introduced operator A^1 on $\mathscr{H}'$ by the same letter A, but introducing the clause that when A stands, *de facto*, for the operator A^1, then it acts on the bra vector *from the right*.

In this notation one writes

$$\phi(Af) = (A^1\phi)(f) = \langle g_\phi | A | f \rangle.$$

According to Theorem 2.7, we have

$$\phi(Af) = \langle A^* g_\phi \mid f \rangle,$$

which implies that, using Dirac notation,

$$\langle A^* g_\phi \mid = \langle g_\phi \mid A.$$

2.6. Closed Operators and the Graph of an Operator

In the case that a linear operator A is unbounded and has an adjoint A^*, we can deduce from the preceding theorem that A^* is also unbounded when $\bar{\mathscr{D}}_{A^*} = \mathscr{H}$ [so that $(A^*)^* = A^{**}$ exists]. In fact, if A^* were bounded, then A^{**} would also be bounded, and since obviously $A \subseteq A^{**}$, this would imply that A is bounded. However, A^* has always the property of being closed, which is a somewhat weaker property then boundedness.

Definition 2.4. The linear operator A in the Hilbert space $\mathscr{H}$ is *closed* if it has the property that whenever $f_1, f_2, \ldots \in \mathscr{D}_A$ converges to a vector $f \in \mathscr{H}$ and $Af_1, Af_2, \ldots$ *also* converges to some vector $h \in \mathscr{H}$, then $f \in \mathscr{D}_A$ and $h = Af$.

Theorem 2.8. The adjoint A^* of a linear operator A in $\mathscr{H}$ is closed.

Proof. Let $g_1, g_2, \ldots \in \mathscr{D}_{A^*}$ converge to $g \in \mathscr{H}$ and assume that $A^*g_1, A^*g_2, \ldots$ converges to a vector h. Then for any $f \in \mathscr{D}_A$

$$\langle h \mid f \rangle = \lim_{n\to\infty} \langle A^* g_n \mid f \rangle = \lim_{n\to\infty} \langle g_n \mid Af \rangle = \langle g \mid Af \rangle,$$

which by the definition of A^* implies that $g \in \mathscr{D}_{A^*}$ and that $h = A^*g$.

Q.E.D.

The concept of a closed operator can be readily grasped in terms of the graph of an operator. The graph of an operator is a generalization of the concept of the graph of a real function $f(x)$, $x \in \mathbb{R}^1$, which, we recall, consists of all the points $(x, f(x))$ in $(\mathbb{R}^2) = (\mathbb{R}^1) \oplus (\mathbb{R}^1)$ corresponding to x from the domain of definition of $f(x)$.

Definition 2.5. The *graph* G_A *of an operator* A acting in the Hilbert space $\mathscr{H}$ is the subset

$$\mathsf{G}_A = \{(f, Af) : f \in \mathscr{D}_A\}$$

of the direct sum $\mathscr{H} \oplus \mathscr{H}$ (see Chapter II, Theorems 6.1 and 6.3).

We note that A is closed if and only if whenever

$$(f_1, Af_1), (f_2, Af_2), \ldots \in \mathsf{G}_A$$

is a convergent sequence in $\mathscr{H} \oplus \mathscr{H}$, then its limit (f, g) lies in G_A, i.e., G_A is a closed subset of $\mathscr{H} \oplus \mathscr{H}$ if and only if A is a closed operator.

In Chapters I and II we encountered many operators A which were *Hermitian*, i.e., for which $\langle Af \mid g\rangle = \langle f \mid Ag\rangle$ for all $f, g \in \mathscr{D}_A$. These operators were in addition linear. In mathematical literature such an operator is called symmetric.

Definition 2.6. The linear operator A acting in a Hilbert space is *symmetric* if $\langle Af \mid g\rangle = \langle f \mid Ag\rangle$ for all $f, g \in \mathscr{D}_A$. A symmetric operator A is called *self-adjoint* if $A \equiv A^*$, i.e., $A^* \subseteq A$ and $A \subseteq A^*$.

We note that a self-adjoint operator is always closed (because by Theorem 2.8 A^* is closed) and densely defined (since A^* exists).

All the Hermitian operators which have been introduced while studying different physical problems were symmetric. For instance, we have seen that the Schroedinger operator H_S defined in Chapter II, §5, is linear, Hermitian, and defined—under reasonable requirements on the potential (see Chapter II, Theorem 5.5)—on the dense subspace $\mathscr{C}_b^\infty(\mathbb{R}^{3n})$ (see Chapter II, Theorem 5.6).

It is an unfortunate feature of unbounded linear operators defined on linear subspaces of a Hilbert space that they do not have closed extensions to the entire Hilbert space (see Section 3.5), and for this reason we had to give particular attention in Definition 2.3 to the domain of definition of an operator. This fact makes the study of unbounded operators particularly difficult, especially since it prevents simple definitions of quite fundamental algebraic operations with such operators. For instance, if A and B are two operators for which $\mathscr{D}_A \neq \mathscr{H}$ and $\mathscr{D}_B \neq \mathscr{H}$, then we can define $A + B$ only on $\mathscr{D}_A \cap \mathscr{D}_B$,

$$(A + B)f = Af + Bf, \qquad f \in \mathscr{D}_A \cap \mathscr{D}_B.$$

Since it might happen that $\mathscr{D}_A \cap \mathscr{D}_B = \{\mathbf{0}\}$ even when both $\mathscr{D}_A$ and $\mathscr{D}_B$ are everywhere dense in $\mathscr{H}$, it follows that $A + B$ might not be defined at all except on the zero vector $\mathbf{0}$.

Similarly, in order to define AB, we must assume that $Bf \in \mathscr{D}_A$ whenever we apply AB to some $f \in \mathscr{D}_B$,

$$(AB)f = A(Bf), \qquad f \in \mathscr{D}_B, \qquad Bf \in \mathscr{D}_A.$$

Only in case of multiplication of an operator A by a scalar a can it be stated simply

$$(aA)f = a(Af), \qquad f \in \mathscr{D}_A.$$

Despite these difficulties we shall have to study unbounded operators in some detail since practically all of the operators which appear in physics are unbounded, such as the position and momentum operators or the Schroedinger operator (see Exercise 2.4).

2.7. Nonexistence of Unbounded Everywhere-Defined Self-Adjoint Operators

We shall proceed to prove that no symmetric unbounded operator has a self-adjoint extension to the entire Hilbert space. For this proof we need some preliminary results.

***Lemma 2.2.** If $\phi_1, \phi_2, \ldots$ is a sequence of continuous linear functionals on $\mathscr{H}$, which is bounded for each fixed $f \in \mathscr{H}$, i.e.,

$$|\phi_n(f)| \leqslant C_f, \qquad f \in \mathscr{H},$$

then $\phi_1, \phi_2, \ldots$ are uniformly bounded within the unit sphere $S_1 = \{h: \| h \| \leqslant 1\}$, i.e.,

$$|\phi_n(g)| \leqslant C, \qquad g \in S_1, \qquad n = 1, 2, \ldots,$$

or equivalently

$$|\phi_n(f)| \leqslant C \| f \|, \qquad n = 1, 2, \ldots,$$

for all $f \in \mathscr{H}$.

Proof. We shall show first that $\phi_1(f), \phi_2(f), \ldots$ is uniformly bounded inside at least one sphere $\{h: \| h - h_0 \| \leqslant r\}$ in $\mathscr{H}$.

Assume that the above statement were not true! Then we would have $|\phi_{n_1}(f_1)| > 1$ for at least some integer n_1 and some $f_1 \in \mathscr{H}$. By the continuity of $\phi_{n_1}(f)$, we also have

$$|\phi_{n_1}(f)| > 1, \qquad f \in S^{(1)},$$

on a sphere $S^{(1)} \subset \mathscr{H}$ with the centre at f_1. Since $\phi_1(f), \phi_2(f), \ldots$ are not uniformly bounded in any sphere, including $S^{(1)}$, there must be an $f_2 \in S^{(1)}$ such that $|\phi_{n_2}(f_2)| > 2$ for some integer n_2. Due to the continuity of $\phi_{n_2}(f)$,

$$|\phi_{n_2}(f)| > 2, \qquad f \in S^{(2)},$$

for some sphere $S^{(2)}$ with its center at f_2. By continuing this reasoning, we establish the existence of a sequence $S^{(1)} \supset S^{(2)} \supset \cdots$ of spheres in which

$$|\phi_{n_k}(f)| > k, \qquad f \in S^{(k)},$$

for some n_k, $k = 1, 2,\ldots$. Since $\mathcal{H}$ is complete, $\bigcap_{k=1}^{\infty} S^{(k)}$ contains at least one element f_0, for which we have according to the above construction

$$|\phi_{n_k}(f_0)| > k, \qquad k = 1, 2,\ldots .$$

Thus, $\phi_1(f_0), \phi_2(f_0),\ldots$ is not a bounded sequence, contrary to the assumption of the theorem. This means that if $\phi_1, \phi_2,\ldots$ are bounded for each $f \in \mathcal{H}$, then $\phi_1, \phi_2,\ldots$ are uniformly bounded on at least one sphere $S_0 = \{h: \| h - h_0 \| \leqslant r\}$ in $\mathcal{H}$,

$$|\phi_n(f)| \leqslant C_0, \qquad f \in S_0, \qquad n = 1, 2,\ldots .$$

Consequently, we get by exploiting the linearity of $\phi_n(f)$

$$\phi_n(g) = (1/r)(\phi_n(rg - h_0) + \phi(h_0)).$$

Since for $g \in S_1$ we have $\| g \| \leqslant 1$ and therefore $rg - h_0 \in S_0$ so that

$$|\phi_n(rg - h_0)| \leqslant C_0,$$

we have established that $\phi_1, \phi_2,\ldots$ are uniformly bounded on S_1:

$$|\phi_n(g)| \leqslant \frac{2C_0}{r}. \quad \text{Q.E.D.}$$

***Theorem 2.9.** The adjoint A^* of a linear operator A with domain of definition $\mathcal{D}_A = \mathcal{H}$ is a bounded operator on $\mathcal{H}$.

Proof. Since $\mathcal{D}_A = \mathcal{H}$, A^* exists by Theorem 2.5. Assume that the theorem is not true for A, i.e., that A^* is not bounded. Then there must be a sequence $h_1, h_2,\ldots \in \mathcal{D}_{A^*}$ such that $\| h_1 \|, \| h_2 \|,\ldots$ is bounded while $\| A^*h_n \| \to +\infty$ for $n \to +\infty$. Since $\| h_1 \|, \| h_2 \|,\ldots$ is bounded, we have

$$\lim_{n\to\infty} \| A^*g_n \| = +\infty, \qquad \| g_n \| = 1, \tag{2.4}$$

for the normalized vectors $g_n = \| h_n \|^{-1} h_n$.

The functionals ϕ_n defined by

$$\phi_n(f) = \langle g_n \mid Af \rangle, \qquad f \in \mathcal{H},$$

are obviously linear and defined everywhere on $\mathcal{H}$ because $\mathcal{D}_A = \mathcal{H}$. Moreover, these functionals are bounded,

$$|\phi_n(f)| = |\langle A^*g_n \mid f \rangle| \leqslant \| A^*g_n \| \, \| f \|,$$

and therefore continuous (see Theorem 2.1). Since in addition the sequence $\phi_1, \phi_2, \ldots$ is bounded for every fixed $f \in \mathscr{H}$,

$$|\phi_n(f)| = |\langle g_n \mid Af\rangle| \leqslant \| g_n \| \, \| Af \| = \| Af \|,$$

we can apply Lemma 2.2. According to this lemma the functionals $\phi_1, \phi_2, \ldots$ have a common bound for all $f \in \mathscr{H}$,

$$|\phi_n(f)| \leqslant C \| f \|, \qquad n = 1, 2, \ldots .$$

But then we get by setting $f = A^* g_n$

$$|\phi_n(A^* g_n)| = \| A^* g_n \|^2 \leqslant C \| A^* g_n \|,$$

which implies that

$$\| A^* g_n \| \leqslant C, \qquad n = 1, 2, \ldots,$$

contradicting (2.4). Q.E.D.

From Theorem 2.9 we can immediately deduce that the following theorem is true.

Theorem 2.10 (*Hellinger and Toeplitz*). Any symmetric operator A defined on the entire Hilbert space $\mathscr{H}$ is bounded.

The above theorem is evidently true, since $A \subseteq A^*$ and the fact that $\mathscr{D}_A = \mathscr{H}$ imply that $A \equiv A^*$, while on the other hand A^* is bounded by Theorem 2.9.

We note that in the Hellinger–Toeplitz theorem we have that $A \equiv A^*$, and therefore A is closed by Theorem 2.8. In §3 we shall generalize this theorem to any closed operator defined on a dense domain.

EXERCISES

2.1. Show that the functional

$$\phi(f) = f(0), \qquad f \in \mathscr{C}^0_{(2)}(\mathbb{R}^1)$$

is linear but it is not continuous in the norm

$$\| f \| = \left[\int_{-\infty}^{+\infty} |f(x)|^2 \, dx \right]^{1/2}.$$

2.2. Show that the functional

$$\phi(f) = \int_0^1 f(x) \, dx, \qquad f \in L^2(\mathbb{R}^1),$$

is linear and continuous in $L^2(\mathbb{R}^1)$, and find a $g(x) \in L_{(2)}(\mathbb{R}^1)$ such that

$$\phi(f) = \int_0^1 g^*(x) f(x)\, dx = \langle g \mid f \rangle.$$

2.3. Let $A = \| a_{ik} \|$ be the operator acting on $\| c_k \| \in l^2(n)$ in the following manner:

$$A \| c_i \| = \| b_i \|, \qquad b_i = \sum_{k=1}^{n} a_{ik} c_k .$$

Show that the above operator is linear and bounded, and that its adjoint is represented by the Hermitian conjugate matrix $\| a_{ki}^* \|$,

$$A^* \| c_i \| = \| b_i' \|, \qquad b_i' = \sum_{k=1}^{n} a_{ki}^* c_k .$$

2.4. Prove that the operator

$$(\Delta f)(\mathbf{r}) = \frac{\partial^2 f(\mathbf{r})}{\partial x^2} + \frac{\partial^2 f(\mathbf{r})}{\partial y^2} + \frac{\partial^2 f(\mathbf{r})}{\partial z^2}, \qquad f \in \mathscr{C}_{\mathrm{b}}^{\infty}(\mathbb{R}^3),$$

is an *unbounded* linear operator on $L^2(\mathbb{R}^3)$.

2.5. Let $(f \mid g)$, $f, g \in \mathscr{H}$, be a bilinear form on the Hilbert space $\mathscr{H}$ which is bounded, i.e.,

$$|(f \mid g)| \leqslant C \| f \| \, \| g \|.$$

Prove that there is a bounded linear operator A on $\mathscr{H}$, with $\| A \| \leqslant C$ and $\mathscr{D}_A = \mathscr{H}$, such that

$$(f \mid g) = \langle Af \mid g \rangle.$$

Show that $(f \mid g)$ is Hermitian if and only if $A^* \equiv A$.

2.6. Prove that if A^* exists and λ is any number, then $(\lambda A)^*$ and $(A + \lambda)^*$ exist (in $A + \lambda$, λ stands for the operator which multiplies each vector by the number λ) and

$$(\lambda A)^* = \lambda^* A^*, \qquad (A + \lambda)^* = A^* + \lambda^*.$$

2.7. Show that if A and B are bounded operators with $\mathscr{D}_A = \mathscr{D}_B = \mathscr{H}$, then $(A + B)^* = A^* + B^*$ and $(AB)^* = B^* A^*$.

2.8. Prove that if A^* and B^* exist, then $(A + B)^* \supseteq A^* + B^*$ if $\bar{\mathscr{D}}_{A+B} = \mathscr{H}$, and $(AB)^* \supseteq B^* A^*$ if $\bar{\mathscr{D}}_{AB} = \mathscr{H}$.

2.9. Show that if A^* exists and $A \subseteq B$, then B^* also exists and $B^* \subseteq A^*$.

2.10. Quote the theorems in this and the preceding sections which show that every Hermitian linear operator on a finite-dimensional Hilbert space is self-adjoint.

3. Orthogonal Projection Operators

3.1. Projectors onto Closed Subspaces of a Hilbert Space

The simplest type of operator, which is of essential significance in the spectral theory of self-adjoint operators as well as in quantum mechanical problems, is the *orthogonal projection operator*. As the name suggests, this concept of a projection operator acting on a Hilbert space $\mathscr{H}$ is a generalization of the notion of an orthogonal projection of a vector $\mathbf{r}$ in $(\mathbb{R}^3)$ onto a line or a plane in $(\mathbb{R}^3)$. For instance, if we have a line through the origin of $\mathbb{R}^3$ [i.e., a one-dimensional subspace of $(\mathbb{R}^3)$],

$$\mathsf{M} = \{t\mathbf{e}: -\infty < t < +\infty\}$$

determined by the unit vector $\mathbf{e}$, $\|\mathbf{e}\| = 1$, recall that the orthogonal projection of a vector $\mathbf{r}$ onto this line is the vector

$$\mathbf{r}_{\mathsf{M}} = (\mathbf{e} \cdot \mathbf{r})\, \mathbf{e}.$$

Theorem 3.1. Let M be a *closed* linear subspace of the Hilbert space $\mathscr{H}$. Denote by $\mathsf{M}^\perp$ (or $\mathscr{H} \ominus \mathsf{M}$) the linear space (see Exercise 3.1) of all vectors orthogonal to M:

$$\mathsf{M}^\perp = \{h: \langle h \mid g\rangle = 0,\ g \in \mathsf{M}\}.$$

Then each vector $f \in \mathscr{H}$ can be written uniquely as a sum

$$f = f' + f'', \qquad f' \in \mathsf{M}, \qquad f'' \in \mathsf{M}^\perp, \tag{3.1}$$

and the mapping

$$f \to f' = E_{\mathsf{M}}(f), \qquad f \in \mathscr{H}, \qquad f' \in \mathsf{M}, \tag{3.2}$$

is a linear operator defined on the entire $\mathscr{H}$, called the *projector* (or *orthogonal projection operator*) onto M.

Proof. The possibility of the decomposition (3.1) for each $f \in \mathscr{H}$ is guaranteed by Lemma 2.1. In order to prove the uniqueness of this decomposition, assume that

$$f = f_1' + f_1'', \qquad f_1' \in \mathsf{M}, \qquad f_1'' \in \mathsf{M}^\perp$$

Then we have

$$f' - f_1' = -(f'' - f_1'').$$

Since, on the other hand, $f' - f_1' \in \mathsf{M}$, $f'' - f_1'' \in \mathsf{M}^\perp$, and therefore $f' - f_1' \perp f'' - f_1''$, we get

$$\langle f' - f_1' \mid f' - f_1' \rangle = -\langle f' - f_1' \mid f'' - f_1'' \rangle = 0,$$

which implies $f' - f_1' = \mathbf{0}$ and $f'' - f_1'' = \mathbf{0}$. Thus, the uniqueness of f' and f'' for each $f \in \mathscr{H}$ is established, and consequently (3.2) defines a mapping of $\mathscr{H}$ into M.

To prove that this mapping is a linear operator, note that if

$$g = g' + g'', \qquad g' \in \mathsf{M}, \qquad g'' \in \mathsf{M}^\perp,$$

we have

$$af + bg = (af' + bg') + (af'' + bg''),$$

where, due to the linearity of M and $\mathsf{M}^\perp$,

$$af' + bg' \in \mathsf{M} \qquad \text{and} \qquad af'' + bg'' \in \mathsf{M}^\perp.$$

Since the decomposition (3.1) of any vector in $\mathscr{H}$ is unique, we get

$$(af + bg)' = E_{\mathsf{M}}(af + bg) = af' + bg' = aE_{\mathsf{M}}(f) + bE_{\mathsf{M}}(g),$$

which proves the linearity of E_{M}. Q.E.D.

If M is a closed linear subspace of a Hilbert space $\mathscr{H}$ and E_{M} is the projector on M, then $f' = E_{\mathsf{M}} f$ is called the *projection* of the vector $f \in \mathscr{H}$ onto the subspace M. The following theorem provides a simple recipe for computing projections in practically important cases.

Theorem 3.2. If M is a *separable* closed linear subspace of a (not necessarily separable) Hilbert space $\mathscr{H}$, and if $\{e_1, e_2, \ldots\}$ is an orthonormal basis in M, then

$$E_{\mathsf{M}} f = \sum_k \langle e_k \mid f \rangle\, e_k, \qquad f \in \mathscr{H}, \tag{3.3}$$

where E_{M} is the projector on M.

Proof. In order to prove that $E_{\mathsf{M}} f$ is given by (3.3) when $\{e_1, e_2, \ldots\}$ is a (countable) orthonormal basis in M, note that

$$g' = \sum_k \langle e_k \mid f \rangle\, e_k$$

exists and $g' \in \mathsf{M}$ (Chapter I, Theorem 4.6) for all $f \in \mathscr{H}$. Furthermore, $g'' = f - g' \in \mathsf{M}^\perp$ because for any $h \in \mathsf{M}$ we get

$$\begin{aligned}\langle g'' \mid h\rangle &= \langle f \mid h\rangle - \sum_k \langle f \mid e_k\rangle\langle e_k \mid h\rangle \\ &= \langle f' \mid h\rangle - \sum_k \langle f' \mid e_k\rangle\langle e_k \mid h\rangle = 0,\end{aligned}$$

where in deriving the third expression we used the fact that

$$\langle f'' \mid h\rangle = \langle f'' \mid e_1\rangle = \langle f'' \mid e_2\rangle = \cdots = 0,$$

and then Parseval's relation (4.15) of Chapter I was applied to the orthonormal basis $\{e_1, e_2, \ldots\}$ in M. Thus, we have

$$f = g' + g'', \qquad g' \in \mathsf{M}, \qquad g'' \in \mathsf{M}^\perp,$$

and therefore, due to the uniqueness of the decomposition (3.1), $g' = f' = E_{\mathsf{M}} f$. Q.E.D.

An illustration of the above theorem is provided by the special case of a one-dimensional space M. If e is a normalized vector from M, then we have

$$E_{\mathsf{M}} f = \langle e \mid f\rangle\, e.$$

In Dirac's notation the projector E_{M} is denoted by $| e\rangle \langle e |$, thus providing a handy mnemonic device for writing* the above formula in the ket formalism:

$$E_{\mathsf{M}} f = (| e\rangle\langle e |) \,| f\rangle = | e\rangle\langle e \mid f\rangle = \langle e \mid f\rangle\, e. \tag{3.4}$$

It is immediately noticeable that in defining the projection on a closed subspace M, the spaces M and $\mathsf{M}^\perp$ play a symmetric role. In fact, we see from (3.1) that

$$E_{\mathsf{M}} + E_{\mathsf{M}^\perp} = 1. \tag{3.5}$$

$\mathsf{M}^\perp$ is called the *orthogonal complement* of M in $\mathscr{H}$. We say that $\mathscr{H}$ is the *orthogonal sum* of the subspaces M and $\mathsf{M}^\perp$, and express this statement symbolically by writing

$$\mathscr{H} = \mathsf{M} \oplus \mathsf{M}^\perp.$$

More generally, we define the orthogonal sum of subspaces as follows.

* In mathematical literature, instead of writing $| e\rangle\langle e |$, one would write $\langle e \mid \cdot\rangle\, e$.

Definition 3.1. Let $\mathscr{F}$ be a family of mutually orthogonal closed linear subspaces of a Hilbert space $\mathscr{H}$, i.e., if $\mathsf{M}_1, \mathsf{M}_2 \in \mathscr{F}$, then $\mathsf{M}_1 \perp \mathsf{M}_2$. The closed linear subspace

$$\bigoplus_{\mathsf{M}\in\mathscr{F}} \mathsf{M} = \left[\left\{f: f \in \bigcup_{\mathsf{M}\in\mathscr{F}} \mathsf{M}\right\}\right]$$

spanned by $\bigcup_{\mathsf{M}\in\mathscr{F}} \mathsf{M}$ is called the *orthogonal sum* of the subspaces $\mathsf{M} \in \mathscr{F}$.

In the case where $\mathscr{F}$ can be written as a finite or infinite sequence $\{\mathsf{M}_1, \mathsf{M}_2, \ldots\}$ of closed subspaces of $\mathscr{H}$, their orthogonal sum is also denoted by

$$\bigoplus_k \mathsf{M}_k = \mathsf{M}_1 \oplus \mathsf{M}_2 \oplus \cdots .$$

It is significant to note the similarity between the notation for the *direct* sum of Hilbert spaces and the *orthogonal* sum of subspaces of a certain Hilbert space. It will soon be evident that the direct sum of the closed linear subspaces $\mathsf{M}_1, \mathsf{M}_2, \ldots$ (which are themselves Hilbert spaces) is unitarily equivalent to the orthogonal sum of $\mathsf{M}_1, \mathsf{M}_2, \ldots$.

3.2. Algebraic Properties of Projectors

Theorem 3.3. (a) A linear operator E defined on the entire Hilbert space $\mathscr{H}$ is a projector if and only if it is self-adjoint,

$$E = E^*, \tag{3.6}$$

and *idempotent*,

$$E = E^2. \tag{3.7}$$

(b) The closed linear subspace M on which a projector E defined on $\mathscr{H}$ projects is identical with the set of all vectors $f \in \mathscr{H}$ satisfying the equation

$$Ef = f. \tag{3.8}$$

(c) $\| E \| = 1$ for any projector E which is different from the zero operator 0, i.e., which projects on a subspace $\mathsf{M} \neq \{0\}$.

Proof. Let E_M be the projector on the closed linear subspace M of $\mathscr{H}$. For arbitrary $f, g \in \mathscr{H}$ we get by using the familiar decomposition

$$f = f' + f'', \qquad g = g' + g'', \qquad f', g' \in \mathsf{M}, \qquad f'', g'' \in \mathsf{M}^\perp,$$

that

$$\begin{aligned}\langle f \mid E_\mathsf{M} g\rangle &= \langle f' + f'' \mid g'\rangle = \langle f' \mid g'\rangle = \langle f' \mid g' + g''\rangle \\ &= \langle E_\mathsf{M} f \mid g\rangle,\end{aligned}$$

which proves that $E_M{}^* = E_M$. Furthermore, we see from the above that for any $f \in \mathscr{H}$

$$\langle E_M f \mid g \rangle = \langle f' \mid g' \rangle = \langle E_M f \mid E_M g \rangle = \langle E_M{}^2 f \mid g \rangle$$

for all $g \in \mathscr{H}$, which implies that $E_M f = E_M{}^2 f$, i.e., $E_M = E_M{}^2$. Note that in particular $E_M f = f$ when $f \in M$!

Assume now that a linear operator E, with $\mathscr{D}_E = \mathscr{H}$, is such that $E = E^*$ and $E = E^2$. Denote by R the set of all vectors $f \in \mathscr{H}$ for which

$$Ef = f.$$

R is a linear subspace of $\mathscr{H}$, since if $f, g \in R$, then

$$E(af + bg) = aEf + bEg = af + bg$$

for all scalars a, b. Moreover, R is closed; namely, if the sequence $f_1, f_2, \ldots \in R$ converges to some vector $f \in \mathscr{H}$, we get

$$\lim_{k\to\infty} \| Ef - f_k \| = \lim_{k\to\infty} \| Ef - Ef_k \| = 0$$

by using the fact that E is bounded by Theorem 2.10, and therefore continuous (Theorem 1.3). Since $f_1, f_2, \ldots$ can have only f as their limit, we conclude that $Ef = f$, i.e., $f \in R$.

In order to establish that E is the projector on R, decompose each vector $g \in \mathscr{H}$ in the following way:

$$g = Eg + (g - Eg).$$

But $Eg \in R$, as we can see by using the relation $E^2 = E$,

$$E(Eg) = E^2 g = Eg.$$

On the other hand, $g - Eg \in R^\perp$, since for any $f \in R$

$$\langle g - Eg \mid f \rangle = \langle g - Eg \mid Ef \rangle = \langle E(g - Eg) \mid f \rangle = \langle Eg - E^2 g \mid f \rangle = 0,$$

which shows that $E = E_R$.

The above considerations establish the points (a) and (b) of the theorem. Since by (3.1)

$$\| f \|^2 \geqslant \| f' \|^2 = \| E_M f \|^2,$$

we have $\| E_M \| \leqslant 1$. If $E_M \neq 0$, there must be at least one vector $f_1 \neq 0$ which belongs to M and therefore satisfies (3.8). Thus, $\| E_M f_1 \| = \| f_1 \|$, which shows that $\| E_M \| = 1$. Q.E.D.

Theorem 3.3 provides a very useful criterion for verifying whether a linear operator is a projector, without having to find the closed subspace in which it projects (see Exercise 3.2). It will be used so frequently in the following that no explicit mention will be made of the fact that the theorem has been applied while carrying out a certain step.

Theorem 3.4. Two closed subspaces M and N of $\mathscr{H}$ are orthogonal if and only if $E_\mathsf{M} E_\mathsf{N} = E_\mathsf{N} E_\mathsf{M} = 0$.

Proof. Since $E_\mathsf{M} = E_\mathsf{M}^*$, we have

$$\langle E_\mathsf{M} f \mid E_\mathsf{N} g \rangle = \langle f \mid E_\mathsf{M} E_\mathsf{N} g \rangle \tag{3.9}$$

for all $f, g \in \mathscr{H}$. If $\mathsf{M} \perp \mathsf{N}$, as we have $E_\mathsf{M} f \in \mathsf{M}$ and $E_\mathsf{N} g \in \mathsf{N}$ for all $f, g \in \mathscr{H}$, we get from (3.9) that $\langle f \mid E_\mathsf{M} E_\mathsf{N} g \rangle = 0$. Conversely, if $E_\mathsf{M} E_\mathsf{N} = 0$ and f, g are any vectors from M and N, respectively, due to $E_\mathsf{M} f = f$ and $E_\mathsf{N} g = g$ we obtain from (3.9) $f \perp g$.

Since $E_\mathsf{M} E_\mathsf{N} = 0$ implies that (see Exercise 2.7)

$$E_\mathsf{M} E_\mathsf{N} = 0 = (E_\mathsf{M} E_\mathsf{N})^* = E_\mathsf{N} E_\mathsf{M},$$

we see that the statement $E_\mathsf{M} E_\mathsf{N} = 0$ is equivalent to $E_\mathsf{N} E_\mathsf{M} = 0$. Q.E.D.

3.3. Partial Ordering of Projectors

If M and N are two subspaces and $\mathsf{M} \subset \mathsf{N}$, we shall write $E_\mathsf{M} \leqslant E_\mathsf{N}$. The following theorem gives us two additional criteria for this relation to hold.

Theorem 3.5. The following three statements about the projectors E_M and E_N defined on $\mathscr{H}$ are equivalent:

(a) $\mathsf{M} \subset \mathsf{N}$, i.e., $E_\mathsf{M} \leqslant E_\mathsf{N}$;
(b) $E_\mathsf{M} E_\mathsf{N} = E_\mathsf{N} E_\mathsf{M} = E_\mathsf{M}$;
(c) $\| E_\mathsf{M} f \| \leqslant \| E_\mathsf{N} f \|$ for all $f \in \mathscr{H}$.

Proof. If $\mathsf{M} \subset \mathsf{N}$, then $E_\mathsf{M} f \in \mathsf{N}$ for any $f \in \mathscr{H}$, and consequently, by (3.8),

$$E_\mathsf{N}(E_\mathsf{M} f) = E_\mathsf{M} f$$

for all $f \in \mathscr{H}$. Hence, $E_\mathsf{M} = E_\mathsf{N} E_\mathsf{M}$, which implies

$$\| E_\mathsf{M} f \| = \| E_\mathsf{M} E_\mathsf{N} f \| \leqslant \| E_\mathsf{M} \| \, \| E_\mathsf{N} f \| \leqslant \| E_\mathsf{N} f \|$$

since $\| E_\mathsf{M} \| \leqslant 1$ according to Theorem 3.3(c). Thus, (b) implies (c).

Finally, to prove that (c) implies (a), note for any closed subspace M_1 we can infer from

$$\|f\|^2 = \|E_{\mathsf{M}_1}f\|^2 + \|E_{\mathsf{M}_1^\perp}f\|^2$$

that $f \in \mathsf{M}_1$ (i.e., $f = E_{\mathsf{M}_1}f$) if and only if

$$\|f\| = \|E_{\mathsf{M}_1}f\|.$$

If (c) is true, then whenever $f \in \mathsf{M}$, we have

$$\|f\| = \|E_{\mathsf{M}}f\| \leqslant \|E_{\mathsf{N}}f\|.$$

Combining the above with the fact that $\|E_{\mathsf{N}}f\| \leqslant \|f\|$ for any $f \in \mathscr{H}$, we deduce that

$$\|f\| = \|E_{\mathsf{N}}f\|.$$

This means, according to the aforesaid, that $f \in \mathsf{N}$, thus proving that $\mathsf{M} \subset \mathsf{N}$. Q.E.D.

The reason for the particular notation $\leqslant$ used in designating the relation between any two projectors satisfying the conditions of the preceding theorem is that this relation obviously introduces a partial ordering in the set of all projectors acting on $\mathscr{H}$.

Definition 3.2. A set S is said to be *partially ordered* if a relation $\xi \leqslant \eta$ is defined for some pairs ξ, η of elements of S, which satisfies the following conditions:

(1) $\xi \leqslant \xi$ for all $\xi \in S$;

(2) if $\xi \leqslant \eta$ and $\eta \leqslant \xi$, then $\xi = \eta$;

(3) if $\xi \leqslant \eta$ and $\eta \leqslant \zeta$, then $\xi \leqslant \zeta$.

Other examples of partial ordering of a set are provided by the "equal or smaller" relation $\leqslant$ holding between real numbers, and by the relation $\subset$ of set inclusion holding between the sets in a family of sets.

3.4. Projectors onto Intersections and Orthogonal Sums of Subspaces

Theorem 3.6. The product $E_{\mathsf{M}}E_{\mathsf{N}}$ of two projectors E_{M} and E_{N} in $\mathscr{H}$ is a projector if and only if E_{M} and E_{N} commute,

$$E_{\mathsf{M}}E_{\mathsf{N}} = E_{\mathsf{N}}E_{\mathsf{M}}. \tag{3.10}$$

If this is the case, the set $\mathsf{L} = \mathsf{M} \cap \mathsf{N}$ is a closed linear subspace of $\mathscr{H}$ and

$$E_{\mathsf{L}} = E_{\mathsf{M}} E_{\mathsf{N}} .$$

Proof. If the linear operator $E_{\mathsf{M}} E_{\mathsf{N}}$ is a projector, by Theorem 3.3, it must be Hermitian, and consequently

$$E_{\mathsf{M}} E_{\mathsf{N}} = (E_{\mathsf{M}} E_{\mathsf{N}})^* = E_{\mathsf{N}}^* E_{\mathsf{M}}^* = E_{\mathsf{N}} E_{\mathsf{M}} .$$

Conversely, if E_{M} and E_{N} commute, then $E_{\mathsf{M}} E_{\mathsf{N}}$ is Hermitian,

$$(E_{\mathsf{M}} E_{\mathsf{N}})^* = E_{\mathsf{N}}^* E_{\mathsf{M}}^* = E_{\mathsf{N}} E_{\mathsf{M}} = E_{\mathsf{M}} E_{\mathsf{N}} ,$$

and idempotent,

$$(E_{\mathsf{M}} E_{\mathsf{N}})^2 = E_{\mathsf{M}}(E_{\mathsf{N}} E_{\mathsf{M}})\, E_{\mathsf{N}} = E_{\mathsf{M}}^2 E_{\mathsf{N}}^2 = E_{\mathsf{M}} E_{\mathsf{N}} ,$$

i.e., $E_{\mathsf{M}}\, E_{\mathsf{N}}$ is a projector.

To prove the second part of the theorem, assume that (3.10) is true, and denote by L the closed subspace of $\mathscr{H}$ onto which $E_{\mathsf{M}} E_{\mathsf{N}}$ projects, so that $E_{\mathsf{L}} = E_{\mathsf{M}} E_{\mathsf{N}}$. Then we have

$$\| E_{\mathsf{L}} f \| = \| E_{\mathsf{N}} E_{\mathsf{M}} f \| \leqslant \| E_{\mathsf{M}} f \|,$$

$$\| E_{\mathsf{L}} f \| = \| E_{\mathsf{M}} E_{\mathsf{N}} f \| \leqslant \| E_{\mathsf{N}} f \|,$$

for all $f \in \mathscr{H}$, i.e., according to Theorem 3.5 $E_{\mathsf{L}} \leqslant E_{\mathsf{M}}$ and $E_{\mathsf{L}} \leqslant E_{\mathsf{N}}$, and consequently $\mathsf{L} \subset \mathsf{M} \cap \mathsf{N}$. Conversely, if $f \in \mathsf{M} \cap \mathsf{N}$, then $E_{\mathsf{L}} f = E_{\mathsf{M}}(E_{\mathsf{N}} f) = E_{\mathsf{M}} f = f$ and therefore $f \in \mathsf{L}$, i.e., $\mathsf{L} \supset \mathsf{M} \cap \mathsf{N}$; thus, $\mathsf{L} = \mathsf{M} \cap \mathsf{N}$. Q.E.D.

Theorem 3.7. The sum

$$E_{\mathsf{M}_1} + E_{\mathsf{M}_2} + \cdots + E_{\mathsf{M}_n}$$

of a finite number of projectors defined on $\mathscr{H}$ is a projector if and only if

$$E_{\mathsf{M}_i} E_{\mathsf{M}_j} = 0 \qquad \text{for} \quad i \neq j; \quad i, j = 1, \ldots, n, \tag{3.11}$$

i.e., if and only if M_1 , M_2 ,..., M_n are mutually orthogonal; in that case

$$E_{\mathsf{M}} = E_{\mathsf{M}_1} + E_{\mathsf{M}_2} + \cdots + E_{\mathsf{M}_n} , \tag{3.12}$$

where

$$\mathsf{M} = \mathsf{M}_1 \oplus \mathsf{M}_2 \oplus \cdots \oplus \mathsf{M}_n . \tag{3.13}$$

Proof. The linear operator

$$A = E_{\mathsf{M}_1} + \cdots + E_{\mathsf{M}_n}$$

is Hermitian, $A = A^*$, since it is a sum of bounded Hermitian operators,

$$A^* = E^*_{\mathsf{M}_1} + \cdots + E^*_{\mathsf{M}_n} = E_{\mathsf{M}_1} + \cdots + E_{\mathsf{M}_n} = A.$$

Thus, A is a projector if and only if $A = A^2$. This will be the case whenever (3.11) is satisfied, since then

$$A^2 = \sum_{i,j=1}^{n} E_{\mathsf{M}_i} E_{\mathsf{M}_j} = \sum_{i=1}^{n} E^2_{\mathsf{M}_i} = \sum_{i=1}^{n} E_{\mathsf{M}_i} = A.$$

Conversely, assume that A is a projector. Then for $i \neq j$

(3.14)
$$\langle f \mid E_{\mathsf{M}_i} f\rangle + \langle f \mid E_{\mathsf{M}_j} f\rangle \leqslant \sum_{k=1}^{n} \langle f \mid E_{\mathsf{M}_k} f\rangle$$
$$= \langle f \mid Af\rangle \leqslant \langle f \mid f\rangle,$$

where in deriving the first inequality we have used the fact that for any projector

$$\langle f \mid Ef\rangle = \langle f \mid E^2 f\rangle = \langle Ef \mid Ef\rangle \geqslant 0, \qquad f \in \mathscr{H},$$

and in deriving the second inequality we employed the property that $\| A \| \leqslant 1$ if A is a projector.

From (3.14) we obtain

$$\| E_{\mathsf{M}_i} f \|^2 + \| E_{\mathsf{M}_j} f \|^2 \leqslant \| f \|^2.$$

If we set above $f = E_{\mathsf{M}_j} f$, we get

$$\| E_{\mathsf{M}_i} E_{\mathsf{M}_j} f \|^2 \leqslant 0,$$

which implies $E_{\mathsf{M}_i} E_{\mathsf{M}_j} f = 0$ for all $f \in \mathscr{H}$, i.e., that $\mathsf{M}_1, \ldots, \mathsf{M}_n$ are mutually orthogonal (see Theorem 3.4).

To prove (3.12), assume that (3.11) is satisfied, so that A is a projector. Denote by M the space on which A projects, i.e., $A = E_{\mathsf{M}}$. If $f \in \mathsf{M}$, then

$$f = Af = E_{\mathsf{M}_1} f + \cdots + E_{\mathsf{M}_n} f \in \mathsf{M}_1 \oplus \cdots \oplus \mathsf{M}_n,$$

which shows that

(3.15)
$$\mathsf{M} \subset \mathsf{M}_1 \oplus \cdots \oplus \mathsf{M}_n.$$

Suppose that $g \in \mathsf{M}_1 \oplus \cdots \oplus \mathsf{M}_n$. By Definition 3.1, g is in the closure of the linear manifold spanned by $\mathsf{M}_1 \cup \cdots \cup \mathsf{M}_n$. This means that there is a sequence $g_1, g_2, \ldots \in (\bigcup_{k=1}^n \mathsf{M}_k)$

$$g_k = g_k^{(1)} + \cdots + g_k^{(n)}, \qquad g_k^{(1)} \in \mathsf{M}_1, \ldots, g_k^{(n)} \in \mathsf{M}_n,$$

converging in the norm to g. From the mutual orthogonality of $\mathsf{M}_1, \ldots, \mathsf{M}_n$ we get

$$\| g_k - g_l \|^2 = \sum_{i,j=1}^{n} \langle g_k^{(i)} - g_l^{(i)} \mid g_k^{(j)} - g_l^{(j)} \rangle = \sum_{j=1}^{n} \| g_k^{(j)} - g_l^{(j)} \|^2$$

by using that fact that $g_k^{(i)} - g_l^{(i)} \in \mathsf{M}_i$ for any $k, l = 1, 2, \ldots$. Thus, for any fixed $i = 1, \ldots, n$,

$$\| g_k^{(i)} - g_l^{(i)} \| \leqslant \| g_k - g_l \| \to 0 \qquad \text{for} \quad k, l \to \infty,$$

so that $g_1^{(i)}, g_2^{(i)}, \ldots$ is a Cauchy sequence in M_i. Since M_i is closed, this sequence has a limit $g^{(i)}$. Thus, we can write our arbitrarily chosen element g from $\mathsf{M}_1 \oplus \cdots \oplus \mathsf{M}_n$ in the form

$$g = g^{(1)} + \cdots + g^{(n)}, \qquad g^{(1)} \in \mathsf{M}_1, \ldots, g^{(n)} \in \mathsf{M}_n.$$

Moreover, since $\mathsf{M}_i \perp \mathsf{M}_j$ for $i \neq j$, we have

$$E_{\mathsf{M}_i} g^{(j)} = \delta_{ij} g^{(j)}.$$

It follows that

$$Ag = \sum_{i,j=1}^{n} E_{\mathsf{M}_i} g^{(j)} = \sum_{j=1}^{n} g^{(j)} = g, \tag{3.16}$$

for the arbitrarily chosen $g \in \mathsf{M}_1 \oplus \cdots \oplus \mathsf{M}_n$. From the above relation (3.16) we deduce that $g \in \mathsf{M}$, thus proving that

$$\mathsf{M} \supset \mathsf{M}_1 \oplus \cdots \oplus \mathsf{M}_n. \tag{3.17}$$

The relations (3.15) and (3.17) combined together imply (3.13). Q.E.D.

Theorem 3.8. The difference $E_{\mathsf{M}} - E_{\mathsf{N}}$ of two projectors is a projector if and only if $\mathsf{N} \subset \mathsf{M}$; in that case

$$E_{\mathsf{M}} - E_{\mathsf{N}} = E_{\mathsf{M} \ominus \mathsf{N}},$$

where $\mathsf{M} \ominus \mathsf{N}$ is the orthogonal complement of N with respect to M:

$$\mathsf{M} \ominus \mathsf{N} = \{f : f \in \mathsf{M}, f \perp \mathsf{N}\}. \tag{3.18}$$

Proof. Suppose that $E_{\mathsf{M}} - E_{\mathsf{N}}$ is a projector, and let L be the space on which it projects:

$$E_{\mathsf{M}} - E_{\mathsf{N}} = E_{\mathsf{L}} .$$

Thus, E_{M} is a projector which satisfies the relation

$$E_{\mathsf{M}} = E_{\mathsf{L}} + E_{\mathsf{N}} ,$$

In this case we have according to Theorem 3.7 that $E_{\mathsf{L}}E_{\mathsf{N}} = 0$ and $\mathsf{M} = \mathsf{L} \oplus \mathsf{N}$.

Conversely, if $\mathsf{N} \subset \mathsf{M}$, then by Theorem 3.5

$$E_{\mathsf{N}} = E_{\mathsf{M}}E_{\mathsf{N}} = E_{\mathsf{N}}E_{\mathsf{M}} .$$

This implies that the evidently self-adjoint operator $E_{\mathsf{M}} - E_{\mathsf{N}}$ is idempotent,

$$(E_{\mathsf{M}} - E_{\mathsf{N}})^2 = E_{\mathsf{M}} - E_{\mathsf{M}}E_{\mathsf{N}} - E_{\mathsf{N}}E_{\mathsf{M}} + E_{\mathsf{N}} = E_{\mathsf{M}} - E_{\mathsf{N}} ,$$

i.e., $E_{\mathsf{M}} - E_{\mathsf{N}}$ is a projector. Q.E.D.

We shall now generalize Theorem 3.7 to the case of a direct sum of a countable infinity of closed subspaces.

Theorem 3.9. If $\mathsf{M}_1, \mathsf{M}_2, \ldots$ is a sequence of mutually orthogonal linear subspaces of a Hilbert space $\mathscr{H}$, and M is their orthogonal sum

$$\mathsf{M} = \mathsf{M}_1 \oplus \mathsf{M}_2 \oplus \cdots ,$$

then

$$\lim_{n\to\infty} \left\| \left(E_{\mathsf{M}} - \sum_{k=1}^{n} E_{\mathsf{M}_k} \right) f \right\| = 0 \tag{3.19}$$

for every $f \in \mathscr{H}$.

Proof. By applying Theorem 3.7 to $E_{\mathsf{M}_1} + \cdots + E_{\mathsf{M}_n}$ we easily derive

$$\sum_{k=1}^{n} \| E_{\mathsf{M}_k} f \|^2 = \sum_{k=1}^{n} \langle E_{\mathsf{M}_k} f \mid f \rangle = \left\langle \sum_{k=1}^{n} E_{\mathsf{M}_k} f \mid f \right\rangle$$

$$= \langle E_{\mathsf{M}_1 \oplus \ldots \oplus \mathsf{M}_n} f \mid f \rangle \leqslant \| f \|^2,$$

which shows that the series $\sum_{k=1}^{\infty} \| E_{\mathsf{M}_k} f \|^2$ converges. Consequently, the sequence

$$\left\{ \sum_{k=1}^{n} E_{\mathsf{M}_k} f : \; n = 1, 2, \ldots \right\} \tag{3.20}$$

is a Cauchy sequence, as can be seen from the relation

$$\left\| \sum_{k=m}^{m+r} E_{\mathsf{M}_k} f \right\|^2 = \sum_{k=m}^{m+r} \| E_{\mathsf{M}_k} f \|^2.$$

If we denote the limit of the sequence (3.20) by f', the operator $Af = f'$ is obviously linear and defined everywhere in $\mathscr{H}$. Moreover, we have

$$\begin{aligned} \langle Af \mid g \rangle &= \lim_{n\to\infty} \left\langle \sum_{k=1}^{n} E_{\mathsf{M}_k} f \mid g \right\rangle \\ &= \lim_{n\to\infty} \left\langle f \middle| \sum_{k=1}^{n} E_{\mathsf{M}_k} g \right\rangle = \langle f \mid Ag \rangle \\ &= \lim_{n\to\infty} \left\langle \sum_{k=1}^{n} E_{\mathsf{M}_k} f \middle| \sum_{k=1}^{n} E_{\mathsf{M}_k} g \right\rangle = \langle Af \mid Ag \rangle, \end{aligned}$$

which shows that $A = A^*$ and $A^2 = A$. Thus, A is a projector.

Denote by M the linear subspace of $\mathscr{H}$ on which A projects. If $f \in \mathsf{M}$, then

$$f = Af = E_{\mathsf{M}_1} f + E_{\mathsf{M}_2} f + \cdots \in \mathsf{M}_1 \oplus \mathsf{M}_2 \oplus \cdots,$$

i.e., we must have $\mathsf{M} \subset \mathsf{M}_1 \oplus \mathsf{M}_2 \oplus \cdots$.

If we assume that $\mathsf{M} \neq \mathsf{M}_1 \oplus \mathsf{M}_2 \oplus \cdots$, there would be a vector which belongs to $\mathsf{M}_1 \oplus \mathsf{M}_2 \oplus \cdots$ but not to M. Since $\mathsf{M}_1 \oplus \mathsf{M}_2 \oplus \cdots$ is a closed linear subspace of a Hilbert space and as such is a Hilbert space itself, we could deduce from Lemma 2.1 the existence of a nonzero vector $g \in \mathsf{M}_1 \oplus \mathsf{M}_2 \oplus \cdots$ which is orthogonal to M. This implies that $g \perp \mathsf{M}_k$, $k = 1, 2, \ldots$, because we obviously have $\| E_{\mathsf{M}_k} f \| \leqslant \| Af \|$ for all $f \in \mathscr{H}$, and therefore $\mathsf{M}_1 \oplus \mathsf{M}_2 \oplus \cdots \subset \mathsf{M}$.

On the other hand, by Definition 3.1 of orthogonal sums, g is the limit in the norm of linear combinations of vectors from $\mathsf{M}_1, \mathsf{M}_2, \ldots$, i.e., we have vectors $g_1^k \in \mathsf{M}_1, \ldots, g_k^k \in \mathsf{M}_k$, $k = 1, 2, \ldots$, such that

$$\lim_{k\to\infty} \| g - (g_1^k + \cdots + g_k^k) \| = 0.$$

Consequently, since $g \perp \mathsf{M}_1, \mathsf{M}_2, \ldots$, we have

$$\langle g \mid g \rangle = \lim_{k\to\infty} \langle g \mid g_1^k + \cdots + g_k^k \rangle = 0,$$

which shows that $g = \mathbf{0}$, thus contradicting the assumption that $\mathsf{M} \neq \mathsf{M}_1 \oplus \mathsf{M}_2 \oplus \cdots$. Q.E.D.

Theorem 3.10. If $E_{\mathsf{M}_1}, E_{\mathsf{M}_2}, \ldots$ is a sequence of projectors defined on $\mathscr{H}$ which is *monotonically increasing*

$$E_{\mathsf{M}_1} \leqslant E_{\mathsf{M}_2} \leqslant \cdots,$$

or *monotonically decreasing*

$$E_{\mathsf{M}_1} \geqslant E_{\mathsf{M}_2} \geqslant \cdots,$$

then we have for any $f \in \mathscr{H}$

$$\lim_{n\to\infty} \|(E_{\mathsf{M}} - E_{\mathsf{M}_n})f\| = 0,$$

where E_{M} is the projector onto (see Chapter II, Definition 1.2)

$$\mathsf{M} = \lim_{n\to\infty} \mathsf{M}_n .$$

Proof. In the case of a monotonically increasing sequence, apply Theorem 3.9 to the mutually orthogonal projectors (take $\mathsf{M}_0 = \{0\}$)

$$E_{\mathsf{M}_k \ominus \mathsf{M}_{k-1}} = E_{\mathsf{M}_k} - E_{\mathsf{M}_{k-1}}, \qquad k = 1, 2, \ldots$$

to deduce that

$$\lim_{n\to\infty} \|(E_{\mathsf{M}} - E_{\mathsf{M}_n})f\| = \lim_{n\to\infty} \left\|\left(E_{\mathsf{M}} - \sum_{k=1}^{n} E_{\mathsf{M}_k \ominus \mathsf{M}_{k-1}}\right)f\right\| = 0,$$

where

$$\mathsf{M} = \bigoplus_{k=1}^{\infty} (\mathsf{M}_k \ominus \mathsf{M}_{k-1}) = \overline{\bigcup_{k=1}^{\infty} \mathsf{M}_k} .$$

The case of a monotonically decreasing sequence can be reduced to the above case by considering the projectors $E_{\mathsf{M}_n^\perp} = 1 - E_{\mathsf{M}_n}$. Q.E.D.

*3.5. Appendix: Extensions and Adjoints of Closed Linear Operators

By using the concept of orthogonal complement, we can prove the following theorem which then easily yields the generalization of the Hellinger–Toeplitz theorem (Theorem 2.10) to any closed operators.

***Theorem 3.11.** If A is a closed linear operator in $\mathscr{H}$ and A^* exists, then the domain of A^* is dense in $\mathscr{H}$, and $A^{**} = A$.

Proof. If $\mathscr{D}_{A^*}$ is not dense in $\mathscr{H}$, there is a nonzero vector h orthogonal to $\mathscr{D}_{A^*}$, i.e., $\langle h \mid g \rangle = 0$ for all $g \in \mathscr{D}_{A^*}$. But then we also have

$$\langle 0 \mid A^*g \rangle + \langle h \mid -g \rangle = 0.$$

In terms of the inner product $\langle \cdot \mid \cdot \rangle_2$ in $\mathscr{H} \oplus \mathscr{H}$, the above relation yields

(3.21) $$\langle (0, h) \mid (A^*g, -g) \rangle_2 = 0$$

for all $g \in \mathscr{D}_{A^*}$.

According to Theorem 2.5, which contains the definition of A^*, the set

(3.22) $$\{(A^*g, -g): g \in \mathscr{D}_{A^*}\} \subset \mathscr{H} \oplus \mathscr{H}$$

is the set of all the points $(g^*, -g) \in \mathscr{H} \oplus \mathscr{H}$ satisfying (2.3). This set consists of all the points in $\mathscr{H} \oplus \mathscr{H}$ which satisfy the relation

(3.23) $$\langle (g^*, -g) \mid (f, Af) \rangle_2 = 0$$

for all $f \in \mathscr{D}_A$, i.e., which are orthogonal to the graph G_A of A. The graph G_A is a linear subset of $\mathscr{H} \oplus \mathscr{H}$ because A is linear, and moreover, it is closed since A is closed. Thus, the set (3.22) is the orthogonal complement $\mathsf{G}_A^\perp$ of the closed subspace G_A of $\mathscr{H} \oplus \mathscr{H}$. By (3.21), $(0, h)$ is orthogonal to $\mathsf{G}_A^\perp$ and consequently, $(0, h) \in \mathsf{G}_A$. This implies that $h = A0$, i.e., $h = 0$, thus establishing that $\mathscr{D}_{A^*}$ is dense in $\mathscr{H}$.

Since $\mathscr{D}_{A^*}$ is dense in $\mathscr{H}$, $(A^*)^* = A^{**}$ exists by Theorem 2.5. The graph of A^{**} consists of all the points (f, f^*) of $\mathscr{H} \oplus \mathscr{H}$ satisfying

$$\langle A^*g \mid f \rangle = \langle g \mid f^* \rangle, \qquad g \in \mathscr{D}_{A^*},$$

i.e., for which

(3.24) $$\langle (A^*g, -g) \mid (f, f^*) \rangle_2 = 0.$$

This implies that $\mathsf{G}_{A^{**}} = (\mathsf{G}_A^\perp)^\perp = \mathsf{G}_A$ and consequently $\mathsf{G}_{A^{**}} = \mathsf{G}_A$, which means that $A = A^{**}$. Q.E.D.

Let A be a closed linear operator with $\mathscr{D}_A = \mathscr{H}$. By Theorem 2.9 A^* is a bounded linear operator with $\mathscr{D}_{A^*} = \mathscr{H}$, and again by the same theorem, A^{**} is also a bounded linear operator with $\mathscr{D}_{A^{**}} = \mathscr{H}$. On the other hand, by Theorem 3.11, $A^{**} = A$, and therefore we can state Theorem 3.12.

Theorem 3.12 (*The closed-graph theorem*). All *closed* linear operators defined on the entire Hilbert space $\mathscr{H}$ are bounded.

It is also easy to deduce from Theorem 3.11 the following useful result.

***Theorem 3.13.** The adjoint A^* of the linear operator A has a dense domain of definition in $\mathscr{H}$ if and only if A has a closed linear extension; in that case A^{**} is the smallest closed linear extension of A, i.e., any closed linear extension of A is also an extension of A^{**}.

Proof. According to Theorem 3.11, if A is closed, then $\bar{\mathscr{D}}_{A^*} = \mathscr{H}$ and $A^{**} = A$, so that Theorem 3.13 stands.

Conversely, if $\bar{\mathscr{D}}_{A^*} = \mathscr{H}$ then A^{**} exists. From the relation

$$\langle A^*g \mid f\rangle = \langle g \mid Af\rangle, \qquad g \in \mathscr{D}_{A^*},$$

we obviously get $A^{**} \supseteq A$. Moreover, if B is a closed extension of A, then its graph G_B is closed in $\mathscr{H} \oplus \mathscr{H}$ and it contains G_A. However, as can easily be seen from (3.23) and (3.24), we always have $\mathsf{G}_{A^{**}} = (\mathsf{G}_A^{\perp})^{\perp}$. This implies that $\mathsf{G}_{A^{**}}$ is the closure of G_A, i.e., $\mathsf{G}_{A^{**}} \subset \mathsf{G}_B$ and, therefore, $A^{**} \subseteq B$. Q.E.D.

Exercises

3.1. Let S be a nonempty subset of a Hilbert space $\mathscr{H}$. Prove that the set $S^{\perp}$ of all vectors f in $\mathscr{H}$ which are orthogonal to S

$$S^{\perp} = \{f: \langle f \mid g\rangle = 0, g \in S\}$$

is a closed linear subspace of $\mathscr{H}$.

3.2. Use Theorem 3.3 to show that the linear operator A, acting on the two-dimensional space $l^2(2)$, which is represented (see Theorem 1.1) in some orthonormal basis of $l^2(2)$ by the matrix

$$\frac{1}{2}\begin{pmatrix} 1 & i \\ -i & 1 \end{pmatrix},$$

is a projector.

3.3. Show that any projector E on an n-dimensional Hilbert space $\mathscr{H}$ can be represented by a matrix of the form

$$\begin{bmatrix} 1 & & & & & & 0 \\ & \ddots & & & & & \\ & & 1 & & & & \\ & & & 0 & & & \\ & & & & \ddots & & \\ 0 & & & & & & 0 \end{bmatrix}$$

in a suitably chosen orthonormal basis $\{e_1, \ldots, e_n\}$, where the number of 1's in the diagonal (the rank of E) is equal to the dimension of the space on which E projects.

3.4. Prove that any 2×2 Hermitian matrix either is a multiple λE $(\lambda = \lambda^*)$ of a matrix E representing a projector on $l^2(2)$, or it can be written uniquely as $\lambda_1 E_1 + \lambda_2 E_2$, where E_1 and E_2 represent projectors on $l^2(2)$, and λ_1, λ_2 are real numbers.

4. Isometric and Unitary Transformations

4.1. Isometric Transformations in between Hilbert Spaces

We encountered unitary transformations in Chapter I (Definition 2.4) when defining the concept of isomorphism of two Euclidean spaces. The concept of isometric transformation presents a straightforward generalization of the concept of a unitary transformation, and since it will be of fundamental significance in scattering theory, we shall define it now.

Definition 4.1. A linear transformation T of a Hilbert space $\mathscr{H}_1$ into another Hilbert space $\mathscr{H}_2$ (with inner products $\langle \cdot \mid \cdot \rangle_1$ and $\langle \cdot \mid \cdot \rangle_2$, respectively) which preserves the inner product, i.e.,

$$\langle Tf \mid Tg \rangle_2 = \langle f \mid g \rangle_1$$

for all $f, g \in \mathscr{H}_1$, is called an *isometric transformation.* If in addition the range $\mathscr{R}_T$ of such a transformation is the entire space $\mathscr{H}_2$, i.e., $\mathscr{R}_T = \mathscr{H}_2$, then the transformation is said to be *unitary*.

We have the following weaker condition of isometry.

Theorem 4.1. The linear transformation T of $\mathscr{H}_1$ into $\mathscr{H}_2$ is an isometric operator if

$$\| Tf \|_2 = \| f \|_1 \tag{4.1}$$

for all $f \in \mathscr{H}_1$.

Proof. To prove the above statement, first substitute $f + g$ for f in (4.1) to derive

$$\begin{aligned}\langle T(f+g) \mid T(f+g) \rangle_2 &= \| Tf \|_2^2 + \langle Tf \mid Tg \rangle_2 + \langle Tg \mid Tf \rangle_2 + \| Tg \|_2^2 \\ &= \langle f+g \mid f+g \rangle_1 \\ &= \| f \|_1^2 + \langle f \mid g \rangle_1 + \langle g \mid f \rangle_1 + \| g \|_1^2,\end{aligned}$$

then $f + ig$ to obtain

$$\begin{aligned}\langle T(f + ig) \mid T(f + ig)\rangle_2 &= \| Tf \|_2^2 + i\langle Tf \mid Tg\rangle_2 - i\langle Tg \mid Tf\rangle_2 + \| Tg \|_2^2 \\ &= \langle f + ig \mid f + ig\rangle_1 \\ &= \| f \|_1^2 + i\langle f \mid g\rangle_1 - i\langle g \mid f\rangle_1 + \| g \|_1^2.\end{aligned}$$

After multiplying the above equation by $-i$ and adding it to the one preceding it, we arrive at the result

$$\langle Tf \mid Tg\rangle_2 = \langle f \mid g\rangle_1$$

for all $f, g \in \mathscr{H}_1$. Q.E.D.

It is easy to see that an isometric *operator*, mapping $\mathscr{H}$ into $\mathscr{H}$, is necessarily unitary if $\mathscr{H}$ is finite dimensional (see Exercise 4.1). However, in the case of linear operators on infinite-dimensional Hilbert spaces these two concepts do not coincide. For instance, if $\{e_1, e_2, \ldots\}$ is an orthonormal basis in the Hilbert space $\mathscr{H}$, then the *shift* operator

$$A_l f = A_l\left(\sum_{k=1}^{\infty} \langle e_k \mid f\rangle\, e_k\right) = \sum_{k=1}^{\infty} \langle e_k \mid f\rangle\, e_{k+l}$$

is obviously isometric, but not unitary, for any fixed $l = 1, 2, \ldots$.

The relation that characterizes a unitary operator is $U^* = U^{-1}$, where U^{-1} is the inverse of U. The definition of the inverse T^{-1} of a linear transformation T is contained in the following statement.

Theorem 4.2. Let T be a linear transformation with domain $\mathscr{D}_T$ and range $\mathscr{R}_T$. The inverse mapping T^{-1} of $\mathscr{R}_T$ into $\mathscr{D}_T$ exists if and only if the zero vector $\mathbf{0}_1$ of $\mathscr{D}_T$ is the only vector mapped by T into the zero vector* $\mathbf{0}_2$ of $\mathscr{R}_T$. If T^{-1} exists, then T^{-1} is a linear transformation with domain $\mathscr{D}_{T^{-1}} = \mathscr{R}_T$ and range $\mathscr{R}_{T^{-1}} = \mathscr{D}_T$.

Proof. For the existence of the inverse transformation T^{-1} of the linear transformation

$$f \to f' = Tf, \qquad f \in \mathscr{D}_T, \qquad f' \in \mathscr{R}_T, \tag{4.2}$$

it is necessary and sufficient that $f \neq g$ (i.e., $f - g \neq \mathbf{0}_1$) implies $Tf \neq Tg$ [i.e., $T(f - g) \neq \mathbf{0}_2$]. The existence of the inverse of (4.2) is thus obviously equivalent to the requirement that $Tf = \mathbf{0}_2$ has only the solution $f = \mathbf{0}_1$.

* Naturally, $\mathbf{0}_2 = \mathbf{0}_1$ if $\mathscr{D}_A$ and $\mathscr{R}_A$ belong to the same vector space.

If the inverse transformation T^{-1} is defined on the linear set $\mathscr{R}_T$, it is obviously linear, as can be easily seen in the following way: if we have $f = T^{-1}f'$ and $g = T^{-1}g'$, $f', g' \in \mathscr{D}_{T^{-1}} = \mathscr{R}_T$, it follows that $f' = Tf$ and $g' = Tg$, and consequently for any scalars a, b,

$$af' + bg' = aTf + bTg = T(af + bg),$$

i.e.,

$$af + bg = T^{-1}(af' + bg'). \quad \text{Q.E.D.}$$

4.2. Unitary Operators and the Change of Orthonormal Basis

Theorem 4.3. A linear operator U, defined on the entire Hilbert space $\mathscr{H}$, is unitary if and only if

$$U^*U = UU^* = 1, \tag{4.3}$$

i.e., if and only if $U^{-1} = U^*$.

Proof. A unitary operator U is obviously bounded ($\| U \| = 1$) and consequently its adjoint U exists and is defined everywhere in $\mathscr{H}$. Since

$$\langle f \mid g \rangle = \langle Uf \mid Ug \rangle = \langle f \mid U^*Ug \rangle$$

is true for all $f, g \in \mathscr{H}$, we conclude that $U^*U = 1$. Moreover, from $\mathscr{R}_U = \mathscr{H}$ we deduce that in the relation

$$\langle Uf \mid g \rangle = \langle f \mid U^*g \rangle = \langle Uf \mid UU^*g \rangle$$

Uf assumes all the values in $\mathscr{H}$ when f takes on all the values in $\mathscr{H}$. This implies that $UU^*g = g$ for all $g \in \mathscr{H}$, i.e., $UU^* = 1$.

Conversely, if (4.3) is true for some linear operator U, we have $U^{-1} = U^*$ and therefore

$$\mathscr{R}_U = \mathscr{D}_{U^{-1}} = \mathscr{D}_{U^*} = \mathscr{H}.$$

Furthermore, (4.3) also implies that U is isometric:

$$\langle Uf \mid Ug \rangle = \langle f \mid U^*Ug \rangle = \langle f \mid g \rangle.$$

Thus, we have proved that U is unitary. Q.E.D.

One of the most important features of unitary operators is that they perform the transformation of one orthonormal basis into another.

Theorem 4.4. (a) If $\{e_1, e_2, \ldots\}$ is an orthonormal basis in the Hilbert space $\mathscr{H}$, then $\{Ue_1, Ue_2, \ldots\}$ is also an orthonormal basis.

(b) Conversely, if $\{e_1, e_2, \ldots\}$ and $\{e_1', e_2', \ldots\}$ are two orthonormal bases, there is a unique unitary operator U such that $e_k' = Ue_k$, $k = 1, 2, \ldots$.

Proof. (a) The set $\{Ue_1, Ue_2, \ldots\}$ is orthonormal, since due to the unitarity of U

$$\langle Ue_i \mid Ue_j \rangle = \langle U^* Ue_i \mid e_j \rangle = \langle e_i \mid e_j \rangle = \delta_{ij}.$$

To see that $\{Ue_1, Ue_2, \ldots\}$ is also complete, assume that for some $f \in \mathscr{H}$

$$\langle f \mid Ue_k \rangle = 0, \qquad k = 1, 2, \ldots.$$

Then we also have

$$\langle U^* f \mid e_k \rangle = 0, \qquad k = 1, 2, \ldots.$$

The above relations imply that $U^* f = \mathbf{0}$, due to the completeness of $\{e_1, e_2, \ldots\}$. Since $UU^* = 1$,

$$f = U(U^* f) = \mathbf{0}.$$

(b) Given $\{e_1, e_2, \ldots\}$ and $\{e_1', e_2', \ldots\}$, we can define an operator U, acting on all $f \in \mathscr{H}$ as follows:

$$Uf = U\left(\sum_{k=1}^{\infty} \langle e_k \mid f \rangle e_k\right) = \sum_{k=1}^{\infty} \langle e_k \mid f \rangle e_k'. \tag{4.4}$$

The series on the right-hand side converges in the norm since the sequence

$$h_n = \sum_{k=1}^{n} \langle e_k \mid f \rangle e_k', \qquad n = 1, 2, \ldots,$$

is a Cauchy sequence; this statement is a direct consequence of the relation

$$\| h_m - h_n \|^2 = \sum_{k=n+1}^{m} |\langle e_k \mid f \rangle|^2$$

and the convergence of the series

$$\sum_{k=1}^{\infty} |\langle e_k \mid f \rangle|^2 = \| f \|^2.$$

It is trivial to check that the operator U defined by (4.4) is linear. Furthermore, we have for any $f, g \in \mathscr{H}$,

$$\langle Uf \mid Ug\rangle = \lim_{n\to\infty} \left\langle \sum_{i=1}^{n} \langle e_i \mid f\rangle\, e_i' \,\middle|\, \sum_{j=1}^{n} \langle e_j \mid g\rangle\, e_j' \right\rangle \tag{4.5}$$

$$= \sum_{i,j=1}^{\infty} \langle e_i \mid f\rangle^* \,\delta_{ij} \langle e_j \mid g\rangle = \langle f \mid g\rangle.$$

Hence, the operator U is isometric. From its definition (4.4) it is obvious that the range $\mathscr{R}_U$ of U is the entire Hilbert space $\mathscr{H}$. Therefore, U is a unitary operator. Q.E.D.

4.3. The Fourier–Plancherel Transform

The Fourier transform—which is the best known of all integral transforms and is frequently used in quantum mechanics—can be looked upon as being essentially the restriction of a unitary operator to some linear submanifold of an $L^2(\mathbb{R}^n)$ space. In order to prove this unitarity property of the Fourier transform, we need a few auxiliary lemmas and theorems.

Lemma 4.1 (*The Riemann–Lebesgue lemma*). If $f(x)$ is a complex function which is Lebesgue integrable on the interval I, $I \subset \mathbb{R}^1$, then

$$\lim_{\lambda\to\infty} \int_I f(x) \sin \lambda x \, dx = \lim_{\lambda\to\infty} \int_I f(x) \cos \lambda x \, dx = 0. \tag{4.6}$$

Proof. We shall prove the lemma only for the sine integral, since an analogous proof applies to the cosine integral.

Assume that I is a finite interval, for instance that $I = [a, b]$. Take a function $g(x)$ which is differentiable on $[a, b]$, and for which (see Exercise 4.2)

$$\int_a^b |f(x) - g(x)| \, dx < \epsilon/2.$$

Integrating by parts we derive

$$\int_a^b g(x) \sin \lambda x \, dx = -\frac{g(b) \cos \lambda b - g(a) \cos \lambda a}{\lambda}$$

$$+ \frac{1}{\lambda} \int_a^b g'(x) \cos \lambda x \, dx \to 0 \qquad \text{for} \quad \lambda \to \infty.$$

Consequently, for sufficiently large values of λ,

$$\left| \int_a^b g(x) \sin \lambda x \, dx \right| < \epsilon/2,$$

and therefore for such λ,

$$\left| \int_a^b f(x) \sin \lambda x \, dx \right| \leqslant \left| \int_a^b g(x) \sin \lambda x \, dx \right| + \int_a^b |f(x) - g(x)| \, |\sin \lambda x| \, dx < \epsilon,$$

which proves the theorem.

The case of infinite I can be reduced to the above case by chosing a and b, $a < b$, so that

$$\int_{I-[a,b]} |f(x)| \, dx < \epsilon.$$

In that case

$$\left| \int_I f(x) \sin \lambda x \right| \leqslant \int_{I-[a,b]} |f(x)| \, dx + \left| \int_a^b f(x) \sin \lambda x \, dx \right|$$
$$\leqslant \left| \int_a^b f(x) \sin \lambda x \, dx \right| + \epsilon,$$

and the previous considerations apply. Q.E.D.

Lemma 4.2. If $f(x)$ is once continuously differentiable and Lebesgue integrable on the interval I in $\mathbb{R}^n$, and if $\mathbf{0} \in \mathbb{R}^n$ is an internal point of I, then

$$\lim_{\lambda \to \infty} \frac{1}{\pi^n} \int_I f(u + x) \frac{\sin \lambda x_1}{x_1} \cdots \frac{\sin \lambda x_n}{x_n} \, d^n x = f(u). \tag{4.7}$$

Proof. We shall give the proof for the case $n = 1$. The general case can be easily reduced to this case (see Exercise 4.3).

Since the origin of $\mathbb{R}^1$ is an interior point of I, we can choose $\alpha > 0$ so that $[-\alpha, +\alpha] \subset I$. Then

$$\int_I f(u + x) \frac{\sin \lambda x}{x} \, dx = \int_{-\alpha}^{+\alpha} f(u + x) \frac{\sin \lambda x}{x} \, dx + \int_{I-[-\alpha,+\alpha]} f(u + x) \frac{\sin \lambda x}{x} \, dx, \tag{4.8}$$

where by Lemma 4.1 the second integral approaches zero as $\lambda \to +\infty$.

The function

$$h(x) = \begin{cases} \dfrac{f(u + x) - f(u)}{x}, & x \neq 0 \\ f'(u), & x = 0 \end{cases}$$

is a continuous function of x because $f(x)$ is continuously differentiable. Therefore, after writing

$$\int_{-\alpha}^{+\alpha} f(u + x) \frac{\sin \lambda x}{x} \, dx = f(u) \int_{-\alpha}^{+\alpha} \frac{\sin \lambda x}{x} \, dx + \int_{-\alpha}^{+\alpha} \frac{f(u + x) - f(u)}{x} \sin \lambda x \, dx, \tag{4.9}$$

we can apply Lemma 4.1 to the second integral on the right-hand side of (4.9) to infer that this integral vanishes in the limit $\lambda \to \infty$. Since

$$\lim_{\lambda\to\infty} \int_{-\alpha}^{+\alpha} \frac{\sin \lambda x}{x}\, dx = \lim_{\lambda\to\infty} \int_{-\lambda\alpha}^{+\lambda\alpha} \frac{\sin t}{t}\, dt = \int_{-\infty}^{+\infty} \frac{\sin t}{t}\, dt = \pi,$$

we obtain (4.7) by combining (4.8) and (4.9) and then letting $\lambda \to \infty$.

Q.E.D.

Theorem 4.5. If $f(x) \in \mathscr{C}^1(\mathbb{R}^m)$ is Lebesgue integrable on $\mathbb{R}^m$, then its *Fourier transform**

$$\tilde{f}(p) = (2\pi\hbar)^{-m/2} \int_{\mathbb{R}^m} f(x)\, e^{-(i/\hbar)p\cdot x}\, d^m x \tag{4.10}$$

is bounded and continuous on $\mathbb{R}^m$, and if $D_\lambda = [-\lambda, +\lambda] \times \cdots \times [-\lambda, +\lambda]$,

$$f(x) = (2\pi\hbar)^{-m/2} \lim_{\lambda\to\infty} \int_{D_\lambda} \tilde{f}(p)\, e^{(i/\hbar)x\cdot p}\, d^m p. \tag{4.11}$$

If, in addition, $f(x)$ is of compact support in $\mathbb{R}^m$, then

$$\int_{\mathbb{R}^m} |f(x)|^2\, d^m x = \int_{\mathbb{R}^m} |\tilde{f}(p)|^2\, d^m p. \tag{4.12}$$

Proof. The integral on the right-hand side of (4.10) is uniformly convergent in p due to the fact that $f(x)$ is integrable and

$$|f(x)\, e^{-(i/\hbar)p\cdot x}| \leqslant |f(x)|.$$

Hence, $\tilde{f}(p)$ exists, it is continuous, and by Lemma 4.1 it is bounded.

In order to derive (4.11), first apply Fubini's theorem (Chapter II, Theorem 3.13) to invert the order of integration, and then Lemma 4.2 to obtain upon setting $x' = \hbar^{-1/2}x$, $p' = \hbar^{-1/2}p$, $\lambda' = \hbar^{-1/2}\lambda$,

$$\begin{aligned}
&(2\pi\hbar)^{-m/2} \lim_{\lambda\to\infty} \int_{D_\lambda} \tilde{f}(p)\, e^{(i/\hbar)x\cdot p}\, d^m p \\
&= \lim_{\lambda\to\infty} (2\pi)^{-m} \int_{-\lambda}^{\lambda} \cdots \int_{-\lambda}^{\lambda} d^m p' \int_{\mathbb{R}^m} f(y)\, e^{ip'\cdot(x'-y')}\, d^m y' \\
&= \lim_{\lambda\to\infty} \pi^{-m} \int_{\mathbb{R}^m} f(y) \frac{\sin \lambda(x_1' - y_1')}{x_1' - y_1'} \cdots \frac{\sin \lambda(x_m' - y_m')}{x_m' - y_m'}\, d^m y' \\
&= \lim_{\lambda'\to\infty} \pi^{-m} \int_{\mathbb{R}^m} f(x+u) \frac{\sin \lambda' u_1}{u_1} \cdots \frac{\sin \lambda' u_m}{u_m}\, d^m u = f(x).
\end{aligned}$$

* The definition found in mathematical literature corresponds to the choice $\hbar = 1$.

If $f(x) \in \mathscr{C}_b^1(\mathbb{R}^m)$ then $P(p)\tilde{f}(p)$ is bounded for any polynomial $P(p)$ so that $\tilde{f} \in L^1(\mathbb{R}^m)$, and we can apply Fubini's theorem to obtain

$$\int_{\mathbb{R}^m} |f(x)|^2\, d^m x = (2\pi\hbar)^{-m/2} \int_{\mathbb{R}^m} d^m x\, f^*(x) \int_{\mathbb{R}^m} d^m p\, \tilde{f}(p)\, e^{(i/\hbar)x\cdot p}$$

$$= (2\pi\hbar)^{-m/2} \int_{\mathbb{R}^m} d^m p\, \tilde{f}(p) \int_{\mathbb{R}^m} d^m x\, f^*(x)\, e^{(i/\hbar)p\cdot x}$$

$$= \int_{\mathbb{R}^m} |\tilde{f}(p)|^2\, d^m p. \quad \text{Q.E.D.}$$

Theorem 4.5 constitutes one of the most important results in the theory of the Fourier transform. We are interested, however, in the Fourier transform as an operator on a $L^2(\mathbb{R}^m)$ space. In that context the following result is of greatest interest to us.

Theorem 4.6. There is a unique linear and bounded transformation U_F of $L^2(\mathbb{R}^m)$ into itself, which maps every element $f \in \mathscr{C}_b^\infty(\mathbb{R}^m)$ into

$$\tilde{f} = U_F f, \qquad f \in \mathscr{C}_b^\infty(\mathbb{R}^m), \qquad \tilde{f} \in L^2(\mathbb{R}^m),$$

where $\tilde{f}(p)$ is related to $f(x)$ by the Fourier transform (4.10); U_F is a unitary operator, called the *Fourier–Plancherel transform on* $\mathbb{R}^m$.

Proof. The integral operator defined by (4.10) and having $\mathscr{C}_b^\infty(\mathbb{R}^m)$ as a domain of definition is obviously linear. Since $\mathscr{C}_b^\infty(\mathbb{R}^n)$ is dense in $L^2(\mathbb{R}^m)$ (Chapter II, Theorem 5.6), and since by the relation (4.12) in Theorem 4.5 the present operator is isometric on $\mathscr{C}_b^\infty(\mathbb{R}^m)$, it has according to Theorem 2.6 a unique bounded linear extension to $L^2(\mathbb{R}^m)$. This extension U_F is isometric (see Exercise 4.4).

In order to prove that U_F is unitary, i.e., that $\mathscr{R}_{U_F} = L^2(\mathbb{R}^m)$, it is sufficient to show that it maps a dense set onto another dense set. This last property follows from Theorem 4.5; namely, we easily see from (4.10) and (4.11) that the dense set consisting of all of functions $f(x) \in \mathscr{C}^\infty(\mathbb{R}^m)$ for which $x_1^{i_1} \cdots x_m^{i_m}\, \partial_{x_1}^{j_1} \cdots \partial_{x_m}^{j_m} f(x)$ is bounded on $\mathbb{R}^m$ for any $i_1, \ldots, i_m, j_1, \ldots, j_m = 0, 1, 2, \ldots$ is mapped onto itself by U_F. Q.E.D.

*4.4. Cayley Transforms of Symmetric Operators

An important category of isometric operators are the *Cayley transforms* of symmetric operators A,

$$V = (A - i)(A + i)^{-1}, \tag{4.13}$$

which play a key role in the theory of self-adjoint operators—as we shall see later in this chapter.

***Theorem 4.7.** *If A is a (closed) symmetric* operator in a Hilbert space $\mathscr{H}$, its Cayley transform $V = (A - i)(A + i)^{-1}$ exists as a linear operator in $\mathscr{H}$, and it is an isometric transformation of the closed subspace $\mathscr{R}_{A+i}$ onto the (closed) subspace $\mathscr{R}_{A-i}$.

Proof. The existence of the operators $(A \pm i)^{-1}$ follows from the relation

$$(4.14)\qquad \begin{aligned} \|(A \pm i)f\|^2 &= \langle Af \mid Af\rangle \mp i\langle f \mid Af\rangle \pm i\langle Af \mid f\rangle + \langle f \mid f\rangle \\ &= \|Af\|^2 + \|f\|^2, \qquad f \in \mathscr{D}_A, \end{aligned}$$

and Theorem 4.2, since according to the above relation $(A \pm i)f = \mathbf{0}$ implies $f = \mathbf{0}$.

We shall prove now that the ranges $\mathscr{R}_{A\pm i}$ of the two operators $A \pm i$ are closed linear subspaces of $\mathscr{H}$ if A is closed.

Since $\mathscr{R}_{A\pm i}$ are the ranges of linear operators, they are linear. To show that $\mathscr{R}_{A+i}$ and $\mathscr{R}_{A-i}$ are closed sets, we first derive from (4.14) that

$$\|f\| \leqslant \|(A \pm i)f\|, \qquad f \in \mathscr{D}_A.$$

Thus, if the sequence $f_1' = (A + i)f_1, f_2' = (A + i)f_2, \ldots$ converges to some $f' \in \mathscr{H}$, the sequence $f_1, f_2, \ldots$ also converges to some $f \in \mathscr{H}$. Since $A + i$ is obviously closed when A is closed (see Definition 2.4), we must have $f \in \mathscr{D}_A$ and $f' \in \mathscr{R}_{A+i}$.

A similar argument establishes that $\mathscr{R}_{A-i}$ is closed too.

The operator $(A + i)^{-1}$ is defined on $\mathscr{R}_{A+i}$ and its range is $\mathscr{D}_A$, while $A - i$ is defined on $\mathscr{D}_A$ and its range is $\mathscr{R}_{A-i}$. Consequently, $V = (A - i)(A + i)^{-1}$ maps $\mathscr{R}_{A+i}$ onto $\mathscr{R}_{A-i}$. To establish that V is isometric, note that (4.14) yields

$$\|(A - i)g\| = \|(A + i)g\|, \qquad g \in \mathscr{D}_A.$$

If we insert above $g = (A + i)^{-1}f, f \in \mathscr{R}_{A+i}$, we get

$$\|(A - i)(A + i)^{-1}f\| = \|f\|$$

for all $f \in \mathscr{R}_{A+i}$, i.e., V is isometric. Q.E.D.

In the above Theorem the term "closed" is optional, but it should be noted that each densely defined symmetric operator A has at least one closed extension, namely A^{**}; we see immediately that $A^{**} \supseteq A$ by looking at the relation

$$\langle A^*f \mid g\rangle = \langle f \mid Ag\rangle, \qquad f \in \mathscr{D}_{A^*}, \qquad g \in \mathscr{D}_A,$$

and recalling that since $\mathscr{D}_{A^*} \supset \mathscr{D}_A$ and $\mathscr{D}_A$ is dense in $\mathscr{H}$, $\mathscr{D}_{A^*}$ itself is dense in $\mathscr{H}$, and therefore A^{**} exists and is closed (see Theorem 2.8).

Moreover, as we shall see later, $A^{**} \subseteq A^*$, and since obviously $A^* \subseteq A^{***}$, it follows that A^{**} is symmetric.

***Theorem 4.8.** If the Cayley transform V of the linear operator A exists, then $\mathscr{D}_A = \mathscr{R}_{1-V}$ and

$$A = i(1 + V)(1 - V)^{-1}. \tag{4.15}$$

Proof. According to Definition (4.13) of V, $\mathscr{D}_V$ consists of all vectors g for which there is an $f \in \mathscr{D}_A$ such that

$$g = (A + i)f. \tag{4.16}$$

When V is applied to the above g, we get

$$Vg = (A - i)f. \tag{4.17}$$

By adding and then subtracting (4.16) and (4.17) we obtain

$$(1 + V)g = 2Af, \qquad (1 - V)g = 2if. \tag{4.18}$$

The second of the above relations shows that $(1 - V)^{-1}$ exists; namely, if $(1 - V)g = \mathbf{0}$, the aforementioned relation implies that $f = \mathbf{0}$, which due to (4.16) implies in its turn that $g = \mathbf{0}$, thus establishing, in accordance with Theorem 4.2, the existence of $(1 - V)^{-1}$. Hence, we can write

$$g = 2i(1 - V)^{-1}f,$$

which in conjunction with the first of the relations (4.18) yields

$$Af = i(1 + V)(1 - V)^{-1}f, \qquad f \in \mathscr{D}_A.$$

Moreover, from the second of the relations (4.18) we see that $\mathscr{D}_{(1-V)^{-1}} = \mathscr{D}_A$, so that the domains of definition of A and $i(1 + V)(1 - V)^{-1}$ are identical, i.e., (4.15) is true. Q.E.D.

The following theorem is of great practical importance in establishing the self-adjointness of a symmetric operator.

***Theorem 4.9.** The symmetric and densely defined operator A is self-adjoint if and only if its Cayley transform is a unitary operator.

Proof. Assume that $A = A^*$. We shall prove that the domain of definition $\mathscr{R}_{A+i}$ of V is dense in $\mathscr{H}$.

Assume that $h \perp \mathscr{R}_{A+i}$, i.e., that $\langle h \mid (A + i)f\rangle = 0$ for all $f \in \mathscr{D}_A$, or, equivalently, that $\langle h \mid (A + i)f\rangle = \langle \mathbf{0} \mid f\rangle$ for all $f \in \mathscr{D}_A$. According to Theorem 2.5 we have $h \in \mathscr{D}_{(A+i)^*}$ and

$$(A + i)^* h = (A^* - i)h = (A - i)h = \mathbf{0}.$$

It follows from (4.14) that $(A \pm i)h = \mathbf{0}$ only if $h = \mathbf{0}$. Thus, R_{A+i} is dense in $\mathscr{H}$, and since $\mathscr{R}_{A+i}$ is closed (see Theorem 4.7), we must have $\mathscr{D}_V = \mathscr{R}_{A+i} = \mathscr{H}$.

It can be proved in a completely analogous manner that $\mathscr{R}_V = \mathscr{R}_{A-i} = \mathscr{H}$. The identities $\mathscr{R}_{A\pm i} = \mathscr{H}$ establish, in conjunction with Theorem 4.7, that V is unitary when $A = A^*$.

Assume now that V is unitary and A^* exists. We shall prove that in this case $A = A^*$, where A^* exists since $\mathscr{D}_{A^*}$ is dense in $\mathscr{H}$.

Take $g \in \mathscr{D}_{A^*}$ and set $g^* = A^*g$. Then

$$\langle g \mid Af \rangle = \langle g^* \mid f \rangle$$

for all $f \in \mathscr{D}_A$. According to (4.15), for each $f \in \mathscr{D}_A$ there is an $h \in \mathscr{D}_V = \mathscr{H}$ such that $f = (1 - V)h$ and $Af = i(1 + V)h$. Hence

$$i\langle g \mid (1 + V)h \rangle = \langle g^* \mid (1 - V)h \rangle$$

for all $h \in \mathscr{H}$, or

$$i\langle g \mid h \rangle + i\langle g \mid Vh \rangle = \langle g^* \mid h \rangle - \langle g^* \mid Vh \rangle.$$

Due to the unitarity of V,

$$\langle g \mid h \rangle = \langle Vg \mid Vh \rangle \qquad \text{and} \qquad \langle g^* \mid h \rangle = \langle Vg^* \mid Vh \rangle$$

and the preceding relation yields

$$\langle -ig - iVg - Vg^* + g^* \mid Vh \rangle = 0$$

for all $h \in \mathscr{H}$. Since V is unitary, Vh assumes all the values in $\mathscr{H}$ when h takes on all the values in $\mathscr{H}$, and therefore

$$i(1 + V)g - (1 - V)g^* = \mathbf{0}.$$

Using the above relation we obtain

$$(1 - V)\frac{g - ig^*}{2} = g,$$

which shows that $g \in \mathscr{R}_{(1-V)} = \mathscr{D}_A$; the same relation yields

$$i(1 + V)\frac{g - ig^*}{2} = g^*,$$

which shows that $g^* = Ag$. Thus, $A^* \subseteq A$, and since A is symmetric, $A^* = A$. Q.E.D.

We have already seen in Theorem 4.7 that $\mathscr{R}_{A+i}$ and $\mathscr{R}_{A-i}$ are closed linear subspaces of $\mathscr{H}$ when A is a closed symmetric operator. Let us denote by m and n the dimensions of the ortogonal complements of $\mathscr{R}_{A-i}$ and $\mathscr{R}_{A+i}$, respectively,

$$m = \dim \mathscr{R}^{\perp}_{A-i}, \qquad n = \dim \mathscr{R}^{\perp}_{A+i}.$$

Naturally, m and n can assume any nonnegative integer values, including $+\infty$ when the corresponding space is infinite dimensional. Moreover, if $\mathscr{H}$ is not separable, we have to distinguish between infinite dimensions of separable and nonseparable subspaces.

The two numbers m and n are called the *deficiency indices* of A, and the two subspaces $\mathscr{R}^{\perp}_{A-i}$ and $\mathscr{R}^{\perp}_{A+i}$ are the *deficiency subspaces* of A. According to Theorem 4.9, A is self-adjoint if and only if $m = n = 0$.

If the deficiency indices of A are equal but not zero, then we can easily extend the Cayley transform V of A to a unitary operator V_1 on $\mathscr{H}$. This can be achieved by choosing any two orthonormal bases $\{e_1', e_2',...\}$ and $\{e_1'', e_2'',...\}$ in $\mathscr{R}^{\perp}_{A-i}$ and $\mathscr{R}^{\perp}_{A+i}$, and extending V to $\mathscr{R}^{\perp}_{A+i}$ by writing $V_1 e_k'' = e_k'$. Since this extended V_1 is unitary, the operator

$$A_1 = i(1 + V_1)(1 - V_1)^{-1}$$

is self-adjoint by Theorem 4.9, and it obviously provides an extension of A. Hence, if $m = n > 0$, the closed symmetric operator A will have many self-adjoint extensions. If, however, $m \neq n$, then A has no self-adjoint extensions. As a matter of fact, if A_0 were a self-adjoint extension of A, then the Cayley transform V_0 of A_0 would be unitary, and it would be an extension of the Cayley transform V of A. Since obviously V maps $\mathscr{R}^{\perp}_{A-i}$ isometrically onto $\mathscr{R}^{\perp}_{A+i}$, this would imply that $m = n$, which contradicts the assumption. In summary, we can state the following theorem (see also Lemma 7.1 in Chapter IV).

Theorem 4.10. A closed symmetric operator A has self-adjoint extensions if and only if its deficiency indices $m = \dim \mathscr{R}^{\perp}_{A-i}$ and $n = \dim \mathscr{R}^{\perp}_{A+i}$ are equal; moreover, $A = A^*$ if and only if $m = n = 0$.

Theorems 4.7–4.10, and in particular Theorem 4.10, indicate that the self-adjointness or essential self-adjointness of a Hermitian operator in an infinite-dimensional Hilbert space is by no means a "natural" feature of such an operator. On the contrary, there are many more symmetric operators which have no self-adjoint extensions than symmetric operators which possess self-adjoint extensions; namely, to any isometry which is not the restriction of a unitary operator there corresponds a closed symmetric operator without self-adjoint extensions. The reader should keep this in mind when trying to decide within the context of some quantum mechanical theory whether or not some obviously Hermitian differential operator represents a self-adjoint operator.

4.5. Self-Adjointness of Position and Momentum Operators in Wave Mechanics

As an easy application of the results obtained in this section, we can prove the self-adjointness of the j-coordinate position operator Q_j and the canonically conjugate momentum operator P_j.

We define Q_j as the operator with domain $\mathscr{D}_{Q_j}$ consisting of elements f of $L^2(\mathbb{R}^m)$, representable by functions $f(x_1, \ldots, x_m)$ which are such that $x_j f(x_1, \ldots, x_m)$ are Lebesgue square integrable, and acting on $f \in \mathscr{D}_{Q_i}$ in the following way:

$$(Q_j f)(x_1, \ldots, x_m) = x_j f(x_1, \ldots, x_m), \qquad f \in \mathscr{D}_{Q_j}. \tag{4.19}$$

The above operator is obviously symmetric (see Exercise 4.5).

To prove that $Q_j^* = Q_j$, we will show that the Cayley transform $(Q_j - i)(Q_j + i)^{-1}$ is unitary (see Theorem 4.9).

The ranges of $Q_j \pm i$ are $L^2(\mathbb{R}^m)$; e.g., if $g \in L^2(\mathbb{R}^m)$, we can write

$$g(x_1, \ldots, x_m) = (x_j + i)\, h(x_1, \ldots, x_m),$$
$$h(x_1, \ldots, x_m) = \frac{g(x_1, \ldots, x_m)}{x_j + i},$$

where obviously $h \in \mathscr{D}_{Q_j}$. Moreover, the operator $(Q_j - i)(Q_j + i)^{-1}$ is isometric,

$$\|(Q_j - i)(Q_j + i)^{-1} f\|^2 = \int_{\mathbb{R}^m} \left| \frac{x_j - i}{x_j + i} \right|^2 |f(x_1, \ldots, x_m)|^2 \, d^m x$$
$$= \int_{\mathbb{R}^m} |f(x_1, \ldots, x_m)|^2 \, d^m x = \|f\|^2$$

and since its domain and range are (see Theorem 4.7) $\mathscr{R}_{Q_j+i}$ and $\mathscr{R}_{Q_j-i}$, respectively, this operator is unitary. Hence, Q_j is self-adjoint.

We define P_j by

$$P_j = U_F^{-1} Q_j U_F, \qquad \mathscr{D}_{P_j} = U_F^{-1} \mathscr{D}_{Q_j}, \tag{4.20}$$

where U_F is the Fourier–Plancherel transform introduced in Theorem 4.6; in (4.20) we used a very useful notation, according to which if T is any transformation mapping the set $S_1 \subset \mathscr{D}_T$ onto S_2 we write

$$S_2 = TS_1 = \{Tf : f \in S_1\}. \tag{4.21}$$

The self-adjointness of P_j follows from the unitarity of U_F and the following lemma.

Lemma 4.3. If A is a self-adjoint operator in $\mathscr{H}$, and U is a unitary operator on $\mathscr{H}$, then the operator

$$B = UAU^{-1}, \qquad \mathscr{D}_B = U\mathscr{D}_A$$

is self-adjoint.

Proof. Since obviously $\bar{\mathscr{D}}_B = \mathscr{H}$ because $\bar{\mathscr{D}}_A = \mathscr{H}$ (see Exercise 4.6), it follows that B^* exists. We must have $B^* \supseteq UA^*U^{-1} = B$ due to the following relation:

$$\begin{aligned} \langle f \mid Bg\rangle &= \langle f \mid UA^*U^{-1}g\rangle = \langle U^{-1}f \mid A^*U^{-1}g\rangle \\ &= \langle AU^{-1}f \mid U^{-1}g\rangle = \langle UAU^{-1}f \mid g\rangle = \langle Bf \mid g\rangle, \qquad f, g \in U\mathscr{D}_A. \end{aligned}$$

By inverting the roles of A and B, and noting that U^{-1} is unitary, we get $A^* \supseteq U^{-1}B^*U$, which implies that $B^* \subseteq UA^*U^{-1} = UAU^{-1} = B$. Hence, $B = B^*$. Q.E.D.

As a result of the above considerations we can state Theorem 4.11.

Theorem 4.11. The wave mechanical canonical position and momentum operators Q_j and P_j, defined by (4.19) and (4.20), are self-adjoint.

In order to see that (4.20) is nothing but a precise definition of the symbol $-i\hbar(\partial/\partial x_j)$, note that for f belonging to the set $U_F^{-1}\mathscr{C}_b^1(\mathbb{R}^m)$ [which according to Exercise 4.6 is dense in $L^2(\mathbb{R}^m)$ since $\mathscr{C}_b^1(\mathbb{R}^m)$ is dense] we have

$$P_j f = U_F^{-1} Q_j \tilde{f},$$

where

$$(Q_j\tilde{f})(p_1, \ldots, p_m) = p_j\tilde{f}(p_1, \ldots, p_m) = g(p_1, \ldots, p_m).$$

Hence

$$\begin{aligned} (P_j f)(x_1, \ldots, x_m) &= (U_F^{-1}g)(x_1, \ldots, x_m) \\ &= (2\pi\hbar)^{-m/2} \int_{\mathbb{R}^m} p_j\tilde{f}(p_1, \ldots, p_m)\, e^{(i/\hbar)p\cdot x}\, d^m p \\ &= -i(2\pi\hbar)^{-m/2} \frac{\partial}{\partial x_j} \int_{\mathbb{R}^m} \tilde{f}(p_1, \ldots, p_m)\, e^{(i/\hbar)p\cdot x}\, d^m p \\ &= -i\hbar \frac{\partial f(x_1, \ldots, x_m)}{\partial x_j}, \end{aligned}$$

where the interchanging of the order of differentiation and integration in carrying out the third step in the above derivation is easily deducible from standard theorems of the calculus, when the fact that $\tilde{f} \in \mathscr{C}_b^1(\mathbb{R}^m)$ is taken into account.

EXERCISES

4.1. Prove that an isometric operator V defined on a finite-dimensional Hilbert space $\mathscr{H}$ is necessarily unitary.

4.2. Prove that if $f(x)$ is Lebesgue integrable on $[a, b]$, there is for any given $\epsilon > 0$ an infinitely many times differentiable function $g(x) \in \mathscr{C}^\infty([a, b])$ such that

$$\int_a^b |f(x) - g(x)| \, dx < \epsilon.$$

4.3. Prove in detail Lemma 4.2 for any integer n.

4.4. Prove that if A and B are bounded linear operators satisfying $A \subseteq B$, and if $\mathscr{D}_A$ is dense in $\mathscr{D}_B$, then B is isometric when A is isometric.

4.5. Show in detail that the operator Q_j defined in the subsection 4.5 [see (4.19)] is linear and Hermitian, and that $Q_j^* \supseteq Q_j$.

4.6. Show that under a unitary transformation the image of a dense set is also a dense set.

5. Spectral Measures

5.1. THE POINT SPECTRUM OF A SELF-ADJOINT OPERATOR

We encountered the eigenvalue problem when determining the bound-energy spectrum of a quantum mechanical system from the wave mechanics point of view. In this section we shall formulate the eigenvalue problem for symmetric operators in general.

Definition 5.1. A complex number λ is said to be an *eigenvalue* (*characteristic value*) of the linear operator A, with domain of definition $\mathscr{D}_A$, if there is at least one nonzero vector $f \in \mathscr{D}_A$ satisfying the eigenvalue equation

$$Af = \lambda f. \tag{5.1}$$

Each vector $f \in \mathscr{D}_A$ satisfying the above equation is called an *eigenvector* (or *characteristic vector*) of A with eigenvalue λ. All the eigenvectors with the same eigenvalue λ form a linear subspace M_λ, which is usually referred to as the *characteristic subspace* corresponding to the eigenvalue λ.

If M_λ is infinite dimensional, then M_λ is not necessarily closed. However, it is easy to see that the characteristic subspaces of closed linear operators are always closed.

Theorem 5.1. If the linear operator A is symmetric, all its eigenvalues are real and the characteristic subspaces corresponding to different eigenvalues are mutually orthogonal.

Specialized versions of Theorem 5.1 were proved in Chapter I (Theorem 5.1) and Chapter II (Theorem 5.3). The proof in the general case proceeds along the same lines.

To derive that an eigenvalue λ is real if A is symmetric, choose any nonzero eigenvector f and use its symmetry property to write

$$\lambda\langle f \mid f\rangle = \langle f \mid \lambda f\rangle = \langle f \mid Af\rangle = \langle Af \mid f\rangle = \langle \lambda f \mid f\rangle = \lambda^*\langle f \mid f\rangle,$$

thus showing that $\lambda = \lambda^*$.

The orthogonality of two characteristic subspaces M_{λ_1} and M_{λ_2}, corresponding to two different eigenvalues λ_1 and λ_2, follows from the orthogonality of any two respective eigenvectors $f_1 \in \mathsf{M}_{\lambda_1}$ and $f_2 \in \mathsf{M}_{\lambda_2}$:

$$\lambda_1\langle f_1 \mid f_2\rangle = \langle \lambda_1 f_1 \mid f_2\rangle = \langle Af_1 \mid f_2\rangle = \langle f_1 \mid Af_2\rangle = \langle f_1 \mid \lambda_2 f_2\rangle = \lambda_2\langle f_1 \mid f_2\rangle.$$

The set $\mathsf{S}_{\mathrm{p}}^A$ of all eigenvalues of an operator A is called its *point spectrum* (see also Definition 3.3 in Chapter V).

5.2. Spectral Resolution of Self-Adjoint Operators with Pure Point Spectrum

Consider a closed symmetric operator A for which the closed linear space spanned by its eigenvectors is the entire Hilbert space $\mathscr{H}$ (an operator with a so-called *pure point spectrum*), i.e., for which, by Definition 3.1,

$$\mathscr{H} = \bigoplus_{\lambda \in \mathsf{S}_{\mathrm{p}}^A} \mathsf{M}_\lambda,$$

where $\mathsf{S}_{\mathrm{p}}^A$ denotes the point spectrum of A. We assume that $\mathscr{H}$ is separable, so that $\mathsf{S}_{\mathrm{p}}^A$ is necessarily at most countable, and hence it can be written in the form

$$\mathsf{S}_{\mathrm{p}}^A = \{\lambda_1, \lambda_2, \ldots\}.$$

By using Theorem 3.9 we get

$$\sum_{k=1}^{\infty} E_{\mathsf{M}_{\lambda_k}} f = E_{\mathscr{H}} f = f.$$

If we have for s me f

$$\sum_{k=1}^{\infty} |\lambda_k|^2 \| E_{\mathsf{M}_{\lambda_k}} f \|^2 < +\infty, \tag{5.2}$$

then the sequence $f_1, f_2, \ldots,$

$$f_n = \sum_{k=1}^{n} E_{\mathsf{M}_{\lambda_k}} f \tag{5.3}$$

converges to f, while $Af_1, Af_2, \ldots,$

$$Af_n = \sum_{k=1}^{n} A(E_{\mathsf{M}_{\lambda_k}} f) = \sum_{k=1}^{n} \lambda_k (E_{\mathsf{M}_{\lambda_k}} f) \tag{5.4}$$

is a Cauchy sequence, due to (5.2). Since $\mathscr{H}$ is complete, this sequence has a limit g. Therefore, A being closed, we must have that $f \in \mathscr{D}_A$ and

$$Af = g = \sum_{k=1}^{\infty} \lambda_k (E_{\mathsf{M}_{\lambda_k}} f).$$

Vice versa, if $f \in \mathscr{D}_A$, (5.2) must be satisfied. To see that, note that the restriction of A to $\mathsf{M}_{\lambda_1} \oplus \cdots \oplus \mathsf{M}_{\lambda_n}$ is a bounded operator and therefore, A being closed, it must be defined everywhere on $\mathsf{M}_{\lambda_1} \oplus \cdots \oplus \mathsf{M}_{\lambda_n}$. Hence, the vector f_n, defined in (5.3), belongs to $\mathscr{D}_A$. Moreover, A obviously maps $\mathsf{M}_{\lambda_1} \oplus \cdots \oplus \mathsf{M}_{\lambda_n}$ into itself, and therefore

$$\langle Af_n \mid A(f - f_n) \rangle = \langle A^2 f_n \mid f - f_n \rangle = 0.$$

By using this relation we immediately get

$$\| Af \|^2 = \| A(f - f_n) \|^2 + \| Af_n \|^2 \geqslant \| Af_n \|^2,$$

which shows, in conjunction with (5.4), that (5.2) must be satisfied if $f \in \mathscr{D}_A$.

The above arguments can be applied to A^*, $A^* \supseteq A$, to derive that $\mathscr{D}_{A^*}$ consists of all vectors f satisfying (5.2), so that $\mathscr{D}_{A^*} = \mathscr{D}_A$. Hence, we have proved Theorem 5.2.

Theorem 5.2. If the closed linear manifold spanned by all the eigenvectors of a closed symmetric operator A acting on a separable Hilbert space $\mathscr{H}$ is identical with $\mathscr{H}$, then A is self-adjoint, and $\mathscr{D}_A$ consists of a the vectors f satisfying (5.2).

In the considered case, the complete solution of the eigenvalue problem is accomplished when the characteristic subspaces M_{λ_1}, M_{λ_2},... are found, or equivalently, when the projectors $E_{\mathsf{M}_{\lambda_1}}$, $E_{\mathsf{M}_{\lambda_2}}$,... have been specified. However, this formulation of the eigenvalue problem is not sufficiently general, because it does not cover the cases of self-adjoint operators whose eigenvectors do not span $\mathscr{H}$ (such as the general case of the Schroedinger operator), or do not even have any eigenvalues at all corresponding to Hilbert space eigenvectors (such as the position and momentum operators in quantum mechanics). Fortunately, one can solve the most general case of self-adjoint operators after rephrasing the above offered solution of the eigenvalue problem for operators A with pure point spectrum in the following way: the knowledge of the projectors $E_{\mathsf{M}_{\lambda_1}}$, $E_{\mathsf{M}_{\lambda_2}}$,... on the characteristic subspaces M_{λ_1}, M_{λ_2},... of A is equivalent to the knowledge of the projector-valued set function

$$E(B) = \sum_{\lambda \in B \cap S_p^A} E_{\mathsf{M}_\lambda}, \qquad B \in \mathscr{B}^1, \tag{5.5}$$

for all Borel sets B on the real line. The set function $E(B)$, $B \in \mathscr{B}^1$, whose assumed values are projectors, has certain properties which are common to a class of projector-valued set functions which are known as spectral measures. Moreover, these properties, which we shall formulate and study later in this section, make $E(B)$ unique in the sense of $E(B)$ being the only spectral measure that can be associated with A. When it is thus generalized, the eigenvalue problem becomes the problem of determining the spectral measure of A, and as such it can be applied to any self-adjoint operator—as we shall see in §6.

5.3. Weak, Strong, and Uniform Operator Limits

The above approach to the problem necessitates a generalization of the concept of measure which would also apply to operator-valued set functions. A careful analysis of the concept of limit of sequences of operators is essential for that. As seen in the following definition, we can distinguish three main concepts of limits for operator-valued functions.

Definition 5.2. Let $A(t)$ denote an operator-valued function on some subset S of $\mathbb{R}^1$, assigning to each $t \in S$ a bounded operator $A(t)$ acting on a Hilbert space $\mathscr{H}$. We can define A to be a limit of $A(t)$ when t tends to t_0 (where t_0 is a real number or $\pm\infty$) in the following three ways:

(a) If for all $f, g \in \mathscr{H}$,

$$\langle f \mid Ag \rangle = \lim_{t \to t_0} \langle f \mid A(t)\, g \rangle, \tag{5.6}$$

A is the *weak limit* of $A(t)$ when $t \to t_0$, and we write

$$A = \operatorname*{w-lim}_{t \to t_0} A(t) \qquad \text{or} \qquad A(t) \xrightarrow[t \to t_0]{} A.$$

(b) If for all $f \in \mathscr{H}$

$$\lim_{t \to t_0} \|(A(t) - A)f\| = 0, \tag{5.7}$$

A is the *strong limit* of $A(t)$ for $t \to t_0$:

$$A = \operatorname*{s-lim}_{t \to t_0} A(t) \qquad \text{or} \qquad A(t) \xrightarrow[t \to t_0]{} A.$$

(c) Finally, if $\| A - A(t)\| \to 0$ when $t \to t_0$, we call A the *uniform limit* of $A(t)$ and write

$$A = \operatorname*{u-lim}_{t \to t_0} A(t) \qquad \text{or} \qquad A(t) \underset{t \to t_0}{\Longrightarrow} A.$$

These three definitions of limits are related in the sense that a uniform limit is also a strong limit, and a strong limit is also a weak limit, though the converses of these statements are by no means true. Thus, from the inequality

$$\|(A - A(t))f\| \leqslant \| A - A(t)\| \, \|f\|$$

we derive that when $A = \text{u-lim}_{t \to t_0} A(t)$, then (5.7) is also true. By means of the Schwarz–Cauchy inequality we get

$$|\langle f \mid (A - A(t))\, g\rangle| \leqslant \|f\| \, \|(A - A(t))\, g\|,$$

which shows that (5.7) implies (5.6).

As an example of strong convergence which does not imply uniform convergence, consider a sequence $E_1, E_2, \ldots$ of mutually orthogonal nonzero projectors. According to Theorem 3.9, there is such a projector E that

$$E = \operatorname*{s-lim}_{n \to \infty} \sum_{k=1}^{n} E_k .$$

However, the sequence $E^{(1)}, E^{(2)}, \ldots,$

$$E^{(n)} = \sum_{k=1}^{n} E_k, \qquad n = 1, 2, \ldots,$$

is not uniformly convergent, since it does not satisfy the necessary condition (see Exercise 5.3) of uniform convergence:

$$\| E^{(n+1)} - E^{(n)}\| \to 0 \qquad \text{as} \quad n \to \infty;$$

in fact, since $E_{n+1} \neq 0$, we have

$$\| E^{(n+1)} - E^{(n)} \| = \| E_{n+1} \| = 1.$$

Definition 5.2 suggests that there are two different kinds of convergence of sequences of vectors in a Euclidean space.

Definition 5.3. If $f(t)$ is a vector-valued function assigning a vector $f(t)$ from the Euclidean space $\mathscr{E}$ to each value t from the set S, $S \subset \mathbb{R}^1$, we call f a *weak limit* of $f(t)$ for $t \to t_0$,

$$f = \operatorname*{w-lim}_{t \to t_0} f(t) \qquad \text{or} \qquad f(t) \xrightarrow[t \to t_0]{} f,$$

if $\langle g \mid f \rangle = \lim_{t \to t_0} \langle g \mid f(t) \rangle$ for all $g \in \mathscr{E}$; we say that f is a *strong limit* of $f(t)$ for $t \to t_0$,

$$f = \operatorname*{s-lim}_{t \to t_0} f(t) \qquad \text{or} \qquad f(t) \xrightarrow[t \to t_0]{} f,$$

if $\lim_{t \to t_0} \| f - f(t) \| = 0$.

From the inequality

$$|\langle g \mid f - f(t) \rangle| \leqslant \| g \| \, \| f - f(t) \|,$$

we conclude that if f is a strong limit of $f(t)$, then it is also a weak limit. The converse is by no means generally true.

It has to be emphasised that we have been considering until now only limits in the norm for sequences of vectors. The concept of the limit in the norm is obviously identical with the concept of strong limit.

5.4. Spectral Measures and Complex Measures

We are now ready to define the concept of a spectral measure. The reader should review Definition 2.1 in Chapter II of an ordinary measure, and note the formal analogies.

Definition 5.4. A *spectral measure* $E(S)$ on the measurable space $(\mathscr{X}, \mathscr{A})$ is a projector-valued function which assigns a projector $E(S)$ to each element S of the Boolean σ algebra $\mathscr{A}$ in such a manner that

(1) $E(\mathscr{X}) = 1$;

(2) $E\left(\bigcup_{k=1}^{\infty} S_k\right) = \sum_{k=1}^{\infty} E(S_k)$, if $S_1, S_2, \ldots,$ is any sequence of disjoint sets from $\mathscr{A}$.

In the last relation above, we define

$$\sum_{k=1}^{\infty} E(S_k) = \operatorname*{s-lim}_{n \to \infty} \sum_{k=1}^{n} E(S_k).$$

We agree, in general, that given an infinite sequence $E_1, E_2, \ldots$ of projectors, the sum of their infinite series shall always mean, if not otherwise stated, the *strong* limit of the partial sums:

$$\sum_{k=1}^{\infty} E_k = \text{s-}\lim_{n\to\infty} \sum_{k=1}^{n} E_k .$$

If we choose in the relation

$$E\left(\bigcup_{k=1}^{\infty} S_k\right) = \sum_{k=1}^{\infty} E(S_k), \qquad S_i \cap S_j = \varnothing \quad \text{for} \quad i \neq j, \tag{5.8}$$

$S_2 = S_3 = \cdots = \varnothing$ we get $E(\varnothing) = 0$. By taking above instead just $S_3 = S_4 = \cdots = \varnothing$, we obtain $E(S_1 \cup S_2) = E(S_1) + E(S_2)$ if $S_1 \cap S_2 = \varnothing$, so that $E(S_1)E(S_2) = 0$ by Theorem 3.7. More generally, setting $B = R \cap S$, $S_1 = R - B$, $S_2 = S - B$ for $R, S \in \mathscr{A}$, we get $E(R)E(S) = [E(S_1) + E(B)][E(S_2) + E(B)] = E^2(B) = E(B)$ since $S_1 \cap S_2 = S_1 \cap B = S_2 \cap B = \varnothing$. Hence we can state the following theorem.

Theorem 5.3. If $E(S)$ is a spectral measure on $(\mathscr{X}, \mathscr{A})$, then $E(\varnothing) = 0$ and $E(R)E(S) = E(R \cap S)$ whenever $R, S \in \mathscr{A}$.

The concept of a spectral measure can be related to that of a measure most conveniently by introducing the concept of a complex measure. For our purposes the following definition of a complex measure is most adequate, though it is not the customary one.

Definition 5.5. A *complex measure* $\mu(S)$ on the measurable space $(\mathscr{X}, \mathscr{A})$ is a finite linear combination, with complex coefficients $c_1, \ldots, c_\nu$

$$\mu(S) = \sum_{\alpha=1}^{\nu} c_\alpha \mu_\alpha(S)$$

of a given number ν of *finite* measures $\mu_1(S), \ldots, \mu_\nu(S)$ on $(\mathscr{X}, \mathscr{A})$. A measurable function $f(\xi)$, $\xi \in \mathscr{X}$, is integrable on $R \in \mathscr{A}$ with respect to μ if and only if it is integrable on R with respect to each of the measures $\mu_1, \ldots, \mu_\nu$; in that case, by definition,

$$\int_R f(\xi)\, d\mu(\xi) = \sum_{\alpha=1}^{\nu} c_\alpha \int_R f(\xi)\, d\mu_\alpha(\xi). \tag{5.9}$$

With the above definition of integration most of the results of Chapter II, §3 on integration can be easily generalized directly to complex measures. Moreover, it is easy to see that the above integral (5.9) is independent of the way the complex measure μ is written as a sum of measures. That is, if

$$\mu(S) = \sum_{\alpha=1}^{\nu} c_\alpha \mu_\alpha(S) = \sum_{\beta=1}^{\nu'} c_\beta' \mu_\beta'(S), \tag{5.10}$$

any function $f(\xi)$ integrable on R with respect to $\mu_1, \ldots, \mu_\nu$ is also integrable with respect to $\mu_1', \ldots, \mu_{\nu'}'$ and

$$\sum_{\alpha=1}^{\nu} c_\alpha \int_R f(\xi)\, d\mu_\alpha(\xi) = \sum_{\beta=1}^{\nu'} c_\beta' \int_R f(\xi)\, d\mu_\beta'(\xi). \tag{5.11}$$

The simplest way to verify (5.11) is by noting that for $f(\xi) = \chi_S(\xi)$, $S \subset R$, (5.11) becomes (5.10) and is therefore certainly true. Consequently, (5.11) is also true when $f(\xi)$ is a simple function. Following the usual procedure, indicated by Definitions 3.4 and 3.5 of Chapter II, one can arrive at the verification of (5.11) for the most general case when $f(\xi)$ is an extended complex-valued function on R.

The reader can easily convince himself that a complex measure is a complex-valued σ-additive set function on the measurable space $(\mathscr{X}, \mathscr{A})$. It is worthwhile to mention that due to a fundamental theorem of measure theory* the converse is also true: any finite complex-valued σ-additive set function on a measurable space $(\mathscr{X}, \mathscr{A})$ is a complex measure (see Theorem 5.8).

***Theorem 5.4.** A projector-valued set function $E(S)$, $S \in \mathscr{A}$, defined on the measurable space $(\mathscr{X}, \mathscr{A})$ and such that $E(\mathscr{X}) = 1$, is a spectral measure if and only if for any two given vectors f, g

$$\mu_{f,g}(S) = \langle f \mid E(S) g \rangle, \qquad S \in \mathscr{A}, \tag{5.12}$$

is a complex measure on $(\mathscr{X}, \mathscr{A})$.

Proof. From the easily verifiable relation

$$\begin{aligned} \langle f \mid E(S) g \rangle = {} & \frac{1}{2i} \langle f + ig \mid E(S)(f + ig) \rangle + \frac{1}{2} \langle f + g \mid E(S)(f + g) \rangle \\ & - \frac{1 - i}{2} (\langle f \mid E(S) f \rangle + \langle g \mid E(S) g \rangle) \end{aligned} \tag{5.13}$$

containing on the right-hand side only set functions of the form $\langle h \mid E(S) h \rangle$, which are obviously measures, we immediately deduce that $\mu_{f,g}(S)$ is a complex measure.

* Hahn's theorem: see Theorem 5.7 in Appendix 5.6.

Assume now that, conversely, (5.12) is a complex measure for any two given $f, g \in \mathscr{H}$. This implies that $\mu_{f,g}(S)$ is σ additive, i.e., for any sequence $S_1, S_2, \ldots$ of disjoint sets from $\mathscr{A}$

$$(5.14)\quad \langle f \mid E(S_1 \cup S_2 \cup \cdots) g\rangle = \mu_{f,g}(S_1 \cup S_2 \cup \cdots) = \sum_{k=1}^{\infty} \mu_{f,g}(S_k) = \sum_{k=1}^{\infty} \langle f \mid E(S_k) g\rangle = \lim_{n\to\infty} \langle f \mid (E(S_1) + \cdots + E(S_n)) g\rangle$$

By taking in the above relation $S_3 = S_4 = \cdots = \varnothing$, we derive

$$E(S_1 \cup S_2) = E(S_1) + E(S_2)$$

for any two disjoint $S_1, S_2 \in \mathscr{A}$. This implies, according to Theorem 3.7, that

$$E(S_1)\, E(S_2) = 0$$

for any disjoint sets $S_1, S_2 \in \mathscr{A}$. Thus, when, in general, the sets in $S_1, S_2, \ldots \in \mathscr{A}$ are disjoint,

$$\sum_{n=1}^{\infty} E(S_n) = \operatorname{s-lim}_{n\to\infty} \sum_{k=1}^{n} E(S_k)$$

exists by Theorem 3.9, and due to (5.14) we have

$$\sum_{n=1}^{\infty} E(S_n) = E\left(\bigcup_{n=1}^{\infty} S_n\right).$$

This establishes that $E(S)$ is a spectral measure. Q.E.D.

In defining the concept of a spectral measure $E(S)$ we required that S should assume values from a Boolean σ algebra instead of an algebra. This requirement does not entail any loss of generality, as can be seen from the following analogue of Theorem 2.3 in Chapter II.

***Theorem 5.5.** Let $E_0(S)$ be a projector-valued set function defined on a Boolean algebra $\mathscr{A}_0$ of subsets of a set $\mathscr{X}$. Suppose that $E_0(\mathscr{X}) = 1$ and

$$E_0\left(\bigcup_{k=1}^{\infty} S_k\right) = \sum_{k=1}^{\infty} E_0(S_k)$$

for any sequence $S_1, S_2, \ldots$ of disjoint sets from $\mathscr{A}_0$, for which $\bigcup_{k=1}^{\infty} S_k \in \mathscr{A}_0$. In that case there is a unique spectral measure $E(S)$,

$S \in \mathscr{A}$, defined on the Boolean σ algebra $\mathscr{A}$ generated by $\mathscr{A}_0$, which is an *extension* of $E_0(S)$, i.e., $E_0(S) = E(S)$ for all $S \in \mathscr{A}_0$.

Proof. To prove the theorem, we shall follow the lead suggested by the proof of Theorem 2.3 in Chapter II, and introduce for each subset R of $\mathscr{X}$ the equivalent of an infimum of the family of all projectors

$$(5.15) \quad \sum_{k=1}^{\infty} E_0(S_k), \qquad \bigcup_{k=1}^{\infty} S_k \supset R, \qquad S_k \in \mathscr{A}_0, \qquad S_i \cap S_k = \varnothing \qquad \text{for } i \neq k.$$

This "infimum" $E(R)$ can be defined as the projector onto the closed subspace $\mathsf{M}(R)$ which is the intersection of all subspaces on which the projectors (5.15) project (see Exercise 5.7). In this way, we have defined a projector-valued set function on the family of all subsets R of $\mathscr{X}$, which obviously has the property that

$$\langle f \mid E(R)f\rangle = \inf \left\{ \left\langle f \mid \sum_{k=1}^{\infty} E_0(S_k)f \right\rangle : \bigcup_{k=1}^{\infty} S_k \supset R,\ S_k \in \mathscr{A}_0 \right\}.$$

According to Chapter II, Theorem 2.3, $\langle f \mid E(S)f\rangle$, $S \in \mathscr{A}$, is a measure on $\mathscr{A}$. Hence, as we see from (5.13), it follows that $\langle f \mid E(S)g\rangle$, $S \in \mathscr{A}$, is a complex measure on $\mathscr{A}$ for any vectors f and g. Thus, $E(S)$, $S \in \mathscr{A}$, is by Theorem 5.4 a projector-valued measure on $\mathscr{A}$, which evidently coincides with $E_0(S)$ for $S \in \mathscr{A}_0$. Q.E.D.

5.5. Spectral Functions

A concept very closely related to that of spectral measures, and also very useful when formulating the spectral theorem in §6, is that of spectral function. We shall give a definition of spectral function in $\mathbb{R}^n$, rather than limiting ourselves to the more conventional special case of $n = 1$.

Definition 5.6. A projector-valued function $E_\lambda = E_{\lambda_1,\ldots,\lambda_n}$ on $\mathbb{R}^n$ is called a *spectral function* (or *spectral family*, or *decomposition* [*resolution*] *of the identity*) if it satisfies the following requirements:

(1) $E_\lambda \leqslant E_{\lambda'}$ whenever $\lambda \leqslant \lambda'$,

(2) $E_\lambda = E_{\lambda+0} = \underset{\substack{\lambda' \to \lambda \\ \lambda' \geqslant \lambda}}{\text{s-lim}}\, E_{\lambda'}$,

(3) $E_{-\infty} = \underset{\lambda \to -\infty}{\text{s-lim}}\, E_\lambda = 0$ and $E_{+\infty} = \underset{\lambda \to +\infty}{\text{s-lim}}\, E_\lambda = 1$,

where $\lambda' \geqslant \lambda$ for $\lambda = (\lambda_1, \ldots, \lambda_n)$ and $\lambda' = (\lambda_1', \ldots, \lambda_n')$ means that $\lambda_1 \leqslant \lambda_1', \ldots, \lambda_n \leqslant \lambda_n'$.

***Theorem 5.6.** To every spectral measure $E(B)$, $B \in \mathscr{B}^n$, on $(\mathbb{R}^n, \mathscr{B}^n)$ corresponds a unique spectral function E_λ, $\lambda \in \mathbb{R}^n$, satisfying

$$E_{\lambda_1,\dots,\lambda_n} = E(I_{\lambda_1,\dots,\lambda_n}),$$
(5.16)
$$I_{\lambda_1,\dots,\lambda_n} = (-\infty, \lambda_1] \times \cdots \times (-\infty, \lambda_n],$$

and, conversely, to each spectral function on $\mathbb{R}^n$ corresponds one and only one spectral measure on $(\mathbb{R}^n, \mathscr{B}^n)$ satisfying (5.16).

Proof. If $E(B)$, $B \in \mathscr{B}^n$, is a spectral measure, the projector-valued function E_λ, $\lambda \in \mathbb{R}^n$, defined by (5.16) satisfies the first requirement of Definition 5.6 since for $\lambda' \geqslant \lambda$,

$$E_{\lambda'} = E(I_{\lambda'}) = E(I_\lambda) + E(I_{\lambda'} - I_\lambda) \geqslant E(I_\lambda) = E_\lambda,$$

while the other two requirements of Definition 5.5 are satisfied by E_λ due to the continuity from above of the spectral measure $E(B)$ (see Exercise 5.8); for instance,

$$\operatorname*{s-lim}_{\substack{\lambda' \geqslant \lambda \\ \lambda' \to \lambda}} E_{\lambda'} = \operatorname*{s-lim}_{\substack{\lambda' \geqslant \lambda \\ \lambda' \to \lambda}} E(I_{\lambda'}) = E(I_\lambda) = E_\lambda .$$

To prove the converse, define a projector-valued set function $E(B)$ on the Borel algebra $\mathscr{B}_0^n$ in the following way. For intervals I_λ defined in (5.16), the set function $E(I_\lambda)$ is defined by (5.16). Assuming now that the $E(B)$ is defined for two sets B_1 and B_2, write

$$\begin{aligned} E(B_1 \cup B_2) &= E(B_1) + E(B_2) \qquad \text{if} \quad B_1 \cap B_2 = \varnothing, \\ E(B_1 \cap B_2) &= E(B_1)\, E(B_2), \\ E(B_1') &= 1 - E(B_1). \end{aligned}$$

The reader can easily convince himself that by the above indicated operations one can extend the domain of definition of $E(B)$ from the intervals I_λ to the family $\mathscr{I}^n$ of all intervals in $\mathbb{R}^n$, and then to $\mathscr{B}_0^n$, and that the derived function $E(B)$, $B \in \mathscr{B}_0^n$, is σ additive on $\mathscr{B}_0^n$. Since

$$E(\mathbb{R}^n) = \operatorname*{s-lim}_{\lambda_1,\dots,\lambda_n \to \infty} E(I_{\lambda_1,\dots,\lambda_n}) = 1,$$

$E(B)$, $B \in \mathscr{B}_0^n$, satisfies the condition of Theorem 5.5. Therefore, it has a unique extension to $\mathscr{B}^n$, and this extension is a spectral measure.

Q.E.D.

*5.6 Appendix: Signed Measures

The real and imaginary parts of a complex measure are σ additive but, in general, not nonnegative set functions. These set functions belong to the class of signed measures.

Definition 5.7. A countably additive set function $\mu(S)$ on a measurable space $(\mathscr{X}, \mathscr{A})$, for which $\mu(\varnothing) = 0$, is called a *signed measure* if it assumes values only from the set $\mathbb{R}^1 \cup \{+\infty\}$ or only from the set $\mathbb{R}^1 \cup \{-\infty\}$.

The reason for not allowing a signed measure to assume $+\infty$ as well as $-\infty$ values is to avoid the ambiguities inherent in the addition of $+\infty$ and $-\infty$.

We recall that, in general, a measure was not a special case of a complex measure since a complex measure is, by definition, finite. Hence, only the class of finite measures is a subset of the family of complex measures. However, we note that all measures are certainly special cases of signed measures.

One could ask the question whether a signed measure could be always written as a linear combination of measures. The following theorem provides an affirmative answer to this question.

***Theorem 5.7** (*Hahn's theorem*). If μ is a signed measure on the measurable space $(\mathscr{X}, \mathscr{A})$, then there are two measurable disjoint sets $S^{(+)}$ and $S^{(-)}$ whose union is $\mathscr{X}$, such that $\mu(S \cap S^{(+)}) \geqslant 0$ and $\mu(S \cap S^{(-)}) \leqslant 0$ for all $S \in \mathscr{A}$.

Proof. It is sufficient to consider only the case when $\mu(S)$ assumes values only from $\mathbb{R}^1 \cup \{+\infty\}$.

Denote by $\mathscr{F}^{(-)}$ the family of all sets R such that $\mu(R \cap S) \leqslant 0$ for all $S \in \mathscr{A}$. If

$$\gamma = \inf_{R \in \mathscr{F}^{(-)}} \mu(R), \tag{5.17}$$

choose a sequence $R_1, R_2, \ldots \in \mathscr{F}^{(-)}$ such that

$$\gamma = \lim_{k \to +\infty} \mu(R_k).$$

The sets $S_k = \bigcup_{i=1}^{n} R_i$ also belong to $\mathscr{F}^{(-)}$ since

$$\mu(S \cap S_k) = \sum_{i=1}^{k} \mu\left(S \cap \left(R_i - \bigcup_{j=1}^{i-1} R_j\right)\right) = \sum_{i=1}^{k} \mu\left(R_i \cap S \cap \left(\bigcap_{j=1}^{i-1} R_j'\right)\right) \leqslant 0$$

for all $S \in \mathscr{A}$. Moreover, we also have

$$\mu(S_k) = \mu\left(\bigcup_{i=1}^{k}\left(R_i - \bigcup_{j=i+1}^{k} R_j\right)\right) = \sum_{i=1}^{k} \mu\left(R_i - \bigcup_{j=i+1}^{k} R_j\right)$$

$$= \mu(R_k) + \sum_{i=1}^{k-1} \mu\left(R_i \cap \left(\bigcap_{j=i+1}^{k} R_j'\right)\right) \leqslant \mu(R_k),$$

and consequently

$$\gamma = \lim_{k\to\infty} \mu(S_k).$$

The set

$$S^{(-)} = \bigcup_{k=1}^{\infty} S_k = \bigcup_{k=1}^{\infty} R_k$$

also belongs to $\mathscr{F}^{(-)}$ since $\mu(S \cap S^{(-)}) \leqslant 0$ for any $S \in \mathscr{A}$, as can be seen from the relation

$$\mu(S \cap S^{(-)}) = \sum_{k=1}^{\infty} \mu(S \cap (S_{k+1} - S_k)) = \lim_{k\to\infty} \mu(S \cap S_k).$$

Setting $S = \mathscr{X}$, we also get

$$\mu(S^{(-)}) = \lim_{k\to+\infty} \mu(S_k) = \gamma.$$

We claim now that $\mu(S \cap S^{(+)}) \geqslant 0$ for all $S \in \mathscr{A}$ if $S^{(+)} = \mathscr{X} - S^{(-)}$.

Assume to the contrary that $\mu(\bar{S} \cap S^{(+)}) < 0$ for some $\bar{S} \in \mathscr{A}$. The set $S_0^{(+)} = \bar{S} \cap S^{(+)}$ could not belong to $\mathscr{F}^{(-)}$, since then we would have $S_0^{(+)} \cup S^{(-)} \in \mathscr{F}^{(-)}$ and

$$\mu(S_0^{(+)} \cup S^{(-)}) = \mu(S_0^{(+)}) + \mu(S^{(-)}) < \gamma,$$

which contradicts (5.17). Hence, $S_0^{(+)}$ has measurable subsets of positive measure μ. Let n_1 be the smallest positive integer for which there is a measurable subset $S_1^{(+)}$ of $S_0^{(+)}$ such that $\mu(S_1^{(+)}) > 1/n_1$. Since

$$\mu(S_0^{(+)} - S_1^{(+)}) = \mu(S_0^{(+)}) - \mu(S_1^{(+)}) \leqslant \mu(S_0^{(+)}) < 0,$$

we can apply the same argument to $S_0^{(+)} - S_1^{(+)}$ to infer the existence of a smallest positive integer n_2 such that $S_0^{(+)} - S_1^{(+)}$ contains a measurable set $S_2^{(+)}$ with $\mu(S_2^{(+)}) \geqslant 1/n_2$. We can continue with this procedure, thus obtaining the sets $S_3^{(+)}, S_4^{(+)}, \ldots$. Then

$$\mu\left(\bigcup_{k=1}^{\infty} S_k^{(+)}\right) = \sum_{k=1}^{\infty} \mu(S_k^{(+)}) \geqslant \sum_{k=1}^{\infty} \frac{1}{n_k} \geqslant 0,$$

and consequently

$$\mu\left(S_0^{(+)} - \bigcup_{k=1}^{\infty} S_k^{(+)}\right) < 0.$$

For arbitrary $S \in \mathscr{A}$ we have

$$\mu\left(S \cap \left(S_0^{(+)} - \bigcup_{k=1}^{l-1} S_k^{(+)}\right)\right) \leqslant \frac{1}{n_l - 1}. \tag{5.18}$$

In fact, if that were not so for some $l = k_0$, then n_{k_0} would not fulfill the conditions of its definition since we could choose $n'_{k_0} < n_{k_0}$ so that

$$\mu\left(S \cap \left(S^{(+)} - \bigcup_{k=1}^{k_0-1} S_k^{(+)}\right)\right) > \frac{1}{n'_{k_0}}$$

and replace $S_{k_0}^{(+)}$ by

$$S \cap \left(S^{(+)} - \bigcup_{k=1}^{k_0-1} S_k^{(+)}\right) \subset S^{(+)} - \bigcup_{k=1}^{k_0-1} S_k^{(+)}.$$

The sequence $n_1, n_2, \ldots$ diverges to infinity since

$$\sum_{k=1}^{\infty} \frac{1}{n_k} \leqslant \sum_{k=1}^{\infty} \mu(S_k^{(+)}) = \mu(S_0^{(+)}) - \mu\left(S_0^{(+)} - \bigcup_{k=1}^{\infty} S_k^{(+)}\right),$$

and the measures of $S_0^{(+)}$ and $S_0^{(+)} - \bigcup_{k=1}^{\infty} S_k^{(+)}$ are nonpositive and therefore finite. By combining this result with (5.18) we obtain

$$\mu\left(S \cap \left(S_0^{(+)} - \bigcup_{k=1}^{\infty} S_k^{(+)}\right)\right) \leqslant 0$$

for all $S \in \mathscr{A}$. This would imply that $S_0^{(+)} - \bigcup_{k=1}^{\infty} S_k^{(+)}$ is a set which belongs to $\mathscr{F}^{(-)}$, is of negative measure, and is disjoint from $S^{(-)}$. Hence, we would have

$$\mu\left(S^{(-)} \cup \left(S_0^{(+)} - \bigcup_{k=1}^{\infty} S_k^{(+)}\right)\right) < \mu(S^{(-)}) = \gamma,$$

which contradicts (5.17). Q.E.D.

It is easy to see that if $\mu(S)$ is a signed measure, then the set functions $\mu^{(+)}(S) = \mu(S \cap S^{(+)})$ and $\mu^{(-)}(S) = -\mu(S \cap S^{(-)})$ are measures. These measures are called the *upper variation* and the *lower variation* of μ, respectively. According to the above theorem, we can always write

$$\mu(S) = \mu^{(+)}(S) - \mu^{(-)}(S), \qquad S \in \mathscr{A}.$$

Using this result we arrive at the following conclusion.

Theorem 5.8. Any countably additive set function $\mu(S)$ on a measure space $(\mathscr{X}, \mathscr{A})$ for which $\mu(\varnothing) = 0$ and which assumes values from $\mathbb{C}^1$ is a complex measure (in the sense of Definition 5.5).

In order to establish the above theorem we only have to note that $\operatorname{Re} \mu(S)$ and $\operatorname{Im} \mu(S)$ are signed measures and therefore they can be written as the difference of their respective upper and lower variations.

Exercises

5.1. Show that in an n-dimensional Hilbert space $\mathscr{H}$ the eigenvalue problem

$$Af = \lambda f$$

can be reformulated as a matrix eigenvalue problem

$$\sum_{k=1}^{n} A_{ik} x_k = \lambda x_i\,, \qquad A_{ik} = \langle e_i \mid Ae_k \rangle, \qquad x_i = \langle e_i \mid f \rangle,$$

in some orthonormal basis $\{e_1, \ldots, e_n\}$.

5.2. Show that every linear Hermitian operator A acting on an n-dimensional Hilbert space $\mathscr{H}$ has a pure point spectrum containing at most n points.

5.3. Prove that if $A_1, A_2, \ldots,$ is a uniformly convergent sequence of operators on $\mathscr{H}$ then for each ϵ there is an $N(\epsilon)$ such that $\| A_m - A_n \| < \epsilon$ for all $m, n > N(\epsilon)$.

5.4. Let $\{e_1, e_2, \ldots\}$ be an orthonormal basis in an infinite-dimensional Hilbert space $\mathscr{H}$. Denote by $| e_k \rangle \langle e_k |$ the projector (3.4) onto e_k. Show that on account of Theorem 4.6(b) in Chapter I,

$$1 = \operatorname*{s-lim}_{n\to\infty} \sum_{k=1}^{n} | e_k \rangle \langle e_k | = \sum_{k=1}^{\infty} | e_k \rangle \langle e_k |.$$

5.5. Show that when $\operatorname{s-lim}_{t\to t_0} A(t)$ exists, then

$$\lim_{t\to t_0} \| A(t) f \| = \| \operatorname*{s-lim}_{t\to t_0} A(t) f \| \qquad \text{for all} \quad f \in \mathscr{H}.$$

5.6. Show that if $A = \operatorname{s-lim}_{t\to t_0} A(t)$ and $B = \operatorname{s-lim}_{t\to t_0} B(t)$, then $\operatorname{s-lim}_{t\to t_0} A(t)\, B(t)$ exists and is equal to AB in the case that $\| A(t) \| < c$ for all t.

5.7. Show that the intersection of all the subspaces in any given family of closed linear subspaces of a Hilbert space $\mathscr{H}$ is a closed linear subspace of $\mathscr{H}$.

5.8. Prove that any spectral measure $E(B)$, $B \in \mathscr{A}$, is *continuous from above* and *from below*, i.e., if $B = \lim_{n\to\infty} B_n$, then

$$E(B) = \operatorname*{s-lim}_{n\to\infty} E(B_n),$$

where $B_1 \supset B_2 \supset \cdots \supset B$ and $B_1 \subset B_2 \subset \cdots \subset B$, respectively.

5.9. Prove that $\| A^* \| = \| A \|$.

5.10. Show that if $A = \text{u-lim}_{t \to t_0} A_t$ and $B = \text{u-lim}_{t \to t_0} B_t$, then

$$A + B = \underset{t \to t_0}{\text{u-lim}} (A_t + B_t),$$

$$AB = \underset{t \to t_0}{\text{u-lim}} A_t B_t,$$

$$A^* = \underset{t \to t_0}{\text{u-lim}} A_t^*.$$

5.11. Explain why any infinite orthonormal sequence $e_1, e_2, \ldots$ in a Hilbert space $\mathscr{H}$ is weakly convergent to $\mathbf{0}$, but it does not converge strongly to any vector.

*6. The Spectral Theorem for Unitary and Self-Adjoint Operators

6.1. Spectral Decomposition of a Unitary Operator

We shall first state the spectral theorem for unitary operators, and later on proceed to prove it in a few stages.

***Theorem 6.1.** To every unitary operator U on a Hilbert space $\mathscr{H}$ corresponds a unique spectral function E_λ having the properties that $E_0 = 0$, $E_{2\pi} = 1$, which is such that

$$U = \underset{\substack{n \to \infty \\ \epsilon \to 0}}{\text{u-lim}} \sum_{k=1}^{n} e^{i\lambda_k'}(E_{\lambda_k} - E_{\lambda_{k-1}}), \tag{6.1}$$

$$\lambda_{k-1} < \lambda_k' \leqslant \lambda_k, \qquad \epsilon = \max_{k=1,\ldots,n} (\lambda_k - \lambda_{k-1}),$$

where the limit is taken over finer and finer partitions of $[0, 2\pi]$:

$$0 = \lambda_0 < \lambda_1 < \cdots < \lambda_n = 2\pi. \tag{6.2}$$

Since uniform convergence implies weak convergence, we can write

$$\langle f \mid Ug \rangle = \int_{[0,2\pi]} e^{i\lambda}\, d\langle f \mid E_\lambda g \rangle \tag{6.3}$$

for any two vectors $f, g \in \mathscr{H}$.

It should be noted that since $E_{\lambda_1} \leqslant E_{\lambda_2}$ for $\lambda_1 \leqslant \lambda_2$, we have

$$E_\lambda = \begin{cases} 0 & \text{for } \lambda \leqslant 0 \\ 1 & \text{for } \lambda \geqslant 2\pi. \end{cases} \tag{6.4}$$

Thus, the support of $E(B)$ is within the set $[0, 2\pi]$, i.e.,

$$E(B) = 0 \qquad \text{if} \quad B \cap [0, 2\pi] = \varnothing .$$

Consequently, the integration in (6.3) can be extended over $\mathbb{R}^1$.

The formula (6.1) is customarily written in the symbolic form

$$U = \int_{\mathbb{R}^1} e^{i\lambda}\, dE_\lambda\,, \tag{6.5}$$

and as such is known as the *spectral decomposition* of U.

The basic idea of the construction of the spectral function E_λ consists in defining E_λ as the strong limit of monotonically decreasing sequences of polynomials in U and U^{-1}. We define the yet unfamiliar concept of a monotonic sequence of bounded operators so as to obtain a generalization of the already familiar case for projectors (see Theorem 3.10).

6.2. Monotonic Sequences of Linear Operators

Definition 6.1. If A is a symmetric operator in $\mathscr{H}$ and c is a real number, we write $A \leqslant c$ or $A \geqslant c$ if, respectively, $\langle f \mid Af\rangle \leqslant c\langle f \mid f\rangle$ or $\langle f \mid Af\rangle \geqslant c\langle f \mid f\rangle$ for all $f \in \mathscr{D}_A$. In that case A is said to be *bounded from above or below*, respectively; A is said to be *positive* if $A \geqslant 0$.

If A and B are two symmetric operators on $\mathscr{H}$ and $A - B \geqslant 0$, then we write $A \geqslant B$. It is very easy to check that the above relation $\geqslant$ is a partial-ordering relation (see Exercise 6.2).

Theorem 6.2. If $A_1, A_2, \ldots$ is a sequence of symmetric operators on $\mathscr{H}$, which is *monotonically increasing* (i.e., $A_1 \leqslant A_2 \leqslant \cdots$) or *monotonically decreasing* (i.e., $A_1 \geqslant A_2 \geqslant \cdots$), and bounded, respectively, from above (i.e., $A_k \leqslant c$ for all k and some constant c) or from below (i.e., $A_k \geqslant c$), then $A_1, A_2, \ldots$ converges strongly to a limit A, which is a symmetric operator on $\mathscr{H}$.

Proof. Consider the case $A_1 \leqslant A_2 \leqslant \cdots$. Since the operators A_1, $A_2, \ldots$ are symmetric and defined everywhere on $\mathscr{H}$, they are bounded (see Theorem 2.10).

We shall assume that all the operators $A_1, A_2, \ldots$ are positive; the general case $B_1 \leqslant B_2 \leqslant \cdots$ can be easily reduced to this case by setting $A_k = B_k + \| B_1 \|$:

$$\langle f \mid A_k f\rangle = \langle f \mid B_k f\rangle + \langle f \mid \| B_1 \| f\rangle \geqslant \langle f \mid B_1 f\rangle + \| B_1 \| \langle f \mid f\rangle \geqslant 0.$$

To prove the strong convergence of $A_1 f, A_2 f, \ldots$, note that $A_i - A_j \geqslant 0$

if $i \geqslant j$. For any positive symmetric operator A we have the generalized Schwarz–Cauchy inequality (see Exercise 6.1):

$$|\langle g \mid Af \rangle|^2 \leqslant \langle f \mid Af \rangle \langle g \mid Ag \rangle. \tag{6.6}$$

By inserting above $A = A_i - A_j$ and $g = (A_i - A_j)f$, we get

$$\langle (A_i - A_j)f \mid (A_i - A_j)f \rangle^2 \leqslant \langle f \mid (A_i - A_j)f \rangle \langle (A_i - A_j)f \mid (A_i - A_j)^2 f \rangle. \tag{6.7}$$

Let c be an upper bound of the sequence $A_1 \leqslant A_2 \leqslant \cdots$, so that $\langle f \mid A_i f \rangle \leqslant c \|f\|^2$ for all $i = 1, 2, \ldots$. From the inequality

$$\langle f \mid (A_i - A_j)f \rangle \leqslant \langle f \mid A_i f \rangle \leqslant c \|f\|^2$$

we get by using the relation (6.6) with $A = A_i - A_j$,

$$|\langle g \mid (A_i - A_j)f \rangle|^2 \leqslant c^2 \|f\|^2 \|g\|^2, \qquad f, g \in \mathscr{H}.$$

The above inequality implies that $\| A_i - A_j \| \leqslant c$. Using this result in (6.7) we obtain

$$\|(A_i - A_j)f\|^4 \leqslant c^3 \|f\|^2 |\langle f \mid A_i f \rangle - \langle f \mid A_j f \rangle|. \tag{6.8}$$

Since

$$\langle f \mid A_1 f \rangle \leqslant \langle f \mid A_2 f \rangle \leqslant \cdots \leqslant c \|f\|^2,$$

the above sequence of numbers is convergent, so that the term $|\langle f \mid A_i f \rangle - \langle f \mid A_j f \rangle|$ can be made arbitrarily small for sufficiently large i and j. In the light of this remark, (6.8) yields the strong convergence of $A_1 f, A_2 f, \ldots$.

The linearity and Hermiticity of A, where

$$Af = \operatorname*{s-lim}_{n \to \infty} A_n f,$$

is easily derivable from the corresponding properties of each of the operators $A_1, A_2, \ldots$ (see Exercise 6.3). Q.E.D.

6.3. Construction of Spectral Families for Unitary Operators

Assign to every trigonometric polynomial

$$p(e^{i\varphi}) = \sum_{k=-n}^{+n} c_k e^{ik\varphi} \tag{6.9}$$

the operator $p(U)$, where

$$p(U) = \sum_{k=-n}^{+n} c_k U^k. \tag{6.10}$$

Then we can state the following lemma.

Lemma 6.1. The mapping

$$p(e^{i\varphi}) \to p(U) \tag{6.11}$$

of the family of all trigonometric polynomials (6.9) into the family of linear operators on $\mathscr{H}$ has the following properties:

$$a_1 p_1(e^{i\varphi}) + a_2 p_2(e^{i\varphi}) \to a_1 p_1(U) + a_2 p_2(U), \tag{6.12}$$

$$p_1(e^{i\varphi})\, p_2(e^{i\varphi}) \to p_1(U)\, p_2(U), \tag{6.13}$$

$$(p(e^{i\varphi}))^* \to (p(U))^*. \tag{6.14}$$

Moreover, if $p(e^{i\varphi}) \geqslant 0$ for all $\varphi \in \mathbb{R}^1$, then $p(U)$ is a positive operator.

Proof. The verification of the relations (6.12)–(6.14) is straightforward algebra and is left to the reader (Exercise 6.4).

In order to prove that $p(U) \geqslant 0$ if $p(e^{i\varphi}) \geqslant 0$, we use a lemma by Fejér and Riesz (see Exercise 6.5) which states that if $p(e^{i\varphi}) \geqslant 0$, then there is another trigonometric polynomial $q(e^{i\varphi})$ such that

$$p(e^{i\varphi}) = |\, q(e^{i\varphi})|^2 = [q(e^{i\varphi})]^*\, q(e^{i\varphi}).$$

Using (6.13) we get

$$\begin{aligned} \langle f \,|\, p(U) f\rangle &= \langle f \,|\, (q(U))^*\, q(U) f\rangle \\ &= \langle q(U) f \,|\, q(U) f\rangle \geqslant 0, \end{aligned}$$

which shows that $p(U) \geqslant 0$. Q.E.D.

We can extend the mapping (6.11) to the family $\mathscr{F}_{\mathrm{p}}$ of all functions $w(\varphi)$, $\varphi \in \mathbb{R}^1$, which are the limits

$$w(\varphi) = \lim_{n\to\infty} p_n(e^{i\varphi}) \tag{6.15}$$

of monotonically decreasing sequences

$$p_1(e^{i\varphi}) \geqslant p_2(e^{i\varphi}) \geqslant \cdots \geqslant 0 \tag{6.16}$$

of positive trigonometric polynomials. This result constitutes the following lemma.

Lemma 6.2. To each function $w(\varphi) \in \mathscr{F}_p$ corresponds a symmetric positive operator A_w on $\mathscr{H}$

$$A_w = \operatorname*{s\text{-}lim}_{n\to\infty} p_n(U),$$

which is independent of the chosen monotonic sequence (6.16). The mapping

$$w(\varphi) \to A_w \tag{6.17}$$

is such that

$$w_1(\varphi)\, w_2(\varphi) \to A_{w_1 w_2} = A_{w_1} A_{w_2}\,. \tag{6.18}$$

Proof. Due to the last statement in Lemma 6.1, we have for any sequence (6.16) that

$$p_1(U) \geqslant p_2(U) \geqslant \cdots \geqslant 0.$$

Hence, by Theorem 6.2, the above sequence has a strong limit A, which is a symmetric operator on $\mathscr{H}$. Moreover, $A \geqslant 0$,

$$\langle f \mid Af \rangle = \lim_{n\to\infty} \langle f \mid p_n(U) f \rangle \geqslant 0.$$

In order to prove that the limit A is independent of the chosen sequence (6.16), take two such monotonically decreasing sequences $\{p_n^{(1)}(e^{i\varphi})\}$ and $\{p_n^{(2)}(e^{i\varphi})\}$, for which

$$\lim_{n\to\infty} p_n^{(1)}(e^{i\varphi}) = \lim_{n\to\infty} p_n^{(2)}(e^{i\varphi}).$$

The above assumption implies (see Exercise 6.6) that for every integer n there is some $p_{k_n}^{(1)}(e^{i\varphi})$ such that

$$p_{k_n}^{(1)}(e^{i\varphi}) \leqslant p_n^{(2)}(e^{i\varphi}) + 1/n.$$

By using again the last statement of Lemma 6.1 we infer that

$$p_{k_n}^{(1)}(U) \leqslant p_n^{(2)}(U) + 1/n,$$

which implies that

$$\operatorname*{s\text{-}lim}_{n\to\infty} p_n^{(1)}(U) \leqslant \operatorname*{s\text{-}lim}_{n\to\infty} p_n^{(2)}(U).$$

By reversing in the above argument the roles of $p_n^{(1)}(e^{i\varphi})$ and $p_n^{(2)}(e^{i\varphi})$ we can deduce that

$$\operatorname*{s\text{-}lim}_{n\to\infty} p_n^{(2)}(U) \leqslant \operatorname*{s\text{-}lim}_{n\to\infty} p_n^{(1)}(U),$$

which implies (see Exercise 6.2) that the above two limits are equal.

Finally, the property (6.18) of the mapping (6.17) follows from (6.13) by using the result of Exercise 5.6 on strong limits. Q.E.D.

Consider the following family of periodic functions $\Pi_\lambda(\varphi)$ defined by

$$\Pi_\lambda(\varphi) \equiv 0 \qquad \text{for} \quad \lambda < 0,$$
$$\Pi_\lambda(\varphi) \equiv 1 \qquad \text{for} \quad \lambda > 2\pi,$$

$$(6.19) \quad \Pi_\lambda(\varphi) = \begin{cases} 1 & \text{when} \quad 2n\pi \leqslant \varphi \leqslant 2n\pi + \lambda \\ 0 & \text{when} \quad 2n\pi + \lambda < \varphi \leqslant 2(n+1)\pi, \end{cases}$$
$$0 \leqslant \lambda \leqslant 2\pi, \quad n = 0, \pm 1, \pm 2, \ldots .$$

We shall first show that the operators E_λ assigned to the above functions by the mapping (6.17) are projectors, and later on we shall prove that these projectors constitute the spectral family of Theorem 6.1.

Lemma 6.3. The functions (6.19) belong to the family $\mathscr{F}_{\mathrm{p}}$. The operator E_λ assigned to $\Pi_\lambda(\varphi)$ by the mapping (6.17) is a projector in $\mathscr{H}$.

Proof. It is easy to construct* for any given function $\Pi_\lambda(\varphi)$ a monotonically decreasing sequence

$$u_1(\varphi) > u_2(\varphi) > \cdots$$

of once continuously differentiable and periodic functions $u_n(\varphi)$

$$u_n(\varphi) = u_n(\varphi + 2k\pi), \qquad k = \pm 1, \pm 2, \ldots,$$

converging to $\Pi_\lambda(\varphi)$. Choose a sequence $\delta_1, \delta_2, \ldots \to 0$, where

$$(6.20) \qquad 0 < \delta_n \leqslant \tfrac{1}{2} \inf_{\varphi \in \mathbb{R}^1} [\min\{u_n(\varphi) - u_{n-1}(\varphi), u_{n+1}(\varphi) - u_n(\varphi)\}].$$

According to Chapter II, Theorem 7.3, there can be found for each $u_n(\varphi)$ satisfying the above imposed conditions a trigonometric polynomial $p_n(e^{i\varphi})$ for which

$$| u_n(\varphi) - p_n(e^{i\varphi})| < \delta_n , \qquad \varphi \in \mathbb{R}^1.$$

Due to the conditions (6.20), we have that $p_1(e^{i\varphi}) \geqslant p_2(e^{i\varphi}) \geqslant \cdots$, and obviously

$$\lim_{n\to\infty} p_n(e^{i\varphi}) = \lim_{n\to\infty} u_n(\varphi) = \Pi_\lambda(\varphi).$$

This proves that $\Pi_\lambda(\varphi) \in \mathscr{F}_{\mathrm{p}}$.

* See the construction of the functions $h_n(x)$ in the course of the proof of Theorem 5.6 in Chapter II.

If E_λ denotes the symmetric operator in $\mathscr{H}$ assigned to $\Pi_\lambda(\varphi)$, by (6.18) we shall obtain $E_\lambda^2 = E_\lambda$, due to the fact that

$$\Pi_\lambda^2(\varphi) = \Pi_\lambda(\varphi).$$

Consequently, E_λ is a projector. Q.E.D.

6.4. Uniqueness of the Spectral Family of a Unitary Operator

We have to prove now that the projectors E_λ defined in Lemma 6.3 constitute a spectral family.

From the obvious relation

$$\Pi_\lambda(\varphi)\, \Pi_{\lambda'}(\varphi) = \Pi_\lambda(\varphi), \qquad \lambda < \lambda',$$

and (6.17) we get

$$E_\lambda E_{\lambda'} = E_\lambda, \qquad \lambda < \lambda'. \tag{6.21}$$

Thus, in order to establish that E_λ, $\lambda \in \mathbb{R}^1$, is a spectral family, we are left with the task of demonstrating that

$$E_\lambda = E_{\lambda+0} = \operatorname*{s-lim}_{\lambda' \to \lambda+0} E_{\lambda'}. \tag{6.22}$$

First, we note (as can be easily and explicitly proven in the same fashion as in Lemma 6.3) that a monotonically decreasing sequence $\{p_n(e^{i\varphi})\}$ of trigonometric polynomials can be found which is such that

$$\lim_{n\to\infty} p_n(e^{i\varphi}) = \Pi_\lambda(e^{i\varphi}),$$

and in addition

$$p_n(e^{i\varphi}) \geqslant \Pi_{\lambda+1/n}(\varphi), \qquad n = 1, 2, \ldots .$$

Consequently

$$p_n(U) \geqslant E_{\lambda+1/n},$$

and since $E_\lambda \leqslant E_{\lambda+1/n}$,

$$E_\lambda \leqslant \operatorname*{s-lim}_{n\to\infty} E_{\lambda+1/n} = \operatorname*{s-lim}_{n\to\infty} p_n(U) = E_\lambda,$$

i.e.,

$$\operatorname*{s-lim}_{n\to\infty} E_{\lambda+1/n} = E_\lambda .$$

As any sequence $\lambda_1, \lambda_2, \ldots \to \lambda + 0$ can be majorized by a sequence containing only the numbers $\lambda + 1/n$, $n = 1, 2, \ldots$, and converging to λ, we see that (6.22) is true.

To establish now that E_λ, $\lambda \in \mathbb{R}^1$, of Lemma 6.3 is the desired spectral family, note that the definition (6.19) of $\Pi_\lambda(\varphi)$ yields immediately that (6.4) is satisfied. Thus, we have to verify that (6.1) is true.

If we keep the notation of (6.1) and (6.2), we see that for $\lambda_{k-1} \leqslant \varphi \leqslant \lambda_k$

$$\left| e^{i\varphi} - \sum_{l=1}^{n} e^{i\lambda_l'}(\Pi_{\lambda_l}(\varphi) - \Pi_{\lambda_{l-1}}(\varphi)) \right| \leqslant | e^{i\varphi} - e^{i\lambda_k'} | \leqslant | \varphi - \lambda_k' | \leqslant \epsilon.$$

Since the above holds for $k = 1,..., n$, we have

$$(6.23) \qquad 0 \leqslant \left[e^{i\varphi} - \sum_{k=1}^{n} e^{i\lambda_k'}(\Pi_{\lambda_k}(\varphi) - \Pi_{\lambda_{k-1}}(\varphi))\right]^* \times \left[e^{i\varphi} - \sum_{k=1}^{n} e^{i\lambda_k'}(\Pi_{\lambda_k}(\varphi) - \Pi_{\lambda_{k-1}}(\varphi))\right] \leqslant \epsilon^2$$

for any $\varphi \in [0, 2\pi]$. From the last statement in Lemma 6.1 about the mapping (6.17), alias (6.11), it can be easily derived that (6.23) implies

$$0 \leqslant \left[U - \sum_{k=1}^{n} e^{i\lambda_k'}(E_{\lambda_k} - E_{\lambda_{k-1}})\right]^* \left[U - \sum_{k=1}^{n} e^{i\lambda_k'}(E_{\lambda_k} - E_{\lambda_{k-1}})\right] \leqslant \epsilon^2.$$

Thus, we have

$$\left\| \left(U - \sum_{k=1}^{n} e^{i\lambda_k'}(E_{\lambda_k} - E_{\lambda_{k-1}})\right) f \right\|^2 \leqslant \epsilon^2 \|f\|^2,$$

for all $f \in \mathscr{H}$, and consequently

$$\left\| U - \sum_{k=1}^{n} e^{i\lambda_k'}(E_{\lambda_k} - E_{\lambda_{k-1}}) \right\| \leqslant \epsilon,$$

which proves (6.1).

To establish the uniqueness of the spectral decomposition of U, assume that E_λ' is a spectral function satisfying (6.1) and (6.4). Because of the pairwise orthogonality of $E'_{\lambda_k} - E'_{\lambda_{k-1}}$ for different values of k, we easily derive (using the results of Exercise 5.10) that for any nonnegative integer m

$$\underset{\substack{n\to\infty \\ \epsilon\to 0}}{\text{u-lim}} \sum_{k=1}^{n} e^{im\lambda_k''}(E'_{\lambda_k} - E'_{\lambda_{k-1}}) = \underset{\substack{n\to\infty \\ \epsilon\to 0}}{\text{u-lim}} \left[\sum_{k=1}^{n} e^{i\lambda_k''}(E'_{\lambda_k} - E'_{\lambda_{k-1}})\right]^m = U^m,$$

$$\operatorname*{u-lim}_{\substack{n\to\infty\\ \epsilon\to 0}} \sum_{k=1}^{n} e^{-im\lambda_k''}(E'_{\lambda_k} - E'_{\lambda_{k-1}})$$

$$= \operatorname*{u-lim}_{\substack{n\to\infty\\ \epsilon\to 0}} \left[\sum_{k=1}^{n} e^{-i\lambda_k''}(E'_{\lambda_k} - E'_{\lambda_{k-1}})\right]^m = (U^*)^m = U^{-m},$$

i.e., in symbolic notation,

$$\int_{\mathbb{R}^1} e^{im\varphi}\, dE_\varphi' = U^m, \qquad m = 0, \pm 1, \ldots .$$

Due to linearity of the above integration procedure, we easily establish that for all trigonometric polynomials $p(e^{i\varphi})$ we have

$$\int_{\mathbb{R}^1} p(e^{i\varphi})\, d\langle f \mid E_\varphi' g\rangle = \langle f \mid p(U)\, g\rangle$$

for any $f, g \in \mathscr{H}$. With the help of Theorem 3.10, Chapter II, which remains valid for complex measures, we can generalize the above formula to any function $w(\varphi) \in \mathscr{F}_\mathrm{p}$

$$\int_{\mathbb{R}^1} w(\varphi)\, d\langle f \mid E_\varphi' g\rangle = \langle f \mid A_w g\rangle,$$

by approximating $w(\varphi)$ by a sequence (6.16) and then going to the limit. Thus, in particular, for $w(\varphi) = \Pi_\lambda(\varphi)$ we get

$$\int_{\mathbb{R}^1} \Pi_\lambda(\varphi)\, d\langle f \mid E_\varphi' g\rangle = \langle f \mid E_\lambda g\rangle,$$

i.e., after carrying out the above integration,

$$\langle f \mid E_\lambda' g\rangle - \langle f \mid E_0' g\rangle = \langle f \mid E_\lambda g\rangle.$$

Since $E_0' = 0$, and the above identity is true for any $f, g \in \mathscr{H}$, we get $E_\lambda' = E_\lambda$ for all $\lambda \in \mathbb{R}^1$. This establishes the uniqueness of the spectral decomposition of U and completes the proof of Theorem 6.1.

6.5. Spectral Decomposition of a Self-Adjoint Operator

The spectral theorem for self-adjoint operators on a Hilbert space can now be stated and proved.

Theorem 6.3. (*The spectral theorem*). To each self-adjoint operator A in a Hilbert space $\mathscr{H}$ corresponds a unique spectral function E_λ, $\lambda \in \mathbb{R}^1$, such that the domain $\mathscr{D}_A$ of A consists of all the vectors f for which

$$\int_{\mathbb{R}^1} \lambda^2 \, d\langle f \mid E_\lambda f\rangle < +\infty, \tag{6.24}$$

and for any $f \in \mathscr{D}_A$ and $g \in \mathscr{H}$

$$\langle g \mid Af\rangle = \int_{\mathbb{R}^1} \lambda \, d\langle g \mid E_\lambda f\rangle. \tag{6.25}$$

Proof. We shall derive the existence and uniqueness of the spectral function of A from the existence and uniqueness of the spectral function of the Cayley transform V of A.

Inasmuch as A is self-adjoint, its Cayley transform V is a unitary operator (see Theorem 4.9), which, therefore, has a spectral decomposition

$$V = \int_{[0,2\pi]} e^{i\varphi} \, d\tilde{E}_\varphi .$$

According to Theorem 4.8, if $f \in \mathscr{D}_A$ we have

$$Af = i(1 + V)h, \qquad h = (1 - V)^{-1}f. \tag{6.26}$$

Consequently, for any $g \in \mathscr{H}$ we get by applying in the last step of the following derivation the result of Exercise 6.7:

$$\begin{aligned}\langle g \mid Af\rangle &= i\langle g \mid h\rangle + i\langle g \mid Vh\rangle \\ &= i \int_{[0,2\pi]} d\langle g \mid \tilde{E}_\varphi h\rangle + i \int_{[0,2\pi]} e^{i\varphi} \, d\langle g \mid \tilde{E}_\varphi h\rangle \\ &= i \int_{[0,2\pi]} \frac{1 + e^{i\varphi}}{1 - e^{i\varphi}} (1 - e^{i\varphi}) \, d\langle g \mid \tilde{E}_\varphi h\rangle \\ &= \int_{[0,2\pi]} \left(- \cot \frac{\varphi}{2}\right) d\mu_{g,h}(\varphi),\end{aligned}$$

where the complex measure $\mu_{g,h}(B)$, $B \in \mathscr{B}^1$, is defined as

$$\mu_{g,h}(B) = \int_B (1 - e^{i\varphi}) \, d\langle g \mid \tilde{E}_\varphi h\rangle.$$

On the other hand, by making use of the fact that $f = (1 - V)h$ and $\tilde{E}(B)V = V\tilde{E}(B)$ (see Exercise 6.8), we easily derive

$$(6.27)\quad \langle g \mid \tilde{E}(B)f\rangle = \langle g \mid \tilde{E}(B)\, h\rangle - \langle g \mid V\tilde{E}(B)\, h\rangle$$

$$= \int_{[0,2\pi]} d\langle g \mid \tilde{E}_\varphi \tilde{E}(B)\, h\rangle - \int_{[0,2\pi]} e^{i\varphi}\, d\langle g \mid \tilde{E}_\varphi \tilde{E}(B)\, h\rangle$$

$$= \int_B (1 - e^{i\varphi})\, d\langle g \mid \tilde{E}_\varphi h\rangle = \mu_{g,h}(B),$$

by noting that by (5.16) and Theorem 5.3

$$\tilde{E}_\varphi \tilde{E}(B) = \tilde{E}(B \cap [0, \varphi]), \qquad B \subset [0, 2\pi].$$

Thus, we obtain

$$\langle g \mid Af\rangle = -\int_{[0,2\pi]} \cot \tfrac{1}{2}\varphi\, d\langle g \mid \tilde{E}_\varphi f\rangle = \int_{\mathbb{R}^1} \lambda\, d\langle g \mid E_\lambda f\rangle,$$

where in the last step we had introduced the new variable

$$\lambda = -\cot \tfrac{1}{2}\varphi, \qquad \varphi \in (0, 2\pi),$$

and in terms of it the new spectral function

$$E_\lambda = \tilde{E}_{(-2\,\mathrm{arccot}\,\lambda)} = \tilde{E}_\varphi\, .$$

For $f \in \mathscr{D}_A$ we get from (6.27)

$$\langle f \mid \tilde{E}(B) f\rangle = \int_B (1 - e^{i\varphi})\, d\langle f \mid \tilde{E}_\varphi h\rangle$$

$$= \int_B (1 - e^{i\varphi})\, d\langle h \mid \tilde{E}_\varphi f\rangle^*$$

$$= \int_B (1 - e^{i\varphi})\, d\mu^*_{h,h}(\varphi)$$

$$= \int_B |\, 1 - e^{i\varphi}\,|^2\, d\langle h \mid \tilde{E}_\varphi h\rangle$$

$$= 4 \int_B \sin^2 \tfrac{1}{2}\varphi\, d\langle h \mid \tilde{E}_\varphi h\rangle,$$

and consequently (see Exercise 6.7)

$$\int_{\mathbb{R}^1} \lambda^2\, d\langle f \mid E_\lambda f\rangle = \int_{[0,2\pi]} \cot^2 \tfrac{1}{2}\varphi\, d\langle f \mid \tilde{E}_\varphi f\rangle$$

$$= 4 \int_{[0,2\pi]} \cot^2 \tfrac{1}{2}\varphi \sin^2 \tfrac{1}{2}\varphi\, d\langle h \mid \tilde{E}_\varphi h\rangle$$

$$= 4 \int_{[0,2\pi]} \cos^2 \tfrac{1}{2}\varphi\, d\langle h \mid \tilde{E}_\varphi h\rangle$$

$$\leqslant 4 \int_{[0,2\pi]} d\langle h \mid \tilde{E}_\varphi h\rangle = 4\langle h \mid h\rangle < +\infty.$$

The above consideration proves the existence of a spectral function E_λ satisfying (6.24) and (6.25) for all $f \in \mathscr{D}_A$. We now want to show that there is only one such spectral function.

Suppose E_λ', $\lambda \in \mathbb{R}^1$, is a spectral function satisfying (6.24) and (6.25) for all $f \in \mathscr{D}_A$. It is easy to see (see also Exercise 6.9) that the operator A' defined on the set

$$\left\{f: \int_{\mathbb{R}^1} \lambda^2 \, d\langle f \mid E_\lambda f\rangle < +\infty,\ f \in \mathscr{H}\right\} \tag{6.28}$$

as the operator for which $A'f$ satisfies

$$\langle g \mid A'f\rangle = \int_{\mathbb{R}^1} \lambda \, d\langle g \mid E_\lambda' f\rangle$$

for all $g \in \mathscr{H}$, is linear. By the original requirement, we have $A' \supseteq A$ and consequently $A'^* \subseteq A^* = A$ (see Exercise 2.9). On the other hand, A' is symmetric, since for $g \in \mathscr{D}_{A'}$,

$$\begin{aligned}\langle g \mid A'f\rangle &= \int_{\mathbb{R}^1} \lambda \, d\langle E_\lambda' g \mid f\rangle \\ &= \int_{\mathbb{R}^1} \lambda \, d\langle f \mid E_\lambda' g\rangle^* = \langle f \mid A'g\rangle^*.\end{aligned}$$

Thus we also have $A' \subseteq A'^* \subseteq A$, i.e., $A = A'$.

Now we can reverse the earlier procedure of going over from the spectral function of V to the spectral function of A, and derive that

$$\tilde{E}_\varphi' = E'_{-\cot \frac{1}{2}\varphi}$$

is the spectral function of the Cayley transform of A. Due to the uniqueness of $\tilde{E}_\varphi$ (see Theorem 6.1), it follows that $\tilde{E}_\varphi = \tilde{E}_\varphi'$, i.e., $E_\lambda = E_\lambda'$.

The above argument establishes at the same time that (6.28) coincides with the domain of definition $\mathscr{D}_A$ of A. Q.E.D.

The spectral function E_λ and the corresponding spectral measure $E(B)$ satisfying the conditions of Theorem 6.3 are said to belong to the operator A. In that case it is customary to write symbolically

$$A = \int_{\mathbb{R}^1} \lambda \, dE_\lambda . \tag{6.29}$$

The integral on the right-hand side is usually referred to as the *spectral decomposition* of A.

The integration in (6.25) extends effectively only over the spectrum S^A of the operator A, where the spectrum can be defined as follows.

Definition 6.2. A point $\lambda \in \mathbb{R}^1$ belongs to the *spectrum* S^A of the self-adjoint operator A with the spectral measure $E(B)$ if $E(I)$ is nonzero for every open interval I containing λ.

If $\lambda \notin \mathsf{S}^A$, then $E(I_0) = 0$ for some open interval I_0 containing λ. Hence, all the points in I_0 do not belong to the spectrum. Thus, all the points not belonging to the spectrum constitute an open set, and therefore the spectrum is always a closed set.

Definition 6.3. A point $\lambda \in \mathsf{S}^A$ is said to belong to the *point spectrum* $\mathsf{S}_{\mathrm{p}}^A$ of A if the spectral measure of A is different from zero at λ, i.e., $E(\{\lambda\}) \neq 0$. The set $\mathsf{S}_{\mathrm{c}}^A = \mathsf{S}^A - \mathsf{S}_{\mathrm{p}}^A$ is called the *continuous spectrum* of A.

One of the most important properties of the spectrum of a self-adjoint operator A is that it is the support of the spectral measure of A, i.e.,

$$E^A(\mathsf{S}^A) = 1. \tag{6.30}$$

In order to establish the above relation, let us write

$$\mathsf{S}^A = \bigcap_{k=1}^{\infty} S_k, \qquad S_1 \supset S_2 \supset \cdots,$$

where S_k is defined as the union of all intervals $[n/2^k, (n+1)/2^k]$ which contain at least one point of the spectrum of A. Now, if a closed interval I does not contain any point of the spectrum then $E^A(I) = 0$. In fact, if that were not so, we could split the interval in two halves and then at least on one of the halves the spectral measure would be different from zero. By continuing the process we would obtain a monotonic sequence of intervals which shrink to a point in I, and that point would obviously belong to the spectrum. Using this result, we can immediately infer that $E^A(\mathsf{S}^A) = 1$ and consequently, due to the continuity from above of spectral measures (see Exercise 5.8), we conclude that

$$E^A(\mathsf{S}^A) = \operatorname*{s-lim}_{k\to+\infty} E^A(S_k) = 1,$$

i.e., (6.30) is true.

6.6. The Spectral Theorem for Bounded Self-Adjoint Operators

We conclude this section by showing that for bounded self-adjoint operators the integral in (6.29) can be proven to converge uniformly, as was the case with the similar relation (6.1) for unitary operators. To prove this statement we need the following lemma.

Lemma 6.4. If A is a bounded self-adjoint operator on $\mathscr{H}$, its spectrum S^A lies within the interval $[-\| A \|, +\| A \|]$, and at least one of the two numbers $-\| A \|$ and $+\| A \|$ belongs to the spectrum of A.

Proof. If $\lambda \in S^A$ and if $E(B)$, $B \in \mathscr{B}^1$, is the spectral measure of A, then $E(I_\epsilon) \neq 0$, $I_\epsilon = (\lambda - \epsilon, \lambda + \epsilon)$ for any $\epsilon > 0$. Hence, there is a nonzero $f_\epsilon \in \mathscr{H}$ for which $E(I_\epsilon)f_\epsilon = f_\epsilon$. Consequently (see Exercise 6.10)

$$\| Af_\epsilon \|^2 = \int_{I_\epsilon} \lambda^2 \, d \| E_\lambda f_\epsilon \|^2 \geqslant (| \lambda | - \epsilon)^2 \| f_\epsilon \|^2, \qquad 0 < \epsilon < | \lambda |,$$

so that $| \, | \lambda | - \epsilon \, | \leqslant \| A \|$ for any of the above $\epsilon > 0$, i.e., $| \lambda | \leqslant \| A \|$. This establishes that $S^A \subset [-\| A \|, \| +A \|]$.

If $\alpha = \sup\{| \lambda |: \lambda \in S^A\}$, then by (6.30) and Exercise 6.10

$$\| Af \|^2 = \int_{[-\alpha, +\alpha]} \lambda^2 \, d \| E_\lambda f \|^2 \leqslant \alpha^2 \| f \|^2, \qquad f \in \mathscr{D}_A = \mathscr{H},$$

i.e., $\| A \| \leqslant \alpha$. On the other hand, $\alpha \leqslant \| A \|$ since $S^A \subset [-\| A \|, +\| A \|]$. Hence $\alpha = \| A \|$, and since S^A is closed we must have that either $\alpha \in S^A$ or $-\alpha \in S^A$, or both. Q.E.D.

We are now ready to prove Theorem 6.4.

Theorem 6.4. Let $-\| A \| = \lambda_0 < \lambda_1 < \cdots < \lambda_n = +\| A \|$ be a partition of the interval $[-\| A \|, +\| A \|]$ and $\lambda_1', \ldots, \lambda_n'$ any points for which $\lambda_{k-1} \leqslant \lambda_k' < \lambda_k$, $k = 1, \ldots, n$. Then in the limit of finer and finer partitions

$$A = \operatorname*{u-lim}_{\substack{n \to +\infty \\ \omega \to 0}} \sum_{k=1}^{n} \lambda_k'(E_{\lambda_k} - E_{\lambda_{k-1}}), \qquad \omega = \max_{k=1,\ldots,n} | \lambda_k - \lambda_{k-1} |, \tag{6.31}$$

where E_λ is the spectral function of A.

We shall only sketch the proof of this theorem.

By following the procedure used in proving Lemma 6.1, it can be easily established that the mapping $p(x) \to p(A)$ of the set of polynomials in $x \in \mathbb{R}^1$ into the set of linear operators in $\mathscr{H}$ has the following properties:

$$p_1(x) + p_2(x) \to p_1(A) + p_2(A),$$

$$p_1(x) \, p_2(x) \to p_1(A) \, p_2(A),$$

$$p^*(x) \to [p(A)]^*,$$

$$p(A) \geqslant 0 \qquad \text{if} \quad p(x) \geqslant 0.$$

Then, by using the method employed in proving Lemma 6.2 one can show that the characteristic function $\chi_{I_\lambda}(x)$ of the closed interval $I_\lambda = [-\| A \|, \lambda]$ is on the interval $[-\| A \|, +\| A \|]$ the limit of a monotonic sequence $p_1(x) \geqslant p_2(x) \geqslant \cdots$ of positive polynomials. Combining these two results one can easily show that

$$E_\lambda = \operatorname*{s-lim}_{n\to+\infty} p_n(A)$$

is a spectral function with $E_{\| A \|} = 1$.

From the construction of this spectral function it follows that, given any partition $-\| A \| = \lambda_0 < \lambda_1 < \cdots < \lambda_n = +\| A \|$ of the interval $[-\| A \|, +\| A \|]$, we have

$$\lambda_{k-1}(E_{\lambda_k} - E_{\lambda_{k-1}}) \leqslant A(E_{\lambda_k} - E_{\lambda_{k-1}}) \leqslant \lambda_k(E_{\lambda_k} - E_{\lambda_{k-1}}).$$

A summation in $k = 1,\ldots, n$ obviously yields

$$\sum_{k=1}^{n} \lambda_{k-1}(E_{\lambda_k} - E_{\lambda_{k-1}}) \leqslant A \leqslant \sum_{k=1}^{n} \lambda_k(E_{\lambda_k} - E_{\lambda_{k-1}}). \tag{6.32}$$

Furthermore, from the inequality

$$\sum_{k=1}^{n} \lambda_k(E_{\lambda_k} - E_{\lambda_{k-1}}) - \sum_{k=1}^{n} \lambda_{k-1}(E_{\lambda_k} - E_{\lambda_{k-1}})$$

$$= \sum_{k=1}^{n} (\lambda_k - \lambda_{k-1})(E_{\lambda_k} - E_{\lambda_{k-1}}) \leqslant \omega \sum_{k=1}^{n} (E_{\lambda_k} - E_{\lambda_{k-1}}) = \omega$$

we see that both sums in (6.32) have to converge to A in the limit $\omega \to 0$. Hence, (6.31) is established.

From (6.31) we easily deduce that

$$\langle f \mid Ag \rangle = \int \lambda \, d\langle f \mid E_\lambda g \rangle$$

for all $f, g \in \mathscr{H}$. On account of the uniqueness of the spectral decomposition of A, we conclude that E_λ is the spectral function of A. This completes the proof of the theorem.

Exercises

6.1. Prove the generalized Schwarz–Cauchy inequality

$$|\langle f \mid Ag \rangle|^2 \leqslant \langle f \mid Af \rangle \langle g \mid Ag \rangle,$$

which is valid for any positive symmetric operator A defined on $\mathscr{H}$.

6.2. Show that if A and B are two symmetric operators on $\mathscr{H}$, then $A \geqslant B$ and $B \geqslant A$ implies $A = B$.

6.3. Show that if $A_1, A_2, \ldots$ is a sequence of self-adjoint operators on $\mathscr{H}$ and $\text{s-lim}_{n\to\infty} A_n f = f'$ for each $f \in \mathscr{H}$, then $A(f) = f'$ is also a self-adjoint operator.

6.4. Verify explicitly that Lemma 6.1 is true.

6.5. Prove the *lemma of Fejér and Riesz*, which states the following: if $p(e^{i\varphi})$ is a positive trigonometric polynomial, $p(e^{i\varphi}) \geqslant 0$, there is a trigonometric polynomial $q(e^{i\varphi})$ such that $p(e^{i\varphi}) = |q(e^{i\varphi})|^2$.

6.6. Assume that $u_1(x), u_2(x), \ldots,$ and $v_1(x), v_2(x), \ldots,$ are two monotonically decreasing sequences of continuous functions on the finite interval I, for which $\lim_{n\to\infty} u_n(x) = \lim_{n\to\infty} v_n(x)$. Prove that for each integer n there is some other integer k_n such that

$$u_{k_n}(x) \leqslant v_n(x) + 1/n.$$

6.7. Show that if $f(\xi)g(\xi)$ and $g(\xi)$ are μ-integrable complex functions on R, and $\mu_1(S)$ denotes the complex measure

$$\mu_1(S) = \int_S g(\xi)\, d\mu(\xi), \qquad S \subset R,$$

then $f(\xi)$ is μ_1 integrable on R and

$$\int_R f(\xi)\, d\mu_1(\xi) = \int_R f(\xi)\, g(\xi)\, d\mu(\xi).$$

6.8. Show that if U is a unitary operator and $E(B)$, $B \in \mathscr{B}^1$, is its spectral measure, then $UE(B) = E(B)U$.

6.9. Show that if E_λ, $\lambda \in \mathbb{R}^1$, is a spectral function, the set

$$\left\{f\colon \int_{\mathbb{R}^1} \lambda^2\, d\langle f \mid E_\lambda f\rangle < +\infty\right\}$$

is a linear subspace of $\mathscr{H}$. (*Hint*: $\frac{1}{2}\|E_\lambda(f+g)\|^2 \leqslant \|E_\lambda f\|^2 + \|E_\lambda g\|^2$).

6.10. Prove that if $A = A^*$, then for $f \in \mathscr{D}_A$

$$\|Af\|^2 = \int_{\mathbb{R}^1} \lambda^2\, d\langle f \mid E_\lambda f\rangle.$$

References for Further Study

Consult any text on functional analysis, such as Akhiezer and Glazman [1961], Kato [1966], Riesz and Sz. Nagy [1955], Schechter [1971], Yosida [1974].

CHAPTER IV

The Axiomatic Structure of Quantum Mechanics

In the quantum mechanical problems encountered thus far we have had to limit ourselves basically to the special version of quantum mechanics known as wave mechanics, in which we could characterize the state as a wave function. The knowledge of the theory of linear operators in Hilbert spaces acquired in Chapter III will enable us to study the structure of quantum mechanics from an abstract and very general point of view, in which the formalism of wave mechanics becomes a particular realization of the general formalism.

We start this chapter with a section on the theory of measurement in quantum mechanics, which investigates the relation of the basic theoretical constructs to experimental material. We proceed then to explore the basic mathematical concepts of quantum mechanics, formulating in the process the fundamental axioms which these constructs have to obey. This entire exposition will lead in a natural way to specialized versions of quantum mechanics, such as the already familiar case of wave mechanics.

1. Basic Concepts in the Quantum Theory of Measurement

1.1. Observables and States in Quantum Mechanics

The theory of linear operators in Hilbert space developed in Chapter III has equipped us with all the necessary tools for a very general formulation of quantum mechanics.

As we saw in the Introduction, the basic categories of primitive concepts of quantum mechanics are, as in the case of classical mechanics, those of *state* and *observable*. In order to be able to give a mathematical description of these objects, we associate in classical mechanics with any given system a phase space; e.g., with an n-particle system we can associate a $6n$-dimensional vector space. For the same reason, we associate a Hilbert space with any given system which we intend to describe quantum mechanically. For instance, as we have seen in §5 of Chapter II, if the system consists of n particles, which are of different kinds, without spin and moving in three dimensions, then in wave mechanics (which is a special version of quantum mechanics) we associate with that system the Hilbert space $L^2(\mathbb{R}^{3n})$.

Let us assume now that we are dealing with a particular quantum mechanical problem in which a certain system has been specified (e.g., a hydrogen atom) and with which a certain Hilbert space $\mathscr{H}$ is associated [e.g., $L^2(\mathbb{R}^6)$]. It is then postulated that to each observable corresponds in the formalism a unique self-adjoint operator acting in $\mathscr{H}$ —though the converse is by no means required, i.e., there might be self-adjoint operators in $\mathscr{H}$ which do not correspond* to any observables.

As an example of operators corresponding to observables related to a two-particle system (e.g., the hydrogen atom) described wave mechanically in $L^2(\mathbb{R}^6)$, consider the self-adjoint operators $Q_i^{(x)}$, $Q_i^{(y)}$, $Q_i^{(z)}$, $P_i^{(x)}$, $P_i^{(y)}$, $P_i^{(z)}$, $i = 1, 2$, defined as follows.

$Q_i^{(x)}$ is the self-adjoint operator [see (4.19) of Chapter III] with a domain of definition

$$\mathscr{D}_i^{(x)} = \left\{\psi: \psi \in L^2(\mathbb{R}^6), \int_{\mathbb{R}^6} |x_i \psi(x)|^2 \, d^6x < +\infty\right\}, \tag{1.1}$$

where

$$x = (x_1, y_1, z_1, x_2, y_2, z_2), \qquad d^6x = dx_1 \cdots dz_2, \tag{1.2}$$

which acts on a $\psi \in \mathscr{D}_i^{(x)}$ in the following way:

$$(Q_i^{(x)}\psi)(x) = x_i \psi(x). \tag{1.3}$$

An analogous definition holds for $Q_i^{(y)}$ and $Q_i^{(z)}$.

$P_i^{(x)}$ is the self-adjoint operator [see (4.20) of Chapter III] with a domain of definition

$$\tilde{\mathscr{D}}_i^{(x)} = \left\{\psi: \psi \in L^2(\mathbb{R}^6), \int_{\mathbb{R}^6} |p_i^{(x)} \tilde{\psi}(p)|^2 \, d^6p < +\infty\right\}, \tag{1.4}$$

* In the early days of quantum mechanics there was a tendency to assume that each self-adjoint operator represents an observable (see von Neumann [1955]). However, the existence of so-called superselection rules (which will be considered later in this chapter) indicated for the first time that this requirement lacks experimental support.

where $\tilde{\psi}(p)$ is the Fourier–Plancherel transform of $\psi(x)$, and

$$(1.5)\qquad p = (p_1^{(x)}, p_1^{(y)}, p_1^{(z)}, p_2^{(x)}, p_2^{(y)}, p_2^{(z)}), \qquad d^6p = dp_1^{(x)} \cdots dp_2^{(z)};$$

$P_i^{(x)}$ acts on any $\psi \in \mathscr{C}_b^\infty(\mathbb{R}^6) \subset \mathscr{D}_i^{(x)}(\mathbb{R}^6)$ according to the formula

$$(1.6)\qquad (P_i^{(x)}\psi)(x) = -i\hbar \frac{\partial\psi(x)}{\partial x_i};$$

$P_i^{(y)}$ and $P_i^{(z)}$ are defined analogously. The operators $Q_i^{(x)}$, $Q_i^{(y)}$, $Q_i^{(z)}$ correspond then to observables which from the experimental point of view represent measurements of the x, y, and z coordinate, respectively, of the ith particle in a Cartesian inertial reference frame. Similarly, $P_i^{(x)}$, $P_i^{(y)}$, $P_i^{(z)}$ correspond to the measurements of the x, y, and z components, respectively, of the momentum of the ith particle with respect to the same frame.*

As another example of an observable for the same system, consider the operator H' with domain of definition $\mathscr{D}_{H'} = \mathscr{C}_b^\infty(\mathbb{R}^6)$, which maps a $\psi \in \mathscr{D}_{H'}$ into

$$(1.7)\qquad (H'\psi)(x) = \left(-\frac{\hbar^2}{2m_1}\Delta_1 - \frac{\hbar^2}{2m_2}\Delta_2 + V(x)\right)\psi(x).$$

For a very wide class of suitable potentials $V(x)$, the above operator H' is essentially self-adjoint, i.e., its adjoint $(H')^* \supseteq H'$ is a self-adjoint operator. Its self-adjoint extension $(H')^*$, customarily denoted by H, is called the *Hamiltonian* of the system, and is an observable corresponding to measurements of the *total* energy of the system.

In general, for any given system, there are *correspondence rules specifying which self-adjoint operator corresponds to each particular observable* related to the system. *These rules are part of the set of correspondence rules relating the formalism to the experimental framework.* We shall meet some of the most important of these rules, which are concerned with the most fundamental observables, such as energy, position, momentum, angular momentum, spin, and charge, as we proceed in our study of quantum mechanics.

Let us turn now to the concept of state. As a generalization of the consideration in §5 of Chapter I and §5 of Chapter II, it is postulated, in general, that a state† is specified by a vector-valued function $\Psi(t)$,

* See Exercises 1.5 and 1.6 for the spectral measures of these operators.

† We are adopting here the *Schroedinger picture* of quantum mechanics. In §3 we shall encounter *other physically equivalent* pictures.

$\Psi(t) \in \mathscr{H}$, of the time parameter t; $\Psi(t)$ is never zero, i.e., $\| \Psi(t)\| > 0$ for all $t \in \mathbb{R}^1$. We have, again, as in the aforementioned sections, that if $\Psi(t)$ describes a state, then $c\Psi(t)$, for any nonzero constant c, describes the same state. The reason for this assumption lies in the nature of Born's correspondence rule, which is the fundamental principle on which the accepted interpretation of the quantum mechanical formalism lies. We shall now focus our attention on this rule.

1.2. The Concept of Compatible Observables

Let A be a self-adjoint operator corresponding to an observable (also referred to as A), and denote by $E(B)$, $B \in \mathscr{B}^1$, the spectral measure of A

$$A = \int_{-\infty}^{+\infty} \lambda \, dE_\lambda .$$

Consider a system which is known to be at time t in a state described by the normalized vector $\Psi(t)$. In keeping with a suggestion first forwarded by Born [1926a, b] for the case when A is a position observable [see (5.2) of Chapter II], we interpret the measure

$$P^A_{\Psi(t)}(B) = \langle \Psi(t) \mid E(B)\, \Psi(t)\rangle, \qquad B \in \mathscr{B}^1, \tag{1.8}$$

as a probability measure in the following way: a measurement to determine the value of A, carried out on the system at the time t, has the probability $P^A_{\Psi(t)}(B)$ to have as outcome a value within the set B.

A generalization of this interpretation can be applied to any finite set of *compatible observables*. This necessitates an understanding of the concept of compatibility, which is empirically rooted and to which we give the following meaning.

Definition 1.1. The observables $A_1 ,..., A_n$ related to a certain system are *compatible* if, in principle, *arbitrarily accurate* measurements of the simultaneous values of these observables can be carried out; i.e., if for any *prepared* $\alpha \in \mathrm{S}$ from the set $\mathrm{S} \subset \mathbb{R}^n$ of feasible (cf. Definition 5.2) simultaneous values of $A_1 ,..., A_n$, and for any open confidence* interval I containing α, the confidence probability p of actual readings falling within I can be brought arbitrarily close to one by a suitable design of apparatus for the simultaneous measurement of $A_1 ,..., A_n$.

Denote the self-adjoint operators corresponding to the observables $A_1 ,..., A_n$ also by $A_1 ,..., A_n$. It is a basic feature of all quantum mechan-

* See Sections 2.49 and 5.21 of Dietrich [1973].

ical theories that if the n observables $A_1, \ldots, A_n$ are compatible, then the corresponding operators must commute with each other. Note, however, that in general $A_1, \ldots, A_n$ can be unbounded operators which are not defined on the entire Hilbert space $\mathscr{H}$ and, therefore, for which we do not have an *a priori* guarantee that the commutator

$$[A_i, A_k] = A_i A_k - A_k A_i$$

is defined on a domain which is dense in $\mathscr{H}$. We bypass this difficulty by introducing the following concept of commutativity.

Definition 1.2. Two self-adjoint operators A_1 and A_2 are said to *commute* if their respective spectral measures $E^{(1)}(B)$ and $E^{(2)}(B)$ commute, i.e.,

$$[E^{(1)}(B_1), E^{(2)}(B_2)] = 0 \tag{1.9}$$

for all $B_1, B_2 \in \mathscr{B}^1$.

In the case when A_1 and A_2 are bounded self-adjoint operators with $\mathscr{D}_{A_1} \equiv \mathscr{D}_{A_2} \equiv \mathscr{H}$, (1.9) is a sufficient and necessary condition (see Exercises 1.1 and 1.2) for $[A_1, A_2] = 0$.

1.3. Born's Correspondence Rule for Determinative Measurements

The above considerations show that when $A_1, \ldots, A_n$ are self-adjoint operators corresponding to compatible observables and $E^{(1)}(B), \ldots, E^{(n)}(B)$ are their respective spectral measures, then $E^{(1)}(B_1), \ldots, E^{(n)}(B_n)$ commute, and consequently the operator

$$E(B_1 \times \cdots \times B_n) = E^{(1)}(B_1) \cdots E^{(n)}(B_n) \tag{1.10}$$

is a projector (see Chapter III, Theorem 3.6). If $\Psi(t)$ is a normalized vector, $\| \Psi(t) \| = 1$, representing the state of a system at the time t, then the generalization

$$P_{\Psi(t)}^{A_1,\ldots,A_n}(B_1 \times \cdots \times B_n) = \langle \Psi(t) \mid E^{(1)}(B_1) \cdots E^{(n)}(B_n) \Psi(t) \rangle \tag{1.11}$$

of (1.8) is interpreted as being the probability that a measurement designed to determine the simultaneous values $\lambda_1, \ldots, \lambda_n$ of $A_1, \ldots, A_n$, respectively, will yield a result

$$\lambda_1 \in B_1, \ldots, \lambda_n \in B_n, \qquad B_1, \ldots, B_n \in \mathscr{B}^1,$$

when carried out on the system at time t.

We can generalize the above statement further by noting that the set function (1.11), defined by (1.11) only on the Borel sets of $\mathbb{R}^n$ which belong to $(\mathscr{B}^1)^n = \mathscr{B}^1 \times \cdots \times \mathscr{B}^1$, can be extended to a measure $P_{\Psi(t)}^{A_1,\ldots,A_n}(B)$, $B \in \mathscr{B}^n$, on the family $\mathscr{B}^n$ of all the Borel sets in $\mathbb{R}^n$. Moreover, by Theorem 5.5 in Chapter III, the projector-valued set function (1.10) can be extended to a spectral measure $E^{A_1,\ldots,A_n}(B)$ defined for all $B \in \mathscr{B}^n$, and such that

$$P_{\Psi(t)}^{A_1,\ldots,A_n}(B) = \langle \Psi(t) \mid E^{A_1,\ldots,A_n}(B)\, \Psi(t)\rangle.$$

This suggests the following interpretation, which applies to measurements which are designed to determine (i.e., *determinative* measurements; see the Introduction) the simultaneous values of any finite number of given *compatible** observables (see Section 8.9 for incompatible ones).

Definition 1.3 *(Born's correspondence rule for determinative measurements).* Let $A_1, \ldots, A_n$ be n commuting self-adjoint operators acting in $\mathscr{H}$, with spectral measures $E^{A_1}(B), \ldots, E^{A_n}(B)$, respectively, and representing n compatible observables. Denote by

$$E^{A_1,\ldots,A_n}(B), \qquad B \in \mathscr{B}^n, \tag{1.12}$$

the unique spectral measure which satisfies the equalities

$$E^{A_1,\ldots,A_n}(B_1 \times \cdots \times B_n) = E^{A_1}(B_1) \cdots E^{A_n}(B_n) \tag{1.13}$$

for all choices of $B_1, \ldots, B_n \in \mathscr{B}^1$. If the vector-valued function $\Psi(t)$

$$\Psi(t) \in \mathscr{H}, \qquad -\infty < t < +\infty,$$

represents an already prepared state of a given system, then the measure

$$P_{\Psi(t)}^{A_1,\ldots,A_n}(B) = \| \Psi(t)\|^{-2}\langle \Psi(t) \mid E^{A_1,\ldots,A_n}(B)\, \Psi(t)\rangle \tag{1.14}$$

is a probability measure: for each $B \in \mathscr{B}^n$ the number $P_{\Psi(t)}^{A_1,\ldots,A_n}(B)$ is the theoretically predicted probability that a measurement carried out at time t to *determine* the *simultaneous* values $\lambda_1, \ldots, \lambda_n$ of $A_1, \ldots, A_n$, respectively, will yield a result within B, i.e., $(\lambda_1, \ldots, \lambda_n) \in B$ at t.

* The first attempts of treating incompatible observables were made in the 1960s (She and Hefner [1966], Prugovečki [1966. 1967], Park and Margenau [1968]). A consistent extrapolation of Born's correspondence rule turns out to be indeed feasible (Prugovečki [1976], Ali and Prugovečki [1977], Schroeck [1978]), and in fact quite essential for *relativistic* quantum mechanics (Prugovečki [1978d, 1981a-c]).

According to the above interpretation

$$(1.15)\quad \langle A\rangle_{\Psi(t)} = \langle \Psi(t) \mid A\Psi(t)\rangle = \int_{-\infty}^{+\infty} \lambda \, dP^A_{\Psi(t)}(\lambda), \qquad \| \Psi(t)\| = 1,$$

represents the *mean value* (also called *expectation value*) of the observable A for the system in the state $\Psi(t)$ at time t. If A is an unbounded operator, then (1.15) is not necessarily defined for all vectors $\Psi(t) \in \mathscr{H}$, since the integral on the right-hand side might diverge for some such vectors. However, if $\Psi(t) \in \mathscr{D}_A$, then by the spectral theorem

$$(1.16)\qquad \int_{-\infty}^{+\infty} \lambda^2 \, dP^A_{\Psi(t)}(\lambda) < +\infty,$$

and since $|\lambda| \leqslant \lambda^2$ for $|\lambda| \geqslant 1$, the integral in (1.15) converges and $\langle A\rangle_{\Psi(t)}$ exists.

As an example of compatible observables, take the six position observable $Q_i^{(x)},\ldots$, $i = 1, 2$, defined by (1.1) and (1.3). The spectral measure (1.12) is (see Exercises 1.4 and 1.5)

$$(E^{\mathbf{Q}_1,\mathbf{Q}_2}(B)\psi)(x, t) = \chi_B(x)\, \psi(x, t),$$

and the probability of discovering the first particle at $\mathbf{r}_1$ and the second particle at $\mathbf{r}_2$, where $x = (\mathbf{r}_1, \mathbf{r}_2) \in B$, is given according to (1.14) by

$$(1.17)\qquad P^{\mathbf{Q}_1,\mathbf{Q}_2}_{\Psi(t)}(B) = \int_B |\psi(x, t)|^2 \, d^6x.$$

This agrees with (5.2) of Chapter II.

The six operators $P_i^{(x)},\ldots$, $i = 1, 2$, representing momentum and defined by (1.4) and (1.6), also correspond to compatible observables. In this case we have (see also Exercise 1.6)

$$(1.18)\qquad (E^{\mathbf{P}_1,\mathbf{P}_2}(B)\psi)(x, t) = (U_{\mathbf{F}}^{-1}[\chi_B \cdot \tilde{\psi}])(x, t).$$

Thus, in terms of the Fourier–Plancherel transform $\tilde{\psi}(p)$ of $\psi(x)$,

$$(1.19)\qquad P^{\mathbf{P}_1,\mathbf{P}_2}_{\Psi(t)}(B) = \int_B |\tilde{\psi}(p, t)|^2 \, d^6p.$$

The operators $Q_i^{(x)}$ and $P_i^{(x)}$ do not commute. This feature corresponds to the experimental fact that the position and momentum are incompatible observables.

1.4. Born's Correspondence Rule for Preparatory Measurements

The above correspondence rule (Definition 1.3) and discussion of the process of measurement is related to the subject of determinative measurement. In this case, we have a system about which it is *already known* to be in a state described by the vector-valued function $\Psi(t)$. Thus, we have a situation in which, on the one hand, the correspondence rule in Definition 1.3 predicts the probability that different observables of the system have certain values at any given time t, while, on the other hand, a procedure of measurement (such as those described in the Introduction for position and momentum) enable us to experimentally determine the values of these same observables. These considerations do not tell us, however, how to make sure that a system is indeed in a state $\Psi(t)$. This last question involves *preparatory measurements* or *preparation-of-state procedures.*

Let us assume that we have (for the considered system) n compatible observables represented by the operators $A_1, \ldots, A_n$, which are *complete with respect to some simultaneous eigenvector* Ψ_0,

$$A_1\Psi_0 = \lambda_1^{(0)}\Psi_0, \ldots, A_n\Psi_0 = \lambda_n^{(0)}\Psi_0, \tag{1.20}$$

i.e., for which Ψ_0 is, up to a multiplicative constant, the only vector satisfying the n eigenvalue equations in (1.20) with the respective eigenvalues $\lambda_1^{(0)}, \ldots, \lambda_n^{(0)}$. According to Definition 1.1 of compatible observables, there must be not only determinative but also preparatory procedures which would enable us to prepare the system so as to have at some time t_0 a value of A_k in the ϵ neighborhood of $\lambda_k^{(0)}$. If $\lambda_1^{(0)}, \ldots, \lambda_n^{(0)}$ are not accumulation points of the spectra of $A_1, \ldots, A_n$, respectively, then for some ϵ_0 there are in addition to $\lambda_k^{(0)}$ no other points λ_k in the spectrum of A_k which lie in the ϵ_0 neighborhood of $\lambda_k^{(0)}$. Thus, there should be, in principle, a sufficiently accurate (ϵ_0-accurate) preparatory measurement which would tell us that the considered system had at time t_0 the value $\lambda_k^{(0)}$ for the observable A_k, $k = 1, \ldots, n$. Then the accepted framework of quantum mechanics says that the state $\Psi(t)$, $-\infty < t < +\infty$, of the system immediately after the preparatory measurement is such that

$$\Psi(t_0) = \Psi_0, \tag{1.21}$$

where Ψ_0 satisfies (1.20). As we shall see in §3.2, the dynamical law of quantum mechanics assures us that in this case there is only one state $\Psi(t)$ satisfying (1.21).

An ideal preparatory measurement, like that above, is rarely the case in practice. Usually we can establish that the simultaneous values $\lambda_1, \ldots, \lambda_n$ of the n compatible observables $A_1, \ldots, A_n$ are within an n-dimensional interval I, or, more generally, within a Borel set $B \in \mathscr{B}^n$ at some instant t_0. For instance, we have seen in the Introduction that the experimental arrangement on Fig. 2 does not offer us an absolutely precise knowledge of the momentum of a charged particle which emerges from the apparatus in the interaction region, but it prepares a range Δ of momenta.

We see that we cannot limit ourselves to the very idealized case treated by (1.20). Instead, we should be able to cope with the most general imaginable case. To arrive at such a generalization of the case treated by (1.20), we shall recast (1.20) in a form which yields itself to straightforward generalization.

The vector Ψ_0 satisfying (1.20) can be characterized as the unique vector (up to a multiplicative constant) satisfying

$$(1.22) \qquad E^{A_1,\ldots,A_n}(\{\lambda_1^{(0)}\} \times \cdots \times \{\lambda_n^{(0)}\})\, \Psi_0 = \Psi_0\,,$$

where the above projector is given by the formula (1.13) with $B_1 = \{\lambda_1^{(0)}\}, \ldots, B_n = \{\lambda_n^{(0)}\}$. Thus, by Theorem 3.3 in Chapter III, Ψ_0 is the unique vector (up to a multiplicative constant) in $\mathscr{H}$ for which

$$(1.23) \qquad \langle \Psi_0 \mid E^{A_1,\ldots,A_n}(\{\lambda_1^{(0)}\} \times \cdots \times \{\lambda_n^{(0)}\})\, \Psi_0\rangle = \| \Psi_0 \|^2.$$

The characterization of Ψ_0 by (1.22), or equivalently by (1.23), can be applied to the problem of characterizing the state of a particle prepared by the apparatus described in Fig. 2 (see the Introduction) to have at time t_0 its momentum within Δ.

We achieve this generalization by introducing the projector $E^{\mathbf{P}_1}(\Delta)$, which, for instance, in the case of the two-particle system considered earlier, can be written as

$$E^{\mathbf{P}_1}(\Delta) = E^{\mathbf{P}_1,\mathbf{P}_2}(\Delta \times \mathbb{R}^3),$$

with $E^{\mathbf{P}_1,\mathbf{P}_2}$ given by (1.18). According to (1.22) we can state that this two-particle system, with the first particle emerging from the aperture O_3 (see Fig. 2) at time t_0, can be afterwards in any state $\Psi(t)$ which at t_0 satisfies the relation

$$(1.24) \qquad \langle \Psi(t_0) \mid E^{\mathbf{P}_1}(\Delta)\, \Psi(t_0)\rangle = \| \Psi(t_0)\|^2.$$

It should be noted that the outlined preparatory measurement obviously does not single out a unique state $\Psi(t)$, but rather a whole set of states. This state of affairs is by no means peculiar to quantum mechanics, but is shared by other physical theories. It is due to the ever-present "errors" of measurement. Thus, when trying to determine* in classical physics the state $\mathbf{r}(t)$, $-\infty < t < +\infty$, of a particle of mass m moving in a known force field by measuring its momentum $\mathbf{p}$ at some instant t_0, we determine, in fact, only some range Δ of momenta, since "errors" of measurement are always present. After this measurement the particle could be in any of the states $\mathbf{r}(t)$ for which

$$m\dot{\mathbf{r}}(t_0) \in \Delta. \tag{1.25}$$

A simultaneous measurement of position of the particle, which would locate it at t_0 within the space region Δ', would reduce the selectable set of states to those which in addition to (1.25) also satisfy the relation

$$\mathbf{r}(t_0) \in \Delta'.$$

However, even this measurement will not, strictly speaking, single out a unique state $\mathbf{r}(t)$ since in practice Δ and Δ' are never one-point sets.

If we return to the quantum mechanical case, it is interesting to note that the experimental arrangement in Fig. 2 prepares not only a range Δ of momentum values of the particle, but also a range Δ' of the position coordinates; namely, if a particle has emerged in the interaction region during the time interval t_0 to $t_0 + \Delta t$ while the aperture O_3 was open, the particle must have been somewhere within the confines of that aperture at the instant t_0. However, the accepted interpretation of quantum mechanics deals only with the measurements of compatible observables, and cannot take advantage of the additional information that is derived from simultaneous measurements of incompatible observables.

We get the general form of this interpretation for the case of preparatory measurements by straightforward generalization from the special case of momentum measurement discussed above.

Definition 1.4 (*Born's correspondence rule for preparatory measurements*). Suppose that a preparatory measurement of n compatible observables, represented by the self-adjoint operators $A_1, \ldots, A_n$, has been carried out on the system at time t_0, and that this measurement establishes that the simultaneous values $\lambda_1, \ldots, \lambda_n$ of these observables

* Since the disturbances of macroscopic systems caused by the impact of photons used in the measurement process are negligible, the distinction between preparatory and determinative measurement is of no practical consequence in classical mechanics.

are within the set $B \in \mathscr{B}^n$ at the instant t_0. Then, after the instant t_0, the system can only be in a state $\Psi(t)$ satisfying at t_0 the relation

$$P^{A_1,\ldots,A_n}_{\Psi(t_0)}(B) = \| \Psi(t_0)\|^{-2}\langle\Psi(t_0) \mid E^{A_1,\ldots,A_n}(B)\, \Psi(t_0)\rangle = 1. \tag{1.26}$$

According to the above, a preparatory measurement assigns to the system the entire family of all states satisfying (1.26). Since, in general, this family contains more than one state, the theory will not predict a sharp probality $P^{A_1',\ldots,A_m'}_{\Psi(t_1)}(B')$ for the outcome within $B' \in \mathscr{B}^m$ of a determinative simultaneous measurement of $A_1',\ldots, A_m'$ at time t_1 ($t_1 > t_0$), but rather a whole range of such values. Statistical considerations (to be mentioned in §8 in connection with the "statistical operator") can replace this range by a representative "mean" value.

1.5. The Stochastic Nature of the Quantum Theory of Measurement

As we see from Definitions 1.3 and 1.4, the only link between the formalism and the physical reality is via the probability measures (1.14). An experiment consists in first having the system submitted to a preparatory procedure at some instant t_0 and then carrying out a determinative measurement at some later instant t_1. The Born correspondence rule in Definition 1.4 enables us to specify a set of states $\Psi(t)$ as being the states in which the system can "find itself" after the preparatory procedure. We note that even in the most ideal case we cannot assign to the system at time t_0 a single vector Ψ_0, but rather an entire family $\{\Psi\colon \Psi = c\Psi_0, c \in \mathbb{C}^1, c \neq 0\}$ of vectors. This explains the reason for the statement that if $\Psi(t)$ represents a state, then $c\Psi(t)$, $|c| > 0$, represents the same state.

From Born's correspondence rule for determinative measurements (Definition 1.3) we note that the quantum mechanical theory, as opposed to the classical theory, does not predict the outcome of a *single* measurement carried out at some instant $t_1 > t_0$. The prediction is statistical, and as such it refers to a large ensemble of identical measurements rather than a single measurement. This means that the preparatory procedure should be carried out at the time t_0 on a large sample of N identical and independent systems. Then the theory predicts the frequency with which a determinative simultaneous measurement of certain compatible observables $A_1,\ldots, A_n$, carried out at time t, will have an outcome within the set B; i.e., if out of the N systems a number $\nu(B)$ will have at t values within B for $A_1,\ldots, A_n$, then, if the theory is correct, we must have the following approximate relationship,

$$\frac{\nu(B)}{N} \approx P^{A_1,\ldots,A_n}_{\Psi(t)}(B), \tag{1.27}$$

with an accuracy which increases with an increasing number N of systems in the ensemble.

It is conceivable that in some exceptional* cases a measurement on the set $A_1, \ldots, A_n$ of compatible observables, which is complete with respect to Ψ_0 and satisfies (1.20), can serve not only to determine whether the values of $A_1, \ldots, A_n$ are $\lambda_1^{(0)}, \ldots, \lambda_n^{(0)}$, but also as a preparatory procedure which leaves the system with the values $\lambda_1^{(0)}, \ldots, \lambda_n^{(0)}$ for $A_1, \ldots, A_n$, respectively—at least for a short time after the measurement. If the system was prior to this measurement in a state $\Psi(t)$, then after the measurement at time t_1 it will be in a state $\Psi'(t)$ for which $\Psi'(t_0) = \Psi_0$. In this case, the probability

$$P_{\Psi(t_0)}^{A_1,\ldots,A_n}(\{\lambda_1^{(0)}\} \times \cdots \times \{\lambda_n^{(0)}\}) = |\langle \Psi_0 \mid \Psi(t_0)\rangle|^2 \tag{1.28}$$

can be called the transition probability of having the interference introduced by the process of measurement cause the transition of the system from the state $\Psi(t)$ to the state $\Psi'(t)$ which is equal to Ψ_0 at t_0. For this reason the expression $|\langle \Psi_1 \mid \Psi_2\rangle|^2$ for two normalized vectors representing two states at some instant is sometimes called a *transition probability*.

Exercises

1.1. Assume that A_i, $i = 1, 2$, are symmetric operators on $\mathscr{H} = \mathscr{D}_{A_1} = \mathscr{D}_{A_2}$, with spectral measures $E^{(i)}(B)$, $i = 1, 2$. Show that if $[E^{(1)}(B_1), E^{(2)}(B_2)] = 0$ for all $B_1, B_2 \in \mathscr{B}^1$, then $[A_1, A_2] = 0$.

1.2. Show that if A_k, $k = 1, 2$, are commuting bounded symmetric operators on $\mathscr{H}$, then their spectral measures commute.

Note. A corresponding statement is by no means true when A_1, A_2 are unbounded operators, i.e., when $\mathscr{D}_{A_1} \neq \mathscr{H}$ and $\mathscr{D}_{A_2} \neq \mathscr{H}$. In fact, it has been shown by Nelson [1959] that there are self-adjoint operators with a common dense domain which they leave invariant and on which they even commute (being essentially self-adjoint; see §7) but possessing noncommuting spectral measures (see also Exercise 5.4).

1.3. Suppose $\rho(x) \geqslant 0$ is a Lebesgue integrable function on $\mathbb{R}^n$. Introduce the Lebesgue–Stieltjes measure $\mu_\rho(B) = \int_B \rho(x)\, d^n x$, $B \in \mathscr{B}^n$.

* That this is not always so can be seen from the extreme case when the system itself is destroyed by the determinative measurement, e.g., by becoming a part of the detector (for instance, when the detector is a photographic plate).

Show that a Borel measurable function $f(x)$, $x \in \mathbb{R}^n$, is μ_ρ integrable on $\mathbb{R}^n$ if and only if $f(x)\,\rho(x)$ is Lebesgue integrable on $\mathbb{R}^n$, and in this case*

$$\int_{\mathbb{R}^n} f(x)\,\rho(x)\,d^n x = \int_{\mathbb{R}^n} f(x)\,d\mu_\rho(x).$$

1.4. Use the result of Exercise 1.3 to establish that the spectral measure $(E^{(k)}(B)\psi)(x) = \chi_B(x_k)\,\psi(x)$, $x \in \mathbb{R}^n$, belongs to the operator $(Q_k\psi)(x) = x_k\,\psi(x)$, $x = (x_1, ..., x_n)$, which has the domain of definition $\mathscr{D}_{Q_k} = \{\psi: \int_{\mathbb{R}^n} | x_k\psi(x)|^2\,d^n x < +\infty\}$.

1.5. The spectral measures of $Q_1, ..., Q_n$ are $(E^{Q_k}(B_k)\psi)(x) = \chi_{B_k}(x_k)\,\psi(x)$. Prove that the spectral measure $E(B)$ defined by $(E(B)\psi)(x) = \chi_B(x)\,\psi(x)$ coincides with $E^{Q_1,...,Q_n}(B)$.

1.6. Prove that if U is a unitary operator and A is a self-adjoint operator with the spectral measure $E^A(B)$, then the spectral measure of the self-adjoint operator $A_1 = U^{-1}AU$ is $E^{A_1}(B) = U^{-1}E^A(B)U$.

2. Functions of Compatible Observables

2.1. Fundamental and Nonfundamental Observables

In §1 we had postulated that in quantum mechanics an observable is represented by a self-adjoint operator in a Hilbert space and that a state is represented by a vector-valued function $\Psi(t)$, $t \in \mathbb{R}^1$. Thus, as the most basic requirement for a quantum mechanical description of a system we have the following.

Axiom O. To any quantum mechanical system belongs at least one complex Hilbert space $\mathscr{H}$ in which the quantum mechanical theory of that system can be formulated.

If h is a normalized vector in the complex Hilbert space $\mathscr{H}$, the set

$$\underset{\sim}{h} = \{ah: a \in \mathbb{C}^1, | a | = 1\}$$

is called a *ray* in $\mathscr{H}$. We note that if Ψ_0 is a normalized vector representing a state at an instant t_0, then for any complex number a, $| a | = 1$, $a\Psi_0$ is another normalized vector representing the same state. These

* This result can be deduced also from Radon–Nikodym's theorem (see Halmos [1950]) once it is observed that $\rho(x)$ equals the Radon–Nikodym derivative $d\mu_\rho(x)/d^n x$.

remarks justify the frequently made statement that "a state of a quantum system is (at any given instant) a ray in the Hilbert space."

In analyzing the concept of observables we have demanded until now that an observable play the dual role of being related to a symbol of the formalism (a self-adjoint operator, in case of quantum mechanics) and at the same time be anchored directly in the experiment by being related to some empirical procedure (or procedures) for measuring it. We shall call observables satisfying both these conditions *fundamental observables*, to distinguish them as a subfamily of the wider family of observables in general, which will be introduced next.

In order to realize the necessity of enlarging the concept of observables, consider the concept of a function of observables in classical mechanics. Take, for instance, a one-particle system. If we know the position $\mathbf{r}$ and momentum $\mathbf{p}$ of the particle at some instant, then we can compute the values of other observables—such as angular momentum $\mathbf{r} \times \mathbf{p}$, kinetic energy* $\mathbf{p}^2/2m$, potential energy $V(\mathbf{r})$—since all these other observables are functions of $\mathbf{r}$ and $\mathbf{p}$. More generally, if we take any real-valued function $F(\mathbf{r}, \mathbf{p})$ in the six variables of $\mathbf{r}, \mathbf{p}$, we can think of it as an observable, because an indirect measurement of $F(\mathbf{r}, \mathbf{p})$ can be carried out by measuring the position and momentum of the particle and then computing $F(\mathbf{r}, \mathbf{p})$.

The above argument indicates that we can generalize the concept of an observable in quantum mechanics by introducing functions of one or more already given fundamental observables. However, while in classical physics all observables are assumed to be compatible, this is not the case in quantum physics. Thus, in quantum mechanics we shall be able to give an adequate definition only of functions of a single observable or, at most, of a set of compatible observables.

2.2. Bounded Functions of Commuting Self-Adjoint Operators

The mathematical basis of the concept of a function of observables is contained mainly in Theorems 2.1 and 2.3 of the present section.

Theorem 2.1. Let $E^{A_1}(B),\ldots, E^{A_n}(B)$ be the spectral measures of n commuting self-adjoint operators $A_1,\ldots, A_n$ in $\mathscr{H}$. Denote by $E^{A_1,\ldots,A_n}(B)$, $B \in \mathscr{B}^n$, the spectral measure on $\mathscr{B}^n$ which satisfies the relation

$$E^{A_1,\ldots,A_n}(B_1 \times \cdots \times B_n) = E^{A_1}(B_1) \cdots E^{A_n}(B_n) \tag{2.1}$$

* The mass m and the force field characterizing the problem are assumed to be given.

for all $B_1, \dots, B_n \in \mathscr{B}^1$. If $F(\lambda_1, \dots, \lambda_n)$ is a bounded Borel measurable function on $\mathbb{R}^n$,

$$|F(\lambda_1, \dots, \lambda_n)| \leqslant M, \qquad \lambda_1, \dots, \lambda_n \in \mathbb{R}^1, \tag{2.2}$$

there is a unique bounded linear operator A satisfying the relation

$$\langle f \mid Ag \rangle = \int_{\mathbb{R}^n} F(\lambda_1, \dots, \lambda_n)\, d\langle f \mid E^{A_1, \dots, A_n}_{\lambda_1, \dots, \lambda_n} g \rangle \tag{2.3}$$

for all vectors $f, g \in \mathscr{H}$. This operator A, which is conventionally denoted by $F(A_1, \dots, A_n)$, is bounded and $\|F(A_1, \dots, A_n)\| \leqslant M$.

Proof. Since the complex measure

$$\mu_{f,g}(B) = \langle f \mid E^{A_1, \dots, A_n}(B) g \rangle, \qquad B \in \mathscr{B}^n,$$

is equal to a sum [see (5.13) of Chapter III] of finite measures, any bounded measurable function on $\mathbb{R}^n$ is integrable (see Chapter II, Theorem 3.9). Hence, the functional

$$(f \mid g) = \int_{\mathbb{R}^n} F(\lambda_1, \dots, \lambda_n)\, d\mu_{f,g}(\lambda_1, \dots, \lambda_n)$$

is defined for all $f, g \in \mathscr{H}$. This functional is obviously a bilinear form. Let us show that $(f \mid g)$ is a bounded bilinear form.

Construct a sequence $s_1, s_2, \dots$ of complex simple functions on $\mathbb{R}^n$

$$s_k(\lambda_1, \dots, \lambda_n) = \sum_{i=1}^{n_k} a_{ik} \chi_{B_{ik}}(\lambda_1, \dots, \lambda_n), \tag{2.4}$$

$$B_{ik} \cap B_{i'k} = \varnothing \qquad \text{for} \quad i \neq i',$$

for which (see Exercise 2.1)

$$|s_k(\lambda_1, \dots, \lambda_n)| \leqslant |F(\lambda_1, \dots, \lambda_n)| \leqslant M, \tag{2.5}$$

and for any $f, g \in \mathscr{H}$

$$\lim_{k \to \infty} \int_{\mathbb{R}^n} s_k(\lambda)\, d\mu_{f,g}(\lambda) = \int_{\mathbb{R}^n} F(\lambda)\, d\mu_{f,g}(\lambda). \tag{2.6}$$

An application of (2.5) and of the Schwarz–Cauchy inequality, first in $\mathscr{H}$ and then in $l^2(n_k)$, yields

$$\begin{aligned}
&\left| \int_{\mathbb{R}^n} s_k(\lambda)\, d\mu_{f,g}(\lambda) \right| \\
&\quad = \left| \sum_{i=1}^{n_k} a_{ik}\, \mu_{f,g}(B_{ik}) \right| \\
&\quad \leqslant \sum_{i=1}^{n_k} | a_{ik} |\, |\langle f \mid E^{A_1,\ldots,A_n}(B_{ik})g\rangle| \\
&\quad \leqslant M \sum_{i=1}^{n_k} \| E^{A_1,\ldots,A_n}(B_{ik})f \|\, \| E^{A_1,\ldots,A_n}(B_{ik})g \| \\
&\quad \leqslant M \left[\sum_{i=1}^{n_k} \| E^{A_1,\ldots,A_n}(B_{ik})f \|^2\right]^{1/2} \left[\sum_{i=1}^{n_k} \| E^{A_1,\ldots,A_n}(B_{ik})g \|^2\right]^{1/2} \\
&\quad = M \left|\left\langle f \,\Big|\, \sum_{i=1}^{n_k} E^{A_1,\ldots,A_n}(B_{ik})f \right\rangle\right|^{1/2} \left|\left\langle g \,\Big|\, \sum_{i=1}^{n_k} E^{A_1,\ldots,A_n}(B_{ik})g \right\rangle\right|^{1/2} \\
&\quad \leqslant M \| f \|\, \| g \|.
\end{aligned}$$

The above inequality implies, when combined with (2.6), that

$$|(f \mid g)| \leqslant M \| f \|\, \| g \|.$$

Thus, $(f \mid g)$ is a bounded bilinear form, and therefore by a straightforward application of the Riesz theorem (Chapter III, Theorem 2.3; see Chapter III, Exercise 2.5), we infer the existence of a linear operator A for which

$$\langle f \mid Ag\rangle = (f \mid g)$$

and $\| A \| \leqslant M$. Q.E.D.

An operator A for which a Borel measurable, bounded* function $F(\lambda_1, \ldots, \lambda_n)$ and commuting self-adjoint operators $A_1, \ldots, A_n$ can be found such that $A = F(A_1, \ldots, A_n)$ is said to be a *function of the operators* $A_1, \ldots, A_n$. It is important to note that for two different functions $F_1(\lambda_1, \ldots, \lambda_n)$ and $F_2(\lambda_1, \ldots, \lambda_n)$ we shall have $F_1(A_1, \ldots, A_n) =$

* The case of unbounded functions is going to be treated in Section 2.4.

$F_2(A_1, \ldots, A_n)$ in the case that these two functions coincide when $\lambda_1 \in \mathsf{S}^{A_1}, \ldots, \lambda_n \in \mathsf{S}^{A_n}$ (see Exercise 2.2). Thus, we can define $F(A_1, \ldots, A_n)$ even in the case when the function $F(\lambda_1, \ldots, \lambda_n)$ is defined only on the set $\mathsf{S}^{A_1} \times \cdots \times \mathsf{S}^{A_n}$, and is bounded and Borel measurable on $\mathsf{S}^{A_1} \times \cdots \times \mathsf{S}^{A_n}$. We can do that by extending $F(\lambda_1, \ldots, \lambda_n)$ to a bounded function on $\mathbb{R}^n$; this can be done, for instance, by setting $F(\lambda_1, \ldots, \lambda_n) = 0$ when $(\lambda_1, \ldots, \lambda_n) \notin \mathsf{S}^{A_1} \times \cdots \times \mathsf{S}^{A_n}$.

In the above considerations we have allowed $F(\lambda_1, \ldots, \lambda_n)$ to be a complex function. We shall establish that $F(A_1, \ldots, A_n)$ is self-adjoint when $F(\lambda_1, \ldots, \lambda_n)$ is real by proving Theorems 2.2 and 2.3.

Theorem 2.2. Let $\{A_1, \ldots, A_n\}$ be a set of commuting self-adjoint operators. If $F(\lambda_1, \ldots, \lambda_n)$ is a complex-valued, bounded, Borel measurable function on $\mathbb{R}^n$ and $F^*(\lambda_1, \ldots, \lambda_n)$ is its complex conjugate, then $F^*(A_1, \ldots, A_n)$ is the adjoint of $F(A_1, \ldots, A_n)$.

Proof. According to Theorem 2.1

$$\langle f \mid F(A_1, \ldots, A_n)g\rangle = \int_{\mathbb{R}^n} F(\lambda)\, d\langle f \mid E_\lambda^{A_1,\ldots,A_n} g\rangle.$$

The same theorem applied to $F^*(A_1, \ldots, A_n)$ yields

$$\langle g \mid F^*(A_1, \ldots, A_n)f\rangle = \int_{\mathbb{R}^n} F^*(\lambda)\, d\langle g \mid E_\lambda^{A_1,\ldots,A_n} f\rangle.$$

Thus, we get

$$\begin{aligned}
&\langle F^*(A_1, \ldots, A_n) f \mid g\rangle \\
&= \int_{\mathbb{R}^n} F(\lambda)\, d\langle g \mid E_\lambda^{A_1,\ldots,A_n} f\rangle^* \\
&= \int_{\mathbb{R}^n} F(\lambda)\, d\langle E_\lambda^{A_1,\ldots,A_n} f \mid g\rangle \\
&= \int_{\mathbb{R}^n} F(\lambda)\, d\langle f \mid E_\lambda^{A_1,\ldots,A_n} g\rangle \\
&= \langle f \mid F(A_1, \ldots, A_n) g\rangle
\end{aligned}$$

for all $f, g \in \mathscr{H}$. This shows that $F^*(A_1, \ldots, A_n) = [F(A_1, \ldots, A_n)]^*$.

Q.E.D.

Theorem 2.3. If $F(\lambda_1, \ldots, \lambda_n)$ is a real bounded Borel measurable function on $\mathbb{R}^n$ and $A_1, \ldots, A_n$ are n commuting self-adjoint operators, then the operator $A = F(A_1, \ldots, A_n)$ is self-adjoint and its spectral measure $E^A(B)$ satisfies the relation

$$E^A(B) = E^{A_1,\ldots,A_n}(F^{-1}(B)) \tag{2.7}$$

for all $B \in \mathscr{B}^1$.

Proof. If $F(\lambda_1, \ldots, \lambda_n) = F^*(\lambda_1, \ldots, \lambda_n)$, we have according to the preceding theorem

$$[F(A_1, \ldots, A_n)]^* = F^*(A_1, \ldots, A_n) = F(A_1, \ldots, A_n).$$

Thus, the operator $A = F(A_1, \ldots, A_n)$ is self-adjoint.

The projector-valued set function

$$E^{(F)}(B) = E^{A_1,\ldots,A_n}(F^{-1}(B)) \tag{2.8}$$

is obviously a spectral measure on $\mathscr{B}^1$. We shall prove that for all f, g

$$\langle f \mid Ag \rangle = \int_{\mathbb{R}^1} \lambda \, d\langle f \mid E_\lambda^{(F)} g \rangle, \tag{2.9}$$

and that consequently the spectral measure $E^{(F)}(B)$ defined in (2.8) is indeed the spectral measure of A.

From (2.3) we see that the relation (2.9) is equivalent to the relation

$$\int_{\mathbb{R}^n} F(\lambda_1, \ldots, \lambda_n) \, d\langle f \mid E_{\lambda_1,\ldots,\lambda_n}^{A_1,\ldots,A_n} g \rangle = \int_{\mathbb{R}^1} \lambda \, d\langle f \mid E_\lambda^{(F)} g \rangle. \tag{2.10}$$

In the case when $F(\lambda_1, \ldots, \lambda_n) = \chi_{B_1}(\lambda_1, \ldots, \lambda_n)$, $B_1 \in \mathscr{B}^n$, an easy computation shows that (2.10) reduces to the relation (2.8), which is true by definition. Hence, we can immediately conclude that (2.10) is also true when $F(\lambda_1, \ldots, \lambda_n)$ is a simple function. From this result, (2.10) can be established straightforwardly first for $F(\lambda_1, \ldots, \lambda_n)$ nonnegative, and afterward for any real function $F(\lambda_1, \ldots, \lambda_n)$, by employing Definitions 3.4 and 3.5 in Chapter II for integration on measure spaces. Q.E.D.

2.3. Algebras of Compatible Observables

Suppose that the self-adjoint operators $A_1, \ldots, A_n$ represent compatible observables. If we carry out at some instant t a simultaneous measurement of these observables, we will determine, ideally, a measure $P_{\Psi(t)}^{A_1,\ldots,A_n}(B)$—in the case that the system was in the state $\Psi(t)$ at that instant. Thus, we can also compute the measure

$$P_{\Psi(t)}^{A}(B) = P_{\Psi(t)}^{A_1,\ldots,A_n}(F^{-1}(B)). \tag{2.11}$$

It is then natural to say that the measure (2.11) provides the distribution of values of the observable represented by the operator $A = F(A_1, \ldots, A_n)$. This ruling is in agreement with (1.14), since by Theorem 2.3 we find that $P^A_{\Psi(t)}(B)$, defined by (2.11), satisfies indeed the following relation:

$$P^A_{\Psi(t)}(B) = \| \Psi(t) \|^{-2} \langle \Psi(t) \mid E^A(B)\, \Psi(t) \rangle.$$

We shall say that A represents a bounded observable which is a function of $A_1, \ldots, A_n$. The family of all such observables has an algebraic structure, which can be established on the basis of Theorem 2.4.

Theorem 2.4. If $F_i(\lambda_1, \ldots, \lambda_n)$, $i = 1, 2$, are two bounded Borel measurable functions on $\mathbb{R}^n$ and $A_1, \ldots, A_n$ are commuting self-adjoint operators, then

$$(F_1 + F_2)(A_1, \ldots, A_n) = F_1(A_1, \ldots, A_n) + F_2(A_1, \ldots, A_n), \tag{2.12}$$

$$(F_1 \cdot F_2)(A_1, \ldots, A_n) = F_1(A_1, \ldots, A_n) F_2(A_1, \ldots, A_n). \tag{2.13}$$

Proof. Since F_1 and F_2 are bounded functions, their sum $F_1 + F_2$ and product $F_1 \cdot F_2$ are also bounded. In addition, $F_1 + F_2$ and $F_1 \cdot F_2$ are Borel measurable according to Theorem 3.5 of Chapter II so that the operators

$$A = (F_1 + F_2)(A_1, \ldots, A_n),$$

$$B = (F_1 \cdot F_2)(A_1, \ldots, A_n),$$

are defined by Theorem 2.1.

We easily obtain (2.12) by using the definition of A as a function of $A_1, \ldots, A_n$ to derive that for all $f, g \in \mathscr{H}$,

$$\begin{aligned}\langle f \mid Ag \rangle &= \int_{\mathbb{R}^n} (F_1(\lambda_1, \ldots, \lambda_n) + F_2(\lambda_1, \ldots, \lambda_n))\, d\langle f \mid E^{A_1, \ldots, A_n}_{\lambda_1, \ldots, \lambda_n} g \rangle \\ &= \langle f \mid F_1(A_1, \ldots, A_n) g \rangle + \langle f \mid F_2(A_1, \ldots, A_n) g \rangle.\end{aligned}$$

In order to prove (2.13), write first

$$\begin{aligned}&\langle f \mid F_1(A_1, \ldots, A_n) F_2(A_1, \ldots, A_n) g \rangle \\ &\quad = \int_{\mathbb{R}^n} F_1(\lambda_1, \ldots, \lambda_n)\, d\langle f \mid E^{A_1, \ldots, A_n}_{\lambda_1, \ldots, \lambda_n} F_2(A_1, \ldots, A_n) g \rangle.\end{aligned} \tag{2.14}$$

Now we have for any $B \in \mathscr{B}^n$,

$$\begin{aligned}
&\langle f \mid E^{A_1,\ldots,A_n}(B) F_2(A_1, \ldots, A_n) g\rangle \\
&= \langle E^{A_1,\ldots,A_n}(B) f \mid F_2(A_1, \ldots, A_n) g\rangle \\
&= \int_{\mathbb{R}^n} F_2(\lambda_1, \ldots, \lambda_n)\, d\langle E^{A_1,\ldots,A_n}(B) f \mid E^{A_1,\ldots,A_n}_{\lambda_1,\ldots,\lambda_n} g\rangle \\
&= \int_B F_2(\lambda_1, \ldots, \lambda_n)\, d\langle f \mid E^{A_1,\ldots,A_n}_{\lambda_1,\ldots,\lambda_n} g\rangle
\end{aligned}$$

because the complex measure

$$\mu_{f,g}(B_1) = \langle E^{A_1,\ldots,A_n}(B) f \mid E^{A_1,\ldots,A_n}(B_1) g\rangle$$

obviously vanishes on sets B_1 lying outside B. Consequently (see Chapter III, Exercise 6.7)

$$\begin{aligned}
&\int_{\mathbb{R}^n} F_1(\lambda_1, \ldots, \lambda_n)\, d\langle f \mid E^{A_1,\ldots,A_n}_{\lambda_1,\ldots,\lambda_n} F_2(A_1, \ldots, A_n) g\rangle \\
&= \int_{\mathbb{R}^n} F_1(\lambda_1, \ldots, \lambda_n) F_2(\lambda_1, \ldots, \lambda_n)\, d\langle f \mid E^{A_1,\ldots,A_n}_{\lambda_1,\ldots,\lambda_n} g\rangle \\
&= \langle f \mid (F_1 \cdot F_2)(A_1, \ldots, A_n) g\rangle,
\end{aligned}$$

which, when combined with (2.14), shows that (2.13) is true. Q.E.D.

We note that albeit in general we cannot add or multiply unbounded operators without paying close attention to domain questions (cf. Chapter III, §2.6), we do not run into similar problems with bounded functions of self-adjoint operators $A_1, \ldots, A_n$ representing compatible observables. Denote by $\mathcal{O}_b(A_1, \ldots, A_n)$ the family of all bounded self-adjoint functions of $A_1, \ldots, A_n$. If a is a real number and $F(A_1, \ldots, A_n) \in \mathcal{O}_b(A_1, \ldots, A_n)$, then obviously also $(aF)(A_1, \ldots, A_n) \in \mathcal{O}_b(A_1, \ldots, A_n)$. Moreover, if $F_1(A_1, \ldots, A_n)$, $F_2(A_1, \ldots, A_n) \in \mathcal{O}_b(A_1, \ldots, A_n)$, then on account of Theorem 2.4,

$$F_1(A_1, \ldots, A_n) + F_2(A_1, \ldots, A_n) = (F_1 + F_2)(A_1, \ldots, A_n) \in \mathcal{O}_b(A_1, \ldots, A_n),$$

$$F_1(A_1, \ldots, A_n) F_2(A_1, \ldots, A_n) = (F_1 \cdot F_2)(A_1, \ldots, A_n) \in \mathcal{O}_b(A_1, \ldots, A_n).$$

Thus, $\mathcal{O}_b(A_1, \ldots, A_n)$ is a real algebra, usually referred to as the *algebra generated by the observables* $A_1, \ldots, A_n$. Furthermore, $\mathcal{O}_b(A_1, \ldots, A_n)$ is obviously a commutative algebra.

In studying the relation between theory and experiment, it is often convenient to consider instead of the original observables $A_1, ..., A_n$ the algebra $\mathcal{O}_b(A_1, ..., A_n)$. The reason for this is that, while some of the operators $A_1, ..., A_n$ will be, generally speaking, unbounded and therefore not defined on the entire Hilbert space, the operators in $\mathcal{O}_b(A_1, ..., A_n)$ are bounded and defined on the entire Hilbert space. On the other hand, $\mathcal{O}_b(A_1, ..., A_n)$ supplies all the *physical* information that $A_1, ..., A_n$ do. The reason for this is that this information is contained exclusively in the expressions

$$\langle \Psi(t) \mid E^{A_1, ..., A_n}(B)\, \Psi(t)\rangle, \tag{2.15}$$

where $E^{A_1, ..., A_n}(B)$ are bounded functions of $A_1, ..., A_n$,

$$E^{A_1, ..., A_n}(B) = \chi_B(A_1, ..., A_n), \tag{2.16}$$

as can be seen by taking for $F(\lambda_1, ..., \lambda_n)$ in (2.3) the characteristic function $\chi_B(\lambda_1, ..., \lambda_n)$ of the set $B \in \mathscr{B}^n$.

If A is an operator representing an observable and $\Psi(t) \in \mathscr{D}_A$, then the expression

$$\langle A\rangle_{\Psi(t)} = \langle \Psi(t) \mid A\Psi(t)\rangle, \qquad \|\Psi(t)\| = 1, \tag{2.17}$$

is called the *expectation value* of A at time t. In §1 we saw [see (1.15)] that in the case of a determinative measurement of A at time t, (2.17) is the mean value of the statistical sample obtained as an outcome of that measurement if the system was in the state $\Psi(t)$ at time t.

According to (2.15) and (2.16) we always have

$$P_{\Psi(t)}^{A_1, ..., A_n}(B) = \langle \chi_B(A_1, ..., A_n)\rangle_{\Psi(t)}, \qquad \|\Psi(t)\| = 1. \tag{2.18}$$

This prompts us to state that a knowledge of the expectation values of all self-adjoint bounded functions of the observables $A_1, ..., A_n$ is equivalent to a knowledge of the statistical distribution of the simultaneous value of $A_1, ..., A_n$. Hence we shall call such functions also observables.

*2.4. Unbounded Functions of Commuting Self-Adjoint Operators

We shall study now the general concept of a function $F(A_1, ..., A_n)$ of n self-adjoint commuting operators $A_1, ..., A_n$ for the case when $F(\lambda)$, $\lambda \in \mathbb{R}^n$, is any Borel measurable function on $\mathbb{R}^n$.

Theorem 2.5. Let $A_1, \ldots, A_n$ be n commuting self-adjoint operators in $\mathscr{H}$, and $E^{A_1,\ldots,A_n}(B)$, $B \in \mathscr{B}^n$, the spectral measure defined by (2.1). If $F(\lambda)$, $\lambda \in \mathbb{R}^n$, is a Borel measurable function, then there is a unique linear operator A which satisfies

$$\langle g \mid Af \rangle = \int_{\mathbb{R}^n} F(\lambda)\, d\langle g \mid E_\lambda^{A_1,\ldots,A_n} f \rangle \tag{2.19}$$

for all $g \in \mathscr{H}$. This operator is defined on the domain

$$\mathscr{D}_A = \left\{ f\colon \int_{\mathbb{R}^n} |F(\lambda)|^2\, d\, \| E_\lambda^{A_1,\ldots,A_n} f \|^2 < \infty \right\}; \tag{2.20}$$

A is usually denoted by $F(A_1, \ldots, A_n)$.

Proof. Denote by R the set of all vectors $f \in \mathscr{H}$ for which the linear functional

$$\varphi_f(g) = \int_{\mathbb{R}^n} F^*(\lambda)\, d\langle E_\lambda^{A_1,\ldots,A_n} f \mid g \rangle$$

exists for all $g \in \mathscr{H}$ and is continuous. For each such $f \in R$ there is by the theorem of Riesz (Chapter III, Theorem 2.3) a vector h such that

$$\langle h \mid g \rangle = \varphi_f(g),$$

or equivalently, and with the notation $h = A(f)$,

$$\langle g \mid A(f) \rangle = \int_{\mathbb{R}^n} F(\lambda)\, d\langle g \mid E^{A_1,\ldots,A_n} f \rangle.$$

The set R is linear, because if $f_1, f_2 \in R$, we can derive for any $g \in \mathscr{H}$, by using the rules of integration,

$$\begin{aligned} a_1 \int_{\mathbb{R}^n} F(\lambda)\, d\langle g \mid E_\lambda^{A_1,\ldots,A_n} f_1 \rangle &+ a_2 \int_{\mathbb{R}^n} F(\lambda)\, d\langle g \mid E_\lambda^{A_1,\ldots,A_n} f_2 \rangle \\ &= \int_{\mathbb{R}^n} F(\lambda)\, d\langle g \mid E_\lambda^{A_1,\ldots,A_n}(a_1 f_1 + a_2 f_2) \rangle, \end{aligned}$$

i.e., $a_1 f_1 + a_2 f_2 \in R$; the above relation can also be written as

$$\begin{aligned} a_1\langle g \mid A(f_1) \rangle &+ a_2\langle g \mid A(f_2) \rangle \\ &= \langle g \mid a_1 A(f_1) + a_2 A(f_2) \rangle \\ &= \langle g \mid A(a_1 f_1 + a_2 f_2) \rangle \end{aligned}$$

for all $g \in \mathscr{H}$, which implies that A is linear.

In order to show that R coincides with the set $\mathscr{D}_A$ defined in (2.20), let us show first that $f \in R$ implies that $f \in \mathscr{D}_A$. Taking in (2.19) $g = Af$, we get

$$\| Af \|^2 = \int_{\mathbb{R}^n} F(\lambda)\, d\langle Af \mid E_\lambda^{A_1,\ldots,A_n} f\rangle$$

$$= \int_{\mathbb{R}^n} F(\lambda)\, d\langle E_\lambda^{A_1,\ldots,A_n} f \mid Af\rangle^*,$$

and since we have (see the proof of Theorem 2.4)

$$\langle E^{A_1,\ldots,A_n}(B) f \mid Af\rangle = \int_{\mathbb{R}^n} F(\lambda')\, d\langle E^{A_1,\ldots,A_n}(B) f \mid E_{\lambda'}^{A_1,\ldots,A_n} f\rangle$$

$$= \int_B F(\lambda')\, d\langle f \mid E_{\lambda'}^{A_1,\ldots,A_n} f\rangle,$$

we obtain $f \in \mathscr{D}_A$ (see Chapter III, Exercise 6.7):

$$\| Af \|^2 = \int_{\mathbb{R}^n} | F(\lambda)|^2\, d \| E_\lambda^{A_1,\ldots,A_n} f \|^2 < +\infty.$$

Conversely, let us assume that the above relation holds for some $f \in \mathscr{H}$, and let us show that $f \in R$. Choose a sequence of simple functions $s_1(\lambda)$, $s_2(\lambda),\ldots$ such that (see Exercise 2.1)

$$\text{(2.21)} \qquad \begin{gathered} |\operatorname{Re} s_1(\lambda)| \leqslant |\operatorname{Re} s_2(\lambda)| \leqslant \cdots \leqslant |\operatorname{Re} F(\lambda)|, \\ |\operatorname{Im} s_1(\lambda)| \leqslant |\operatorname{Im} s_2(\lambda)| \leqslant \cdots \leqslant |\operatorname{Im} F(\lambda)|, \\ \lim_{k\to\infty} s_k(\lambda) = F(\lambda), \qquad \lambda \in \mathbb{R}^n. \end{gathered}$$

If we write $s_k(\lambda)$ in the form (2.4), we can derive for any $g \in \mathscr{H}$

$$\text{(2.22)} \qquad \sum_{i=1}^{n_k} |\operatorname{Re} a_{ik}|\, |\langle g \mid E^{A_1,\ldots,A_n}(B_{ik}) f\rangle|$$

$$\leqslant \sum_{i=1}^{n_k} | a_{ik} |\, \| E^{A_1,\ldots,A_n}(B_{ik}) f \|\, \| E^{A_1,\ldots,A_n}(B_{ik}) g \|$$

$$\leqslant \left(\sum_{i=1}^{n_k} | a_{ik} |^2 \| E^{A_1,\ldots,A_n}(B_{ik}) f \|^2\right)^{1/2} \left(\sum_{i=1}^{n_k} \| E^{A_1,\ldots,A_n}(B_{ik}) g \|^2\right)^{1/2}.$$

The expressions on the left-hand side of (2.22) constitute a convergent sequence since they are monotonically increasing for $k = 1, 2, \ldots$, while the right-hand side is bounded by

$$\left(\int_{\mathbb{R}^n} |F(\lambda)|^2 \, d \| E_\lambda^{A_1,\ldots,A_n} f \|^2\right)^{1/2} \cdot \left(\int_{\mathbb{R}^n} d \| E_\lambda^{A_1,\ldots,A_n} g \|^2\right)^{1/2}$$
$$= \| g \| \left(\int_{\mathbb{R}^n} |F(\lambda)|^2 \, d \| E^{A_1,\ldots,A_n} f \|^2\right)^{1/2}.$$

As this observation stays true if Re a_{ik} is replaced in (2.22) by Im a_{ik}, we arrive at the existence of the limit

$$\lim_{k\to\infty} \sum_{i=1}^{n_k} a_{ik} \langle g \mid E^{A_1,\ldots,A_n}(B_{ik}) f\rangle = \int_{\mathbb{R}^n} F(\lambda) \, d\langle g \mid E_\lambda^{A_1,\ldots,A_n} f\rangle.$$

This argument establishes, incidentally, the following inequality:

$$\left| \int_{\mathbb{R}^n} F(\lambda) \, d\langle g \mid E^{A_1,\ldots,A_n} f\rangle \right| \leqslant \| g \| \left(\int_{\mathbb{R}^n} |F(\lambda)|^2 \, d \| E_\lambda^{A_1,\ldots,A_n} f \|^2\right)^{1/2}.$$

Since the above argument stands for any $g \in \mathscr{H}$, it follows that $f \in R$. Q.E.D.

It should be noted that, in general, the domain of definition of $F(A_1, \ldots, A_n)$ is not dense in $\mathscr{H}$. However, in the case that this domain is dense and $F(\lambda)$ is real, it follows from Theorem 2.6 that $F(A_1, \ldots, A_n)$ is self-adjoint.

***Theorem 2.6.** If $A_1, \ldots, A_n$ commute and $A = F(A_1, \ldots, A_n)$ has a dense domain of definition, then $A^* = F^*(A_1, \ldots, A_n)$, where the function $F^*(\lambda)$ is the complex conjugate of $F(\lambda)$.

Proof. Consider the operator D, which is defined as that operator which satisfies

$$\langle f \mid Dg\rangle = \int_{\mathbb{R}^n} F^*(\lambda) \, d\langle f \mid E_\lambda^{A_1,\ldots,A_n} g\rangle$$

for all $f \in \mathscr{H}$. It is obvious from Theorem 2.5, especially from (2.20), that the domain $\mathscr{D}_D$ of D is identical to $\mathscr{D}_A$. If we take $f, g \in \mathscr{D}_A \equiv \mathscr{D}_D$, we get

$$\langle f \mid Dg\rangle = \left(\int_{\mathbb{R}^n} F(\lambda) \, d\langle g \mid E_\lambda^{A_1,\ldots,A_n} f\rangle\right)^*$$
$$= \langle g \mid Af\rangle^* = \langle Af \mid g\rangle,$$

which shows that $D \subseteq A^*$. We shall prove that $D \equiv A^*$ by showing that $\mathscr{D}_{A^*} \subset \mathscr{D}_D$.

Assume that $h \in \mathscr{D}_{A^*}$ and $h^* = A^*h$, so that the relation

$$\langle h \mid Af \rangle = \langle h^* \mid f \rangle \tag{2.23}$$

is valid for all $f \in \mathscr{D}_A$. Consider the bounded and Borel measurable function

$$F_M(\lambda) = \begin{cases} F(\lambda) & \text{if } |F(\lambda)| \leqslant M \\ 0 & \text{if } |F(\lambda)| > M. \end{cases}$$

According to Theorem 2.1, the operator $A_M^* = F_M^*(A_1, \ldots, A_n)$ is bounded. We easily see that $A_M^*h \in \mathscr{D}_A$,

$$\begin{aligned}
&\int_{\mathbb{R}^n} |F(\lambda)|^2 \, d \| E_\lambda^{A_1,\ldots,A_n} A_M^* h \|^2 \\
&\quad = \int_{\mathbb{R}^n} |F(\lambda)|^2 \, d \| A_M^* E_\lambda^{A_1,\ldots,A_n} h \|^2 \\
&\quad = \int_{\mathbb{R}^n} |F(\lambda)|^2 \, d_\lambda \left[\int_{\mathbb{R}^n} |F_M(\mu)|^2 \, d_\mu \| E_\mu^{A_1,\ldots,A_n} E_\lambda^{A_1,\ldots,A_n} h \|^2 \right] \\
&\quad = \int_{\mathbb{R}^n} |F(\lambda) F_M(\lambda)|^2 \, d \| E_\lambda^{A_1,\ldots,A_n} h \|^2 \\
&\quad = \int_{\mathbb{R}^n} | F_M(\lambda)|^4 \, d \| E_\lambda^{A_1,\ldots,A_n} h \|^2 < +\infty.
\end{aligned}$$

Thus, we can insert in (2.23) $f = A_M^*h$, and by taking into account that A_M^* and $E^{A_1,\ldots,A_n}$ commute (see Exercise 2.4), we obtain

$$\begin{aligned}
&\langle h \mid A A_M^* h \rangle \\
&\quad = \int_{\mathbb{R}^n} F(\lambda) \, d\langle h \mid E^{A_1,\ldots,A_n} A_M^* h \rangle \\
&\quad = \int_{\mathbb{R}^n} F(\lambda) \, d\langle h \mid A_M^* E_\lambda^{A_1,\ldots,A_n} h \rangle \\
&\quad = \int_{\mathbb{R}^n} F(\lambda) \, d_\lambda \left[\int_{\mathbb{R}^n} F_M^*(\lambda') \, d_{\lambda'} \langle h \mid E_{\lambda'}^{A_1,\ldots,A_n} E_\lambda^{A_1,\ldots,A_n} h \rangle \right] \\
&\quad = \int_{\mathbb{R}^n} F(\lambda) F_M^*(\lambda) \, d \| E_\lambda^{A_1,\ldots,A_n} h \|^2 \\
&\quad = \int_{\mathbb{R}^n} | F_M(\lambda)|^2 \, d \| E_\lambda^{A_1,\ldots,A_n} h \|^2 = \| A_M^* h \|^2.
\end{aligned}$$

From (2.23) and the above equation we get

$$\| A_M{}^*h \|^2 = \langle h^* \mid A_M{}^*h \rangle \leqslant \| h^* \| \, \| A_M{}^*h \|,$$

which implies that

$$\left(\int_{\mathbb{R}^n} | F_M(\lambda) |^2 \, d \, \| E_\lambda^{A_1,\ldots,A_n} h \|^2 \right)^{1/2} = \| A_M{}^*h \| \leqslant \| h^* \|.$$

If we let $M \to +\infty$, then the left-hand side in the above relation stays bounded, and since

$$\lim_{M \to \infty} F_M(\lambda) = F(\lambda),$$

the integral

$$\int_{\mathbb{R}^n} | F(\lambda) |^2 \, d \, \| E_\lambda^{A_1,\ldots,A_n} h \|^2$$

exists and is finite (see Chapter II, Theorem 3.10). Thus, $h \in \mathscr{D}_D$, which establishes that $\mathscr{D}_{A^*} \subset \mathscr{D}_D$. Q.E.D.

We can generalize the concept of a function of compatible observables $A_1, \ldots, A_n$ by including all real Borel measurable functions $F(\lambda)$ which are such that the set (2.20) is dense in $\mathscr{H}$. Theorem 2.6 guarantees that the operator $F(A_1, \ldots, A_n)$ representing such a function will be self-adjoint, and therefore we can call it observable. As a generalization of Theorem 2.4 to the unbounded case we prove Theorem 2.7.

Theorem 2.7. If $F_i(A_1, \ldots, A_n)$, $i = 1, 2$, are any two functions of the commuting self-adjoint operators $A_1, \ldots, A_n$, then

$$F_1(A_1, \ldots, A_n) + F_2(A_1, \ldots, A_n) \subseteq (F_1 + F_2)(A_1, \ldots, A_n), \tag{2.24}$$

$$F_1(A_1, \ldots, A_n) F_2(A_1, \ldots, A_n) \subseteq (F_1 \cdot F_2)(A_1, \ldots, A_n). \tag{2.25}$$

For any function $F(A_1, \ldots, A_n)$ and any integer $m \geqslant 0$ we have

$$[F(A_1, \ldots, A_n)]^m = (F^m)(A_1, \ldots, A_n). \tag{2.26}$$

Proof. It is easy to see that

$$\mathscr{D}_{F_1} \cap \mathscr{D}_{F_2} \subseteq \mathscr{D}_{F_1+F_2}$$

since whenever $F_1(\lambda)$ and $F_2(\lambda)$ are square integrable with respect to a measure

$$\mu_f(B) = \| E^{A_1,\ldots,A_n}(B) f \|^2,$$

the function $(F_1 + F_2)(\lambda)$ is also square integrable (see Chapter II, Theorem 4.2). If we introduce the functions

$$F_{iM}(\lambda) = \begin{cases} F_i(\lambda) & \text{if } |F_i(\lambda)| \leqslant M \\ 0 & \text{if } |F_i(\lambda)| > M, \end{cases}$$

then we have by Theorem 2.4

$$F_{1M}(A_1, \ldots, A_n)f + F_{2M}(A_1, \ldots, A_n)f = (F_{1M} + F_{2M})(A_1, \ldots, A_n)f.$$

By letting $M \to \infty$ and noting that for $f \in \mathscr{D}_{F_1} \cap \mathscr{D}_{F_2}$ we have

$$F_i(A_1, \ldots, A_n)f = \operatorname*{s-lim}_{M\to\infty} F_{iM}(A_1, \ldots, A_n)f, \qquad i = 1, 2,$$

$$(F_1 + F_2)(A_1, \ldots, A_n)f = \operatorname*{s-lim}_{M\to\infty} (F_{1M} + F_{2M})(A_1, \ldots, A_n)f,$$

we get in the limit

$$F_1(A_1, \ldots, A_n)f + F_2(A_1, \ldots, A_n)f = (F_1 + F_2)(A_1, \ldots, A_n)f.$$

We turn now to proving (2.25). By the already familiar procedure of dealing with spectral-resolution integrals, we get

$$\begin{aligned} &\int_{\mathbb{R}^n} |F_1(\lambda)|^2 \, d \,\| E_\lambda^{A_1,\ldots,A_n} F_2(A_1, \ldots, A_n)f \|^2 \\ &= \int_{\mathbb{R}^n} |F_1(\lambda)|^2 \, d_\lambda \left[\int_{\mathbb{R}^n} |F_2(\mu)|^2 \, d_\mu \,\| E_\mu^{A_1,\ldots,A_n} E_\lambda^{A_1,\ldots,A_n} f \|^2 \right. \\ &= \int_{\mathbb{R}^n} |F_1(\lambda) F_2(\lambda)|^2 \, d_\lambda \,\| E_\lambda^{A_1,\ldots,A_n} f \|^2. \end{aligned}$$

The above relation shows that whenever $F_2(A_1, \ldots, A_n)f$ belongs to the domain of definition of $F_1(A_1, \ldots, A_n)$, so that $F_1(A_1, \ldots, A_n) F_2(A_1, \ldots, A_n)f$ is defined, then f belongs to the domain of $(F_1 \cdot F_2)(A_1, \ldots, A_n)$. If we take such a vector f, and apply Theorem 2.4 to $F_{1M_1}(A_1, \ldots, A_n)$ and $F_{2M_2}(A_1, \ldots, A_n)$ to obtain

$$(2.27) \qquad F_{1M_1}(A_1, \ldots, A_n) F_{2M_2}(A_1, \ldots, A_n)f = (F_{1M_1} \cdot F_{2M_2})(A_1, \ldots, A_n)f,$$

we get by taking the strong limits, first for $M_2 \to \infty$ [note that the operator $F_{1M_1}(A_1, \ldots, A_n)$ is bounded and therefore continuous] and then for $M_1 \to \infty$,

$$F_1(A_1, \ldots, A_n) F_2(A_1, \ldots, A_n)f = (F_1 \cdot F_2)(A_1, \ldots, A_n)f.$$

Thus, (2.25) is established.

From (2.25) it follows directly that

$$[F(A_1,..., A_n)]^m \subseteq (F^m)(A_1,..., A_n)$$

for any $m = 0, 1, 2,...$. However, when $m \geqslant 2$, the square integrability of $(F(\lambda))^m$ for some measure μ_f implies the square integrability of $(F(\lambda))^{m-1}$, since $|F(\lambda)|^{2(m-1)} \leqslant |F(\lambda)|^{2m}$ at any point λ where $|F(\lambda)| \geqslant 1$. This means that $\mathscr{D}_{F^m} \subset \mathscr{D}_{F^{m-1}}$. Hence, by taking in (2.27) $F_1 = F$ and $F_2 = F^{m-1}$, we get for $f \in \mathscr{D}_{F^m} \subset \mathscr{D}_{F^{m-1}}$ in the limit $M_2 \to +\infty$

$$F_{M_1}(A_1,..., A_n)F^{m-1}(A_1,..., A_n)\, f = (F^m_{M_1})(A_1,..., A_n)\, f.$$

By letting $M_1 \to +\infty$ we arrive at the result that $FF^{m-1}f = F^m f$ and that $\mathscr{D}_{FF^{m-1}} \supset \mathscr{D}_{F^m}$, i.e., $FF^{m-1} \supseteq F^m$. Thus, by induction in m

$$[F(A_1,..., A_n)]^m \supseteq (F^m)(A_1,..., A_n),$$

i.e., (2.26) holds. Q.E.D.

The relation (2.26) indicates that the definition of functions of observables given in Theorem 2.5 coincides with the algebraic definition whenever the algebraic definition is feasible. In fact, the function of n commuting operators $A_1,..., A_n$ corresponding to a polynomial of n real variables $\lambda_1,..., \lambda_n$ is equal to the operator obtained by replacing λ_i by A_i for $i = 1,..., n$ everywhere in the polynomial. In particular, if

$$p(\lambda) = a_n\lambda^n + a_{n-1}\lambda^{n-1} + \cdots + a_1\lambda + a_0$$

and $A^* = A$, and if $p(A)$ is defined by

$$\text{(2.28)} \qquad \langle f \mid p(A)g\rangle = \int_{\mathbb{R}^1} p(\lambda)\, d\langle f \mid E^A_\lambda g\rangle, \qquad f \in \mathscr{H}, \qquad g \in \mathscr{D}_{A^n},$$

it is easy to establish that

$$p(A) = a_nA^n + a_{n-1}A^{n-1} + \cdots + a_1A + a_0$$

by using the linearity property of the integral in (2.28) and the relation (2.26) with $F(\lambda) = \lambda$.

Exercises

2.1. Prove that if $F(\lambda)$, $\lambda \in \mathbb{R}^n$, is Borel measurable on $\mathbb{R}^n$ and $\mu(B)$, $B \in \mathscr{B}^n$, is a complex measure, then there is a sequence of simple functions that satisfy (2.21) with sign Re $F(\lambda)$ = sign Re $s_k(\lambda)$, sign Im $F(\lambda)$ = sign Im $s_k(\lambda)$, and $F(\lambda)$ is μ integrable if and only if

$$\lim_{k\to\infty} \int_{\mathbb{R}^n} s_k(\lambda)\, d\mu(\lambda)$$

exists, in which case the above limit is equal to

$$\int_{\mathbb{R}^n} F(\lambda)\, d\mu(\lambda).$$

2.2. Let $A_1, \ldots, A_n$ be n commuting self-adjoint operators and $F(\lambda)$, $\lambda \in \mathbb{R}^n$, a function which vanishes on $S^{A_1} \times \cdots \times S^{A_n}$. Show that $F(A_1, \ldots, A_n) = 0$.

2.3. The proof of Theorem 2.5 is independent of Theorem 2.1. Derive Theorem 2.1 from Theorem 2.5.

2.4. Show that in general the symbol $\subseteq$ cannot be replaced in (2.24) and (2.25) by the symbol $\equiv$.

2.5. Show that any two bounded functions $F_1(A_1, \ldots, A_n)$ and $F_2(A_1, \ldots, A_n)$ of $A_1, \ldots, A_n$ commute.

2.6. Prove that to each spectral function E_λ, $\lambda \in \mathbb{R}^1$, corresponds a unique self-adjoint operator A such that

$$\langle g \mid Af \rangle = \int_{\mathbb{R}^1} \lambda d \langle g \mid E_\lambda f \rangle$$

for all $g \in \mathscr{H}$ and for all $f \in \mathscr{D}_A$.

2.7. Show that if $Q_1, \ldots, Q_n$ are the self-adjoint multiplication operators $(Q_k\psi)(x_1, \ldots, x_n) = x_k\psi(x_1, \ldots, x_n)$ in $L^2(\mathbb{R}^n)$, then $F(Q_1, \ldots, Q_n)$ coincides with the operator A defined by

$$(A\psi)(x_1, \ldots, x_n) = F(x_1, \ldots, x_n)\, \psi(x_1, \ldots, x_n),$$
$$\mathscr{D}_A = \{\psi : F \cdot \psi \in L^2(\mathbb{R}^n)\}.$$

3. The Schroedinger, Heisenberg, and Interaction Pictures

3.1. The General Form of the Schroedinger Equation

We have seen in §§1 and 2 that in order to "describe" a system quantum mechanically, we have to associate with it a Hilbert space $\mathscr{H}$ (§2, Axiom 0). The observables are represented by self-adjoint operators in $\mathscr{H}$, while the physical states are vector-valued functions $\Psi(t)$ in the time variable $t \in \mathbb{R}^1$. Clearly, not every vector-valued function can be a state, because in that case—as can be easily seen from the Born correspondence rules in §1—the theory would lack predictive power. We must require that a function $\Psi(t)$ which represents a state should satisfy a dynamical law (see the Introduction). This law is given in quantum mechanics by an equation of motion, called the Schroedinger equation.

We can arrive at the general form of this equation by generalizing the Schroedinger equations of Chapter I, §5 and Chapter II, §5. The essential feature of those special cases was the existence of an operator H, which in those cases was the Schroedinger operator [essentially given by the Schroedinger operator form in (5.8) of Chapter II] and it was also related to the total energy observable.

More generally, we postulate for each system which can be described quantum mechanically, the existence of a self-adjoint operator H, commonly called the *Hamiltonian of the system*. This operator is assumed to play a double role in the theory. On one hand, it represents the total energy observable, and it is therefore related to measurements of the energy of the entire system. On the other hand, it plays a key role in the equation of motion; namely, as a generalization of (5.5) in Chapter II, we postulate that a state satisfies the *Schroedinger equation*

$$i\hbar \frac{d\Psi(t)}{dt} = H\Psi(t). \tag{3.1}$$

There are, however, two mathematical questions which a careful look at the above equation immediately brings to mind.

The first question is concerned with the meaning of the derivative (3.1). With some foresight (see also Exercise 3.1), we answer that question by defining

$$\frac{d\Psi(t)}{dt} = \operatorname*{s-lim}_{\Delta t \to 0} \frac{1}{\Delta t}(\Psi(t + \Delta t) - \Psi(t)).$$

The second question is related to the fact that since H is usually an unbounded operator, we cannot find a solution of (3.1) for every state Ψ_0 prepared at some instant t, since we have to require that $\Psi_0 \in \mathscr{D}_H$, while $\mathscr{D}_H \neq \mathscr{H}$. We can solve this last difficulty by formulating the dynamical law by other means than the equation of motion (3.1).

3.2. The Evolution Operator

Consider the function $\exp(-(i/\hbar)\,\lambda(t - t_0))$, which is continuous (and therefore Borel measurable) as well as bounded for $\lambda \in \mathbb{R}^1$. Due to these properties, we can define a bounded operator

$$U(t, t_0) = \exp[-(i/\hbar)\, H(t - t_0)], \tag{3.2}$$

customarily called the *time-evolution operator*. In accordance with Theorem 2.1, $U(t, t_0)$ is the operator satisfying

$$\langle f \mid U(t, t_0)g\rangle = \int_{\mathbb{R}^1} \exp(-(i/\hbar)\,\lambda(t - t_0))\, d\langle f \mid E_\lambda g\rangle \tag{3.3}$$

for all $f, g \in \mathscr{H}$, where E_λ is the spectral function of H,

$$H = \int_{\mathbb{R}^1} \lambda\, dE_\lambda\,. \tag{3.4}$$

We shall prove a theorem from which the main properties of $U(t, t_0)$ can be derived immediately. In proving that theorem we need two lemmas that will find many later applications.

***Lemma 3.1.** (*Lebesgue's dominated, or bounded, convergence theorem*). Let $f_1(\xi)$, $f_2(\xi)$,... be a sequence of nonnegative functions defined on the measurable set R of a measure space $(\mathscr{X}, \mathscr{A}, \mu)$ and integrable on R. Suppose that $\lim_{n\to\infty} f_n(\xi) = f(\xi)$ almost everywhere. If there is a function $g(\xi)$ integrable on R and such that $|f_n(\xi)| \leqslant g(\xi)$ for all $n = 1, 2,\ldots$, then $f(\xi)$ is integrable on R and

$$\lim_{n\to\infty} \int_R f_n(\xi)\, d\mu(\xi) = \int_R f(\xi)\, d\mu(\xi).$$

Proof. By Theorem 3.4 in Chapter II, $f(\xi)$ is measurable so that

$$g_n(\xi) = \{|f_k(\xi) - f(\xi)| : k = n, n+1,\ldots\}, \qquad n = 1, 2,\ldots,$$

are also measurable (see Chapter II, Exercise 3.3). Since $g_n(\xi) \leqslant 2g(\xi)$, the functions $g_n(\xi)$ are integrable on R (see Chapter II, Theorem 3.9). Furthermore, $\lim_{n\to\infty} g_n(\xi) = 0$ almost everywhere on R. Since $g_1(\xi) \geqslant g_2(\xi) \geqslant \cdots$, we can apply Theorem 3.10 of Chapter II to derive

$$\lim_{n\to\infty} \int_R g_n(\xi)\, d\mu(\xi) = 0.$$

We get the desired result by observing that $|f_n(\xi) - f(\xi)| \leqslant g_n(\xi)$. Q.E.D.

***Lemma 3.2** (*Fatou's lemma*). If $f_1(\xi)$, $f_2(\xi)$,... is a sequence of nonnegative functions, integrable on the subset R of measure space $(\mathscr{X}, \mathscr{A}, \mu)$ and converging almost everywhere to a function $f(\xi)$, and if

$$\int_R f_n(\xi)\, d\mu(\xi) \leqslant M$$

for all n, then $f(\xi)$ is integrable on R and

$$\int_R f(\xi)\, d\mu(\xi) \leqslant M.$$

Proof. Consider the functions

$$g_n(\xi) = \inf\{f_n(\xi), f_{n+1}(\xi),...\}$$

which are integrable on R (see Chapter II, Exercise 3.3) and note that $f(\xi) = \lim_{n\to\infty} g_n(\xi)$ almost everywhere in R. By applying Theorem 3.10 in Chapter II to the sequence $g_1(\xi), g_2(\xi),...$ we infer that $f(\xi)$ is integrable on R and

$$\int_R f(\xi)\, d\mu(\xi) = \lim_{n\to\infty} \int_R g_n(\xi)\, d\mu(\xi).$$

Since $g_n(\xi) \leqslant f_n(\xi)$ for all values of n, we have

$$\int_R g_n(\xi)\, d\mu(\xi) \leqslant M,$$

which establishes that M is a bound for the integral of $f(\xi)$ over R.

Q.E.D.

Theorem 3.1. If A is a self-adjoint operator in $\mathscr{H}$, the operator-valued function

$$U_t = e^{iAt}, \qquad t \in \mathbb{R}^1, \tag{3.5}$$

has the following properties:

(a) $U_0 = 1$;

(b) U_t is a unitary operator on $\mathscr{H}$ for all $t \in \mathbb{R}^1$;

(c) $U_{t_1} U_{t_2} = U_{t_1+t_2}$ for all $t_1, t_2 \in \mathbb{R}^1$;

(d) if $f \in \mathscr{D}_A$,

$$iAf = \operatorname*{s-lim}_{t\to 0} \frac{U_t f - f}{t} \tag{3.6}$$

and, conversely, the above strong limit exists only if $f \in \mathscr{D}_A$.

Proof. (a) If E_λ is the spectral function of A

$$A = \int_{\mathbb{R}^1} \lambda\, dE_\lambda,$$

then by Theorem 2.1

$$\langle f \mid U_t g\rangle = \int_{\mathbb{R}^1} e^{i\lambda t}\, d\langle f \mid E_\lambda g\rangle. \tag{3.7}$$

When we insert above $t = 0$, we obtain

$$\langle f \mid U_0 g\rangle = \int_{\mathbb{R}^1} d\langle f \mid E_\lambda g\rangle = \langle f \mid g\rangle$$

for all $f, g \in \mathscr{H}$, which proves that $U_0 = 1$.

(b) According to Theorem 4.3 in Chapter III, U_t is unitary if it satisfies the relations

$$U_t^* U_t = U_t U_t^* \equiv 1. \tag{3.8}$$

The above relation can be established immediately by noting that

$$e^{i\lambda t}(e^{i\lambda t})^* = 1$$

and using Theorems 2.2 and 2.4.

(c) By using Theorem 2.4 we get

$$\begin{aligned}\langle f \mid U_{t_1} U_{t_2} g\rangle &= \int_{\mathbb{R}^1} e^{i\lambda t_1} e^{i\lambda t_2}\, d\langle f \mid E_\lambda g\rangle \\ &= \int_{\mathbb{R}^1} e^{i\lambda(t_1+t_2)}\, d\langle f \mid E_\lambda g\rangle \\ &= \langle f \mid U_{t_1+t_2} g\rangle\end{aligned}$$

for all $f, g \in \mathscr{H}$, which proves that $U_{t_1} U_{t_2} = U_{t_1+t_2}$.

(d) The domain of definition of the operator

$$(1/t)(U_t - 1) - iA \tag{3.9}$$

is $\mathscr{D}_A$ since U_t is defined everywhere on $\mathscr{H}$. The operator (3.9) is a function of A, corresponding to the complex function

$$F_t(\lambda) = (1/t)(e^{i\lambda t} - 1) - i\lambda. \tag{3.10}$$

For $f \in \mathscr{D}_A$ we have by Theorems 2.5 and 2.7

$$\begin{aligned}&\left\| \frac{1}{t}(U_t - 1)f - iAf \right\|^2 \\ &= \left\langle f \,\middle|\, \left(\frac{1}{t}(U_t - 1) - iA\right)^* \left(\frac{1}{t}(U_t - 1) - iA\right) f \right\rangle \\ &= \int_{\mathbb{R}^1} \left| \frac{1}{t}(e^{i\lambda t} - 1) - i\lambda \right|^2 d\langle f \mid E_\lambda f\rangle.\end{aligned} \tag{3.11}$$

Since we have by the mean value theorem of the differential calculus

$$(1/t)(e^{i\lambda t}-1)=i\lambda(\cos\lambda\theta_1 t+i\sin\lambda\theta_2 t),\qquad 0\leqslant\theta_{1,2}\leqslant 1,$$

we get for all $t\neq 0$ and $\lambda\in\mathbb{R}^1$,

$$|F_t(\lambda)|^2\leqslant|\lambda|^2\,|\cos\lambda\theta_1 t+i\sin\lambda\theta_2 t-1|^2\leqslant 4\,|\lambda|^2.$$

Thus, $|F_t(\lambda)|^2$ is a function majorized for $t\neq 0$ by the function $4\,|\lambda|^2$, which is integrable in the measure $\mu_f(B)=\langle f\mid E(B)f\rangle$, and for which $\lim_{t\to 0}|F_t(\lambda)|^2=0$. Consequently, by Lemma 3.1 we have

$$\lim_{t\to 0}\int_{\mathbb{R}^1}|F_t(\lambda)|^2\,d\langle f\mid E_\lambda f\rangle=0,$$

and we can conclude from (3.11) that (3.6) is true.

Assume now that

$$\text{(3.12)}\qquad \operatorname*{s-lim}_{t\to 0}(1/t)(U_t-1)g$$

exists for some $g\in\mathscr{H}$. Due to Theorems 2.1, 2.2, and 2.4 we can write

$$\text{(3.13)}\qquad \left\|\frac{1}{t}(U_t-1)g\right\|^2=\left\langle g\,\middle|\,\frac{1}{t^2}(U_t-1)^*(U_t-1)g\right\rangle=\int_{\mathbb{R}^1}\left|\frac{1}{t}(e^{i\lambda t}-1)\right|^2 d\langle g\mid E_\lambda g\rangle.$$

The existence of the strong limit (3.12) implies (see Chapter III, Exercise 5.5) the existence of a constant M such that

$$\|(1/t)(U_t-1)f\|^2\leqslant M$$

for all $t\in\mathbb{R}^1$. Since in (3.13)

$$\lim_{t\to 0}|(1/t)(e^{i\lambda t}-1)|^2=\lambda^2,$$

we get by applying Fatou's lemma (Lemma 3.2) that λ^2 is integrable with respect to the measure $\mu_g(B)=\langle g\mid E(B)g\rangle$. According to the spectral theorem in Chapter III, §6, this implies that $g\in\mathscr{D}_A$. Q.E.D.

Theorem 3.1 shows that the operator $U(t,t_0)$ in (3.2) is unitary. Furthermore, from (3.6) it follows that if $f\in\mathscr{D}_H$ then for any $t\in\mathbb{R}^1$

$$\text{(3.14)}\qquad \operatorname*{s-lim}_{t\to t_0}\frac{U(t,t_0)-1}{t-t_0}f=\frac{i}{\hbar}Hf.$$

The operator $U(t, t_0)$ obviously leaves $\mathscr{D}_H$ invariant, i.e., if $f \in \mathscr{D}_H$, then $U(t, t_0)f \in \mathscr{D}_H$, for all $t, t_0 \in \mathbb{R}^1$. This can be seen by applying the spectral theorem, and noting that $[E_\lambda, U(t, t_0)] = 0$, to obtain

$$\int_{\mathbb{R}^1} \lambda^2 \, d \| E_\lambda U(t, t_0) f \|^2 = \int_{\mathbb{R}^1} \lambda^2 \, d \| U(t, t_0) E_\lambda f \|^2$$
$$= \int_{\mathbb{R}^1} \lambda^2 \, d \| E_\lambda f \|^2.$$

From Theorem 3.1c we easily get

$$U(t, t_0) \, U(t_0, t_1) = U(t, t_1). \tag{3.15}$$

Combining (3.14) and (3.15), we derive that for any $f \in \mathscr{D}_H$

$$i\hbar \frac{d}{dt} U(t, t_0) f = i\hbar \operatorname*{s-lim}_{\Delta t \to 0} \frac{U(t + \Delta t, t_0) - U(t, t_0)}{\Delta t} f$$
$$= i\hbar \operatorname*{s-lim}_{\Delta t \to 0} \frac{U(t + \Delta t, t) - 1}{\Delta t} U(t, t_0) f$$
$$= HU(t, t_0) f.$$

Thus, if $\Psi_0 \in \mathscr{D}_H$, then

$$\Psi(t) = U(t, t_0) \, \Psi_0$$

satisfies the Schroedinger equation (3.1), and since $U(t_0, t_0) = 1$, it also satisfies the initial condition

$$\Psi(t_0) = U(t_0, t_0) \, \Psi_0 = \Psi_0 .$$

We shall see in a later section that $\Psi(t)$ is, as a matter of fact, the only vector-valued function of t satisfying these conditions and the requirement that $\| \Psi(t) \| = \| \Psi_0 \|$ for all $t \in \mathbb{R}^1$.

3.3. The Schroedinger Picture

The formalism of quantum mechanics espoused in the last two sections has the striking feature that a state is represented by a vector-valued function $\Psi(t)$ in the time variable t, while the observables are represented by self-adjoint operators which do not depend on t. This formalism for quantum mechanics is called the *Schroedinger picture.*

The statement that in the Schroedinger picture the observables are represented by operators which do not depend on t has to be accepted with one qualification; namely, there are cases when the Hamiltonian of the system depends on t, i.e., is given by $H(t)$. Such instances will occur when a system $\mathfrak{S}$ interacts with another system $\mathfrak{S}_1$, and we know how the interaction term in the Hamiltonian depends on t. In such a case it is very often impractical to study the behavior of the compound system $\mathfrak{S} + \mathfrak{S}_1$, when, for instance, $\mathfrak{S}_1$ is very large* and complicated.

We shall summarize the characteristic features of the Schroedinger picture in the following axioms, in such a way as to include the possibility of the most general case when the Hamiltonian is time dependent.

Axiom S1. A Hamiltonian $H(t)$ is assigned to every quantum mechanical system; $H(t)$ is an operator-valued function of the time parameter t. For each value of $t \in \mathbb{R}^1$, $H(t)$ is a self-adjoint operator in the Hilbert space $\mathscr{H}$ associated with the system (see §2, Axiom O) and $H(t)$ represents the total energy of the system.

Axiom S2. All the observables of the system that are not functions of the energy are represented by self-adjoint operators in $\mathscr{H}$ which do not depend on t. Two or more observables are compatible if and only if their corresponding operators commute.

Axiom S3. There is a unitary operator-valued function $U(t, t_0)$, $t, t_0 \in \mathbb{R}^1$, called the *time-evolution operator*, which is such that

$$U(t_0, t_0) = 1, \qquad t_0 \in \mathbb{R}^1, \tag{3.16}$$

$$U(t, t_1)\, U(t_1, t_0) = U(t, t_0), \qquad t, t_1, t_0 \in \mathbb{R}^1, \tag{3.17}$$

and is strongly continuous in t, $t_0 \in \mathbb{R}^1$, and for all $f \in \mathscr{D}_{H(t)}$,

$$i\hbar \operatorname*{s-lim}_{t_1 \to t} \frac{U(t_1, t) - 1}{t_1 - t} f = i\hbar \operatorname*{s-lim}_{t_2 \to t} \frac{U(t, t_2) - 1}{t - t_2} f = H(t) f. \tag{3.18}$$

Axiom S4. A state of the system is represented by vector-valued functions $\Psi(t)$ which satisfy the relation

$$\Psi(t) = U(t, t_0)\, \Psi(t_0), \qquad \| \Psi(t_0) \| > 0,$$

for all t, $t_0 \in \mathbb{R}^1$. If c is a nonzero complex number, $\Psi(t)$ and $c\Psi(t)$ represent the same state.

* Such a case occurs when $\mathfrak{S}_1$ is an electromagnet producing a magnetic field which varies with time. It is true that in principle we could study the system $\mathfrak{S} + \mathfrak{S}_1$ by considering the electromagnet itself to be a many-body system consisting of many atoms, electrons, ions, etc., but such an approach would be hopeless to pursue from the computational point of view. Instead, we could measure the microscopic electromagnetic field produced by the electromagnet, and treat the problem by introducing in the Hamiltonian of the system $\mathfrak{S}$ a term which is a function of that field, and represents, to a good approximation, the interaction of the system and the electromagnet.

We have seen that the conditions (3.16)–(3.18) are satisfied by the operator-valued function (3.2) when $H(t)$ is the same for all $t \in \mathbb{R}^1$. A certain generalization of (3.2) can be obtained when $H(t)$ depends on time in some regular manner, but the considerations become quite involved if the operators $H(t_1)$ and $H(t_2)$ do not commute for $t_1 \neq t_2$.

In the time-independent case it is easy to see that (see Exercise 2.5)

$$P^H_{\Psi(t)}(B) = \langle \Psi(t) \mid E^H(B)\, \Psi(t)\rangle, \qquad \|\Psi(t)\| = 1, \tag{3.19}$$

is time independent. The physical meaning of this result is that the probability of measuring a certain value of the energy of the system in state $\Psi(t)$ is the same at all times, i.e., that the energy is conserved. Naturally, if $H(t)$ is intrinsically time dependent, this statement does not stay true in general. Physically, this is due to the interaction of the system with an external source in the course of which energy can be transferred to or from the source (see also Section 7.9).

3.4. The Heisenberg Picture and Physical Equivalence of Formalisms

Let A be a bounded operator representing an observable, and let

$$\langle A\rangle_{\Psi(t)} = \langle U(t,0)\,\hat{\Psi} \mid A U(t,0)\hat{\Psi}\rangle \tag{3.20}$$

be the expectation value of that observable in the state

$$\Psi(t) = U(t,0)\hat{\Psi}, \qquad \hat{\Psi} = \Psi(0). \tag{3.21}$$

As we have seen in §2, the link between the quantum mechanical formalism and experiment is contained solely in the expectation values (3.20) of all possible bounded observables, so that two formalisms in which all expectation values are the same are *physically equivalent*. Since we can write (3.20) in the form

$$\langle A\rangle_{\Psi(t)} = \langle \hat{\Psi} \mid \hat{A}(t)\hat{\Psi}\rangle, \tag{3.22}$$

$$\hat{A}(t) = U^{-1}(t,0)\, A U(t,0), \tag{3.23}$$

we could describe the same *physical* state of the system by giving the vector $\hat{\Psi}$ and the operator-valued functions $A(t)$ in (3.23) for all bounded observables of the system. This formalism of quantum mechanics is known as the *Heisenberg picture*.

If D is an unbounded operator representing an observable, the operators

$$\hat{D}(t) = U^{-1}(t, 0)\, DU(t, 0)$$

are self-adjoint for all $t \in \mathbb{R}^1$ because $U(t, 0)$ is unitary (see Chapter III, Lemma 4.3), and represent the same observable in the Heisenberg picture as D did in the Schroedinger picture.

If the Hamiltonian H of the system is time independent, we have

$$\hat{H}(t) = e^{+(i/\hbar)Ht}\, H\, e^{-(i/\hbar)Ht} = H,$$

since $e^{-(i/\hbar)Ht}$ leaves $\mathscr{D}_H$ invariant and commutes with H. Thus, in that case $\hat{H}(t)$ will be time independent (in physicists' language, the energy is conserved) and given by the same operator as in Schroedinger picture.

In the more general case of time-dependent Hamiltonian $H(t)$, we have in the Heisenberg picture

$$\hat{H}(t) = U^{-1}(t, 0)\, H(t)\, U(t, 0). \tag{3.24}$$

In the Schroedinger picture the dynamical law was the Schroedinger equation (3.1), which had to be satisfied by the state $\Psi(t)$. Analogously, in the Heisenberg picture the dynamical law is embodied in an equation of motion, which has to be satisfied by any operator function $\hat{A}(t)$ representing an observable. Since the link between the representations of states and observables in the two pictures is given by (3.21), (3.23), and (3.24), we can derive the equations of motion in the Heisenberg picture from the equations of motion in the Schroedinger picture.

Theorem 3.2. Let $\hat{A}(t)$ represent an observable in the Heisenberg picture, and assume that the same observable is represented by a bounded operator A in the Schroedinger picture. If f and $\hat{A}(t)f$ are in the domain of definition of $\hat{H}(t)$ for all $t \in \mathbb{R}^1$, then $\hat{A}(t)f$ satisfies the *Heisenberg equation*

$$i\hbar \frac{d\hat{A}(t)}{dt} f = [\hat{A}(t), \hat{H}(t)]f. \tag{3.25}$$

Proof. By definition,

$$\frac{d\hat{A}(t)}{dt} f = \operatorname*{s-lim}_{\Delta t \to 0} \frac{\hat{A}(t + \Delta t) - \hat{A}(t)}{\Delta t} f. \tag{3.26}$$

Straightforward algebra yields

$$(3.27)\qquad \frac{\hat{A}(t+\Delta t)-\hat{A}(t)}{\Delta t}$$
$$= \frac{U^{-1}(t+\Delta t, 0)\, AU(t+\Delta t, 0) - U^{-1}(t, 0)\, AU(t, 0)}{\Delta t}$$
$$= U^{-1}(t+\Delta t, 0)\, A\, \frac{U(t+\Delta t, 0) - U(t, 0)}{\Delta t}$$
$$+ \frac{U^{-1}(t+\Delta t, 0) - U^{-1}(t, 0)}{\Delta t}\, AU(t, 0).$$

When $f \in \mathscr{D}_{\hat{H}(t)}$, we get from (3.18)

$$(3.28)\qquad i\hbar \operatorname*{s-lim}_{\Delta t\to 0} \frac{U(t+\Delta t, 0) - U(t, 0)}{\Delta t} f = H(t)\, U(t, 0) f.$$

By using the unitarity of $U^{-1}(t+\Delta t, 0)$ we obtain

(3.29)
$$\left\| \left(U^{-1}(t+\Delta t, 0)\, A\, \frac{U(t+\Delta t, 0) - U(t, 0)}{\Delta t} + \frac{i}{\hbar} U^{-1}(t, 0)\, AH(t)\, U(t, 0) \right) f \right\|$$
$$\leqslant \left\| A \left(\frac{U(t+\Delta t, 0) - U(t, 0)}{\Delta t} + \frac{i}{\hbar} H(t)\, U(t, 0) \right) f \right\|$$
$$+ \left\| \left(\frac{i}{\hbar} U^{-1}(t+\Delta t, 0) - \frac{i}{\hbar} U^{-1}(t, 0) \right) AH(t)\, U(t, 0) f \right\|.$$

The first term on the right-hand side of (3.29) converges to zero due to (3.28) and the fact that A is bounded and therefore continuous (see Chapter III, Theorem 1.3), while the second term converges to zero since, due to (3.17) and (3.16),

$$(3.30)\qquad U^{-1}(t_1, t_2) = U(t_2, t_1)$$

for any $t_1, t_2 \in \mathbb{R}^1$, and therefore

$$(3.31)\qquad \operatorname*{s-lim}_{\Delta t\to 0}(U^{-1}(t+\Delta t, 0) - U^{-1}(t, 0))$$
$$= \operatorname*{s-lim}_{\Delta t\to 0}(U(0, t+\Delta t) - U(0, t)) = 0.$$

Thus, we see that

$$(3.32)\qquad \operatorname*{s-lim}_{\Delta t\to 0} U^{-1}(t+\Delta t, 0)\, A\, \frac{U(t+\Delta t, 0) - U(t, 0)}{\Delta t} f$$
$$= -\frac{i}{\hbar} U^{-1}(t, 0)\, AH(t)\, U(t, 0) f = -\frac{i}{\hbar} \hat{A}(t)\, \hat{H}(t) f.$$

If $AU(t, 0)f \in \mathscr{D}_{H(t)}$, or equivalently, $\hat{A}(t)f \in \mathscr{D}_{\hat{H}(t)}$, we derive by employing (3.18) and (3.30)

$$\text{(3.33)} \qquad \begin{aligned} &\underset{\Delta t \to 0}{\text{s-lim}} \frac{U^{-1}(t + \Delta t, 0) - U^{-1}(t, 0)}{\Delta t} AU(t, 0)f \\ &= \underset{\Delta t \to 0}{\text{s-lim}} \frac{U(0, t + \Delta t) - U(0, t)}{\Delta t} AU(t, 0)f \\ &= \frac{i}{\hbar} U(0, t) H(t) AU(t, 0)f = \frac{i}{\hbar} \hat{H}(t) \hat{A}(t)f. \end{aligned}$$

Now we can obtain (3.25) by applying (3.27) to f and using (3.32) and (3.33) when taking the strong limits for $\Delta t \to 0$. Q.E.D.

The Heisenberg picture is a formulation of quantum mechanics which is physically equivalent* to that of the Schroedinger picture. Consequently, its main features are derivable from Axioms S1–S4.

In order to formulate the dynamics in the Heisenberg picture independently of the Schroedinger picture, we introduce the unitary operators

$$\text{(3.34)} \qquad \hat{U}(t) = U(t, 0).$$

For $\hat{U}(t)f \in \mathscr{D}_{H(t)}$, i.e., when $f \in \mathscr{D}_{\hat{H}(t)}$, (3.17) and (3.18) yield

$$i\hbar \underset{\Delta t \to 0}{\text{s-lim}} \frac{\hat{U}(t + \Delta t) - \hat{U}(t)}{\Delta t} f = H(t) \hat{U}(t)f = \hat{U}(t) \hat{H}(t)f.$$

Furthermore, the relation (3.23) can be recast in the form

$$\text{(3.35)} \qquad \hat{A}(t) = \hat{U}^{-1}(t) \hat{A}(0) \hat{U}(t).$$

The above relations do not contain any direct reference to the Schroedinger picture. Thus, we can now characterize the Heisenberg picture as the formalism fulfilling the following Axioms H1–H4.

Axiom H1. A Hamiltonian $\hat{H}(t)$ is assigned to every quantum mechanical system. $\hat{H}(t)$ is an operator-valued function of the time parameter t. For each value of $t \in \mathbb{R}^1$, $\hat{H}(t)$ is a self-adjoint operator in the Hilbert space $\mathscr{H}$ associated with the system. $\hat{H}(t)$ represents the total energy of the system at time t.

* The physical equivalence of these two pictures was established by Schroedinger [1926b] only after both these pictures had been independently developed by Heisenberg [1925], and Schroedinger [1926a], in the matrix and wave mechanics versions, respectively.

Axiom H2. A state of the system is represented by a time-independent nonzero vector $\hat{\Psi} \in \mathscr{H}$; if $c \neq 0$, $c\hat{\Psi}$ represents the same state.

Axiom H3. There is a unitary and strongly continuous operator-valued function $\hat{U}(t)$ such that

$$\hat{U}(0) = 1, \tag{3.36}$$

and such that for all $f \in \mathscr{D}_{\hat{H}(t)}$

$$i\hbar \operatorname*{s-lim}_{\Delta t \to 0} \frac{\hat{U}(t + \Delta t) - \hat{U}(t)}{\Delta t} f = \hat{U}(t)\, \hat{H}(t) f. \tag{3.37}$$

Axiom H4. An observable is represented by an operator-valued function $\hat{A}(t)$, where for each $t \in \mathbb{R}^1$, $\hat{A}(t)$ is a self-adjoint operator in $\mathscr{H}$. Moreover, $\hat{A}(t)$ satisfies the equation

$$\hat{A}(t) = \hat{U}^{-1}(t)\, \hat{A}(0)\, \hat{U}(t). \tag{3.38}$$

3.5. The Formalism of Matrix Mechanics

The Heisenberg picture and the Heisenberg equation (3.25) were first introduced by Heisenberg in the matrix mechanics version of quantum mechanics. We can obtain the matrix formulation of quantum mechanics for any particular system by choosing in the Hilbert space $\mathscr{H}$ assigned to that system an orthonormal basis* $\{e_1, e_2, \ldots\}$ which is contained in the domain of definition $\mathscr{D}_{\hat{A}(t)}$ of all operator functions $\hat{A}(t)$ representing observables. To each $\hat{A}(t)$ we assign an infinite matrix $\| \hat{A}_{ik}(t) \|$ (see also Exercise 3.2), where

$$\hat{A}_{ik}(t) = \langle e_i \mid \hat{A}(t)\, e_k \rangle, \qquad i, k = 1, 2, \ldots .$$

To a state represented by the normalized vector $\hat{\Psi}$, we assign an infinite one-column matrix $\| c_k \|$, where

$$c_k = \langle e_k \mid \hat{\Psi} \rangle, \qquad k = 1, 2, \ldots .$$

Note that $\| c_k \| \in l^2(\infty)$ and that

$$\begin{aligned}(\hat{A}(t)\hat{\Psi})_i &= \langle e_i \mid \hat{A}(t)\hat{\Psi} \rangle = \langle \hat{A}(t)\, e_i \mid \hat{\Psi} \rangle \\ &= \sum_{k=1}^{\infty} \langle \hat{A}(t)\, e_i \mid e_k \rangle \langle e_k \mid \hat{\Psi} \rangle \\ &= \sum_{k=1}^{\infty} \langle e_i \mid \hat{A}(t)\, e_k \rangle \langle e_k \mid \hat{\Psi} \rangle = \sum_{k=1}^{\infty} \hat{A}_{ik}(t) c_k .\end{aligned}$$

* $\mathscr{H}$ is in practice infinite dimensional and separable.

Thus, $\| \hat{A}_{ik}(t) \|$ can be considered to be an operator in $l^2(\infty)$, which acts on the elements of $l^2(\infty)$ according to the rules of matrix multiplication.

The Heisenberg matrix picture can be useful when the Hamiltonian $\hat{H}$ of the system is time independent and the elements $\{e_1, e_2, \ldots\}$ can be chosen to be eigenvectors of $\hat{H}$, i.e., if we can write $e_k = \Psi_k$ where

$$\hat{H}\Psi_k = E_k \Psi_k, \qquad k = 1, 2, \ldots,$$

(an example of such a system is the quantum mechanical harmonic oscillator; see Exercise 3.3). In that case $\hat{U}(t)$ is represented by the matrix with components

$$\hat{U}_{ik}(t) = \langle \Psi_i \mid e^{-(i/\hbar)\hat{H}t} \Psi_k \rangle = e^{-(i/\hbar)E_i t} \delta_{ik}.$$

Consequently, noting that $\hat{U}^{-1}(t) = \hat{U}^*(t)$, we get from (3.38)

$$\hat{A}_{ik}(t) = \sum_{m,n=1}^{\infty} \hat{U}_{im}^{-1}(t) \hat{A}_{mn}(0) \hat{U}_{nk}(t) = \hat{A}_{ik}(0) e^{(i/\hbar)(E_i - E_k)t}.$$

3.6. The Interaction Picture

The physical equivalence of the Schroedinger and Heisenberg pictures is based on the fact that the physically significant quantities, i.e., the expectation values (3.20), are the same in both pictures. When the Hamiltonian is time independent, the Schroedinger picture is characterized by the property that the vectors representing states are time dependent, while the operators representing observables are time independent; in the Heisenberg picture the opposite is true regarding time dependence.

It is clear that one can invent a large number of "intermediate" pictures "between" these two pictures, which would all preserve the expectation values, but in which the time dependence will be partially assigned to vectors representing states, and partially to operators representing observables.

The most significant example of such an intermediate picture is the interaction picture, which is useful when the Hamiltonian $H(t)$ of the system can be written in the form

$$H = H_0 + V(t), \tag{3.39}$$

where H_0 is the "*free*" Hamiltonian of the system (usually) describing a system without interactions; for instance, in the case of the n-particle system of Chapter II, §5, H_0 is the self-adjoint operator defined by the differential operator form

$$-\sum_{k=1}^{n} \frac{\hbar^2}{2m_k} \Delta_k ,$$

while V is the multiplication operator defined by the potential $V(\mathbf{r}_1, \ldots, \mathbf{r}_n)$.

In order to arrive at the interaction representation for Hamiltonians (3.39), we write the expectation value (3.20) in the form

$$\langle A \rangle_{\Psi(t)} = \langle \tilde{\Psi}(t) \mid \tilde{A}(t) \tilde{\Psi}(t) \rangle, \tag{3.40}$$

where

$$\tilde{A}(t) = e^{(i/\hbar)H_0 t} A \, e^{-(i/\hbar)H_0 t}, \tag{3.41}$$

$$\tilde{\Psi}(t) = e^{(i/\hbar)H_0 t} U(t, 0)\Psi; \tag{3.42}$$

$\tilde{A}(t)$ and $\tilde{\Psi}(t)$ are taken to represent in the interaction picture the observable and the state which were represented in the Schroedinger picture by A and $\Psi(t) = U(t, 0)\Psi$, respectively. In the case where $V(t)$ is time independent, (3.42) becomes

$$\tilde{\Psi}(t) = e^{(i/\hbar)H_0 t} e^{-(i/\hbar)Ht}\Psi. \tag{3.43}$$

If we apply Theorem 3.2 to (3.41), we obtain that for any $f \in \mathscr{D}_{H_0}$

$$i\hbar \frac{d\tilde{A}(t)}{dt} f = [\tilde{A}(t), H_0]f; \tag{3.44}$$

the above relation is the Heisenberg equation for the "free" system.

Let us introduce the operator-valued function

$$H_{\mathrm{I}}(t) = e^{(i/\hbar)H_0 t} V(t) \, e^{-(i/\hbar)H_0 t}. \tag{3.45}$$

The vector function $\tilde{\Psi}(t)$ representing a state in the interaction picture satisfies a Schroedinger equation with the Hamiltonian $H_{\mathrm{I}}(t)$,

$$i\hbar \frac{d\tilde{\Psi}(t)}{dt} = H_{\mathrm{I}}(t) \tilde{\Psi}(t) \tag{3.46}$$

if $\tilde{\Psi}(t) \in \mathscr{D}_{H_{\mathrm{I}}}(t)$ (see Exercise 3.4).

EXERCISES

3.1. In formulating the Schroedinger equation in wave mechanics (see Chapter II, §5) for a state of n particles given by a wave function $\psi(x, t)$, $x \in \mathbb{R}^{3n}$, we have been using the derivative $\psi_t(x, t) = \partial_t\psi(x, t)$,

$$i\hbar\psi_t(x, t) = (H_S\psi)(x, t),$$

instead of the strong limit of (3.1). Show that

$$\underset{\Delta t \to 0}{\text{s-lim}} \frac{\psi(x, t + \Delta t) - \psi(x, t)}{\Delta t} = \psi_t(x, t)$$

in the case when $\psi_t(x, t)$ is continuous in t and Lebesgue square integrable in x.

3.2. Assume that $\{e_1, e_2, ...\}$ is an orthonormal basis in the infinite-dimensional Hilbert space $\mathscr{H}$, and that $A, B,...$ are all the linear operators in $\mathscr{H}$, which are such that their domains of definition as well as the domains of definition of their adjoints contain the set $\{e_1, e_2, ...\}$. Assign to each of the operator $A, B,...$ the infinite matrices

$$\| A_{ik} \|, \| B_{ik} \|,..., \qquad i, k = 1, 2,...,$$

where $A_{ik} = \langle e_i \mid Ae_k \rangle$, etc. Show that if $e_1, e_2, ...$ are in the domains of $A + B$ and AB, then $(A + B)_{ik} = A_{ik} + B_{ik}$ and $(AB)_{ik} = \sum_{i=1}^{\infty} A_{ij}B_{jk}$.

3.3. Use the results of Chapter I, Exercise 5.4 to compute the matrix elements $\hat{X}_{ik}(t)$ and $\hat{P}_{ik}(t)$ of the harmonic-oscillator position operators $\hat{X}(t)$ and momentum operators $\hat{P}(t)$ in the basis $\{\Psi_1, \Psi_2, ...\}$ of energy eigenfunctions.

3.4. Prove that the vector function $\tilde{\Psi}(t)$ representing a state in the interaction picture and satisfying the relation $\tilde{\Psi}(t) = e^{(i/\hbar)H_0 t} U(t, 0)\, \tilde{\Psi}(0)$ also satisfies the equation

$$i\hbar \frac{d\tilde{\Psi}(t)}{dt} = H_I(t)\, \tilde{\Psi}(t)$$

if $\tilde{\Psi}(t) \in \mathscr{D}_{H_0} \cap \mathscr{D}_{H(t)}$ and $H_I(t)$ is given by (3.45).

3.5. Show that if A is a self-adjoint operator in $\mathscr{H}$, and if $E([-\alpha, +\alpha])f = f$ for some real $\alpha > 0$, then

$$e^{iAt}f = \underset{n \to +\infty}{\text{s-lim}} \sum_{k=0}^{n} \frac{(iAt)^k}{k!} f = \sum_{k=0}^{\infty} \frac{(iAt)^k}{k!} f.$$

3.6. A well-known phenomenon in wave mechanics is the *evanescence of wave packets*. For example, if H_0 is the kinetic energy Schroedinger operator in $L^2(\mathbb{R}^3)$, given by $(H_0\psi)^{\sim}(\mathbf{p}) = (\mathbf{p}^2/2m)\,\tilde{\psi}(\mathbf{p})$, then it can be proved* that

$$\lim_{t\to+\infty} \int_{B_0} |(e^{-(i/\hbar)H_0t}\psi)(\mathbf{r})|^2\, d\mathbf{r} = 0$$

if B_0 is any bounded Borel set in $\mathbb{R}^3$. Prove that for any open bounded set B_0 the expansion

$$e^{-(i/\hbar)H_0t}\,\psi = \sum_{k=0}^{\infty} \frac{(-(i/\hbar)\,H_0t)^n}{n!}\psi, \qquad \psi(\mathbf{r}) \in \mathscr{C}_{\mathrm{b}}^{\infty}(B_0),$$

is incorrect [though $(H_0{}^n\psi)(\mathbf{r}) = (-(\hbar^2/2m)\Delta)^n\,\psi(\mathbf{r})$ is well defined, since $\psi(\mathbf{r})$ is infinitely many times differentiable and of bounded support] because it contradicts the evanescence property of $e^{-(i/\hbar)H_0t}\,\psi$.

4. State Vectors and Observables of Compound Systems

4.1. Superselection Rules and State Vectors

Until now we have been talking mainly in terms of the Schroedinger picture of quantum mechanics, in which the observables are represented by time-independent self-adjoint operators, while the states are vector-valued functions of t. In §3 we also encountered, however, the Heisenberg picture, in which the observables are represented by time-dependent operator-valued functions, while the states are time-independent vectors, as well as other pictures of quantum mechanics. In each of these pictures we can define the concept of *state vector* as a vector Ψ which can represent, at least at one instant t, *some* state of the system. The natural question to ask is whether all the vectors in the Hilbert space $\mathscr{H}$ assigned to the system (§2, Axiom O) are state vectors.

It is immediately clear from the last section that the zero vector $\mathbf{0}$ can never be a state vector. Inasmuch as we do not adopt the requirement that a state vector be normalized (a request which is a matter of mere convenience), we can say that it was generally believed until the early 1950s that all the other vectors in $\mathscr{H}$ are state vectors in any quantum mechanical theory. A paper by Wick *et al.* [1952] dispelled that belief by pointing out the existence of superselection rules in quantum field theories.

* See Exercise 7. 7 in Chapter V.

The presently known *superselection rules* are those of charge, baryon number, and univalence. These superselection rules state essentially that the Hilbert space $\mathscr{H}$ is a (in general infinite) direct sum of so-called *coherent subspaces* M_1, M_2,..., and that only the nonzero vectors belonging to these subspaces are state vectors. However, these superselection rules are primarily of significance only in theories treating systems with a variable number of particles, for instance in field theories. Thus, the charge superselection rule states that if Ψ_1 and Ψ_2 are eigenvectors of the charge operator with eigenvalues e_1 and e_2, respectively, where $e_1 \neq e_2$, then $a_1\Psi_1 + a_2\Psi_2$ is not a state vector if $a_1 \neq 0$ and $a_2 \neq 0$. Since we are primarily concerned with nonrelativistic quantum mechanics, in which the nature of the particles and their number does not change, the existence of the above-mentioned superselection rules will not be of any immediate concern to us.

4.2. The Hilbert Space of Compound Systems

The main task of quantum mechanics is to study the interaction of different parts of a system. Nonrelativistic quantum mechanics deals with the case when the nature of these parts remains unchanged during the entire duration of the experiment, so that each one of these parts can be thought of as a stable "particle."* The question which naturally arises is whether we know how to construct the formalism for a more complex system if we know which are the formalisms of the constituent particles.

In the case that a system $\mathfrak{S}$ consists of *n different* other systems (e.g., "particles") $\mathfrak{S}_1$,..., $\mathfrak{S}_n$ with which the Hilbert spaces $\mathscr{H}_1$,..., $\mathscr{H}_n$ are associated, respectively, then the rule is to associate with $\mathfrak{S}$ the Hilbert tensor product $\mathscr{H} = \mathscr{H}_1 \otimes \cdots \otimes \mathscr{H}_n$. Let us assume, in addition, that at a certain instant t we know that the n systems $\mathfrak{S}_1$,..., $\mathfrak{S}_n$ are in states represented at that instant by the state vectors $\Psi_1 \in \mathscr{H}_1$,..., $\Psi_n \in \mathscr{H}_n$, respectively. Then quantum mechanics rules that the entire system $\mathfrak{S}$ is at that instant in a state represented by the state vector $\Psi_1 \otimes \cdots \otimes \Psi_n$. Since in this case we can talk about the state vectors of the individual systems $\mathfrak{S}_1$,..., $\mathfrak{S}_n$ separately, we say that $\mathfrak{S}_1$,..., $\mathfrak{S}_n$ move independently at time t. However, in general, the state of $\mathfrak{S}$ will

* It should be noted that the use of the word "particle" does not preclude the possibility of an intricate internal structure. Thus, an α-particle consists of two protons and two neutrons, a deuteron consists of one proton and one neutron, etc. In fact, there is a tendency in modern elementary-particle physics to treat all particles on an equal footing, and talk about the transformation or decomposition of one particle into other particles rather than of breaking up of a particle into constituent parts.

not be representable at later times by a tensor product of n vectors from $\mathscr{H}_1, \ldots, \mathscr{H}_n$, respectively. Due to the interaction of $\mathfrak{S}_1, \ldots, \mathfrak{S}_n$, the state vectors representing the state of $\mathfrak{S}$ at later times can be vectors of a more general form in $\mathscr{H}$.

4.3. Tensor Products of Linear Operators

Let the operator A_k be an observable of the system $\mathfrak{S}_k$. In the case where $\mathfrak{S}_k$ is an independent part of the system $\mathfrak{S}$, we can certainly measure A_k. Hence, this question arises: which mathematical entity related to the Hilbert space $\mathscr{H} = \mathscr{H}_1 \otimes \cdots \otimes \mathscr{H}_n$ represents this observable. In order to answer this question, we have to define the following concept.

Definition 4.1. Let $A_1, \ldots, A_n$ be n bounded linear operators acting on the Hilbert spaces $\mathscr{H}_1, \ldots, \mathscr{H}_n$, respectively. The tensor product $A_1 \otimes \cdots \otimes A_n$ of these n operators is that bounded linear operator on $\mathscr{H}_1 \otimes \cdots \otimes \mathscr{H}_n$ which acts on a vector $f_1 \otimes \cdots \otimes f_n$, $f_1 \in \mathscr{H}_1, \ldots, f_n \in \mathscr{H}_n$, in the following way:

$$(4.1) \qquad (A_1 \otimes \cdots \otimes A_n)(f_1 \otimes \cdots \otimes f_n) = A_1 f_1 \otimes \cdots \otimes A_n f_n .$$

The above relation (4.1) determines the operator $A_1 \otimes \cdots \otimes A_n$ on the set of all vectors of the form $f_1 \otimes \cdots \otimes f_n$, and therefore, due to the presupposed linearity of $A_1 \otimes \cdots \otimes A_n$, on the linear manifold spanned by all such vectors. Since this linear manifold is dense in $\mathscr{H}_1 \otimes \cdots \otimes \mathscr{H}_n$ and the operator defined by (4.1) in this manifold is bounded (see Exercise 4.1), it has a unique extension to $\mathscr{H}_1 \otimes \cdots \otimes \mathscr{H}_n$. Consequently, the above definition is consistent.

Coming back now to the earlier posed question of how to represent in $\mathscr{H} = \mathscr{H}_1 \otimes \cdots \otimes \mathscr{H}_n$ an observable of $\mathfrak{S}_k$ representable in $\mathscr{H}_k$ by the bounded operator A_k, we see that the natural candidate is

$$(4.2) \qquad \tilde{A}_k = 1 \otimes \cdots \otimes 1 \otimes A_k \otimes 1 \otimes \cdots \otimes 1.$$

The fact that this is the right choice can be established by verifying that the physically meaningful quantities, i.e., the expectation values (2.17), are the same for A_k when $\mathfrak{S}_k$ is an isolated system in the state Ψ_k at t, or when $\mathfrak{S}_k$ is at the instant t an independent part of $\mathfrak{S}$, which is at t in the state $\Psi_1 \otimes \cdots \otimes \Psi_k \otimes \cdots \otimes \Psi_n$; namely, if we take the normalized state vectors $\Psi_1, \ldots, \Psi_n$, then

$$\begin{aligned} &\langle \Psi_1 \otimes \cdots \otimes \Psi_n \mid (1 \otimes \cdots \otimes A_k \otimes \cdots \otimes 1)\, \Psi_1 \otimes \cdots \otimes \Psi_n \rangle \\ &\quad = \langle \Psi_1 \mid \Psi_1 \rangle_1 \cdots \langle \Psi_k \mid A_k \Psi_k \rangle_k \cdots \langle \Psi_n \mid \Psi_n \rangle = \langle \Psi_k \mid A_k \Psi_k \rangle_k . \end{aligned}$$

We note that in Definition 4.1 we defined $A_1 \otimes \cdots \otimes A_n$ only in the case when $A_1, \ldots, A_n$ are bounded. In case of unbounded operators difficulties appear with the domain of definition of $A_1, \ldots, A_n$. If the domains of definition of $A_1, \ldots, A_n$ are $\mathscr{D}_{A_1}, \ldots, \mathscr{D}_{A_n}$, respectively, then we can define $A_1 \otimes \cdots \otimes A_n$ uniquely by (4.1) at least on the algebraic tensor product $\mathscr{D}_{A_1} \otimes_a \cdots \otimes_a \mathscr{D}_{A_n}$. However, if $A_1, \ldots, A_n$ were self-adjoint unbounded operators, then the $A_1 \otimes \cdots \otimes A_n$ so defined might not be self-adjoint when defined on such a domain.

When we are trying to define the operator $\tilde{A}_k$ which would be a generalization of (4.2) to the case when A_k is unbounded, we can circumvent the above-mentioned difficulty by considering the spectral measure $E^{A_k}(B)$ of A_k. Since

$$\tilde{E}_k(B) = 1 \otimes \cdots \otimes E^{A_k}(B) \otimes \cdots \otimes 1, \qquad B \in \mathscr{B}^1, \tag{4.3}$$

is also a spectral measure of projectors in $\mathscr{H}$ (see Exercise 4.5) we can define $\tilde{A}_k$ as the self-adjoint operator with the spectral measure (4.3) (see Exercise 2.6).

4.4. The Observables of a System of Distinct Particles

Let $\mathcal{O}_1, \ldots, \mathcal{O}_n$ be the sets of observables corresponding to the systems $\mathfrak{S}_1, \ldots, \mathfrak{S}_n$, respectively, and denote by $\tilde{\mathcal{O}}_1, \ldots, \tilde{\mathcal{O}}_n$ the same observables represented by operators in $\mathscr{H} = \mathscr{H}_1 \otimes \cdots \otimes \mathscr{H}_n$, such as the operators (4.2) in the case of bounded observables. Naturally, it is true that the set $\mathcal{O}$ of observables of the compound system $\mathfrak{S}$ contains $\tilde{\mathcal{O}}_1 \cup \cdots \cup \tilde{\mathcal{O}}_n$. However, in general it contains additional observables.

One way of obtaining additional observables is by taking functions of compatible observables from $\tilde{\mathcal{O}}_1 \cup \cdots \cup \tilde{\mathcal{O}}_n$. For instance, let H_k be the operator in $\mathscr{H}_k$ representing the energy observable of the system $\mathfrak{S}_k$, i.e., the Hamiltonian of $\mathfrak{S}_k$, and let $E^{H_k}(B)$ be its spectral measure. Denote by $\tilde{H}_k$ the energy operator of $\mathfrak{S}_k$ in $\mathscr{H}$, i.e., the self-adjoint operator with a spectral measure built from $E^{H_k}(B)$ according to (4.3). It is very easy to check that the spectral measures of $\tilde{H}_i$ and $\tilde{H}_j$ commute, i.e., that $\tilde{H}_i$ and $\tilde{H}_j$ are compatible for any $i, j = 1, \ldots, n$. Hence, according to Theorem 2.7, the operator $\tilde{H}_1 + \cdots + \tilde{H}_n$ is self-adjoint. This operator obviously represents the total energy of the system $\mathfrak{S}$ in the case when $\mathfrak{S}_1, \ldots, \mathfrak{S}_n$ are independent, since it gives us the probability of obtaining energy values $E_1 + \cdots + E_n$ for $\mathfrak{S}$ when the probabilities for the energies $E_1, \ldots, E_n$ of $\mathfrak{S}_1, \ldots, \mathfrak{S}_n$, respectively, are known.

The operator H which represents the total energy of the system $\mathfrak{S} = \{\mathfrak{S}_1, \ldots, \mathfrak{S}_n\}$ is usually taken to be of the form

$$H \supseteq \tilde{H}_1 + \cdots + \tilde{H}_n + V,$$

where V is a self-adjoint operator referred to as the interaction part of H. Indeed, the presence of V reflects, from the physical point of view, the existence of interaction between the n parts $\mathfrak{S}_1, \ldots, \mathfrak{S}_n$ of $\mathfrak{S}$.

Let us illustrate the above consideration with the case of n different particles without spin, which was treated in §5 of Chapter II in the context of wave mechanics. In that case $\mathscr{H}$ was taken to be $L^2(\mathbb{R}^{3n})$, and we can take $\mathscr{H}_1 = \cdots = \mathscr{H}_n = L^2(\mathbb{R}^3)$. The space $L^2(\mathbb{R}^{3n})$ is isomorphic to $L^2(\mathbb{R}^3) \otimes \cdots \otimes L^2(\mathbb{R}^3)$ (n factors), and according to Theorem 6.9 of Chapter II, the mapping

$$\psi_1(\mathbf{r}_1) \otimes \cdots \otimes \psi_n(\mathbf{r}_n) \to \psi_1(\mathbf{r}_1) \cdots \psi_n(\mathbf{r}_n) \tag{4.4}$$

induces a unitary transformation U between these two Hilbert spaces. The operator $\tilde{H}_k$ is in this case given essentially by the differential operator form $-(\hbar^2/2m_k)\,\Delta_k$, and therefore $\tilde{H}_1 + \cdots + \tilde{H}_n$ is essentially $-((\hbar^2/2m_1)\,\Delta_1 + \cdots + (\hbar^2/2m_n)\,\Delta_n)$, and represents in this case the kinetic energy of the system $\mathfrak{S}$. The interaction term V is determined by the potential $V(\mathbf{r}_1, \ldots, \mathbf{r}_n)$,

$$(V\psi)(\mathbf{r}_1, \ldots, \mathbf{r}_n) = V(\mathbf{r}_1, \ldots, \mathbf{r}_n)\,\psi(\mathbf{r}, \ldots, \mathbf{r}_n).$$

We note that H corresponds to the Schroedinger operator form (5.8) in Chapter II.

4.5. Symmetric and Antisymmetric Tensor Products of Hilbert Spaces

We have emphasized many times that our considerations on compound systems did not apply, up to this point, to systems containing identical particles. A widening of the scope of our considerations can be achieved only after the introduction of some additional mathematical material on tensor products.

Definition 4.2. The closed linear subspace of the Hilbert space $\mathscr{H} = \mathscr{H}_1 \otimes \cdots \otimes \mathscr{H}_n$, $\mathscr{H}_1 \equiv \cdots \equiv \mathscr{H}_n$, spanned by all the vectors of the form

$$f_1 \otimes^{\mathrm{S}} \cdots \otimes^{\mathrm{S}} f_n = \frac{1}{\sqrt{n!}} \sum_{(k_1, \ldots, k_n)} f_{k_1} \otimes \cdots \otimes f_{k_n}, \qquad f_1, \ldots, f_n \in \mathscr{H}_1, \tag{4.5}$$

where the sum is taken over all the permutations $(k_1, \dots, k_n)$ of $(1, \dots, n)$, is called the *symmetric tensor product* of $\mathscr{H}_1, \dots, \mathscr{H}_n$, and it is denoted by

$$\mathscr{H}_1 \otimes^{\mathrm{S}} \cdots \otimes^{\mathrm{S}} \mathscr{H}_n \quad \text{or} \quad \mathscr{H}_1^{\otimes^{\mathrm{S}} n}.$$

Similarly, in the case of the set of all vectors in $\mathscr{H}$ of the form

$$f_1 \otimes^{\mathrm{A}} \cdots \otimes^{\mathrm{A}} f_n = \frac{1}{\sqrt{n!}} \sum_{(k_1, \dots, k_n)} \pi(k_1, \dots, k_n) f_{k_1} \otimes \cdots \otimes f_{k_n}, \tag{4.6}$$

$$f_1, \dots, f_n \in \mathscr{H}_1,$$

where

$$\pi(k_1, \dots, k_n) = \begin{cases} +1 & \text{if } (k_1, \dots, k_n) \text{ is even} \\ -1 & \text{if } (k_1, \dots, k_n) \text{ is odd,} \end{cases} \tag{4.7}$$

the closed linear subspace spanned by this set is called the *antisymmetric tensor product* of $\mathscr{H}_1, \dots, \mathscr{H}_n$, and it is denoted by

$$\mathscr{H}_1 \otimes^{\mathrm{A}} \cdots \otimes^{\mathrm{A}} \mathscr{H}_n \quad \text{or} \quad \mathscr{H}_1^{\otimes^{\mathrm{A}} n}.$$

The factor $(n!)^{-1/2}$ in (4.5) and (4.6) has been introduced for the sake of convenience in dealing with orthonormal bases (see Exercise 4.2). In case that $f_1, \dots, f_n$ are orthogonal to each other and normalized in $\mathscr{H}_1$, then due to this factor, the symmetric and antisymmetric tensor products of these vectors will be also normalized.

If we take the inner product of a vector (4.5) with a vector (4.6) we get zero. This implies that $\mathscr{H}_1^{\otimes^{\mathrm{S}} n}$ is orthogonal to $\mathscr{H}_1^{\otimes^{\mathrm{A}} n}$ (see Exercise 4.3), when these spaces are treated as subspaces of $\mathscr{H}_1^{\otimes n}$.

If $A_1, \dots, A_n$ are bounded linear operators on $\mathscr{H}_1$, then $A_1 \otimes \cdots \otimes A_n$ will not leave, in general, $\mathscr{H}_1^{\otimes^{\mathrm{S}} n}$ or $\mathscr{H}_1^{\otimes^{\mathrm{A}} n}$ invariant. We can define, however, the symmetric tensor product

$$A_1 \otimes^{\mathrm{S}} \cdots \otimes^{\mathrm{S}} A_n = \sum_{(k_1, \dots, k_n)} A_{k_1} \otimes \cdots \otimes A_{k_n} \tag{4.8}$$

which leaves $\mathscr{H}_1^{\otimes^{\mathrm{A}} n}$ and $\mathscr{H}_1^{\otimes^{\mathrm{S}} n}$ invariant, and which can be considered therefore as defining operators on these respective spaces, by restricting the domain of definition of this operator to these spaces.

4.6. The Connection between Spin and Statistics

Let us consider now a system $\mathfrak{S}$ of n identical particles of spin σ. We recall from the Introduction that the spin projection in some direction can assume only integer and half-integer values. If $\mathscr{H}_1$ is the Hilbert space associated with each particle in this system, then not every nonzero vector in $\mathscr{H}_1^{\otimes n}$ is a state vector. It is an assumption of quantum mechanics, which is consistent with the known experimental material, that the Hilbert space $\mathscr{H}$ of state vectors associated with $\mathfrak{S}$ is the symmetric tensor product $\mathscr{H}_1^{\otimes^S n}$, in the case that the particles have integer spin, and the antisymmetric tensor product $\mathscr{H}_1^{\otimes^A n}$, in the case that the particles have half-integer spin.

The above assumption is based, primarily, on the statistics of gases. A given macroscopic volume of gas can be considered to be a system $\mathfrak{S}$ with a large number n of identicle particles which are, in the first approximation, independent. Well-known statistical consideration, starting with a basic assumption of equal *a priori* probability for each particle to have its energy within a certain range, lead to a certain distribution of particles (energy distribution) along the energy spectrum. This procedure involves counting the number of all possible states of the system which are compatible with a certain energy and some other exterior conditions. It has been observed that the assumption that the state Ψ of the system (e.g., gas) $\mathfrak{S}$ in which particle $\mathfrak{S}_k$ is in the state Ψ_k and particle $\mathfrak{S}_l$ in the state Ψ_l is identical to the state of $\mathfrak{S}$ in which $\mathfrak{S}_k$ is in the state Ψ_l and $\mathfrak{S}_l$ in the state Ψ_k leads to a correct energy distribution in the case of particles of integer spin. This assumption is usually referred to as the *principle of indistinguishability of identical particles* in quantum mechanics, and the statistical considerations based on it are referred to as the *Bose–Einstein statistics*. It reflects the fact that the symmetric tensor product $\Psi_1 \otimes^S \cdots \otimes^S \Psi_n$ does not change under a permutation of Ψ_k and Ψ_l.

In the case of a gas consisting of particles of half-integer spin, it has been observed that one must add to the assumption of indistinguishability of particles the assumption that no two particles can be in the same state. This last assumption is referred to as the *Pauli exclusion principle* (which we have met already in Chapter II, §7) and the statistics based on it is called the *Fermi–Dirac statistics*. The Pauli exclusion principle reflects the mathematical feature of the antisymmetric tensor product $\Psi_1 \otimes^A \cdots \otimes^A \Psi_n$ that it is obviously equal to the zero vector if two vectors Ψ_k and Ψ_l in it are identical for some $k \neq l$.

The indistinguishability of identical particles is reflected in the further natural assumption that those observables A of the system $\mathfrak{S}$ which are built from observables of individual free particles, are symmetric tensor products (4.8) of these. Thus, in measuring the expectation values $\langle A \rangle_\Psi$ of some particle, we cannot say whether it refers to the first, second,..., or nth particle, since obviously for any such A

$$\langle A \rangle_{\cdots \otimes^S \Psi_k \otimes^S \cdots \otimes^S \Psi_l \otimes^S \cdots} = \langle A \rangle_{\cdots \otimes^S \Psi_l \otimes^S \cdots \otimes^S \Psi_k \otimes^S \cdots},$$

$$\langle A \rangle_{\cdots \otimes^A \Psi_k \otimes^A \cdots \otimes^A \Psi_l \otimes^A \cdots} = \langle A \rangle_{\cdots \otimes^A \Psi_l \otimes^A \cdots \otimes^A \Psi_k \otimes^A \cdots}.$$

Let $E(B)$, $B \in \mathscr{B}^1$, be the spectral measure of some observable of a single particle from our system of identical particles. Then

$$\frac{1}{n!} E(B) \otimes^S \cdots \otimes^S E(B) \qquad (n \text{ factors})$$

is a projector, as follows from the fact that it is self-adjoint (see Exercise 4.4) and idempotent (see Exercise 4.5),

$$(E(B) \otimes \cdots \otimes E(B))^2 = (E(B))^2 \otimes \cdots \otimes (E(B))^2 = E(B) \otimes \cdots \otimes E(B).$$

Similarly, by using the definition (4.8) and Exercise 4.5, it is easy to check that

$$E(B_1) \otimes^S \cdots \otimes^S E(B_n), \qquad B_i \cap B_j = \varnothing \qquad \text{for} \quad i \neq j, \tag{4.9}$$

is self-adjoint and idempotent, and therefore it is a projector. More generally, if in the family

$$B_{11}, \ldots, B_{1n_1}, \ldots, B_{k1}, \ldots, B_{kn_k}, \qquad n_1 + \cdots + n_k = n, \tag{4.10}$$

of n Borel sets, any two sets with the same first index are identical while any two sets with different first indices are disjoint, then

$$\frac{1}{n_1! \cdots n_k!} [E(B_{11}) \otimes^S \cdots \otimes^S E(B_{1n_1})] \otimes^S \cdots \otimes^S [E(B_{k1}) \otimes^S \cdots \otimes^S E(B_{kn_k})] \tag{4.11}$$

is a projector (see Exercise 4.6). Thus, the expectation value of (4.11) for some state $\Psi(t)$ in the Schroedinger picture gives us the probability of having n_i particles within the sets $B_{i1} \equiv \cdots \equiv B_{in_i}$ for each $i = 1, \ldots, k$. However, if $B_1, \ldots, B_n$ are not pairwise disjoint, then the operator (4.9) is not, in general, a projector, and its expectation value does not yield the probability of finding at some instant one particle in B_1, one particle in B_2, etc. This probability is given by the expectation value of a projector which is the sum of projectors of the form (4.11), obtained by writing $B_1 \otimes \cdots \otimes B_n$ as a union of disjoint sets of the form (4.10) and adding the operators (4.11) corresponding to these sets.

4.7. Spin and Statistics for the n-Body Problem

As an illustration of the above consideration on the connection between spin and statistics we shall formulate the wave mechanics for n identical particles of spin σ. Wave mechanics postulates that a state vector will be represented in this case by a wavefunction

$$\psi(\mathbf{r}_1, s_1, \ldots, \mathbf{r}_n, s_n), \qquad \mathbf{r}_k \in \mathbb{R}^3, \qquad s_k = -\sigma, -\sigma + 1, \ldots, \sigma, \tag{4.12}$$

which is Lebesgue square integrable and symmetric

$$\psi(\ldots, \mathbf{r}_i, s_i, \ldots, \mathbf{r}_j, s_j, \ldots) = \psi(\ldots, \mathbf{r}_j, s_j, \ldots, \mathbf{r}_i, s_i, \ldots) \tag{4.13}$$

if σ is an integer, or antisymmetric

$$\psi(\ldots, \mathbf{r}_i, s_i, \ldots, \mathbf{r}_j, s_j, \ldots) = -\psi(\ldots, \mathbf{r}_j, s_j, \ldots, \mathbf{r}_i, s_i, \ldots) \tag{4.14}$$

if σ is half-integer. It is easy to see that the Hilbert spaces of functions (4.13) and (4.14) in which the inner product is taken to be

$$\langle f \mid g \rangle = \sum_{s_1=-\sigma}^{+\sigma} \cdots \sum_{s_n=-\sigma}^{+\sigma} \int_{\mathbb{R}^{3n}} f^*(\mathbf{r}_1, s_1, \ldots, \mathbf{r}_n, s_n) \times g(\mathbf{r}_1, s_1, \ldots, \mathbf{r}_n, s_n)\, d\mathbf{r}_1 \ldots d\mathbf{r}_n, \tag{4.15}$$

are unitarily equivalent under the unitary transformation induced by the mapping

$$\psi(\mathbf{r}_1, s_1) \otimes \cdots \otimes \psi(\mathbf{r}_n, s_n) \to \psi(\mathbf{r}_1, s_1) \cdots \psi(\mathbf{r}_n, s_n) \tag{4.16}$$

to the spaces $\mathscr{H}_1^{\otimes_{\mathrm{S}} n}$ and $\mathscr{H}_1^{\otimes_{\mathrm{A}} n}$, respectively, where (see Exercise 4.7)

$$\mathscr{H}_1 = L^2(\mathbb{R}^3) \oplus \cdots \oplus L^2(\mathbb{R}^3) \qquad (2\sigma + 1 \text{ terms}). \tag{4.17}$$

The Hamiltonian of the system is usually taken to be of the form

$$H \supseteq T_{\mathrm{S}} + V,$$

where T_{S} is the kinetic energy operator essentially given (see Theorem 7.3) by $(-\hbar^2/2m)(\Delta_1 + \cdots + \Delta_n)$. Thus, T_{S} is already symmetric with respect to the n particles. If V is the potential energy given by a potential $V(\mathbf{r}_1, s_1, \ldots, \mathbf{r}_n, s_n)$, then the principle of indistinguishability of identical particles requires that this function is symmetric under any permutation of the indices $1, \ldots, n$.

We shall now summarize the main conclusions reached in this section in the following axioms.

Axiom P1. Let $\mathfrak{S}$ by a system consisting of $\kappa \cdot n$ particles $\mathfrak{S}_1^{(1)}, \ldots, \mathfrak{S}_{n_1}^{(1)}, \ldots, \mathfrak{S}_1^{(\kappa)}, \ldots, \mathfrak{S}_{n_\kappa}^{(\kappa)}$, where the particles carrying the same superscript are of the same kind, while those carrying different superscripts belong to distinct species. If $\mathscr{H}^{(\nu)}$ is the Hilbert space associated with each one of the particles $\mathfrak{S}_1^{(\nu)}, \ldots, \mathfrak{S}_{n_\nu}^{(\nu)}$, $\nu = 1, \ldots, \kappa$, then the Hilbert space for $\mathfrak{S}$ is

$$\mathscr{H} = (\mathscr{H}^{(1)})^{\hat{\otimes} n_1} \otimes \cdots \otimes (\mathscr{H}^{(\kappa)})^{\hat{\otimes} n_\kappa}, \tag{4.18}$$

where $\widehat{\otimes}$ should be taken to be the symmetric tensor product if the corresponding systems are of integer spin, or the antisymmetric tensor product if the corresponding systems are of half-integral spin.

Axiom P2. In the absence of superselection rules, every nonzero vector in $\mathscr{H}$ is a state vector.

Axiom P3. If at some instant t the particles $\mathfrak{S}_1^{(1)},\ldots,\mathfrak{S}_{n_\kappa}^{(\kappa)}$ are free and in the states $\Psi_1^{(1)},\ldots,\Psi_{n_\kappa}^{(\kappa)}$, respectively, then the state of $\mathfrak{S}$ at that instant is represented in the Schroedinger picture by

$$(\Psi_1^{(1)} \widehat{\otimes} \cdots \widehat{\otimes} \Psi_{n_1}^{(1)}) \otimes \cdots \otimes (\Psi_1^{(\kappa)} \widehat{\otimes} \cdots \widehat{\otimes} \Psi_{n_\kappa}^{(\kappa)}). \tag{4.19}$$

Axiom P4. If E is a projector in $\mathscr{D}^{(\nu)}$ representing in the Schroedinger picture an observable of the particle $\mathfrak{S}_1^{(\nu)}$ when that system is free, then the same observable is represented in $\mathscr{H}$ by the projector

$$\begin{aligned} 1 \otimes \cdots \otimes 1 \otimes \Big[& \underbrace{E \otimes \cdots \otimes E}_{n_\nu \text{ factors}} \\ & + \sum_{r=1}^{n_\nu - 1} \binom{n_\nu}{r}^{-1} \underbrace{E \otimes^{\mathrm{S}} \cdots \otimes^{\mathrm{S}} E}_{n_\nu - r \text{ factors}} \otimes^{\mathrm{S}} \underbrace{(1-E) \otimes^{\mathrm{S}} \cdots \otimes^{\mathrm{S}} (1-E)}_{r \text{ factors}} \Big] \otimes 1 \otimes \cdots \otimes 1, \end{aligned} \tag{4.20}$$

where the operator in brackets acts in $(\mathscr{H}^{(\nu)})^{\widehat{\otimes} n_\nu}$.

It is easy to derive from (4.20) the expression for any projector related to measurements on the system $\mathfrak{S}$, when the constituent parts of $\mathfrak{S}$ are independent (see Exercise 4.8). In particular, we can arrive at the expression (4.11) for measurements on systems of identical particles.

Exercises

4.1. Prove that the linear operator $A_1 \otimes \cdots \otimes A_n$ on the separable Hilbert space $\mathscr{H}_1 \otimes \cdots \otimes \mathscr{H}_n$ has the bound $\| A_1 \| \cdots \| A_n \|$.

4.2. Show that if $e_1, e_2, \ldots$ is an orthonormal basis in $\mathscr{H}_1$, then $\{e_{k_1} \otimes^{\mathrm{S}} \cdots \otimes^{\mathrm{S}} e_{k_n} : k_1, \ldots, k_n = 1, 2, \ldots\}$ is an orthogonal basis in $\mathscr{H}_1^{\otimes^{\mathrm{S}} n}$, whereas $\{e_{k_1} \otimes^{\mathrm{A}} \cdots \otimes^{\mathrm{A}} e_{k_n} : k_1, \ldots, k_n = 1, 2, \ldots\}$ is an orthonormal basis in $\mathscr{H}_1^{\otimes^{\mathrm{A}} n}$.

4.3. Prove that the subspaces $\mathscr{H}_1^{\otimes^{\mathrm{S}} n}$ and $\mathscr{H}_1^{\otimes^{\mathrm{A}} n}$ of $\mathscr{H}_1^{\otimes n}$ are mutually orthogonal.

4.4. Show that if $A_1, \ldots, A_n$ are self-adjoint bounded operators on $\mathscr{H}_1, \ldots, \mathscr{H}_n$, respectively, then $A_1 \otimes \cdots \otimes A_n$ is also self-adjoint and bounded.

4.5. Prove that for any bounded linear operators $A_1, \ldots, A_n$ and $B_1, \ldots, B_n$, and any scalar a

$$a(A_1 \otimes \cdots \otimes A_n) = (aA_1) \otimes \cdots \otimes A_n = \cdots = A_1 \otimes \cdots \otimes (aA_n),$$
$$(A_1 + B_1) \otimes A_2 \otimes \cdots \otimes A_n$$
$$= A_1 \otimes A_2 \otimes \cdots \otimes A_n + B_1 \otimes A_2 \otimes \cdots \otimes A_n,$$
$$(A_1 \otimes \cdots \otimes A_n)(B_1 \otimes \cdots \otimes B_n) = (A_1B_1) \otimes \cdots \otimes (A_nB_n).$$

Show that the first two of the above relations are true also for symmetric tensor products, while the third one is not true, in general, for symmetric tensor products.

4.6. Use (4.8) and the results of Exercises 4.4 and 4.5 to prove that (4.11) is a projector.

4.7. Prove that the spaces $\hat{L}_\sigma^2(\mathbb{R}^{3n})^{\text{sym}}$ and $\hat{L}_\sigma^2(\mathbb{R}^{3n})^{\text{anti}}$ consisting, respectively, of all the symmetric (4.13) and antisymmetric (4.14) functions which are Lebesgue square integrable in the variables $\mathbf{r}_1, \ldots, \mathbf{r}_n$ are unitarily equivalent to $\mathscr{H}_1^{\otimes S n}$ and $\mathscr{H}_1^{\otimes A n}$, respectively [with $\mathscr{H}_1$ given in (4.17)], under the linear transformation induced by (4.16).

4.8. If $E^{(\nu)}(B)$ is the spectral measure of some observable of the system $\mathfrak{S}_1^{(\nu)} \equiv \cdots \equiv \mathfrak{S}_{n_\nu}^{(\nu)}$ (see Axiom P1), use Axiom P4 to build the projector whose expectation value gives the probability of observing simultaneously one particle in $B_1^{(\nu)}$, one in $B_2^{(\nu)}$, etc. for all $\nu = 1, \ldots, k$, when the particles in $\mathfrak{S}$ are independent.

4.9. Let $U_1, \ldots, U_n$ be unitary transformations of the Hilbert spaces $\mathscr{H}_1', \ldots, \mathscr{H}_n'$ onto the Hilbert spaces $\mathscr{H}_1'', \ldots, \mathscr{H}_n''$. Define $U_1 \otimes \cdots \otimes U_n$ as that linear transformation of $\mathscr{H}_1 = \mathscr{H}_1' \otimes \cdots \otimes \mathscr{H}_n'$ into $\mathscr{H}_2 = \mathscr{H}_1'' \otimes \cdots \otimes \mathscr{H}_n''$ which maps $f_1 \otimes \cdots \otimes f_n$ into $U_1 f_1 \otimes \cdots \otimes U_n f_n$ (see Definition 4.1). Prove that $U_1 \otimes \cdots \otimes U_n$ is a unitary transformation of $\mathscr{H}_1$ *onto* $\mathscr{H}_2$.

*5. Complete Sets of Observables

5.1. The Concept of a Complete Set of Operators

The concept of a complete set of observables was first introduced by Dirac [1930] and it has remained in extensive use ever since. The original formulation is, however, merely heuristic in the general case, and it becomes rigorous only in the special case of a set of self-adjoint operators $A_1, \ldots, A_n$ acting in a separable Hilbert space $\tilde{\mathscr{H}}$ and having pure point spectra (which must be countable since $\tilde{\mathscr{H}}$ is separable),

$$\mathsf{S}^{A_k} = \{\lambda_k^{(1)}, \lambda_k^{(2)}, \ldots\}, \qquad k = 1, \ldots, n.$$

Under these circumstances $\{A_1, \ldots, A_n\}$ is called a complete set of operators if the following statements are true for some $\mathsf{S} \subset \mathbb{R}^n$:

(a) To each n-tuple $(\lambda_1, \ldots, \lambda_n) \in \mathsf{S} \subset \mathsf{S}^{A_1} \times \cdots \times \mathsf{S}^{A_n}$ can be assigned a vector $\Psi_{\lambda_1, \ldots, \lambda_n}$ from the common domain of definition of $A_1, \ldots, A_n$ which satisfies

$$A_k \Psi_{\lambda_1, \ldots, \lambda_n} = \lambda_k \Psi_{\lambda_1, \ldots, \lambda_n}, \qquad k = 1, \ldots, n. \tag{5.1}$$

(b) The family of all vectors

$$\Psi_{\lambda_1, \ldots, \lambda_n}, \qquad (\lambda_1, \ldots, \lambda_n) \in \mathsf{S}, \tag{5.2}$$

is an orthonormal basis in $\tilde{\mathscr{H}}$.

Note that the completeness of the set $\{A_1, \ldots, A_n\}$ implies that the operators $A_1, \ldots, A_n$ have a common dense domain of definition containing the linear manifold spanned by all the vectors $\Psi_{\lambda_1, \ldots, \lambda_n}$, and that they commute with one another on that domain of definition.

Another fact which has to be borne in mind is that, due to Bose–Einstein or Fermi–Dirac statistics, not all vectors (5.2) represent physical states, but only those vectors which are appropriately symmetrized or antisymmetrized with respect to the indices $\lambda_1, \ldots, \lambda_n$ referring to the same observables of identical particles of integer or half-integer spin, respectively. In the presence of identical particles, the Hilbert space of physical states will not be $\tilde{\mathscr{H}}$, but rather a closed subspace $\mathscr{H}$ of $\tilde{\mathscr{H}}$, constructed in accordance with Axiom P1 in §4, in which the appropriately symmetrized vectors in (5.2) will constitute an orthonormal basis. The operators $A_1, \ldots, A_n$ will not, in general, leave $\mathscr{H}$ invariant. (Consider, e.g., the spin operators of identical particles in wave mechanics.) However, appropriately symmetrized polynomials $p(A_1, \ldots, A_n)$ in $A_1, \ldots, A_n$ leave $\mathscr{H}$ invariant.

Definition 5.1. Let $\mathfrak{S}$ be a system consisting of a finite number of particles, and $A_1, \ldots, A_n$ obvervables of single particles from that system. We say that the function $f(x_1, \ldots, x_n)$, $x_1, \ldots, x_n \in \mathbb{R}^1$, is *appropriately symmetrized* with respect to $\{A_1, \ldots, A_n\}$ if it is symmetric (antisymmetric)

$$f(\ldots, x_i, \ldots, x_j, \ldots) = \pm f(\ldots, x_j, \ldots, x_i, \ldots)$$

with respect to any two of its variables x_i and x_j, whenever A_i and A_j represent the same observable quantity (e.g., energy, spin projection, momentum component) of two identical particles of integer (half-integer) spin.

Since the set (5.2) is an orthonormal basis in $\mathscr{H}$, we can expand any vector $\Psi \in \mathscr{H}$ in terms of that basis,

$$\Psi = \sum_{(\lambda_1,\ldots,\lambda_n)\in \mathsf{S}} \phi(\lambda_1,\ldots,\lambda_n)\, \Psi_{\lambda_1,\ldots,\lambda_n}, \tag{5.3}$$

$$\phi(\lambda_1,\ldots,\lambda_n) = \langle \Psi_{\lambda_1,\ldots,\lambda_n} \mid \Psi \rangle, \tag{5.4}$$

where we have

$$\sum_{(\lambda_1,\ldots,\lambda_n)\in \mathsf{S}} |\, \phi(\lambda_1,\ldots,\lambda_n)|^2 < +\infty. \tag{5.5}$$

Denote by $l^2(\mathsf{S})$, where $\mathsf{S} \subset \mathsf{S}^{A_1} \times \cdots \times \mathsf{S}^{A_n}$, the space of all functions $\phi(\lambda_1,\ldots,\lambda_n)$ defined on S and satisfying (5.5). It is easy to see that $l^2(\mathsf{S})$ is a separable Hilbert space and that the one-to-one mapping

$$\Psi \leftrightarrow \phi(\lambda_1,\ldots,\lambda_n) \tag{5.6}$$

is a unitary mapping of $\mathscr{H}$ onto $l^2(\mathsf{S})$.

On the other hand, the space $l^2(\mathsf{S})$ is isomorphic to the space $\hat{L}^2(\mathbb{R}^n, \mu)$ of all properly symmetrized and antisymmetrized (in the case where some of the particles in the system are identical) complex functions $\psi(\lambda_1,\ldots,\lambda_n)$ which are square integrable on $\mathbb{R}^n$ in the measure μ

$$\int_{\mathbb{R}^n} |\, \psi(\lambda_1,\ldots,\lambda_n)|^2\, d\mu(\lambda_1,\ldots,\lambda_n) < +\infty, \tag{5.7}$$

where μ denotes the measure on the Borel sets of $\mathbb{R}^n$, having support S, and such that

$$\mu(\{\lambda\}) = 1, \qquad \lambda = (\lambda_1,\ldots,\lambda_n) \in \mathsf{S}, \tag{5.8}$$

$$\mathsf{S} \subset \mathsf{S}^{A_1} \times \cdots \times \mathsf{S}^{A_n}. \tag{5.9}$$

In other words, for an arbitrary Borel set B in $\mathbb{R}^n$, $\mu(B)$ is equal to the number of the points in the set $B \cap \mathsf{S}$.

The elements of the Hilbert space $\hat{L}^2(\mathbb{R}^n, \mu)$ are equivalence classes of almost everywhere equal functions, i.e., the square-integrable (in the measure μ) functions $\psi_1(x)$ and $\psi_2(x)$, $x \in \mathbb{R}^n$, represent the same element of $\hat{L}^2(\mathbb{R}^n, \mu)$ if $\psi_1(\lambda) = \psi_2(\lambda)$ for all $\lambda \in \mathsf{S}$. The wave functions $\psi(x)$ are such that if x_1 and x_2 are variables corresponding to the same observables of two identical particles, then

$$\psi(\dots, x_1, \dots, x_2, \dots) = \pm\psi(\dots, x_2, \dots, x_1, \dots),$$

where the plus sign has to be taken in the case that the particles have integer spin, and the minus sign for particles of half-integral spin. Thus, $\hat{L}^2(\mathbb{R}^n, \mu)$ is, in general, a closed linear subspace of the space $L^2(\mathbb{R}^n, \mu)$ of all functions on $\mathbb{R}^n$ which are square integrable in the measure μ. In the case where all particles are different, $L^2(\mathbb{R}^n, \mu)$ is identical to $\hat{L}^2(\mathbb{R}^n, \mu)$.

The mapping

$$\text{(5.10)} \qquad \phi(\lambda) \leftrightarrow \psi(x), \qquad \phi \in l^2(\mathrm{S}), \qquad \psi \in L^2(\mathbb{R}^n, \mu),$$

which is such that $\phi(\lambda) = \psi(\lambda)$ for all $\lambda \in \mathrm{S}$, is a unitary transformation of $l^2(\mathrm{S})$ onto $\hat{L}^2(\mathbb{R}^n, \mu)$. Thus, (5.6) and (5.10) determine a unitary transformation

$$\text{(5.11)} \qquad \Psi \leftrightarrow \psi(x) = U\Psi, \qquad \Psi \in \mathscr{H}, \qquad \psi \in L^2(\mathbb{R}^n, \mu),$$

of $\mathscr{H}$ onto $\hat{L}^2(\mathbb{R}^n, \mu)$.

If we denote, in general, by $\mathscr{D}_A$ the domain of definition of an operator A, then we have that for $\Psi \in \mathscr{D}_{A_k}$

$$A_k\Psi = A_k \sum_{\lambda \in \mathrm{S}} \phi(\lambda)\, \Psi_\lambda = \sum_{\lambda \in \mathrm{S}} \lambda_k \phi(\lambda)\, \Psi_\lambda,$$

where $\lambda_k\phi(\lambda) \in l^2(\mathrm{S})$. Thus, $\Psi \in \mathscr{D}_A$ if and only if $\psi(x) = U\Psi$ is from the domain of definition $\mathscr{D}_{A_k'}$ of the operator A_k' acting on $L^2(\mathbb{R}^n, \mu)$ as follows:

$$\text{(5.12)} \qquad (A_k'\psi)(x) = x_k\psi(x), \qquad \psi \in \mathscr{D}_{A_k'};$$

$\mathscr{D}_{A_k'}$ obviously consists of all $\psi \in L^2(\mathbb{R}^n, \mu)$ for which $x_k\, \psi(x)$ is square integrable, i.e.,

$$\text{(5.13)} \qquad \mathscr{D}_{A_k'} = \left\{\psi: \psi \in L^2(\mathbb{R}^n, \mu), \int_{\mathbb{R}^n} x_k^2\, |\psi(x)|^2\, d\mu(x) < +\infty\right\}.$$

Moreover, we obviously have

$$\text{(5.14)} \qquad A_k' = UA_kU^{-1}.$$

The existence of a unitary transformation (5.11) of $\mathscr{H}$ onto $\hat{L}^2(\mathbb{R}^n, \mu)$ which is such that the operators $A_1', \dots, A_n'$, defined in (5.14), satisfy (5.12) and (5.13) is evidently a necessary and sufficient condition for

$A_1, \ldots, A_n$ to be complete in the sense of the definition given at the beginning of this section.

This approach to the concept of a complete set of self-adjoint operators $A_1, \ldots, A_n$ with discrete spectra leads to a straightforward and natural generalization to the case of self-adjoint operators $A_1, \ldots, A_n$ with arbitrary spectra.

Definition 5.2. The self-adjoint operators $A_1, \ldots, A_n$, acting in the Hilbert space $\tilde{\mathscr{H}}$, constitute a *complete set of operators* in the Hilbert space $\mathscr{H}$, which is a subspace of $\tilde{\mathscr{H}}$, if the following three conditions are met:

(a) There is a measure $\mu(B)$ in the Borel sets B of the n-dimensional Euclidean space $\mathbb{R}^n$ with support $\mathsf{S} \subset \mathsf{S}^{A_1} \times \cdots \times \mathsf{S}^{A_n}$, where $\mathsf{S}^{A_1}, \ldots, \mathsf{S}^{A_n}$ are the spectra of $A_1, \ldots, A_n$, respectively.

(b) There is a unitary transformation U of $\tilde{\mathscr{H}}$ onto the Hilbert space $L^2(\mathbb{R}^n, \mu)$, such that the operators

$$A_k' = UA_kU^{-1}, \qquad k = 1, \ldots, n,$$

are the multiplication operators

$$(A_k'\psi)(x) = x_k\psi(x), \qquad \psi \in \mathscr{D}_{A_k'},$$

with domains of definition

$$\mathscr{D}_{A_k'} = \left\{\psi: \int_{\mathbb{R}^n} x_k^2 \mid \psi(x)|^2 \, d\mu(x) < +\infty, \psi \in L^2(\mathbb{R}^n, \mu)\right\}.$$

(c) U maps $\mathscr{H}$ onto the subspace $\hat{L}^2(\mathbb{R}^n, \mu)$ of $L^2(\mathbb{R}^n, \mu)$ consisting of all appropriately symmetrized functions from $L^2(\mathbb{R}^n, \mu)$.

If the first two of the above requirements are fulfilled, the Hilbert space $L^2(\mathbb{R}^n, \mu)$ is called a *spectral representation space* of the operators $A_1, \ldots, A_n$, and $A_1', \ldots, A_n'$ the *spectral representation* (or *canonical form*) of the operators $A_1, \ldots, A_n$.

If A_1 is an operator such that $\{A_1\}$ is complete (i.e., $n = 1$ in the above definition) then A_1 is said to be an operator with a *simple spectrum.*

5.2. Cyclic Vectors and Complete Sets of Operators

The significance of the concept of a complete set of operators in quantum mechanics is reflected in the following assumption, which is ordinarily tacitly made in nonrelativistic quantum mechanics.

Axiom C. For every quantum mechanical system $\mathfrak{S}$ there is a finite set $\{A_1, \ldots, A_n\}$ of operators in $\tilde{\mathscr{H}}$, which is complete in the Hilbert space $\mathscr{H}$, $\mathscr{H} \subset \tilde{\mathscr{H}}$, associated with that system, and such that, in the absence of superselection rules, every function $F(A_1, \ldots, A_n)$ which leaves $\mathscr{H}$ invariant and, when restricted to $\mathscr{H}$, is a self-adjoint operator in $\mathscr{H}$, represents an observable of $\mathfrak{S}$.

To give an example of a complete set of operators of the kind mentioned in the above axiom, consider a system $\mathfrak{S}$ of n particles without spin. We have seen in the preceding section that if all the particles in the system are different, then in wave mechanics we associate with $\mathfrak{S}$ the Hilbert space $L^2(\mathbb{R}^{3n})$. In this case $\mathscr{H} \equiv \tilde{\mathscr{H}} \equiv L^2(\mathbb{R}^{3n})$ and the position operators

$$(X_i\psi)(\mathbf{r}_1, \ldots, \mathbf{r}_n) = x_i\psi(\mathbf{r}_1, \ldots, \mathbf{r}_n),$$
$$(Y_i\psi)(\mathbf{r}_1, \ldots, \mathbf{r}_n) = y_i\psi(\mathbf{r}_1, \ldots, \mathbf{r}_n),$$
$$(Z_i\psi)(\mathbf{r}_1, \ldots, \mathbf{r}_n) = z_i\psi(\mathbf{r}_1, \ldots, \mathbf{r}_n), \qquad i = 1, \ldots, n,$$

constitute a complete set of operators, which in this case represent observables. Thus, in this example $L^2(\mathbb{R}^{3n})$ is the spectral representation space of the position observables and the above introduced operators X_i, Y_i, Z_i, $i = 1, \ldots, n$, are the canonical forms of these observables.

In case that some of the particles in the system are identical, we have seen in the preceding section that $L^2(\mathbb{R}^{3n})$ is not the Hilbert space associated with $\mathfrak{S}$; instead, the space $\hat{L}^2(\mathbb{R}^{3n})$ of appropriately symmetrized Lebesgue square-integrable functions is the space $\mathscr{H}$ associated with $\mathfrak{S}$. In this case, the operators X_i, Y_i, Z_i, $i = 1, \ldots, n$, acting in $L^2(\mathbb{R}^{3n}) = \tilde{\mathscr{H}}$ are still a complete set of operators, but they do not represent observables since they do not leave $\mathscr{H} = \hat{L}^2(\mathbb{R}^{3n})$ invariant. However, any operator $A = F(X_1, Y_1, Z_1, \ldots, X_n, Y_n, Z_n)$

$$(A\psi)(\mathbf{r}_1, \ldots, \mathbf{r}_n) = F(\mathbf{r}_1, \ldots, \mathbf{r}_n)\, \psi(\mathbf{r}_1, \ldots, \mathbf{r}_n),$$

where $F(\mathbf{r}_1, \ldots, \mathbf{r}_n)$ is a real appropriately symmetrized and bounded Borel-measurable function, will represent an observable of $\mathfrak{S}$; for instance, the projectors (4.11) and (4.20) are such functions.

It has become customary to refer to any set $\{A_1, \ldots, A_n\}$ obeying Axiom C as a *complete set of observables* even when, strictly speaking, only their appropriately symmetrized functions are observables.

Additional examples of complete sets of observables will be encountered in §7, when the formalism of wave mechanics will be derived from the general quantum mechanical formalism espoused in this chapter (see also Sections 4.1–4.4 in Chapter V).

We shall relate the completeness property of $A_1, \ldots, A_n$ to the existence of a cyclic vector for $A_1, \ldots, A_n$.

Definition 5.3. The vector Ψ_0 of the Hilbert space $\mathscr{H}$ is a *cyclic vector in* $\mathscr{H}$ *with respect to the commuting operators* $A_1, \ldots, A_n$ in the Hilbert space $\tilde{\mathscr{H}}$, where $\tilde{\mathscr{H}} \supset \mathscr{H}$, if the following three conditions are fulfilled:

(a) $\Psi_0 \in \mathscr{D}_{A_1} \cap \cdots \cap \mathscr{D}_{A_n}$.
(b) For any integers $k_1, \ldots, k_n = 0, 1, 2, \ldots,$

(5.15) $$A_1^{k_1} \cdots A_n^{k_n} \Psi_0 \in \mathscr{D}_{A_1} \cap \cdots \cap \mathscr{D}_{A_n}.$$

(c) The linear manifold

(5.16) $$\mathscr{D}(\Psi_0; A_1, \ldots, A_n) = \{p(A_1, \ldots, A_n)\, \Psi_0\}$$

spanned by all vectors of the form $p(A_1, \ldots, A_n)\, \Psi_0$, corresponding to all choices of appropriately symmetrized polynomials p in $A_1, \ldots, A_n$, is everywhere dense in $\mathscr{H}$.

In the remainder of this section as well as the next section, we shall gradually establish the following main theorem.

Theorem 5.1. The set of self-adjoint operators $A_1, \ldots, A_n$ acting in the separable Hilbert space $\mathscr{H}$ and with range in $\tilde{\mathscr{H}} \supset \mathscr{H}$ is complete if and only if there is a vector $\Psi_0 \in \mathscr{H}$ which is cyclic with respect to $A_1, \ldots, A_n$, and $A_1, \ldots, A_n$ commute on the dense linear manifold $\mathscr{D}(\Psi_0; A_1, \ldots, A_n)$ generated by applying to Ψ_0 all possible polynomial forms $p(A_1, \ldots, A_n)$ which are appropriately symmetrized.

To prove that the completeness of $A_1, \ldots, A_n$ on a separable $\mathscr{H}$ implies the existence of a cyclic vector Ψ_0 for $A_1, \ldots, A_n$ we first verify Lemmas 5.1 and 5.2.

Lemma 5.1. If $L^2(\mathbb{R}^n, \mu)$ is unitarily equivalent to a separable Hilbert space $\mathscr{H}$, then the measure μ is σ finite.

Proof. If μ is not σ finite, then there is an uncountable family $\mathscr{F}$ of mutually disjoint measurable sets of nonzero measure. Denote by $\chi_S(x)$ the characteristic function of the set S. The functions

$$e_S(x) = \frac{1}{\sqrt{\mu(S)}} \chi_S(x), \qquad S \in \mathscr{F},$$

constitute an orthonormal system in $L^2(\mathbb{R}^n, \mu)$, since for $S_1, S_2 \in \mathscr{F}$, we have

$$e_{S_1}(x)\, e_{S_2}(x) = \begin{cases} 0 & \text{if } S_1 \neq S_2 \\ \dfrac{1}{\mu(S_1)}\, \chi_{S_1}(x) & \text{if } S_1 = S_2 . \end{cases}$$

Under a unitary transformation of $L^2(\mathbb{R}^n, \mu)$ onto $\mathscr{H}$, the images of all elements from $\mathscr{F}$ would consitute an uncountable orthonormal system in $\mathscr{H}$. This is impossible if $\mathscr{H}$ is separable. Q.E.D.

***Lemma 5.2.** Let μ be a σ-finite measure and $h(x)$ a continuous positive function representing an element of $L^2(\mathbb{R}^n, \mu)$. The function

$$f_0(x) = \exp(-\alpha(|x_1| + \cdots + |x_n|))\, h(x), \qquad \alpha > 0, \tag{5.17}$$

represents an element of $L^2(\mathbb{R}^n, \mu)$, and the family of all functions

$$p(x) f_0(x), \tag{5.18}$$

obtained by letting $p(x)$ range over all polynomials in $x \in \mathbb{R}^n$ is dense in $L^2(\mathbb{R}^n, \mu)$.

Proof. Since we have

$$\begin{aligned} &|x_1^{k_1} \cdots x_n^{k_n} \exp(-\alpha(|x_1| \cdots |x_n|)) | \\ &\quad \leqslant \frac{|x_1|^{k_1} \cdots |x_n|^{k_n}}{\prod_{r=1}^{n} \left(1 + \dfrac{|\alpha x_r|}{1!} + \cdots + \dfrac{|\alpha x_r|^{k_r}}{k_r!}\right)} \leqslant \frac{k_1! \cdots k_n!}{\alpha^{k_1 + \cdots + k_n}}, \end{aligned} \tag{5.19}$$

the functions (5.18), which include the function $f_0(x)$ itself, are square integrable because

$$|p(x) f_0(x)|^2 \leqslant \text{const}\, |h(x)|^2.$$

Due to the σ finiteness of μ, the family of all continuous functions on $\mathbb{R}^n$ which represent elements from $L^2(\mathbb{R}^n, \mu)$ and are of compact support is dense in $L^2(\mathbb{R}^n, \mu)$ (see Exercise 5.1). Hence, it is sufficient to show that any continuous function $g(x)$ of compact support and representing an element of $L^2(\mathbb{R}^n, \mu)$ can be approximated arbitrarily well in the norm $\|\cdot\|$,

$$\|f\|^2 = \int_{\mathbb{R}^n} |f(x)|^2\, d\mu(x),$$

by a function of the form (5.18).

We carry out the change of variables

$$x \to u = \rho(x) = (w(x_1),\dots, w(x_n)),$$

where $w(y)$ is some analytic function on $\mathbb{R}^1$, mapping $\mathbb{R}^1$ in a one-to-one manner into a finite interval I and such that $|d^n w/dy^n| \leqslant 1$ for all $n = 0, 1, 2,\dots$ (such a function can be built from exponential functions). Under this transformation $\mathbb{R}^n$ is mapped onto the n-dimensional rectangle $R = I \times \cdots \times I$.

Since $g(x)$ is of compact support, the continuous function $f_0(x)$ is bounded from below on the support of $g(x)$, and therefore the function $g(x)/f_0(x)$ is integrable and square integrable on $\mathbb{R}^n$.

The function $g[\rho^{-1}(u)]/f_0[\rho^{-1}(u)]$, where $\rho^{-1}(u)$ is the inverse of $\rho(x)$, is continuous and of compact support R. Hence, according to the Weierstrass approximation theorem (see Chapter II, Theorem 7.4), we can approximate that function arbitrarily well on R by a polynomial, i.e., we can find a polynomial

$$q(u) = \sum a_{m_1,\dots,m_n} u^{m_1} \cdots u^{m_n}$$

for which

$$\left| q(u) - \frac{g[\rho^{-1}(u)]}{f_0[\rho^{-1}(u)]} \right| < \frac{\epsilon}{2\,\| f_0 \|}, \qquad u \in R.$$

If we denote by μ_1 the measure in R resulting after the change of variable $x \to u = \rho(x)$, we have

$$\begin{aligned}(5.20)\qquad &\int_{\mathbb{R}^n} | g(x) - q(\rho(x))\, f_0(x)|^2\, d\mu(x) \\ &= \int_R | f_0[\rho^{-1}(u)]|^2 \left| \frac{g[\rho^{-1}(u)]}{f_0[\rho^{-1}(u)]} - q(u) \right|^2 d\mu_1(u) \\ &\leqslant \frac{\epsilon^2}{4\,\| f_0 \|^2} \int_R | f[\rho^{-1}(u)]|^2\, d\mu_1(u) = \frac{\epsilon^2}{4}.\end{aligned}$$

By expanding each $w(x_k)$ according to the Taylor's formula to the order k, we obtain

$$\begin{aligned} q(w(x_1),\dots, w(x_n)) &= \sum a_{m_1\cdots m_n} [w(x_1)]^{m_1} \cdots [w(x_n)]^{m_n} \\ &= p_k(x) + r_k(x), \end{aligned}$$

where $p_k(x)$ is a polynomial of the order $n(k-1)$, and $r_k(x)$ is the remainder. A simple computation shows that this remainder can be majorized as follows:

$$| r_k(x)| \leqslant \sum_{l=0}^{n} \frac{c_l}{[(k-1)!]^{n-l}(k!)^l} \sum_{(i_1,\ldots,i_n)} | x_{i_1}^{k-1} \cdots x_{i_k}^{k-1} x_{i_{k+1}}^{k} \cdots x_{i_n}^{k} |,$$

where the coefficients c_l are constants built from the coefficients of $q(u)$, and can be chosen to be independent of k, due to the requirement that $| d^n w/dy^n | \leqslant 1$.

By employing (5.19) we get

$$(5.21)\quad \left[\int_{\mathbb{R}^n} \left| \frac{x_1^{k_1} \cdots x_n^{k_n}}{k_1! \cdots k_n!} f_0(x) \right|^2 d\mu(x)\right]^{1/2} \leqslant \frac{1}{\alpha^{k_1+\cdots+k_n}} \left[\int_{\mathbb{R}^n} | h(x)|^2 \, d\mu(x)\right]^{1/2}.$$

By taking advantage of the relation

$$(5.22)\quad \left[\int_{\mathbb{R}^n} |f_1(x) + f_2(x)|^2 \, |f_0(x)|^2 \, d\mu(x)\right]^{1/2}$$
$$\leqslant \left[\int_{\mathbb{R}^n} |f_1(x)|^2 \, |f_0(x)|^2 \, d\mu(x)\right]^{1/2} + \left[\int_{\mathbb{R}^n} |f_2(x)|^2 \, |f_0(x)|^2 \, d\mu(x)\right]^{1/2},$$

which is a direct consequence of the triangle inequality whenever $f_1 \cdot f_0$, $f_2 \cdot f_0 \in L^2(\mathbb{R}^n, \mu)$, we obtain by using (5.21)

$$\left[\int_{\mathbb{R}^n} |f_0(x)|^2 \, | p_k(x) - q[\rho(x)]|^2 \, d\mu(x)\right]^{1/2}$$
$$= \left[\int_{\mathbb{R}^n} | f_0(x) \, r_k(x)|^2 \, d\mu(x)\right]^{1/2}$$
$$\leqslant \sum_{l=0}^{n} | c_l | \left[\int_{\mathbb{R}^n} \left| \frac{x_{i_1}^{k-1} \cdots x_{i_n}^{k}}{[(k-1)!]^{n-l}[k!]^l} f_0(x) \right|^2 d\mu(x)\right]^{1/2}$$
$$\leqslant \frac{\sum_{l=0}^{n} | c_l |}{\alpha^{n(k-1)}} < \frac{\epsilon}{2}$$

for sufficiently large values of k when $\alpha > 1$. Combining the above relation and (5.20), we arrive (by using the triangle inequality) at the inequality

$$\int_{\mathbb{R}^n} | g(x) - p_k(x) \, f_0(x)|^2 \, d\mu(x) < \epsilon^2,$$

which is the desired result when $\alpha > 1$.

The case with $0 < \alpha < 1$ can be reduced to the preceding case by substituting the variables x_k by the new variables $x_k' = 2\alpha x_k$. Q.E.D.

As a direct consequence of Lemmas 5.1 and 5.2 we get the following theorem upon appropriate symmetrization of $f_0(x)$ in (5.17).

Theorem 5.2. If the Hilbert space $\hat{L}(\mathbb{R}^n, \mu)$ is separable, it contains a vector which is cyclic with respect to the canonical operators $A_1',..., A_n'$ defined in (5.12).

5.3. The Construction of Spectral Representation Spaces

We shall prove now that if there is a cyclic vector for the self-adjoint operators $A_1 ,..., A_n$, then $\{A_1 ,..., A_n\}$ is a complete set. The proof will be obtained by constructing the representation space $\hat{L}^2(\mathbb{R}^n, \mu)$.

***Lemma 5.3.** Assume that $f \in \mathscr{H}$ is a cyclic vector with respect to the self-adjoint operators $A_1 ,..., A_n$ in the separable Hilbert space $\mathscr{H}$, and that μ is the measure satisfying the relation

$$(5.23) \quad \langle f \mid p(A_1 ,..., A_n) f\rangle = \int_{\mathbb{R}^n} p(x)\, d\mu(x), \qquad \mu(B) = \langle f \mid E^{A_1,...,A_n}(B)\, f\rangle,$$

for all polynomials $p(x)$. Then there is a unique unitary operator U, mapping $\hat{L}^2(\mathbb{R}^n, \mu)$ onto $\mathscr{H}$, such that

$$(5.24) \quad U(\chi_I \cdot p) = p(A_1 ,..., A_n)\, E^{A_1,...,A_n}(I)$$

for all appropriately symmetrized polynomials $p(x)$ and all finite intervals I in $\mathbb{R}^n$.

Proof. Note first that every polynomial $p(x)$ is square integrable on $\mathbb{R}^n$ with respect to the measure μ since

$$\begin{aligned}\int_{\mathbb{R}^n} | p(x)|^2\, d\mu(x) &= \int_{\mathbb{R}^n} p^*(x)\, p(x)\, d\mu(x)\\ &= \langle f \mid (p(A_1 ,..., A_n))^*\, p(A_1 ,..., A_n) f\rangle\\ &= \| p(A_1 ,..., A_n) f \|^2 < +\infty.\end{aligned}$$

Define an operator U_0 from $\hat{L}^2(\mathbb{R}^n, \mu)$ into $\mathscr{H}$ by

$$U_0(p \cdot \chi_I) = p(A_1 ,..., A_n)\, E^{A_1,...,A_n}(I),$$

where $p(x)$ is any appropriately symmetrized polynomial. The so-defined operator U_0 is isometric,

$$\begin{aligned} \| U_0(p \cdot \chi_I)\|^2 &= \| p(A_1 ,..., A_n) E^{A_1,...,A_n}(I)f \|^2 \\ &= \langle E^{A_1,...,A_n}(I)f \mid (p(A_1 ,..., A_n))^* p(A_1 ,..., A_n) E^{A_1,...,A_n}(I)f \rangle \\ &= \langle E^{A_1,...,A_n}(I)f \mid p^*(A_1 ,..., A_n) p(A_1 ,..., A_n) E^{A_1,...,A_n}(I)f \rangle \\ &= \int_{\mathbb{R}^n} \chi_I^*(x)(p^*(x) p(x) \chi_I(x)) \, d\mu(x) = \int_{\mathbb{R}^n} | p(x) \chi_I(x)|^2 \, d\mu(x); \end{aligned}$$

in deriving the second step we have made use of the fact that $p^*(A_1 ,..., A_n) p(A_1 ,..., A_n)$ is a polynomial in $A_1 ,..., A_n$.

The measure μ is finite and, therefore, the family of all continuous functions in $\hat{L}^2(\mathbb{R}^n, \mu)$ of compact support is dense in $\hat{L}^2(\mathbb{R}^n, \mu)$ (see Exercise 5.1). Since, according to the Weierstrass theorem, each such function can be approximated uniformly and arbitrarily well by a function of the form $\chi_I(x) p(x)$, where I is a finite interval and $p(x)$ is an appropriately symmetrized polynomial, it follows that the family of all such functions is dense in $\hat{L}^2(\mathbb{R}^n, \mu)$. Thus, the domain of definition of the operator U_0 can be extended to $\hat{L}^2(\mathbb{R}^n, \mu)$ in a unique manner, and the operator U obtained in this way is isometric. Moreover, since f is cyclic with respect to $A_1 ,..., A_n$, the family of all vectors $p(A_1 ,..., A_n) f$ is dense in $\mathscr{H}$, and, therefore, the family of all vectors

$$E^{A_1,...,A_n}(I) \, p(A_1 ,..., A_n)f,$$

is also dense in $\mathscr{H}$. Hence, the range of U_0 is dense in $\mathscr{H}$, i.e., the range of U has to be $\mathscr{H}$. Thus, U is unitary. Q.E.D.

Note that if we take a sequence $I_1 \subset I_2 \subset \cdots$ of intervals which are such that $I_1 \cup I_2 \cup \cdots = \mathbb{R}^n$, then $\chi_{I_n}(x) \, p(x)$ converges strongly to $p(x)$ and $E^{A_1,...,A_n}(I) \, p(A_1 ,..., A_n) f$ converges strongly to $p(A_1 ,..., A_n) f$. Since

$$U(p \cdot \chi_{I_k}) = E^{A_1,...,A_n}(I_k) \, p(A_1 ,..., A_n)f$$

and U is bounded and therefore continuous, we get in the limit

$$Up = p(A_1 ,..., A_n)f.$$

By using Lemma 5.3 we can easily prove Theorem 5.3.

Theorem 5.3. Let $f \in \mathscr{H}$ be a cyclic vector with respect to the self-adjoint operators $A_1 ,..., A_n$ in the Hilbert space $\mathscr{H}$, let μ be the finite measure defined in (5.23) and U the unitary operator mapping $L^2(\mathbb{R}^n, \mu)$ onto $\mathscr{H}$ and satisfying (5.24). Then the self-adjoint operators

$$U^{-1}A_kU, \qquad k = 1,\ldots, n,$$

in $L^2(\mathbb{R}^n, \mu)$ are the (μ-almost everywhere) multiplication operators

$$(U^{-1}A_kU\psi)(x) = x_k\, \psi(x_k), \qquad \psi \in \mathscr{D}_{U^{-1}A_kU}\,,$$

and therefore for any of the operators A in Theorem 2.6,

$$(U^{-1}AU\psi)(x) = F(x)\, \psi(x), \qquad \psi \in \mathscr{D}_{U^{-1}AU}\,.$$

Proof. We shall prove the theorem by showing that $U^{-1}A_kU$ coincides on functions of the form $\chi_I(x)\, p(x)$ (I is a finite interval) with the operator A_k',

$$(A_k'\psi)(x) = x_k\, \psi(x), \qquad \psi \in \mathscr{D}_{A_k'}\,,$$

with domain

$$\mathscr{D}_{A_k'} = \left\{\psi\colon \int_{\mathbb{R}^n} x_k^2\, |\psi(x)|^2\, d\mu(x) < +\infty\right\}.$$

Since $x_k\, p(x)$ is a polynomial when $p(x)$ is a polynomial, we have

$$\begin{aligned}[][U^{-1}A_kU(p\cdot x_I)](x) &= [U^{-1}A_k p(A_1,\ldots, A_n)\, E^{A_1,\ldots,A_n}(I)](x)\\ &= x_k p(x)\, \chi_I(x).\end{aligned}$$

Now, $U^{-1}A_kU$ is self-adjoint because A_k is self-adjoint and U is unitary, while A_k' is known to be self-adjoint. Thus, the operators $U^{-1}A_kU$ and A_k' must be identical because they coincide on the set of all functions $h(x) = \chi_I(x)\, p(x)$, which is dense in $L^2(\mathbb{R}^n, \mu)$. In fact, since $(U^{-1}A_kU)^* = U^{-1}A_kU$ and ψ is any vector in $\mathscr{D}_{U^{-1}A_kU}$, we have

$$\begin{aligned}\langle U^{-1}A_kU\psi \mid h\rangle = \langle\psi \mid U^{-1}A_kUh\rangle &= \int_{\mathbb{R}^n} \psi^*(x)\, x_kh(x)\, d\mu(x)\\ &= \int_{\mathbb{R}^n} (x_k\psi(x))^*\, h(x)\, d\mu(x)\end{aligned}$$

for all $h(x) = p(x)\, \chi_I(x)$, which implies that $(U^{-1}A_kU\psi)(x) = x_k\psi(x)$; the representation of $U^{-1}AU$ follows in a similar manner. Q.E.D.

It is evident that Theorems 5.2 and 5.3 imply Theorem 5.1.

It should be noted that a complete set of operators can have many spectral representation spaces corresponding to different measures μ. For instance, if we take the case of the single operator X in $L^2(\mathbb{R}^1)$,

$$(X\psi)(x) = x\psi(x),$$

which has a simple spectrum, then e^{-x^2} is an example of a cyclic vector for this operator, since the family of all functions $p(x)\, e^{-x^2/2}$ [where $p(x)$ is a polynomial] is dense in $L^2(\mathbb{R}^1)$ (see Chapter II, Exercise 7.9). For a polynomial $p(x)$, $x \in \mathbb{R}^1$, we shall have

$$\langle e^{-x^2} \mid p(X)\, e^{-x^2} \rangle = \int_{\mathbb{R}^1} p(x)\, e^{-2x^2}\, dx = \int_{\mathbb{R}^1} p(x)\, d\mu(x),$$

where

$$d\mu(x) = e^{-2x^2}\, dx.$$

Thus, $L^2(\mathbb{R}^1, \mu)$ will be another spectral representation space of X.

5.4. Cyclicity and Maximality

We have seen in the previous section that in the case that n identical particles are present, it can happen that the self-adjoint operators $\tilde{A}_1, \ldots, \tilde{A}_n$ corresponding to observables of individual particles will not, in general, represent observables of the entire n-particle system. Instead, appropriately symmetrized functions of $\tilde{A}_1, \ldots, \tilde{A}_n$ will represent observables, provided we are working in the Schroedinger picture. A special case of such functions will be the projectors $E^{\tilde{A}_1,\ldots,\tilde{A}_n}(B)$ corresponding to Borel sets B whose characteristic functions $\chi_B(x)$ are appropriately symmetrized.

It is therefore convenient to express the property of f_0 being cyclic in terms of projection operators. The next two theorems are related to this problem, and lead to the important Theorem 5.6.

Theorem 5.4. If f_0 is a cyclic vector in $\mathscr{H}$ with respect to the self-adjoint commuting operators $A_1, \ldots, A_n$ in $\mathscr{H}$, then the linear manifold

$$(E_\lambda^{A_1,\ldots,A_n} f_0 \colon \lambda \in \mathbb{R}^n) \tag{5.25}$$

spanned by all vectors $E_\lambda^{A_1,\ldots,A_n} f_0$, $\lambda \in \mathbb{R}^n$, is dense in $\mathscr{H}$.

Proof. Assume that the set (5.25) is not dense in $\mathscr{H}$. Then there exists a nonzero vector g which is orthogonal to all the vectors in this set, i.e.,

$$\langle g \mid E_\lambda^{A_1,\ldots,A_n} f_0 \rangle \equiv 0, \qquad \lambda \in \mathbb{R}^n.$$

Using Theorem 2.7 [see also (2.28)], we can write for any polynomial $p(A_1, \ldots, A_n)$

$$\langle g \mid p(A_1, \ldots, A_n) f_0 \rangle = \int_{\mathbb{R}^n} p(\lambda)\, d\langle g \mid E_\lambda^{A_1,\ldots,A_n} f_0 \rangle.$$

Then the identity $\langle g \mid E_\lambda^{A_1,\ldots,A_n} f_0 \rangle \equiv 0$ implies that

$$\langle g \mid p(A_1, \ldots, A_n) f_0 \rangle = 0$$

for any polynomial $p(A_1, \ldots, A_n)$, which contradicts the assumption that f_0 is cyclic with respect to $A_1, \ldots, A_n$, i.e., that the set of all vectors $p(A_1, \ldots, A_n) f_0$, corresponding to all polynomials $p(\lambda)$, is dense in $\mathscr{H}$.

Q.E.D

The following result represents a converse, in a limited sense, of the above theorem.

***Theorem 5.5.** If g_0 is a vector in the Hilbert space $\mathscr{H}$ which is cyclic with respect to $\{E_\lambda^{A_1,\ldots,A_n} : \lambda \in \mathbb{R}^n\}$ so that

(5.26) $$(E_\lambda^{A_1,\ldots,A_n} g_0 : \lambda \in \mathbb{R}^n)$$

is dense in $\mathscr{H}$, then any vector f_0 of the form

(5.27) $$f_0 = \exp[-\alpha(|A_1| + \cdots + |A_n|)]\, g_0, \qquad \alpha > 0,$$

is a cyclic vector with respect to the self-adjoint commuting operators $A_1, \ldots, A_n$.

Proof. The vector f_0 in (5.27) is in the domain of definition of any polynomial $p(A_1, \ldots, A_n)$, since $p(\lambda_1, \ldots, \lambda_n) \exp[-\alpha(|\lambda_1| + \cdots + |\lambda_n|)]$, $\lambda_1, \ldots, \lambda_n \in \mathbb{R}^1$, is a bounded function and therefore

$$\int_{\mathbb{R}^n} | p(\lambda_1, \ldots, \lambda_n) \exp[-\alpha(|\lambda_1| + \cdots + |\lambda_n|)]|^2\, d\| E_\lambda^{A_1,\ldots,A_n} g_0 \|^2 < +\infty.$$

If f_0 were not cyclic with respect to $A_1, \ldots, A_n$, then the linear manifold

(5.28) $$(E_\lambda^{A_1,\ldots,A_n} \exp[-\alpha(|A_1| + \cdots + |A_n|)]\, g_0 : \lambda \in \mathbb{R}^n)$$

would not be dense in $\hat{\mathscr{H}}$, i.e., there would be a nonzero vector f orthogonal to (5.28),

$$\langle f \mid E_\lambda^{A_1,\ldots,A_n} \exp[-\alpha(|A_1| + \cdots + |A_n|)] g_0 \rangle = 0, \qquad \lambda \in \mathbb{R}^n.$$

This implies that

$$(5.29) \quad \int_{\mathbb{R}^n} p(\lambda) \exp[-\alpha(|\lambda_1| + \cdots + |\lambda_n|)] \, d\langle f \mid E_\lambda^{A_1,\ldots,A_n} g_0 \rangle$$

$$= \int_{\mathbb{R}^n} p(\lambda) \, d\langle f \mid E_\lambda^{A_1,\ldots,A_n} \exp[-\alpha(|A_1| + \cdots + |A_n|)] g_0 \rangle = 0.$$

Since the measure

$$(5.30) \qquad \mu(B) = \langle f \mid E^{A_1,\ldots,A_n}(B) g_0 \rangle, \qquad B \in \mathscr{B}^n,$$

is finite, we can take in Lemma 5.2 $h(\lambda) \equiv 1$, and conclude that the set of all functions

$$(5.31) \qquad p(\lambda) \exp[-\alpha(|\lambda_1| + \cdots + |\lambda_n|)]$$

is dense in $L^2(\mathbb{R}^n, \mu)$. Thus, due to the fact that any characteristic function $\chi_B(\lambda)$ of a Borel set $B \in \mathscr{B}^n$ is square integrable in the measure (5.30), it can be approximated arbitrarily well in the mean in the measure (5.30) by a function of the form (5.31). By taking such an approximation to the limit and using (5.29) we obtain

$$\int_{\mathbb{R}^n} \chi_B(\lambda) \, d\langle f \mid E_\lambda^{A_1,\ldots,A_n} g_0 \rangle = \langle f \mid E^{A_1,\ldots,A_n}(B) g_0 \rangle = 0.$$

This implies that f is orthogonal to the set (5.26). But this is impossible if $f \neq 0$ and (5.26) is dense in $\mathscr{H}$. Q.E.D.

In relation to the most general case encompassed in Definition 5.3, Theorems 5.4 and 5.5 refer to the case when $\mathscr{H} \equiv \hat{\mathscr{H}}$. The formulation of the equivalent of these theorems in the more general case when $\mathscr{H}$ is a nontrivial closed subspace of $\hat{\mathscr{H}}$ can be achieved by replacing $\hat{\mathscr{H}}$ by $\mathscr{H}$ in the above two theorems, and substituting the vectors in (5.25) and (5.26) by their projections onto $\mathscr{H}$. This means, *de facto*, replacing the projector-valued function $E_{\lambda_1,\ldots,\lambda_n}^{A_1,\ldots,A_n}$ in the $\lambda_1,\ldots,\lambda_n \in \mathbb{R}^1$ by an appropriately symmetrized operator-valued function

$$\sum_{(\lambda_1,\ldots,\lambda_n)} E_{\lambda_1,\ldots,\lambda_n}^{A_1,\ldots,A_n},$$

where the above sum is carried out only over those permutations of $(\lambda_1, \ldots, \lambda_n)$ which permute the same observable of identical particles. The alterations in the proofs of Theorems 5.4 and 5.5, which are necessary when we are dealing with this more general case, are very easy to carry out; they are left to the reader.

The following theorem states the important ***maximality property*** of the algebra generated by a complete set of observables.

Theorem 5.6. If the bounded operator A, defined on the Hilbert space $\mathscr{H}$, commutes with the operators $A_1, \ldots, A_n$, where $\{A_1, \ldots, A_n\}$ is a complete set of operators, then A is a function of $A_1, \ldots, A_n$.

Proof. Since $\{A_1, \ldots, A_n\}$ is a complete set, there is at least one vector $f_0 \in \mathscr{H}$ which is cyclic with respect to $A_1, \ldots, A_n$. According to Theorem 5.4, the linear manifold (5.26) is dense in $\mathscr{H}$. Consequently, there is a sequence $f_1, f_2, \ldots$

$$f_m = (a_m^{(1)} E_{\lambda_m^{(1)}}^{A_1,\ldots,A_n} + \cdots + a_m^{(k_m)} E_{\lambda_m^{(k_m)}}^{A_1,\ldots,A_n}) f_0$$

which converges strongly to Af_0. If we introduce the function

$$F_m(\lambda) = a_m^{(1)}\, \theta(\lambda - \lambda_m^{(1)}) + \cdots + a_m^{(k_m)} \theta(\lambda - \lambda_m^{(k_m)}),$$

where

$$\theta(\lambda) = \begin{cases} 1 & \text{for } \lambda \leqslant 0 \\ 0 & \text{for } \lambda > 0, \end{cases}$$

then we observe that $f_m = F_m(A_1, \ldots, A_n) f_0$. Since

$$\| f_l - f_m \|^2 = \int_{\mathbb{R}^n} | F_l(\lambda) - F_m(\lambda)|^2 \, d \, \| E_\lambda^{A_1,\ldots,A_n} f_0 \|^2$$

and $f_1, f_2, \ldots$ converges, we deduce that $F_1(\lambda), F_2(\lambda), \ldots$ is a sequence of Borel measurable functions which is fundamental in the mean with respect to the measure $\mu_0(B) = \| E^{A_1,\ldots,A_n}(B) f_0 \|^2$, $B \in \mathscr{B}^n$. According to the Riesz–Fischer theorem (Chapter II, Theorem 4.4), $F_1(\lambda), F_2(\lambda), \ldots$ converges in the mean, with respect to the measure μ_0, to a μ_0-square-integrable function $F(\lambda)$. Hence, by using Theorem 2.5, we arrive at the conclusion that

$$Af_0 = F(A_1, \ldots, A_n) f_0. \tag{5.32}$$

Since A commutes with $E_{\lambda_1}^{A_1}, \ldots, E_{\lambda_n}^{A_n}$, it commutes with any operator

$$D = a_1 E_{\lambda_1}^{A_1,\ldots,A_n} + \cdots + a_m E_{\lambda_m}^{A_1,\ldots,A_n}.$$

It is easy to check that $Df_0 \in \mathscr{D}_{F(A_1,\ldots,A_n)}$, and

$$ADf_0 = DAf_0 = DF(A_1, \ldots, A_n) f_0 = F(A_1, \ldots, A_n) Df_0 .$$

Thus, A and $F(A_1, \ldots, A_n)$ are operators which coincide on the set of all vectors Df_0, i.e., on the dense linear manifold (5.26). Now, A is bounded and $\mathscr{D}_A \equiv \mathscr{H}$, while $F(A_1, \ldots, A_n)$ is closed, since by Theorem 2.6, $F(A_1, \ldots, A_n)$ also has to be the adjoint of the densely defined operator $F^*(A_1, \ldots, A_n)$. Hence, we conclude* that $A = F(A_1, \ldots, A_n)$. Q.E.D.

Exercises

5.1. Prove that if μ is a σ-finite measure, then the family of all infinitely many-times differentiable functions which are of compact support and belonging to $\hat{L}^2(\mathbb{R}^n, \mu)$ is dense in $\hat{L}^2(\mathbb{R}^n, \mu)$.

5.2. Use Lemma 5.2 to deduce that the set of all functions $p(x)\, e^{-\beta|x|}$, $\beta > 0$, where $p(x)$ are polynomials in $x \in \mathbb{R}^1$, is dense in $L^2(\mathbb{R}^1)$.

5.3. Let $\{A_1^{(1)}, \ldots, A_{k_1}^{(1)}\}, \ldots, \{A_1^{(m)}, \ldots, A_{k_m}^{(m)}\}$ be complete sets of observables of the systems $\mathfrak{S}_1, \ldots, \mathfrak{S}_n$, respectively, and let these operators act in the spaces $\tilde{\mathscr{H}}_1, \ldots, \tilde{\mathscr{H}}_m$ (which have as subspaces the Hilbert spaces $\mathscr{H}_1, \ldots, \mathscr{H}_m$ associated with $\mathfrak{S}_1, \ldots, \mathfrak{S}_m$). Denote by $\tilde{A}_r^{(\nu)}$ the self-adjoint operator with the spectral measure $1 \otimes \cdots \otimes E_r^{(\nu)}(B) \otimes \cdots \otimes 1$ [see (4.3)], where $E_r^{(\nu)}(B)$ is the spectral measure of $A_r^{(\nu)}$. Show that $\{\tilde{A}_1^{(1)}, \ldots, \tilde{A}_{k_1}^{(1)}, \ldots, \tilde{A}_1^{(m)}, \ldots, \tilde{A}_{k_m}^{(m)}\}$ is a complete set of observables of the system $\mathfrak{S} = \{\mathfrak{S}_1, \ldots, \mathfrak{S}_m\}$.

5.4. Show by making use of the definition of a spectral measure, that if $[E^{(1)}(B_1), E^{(2)}(B_2)] = 0$, the set S of all vectors f satisfying $E^{(1)}(B_1)E^{(2)}(B_2)f = f$, where B_1, B_2 are any *bounded* Borel sets in $\mathbb{R}^1$, is dense in $\mathscr{H}$. If A_1 and A_2 are the self-adjoint operators corresponding to $E^{(1)}(B)$ and $E^{(2)}(B)$, respectively, prove that $S \subset \mathscr{D}_{A_1A_2} \cap \mathscr{D}_{A_2A_1}$ and that $A_1A_2f = A_2A_1f$ for all $f \in S$.

5.5. Show that if A is any self-adjoint operator and $E(B)$, $B \in \mathscr{B}^1$, is its spectral measure, then for any bounded Borel set R we have $\mathscr{D}_{AE(R)} = \mathscr{H}$, $\mathscr{D}_{E(R)A} = \mathscr{D}_A$, and $AE(R)f = E(R)Af$ for all $f \in \mathscr{D}_A$, i.e., $AE(R)$ is the extension of $E(R)A$ to the entire Hilbert space.

* In other words, A necessarily belongs to the Abelian algebra generated by $A_1, \ldots, A_n$, and therefore $\{A_1, \ldots, A_n\}$ is complete if and only if that algebra is *maximally* Abelian. This observation is especially significant when dealing with infinite sets of commuting observables (de Dormale and Gautrin [1975]).

6. Canonical Commutation Relations

6.1. The Empirical Significance of Commutation Relations

One of the most striking features of the position and momentum observables defined in (1.1)–(1.6) (where, in general, we can take $i = 1,..., n$) is that all position operators as well as all momentum operators commute among themselves, while this is not true for each pair consisting of a position operator and a momentum operator. As a matter of fact, it is very easy to see that a position operator corresponding to the position coordinate of a certain particle in the system commutes with any momentum operator except the one belonging to the same momentum component of the same particle; for instance, $Q_1^{(x)}$ commutes with $P_i^{(y)}$ and $P_i^{(z)}$, $i = 1,..., n$, as well as with $P_i^{(x)}$ for $i = 2,..., n$, but we have

$$(6.1)\qquad (Q_1^{(x)}P_1^{(x)} - P_1^{(x)}Q_1^{(x)})\psi = i\hbar\psi, \qquad \psi \in \mathscr{D}_{Q_1^{(x)}P_1^{(x)}} \cap \mathscr{D}_{P_1^{(x)}Q_1^{(x)}} \,.$$

This can be easily verified by taking for ψ differentiable functions:

$$x_1\left(-i\hbar\frac{\partial}{\partial x_1}\right)\psi(x_1,..., x_n) - \left(-i\hbar\frac{\partial}{\partial x_1}\right)(x_1\psi(x_1,..., x_n)) = i\hbar\psi(x_1,..., x_n).$$

The commutation relations satisfied by the position and momentum operators in wave mechanics are called canonical commutation relations. In general, given $2n$ self-adjoint operators $Q_1,..., Q_n$, $P_1,..., P_n$ in a Hilbert space $\mathscr{H}$, we say that they satisfy *canonical commutation relations* in which Q_k is *canonically conjugate* to P_k, $k = 1,..., n$, if

$$(6.2)\qquad (Q_kP_k - P_kQ_k)f = i\hbar f, \qquad k = 1,..., n,$$

for any f belonging to the domain of definition of

$$[Q_k, P_k] = Q_kP_k - P_kQ_k,$$

while any other two operators from $\{Q_1,..., Q_n, P_1,..., P_n\}$ commute.

In the first section of this chapter we identified commutativity of operators representing observables with compatibility of those observables. The empirical meaning of compatibility was given in Definition 1.1. From the discussion in §1 we must draw the conclusion that canonically conjugate observables are incompatible, i.e., that there is reason to believe that there is a lower limit to the accuracy with which simultaneous measurements of such observables can be carried out on the *same* system. Such is the case with canonically conjugate position and momentum observables, as revealed by an extensive analysis of different "*gedanken*" experiments (see Messiah [1962, Vol. I]).

It is important to realize that the commutation relations satisfied by canonically conjugate observables also imply that the outcomes of many measurements of those observables on systems in the same state possess certain statistical features. This conclusion can be drawn from the following result.

Lemma 6.1. If f is a vector in the domains of definition of the operators A_1^2, $[A_1, A_2]$, and A_2^2, where A_1 and A_2 are self-adjoint operators in $\mathscr{H}$, then

$$\langle f \mid A_1^2 f\rangle\langle f \mid A_2^2 f\rangle \geqslant \tfrac{1}{4}|\langle f \mid [A_1, A_2] f\rangle|^2. \tag{6.3}$$

Proof. Since for any real value of λ

$$\begin{aligned}\langle f \mid A_1^2 f\rangle - \lambda\langle f \mid i[A_1, A_2] f\rangle + \lambda^2\langle f \mid A_2^2 f\rangle &= \langle f \mid [A_1^2 - i\lambda(A_1A_2 - A_2A_1) + \lambda^2 A_2^2] f\rangle \\ &= \langle f \mid (A_1 + i\lambda A_2)(A_1 - i\lambda A_2) f\rangle \\ &= \langle (A_1 - i\lambda A_2) f \mid (A_1 - i\lambda A_2) f\rangle \geqslant 0,\end{aligned}$$

we deduce that $\operatorname{Im}\langle f \mid i[A_1, A_2] f\rangle = 0$, and the discriminant of the above polynomial of second degree in λ has to be nonpositive. Thus, we have

$$|\langle f \mid [A_1, A_2] f\rangle|^2 - 4\langle f \mid A_1^2 f\rangle\langle f \mid A_2^2 f\rangle \leqslant 0,$$

and (6.3) follows immediately. Q.E.D.

Let us take a normalized state vector Ψ from the domain of Q_k^2, P_k^2, and $[Q_k, P_k]$. Since

$$[A_1, A_2]\Psi = i\hbar\Psi,$$

where

$$A_1 = Q_k - \langle \Psi \mid Q_k\Psi\rangle, \qquad A_2 = P_k - \langle \Psi \mid P_k\Psi\rangle,$$

we get, by applying Lemma 6.1,

$$\langle \Psi \mid (Q_k - \langle \Psi \mid Q_k\Psi\rangle)^2\Psi\rangle\langle \Psi \mid (P_k - \langle \Psi \mid P_k\Psi\rangle)^2\Psi\rangle \geqslant \hbar^2/4. \tag{6.4}$$

The quantities $\sigma_\Psi(Q_k)$ and $\sigma_\Psi(P_k)$, where

$$\sigma_\Psi^2(Q_k) = \int_{\mathbb{R}^1} (\lambda - \langle \Psi \mid Q_k\Psi\rangle)^2 \, d\| E_\lambda^{Q_k}\Psi \|^2 = \langle \Psi \mid (Q_k - \langle \Psi \mid Q_k\Psi\rangle)^2\Psi\rangle,$$

(6.5)

$$\sigma_\Psi^2(P_k) = \int_{\mathbb{R}^1} (\lambda - \langle \Psi \mid P_k\Psi\rangle)^2 \, d\| E_\lambda^{P_k}\Psi \|^2 = \langle \Psi \mid (P_k - \langle \Psi \mid P_k\Psi\rangle)^2\Psi\rangle$$

are the standard deviations of the probability measures $\| E^{Q_k}(B)\, \Psi \|^2$ and $\| E^{P_k}(B)\, \Psi \|^2$, respectively. Thus, inequality (6.4), which can also be written in the form

$$\sigma_\Psi(Q_k)\, \sigma_\Psi(P_k) \geqslant \hbar/2,$$

indicates that we cannot find states for which the standard deviations of canonically conjugate position and momentum are both arbitrarily small.

The standard deviations $\sigma_\Psi(Q_k)$ and $\sigma_\Psi(P_k)$ can be obtained experimentally by preparing an ensemble of systems in the same state, and then carrying out at the same instant, when the systems are in a state represented by Ψ, a measurement of Q_k on the systems in one part of the ensemble, and a measurement of P_k on the remaining systems in the ensemble. This procedure, which is implicit in the theory of measurement of §1, enables us to compute $\| E^{Q_k}(B)\, \Psi \|^2$ and $\| E^{P_k}(B)\, \Psi \|^2$, from which $\sigma_\Psi(Q_k)$ and $\sigma_\Psi(P_k)$ can be computed.

6.2. Representations of Canonical Commutation Relations

If $Q_1, \ldots, Q_n$, $P_1, \ldots, P_n$ are self-adjoint operators satisfying canonical commutation relations in which Q_k and P_k are canonical conjugate pairs, we write, symbolically,

$$(6.6)\quad [Q_j, Q_k] = [P_j, P_k] = 0, \qquad [Q_j, P_k] = i\hbar\delta_{jk}, \qquad j, k = 1, \ldots, n.$$

Any set of $2n$ operators satisfying the above relations is called a *representation* of the canonical commutation relations.

If U is a unitary operator, it is easy to see that the self-adjoint operators

$$(6.7)\qquad Q_j' = UQ_jU^{-1}, \qquad P_j' = UP_jU^{-1}, \qquad j = 1, \ldots, n,$$

constitute another representation of the canonical commutation relation (6.6). In fact, the spectral measures of Q_k' and P_k' are

$$E^{Q_k'}(B) = UE^{Q_k}(B)\, U^{-1}, \qquad E^{P_k'}(B) = UE^{P_k}(B)\, U^{-1},$$

and it is therefore easily verified that

$$[E^{Q_j'}(B_1), E^{P_k'}(B_2)] = U[E^{Q_j}(B_1), E^{P_k}(B_2)]\, U^{-1} = 0, \qquad j \neq k,$$

and that $Q_1', \ldots, Q_n'$, as well as $P_1', \ldots, P_n'$, commute among themselves. Moreover, for $f \in \mathscr{D}_{Q_j'P_j' - P_j'Q_j'} = U\mathscr{D}_{Q_jP_j - P_jQ_j}$ we have

$$[Q_j', P_j']f = U[Q_j, P_j]\, U^{-1}f = i\hbar UU^{-1}f = i\hbar f.$$

Two representations $\{Q_1, ..., P_n\}$ and $\{Q_1', ..., P_n'\}$ of (6.6) which satisfy (6.7) are said to be *unitarily equivalent*. The question arises whether all representations of (6.6) are unitarily equivalent. The answer to this question has physical implications, since *unitarily equivalent representations are physically indistinguishable*. As a matter of fact, assume that there are two quantum mechanical theories $\mathscr{T}$ and $\mathscr{T}'$ describing the same system. If the observables of the two theories are related in a one-to-one fashion,

$$A \leftrightarrow A', \qquad A \in \mathscr{T}, \qquad A' \in \mathscr{T}',$$

and there is a unitary transformation of $\mathscr{H}$ onto $\mathscr{H}'$ such that

$$A' = UAU^{-1},$$

then, by setting $\Psi' = U\Psi$, we map all the states of one theory onto the set of all states of the other theory in a one-to-one fashion. The only physically important quantities, namely the expectation values of bounded observables,

$$\begin{aligned}\langle A' \rangle_{\Psi'} &= \langle \Psi' \mid A'\Psi' \rangle' = \langle U\Psi \mid UAU^{-1}U\Psi \rangle' \\ &= \langle \Psi \mid A\Psi \rangle = \langle A \rangle_{\Psi},\end{aligned}$$

are the same in both theories, and therefore $\mathscr{T}$ and $\mathscr{T}'$ are physically indistinguishable.

From the mathematical point of view, the question of whether all representations of (6.6) are unitarily equivalent is a version of a uniqueness problem, which consists in finding out the manner in which all solutions of (6.6) are related. We already know that there are solutions to the problem, one of them being the *Schroedinger representation*

$$\begin{aligned}(Q_k\psi)(x_1, ..., x_n) &= x_k\, \psi(x_1, ..., x_n), \\ (P_k\psi)(x_1, ..., x_n) &= -i\hbar \frac{\partial}{\partial x_k} \psi(x_1, ..., x_n),\end{aligned} \tag{6.8}$$

in which $\mathscr{H} = L^2(\mathbb{R}^n)$.

Due to difficulties with the domains of definition of unbounded operators, which do not coincide with the entire Hilbert space, the solution of the problem posed above becomes feasible only after the the problem is rephrased in a technically different fashion. This is achieved by the following heuristic procedure.

Assume that P and Q is a canonically conjugate pair

$$[Q, P] = i\hbar. \tag{6.9}$$

If P and Q are treated like bounded operators and (6.9) like a relation valid on the entire Hilbert space, one obtains by induction (see Exercise 6.2, and for rigorous results see Putnam [1967]),

$$[Q, P^n] = i\hbar n P^{n-1}. \tag{6.10}$$

Setting, formally (see Exercises 3.5 and 3.6),

$$e^{iPu} = \sum_{n=0}^{\infty} \frac{(iPu)^n}{n!}, \qquad u \in \mathbb{R}^1,$$

and using (6.10), one derives in the same vein

$$[Q, e^{iPu}] = \sum_{n=0}^{\infty} \frac{(iu)^n}{n!} [Q, P^n] = -\hbar u \; e^{iuP}.$$

Multiplication of the above relation from the left by e^{-iuP} yields

$$e^{-iPu} Q \; e^{iPu} = Q - \hbar u.$$

By taking the nth power of the above equation, one obtains

$$e^{-iPu} Q^n \; e^{iPu} = (Q - \hbar u)^n.$$

Consequently, for $v \in \mathbb{R}^1$

$$e^{-iPu} \, e^{iQv} \, e^{iPu} = \sum_{n=0}^{\infty} e^{-iPu} \frac{(ivQ)^n}{n!} \, e^{iPu}$$

$$= \sum_{n=0}^{\infty} \frac{(iQv - i\hbar uv)^n}{n!} = e^{ivQ - i\hbar uv}.$$

Finally, multiplying the above relation from the left by e^{iPu}, one gets the *Weyl relation*,

$$e^{iQv} \, e^{iPu} = e^{-i\hbar uv} \, e^{iPu} \, e^{iQv}, \qquad u, v \in \mathbb{R}^1. \tag{6.11}$$

The above relation involves only bounded operators defined on the entire Hilbert space. It is advantageous to adopt as a starting point in our study of representations of canonical commutation relations the Weyl relation (6.11), which has a precise meaning, instead of the symbolic relation (6.9). It will be shown that commutation relations, such as $[Q_1, Q_2] = 0$, can also be replaced by relations

$$e^{iQ_1u}\, e^{iQ_2v} = e^{iQ_2v}\, e^{iQ_1u}, \qquad u, v \in \mathbb{R}^1,$$

involving only bounded operators. Such considerations necessitate an understanding of the basic properties of operator families e^{itA}, $t \in \mathbb{R}^1$, where $A = A^*$.

6.3. One-Parameter Abelian Groups of Unitary Operators

The basic properties of the family

$$U_t = e^{iAt}, \qquad t \in \mathbb{R}^1, \tag{6.12}$$

are given in Theorem 3.1. It follows immediately from this theorem that $U_0 = 1$, $U_t^{-1} = U_t^* = U_{-t}$, and $U_{t_1}U_{t_2} = U_{t_2}U_{t_1}$. These three features of the operator family U_t, $t \in \mathbb{R}^1$, characterize it as an Abelian group of unitary operators. Moreover, this family is weakly continuous in t, i.e. (see Exercise 6.3)

$$\operatorname*{w-lim}_{t\to t_0} U_t = U_{t_0}. \tag{6.13}$$

It is a significant result, which constitutes one of the fundamental theorems of functional analysis, called Stone's theorem, that every continuous one-parameter group of operators is of the form (6.12). The key word in this statement is "continuous." The following lemma contributes to the understanding of continuity in functions $U(t)$, $t \in \mathbb{R}^1$, where $U(t)$ are unitary operators.

Lemma 6.2. If $f = \text{w-lim}_{n\to\infty} f_n$ and $\|f_n\| \leqslant \|f\|$, then $f = \text{s-lim}_{n\to\infty} f_n$.

Proof. Since $\langle f \mid f_n\rangle \to \langle f \mid f\rangle = \|f\|^2$, we have

$$\begin{aligned} \|f - f_n\|^2 &= \|f\|^2 - 2\,\mathrm{Re}\langle f \mid f_n\rangle + \|f_n\|^2 \\ &= -\|f\|^2 + \epsilon_n + \|f_n\|^2, \end{aligned}$$

where $\epsilon_n \to 0$ when $n \to +\infty$. On the other hand, $\|f_n\|^2 - \|f\|^2 \leqslant 0$, so that

$$0 \leqslant \|f - f_n\|^2 \leqslant \epsilon_n.$$

Thus, $\|f - f_n\| \to 0$ when $n \to +\infty$. Q.E.D.

Assume that $U(t)$ are unitary operators for all $t \in \mathbb{R}^1$, and

$$U(t_0) = \operatorname*{w-lim}_{t \to t_0} U(t),$$

i.e., $U(t)f$ converges weakly to $U(t_0)f$ for any $f \in \mathscr{H}$. Since

$$\| U(t)f \| = \| f \| = \| U(t_0)f \|,$$

we can apply the above lemma and infer that $U(t)f$ converges strongly to $U(t_0)f$ for any $f \in \mathscr{H}$, i.e.,

$$U(t_0) = \operatorname*{s-lim}_{t \to t_0} U(t).$$

Thus, in the case of one-parameter families of unitary operators, weak continuity implies strong continuity.

***Theorem 6.1** (*Stone's theorem*). To every weakly continuous, one-parameter family $U(t)$, $t \in \mathbb{R}^1$, of unitary operators on a Hilbert space $\mathscr{H}$, obeying

$$U(t_1 + t_2) = U(t_1)\, U(t_2), \qquad t_1, t_2 \in \mathbb{R}^1,$$

corresponds a unique self-adjoint operator

$$A = \int_{\mathbb{R}^1} \lambda \, dE_\lambda^A$$

such that $U(t) = e^{iAt}$ for all $t \in \mathbb{R}^1$.

Proof. We shall prove the theorem by building the spectral function E_λ^A of A from the operators $U(t)$, $t \in \mathbb{R}^1$.

Consider first the special case of families $U_1(t)$ which are periodic in t, with the period 2π, i.e., $U_1(t + 2\pi) = U_1(t)$. We shall prove that there is a family of mutually orthogonal projectors E_n, $n = 0, \pm 1, \pm 2, \ldots$ with a sum equal to the unit operator,

$$\sum_{n=-\infty}^{+\infty} E_n = \operatorname*{s-lim}_{r \to +\infty} \sum_{n=-r}^{+r} E_n = 1, \tag{6.14}$$

such that $U_1(t)$ can be expanded in the following series:

$$U_1(t) = \sum_{n=-\infty}^{+\infty} e^{int} E_n = \operatorname*{s-lim}_{r \to \infty} \sum_{n=-r}^{+r} e^{int} E_n . \tag{6.15}$$

It is easy to verify that

$$(f \mid g)_n = (1/2\pi) \int_0^{2\pi} e^{-int} \langle f \mid U_1(t)\, g \rangle \, dt$$

are bounded bilinear forms in f and g. It follows from Riesz' theorem (see Chapter III, Exercise 2.5) that there is a unique bounded linear operator satisfying $\langle f \mid E_n g \rangle = (f \mid g)_n$ for all $f, g \in \mathscr{H}$. In the symbolic notation

$$E_n = (1/2\pi) \int_0^{2\pi} e^{-int} U_1(t)\, dt \tag{6.16}$$

it becomes obvious that E_n would be the Fourier coefficient of $U_1(t)$ if $U_1(t)$ were an ordinary function.

The operator E_n is self-adjoint, since

$$\begin{aligned}
\langle f \mid E_n{}^* g \rangle = \langle E_n f \mid g \rangle = \langle g \mid E_n f \rangle^* \\
&= (1/2\pi) \int_0^{2\pi} e^{int} \langle g \mid U_1(t) f \rangle^* \, dt \\
&= (1/2\pi) \int_0^{2\pi} e^{int} \langle f \mid U_1{}^*(t) g \rangle \, dt \\
&= (1/2\pi) \int_0^{2\pi} e^{int} \langle f \mid U_1(-t) g \rangle \, dt \\
&= -(1/2\pi) \int_0^{-2\pi} e^{-int_1} \langle f \mid U_1(t_1) g \rangle \, dt_1 \\
&= (1/2\pi) \int_{-2\pi}^{0} e^{-int_1} \langle f \mid U_1(t_1) g \rangle \, dt_1 \\
&= (1/2\pi) \int_0^{2\pi} e^{-int_2} \langle f \mid U_1(t_2) g \rangle \, dt_2 = \langle f \mid E_n g \rangle,
\end{aligned}$$

where the last step was carried out by setting $t_2 = t_1 + 2\pi$ and taking into account that $U_1(t_1 + 2\pi) = U_1(t_1)$. Moreover, using again the periodicity of $U_1(t)$, we obtain (cf. Chapter V, Theorem 3.4)

$$\begin{aligned}
U_1(t)\, E_n &= (1/2\pi) \int_0^{2\pi} e^{-int_1} U_1(t + t_1) \, dt_1 \\
&= (1/2\pi)\, e^{int} \int_t^{t+2\pi} e^{-int_2} U_1(t_2) \, dt_2 \\
&= (1/2\pi)\, e^{int} \int_0^{2\pi} e^{-int_3} U_1(t_3) \, dt_3 = e^{int} E_n ,
\end{aligned} \tag{6.17}$$

and consequently

$$E_m E_n = (1/2\pi) \int_0^{2\pi} e^{-imt} U_1(t)\, E_n\, dt = (1/2\pi) \int_0^{2\pi} e^{-i(m-n)t} E_n\, dt$$

$$= \begin{cases} E_n & \text{for } m = n \\ 0 & \text{for } m \neq n. \end{cases}$$

Thus, we have established that $E_0, E_{\pm 1}, \ldots$ are mutually orthogonal projectors. Hence, it follows from Theorem 3.9 in Chapter III that

$$E = \sum_{n=-\infty}^{\infty} E_n = \operatorname{s-lim}_{r \to +\infty} \sum_{n=-r}^{+r} E_n$$

is also a projector.

It is obvious that $1 - E$ is orthogonal to all projectors E_n. Consequently

$$(1/2\pi) \int_0^{2\pi} e^{-int} \langle f \mid U_1(t)(1 - E)g \rangle\, dt \tag{6.18}$$
$$= \langle f \mid E_n(1 - E)g \rangle = 0, \qquad n = 0, \pm 1, \ldots$$

for all $f, g \in \mathscr{H}$. Since $\langle f \mid U_1(t)(1 - E)g \rangle$ is a continuous periodic function in t, of period 2π, and has vanishing Fourier coefficients (6.18) on the interval $[0, 2\pi]$, it follows that $\langle f \mid U_1(t)(1 - E)g \rangle = 0$ for all $t \in \mathbb{R}^1$ and all $f, g \in \mathscr{H}$. Hence, $U_1(0)(1 - E) = (1 - E) = 0$, i.e., $E = 1$. Thus, (6.14) is established.

Multplying (6.14) from the left by $U(t)$ and using (6.17), we obtain (see also Chapter III, Exercise 5.6)

$$U_1(t) = \operatorname{s-lim}_{n \to \infty} \sum_{n=-r}^{r} U_1(t)\, E_n = \operatorname{s-lim}_{n \to +\infty} \sum_{n=-r}^{+r} e^{int} E_n .$$

This establishes that (6.15) is true when $U(t)$ is periodic in t.

Let us now turn our attention to the general case, when $U(t)$ is not periodic. Introduce the one-parameter family

$$U_1(t) = U(t)(U(2\pi))^{-t/2\pi}, \tag{6.19}$$

where the power in t is defined in terms of the spectral decomposition of the unitary operator $U(2\pi)$ [see (6.5) of Chapter III],

$$U(2\pi) = \int_0^{2\pi} e^{i\lambda}\, d\hat{E}_\lambda ,$$

by the expression

$$(U(2\pi))^t = \int_0^{2\pi} e^{i\lambda t}\, d\hat{E}_\lambda .$$

We obviously have $U_1(2\pi) = 1$. Since $U(t)$ commutes with $U(2\pi)$,

$$U(t)\, U(2\pi) = U(t + 2\pi) = U(2\pi + t) = U(2\pi)\, U(t),$$

it also commutes with $\hat{E}_\lambda$, and consequently with $(U(2\pi))^{-t/2\pi}$. Hence

$$U_1(t_1)\, U_1(t_2) = U(t_1)\, U(t_2)(U(2\pi))^{-t_1/2\pi}\, (U(2\pi))^{-t_2/2\pi} = U_1(t_1 + t_2).$$

In addition, when $t \to t_0$, the weak convergence of $U(t)$ to $U(t_0)$ and the uniform convergence of $(U(2\pi))^{-t/2\pi}$ to $(U(2\pi))^{-t_0/2\pi}$ (see Exercise 6.4) imply that

$$U_1(t_0) = \operatorname*{w-lim}_{t\to t_0} U_1(t).$$

The above results establish that $U_1(t)$ is a periodic weakly continuous one-parameter group, with period 2π, which can be therefore expanded in the series (6.15). Consequently, for any $f, g \in \mathscr{H}$

$$\begin{aligned}
(6.20)\qquad \langle f \mid U(t)g\rangle &= \langle f \mid U_1(t)(U(2\pi))^{t/2\pi} g\rangle \\
&= \left\langle f \middle| \left(\operatorname*{s-lim}_{r\to+\infty} \sum_{n=-r}^{r} e^{int} E_n\right) (U(2\pi))^{t/2\pi} g\right\rangle \\
&= \sum_{n=-\infty}^{+\infty} e^{int} \langle f \mid E_n (U(2\pi))^{t/2\pi} g\rangle \\
&= \sum_{n=-\infty}^{+\infty} e^{int} (1/2\pi) \int_0^{2\pi} e^{i\lambda(t/2\pi)}\, d\langle E_n f \mid \hat{E}_\lambda g\rangle \\
&= \sum_{n=-\infty}^{+\infty} \int_0^{1} e^{i(n+\lambda_1)t}\, d\langle E_n f \mid \hat{E}_{\lambda_1} g\rangle \\
&= \sum_{n=-\infty}^{+\infty} \int_n^{n+1} e^{i\lambda_2 t}\, d\langle f \mid E_n \hat{E}_{\lambda_2 - n} g\rangle .
\end{aligned}$$

Introduce the spectral function (see Exercise 6.5)

$$(6.21)\qquad E_\lambda^A = \sum_{n=-\infty}^{N(\lambda)-1} E_n + E_{N(\lambda)} \hat{E}_{\lambda - N(\lambda)} ,$$

where $N(\lambda)$ is the greatest integer satisfying $N(\lambda) \leqslant \lambda$. In terms of E_λ^A, (6.20) becomes

$$\langle f \mid U(t)g\rangle = \int_{-\infty}^{+\infty} e^{i\lambda t} \langle f \mid E_\lambda^A g\rangle. \tag{6.22}$$

Furthermore, E_λ^A is uniquely determined by $U(t)$. In fact, if I is any finite closed interval, its characteristic function is the pointwise limit of a uniformly bounded sequence of generalized trigonometric polynomials (see Exercise 6.6)

$$p_n(\lambda) = \sum_{k=-r_n}^{r_n} a_{nk}\, e^{(2\pi i/n)k\lambda}.$$

Setting

$$p_n[U(t)] = \sum_{k=-r_n}^{r_n} a_{nk} (U(t))^{(2\pi k/n)}$$

and using (6.22) to obtain

$$|\langle f \mid (p_n[U(t)] - E^A(I))^2 f\rangle| \leqslant \int_{-\infty}^{+\infty} |p_n(\lambda) - \chi_I(\lambda)|^2 \, d\langle f \mid E_\lambda^A f\rangle,$$

we conclude with the help of Lemma 3.1 that the above expression approaches zero when $n \to +\infty$, and consequently

$$E^A(I) = \operatorname*{s\text{-}lim}_{n\to\infty} p_n[U(t)].$$

Thus, $U(t)$ determines $E^A(I)$ completely, and in turn $E^A(I)$ determines (see Chapter III, Theorem 5.6) $E^A(B)$ for any $B \in \mathscr{B}^1$. Q.E.D.

6.4. Representations of Weyl Relations

We are now ready to pose the problem of representations of the commutation relations (6.6) in terms of one-parameter continuous families $e^{iQ_k t}$, $e^{iP_k t}$, $k = 1,\ldots, n$. This reformulation of (6.6) is made possible by Theorems 6.2 and 6.3, which will be proved next.

Theorem 6.2. Two self-adjoint operators A_1 and A_2 in the Hilbert space $\mathscr{H}$ commute (in the sense of Definition 1.2) if and only if

$$e^{iA_1 t_1}\, e^{iA_2 t_2} = e^{iA_2 t_2}\, e^{iA_1 t_1} \tag{6.23}$$

for all $t_1, t_2 \in \mathbb{R}^1$.

Proof. According to Definition 1.2, A_1 and A_2 are said to commute if and only if

$$[E^{A_1}(B_1), E^{A_2}(B_2)] = 0, \tag{6.24}$$

for all $B_1, B_2 \in \mathscr{B}^1$.

If (6.24) is true, we can apply Theorem 2.4 to infer (6.23):

$$e^{iA_1t}\, e^{iA_2t} = e^{i(A_1+A_2)t} = e^{i(A_2+A_1)t} = e^{iA_2t}\, e^{iA_1t}.$$

Conversely, assume that (6.23) is true. To prove that (6.24) is then also true, we resort to the construction of $E_\lambda^{A_1}$ and $E_\lambda^{A_2}$ carried out in the proof of Stone's theorem.

First, we observe that

$$[U_1^{(1)}(t_1), U_1^{(2)}(t_2)] = 0, \qquad t_1, t_2 \in \mathbb{R}^1,$$

where $U_1^{(k)}(t)$ is defined by (6.19) in terms of $U^{(k)}(t) = e^{iA_kt}$, $k = 1, 2$. This implies that all the corresponding projectors $E_n^{(k)}$, $n = 0, \pm 1, \ldots$, in (6.16) and the spectral functions $\hat{E}_\lambda^{(k)}$ of $U^{(k)}(2\pi)$ commute with each other for $k = 1, 2$. From this we deduce that the spectral functions $E_{\lambda_1}^{A_1}$ and $E_{\lambda_2}^{A_2}$ commute. Q.E.D.

We have seen that (6.11) can be derived from (6.9) only in a heuristic manner (but see also Putnam [1967], §§4.6–4.11). However, (6.9) is a rigorous consequence of (6.11), as stated in the following theorem.

Theorem 6.3. If A_1 and A_2 are two self-adjoint operators in $\mathscr{H}$ satisfying

$$e^{iA_1t_1}\, e^{iA_2t_2} = e^{-it_1t_2}\, e^{iA_2t_2}\, e^{iA_1t_1} \tag{6.25}$$

for all $t_1, t_2 \in \mathbb{R}^1$, then

$$A_1A_2f = if + A_2A_1f \tag{6.26}$$

for any $f \in \mathscr{D}_{A_1A_2-A_2A_1} = \mathscr{D}_{A_1A_2} \cap \mathscr{D}_{A_2A_1}$.

Proof. Using Theorem 3.1, Exercise 5.6 in Chapter III, and the fact that $e^{iA_1t_1}$ is a bounded operator, we derive from (6.25)

$$\begin{aligned} e^{iA_1t_1}(iA_2f) &= \operatorname*{s-lim}_{t_2\to 0} \frac{1}{t_2}\, e^{iA_1t_1}(e^{iA_2t_2} - 1)f \\ &= \left.\frac{de^{-it_1t_2}}{dt_2}\right|_{t_2=0} e^{iA_1t_1}f + \operatorname*{s-lim}_{t_2\to 0} \frac{1}{t_2}\,(e^{iA_2t_2} - 1)\, e^{iA_1t_1}f \\ &= -it_1\, e^{iA_1t_1}f + iA_2\, e^{iA_1t_1}f. \end{aligned}$$

Since $A_2 f \in \mathscr{D}_{A_1}$ because $f \in \mathscr{D}_{A_1 A_2}$, we can again employ (3.6) to deduce (note that the existence of the third limit follows from that of the preceding two limits!):

$$\begin{aligned} -A_1 A_2 f &= \operatorname*{s-lim}_{t_1 \to 0} \frac{1}{t_1} (e^{iA_1 t_1} - 1)(iA_2 f) \\ &= -\operatorname*{s-lim}_{t_1 \to 0} \frac{1}{t_1} it_1 e^{iA_1 t_1} f + \operatorname*{s-lim}_{t_1 \to 0} (iA_2) \frac{1}{t_1} (e^{iA_1 t_1} - 1) f \\ &= -if + \operatorname*{s-lim}_{t_1 \to 0} (iA_2) \frac{1}{t_1} (e^{iA_1 t_1} - 1) f. \end{aligned}$$

The above relation implies that for all $g \in \mathscr{D}_{A_2}$

$$\begin{aligned} \langle g \mid A_1 A_2 f \rangle &= i\langle g \mid f \rangle + \lim_{t_1 \to 0} \langle A_2 g \mid (-i/t_1)(e^{iA_1 t_1} - 1) f \rangle \\ &= i\langle g \mid f \rangle + \langle A_2 g \mid A_1 f \rangle = i\langle g \mid f \rangle + \langle g \mid A_2 A_1 f \rangle, \end{aligned} \tag{6.27}$$

where the last step is valid due to the assumption that $f \in \mathscr{D}_{A_2 A_1}$, i.e., $A_1 f \in \mathscr{D}_{A_2}$. Since g assumes all values in $\mathscr{D}_{A_2}$, which has to be dense in $\mathscr{H}$ because A_2^* exists, we infer that (6.26) follows from (6.27). Q.E.D.

The above two theorems indicate that the commutation relations (6.6) are equivalent to the following set of Weyl relations:

$$\begin{gathered} [\exp(iP_j u_j), \exp(iP_k u_k)] = [\exp(iQ_j v_j), \exp(iQ_k v_k)] = 0, \\ \exp(iP_j u_j)\exp(iQ_k v_k) = \exp(i\hbar\delta_{jk} u_j v_k)\exp(iQ_k v_k)\exp(iP_j u_j), \\ u_1, \dots, u_n, v_1, \dots, v_n \in \mathbb{R}^1. \end{gathered} \tag{6.28}$$

However, strictly speaking, we know only that (6.6) follows from (6.28), where the precise meaning of $[Q_k, P_k] = i\hbar$ is given by (6.2). In any case, whenever we say from now on that $Q_1, \dots, Q_n, P_1, \dots, P_n$ are a representation of the canonical commutation relations we mean, according to an established tradition, that the Weyl relations (6.28) are satisfied.

The representations of canonical commutation relations are classified in two classes. The *irreducible representations* are those which leave no nontrivial closed linear subspace invariant, i.e., for which there is no closed linear subspace $\mathsf{M} \neq \{0\}$, $\mathsf{M} \neq \mathscr{H}$, such that

$$\exp(iP_k u_k) f, \ \exp(iQ_k v_k) f \in \mathsf{M}, \qquad k = 1, \dots, n,$$

for all $f \in \mathsf{M}$. Any representation which is not irreducible is called *reducible* [an example is provided by (8.76) in Section 8.10].

The following theorem was proved rigorously for the first time by von Neumann [1931].

Theorem 6.4 (*von Neumann's theorem*). All irreducible representations of the canonical commutation relations (6.28) on a separable Hilbert space are unitarily equivalent. Any reducible representation Q_k, P_k, $k = 1,\ldots, n$, on a separable Hilbert space $\mathscr{H}$ is the direct sum of a finite or infinite sequence of irreducible representations, i.e., $\mathscr{H}$ is the direct sum of a sequence M_1, M_2,... of mutually orthogonal closed linear subspaces and

$$\exp(iP_k u) = \sum_\alpha \exp(iP_k^{(\alpha)} u)\, E_{\mathsf{M}_\alpha}, \qquad \exp(iQ_k v) = \sum_\alpha \exp(iQ_k^{(\alpha)} v)\, E_{\mathsf{M}_\alpha},$$

where $Q_1^{(\alpha)},\ldots, Q_n^{(\alpha)}$, $P_1^{(\alpha)},\ldots, P_n^{(\alpha)}$ are irreducible representations in M_α.

It will be shown at the end of this section that the Schroedinger representation, in which Q_k and P_k are given by the operators (4.19) and (4.20) of Chapter III, is an irreducible representation in $L^2(\mathbb{R}^n)$. This establishes the existence of representations of (6.28) for $n = 1, 2,\ldots$.

In quantum field theory, or in any quantum mechanical theory of systems in which the number of particles can vary and become arbitrarily large, one is concerned with representations of the canonical commutation relations for an infinite number of degrees of freedom, i.e., for an infinite sequence of operators Q_1, P_1, Q_2, P_2,... . It is worth mentioning —though this case lies beyond the scope of the present book—that von Neumann's theorem cannot be generalized to this case. In fact, there is an infinite number of irreducible unitarily nonequivalent representations of an infinite number of canonical commutation relations [Gärding and Wightman, 1954a, b].

*6.5 Appendix: Proof of von Neumann's Theorem

We shall prove in detail von Neumann's theorem only for the case $n = 1$, i.e., when we are dealing with two canonically conjugate operators Q and P which satisfy the Weyl relation

$$e^{iPu}\, e^{iQv} = e^{i\hbar uv}\, e^{iQv}\, e^{iPu}, \qquad u, v \in \mathbb{R}^1. \tag{6.29}$$

This case contains all the essential features of the more general case. The lengthy proof of this theorem, which occupies the rest of this section, can be ignored at the first reading.

Let us introduce the two-parameter family

$$S(u, v) = e^{-[(i/2)\hbar uv]}\, e^{iPu}\, e^{iQv} \tag{6.30}$$

of unitary operators. It is easy to verify that

$$S(u_1, v_1)\, S(u_2, v_2) = \exp[(i/2)\, \hbar(u_1 v_2 - v_1 u_2)]\, S(u_1 + u_2, v_1 + v_2) \tag{6.31}$$

by using (6.29). Taking in the above relation $u = u_1 = -u_2$ and $v = v_1 = -v_2$, and noting that $S(0, 0) = 1$, we obtain the identity

$$S(-u, -v) = S^*(u, v) = S^{-1}(u, v), \qquad u, v \in \mathbb{R}^1. \tag{6.32}$$

If $\rho(u, v)$ is a function which is Lebesgue integrable on $\mathbb{R}^2$, the integral

$$(f \mid g)_\rho = \int_{\mathbb{R}^2} \rho(u, v)\langle f \mid S(u, v)g\rangle \, du \, dv$$

is defined for any $f, g \in \mathscr{H}$ since $\langle f \mid S(u, v)g\rangle$ is continuous and bounded in u, $v \in \mathbb{R}^1$ (see Chapter II, Theorem 3.9). It is easy to check that $(f \mid g)_\rho$ is a bounded bilinear form, and consequently it follows from Riesz' theorem (see Chapter III, Exercise 2.5) that there is a unique bounded linear operator A_ρ satisfying $\langle f \mid A_\rho g\rangle = (f \mid g)_\rho$. The function $\rho(u, v)$ is called the kernel of A_ρ. It is straightforward to verify by direct calculation, employing the relation

$$\langle f \mid A_\rho g\rangle = \int_{\mathbb{R}^2} \rho(u, v)\langle f \mid S(u, v)g\rangle \, du \, dv, \tag{6.33}$$

that the following statements are true:

$$A_\rho^* = A_{\bar\rho}, \qquad \overline{\rho(u, v)} = \rho^*(-u, -v), \qquad A_{\rho_1+\rho_2} = A_{\rho_1} + A_{\rho_2}, \tag{6.34}$$

where ρ, ρ_1, and ρ_2 are any functions which are Lebesgue integrable on $\mathbb{R}^2$. Moreover, we easily deduce that

$$\begin{aligned} &A_{\rho_1\rho_2} = A_{\hat\rho}, \\ &\hat\rho(u, v) = \int_{\mathbb{R}^2} \exp[(i/2)\,\hbar(uv_2 - vu_2)]\, \rho_1(u - u_2, v - v_2)\, \rho_2(u_2, v_2)\, du_2\, dv_2, \end{aligned} \tag{6.35}$$

by the following procedure:

$$\begin{aligned} &\langle f \mid A_{\rho_1}A_{\rho_2}g\rangle \\ &\quad = \langle A_{\rho_1}^* f \mid A_{\rho_2}g\rangle \\ &\quad = \int_{\mathbb{R}^2} \rho_2(u_2, v_2)\langle A_{\rho_1}^* f \mid S(u_2, v_2)g\rangle \, du_2 \, dv_2 \\ &\quad = \int_{\mathbb{R}^2} \rho_2(u_2, v_2)\langle f \mid A_{\rho_1}S(u_2, v_2)g\rangle \, du_2 \, dv_2 \\ &\quad = \int_{\mathbb{R}^2} du_2\, dv_2\, \rho_2(u_2, v_2) \int_{\mathbb{R}^2} du_1\, dv_1\, \rho_1(u_1, v_1)\langle f \mid S(u_1, v_1)\, S(u_2, v_2)g\rangle \end{aligned}$$

(equation continues)

$$= \int_{\mathbb{R}^4} \rho_1(u_1, v_1)\, \rho_2(u_2, v_2) \exp[(i/2)\, \hbar(u_1 v_2 - v_1 u_2)]$$
$$\times \langle f \mid S(u_1 + u_2, v_1 + v_2) g\rangle\, du_1\, dv_1\, du_2\, dv_2$$
$$= \int_{\mathbb{R}^4} \rho_1(u - u_2, v - v_2)\, \rho_2(u_2, v_2) \exp[(i/2)\, \hbar(u v_2 - v u_2)]$$
$$\times \langle f \mid S(u, v) g\rangle\, du\, dv\, du_2\, dv_2$$
$$= \int_{\mathbb{R}^2} du\, dv\, \langle f \mid S(u, v) g\rangle \int_{\mathbb{R}^2} du_2\, dv_2\, \rho_1(u - u_2, v - v_2)$$
$$\times\, \rho_2(u_2, v_2) \exp[(i/2)\, \hbar(u v_2 - v u_2)]$$

in which Fubini's theorem has been used twice.

We need now the following lemma.

Lemma 6.3. $A_\rho = 0$ if and only if its kernel $\rho(u, v)$ vanishes almost everywhere.

Proof. $A_\rho = 0$ implies that $S(-u', -v')\, A_\rho S(u', v') = 0$ for any $u', v' \in \mathbb{R}^1$. We obtain, using (6.31) in the process,

$$\langle f \mid S(-u', -v')\, A_\rho S(u', v') g\rangle$$
$$= \int_{\mathbb{R}^2} \rho(u, v) \langle f \mid S(-u', -v')\, S(u, v)\, S(u', v') g\rangle\, du\, dv$$
$$= \int_{\mathbb{R}^2} \rho(u, v) \exp[(i/\hbar)(u v' - v u')] \langle f \mid S(u, v) g\rangle\, du\, dv.$$

Letting $u'/2\pi l\hbar$ and $v'/2\pi l\hbar$ assume integer values, we infer that

$$\int_{\mathbb{R}^2} F(u, v)\, \rho(u, v) \langle f \mid S(u, v) g\rangle\, du\, dv = 0 \tag{6.36}$$

for any trigonometric polynomial $F(u, v)$ of period l. Since $\rho(u, v)$ is integrable, and $\langle f \mid S(u, v)\, g\rangle$ is integrable and bounded, we can take limits and generalize (6.36) to the case when $F(u, v)$ can be any continuous function of period $l > 0$ (see Exercise 6.6). Thus, we deduce (see Exercise 6.8) that for any fixed $f, g \in \mathscr{H}$, $\langle f \mid \rho(u, v)\, S(u, v) g\rangle = 0$, except on a set $R(f, g)$ of measure zero. If $\{e_n\}$ is a countable orthonormal basis, we conclude that $\langle e_m \mid \rho(u, v)\, S(u, v)\, e_n\rangle = 0$ except on the set $R = \bigcup_{m,n} R(e_m, e_n)$, which is of measure zero. Hence, $\rho(u, v)\, S(u, v) = 0$ outside R, i.e., $\rho(u, v) = 0$ almost everywhere. Q.E.D.

It follows from Lemma 6.3 that the operator $A = A_{\rho_0}$, where $\rho_0(u, v) = \exp[-(\hbar/4)(u^2 + v^2)]$, is different from zero. Moreover, it is easily verified by using (6.31) and (6.33) that $AS(u, v)A$ has the kernel

$$\int_{\mathbb{R}^2} \exp \left\{\frac{\hbar}{2}\Big[i(u_1v_2 - v_1u_2) - \tfrac{1}{2}(u_1 - u_2)^2 - \tfrac{1}{2}(v_1 - v_2)^2 + i(uv_2 - vu_2) - \tfrac{1}{2}(u - u_2)^2 - \tfrac{1}{2}(v - v_2)^2\Big]\right\} du_2\, dv_2$$

$$= \exp\left[-\frac{\hbar}{4}(u^2 + v^2 + u_1^2 + v_1^2)\right] \int_{\mathbb{R}^2} \exp\left[-\frac{\hbar}{2}(u_3^2 + v_3^2)\right] du_3\, dv_3$$

$$= \frac{2\pi}{\hbar} \exp\left[-\frac{\hbar}{4}(u^2 + v^2)\right] \exp\left[-\frac{\hbar}{4}(u_1^2 + v_1^2)\right].$$

This implies that

$$AS(u, v)A = \frac{2\pi}{\hbar} \exp\left[-\frac{\hbar}{4}(u^2 + v^2)\right] A. \tag{6.37}$$

In particular, for $u = v = 0$ we obtain $A^2 = (2\pi/\hbar)A$. In addition, we can derive from (6.33) that $A = A^*$. Thus, $(\hbar/2\pi)A$ is a projector.

Denote by M the closed linear subspace of $\mathscr{H}$ onto which $(\hbar/2\pi)A$ projects. For any $f, g \in \mathsf{M}$ we obtain by using (6.31) and (6.37)

(6.38)

$$\langle S(u_1, v_1)f \mid S(u_2, v_2)g\rangle$$

$$= \frac{\hbar^2}{4\pi^2} \langle S(u_1, v_1)\, Af \mid S(u_2, v_2)\, Ag\rangle$$

$$= \frac{\hbar^2}{4\pi^2} \langle f \mid AS(-u_1, -v_1)\, S(u_2, v_2)\, Ag\rangle$$

$$= \frac{\hbar^2}{4\pi^2} \exp\left[\frac{i\hbar}{2}(v_1u_2 - u_1v_2)\right] \langle f \mid AS(u_2 - u_1, v_2 - v_1)\, Ag\rangle$$

$$= \frac{\hbar}{2\pi} \exp\left[-\frac{\hbar}{4}(u_2 - u_1)^2 - \frac{\hbar}{4}(v_2 - v_1)^2 - \frac{i\hbar}{2}(u_1v_2 - v_1u_2)\right] \langle f \mid Ag\rangle$$

$$= \exp\left[-\frac{\hbar}{4}(u_2 - u_1)^2 - \frac{\hbar}{4}(v_2 - v_1)^2 - \frac{i\hbar}{2}(u_1v_2 - u_1u_2)\right] \langle f \mid g\rangle.$$

Let $\{e_1, e_2, \ldots\}$ be an orthonormal basis in M, and denote by M_α the closed linear subspace spanned by all the vectors $S(u, v)\, e_\alpha$, $u, v \in \mathbb{R}^1$. We easily deduce from (6.38) that the subspaces M_α are mutually orthogonal.

It follows from (6.31) that, for all $u, v \in \mathbb{R}^1$, $S(u, v) f \in \mathsf{M}_\alpha$ if $f \in \mathsf{M}_\alpha$. Denote by $\mathscr{H}_0$ the orthogonal complement of $\bigoplus_\alpha \mathsf{M}_\alpha$. By virtue of (6.32), for any $f \in \mathscr{H}_0^\perp = \bigoplus_\alpha \mathsf{M}_\alpha$ and $g \in \mathscr{H}_0$, we have

$$\langle f \mid S(u, v) g\rangle = \langle S^*(u, v) f \mid g\rangle = \langle S(-u, -v) f \mid g\rangle = 0$$

because $S(-u, -v) f \in \mathscr{H}_0^\perp$ for all $u, v \in \mathbb{R}^1$. On the other hand, it follows from the construction of M_α that $\bigoplus_\alpha \mathsf{M}_\alpha \supset \mathsf{M}$, where M had been defined to be the subspace onto which $(\hbar/2\pi)A$ projects. Consequently, $Af = 0$ for all $f \in \mathscr{H}_0$. But, this implies that $\mathscr{H}_0 = \{0\}$, i.e., $\mathscr{H} = \bigoplus_\alpha \mathsf{M}_\alpha$. As a matter of fact, we have just proven that $S(u, v)$ leaves $\mathscr{H}_0$ invariant, and therefore, if $\mathscr{H}_0 \neq \{0\}$, we could apply Lemma 6.3 to the Hilbert space $\mathscr{H}_0$ and to the restriction of $S(u, v)$ to $\mathscr{H}_0$, and deduce that the restriction of A to $\mathscr{H}_0$ is not zero.

It is clear from the construction of M_α that $S(u, v)$, and therefore e^{iPu} and e^{iQv}, $u, v \in \mathbb{R}^1$, do not leave any nontrivial subspace of M_α invariant. Hence, $e^{iPu}E_{\mathsf{M}_\alpha}$ and $e^{iQv}E_{\mathsf{M}_\alpha}$ are irreducible representations of the canonical commutation relations in M_α. Thus, the first part of von Neumann's theorem is proved.

Consider now two irreducible representations P, Q and P', Q' in $\mathscr{H}$. In view of the above construction, M and M' are spanned by the single vectors e_α and e_α', respectively, and the sets of vectors $f_{u,v} = S(u, v)\, e_\alpha$ and $f'_{u,v} = S'(u, v)\, e_\alpha'$ are dense in $\mathscr{H}$. Introduce the operator U, which is defined on $f_{u,v}$ by $Uf_{u,v} = f'_{u,v}$ and then extended to the entire Hilbert space $\mathscr{H}$. It will be shown that U is unitary.

By virtue of (6.38)

$$\begin{aligned}\langle f_{u_1,v_1} \mid f_{u_2,v_2}\rangle &= \exp\left[-\frac{\hbar}{4}(u_2 - u_1)^2 - \frac{\hbar}{4}(v_2 - v_1)^2 - \frac{i\hbar}{2}(u_1 v_2 - v_1 u_2)\right] \\ &= \langle f'_{u_1,v_1} \mid f'_{u_2,v_2}\rangle,\end{aligned}$$

so that $\| Uf_{u,v} \| = \| f_{u,v} \|$. Thus, U can be extended to the entire Hilbert space by the extension principle (see Chapter III, §2), and the result is an isometric operator. Moreover, since the set $\{f'_{u,v}\colon u, v \in \mathbb{R}^1\}$ is dense in $\mathscr{H}$, the range of U is $\mathscr{H}$, and U is unitary.

It follows from (6.31) that

$$S(u, v) f_{u',v'} = \exp[(i/2)\,\hbar(uv' - vu')]\, f_{u'+u,v'+v},$$

$$S'(u, v) f'_{u',v'} = \exp[i(\hbar/2)(uv' - vu')]\, f'_{u'+u,v'+v}.$$

Consequently, we obtain the relation

$$\begin{aligned}U^{-1}S'(u, v)\, Uf_{u',v'} &= U^{-1}S'(u, v) f'_{u',v'} \\ &= \exp[i(\hbar/2)(uv' - vu')]\, f_{u'+u,v'+v} = S(u, v) f_{u',v'},\end{aligned}$$

which can be extended straightforwardly to $\mathscr{H}$, so that

$$S(u, v) = U^{-1}S'(u, v)U.$$

Since $S(u, 0) = e^{iPu}$, $S(0, v) = e^{iQv}$, etc., we arrive at the result that $e^{iPu} = U^{-1}e^{iP'u}U$ and $e^{iQv} = U^{-1}e^{iQ'v}U$. This concludes the proof of von Neumann's theorem.

The proof of the theorem for the more general case of $2n$ operators $Q_1,..., Q_n$, $P_1,..., P_n$ proceeds along identical lines, $S(u, v)$ being replaced now by

$$\begin{aligned} &S(u_1,..., u_n, v_1,..., v_n) \\ &\quad = \exp[-(i/2)\,\hbar(u_1v_1 + \cdots + u_nv_n)] \exp[i(P_1u_1 + \cdots + P_nu_n)] \\ &\qquad \times \exp[i(Q_1v_1 + \cdots + Q_nv_n)]. \end{aligned}$$

In the case where Q and P constitute the Schroedinger representation in $L^2(\mathbb{R}^1)$, we have (see Exercise 6.9)

$$(e^{iPu}\psi)(x) = \psi(x + \hbar u), \qquad (e^{iQv}\psi)(x) = e^{ixv}\psi(x). \tag{6.39}$$

Consequently, we easily compute that

$$\begin{aligned} (S(u, v)\psi)(x) &= \exp[iv(x + \tfrac{1}{2}\hbar u)]\, \psi(x + \hbar u), \\ (A\psi)(x) &= \int_{\mathbb{R}^2} \exp[-(\hbar/4)(u^2 + v^2)] \exp[iv(x + \tfrac{1}{2}\hbar u)]\, \psi(x + \hbar u)\, du\, dv \\ &= \sqrt{\frac{4\pi}{\hbar}} \int_{\mathbb{R}^1} \exp\left[-\frac{\hbar}{4}u^2 - \frac{1}{\hbar}(x + \tfrac{1}{2}\hbar u)^2\right] \psi(x + \hbar u)\, du \\ &= \sqrt{\frac{4\pi}{\hbar^3}} \int_{\mathbb{R}^1} \exp\left[-\frac{1}{2\hbar}(x^2 + t^2)\right] \psi(t)\, dt. \end{aligned} \tag{6.40}$$

Hence, the equation $A\psi = (2\pi/\hbar)\psi$ has only one linearly independent solution, namely $\psi_0(x) = e^{-(1/2\hbar)x^2}$, i.e., M is one dimensional. This establishes that the Schroedinger representation is irreducible.

Exercises

6.1. Verify that the commutator $[A, B] = AB - BA$ of linear operators defined everywhere on $\mathscr{H}$ has the following properties:

$$\begin{aligned} [A, B] &= -[B, A], \\ [aA, B] &= [A, aB] = a[A, B], \\ [A, BC] &= B[A, C] + [A, B]C. \end{aligned}$$

6.2. Derive (6.10) from (6.9) by induction, using the results of Exercise 6.1.

6.3. Show that $\text{s-lim}_{t \to t_0} e^{iAt} = e^{iAt_0}$.

6.4. Show that if U is a unitary operator, $\text{u-lim}_{t \to t_0} U^t = U^{t_0}$.

6.5. Prove that E_λ^A defined in (6.21) is a spectral function.

6.6. Let $f(\lambda)$, $\lambda \in \mathbb{R}^1$, be a continuous function of compact support. Prove that there is a sequence of trigonometric polynomials $p_n(\lambda) = \sum_k a_{nk} \exp[(2\pi/n)\, ik\lambda]$ which converges pointwise to $f(\lambda)$. Explain also why the same statement is true when $f(\lambda)$ is the characteristic function of a finite interval.

6.7. Verify that the relations (6.34) are true.

6.8. Show that if $\rho(x)$ is integrable on $\mathbb{R}^n$, and $\int_{\mathbb{R}^n} F(x)\, \rho(x)\, dx = 0$ for any continuous function $F(x)$ of compact support, then $\rho(x) = 0$ almost everywhere.

6.9. Derive relations (6.39) and (6.40) for $(Q\psi)(x) = x\psi(x)$, $P = U_F^{-1} Q U_F$, where U_F is the Fourier–Plancherel transform in $L^2(\mathbb{R}^1)$ (see Chapter III, Eq. (4.20) and Theorem 4.6).

7. The General Formalism of Wave Mechanics

7.1. A Derivation of One-Particle Wave Mechanics

In the preceding sections we outlined a very general framework for quantum mechanics, which is contained in Axioms O, S (or H), P, and C. In practice, more specific formulations of this formalism are usually required. Such formulations are achieved by making other assumptions in addition to Axioms O, S, P, and C. These assumptions reflect the peculiarities of the particular class of systems under consideration.

The assumptions most frequently made are those leading to the formalism of wave mechanics. It is believed that these assumptions apply in the nonrelativistic domain to all microscopic systems which can be considered to consist of a finite number n of stable particles (i.e., of particles whose nature does not change during the duration of the experiment).

The fundamental system in nonrelativistic quantum mechanics is the microscopic quantum-mechanical particle (which we have in mind whenever we talk about a "particle"). The following requirement reflects the basic nature of the concept of a particle.

Axiom W1. The set $\{Q^{(x)}, Q^{(y)}, Q^{(z)}, S^{\mathbf{n}}\}$ of the position observables $Q^{(x)}, Q^{(y)}, Q^{(z)}$, and the spin projection $S^{\mathbf{n}}$ in a given direction $\mathbf{n}$, constitute a complete set of observables of any one-particle system.

The above theoretical assumption reflects an interpretation of certain experimental facts which determine the empirical nature of a particle. In other words, a particle can be defined from the experimental point of view as an empirical object to which measurements of position, momentum, and spin can be related.

The observables $Q^{(x)}$, $Q^{(y)}$, and $Q^{(z)}$ correspond, naturally, to the measurement of the x, y, and z coordinates, respectively, of the position of the particle in some inertial Cartesian frame. It has become customary to choose the z axis of this frame to point in the direction $\mathbf{n}$ in which the spin projection is measured (i.e., the direction $\mathbf{H}$ in Fig. 3 in the Introduction). When this is the case, we denote the observable $S^{\mathbf{n}}$ by $S^{(z)}$.

Axiom W1 implies the existence of a spectral representation space $L^2(\mathbb{R}^4, \mu)$ of the complete set $\{Q^{(x)}, Q^{(y)}, Q^{(z)}, S^{\mathbf{n}}\}$, whose elements are represented by μ-square-integrable functions

$$\psi(x, y, z, s), \qquad \int_{\mathbb{R}^4} |\psi(x, y, z, s)|^2 \, d\mu(x, y, z, s) < +\infty. \tag{7.1}$$

It does not tell us anything, however, about the nature of the measure μ. In order to decide which measures μ we have to choose, we have to invoke two additional axioms, which will be given next.

The following axiom reflects the experimental fact that the spin projection of a particle of spin σ in some direction $\mathbf{n}$ can assume only the values $-\sigma, -\sigma + 1, \ldots, +\sigma$ (see the discussion at the end of the Introduction).

Axiom W2. The spin projection observable $S^{\mathbf{n}}$ of a one-particle system of spin σ has the pure point spectrum

$$\mathsf{S}_\sigma = \{-\sigma, -\sigma + 1, -\sigma + 2, \ldots, \sigma\}. \tag{7.2}$$

In the spectral representation space $L^2(\mathbb{R}^4, \mu)$ of $\{Q^{(x)}, Q^{(y)}, Q^{(z)}, S^{\mathbf{n}}\}$, the operator $S^{\mathbf{n}}$ has the spectral representation

$$(S^{\mathbf{n}}\psi)(x, y, z, s) = s\psi(x, y, z, s)$$

so that

$$(E^{S^{\mathbf{n}}}(B)\,\psi)(x, y, z, s) = \chi_B(s)\,\psi(x, y, z, s), \qquad B \in \mathscr{B}^1.$$

Take a function of the form

$$\psi(x, y, z, s) = \chi_{B_1}(x, y, z)\,\chi_{B_2}(s),$$

where B_1 and B_2 are bounded Borel sets in $\mathbb{R}^3$ and $\mathbb{R}^1$, respectively. Such a function obviously belongs to $L^2(\mathbb{R}^4, \mu)$. If $B \subset B_2$, we have

$$\| E^{S^n}(B) \psi \|^2 = \int_{\mathbb{R}^4} \chi_{B_1}(x, y, z) \chi_B(s) \, d\mu(x, y, z, s)$$
$$= \mu(B_1 \times B).$$

Thus, if $B \cap S_\sigma = \varnothing$, we get

$$\mu(B_1 \times B) = \| E^{S^n}(B) \psi \|^2 = 0.$$

Moreover, $E^{S^n}(\{s\}) = 1$ for $s \in S_\sigma$. Hence, we can write

$$\mu(B_1 \times B) = \sum_{s \in B \cap S_\sigma} \mu(B_1 \times \{s\}), \tag{7.3}$$

where the sum is always finite, since by the assumption in (7.2) S_σ is a finite set.

Relation (7.3) has been established by the above argument for the case of bounded $B_1 \in \mathscr{B}^3$ and $B \in \mathscr{B}^1$, but it can be extended also to unbounded Borel sets by using the continuity from below of measures to derive for arbitrary B_1 and B

$$\mu(B_1 \times B) = \lim_{n \to \infty} \mu(B_1^{(n)} \times B^{(n)}),$$

$$B_1^{(n)} = B_1 \cap \{(x, y, z) \colon -n \leqslant x \leqslant n, \ -n \leqslant y \leqslant n, \ -n \leqslant z \leqslant n\},$$

$$B^{(n)} = B \cap [-n, +n].$$

In view of the structure (7.3) of the measure μ, we find that the inner product $\langle \cdot \mid \cdot \rangle$ in $L^2(\mathbb{R}^4, \mu)$ is of the form

$$\langle \psi_1 \mid \psi_2 \rangle = \int_{\mathbb{R}^3 \times S_\sigma} \psi_1^*(x, y, z, s) \psi_2(x, y, z, s) \, d\mu(x, y, z, s)$$
$$= \sum_{s=-\sigma}^{+\sigma} \int_{\mathbb{R}^3 \times \{s\}} \psi_1^*(x, y, z, s) \psi_2(x, y, z, s) \, d\mu(x, y, z, s),$$

due to the fact that

$$\int_{\mathbb{R}^4 - (\mathbb{R}^3 \times S_\sigma)} \psi_1^*(x, y, z, s) \psi_2(x, y, z, s) \, d\mu(x, y, z, s) = 0.$$

If we introduce the measures

$$\mu_s(B) = \mu(B \times \{s\}), \qquad s = -\sigma, -\sigma + 1, \ldots, \sigma,$$

on $\mathscr{B}^3$, we can write

$$(7.4)\qquad \langle\psi_1 \mid \psi_2\rangle = \sum_{s=-\sigma}^{+\sigma} \int_{\mathbb{R}^3} \psi_1^*(x, y, z, s)\, \psi_2(x, y, z, s)\, d\mu_s(x, y, z).$$

However, without any further assumptions we cannot determine in more detail the measures $\mu_s(B)$, $B \in \mathscr{B}^3$ for $s = -\sigma, \ldots, +\sigma$.

The extra postulate which we need for determining μ_s follows.

Axiom W3. The self-adjoint operators $P^{(x)}$, $P^{(y)}$, and $P^{(z)}$ representing the momentum component observables of a one-particle system commute with the spin operator $S^{\mathbf{n}}$. In addition, the operators $Q^{(x)}$, $Q^{(y)}$, $Q^{(z)}$, $P^{(x)}$, $P^{(y)}$, $P^{(z)}$ satisfy canonical commutation relations in which $Q^{(x)}$ and $P^{(x)}$, $Q^{(y)}$ and $P^{(y)}$, $Q^{(z)}$ and $P^{(z)}$ are canonically conjugate.

In order to be able to take advantage of Axiom W3, we note that (7.4) implies that the mapping

$$\psi(x, y, z, s) \to \psi_s(x, y, z), \qquad s = -\sigma, -\sigma + 1, \ldots, \sigma,$$

of $L^2(\mathbb{R}^4, \mu)$ onto $\bigoplus_{s=-\sigma}^{\sigma} L^2(\mathbb{R}^3, \mu_s)$ establishes a unitary equivalence between these two Hilbert spaces. Since we have

$$(Q^{(x)}\, \psi)(x, y, z, s) = x\psi(x, y, z, s) = x\psi_s(x, y, z),$$

and similar relations hold for $Q^{(y)}$ and $Q^{(z)}$, it follows from von Neumann's theorem (Theorem 6.4) that μ_s have to be Lebesgue measures on $\mathscr{B}^3$, and that for differentiable functions $\psi(x, y, z, s)$, $\psi \in \mathscr{D}_{P^{(x)}}$,

$$(P^{(x)}\, \psi)(x, y, z, s) = -i\hbar \frac{\partial\psi(x, y, z, s)}{\partial x},$$

with corresponding relations holding for $P^{(y)}$ and $P^{(z)}$.

7.2. Wave Mechanics of n-Particle Systems

On the basis of the above discussion we have concluded that the spectral representation space of the set $\{Q^{(x)}, Q^{(y)}, Q^{(z)}, S^{\mathbf{n}}\}$ constituting, according to Axiom W1, a complete set of observables of a single particle system of spin σ, is the space $L^2(\mathbb{R}^4, \mu)$, where μ is the measure satisfying

$$\mu(B_1 \times \{s\}) = \mu_l^{(3)}(B_1), \qquad s = -\sigma, -\sigma + 1, \ldots, \sigma,$$

$$\mu(B_1 \times B_2\} = 0, \qquad B_2 \cap \{-\sigma, \ldots, +\sigma\} = \varnothing,$$

and $\mu_l^{(3)}(B_1)$ is the Lebesgue measure of B_1, $B_1 \in \mathscr{B}^3$. We have also seen that a state ψ can be represented at any instant by an element of the direct sum $\bigoplus_s L^2(\mathbb{R}^3)$ of $(2\sigma + 1)\, L^2(\mathbb{R}^3)$ spaces. Wave mechanics is ordinarily formulated in this $\bigoplus_s L^2(\mathbb{R}^3)$ space, in which a state vector is a wave function

$$\psi(x, y, z, s), \qquad x, y, z \in \mathbb{R}^1, \qquad s = -\sigma, -\sigma + 1, \ldots, +\sigma,$$

which is Lebesgue square integrable in the x, y, z variables for each fixed value of s from the set $S_\sigma = \{-\sigma, \ldots, +\sigma\}$.

By means of Axiom P in §4 we can generalize the above results to the n-particle case.

Suppose that we are dealing with a system $\mathfrak{S}$ of n different particles of spins $\sigma_1, \ldots, \sigma_n$, respectively. Then according to Axiom P, we can associate with $\mathfrak{S}$ the Hilbert space

$$\mathscr{H}_1 = \Big[\bigoplus_{s_1} L^2(\mathbb{R}^3)\Big] \otimes \cdots \otimes \Big[\bigoplus_{s_n} L^2(\mathbb{R}^3)\Big], \tag{7.5}$$

where the kth factor, $k = 1, \ldots, n$, is a direct sum of $(2\sigma_k + 1)$ $L^2(\mathbb{R}^3)$ spaces. The linear mapping U_1, which assigns to a vector of the form

$$(\psi^{(1)}_{-\sigma_1} \oplus \cdots \oplus \psi^{(1)}_{+\sigma_1}) \otimes \cdots \otimes (\psi^{(n)}_{-\sigma_n} \oplus \cdots \oplus \psi^{(n)}_{\sigma_n}), \; \psi^{(1)}_{-\sigma_1}, \ldots, \psi^{(n)}_{\sigma_n} \in L^2(\mathbb{R}^3) \tag{7.6}$$

from $\mathscr{H}_1$ the vector

$$\bigoplus_{s_1=-\sigma_1}^{\sigma_1} \cdots \bigoplus_{s_n=-\sigma_n}^{\sigma_n} (\psi^{(1)}_{s_1} \otimes \cdots \otimes \psi^{(n)}_{s_n}) \tag{7.7}$$

from the Hilbert space

$$\mathscr{H}_2 = \bigoplus_{s_1} \cdots \bigoplus_{s_n} [L^2(\mathbb{R}^3)]^{\otimes n}, \tag{7.8}$$

determines a unitary transformation of $\mathscr{H}_1$ onto $\mathscr{H}_2$ (see Exercise 7.1). Moreover, by Theorem 6.9 in Chapter II, there is a unitary transformation of $[L^2(\mathbb{R}^3)]^{\otimes n}$ onto $L^2(\mathbb{R}^{3n})$, which maps $\psi(x_1, y_1, z_1, s_1) \otimes \cdots \otimes \psi(x_n, y_n, z_n, s_n)$ into $\psi(x_1, y_1, z_1, s_1) \cdots \psi(x_n, y_n, z_n, s_n)$. By carrying out this transformation of each one of the terms $[L^2(\mathbb{R}^3)]^{\otimes n}$ of the direct sum (7.8) into $L^2(\mathbb{R}^{3n})$ we arrive at a unitary transformation (see Exercise 7.3) of the Hilbert space $\mathscr{H}_2$ into the Hilbert space

$$\mathscr{H} = \bigoplus_{s_1} \cdots \bigoplus_{s_n} L^2(\mathbb{R}^{3n}). \tag{7.9}$$

The elements of $\mathscr{H}$ (called the *configuration representation* space) are, for fixed $s_1, \ldots, s_n$, square-integrable wave functions

$$\psi(x_1, y_1, z_1, s_1, \ldots, x_n, y_n, z_n, s_n), \qquad s_k = -\sigma_k, \ldots, +\sigma_k. \tag{7.10}$$

Thus, $U = U_1 U_2$ is a unitary transformation of $\mathscr{H}_1$ onto $\mathscr{H}$ mapping a vector of the form (7.6) into

(7.11) $$\psi^{(1)}(x_1, y_1, z_1, s_1) \cdots \psi^{(n)}(x_n, y_n, z_n, s_n).$$

The x-coordinate position operator in $\mathscr{H}_1$ for the kth particle is

$$\tilde{Q}_k^{(x)}(\cdots \otimes \psi_k(x_k, y_k, z_k, s_k) \otimes \cdots) = \cdots \otimes (x_k \psi_k(x_k, y_k, z_k, s_n)) \otimes \cdots,$$

and corresponding expressions hold for $\tilde{Q}_k^{(y)}$ and $\tilde{Q}_k^{(z)}$. It is very easy to check (first for vectors in $\mathscr{H}$ represented by functions of the form (7.7), and then generalize the result) that the x-coordinate position operator in $\mathscr{H}$ of the kth particle

$$Q_k^{(x)} = U \tilde{Q}_k^{(x)} U^{-1}, \qquad k = 1, \ldots, n,$$

is given by (cf. Definition 5.2 of spectral representation spaces)

$$(Q_k^{(x)} \psi)(\ldots, x_k, y_k, z_k, s_k, \ldots) = x_k \psi(\ldots, x_k, y_k, z_k, s_k, \ldots).$$

Naturally, similar expressions hold for $Q_k^{(y)}$ and $Q_k^{(z)}$.

An analogous consideration easily yields that the momentum operator $P_k^{(x)}$ canonically conjugate to $Q_k^{(x)}$ satisfies the relation

$$(P_k^{(x)} \psi)(\ldots, x_k, y_k, z_k, s_k, \ldots) = -i\hbar \frac{\partial \psi(\ldots, x_k, y_k, z_k, s_k, \ldots)}{\partial x_k}$$

when $P_k^{(x)}$ is acting on differentiable functions from the domain of definition of $P_k^{(x)}$.

The above considerations cover the case of n-particle systems containing no identical particles. In the more general case when identical particles are present, the Hilbert space $\mathscr{H}$ of the system will be a closed linear subspace of the Hilbert space in (7.9), consisting of all appropriately symmetrized (see Definition 5.1) functions in (7.9); i.e., the wave functions of the system will be symmetric with respect to position–spin variables $(\mathbf{r}, s)$ corresponding to particles of integral spin (commonly called *bosons*), and antisymmetric with respect to variables corresponding to particles of half-integer spin (commonly called *fermions*).*

The entire discussion of this section can be duplicated with the complete set (see Exercise 7.5) $\{P^{(x)}, P^{(y)}, P^{(z)}, S^{\mathbf{n}}\}$ instead of $\{Q^{(x)}, Q^{(y)}, Q^{(z)}, S^{\mathbf{n}}\}$. The procedure leads then to the same Hilbert space (7.9), except that now the state vectors of this space are wave functions

$$\tilde{\psi}(p_1^{(x)}, p_1^{(y)}, p_1^{(z)}, s_1, \ldots, p_n^{(x)}, p_n^{(y)}, p_n^{(z)}, s_n), \qquad s_k = -\sigma_k, \ldots, \sigma_k,$$

* See also Exercise 7.4.

in momentum variables (so that as the spectral representation space $\mathscr{H}$ of $\{\mathbf{P}_k, S_k^{\mathrm{n}} : k = 1,..., n\}$ is called the *momentum representation* space) and the momentum operators $P_k^{(x)}$ have the representation

$$(P_k^{(x)}\psi)^\sim(..., p_k^{(x)}, p_k^{(y)}, p_k^{(z)}, s_k ,...) = p_k^{(x)}\tilde{\psi}(..., p_k^{(x)}, p_k^{(y)}, p_k^{(z)}, s_k ,...),$$

with corresponding relations holding for $P_k^{(y)}$ and $P_k^{(z)}$. The almost complete symmetry of the problem under the interchange of position and momentum variables (note that $[Q^{(x)}, P^{(x)}] = -[P^{(x)}, Q^{(x)}]$, etc.) makes it obvious that $Q_k^{(x)}$ is represented now by

$$(Q_k^{(x)}\psi)^\sim(..., p_k^{(x)}, p_k^{(y)}, p_k^{(z)}, s_k ,...) = i\hbar \frac{\partial\tilde{\psi}(..., p_k^{(x)}, p_k^{(y)}, p_k^{(z)}, s_k ,...)}{\partial p_k^{(x)}}$$

when acting on differentiable functions from its domain of definition; similar relations hold for $Q_k^{(y)}$ and $Q_k^{(z)}$.

7.3. The Schroedinger Operator

In any picture of quantum mechanics (see §3) we postulate for every quantum mechanical system the existence of a Hamiltonian H, which plays the dual role of determining the dynamics of the system and representing the total energy observable. However, none of the general axioms provides any clue about the more specific form of H.

In nonrelativistic quantum mechanics one normally approaches the problem of finding H of a particular system of particles by looking at that system from the classical-mechanics point of view. Thus, one chooses a classical model with a potential V from which all the forces acting between the particles of the system can be derived. The transition to wave mechanics is made by writing the Schroedinger operator for that system. The hidden assumption contained in this procedure is the following.

Axiom W4. The Hamiltonian H_{s} of a system of n spinless particles is a self-adjoint operator which maps elements $\psi \in \mathscr{D}_{H_{\mathrm{s}}}$, given in the configuration representation space by twice everywhere continuously differentiable functions, into

$$(7.12)\quad \Big[-\sum_{k=1}^{n} \frac{\hbar^2}{2m_k}\Big(\frac{\partial^2}{\partial x_k^{\,2}} + \frac{\partial^2}{\partial y_k^{\,2}} + \frac{\partial^2}{\partial z_k^{\,2}}\Big) + V(x_1, y_1, z_1 ,..., x_n, y_n, z_n)\Big]\, \psi(x_1, y_1, z_1 ,..., x_n, y_n, z_n)$$

whenever the above expression represents an element of $\mathscr{H}$; H_S is the only self-adjoint operator which maps all such elements $\psi \in \mathscr{D}_{H_S}$ into elements representable by (7.12).

An operator H_S satisfying the above conditions is called a *Schroedinger operator.*

For particles with spin, H_s contains in general terms which depend on the spin-component operators. Since spin operators are obviously bounded, the results derived later in this section are easily generalized to that case. Sometimes even "velocity-dependent" potentials, containing first derivatives in the position variables, i.e., depending on the momentum operators, are used. Their treatment might require special attention, and the reader interested in such cases is referred to the literature quoted at the end of this chapter.

We note that Axiom W4 imposes constraints on the functions which can play the role of potentials. A potential which has too many singularities, or some singularities that are too strong, might not satisfy the second requirement of Axiom W4.

7.4. Closures of Linear Operators

The main reason for not treating the Schroedinger operator exclusively as a differential operator is that differential operators given by means of differential operator forms are ill defined, since their domains of definition are ambiguous. However, one hopes that when defining a linear operator on a dense domain by means of the Schroedinger operator form, one will be able to extend that operator uniquely and in a natural way to a self-adjoint operator. This goal is usually achieved by taking the closure of that operator.

Definition 7.1. The linear operator $\bar{A}$ is the *closure of the linear operator* A if it is the minimal closed extension of A, i.e., every closed operator C satisfying $C \supseteq A$ also satisfies $C \supseteq \bar{A}$.

The existence problem of the closure of an operator is treated in Theorems 7.1 and 7.2.

***Theorem 7.1.** If the linear operator A in the Hilbert space $\mathscr{H}$ has an extension which is a closed operator, then it has a minimal extension $\bar{A}$, i.e., its closure exists.

Proof. Let A_0 be some given closed extension of A. Define the operator $\bar{A}$ in the following manner: the vector f is in the domain of definition of $\bar{A}$ if and only if $f \in \mathscr{D}_{A_0}$ and for any $\epsilon > 0$ there is a vector $g \in \mathscr{D}_A$ such that

$$\|f - g\| < \epsilon, \qquad \|A_0 f - Ag\| < \epsilon.$$

When $f \in \mathcal{D}_{\bar{A}}$, we define

$$\bar{A}f = A_0 f.$$

The so-defined operator $\bar{A}$ is certainly linear, since if $f_1, f_2 \in \mathcal{D}_{\bar{A}}$ and $a_1 \neq 0$, $a_2 \neq 0$, there are, for any given $\epsilon > 0$, vectors $g_1, g_2 \in \mathcal{D}_A$ such that

$$\| f_i - g_i \| < \frac{\epsilon}{2 \mid a_i \mid}, \qquad \| A_0 f_i - A g_i \| < \frac{\epsilon}{2 \mid a_i \mid}, \qquad i = 1, 2.$$

Consequently

$$\| a_1 f_1 + a_2 f_2 - (a_1 g_1 + a_2 g_2) \| < \epsilon, \quad \| A_0(a_1 f_1 + a_2 f_2) - A(a_1 g_1 + a_2 g_2) \| < \epsilon,$$

which implies that $a_1 f_1 + a_2 f_2 \in \mathcal{D}_{\bar{A}}$ and

$$\bar{A}(a_1 f_1 + a_2 f_2) = A_0(a_1 f_1 + a_2 f_2).$$

Finally, the operator $\bar{A}$ is closed. To see that, note that if $f_1, f_2, \ldots \in \mathcal{D}_{\bar{A}}$ converges to $f \in \mathcal{H}$ and $\bar{A}f_1, \bar{A}f_2, \ldots$ converges to $\tilde{f}$, then for any integer n there is a $g_n \in \mathcal{D}_A$ such that

$$\| f_n - g_n \| < 1/n, \qquad \| A_0 f_n - A g_n \| < 1/n.$$

Consequently, if $1/n < \epsilon/2$ and

$$\| f - f_n \| < \epsilon/2, \qquad \| \tilde{f} - \bar{A} f_n \| < \epsilon/2,$$

we get by combining the above two inequalities and applying the triangle inequality the following:

$$\| f - g_n \| < \epsilon, \qquad \| \tilde{f} - A g_n \| < \epsilon.$$

This result, in conjugation with the fact that A_0 is closed (and therefore $\tilde{f} = A_0 f$) yields that $\bar{A}$ is closed too. It is easy to infer from the way in which $\bar{A}$ was constructed that any closed extension of A contains $\bar{A}$. Consequently, $\bar{A}$ is the minimal closed extension of A. Q.E.D.

Theorem 7.1 guarantees the existence of the closure $\bar{A}$ of a linear operator A only in those cases when it is known that A has a closed extension. However, symmetric operators A with $\bar{\mathcal{D}}_A = \mathcal{H}$ have A^* as an extension, and since A^* is closed (see Chapter III, Theorem 2.8), $\bar{A}$ exists.

***Theorem 7.2.** If A is a symmetric operator densely defined in $\mathcal{H}$, its closure exists and is identical to A^{**}.

Proof. Since A^{**} is the adjoint of an operator, it is closed. Moreover, $A^{**} \supseteq A$, so that A has a closed extension. Consequently, $\bar{A}$ exists and $\bar{A} \subseteq A^{**}$.

The relation $\bar{A} \subseteq A^{**}$ implies $\bar{A}^* \supseteq A^{***} \equiv A^*$ (see Exercise 7.6), and therefore $\bar{A}^*$ is densely defined and has an adjoint $\bar{A}^{**}$. We shall prove that $\bar{A}^{**} \equiv \bar{A}$.

Both $\bar{A}$ and $\bar{A}^{**}$ are closed linear operators and consequently their respective Cayley transforms V and V_1 are closed also (see Chapter III, Theorem 4.7). Moreover, $\bar{A} \subseteq \bar{A}^{**}$ implies that $V \subseteq V_1$.

We recall that the domain of definition $\mathscr{D}_V$ of V is closed and identical to $\mathscr{R}_{\bar{A}+i}$. Thus, $\mathscr{H} \ominus \mathscr{D}_V$ consists of all vectors $f \in \mathscr{H}$ satisfying

$$\langle (\bar{A} + i) g \mid f \rangle = 0$$

for all $g \in \mathscr{D}_{\bar{A}+i}$. Since the above relation is identical to

$$\langle \bar{A}g \mid f \rangle = \langle g \mid if \rangle, \qquad g \in \mathscr{D}_{\bar{A}},$$

it follows that $f \in \mathscr{H} \ominus \mathscr{D}_V$ if and only if $\bar{A}^* f = if$.

Since $\bar{A}^* = \bar{A}^{***} \equiv (\bar{A}^{**})^*$ (see Exercise 7.6), it follows that $\mathscr{H} \ominus \mathscr{D}_V \equiv \mathscr{H} \ominus \mathscr{D}_{V_1}$, i.e., $\mathscr{D}_V \equiv \mathscr{D}_{V_1}$. Consequently, the relation $V \subseteq V_1$ holds only if $V \equiv V_1$, i.e., $\bar{A} \equiv \bar{A}^{**}$.

Using the result $\bar{A} \equiv \bar{A}^{**}$, we obtain from $A \subseteq \bar{A} \subseteq A^{**}$ that $A^* \supseteq \bar{A}^* \supseteq A^{***} \equiv A^*$. By taking adjoints in the preceding relations we get $A^{**} \subseteq \bar{A} \subseteq A^{**}$. Thus we have arrived at the desired result $A^{**} \equiv \bar{A}$. Q.E.D.

7.5. The Schroedinger Kinetic Energy Operator

By choosing in (7.12) $V(x_1, y_1, z_1, \ldots, x_n, y_n, z_n) \equiv 0$ we obtain the Schroedinger operator form for the kinetic energy operator

$$\begin{aligned} (7.13) \qquad (T_S\psi)(x_1, y_1, z_1, \ldots, x_n, y_n, z_n) \\ = -\frac{\hbar^2}{2} \sum_{k=1}^{n} \frac{1}{m_k} \left(\frac{\partial^2}{\partial x_k{}^2} + \frac{\partial^2}{\partial y_k{}^2} + \frac{\partial^2}{\partial z_k{}^2} \right) \\ \times \psi(x_1, y_1, z_1, \ldots, x_n, y_n, z_n). \end{aligned}$$

In order to be able to treat T_S as a differential operator in $L^2(R^{3n})$, we require that each $\psi \in L^2(\mathbb{R}^{3n})$ to which T_S is applied be representable by a twice differentiable function with square integrable second derivatives. An operator T_S defined in this fashion would be symmetric but not self-adjoint. It will be shown, however, that such an operator is essentially self-adjoint.

Definition 7.2. The symmetric operator A in $\mathscr{H}$ is said to be *essentially self-adjoint* if its closure $\bar{A}$ is self-adjoint.

According to Theorem 7.2, A^{**} is identical to the closure $\bar{A}$ of A. Since $(A^{**})^* \equiv A^* \equiv A^{**}$ (see Exercise 7.6), we see that A is essentially self-adjoint if and only if $A^* = A^{**}$.

***Theorem 7.3.** The symmetric operator T_0, defined on the family $\mathscr{D}_0$ of all elements $f \in L^2(\mathbb{R}^{3n})$ representable by functions

$$f(\mathbf{r}_1, \dots, \mathbf{r}_n) = P(\mathbf{r}_1, \dots, \mathbf{r}_n) \exp\left[-\frac{1}{2\hbar}(\mathbf{r}_1^2 + \cdots + \mathbf{r}_n^2)\right] \tag{7.14}$$

corresponding to all choices of polynomials $P(\mathbf{r}_1, \dots, \mathbf{r}_n)$, and acting on $f \in \mathscr{D}_0$ according to

$$(T_0 f)(\mathbf{r}_1, \dots, \mathbf{r}_n) = -\frac{\hbar^2}{2}\left(\sum_{k=1}^{n} \frac{1}{m_k} \Delta_k\right) f(\mathbf{r}_1, \dots, \mathbf{r}_n),$$

is essentially self-adjoint. Its closure $\bar{T}_0$ is identical to the operator

$$T_S = \sum_{k=1}^{n} \frac{1}{2m_k} (P_k^{(x)2} + P_k^{(y)2} + P_k^{(z)2}). \tag{7.15}$$

Proof. The operator T_S is a function

$$T_S = K(\mathbf{P}_1, \dots, \mathbf{P}_n); \qquad K(\mathbf{P}_1, \dots, \mathbf{P}_n) = \sum_{k=1}^{n} \mathbf{P}_k^2/2m_k, \tag{7.16}$$

in the sense introduced in Theorem 2.5, of the commuting self-adjoint operators $P_k^{(x)}, P_k^{(y)}, P_k^{(z)}$, $k = 1, \dots, n$ (see Chapter III, Theorem 4.11). Therefore, from Theorem 2.7, and in particular (2.26), we deduce that for $f \in \mathscr{D}_{T_S}$, T_S acts on the Fourier–Plancherel transform $\tilde{f}$ of f in the following way (see also Exercise 7.7):

$$(\widetilde{T_S f})(\mathbf{p}_1, \dots, \mathbf{p}_n) = \frac{1}{2} \sum_{k=1}^{n} \frac{1}{m_k} \mathbf{p}_k^2 \tilde{f}(\mathbf{p}_1, \dots, \mathbf{p}_n). \tag{7.17}$$

Now, $\mathscr{D}_0$ is invariant under the Fourier–Plancherel transform U_F (see Exercise 7.7), i.e., $U_F \mathscr{D}_0 = \mathscr{D}_0$, so that if $f \in \mathscr{D}_0$, then $\tilde{f}$ is representable by a function of the form

$$P(\mathbf{p}_1, \dots, \mathbf{p}_n) \exp[-(2\hbar)^{-1}(\mathbf{p}_1^2 + \cdots + \mathbf{p}_n^2)].$$

It follows that $T_S f$ is defined by (7.16) when $f \in \mathscr{D}_0$, i.e., $\mathscr{D}_0 \subset \mathscr{D}_{T_S}$. In view of the fact that $\mathscr{D}_0$ is dense in $L^2(\mathbb{R}^{3n})$ (see Lemma 5.2), the set

$\mathscr{D}_{T_S}$ [consisting of all $f \in L^2(\mathbb{R}^{3n})$ for which the function on the right-hand side of (7.17) is square integrable] is dense in $L^2(\mathbb{R}^{3n})$. Thus, T_S is a real function of self-adjoint operators and a densely defined operator. Therefore, T_S is self-adjoint by Theorem 2.6.

It is easy to see that $T_0 \subseteq T_S$. As a matter of fact, we have

$$\begin{aligned}(T_0 f)(\mathbf{r}_1, \ldots, \mathbf{r}_n) &= \left(-\frac{\hbar^2}{2}\sum_{k=1}^{n}\frac{1}{m_k}\Delta_k\right)\frac{1}{(2\pi\hbar)^{3n/2}}\int_{\mathbb{R}^{3n}}\tilde{f}(\mathbf{p}_1, \ldots, \mathbf{p}_n)\\ &\quad\times \exp[(i/\hbar)(\mathbf{p}_1\cdot\mathbf{r}_1 + \cdots + \mathbf{p}_n\cdot\mathbf{r}_n)]\,d\mathbf{p}_1\cdots d\mathbf{p}_n\\ &= \frac{1}{2}\frac{1}{(2\pi\hbar)^{3n/2}}\int_{\mathbb{R}^{3n}}\sum_{k=1}^{n}\frac{1}{m_k}\mathbf{p}_k^2\tilde{f}(\mathbf{p}_1, \ldots, \mathbf{p}_n)\\ &\quad\times \exp[(i/\hbar)(\mathbf{p}_1\cdot\mathbf{r}_1 + \cdots + \mathbf{p}_n\cdot\mathbf{r}_n)]\,d\mathbf{p}_1\cdots d\mathbf{p}_n\\ &= (T_0 f)(\mathbf{r}_1, \ldots, \mathbf{r}_n)\end{aligned}$$

since for $f \in \mathscr{D}_0 \subset \mathscr{D}_{T_S}$ of the form (7.14) the order of differentiation and integration can be interchanged by Lemma 3.1 and the mean value theorem of differential calculus.

We shall prove now that T_S is the closure of T_0. Let us show first that $(1 + T_0)\,\mathscr{D}_0$ is dense in $L^2(\mathbb{R}^{3n})$.

Assume that $g \in L^2(\mathbb{R}^{3n})$ is orthogonal to $(1 + T_0)\mathscr{D}_0$. Since $U_F\mathscr{D}_0 \equiv \mathscr{D}_0$, this assumption implies that

$$\begin{aligned}&\int_{\mathbb{R}^{3n}}\tilde{g}(\mathbf{p}_1, \ldots, \mathbf{p}_n)[1 + K(\mathbf{p}_1, \ldots, \mathbf{p}_n)]\,P(\mathbf{p}_1, \ldots, \mathbf{p}_n)\\ &\qquad\times \exp[-(2\hbar)^{-1}(\mathbf{p}_1^2 + \cdots + \mathbf{p}_n^2)]\,d\mathbf{p}_1\cdots d\mathbf{p}_n = 0\end{aligned} \tag{7.18}$$

for all choices of polynomials $P(\mathbf{p}_1, \ldots, \mathbf{p}_n)$. The function

$$\begin{aligned}&\tilde{h}(\mathbf{p}_1, \ldots, \mathbf{p}_n)\\ &\quad= \tilde{g}(\mathbf{p}_1, \ldots, \mathbf{p}_n)[1 + K(\mathbf{p}_1, \ldots, \mathbf{p}_n)]\exp[-(4\hbar)^{-1}(\mathbf{p}_1^2 + \cdots + \mathbf{p}_n^2)]\end{aligned} \tag{7.19}$$

is square integrable, and according to (7.18) it is orthogonal to all functions of the form

$$P(\mathbf{p}_1, \ldots, \mathbf{p}_n)\exp[-(4\hbar)^{-1}(\mathbf{p}_1^2 + \cdots + \mathbf{p}_n^2)]. \tag{7.20}$$

By Lemma 5.2 the family of all functions (7.20) is dense in $L^2(\mathbb{R}^{3n})$. Therefore, we must have $\tilde{h}(\mathbf{p}_1, \ldots, \mathbf{p}_n) = 0$ almost everywhere. Since

$$[1 + K(\mathbf{p}_1, \ldots, \mathbf{p}_n)]\exp[-(4\hbar)^{-1}(\mathbf{p}_1^2 + \cdots + \mathbf{p}_n^2)] > 0, \qquad \mathbf{p}_1, \ldots, \mathbf{p}_n \in \mathbb{R}^3,$$

we must have $\tilde{g}(\mathbf{p}_1, \ldots, \mathbf{p}_n) = 0$ almost everywhere, i.e., $g = \mathbf{0}$, and $(1 + T_0)\, \mathscr{D}_0$ is dense in $L^2(\mathbb{R}^{3n})$.

Let us take any $f \in \mathscr{D}_{T_S}$. Since $(1 + T_0)\, \mathscr{D}_0$ is dense in $L^2(\mathbb{R}^{3n})$, there is for any $\epsilon > 0$ a vector $f_\epsilon \in \mathscr{D}_0$ such that

$$\|(1 + T_S) f - (1 + T_0) f_\epsilon\| < \epsilon,$$

i.e., in view of $T_0 \subseteq T_S$,

$$\|(1 + T_S)(f - f_\epsilon)\| < \epsilon.$$

It is easy to see from the relation

$$\|(1 + T_S)(f - f_\epsilon)\|^2 = \int_{\mathbb{R}^{3n}} |[1 + K(\mathbf{p}_1, \ldots, \mathbf{p}_n)](f - f_\epsilon)(\mathbf{p}_1, \ldots, \mathbf{p}_n)|^2 \, d\mathbf{p}_1 \cdots d\mathbf{p}_n$$

that since $K(\mathbf{p}_1, \ldots, \mathbf{p}_n) \geqslant 0$,

$$\|f - f_\epsilon\| \leqslant \|(1 + T_S)(f - f_\epsilon)\| < \epsilon,$$

$$\|T_S f - T_0 f_\epsilon\| = \|T_S(f - f_\epsilon)\| \leqslant \|(1 + T_S)(f - f_\epsilon)\| < \epsilon,$$

and consequently $T_S \subseteq \bar{T}_0$. However, T_S is self-adjoint, and therefore closed, and since $T_0 \subseteq T_S$, we must have $T_S \equiv \bar{T}_0$. Q.E.D.

7.6. The Schroedinger Potential Energy Operator

The operator V representing the potential energy is given in the Schroedinger picture by means of a potential $V(\mathbf{r}_1, \ldots, \mathbf{r}_n)$

$$(V_S \psi)(\mathbf{r}_1, \ldots, \mathbf{r}_n) = V(\mathbf{r}_1, \ldots, \mathbf{r}_n)\, \psi(\mathbf{r}_1, \ldots, \mathbf{r}_n),$$

where $V(\mathbf{r}_1, \ldots, \mathbf{r}_n)$ is a real function. It is quite obvious that V_S is a function of $Q_1^{(x)}, Q_1^{(y)}, Q_1^{(z)}, \ldots$ in the sense of the definition of a function of commuting operators given in Theorem 2.5 (see also Exercise 2.7):

$$V_S = V(Q_1^{(x)}, Q_1^{(y)}, Q_1^{(z)}, \ldots, Q_n^{(x)}, Q_n^{(y)}, Q_n^{(z)}). \tag{7.21}$$

We cannot expect, however, that $V_S \equiv V_S^*$ for any choice of the real function $V(\mathbf{r}_1, \ldots, \mathbf{r}_n)$. The requirements subsequently imposed on $V(\mathbf{r}_1, \ldots, \mathbf{r}_n)$ will enable us to prove next that V_S is self-adjoint and, later on, that $T_S + V_S$ is also self-adjoint by virtue of the fact that V_S is bounded relative to T_S in the sense of the following definition.

Definition 7.3. An operator K is said to be *bounded relative to* another operator A, or *A-bounded*, if $\mathscr{D}_K \supset \mathscr{D}_A$ and if there are constants $a, b \geqslant 0$ such that $\|Kf\| \leqslant a\|Af\| + b\|f\|$ for all $f \in \mathscr{D}_A$. The infimum of all values a for which there is some b such that the preceding condition holds true is called the *relative bound of K with respect to A*, or its *A-bound.*

We note that if K is bounded in the sense of Definition 1.5 in Chapter III, then it is bounded relative to any operator A with domain $\mathscr{D}_A \subset \mathscr{D}_K$, and that its A-bound is zero. Indeed, in case of such a bounded operator K we can satisfy the conditions of Definition 7.3 with the choice $a = 0$ and $b = \| K \|$. However, an operator K can have a relative bound zero with respect to another operator A without actually being bounded. Such examples are provided in the next theorem which, in fact, states that V_S has T_S-bound zero, whereas the conditions on the potentials for which the theorem is valid are sufficiently liberal to allow for the possibility of V_S being an unbounded operator.

***Theorem 7.4.** Assume that the potential $V(\mathbf{r}_1, \ldots, \mathbf{r}_n)$ can be written in the form

$$V(\mathbf{r}_1, \ldots, \mathbf{r}_n) = V_0(\mathbf{r}_1, \ldots, \mathbf{r}_n) + \sum_{\substack{i,j=1 \\ i<j}}^{n} V_{ij}(\mathbf{r}_i - \mathbf{r}_j) + \sum_{i=1}^{n} V_{0i}(\mathbf{r}_i),$$

where $V_0(\mathbf{r}_1, \ldots, \mathbf{r}_n)$ is measurable and bounded on $\mathbb{R}^{3n}$, while $V_{ij}(\mathbf{r})$, $i, j = 0, 1, \ldots, n$ $(i < j)$ are locally square integrable and bounded at infinity,* i.e., there are positive constants R, C such that

$$\int_{r \leqslant R} | V_{ij}(\mathbf{r})|^2 \, d\mathbf{r} < +\infty, \qquad | V_{ij}(\mathbf{r})| \leqslant C, \qquad r > R. \tag{7.22}$$

For such a potential, the operator

$$(V_S f)(\mathbf{r}_1, \ldots, \mathbf{r}_n) = V(\mathbf{r}_1, \ldots, \mathbf{r}_n) f(\mathbf{r}_1, \ldots, \mathbf{r}_n),$$

defined on the family $\mathscr{D}_{V_S}$ of all $f \in L^2(\mathbb{R}^{3n})$ for which the function $V(\mathbf{r}_1, \ldots, r_n) f(\mathbf{r}_1, \ldots, \mathbf{r}_n)$ is square integrable, is self-adjoint, and $\mathscr{D}_{V_S} \supset \mathscr{D}_{T_S}$. Moreover, the operator V_S is such that for any constant $a > 0$ there is another constant $b > 0$ such that

$$\| V_S f \| \leqslant a \| T_S f \| + b \| f \| \tag{7.23}$$

for all $f \in \mathscr{D}_{T_S}$.

* It has become customary to denote the class of all functions $f(\mathbf{r})$ having these two properties by $L^1(\mathbb{R}^3) + L^\infty(\mathbb{R}^3)$, where $L^1(\mathbb{R}^3)$ and $L^\infty(\mathbb{R}^3)$ denote the Banach spaces of functions that are measurable and, respectively, Lebesgue integrable or bounded on $\mathbb{R}^3$.

Proof. It follows from the assumptions on $V(\mathbf{r}_1,\ldots,\mathbf{r}_n)$ that $V(\mathbf{r}_1,\ldots,r_n)f(\mathbf{r}_1,\ldots,\mathbf{r}_n)$ is square integrable on $\mathbb{R}^{3n}$ if $f\in\mathscr{D}_0$. This conclusion can be easily reached by using Theorem 3.9 in Chapter II. This theorem can be immediately applied to $V_0(\mathbf{r}_1,\ldots,\mathbf{r}_n)$, which is bounded $|V_0(\mathbf{r}_1,\ldots,\mathbf{r}_n)|\leqslant b_1$ and therefore

$$|V_0(\mathbf{r}_1,\ldots,\mathbf{r}_n)f(\mathbf{r}_1,\ldots,\mathbf{r}_n)|\leqslant b_1\,|f(\mathbf{r}_1,\ldots,\mathbf{r}_n)|. \tag{7.24}$$

Its application to $V_{ij}(\mathbf{r}_i-\mathbf{r}_j)$ and $V_{0i}(\mathbf{r}_i)$ is somewhat more roundabout (see Exercise 7.8). Since $\mathscr{D}_0$ is dense in $L^2(\mathbb{R}^{3n})$ and, by virtue of (7.21), V_S is a real function of the position observables, it follows from Theorem 2.6 that V_S is self-adjoint.

A glance at (7.24) makes it obvious that V_0 satisfies (7.23) with $a=0$ and $b=b_1$. The proof of (7.23) for V_{ij} is more intricate, and it will be carried out next only for the representative case of V_{12}.

In the new variables

$$\mathbf{r}_1'=(1/\sqrt{2})(\mathbf{r}_1-\mathbf{r}_2),\qquad \mathbf{r}_2'=(1/\sqrt{2})(\mathbf{r}_1+\mathbf{r}_2),\qquad \mathbf{r}_3'=\mathbf{r}_3,\ldots,\mathbf{r}_n'=\mathbf{r}_n,$$

the function

$$g(\mathbf{r}_1',\ldots,\mathbf{r}_n')=f(\mathbf{r}_1,\ldots,\mathbf{r}_n)$$

is of the form (7.14) whenever $f\in\mathscr{D}_0$. This is due to the fact that

$$\mathbf{r}_1'^2+\mathbf{r}_2'^2=\mathbf{r}_1^{\,2}+\mathbf{r}_2^{\,2}.$$

In addition, the Jacobian of the transformation $\mathbf{r}_k\to\mathbf{r}_k'$, $k=1,\ldots,n$, is equal to one,

$$d\mathbf{r}_1'\cdots d\mathbf{r}_n'=d\mathbf{r}_1\cdots d\mathbf{r}_n,$$

so that writing $V_{12}'(\mathbf{r}_1')=V_{12}(\mathbf{r}_1-\mathbf{r}_2)$,

$$\begin{aligned}\|V_{12}f\|^2&=\int_{\mathbb{R}^n}|V_{12}(\mathbf{r}_1-\mathbf{r}_2)f(\mathbf{r}_1,\ldots,\mathbf{r}_n)|^2\,d\mathbf{r}_1\cdots d\mathbf{r}_n\\&=\int_{\mathbb{R}^n}|V_{12}'(\mathbf{r}_1')\,g(\mathbf{r}_1',\ldots,\mathbf{r}_n')|^2\,d\mathbf{r}_1'\cdots d\mathbf{r}_n'.\end{aligned}$$

To be able to estimate the above integral, we split the domain of integration into two parts

$$\int_{\mathbb{R}^n}=\int_{r_1'\leqslant R'}+\int_{r_1'>R'}, \tag{7.25}$$

where $R' = 2^{-1/2}R$ and R is the constant appearing in (7.22). We shall study the two parts separately.

According to Fubini's theorem (Chapter II, Theorem 3.13)

$$(7.26)\quad \int_{\substack{r_1'\leqslant R' \\ \mathbf{r}_2',\ldots,\mathbf{r}_n' \in \mathbb{R}^3}} |V_{12}'(\mathbf{r}_1')\, g(\mathbf{r}_1',\ldots,\mathbf{r}_n')|^2\, d\mathbf{r}_1' \cdots d\mathbf{r}_n'$$

$$= \int_{r_1'\leqslant R'} d\mathbf{r}_1' \, |V(\mathbf{r}_1')|^2 \int_{\mathbb{R}^{3(n-1)}} d\mathbf{r}_2' \cdots d\mathbf{r}_n' \, |g(\mathbf{r}_1',\ldots,\mathbf{r}_n')|^2.$$

If $\tilde{g}(\mathbf{p}_1',\ldots,\mathbf{p}_n')$ denotes the Fourier transform of $g(\mathbf{r}_1',\ldots,\mathbf{r}_n')$, we have, by (4.12) and Theorem 4.6 in Chapter III,

$$(7.27)\quad \int_{\mathbb{R}^{3(n-1)}} |g(\mathbf{r}_1',\ldots,\mathbf{r}_n')|^2\, d\mathbf{r}_2' \cdots d\mathbf{r}_n'$$

$$= (2\pi\hbar)^{-3} \int_{\mathbb{R}^{3(n-1)}} d\mathbf{p}_2' \cdots d\mathbf{p}_n' \left| \int_{\mathbb{R}^3} \exp(i/\hbar)(\mathbf{r}_1' \cdot \mathbf{p}_1')\, \tilde{g}(\mathbf{p}_1',\ldots,\mathbf{p}_n')\, d\mathbf{p}_1' \right|^2$$

$$\leqslant (2\pi\hbar)^{-3} \int_{\mathbb{R}^{3(n-1)}} d\mathbf{p}_2' \cdots d\mathbf{p}_n' \left[\int_{\mathbb{R}^3} |\tilde{g}(\mathbf{p}_1',\ldots,\mathbf{p}_n')|\, d\mathbf{p}_1' \right]^2.$$

Moreover, for an arbitrary positive constant ϵ, $(1 + \epsilon^4(\mathbf{p}_1'^2)^2)^{-1}$ and $|g(\mathbf{p}_1',\ldots,\mathbf{p}_n')|(1 + \epsilon^4(\mathbf{p}_1'^2)^2)^{1/2}$ are square integrable in $\mathbf{p}_1' \in \mathbb{R}^3$. Hence, we can apply the Schwarz–Cauchy inequality in $L^2(\mathbb{R}^3)$ to these two functions, and thus derive

(7.28)

$$\left[\int_{\mathbb{R}^3} |\tilde{g}(\mathbf{p}_1',\ldots,\mathbf{p}_n')|\, d\mathbf{p}_1' \right]^2$$

$$= \left[\int_{\mathbb{R}^3} \frac{|\tilde{g}(\mathbf{p}_1',\ldots,\mathbf{p}_n')|}{[1 + \epsilon^4(\mathbf{p}_1'^2)^2]^{1/2}} [1 + \epsilon^4(\mathbf{p}_1'^2)^2]^{1/2}\, d\mathbf{p}_1' \right]^2$$

$$\leqslant \left[\int_{\mathbb{R}^3} \frac{d\mathbf{p}_1'}{1 + \epsilon^4(\mathbf{p}_1'^2)^2} \right] \int_{\mathbb{R}^3} |\tilde{g}(\mathbf{p}_1',\ldots,\mathbf{p}_n')|^2 (1 + \epsilon^4(\mathbf{p}_1'^2)^2)\, d\mathbf{p}_1'$$

$$= \frac{\pi^2\sqrt{2}}{\epsilon^3} \int_{\mathbb{R}^3} |\tilde{g}(\mathbf{p}_1',\ldots,\mathbf{p}_n')|^2 (1 + \epsilon^4(\mathbf{p}_1'^2)^2)\, d\mathbf{p}_1'$$

$$= \frac{\pi^2\sqrt{2}}{\epsilon^3} \int_{\mathbb{R}^3} |\tilde{g}(\mathbf{p}_1',\ldots,\mathbf{p}_n')|^2\, d\mathbf{p}_1' + \sqrt{2}\,\pi^2\epsilon \int_{\mathbb{R}^3} |\mathbf{p}_1'^2\tilde{g}(\mathbf{p}_1',\ldots,\mathbf{p}_n')|^2\, d\mathbf{p}_1'$$

$$\leqslant \frac{\sqrt{2}\,\pi^2}{\epsilon^3} \int_{\mathbb{R}^3} |\tilde{g}(\mathbf{p}_1',\ldots,\mathbf{p}_n')|^2\, d\mathbf{p}_1'$$

$$+ 4\sqrt{2}\,\pi^2(m_1 + m_2)^2\, \epsilon \int_{\mathbb{R}^3} \left| \left(\frac{\mathbf{p}_1^2}{2m_1} + \cdots + \frac{\mathbf{p}_n^2}{2m_n} \right) \tilde{g}(\mathbf{p}_1',\ldots,\mathbf{p}_n') \right|^2 d\mathbf{p}_1';$$

in the last step of the above derivation we have used the fact that, in terms of the momentum variables $\mathbf{p}_1, \ldots, \mathbf{p}_n$,

$$\mathbf{p}_1' = \frac{1}{\sqrt{2}}(\mathbf{p}_1 - \mathbf{p}_2), \qquad \mathbf{p}_2' = \frac{1}{\sqrt{2}}(\mathbf{p}_1 + \mathbf{p}_2), \qquad \mathbf{p}_3' = \mathbf{p}_3, \ldots, \mathbf{p}_n' = \mathbf{p}_n,$$

so that $\mathbf{r}_1 \cdot \mathbf{p}_1 + \mathbf{r}_2 \cdot \mathbf{p}_2 = \mathbf{r}_1' \cdot \mathbf{p}_1' + \mathbf{r}_2' \cdot \mathbf{p}_2'$, and consequently

$$\begin{aligned}\mathbf{p}_1'^2 &= \tfrac{1}{2}(\mathbf{p}_1^2 - 2\mathbf{p}_1 \cdot \mathbf{p}_2 + \mathbf{p}_2^2) \leqslant \mathbf{p}_1^2 + \mathbf{p}_2^2 \\ &\leqslant 2(m_1 + m_2)\left(\frac{\mathbf{p}_1^2}{2m_1} + \frac{\mathbf{p}_2^2}{2m_2} + \cdots + \frac{\mathbf{p}_n^2}{2m_n}\right).\end{aligned}$$

Inserting the last inequality of (7.28) in (7.27), and noting that

$$\begin{aligned}\int_{\mathbb{R}^{3n}} &|\tilde{g}(\mathbf{p}_1', \ldots, \mathbf{p}_n')|^2 \, d\mathbf{p}_1' \cdots d\mathbf{p}_n' \\ &= \int_{\mathbb{R}^{3n}} |g(\mathbf{r}_1', \ldots, \mathbf{r}_n')|^2 \, d\mathbf{r}_1' \cdots d\mathbf{r}_n' \\ &= \int_{\mathbb{R}^{3n}} |f(\mathbf{r}_1, \ldots, \mathbf{r}_n)|^2 \, d\mathbf{r}_1 \cdots d\mathbf{r}_n = \|f\|^2,\end{aligned}$$

$$\int_{\mathbb{R}^{3n}} \left|\left(\sum_{k=1}^{n} \frac{\mathbf{p}_k^2}{2m_k}\right) \tilde{g}(\mathbf{p}_1', \ldots, \mathbf{p}_n')\right|^2 d\mathbf{p}_1' \cdots d\mathbf{p}_n' = \| T_S f\|^2,$$

we get

$$\begin{aligned}\hbar^3 \int_{\mathbb{R}^{3(n-1)}} &|g(\mathbf{r}_1', \ldots, \mathbf{r}_n')|^2 \, d\mathbf{r}_2' \cdots d\mathbf{r}_n' \\ &\leqslant \frac{1}{4\sqrt{2}\,\pi} \frac{1}{\epsilon^3} \|f\|^2 + \frac{(m_1 + m_2)^2}{\sqrt{2}\,\pi} \epsilon \| T_S f\|^2.\end{aligned}$$

Hence, combining (7.22), (7.25), (7.26), and (7.28) we obtain

$$\begin{aligned}\| V_{12} f\|^2 &\leqslant \left[\frac{\hbar^{-3}}{4\sqrt{2}\pi} \frac{1}{\epsilon^3} \|f\|^2 + \frac{(m_1 + m_2)^2}{\sqrt{2}\pi\hbar^3} \epsilon \| T_0 f\|^2\right] \int_{r_1' \leqslant R'} | V_{12}'(\mathbf{r}_1)|^2 \, d\mathbf{r}_1' \\ &\quad + C^2 \int_{r_1' > R'} d\mathbf{r}_1' \int_{\mathbb{R}^{3(n-1)}} d\mathbf{r}_2' \cdots d\mathbf{r}_n' \, |g(\mathbf{r}_1', \ldots, \mathbf{r}_n')|^2 \\ &\leqslant \left(\frac{c_{12}}{4\sqrt{2}\pi} \frac{1}{\epsilon^3} + C^2\right) \|f\|^2 + \frac{(m_1 + m_2)^2 c_{12}}{\sqrt{2}\pi} \epsilon \| T_0 f\|^2 \\ &\leqslant \left[\left(\frac{c_{12}}{4\sqrt{2}\pi} \frac{1}{\epsilon^3} + C^2\right)^{1/2} \|f\| + (m_1 + m_2)\left(\frac{\epsilon c_{12}}{\sqrt{2}\pi}\right)^{1/2} \| T_0 f\|\right]^2,\end{aligned}$$

where c_{12} is a constant defined in general as follows:

$$c_{ij} = \hbar^{-3} \int_{r_i' \leqslant R'} | V_{ij}'(\mathbf{r}_i')|^2 \, d\mathbf{r}_i', \qquad i < j.$$

Since $\epsilon > 0$ can be chosen arbitrarily small, we have arrived at the result that the inequality (7.23) is true for all $f \in \mathscr{D}_0$ and any *a priori* given $a > 0$, if we choose ϵ such that

$$a \geqslant \sqrt{\epsilon} \sum_{i<j} \left[\frac{(m_i + m_j)^2 c_{ij}}{\sqrt{2}\pi} \right]^{1/2}. \tag{7.29}$$

Take now a vector $h \in \mathscr{D}_{T_S}$, and let $h_1, h_2, \ldots \in \mathscr{D}_0$ be a sequence for which

$$T_S h = \operatorname*{s-lim}_{n\to\infty} T_S h_n .$$

Since (7.23) has been established for vectors from $\mathscr{D}_0$, we can write

$$\| V_S(h_m - h_n)\| \leqslant a \| T_S(h_m - h_n)\| + b \| h_m - h_n \|,$$

which implies that $V_S h_1, V_S h_2, \ldots$ is a Cauchy sequence. Hence

$$h' = \operatorname*{s-lim}_{n\to\infty} V_S h_n$$

exists, and since V_S is self-adjoint, and therefore closed, we have $h \in \mathscr{D}_{V_S}$ and $h' = V_S h$; hence $\mathscr{D}_{T_S} \subset \mathscr{D}_{V_S}$. Furthermore,

$$\| V_S h \| \leqslant \| V_S h_n \| + \| V_S(h - h_n)\| \leqslant a \| T_S h_n \| + b \| h_n \| + \| V_S(h - h_n)\|,$$

and in the limit $n \to +\infty$ we obtain (7.23) for an arbitrary $h \in \mathscr{D}_{T_S}$.

Q.E.D.

The very general type of potential with which Theorem 7.4 deals consists of three distinct parts: V_0, which describes a global interaction in between all n particles; V_{ij}, describing a two-body interaction in between particle i and particle j; and V_{0i}, describing the influence of some sources external to the system (such as the external fields discussed in Section 7.9) onto particle i. As examples of particular practical importance we can mention the case of Coulomb and Yukawa two-body interactions corresponding to the choice

$$\begin{gathered} V_0(\mathbf{r}_1, \ldots, \mathbf{r}_n) \equiv V_{0i}(\mathbf{r}_i) \equiv 0, \\ V_{ij}(\mathbf{r}_i - \mathbf{r}_j) = \frac{\lambda_{ij} \exp(-\kappa_{ij}(\mathbf{r}_i - \mathbf{r}_j))}{| \mathbf{r}_i - \mathbf{r}_j |}, \qquad i < j, \end{gathered} \tag{7.30}$$

with $\kappa_{ij} = 0$ and $\kappa_{ij} > 0$, respectively, where λ_{ij} are coupling constants (which in the Coulomb case equal the product $e_i e_j$ of the respective charges of the ith and jth particle—see Chapter II, Section 7.7).

7.7. The Self-Adjointness of the Schroedinger Operator

The standard method of proving the self-adjointness of an operator of the form $A + K$, where A is already known to be self-adjoint, is based on a result first proven by Rellich [1939] (see Lemma 7.2 later in this section), in which the useful notion of operator core figures prominently.

Definition 7.4. Let A be a closed operator in $\mathscr{H}$, D a subset of $\mathscr{D}_A$, and A_D the restriction of A to D, so that $A_D \subseteq A$. The set D is called a *core* of A if $\overline{A_D} = A$.

By Theorem 7.3 the set $\mathscr{D}_0$ is a core for the self-adjoint operator T_S. As has been illustrated so many time on the differential operator forms we have encountered in preceding sections, such forms define operators on L^2 spaces only when applied to suitably chosen sets D of functions. In order that such forms unambiguously lead to self-adjoint operators, it is essential that these sets D represent cores (i.e., that the action of the differential operator form on all elements of D represent an essentially self-adjoint operator). It should be realized, however, that one and the same operator can have many cores. In fact, if we reconsider the proof of Theorem 4.10 in Chapter III, we see that the argument used there leads to the following more general conclusion.

Lemma 7.1. A densely defined symmetric operator is essentially self-adjoint if and only if both its deficiency indices are zero.

Thus, as long as for some $A = A^*$ a set $D \subset \mathscr{D}_A$ is such that the restriction A_D of A to D has a Cayley transform whose domain $\mathscr{R}_{A_D+i}$ and range $\mathscr{R}_{A_D-i}$ are dense in $\mathscr{H}$, that set D is going to represent a core for A. For example, we might have used Lemma 5.2 to build dense sets D of functions (5.18) with a choice of $f_0(\mathbf{r}_1, ..., \mathbf{r}_n)$ that would be different from the exponential function in (7.14) and thus arrive at other* cores for T_S, but then the computations of the Fourier transform of such functions needed to establish Theorem 7.3 would not have been as straightforward.

Lemma 7.2. (*Rellich's theorem*). If $A = A^*$ and K is symmetric and relatively bounded with respect to A with an A-bound smaller than 1, then $A + K$ is a self-adjoint operator with domain $\mathscr{D}_{A+K} = \mathscr{D}_A$; furthermore, $A + K$ is essentially self-adjoint on any core of A.

Proof. We shall prove first that $A + K$ is self-adjoint by showing that its Cayley transform V is unitary (see Chapter III, Theorem 4.9), i.e., that $\mathscr{R}_{A+K+i} = \mathscr{R}_{A+K-i} = \mathscr{H}$.

* The set $D = \mathscr{C}_0(\mathbb{R}^{3n})$ is a very popular alternative choice (e.g., see Reed and Simon [1975], Sections X.1–X.4]).

Since for any $\lambda > 0$ the operator $\lambda^{-1}A$ is self-adjoint, we infer from Theorems 4.7 and 4.9 of Chapter III that the ranges of the operators $A \pm i\lambda = \lambda(\lambda^{-1}A \pm i)$ coincide with $\mathscr{H}$ and that $(A \pm i\lambda)^{-1}$ exist. Moreover, $(A \pm i\lambda)^{-1} g \in \mathscr{D}_A \subset \mathscr{D}_K$ for all $g \in \mathscr{H}$, and in accordance with the Definition 7.3 of relative boundedness,

$$\| K(A \pm i\lambda)^{-1}g \| \leqslant a \| A(A \pm i\lambda)^{-1}g \| + b \|(A \pm i\lambda)^{-1}g \|, \tag{7.30}$$

where we can choose $a, b \geqslant 0$ with $a < 1$, since the A-bound of K is assumed smaller than 1. On the other hand, as in (4.14) of Chapter III,

$$\|(A \pm i\lambda) f \|^2 = \| Af \|^2 + \lambda^2 \| f \|^2, \qquad f \in \mathscr{D}_A, \tag{7.31}$$

so that by setting above $f = (A \pm i\lambda)^{-1} g$ we get

$$\| A(A \pm i\lambda)^{-1}g \| \leqslant \|(A \pm i\lambda)(A \pm i\lambda)^{-1}g \| = \| g \|, \qquad g \in \mathscr{H},$$

$$\lambda \|(A \pm i\lambda)^{-1}g \| \leqslant \|(A \pm i\lambda)(A \pm i\lambda)^{-1}g \| = \| g \|, \qquad g \in \mathscr{H}.$$

Consequently, chossing in (7.30) at fixed $0 \leqslant a < 1$ and $b \geqslant 0$ a sufficiently large value of $\lambda > 0$, we obtain

$$\| K(A \pm i\lambda)^{-1}g \| \leqslant (a + b/\lambda) \| g \| < \| g \|, \qquad g \in \mathscr{H}.$$

This implies (see Exercise 7.9) that the range of $1 + K(A \pm i\lambda)^{-1}$ equals $\mathscr{H}$. Since we have seen that the ranges of $A \pm i\lambda$ equal $\mathscr{H}$, we conclude that the ranges of

$$A + K \pm i\lambda = [1 + K(A \pm i\lambda)^{-1}](A \pm i\lambda)$$

also equal $\mathscr{H}$. By Theorem 4.9 in Chapter III this implies that $\lambda^{-1}(A + K)$ is self-adjoint, and therefore so is $A + K$.

If D is a core of A then for any $f \in \mathscr{D}_A$ there must be a sequence $f_1, f_2, \ldots \in D$ such that $f_n \to f$ and $Af_n \to Af$ as $n \to \infty$. But then

$$\| K(f - f_n)\| \leqslant a \| A(f - f_n)\| + b \| f - f_n \| \to 0$$

so that the domain of the closure $\overline{(A + K)_D}$ of $A + K$ restricted to D contains the domain of the closure $\overline{A_D} = A$. Since $\mathscr{D}_{A+K} = \mathscr{D}_A$ and $A + K$ is self-adjoint, we conclude that $\overline{(A + K)_D} = A + K$, i.e., D is a core of $A + K$. Q.E.D.

Since by Theorem 7.3 the T_S-bound of V_S is zero, we can immediately apply the above lemma to $T_S + V_S$, and thus state the following result.

Theorem 7.5. (*Kato's theorem*). The linear operator $H_{\mathscr{D}_0}$ acting on any function of the form (7.14) as follows,

$$(7.32)\qquad (H_{\mathscr{D}_0}f)(\mathbf{r}_1,\ldots,\mathbf{r}_n)$$
$$= \left[-\tfrac{1}{2}(\hbar^2)\sum_{k=1}^{n}(m_k)^{-1}\Delta_k + V(\mathbf{r}_1,\ldots,\mathbf{r}_n)\right] f(\mathbf{r}_1,\ldots,\mathbf{r}_n),$$

is essentially self-adjoint on $\mathscr{D}_0$ if the potential satisfies the conditions of Theorem 7.4, and its closure $\overline{H_{\mathscr{D}_0}}$ coincides with the Schroedinger operator $H_S = T_S + V_S$, which is self-adjoint and has the same domain as T_S in Theorem 7.3.

Theorem 7.5 was first established in essentially the above form by T. Kato [1951]. As discussed earlier, there is a great latitude in the choice of core, and a rather popular alternative choice is $\mathscr{C}_b^{\infty}(\mathbb{R}^{3n})$.

Although Theorem 7.5 is applicable to all the specific cases of interactions treated in the next chapter, it does not cover all cases of physical interest. However, extensive studies of essential self-adjointness of other types of Hamiltonians appearing in wave mechanics have been carried out by a host of researchers,* so that all cases of practical interest in nonrelativistic quantum mechanics have by now received suitable treatment. From a purely mathematical point of view, this field of study is a subdiscipline of the theory of elliptic differential operators, and considering the problem in the abstract one cannot expect to have self-adjointness for arbitrary choices of potentials—such as potentials more singular than r^{-2} at the origin.

When applied to Schroedinger operators the general theory of elliptic differential operators (see Hörmander [1969]) leads to other physically important results, such as the following theorem which we state without proof.

Theorem 7.6. Any eigenvector ψ of the Schroedinger operator in Theorem 7.5 can be represented by a function $\psi(\mathbf{r}_1,\ldots,\mathbf{r}_n)$ that has continuous second derivatives in the region where the potential has continuous first derivatives, and it satisfies there the time-independent Schroedinger equation

$$(7.33)\qquad \left\{\sum_{k=1}^{n}\frac{\hbar^2}{2m_k}\Delta_k + [\lambda - V(\mathbf{r}_1,\ldots,\mathbf{r}_n)]\right\}\psi(\mathbf{r}_1,\ldots,\mathbf{r}_n) = 0.$$

This theorem assures us that by completely solving (7.33) (as we have done in §7.7 of Chapter II for a particle moving in a Coulomb potential), we indeed obtain all the eigenvectors and eigenvalues of the Schroedinger operator H_S, i.e., that there are no additional eigenvectors $f \in \mathscr{D}_{H_S}$

* Stummel [1956], Wienholtz [1958], Jörgens [1964], Schminke [1972], and many others; see also Sections X2, X4, and X5 of Reed and Simon [1975] for a review of some of these results.

that we might not have reached through the differential equation (7.33) because they would not have been representable by differentiable functions.

7.8. The Angular Momentum Operators

In classical mechanics angular momentum is next in importance to energy, position, and momentum. This marked significance of angular momentum observables is retained in quantum mechanics.

Following the analogy with classical mechanics, we postulate that the components of the angular momentum of a single spinless particle along the x, y, and z axes, respectively, are represented in quantum mechanics by the following operators in $L^2(\mathbb{R}^3)$:

$$(7.34)\qquad \begin{gathered} L^{(x)} \supseteq Q^{(y)}P^{(z)} - P^{(y)}Q^{(z)}, \qquad L^{(y)} \supseteq Q^{(z)}P^{(x)} - P^{(z)}Q^{(y)}, \\ L^{(z)} \supseteq Q^{(x)}P^{(y)} - P^{(x)}Q^{(y)}. \end{gathered}$$

The expressions on the right-hand side of these relations are obviously symmetric operators. Indeed, consider for example $L^{(z)}$: $Q^{(x)}$ and $P^{(y)}$, as well as $P^{(x)}$ and $Q^{(y)}$, are commuting self-adjoint operators, so that $Q^{(x)}P^{(y)}$ and $P^{(x)}Q^{(y)}$ are self-adjoint by Theorem 2.6. Hence $Q^{(x)}P^{(y)} - P^{(x)}Q^{(y)}$ is certainly symmetric, but in view of Nelson's counter-example,* we cannot claim immediately that it is self-adjoint, despite the fact that

$$(7.35)\qquad (Q^{(x)}P^{(y)}\psi)(\mathbf{r}) - (Q^{(y)}P^{(x)}\psi)(\mathbf{r}) = -i\hbar\left(x\frac{\partial}{\partial y} - y\frac{\partial}{\partial x}\right)\psi(\mathbf{r})$$

is well-defined on certain dense linear sets, such as $\mathscr{C}_b^{\,1}(\mathbb{R}^3)$.

A simple way of bypassing this technical difficulty is through the use of Stone's theorem (Theorem 6.1). We consider therefore rotations $R_\phi^{(z)}$ of the Cartesian inertial frame of reference by an angle ϕ around the z axis. For a fixed spatial point, such a rotation gives rise to a change in its coordinates (x, y, z) with respect to the original frame to coordinates

$$(7.36)\quad x' = x\cos\phi + y\sin\phi, \qquad y' = -x\sin\phi + y\cos\phi, \qquad z' = z,$$

with respect to the new frame. We surmise† that the configuration space wave function $\psi(\mathbf{r}) = \psi(x, y, z)$ will be represented in this new frame by

$$\psi'(\mathbf{r}') = \psi(\mathbf{r}) = \psi(R_\phi^{(z)^{-1}}\mathbf{r}'), \qquad R_\phi^{(z)-1} = R_{-\phi}^{(z)},$$

* See the note following Exercise 1.2 on p. 268. A concise discussion of this counter-example can be found in Section VIII.5 of Reed and Simon [1972].

† Since by Born's correspondence rule (Definition 1.3) only expectation values (1.14) and (1.15) and not state vectors themselves are measurable, a rigorous argument has to be based on group-theoretical approaches to quantum mechanics (see, e.g., Miller [1972], Barut and Rączka [1977]).

where, in general, if R denotes a given rotation of the coordinates system, R^{-1} stands for the inverse rotation, so that $R^{-1}R = RR^{-1}$ is the identity transformation [in fact, note that a rotation is an isometric transformation in the real Euclidean space ($\mathbb{R}^3$) and as such it can be viewed as a unitary operator on ($\mathbb{R}^3$)—cf. Chapter III, Exercise 4.1]. These considerations suggest the introduction in $L^2(\mathbb{R}^3)$ of the linear operators

$$(U_\phi^{(z)}\psi)(\mathbf{r}) = \psi(R_{-\phi}^{(z)}\mathbf{r}) \tag{7.37}$$

that are unitary since the transformation (7.36) is obviously linear and one-to-one, and its Jacobian equals one. Furthermore, it is easy to check that

$$U_{\phi_1}^{(z)}U_{\phi_2}^{(z)} = U_{\phi_1+\phi_2}^{(z)}, \qquad \phi_1, \phi_2 \in \mathbb{R}^1,$$

so that by Stone's theorem there is a unique self-adjoint operator $L^{(z)}$ for which

$$U_\phi^{(z)} = \exp[(i/\hbar)L^{(z)}\phi], \qquad \phi \in \mathbb{R}^1. \tag{7.38}$$

On the other hand, by Theorem 3.1,

$$(i/\hbar)\, L^{(z)}\psi = \operatorname*{s-lim}_{\phi\to 0}[(U_\phi^{(z)}\psi - \psi)/\phi], \qquad \psi \in \mathscr{D}_{L^{(z)}} \subset L^2(\mathbb{R}^3), \tag{7.39}$$

and for $\psi(\mathbf{r}) \in \mathscr{C}_b^{\,1}(\mathbb{R}^3)$, the outcome of taking the above limit coincides with (7.35) multiplied by $i/\hbar$, as can be easily verified by inserting the expressions (7.36) for $R_\phi^{(z)}\mathbf{r}$ into (7.37) and explicitly computing the limit by elementary calculus techniques.

By also considering rotations around the x and y axes we arrive in a similar manner at definitions of $L^{(x)}$ and $L^{(y)}$ as self-adjoint operators that are the extensions of the symmetric operators appearing on the right-hand side of the relations in (7.34). The so-obtained operators, however, do not commute, and by (6.6) (see Axiom $W3$) we easily obtain

$$[L^{(x)}, L^{(y)}] \subseteq i\hbar L^{(z)}, \qquad [L^{(y)}, L^{(z)}] \subseteq i\hbar L^{(x)}, \qquad [L^{(z)}, L^{(x)}] \subseteq i\hbar L^{(y)}. \tag{7.40}$$

Thus in defining the square of the angular momentum operator by analogy with classical mechanics conventions,

$$\mathbf{L}^2 = (L^{(x)})^2 + (L^{(y)})^2 + (L^{(z)})^2, \tag{7.41}$$

we gain run into the technical problem of establishing self-adjointness—as opposed to mere symmetry properties.

General (Casimir operator) techniques for dealing with problems of this nature are made available by the group theoretical approach to symmetry problems in quantum mechanics. These techniques are, however, beyond the scope of this book, but fortunately we can handle this particular case by elementary techniques based on the results of

§7.6 in Chapter II. Indeed, according to those results, $L^2(\Omega_p, \mu_p) = L^2(\Omega_r, \mu_r) \otimes L^2(\Omega_s, \mu_s)$, and (Chapter II, Theorem 7.1) the spherical harmonics Y_l^m constitute an orthonormal basis in $L^2(\Omega_s, \mu_s)$. Let us therefore set by definition

$$(7.42)\quad (\mathbf{L}^2\psi)(r, \theta, \phi) = \hbar^2 \sum_{l=0}^{\infty} l(l+1) \sum_{m=-l}^{+l} Y_l^m(\theta, \phi) \int_{\Omega_s} \psi^*(r, \theta', \phi')\, Y_l^m(\theta', \phi')\, d\mu_s(\theta', \phi')$$

whenever the infinite series on the right-hand side of (7.42) converges in the mean to an element of $L^2(\Omega_p, \mu_p)$. The resulting operator is self-adjoint (see Chapter III, Theorem 5.2) and by Theorem 7.7 in Chapter II its action on $\psi(r, \theta, \phi) \in \mathscr{C}_b^{\,2}(\Omega_r) \otimes \mathscr{C}_p^{\,2}(\Omega_s)$ coincides with that of the differential operator form obtained by inserting (7.34) into (7.41).

One of the useful by-products of these considerations is the joint spectral measure $E^{|\mathbf{Q}|, |\mathbf{L}|, L^{(z)}}$ of $|\mathbf{Q}|$, $|\mathbf{L}|$, and $L^{(z)}$, which for $B = B_1 \times B_2 \times B_3 \in \mathscr{B}^3$ is defined by

$$(7.43)\quad (E^{|\mathbf{Q}|, |\mathbf{L}|, L^{(z)}}(B_1 \times B_2 \times B_3)\psi)(r, \theta, \phi) = \chi_{B_1}(r) \sum_{\hbar l \in B_2} \sum_{\hbar m \in B_3} Y_l^m(\theta, \phi) \times \int_{\Omega_s} \psi^*(r, \theta', \phi')\, Y_l^m(\theta', \phi')\, d\mu_s(\theta', \phi')$$

where, naturally, $l \in \{0, 1, \ldots\}$, $m \in \{-l, \ldots, +l\}$. Clearly, the existence of this measure extablishes that $\{|\mathbf{Q}|, |\mathbf{L}|, L_z\}$ constitutes a complete set of observables in $L^2(\Omega_p, \mu_p)$. We note that, strictly speaking, this measure corresponds in fact to $\{|\mathbf{Q}|, L', L_z\}$, where

$$(7.44)\quad L' = \tfrac{1}{2}[-1 + (1 + 4\mathbf{L}^2)^{1/2}],$$

but in accordance with a tradition dating back to old quantum theory (see Messiah [1962]) the variable $\hbar l$ is thought of as being related directly to $|\mathbf{L}| = (\mathbf{L}^2)^{\frac{1}{2}}$, albeit the eigenvalues of $\mathbf{L}^2$ actually equal $\hbar^2 l(l+1)$.

An even more frequently used complete set of observables is $\{|\mathbf{P}|, |\mathbf{L}|, \mathbf{L}_z\}$. If $\tilde{\psi}(k, \theta, \phi)$ equals the Fourier–Plancherel transform of $\psi(\mathbf{r})$ expressed in spherical coordinates, then

$$(7.45)\quad (E^{|\mathbf{P}|, |\mathbf{L}|, L^{(z)}}(B_1 \times B_2 \times B_3)\psi)^{\sim}(k, \theta, \phi) = \chi_{B_1}(k) \sum_{\hbar l \in B_2} \sum_{\hbar m \in B_3} Y_l^m(\theta, \phi) \times \int_{\Omega_s} \tilde{\psi}^*(k, \theta', \phi')\, Y_l^m(\theta', \phi') \sin\theta'\, d\theta'\, d\phi'.$$

In the presence of spin the operators (7.34) retain the meaning of orbital angular momentum components, but spin components have to be included to arrive at total angular momentum operators. Naturally, as seen in the discussion in Section 7.1, in that case $S^{(z)}$ has to be added to the sets consisting of $|\mathbf{Q}|$ (or $|\mathbf{P}|$), $|\mathbf{L}|$, and $L^{(z)}$ to arrive at complete sets of observables.

**7.9. Time-Dependent Hamiltonians

In §3 we had formulated the Schroedinger picture for the general case of time-dependent Hamiltonians $H(t)$. Such time-dependence would be encountered only if the system under consideration is not isolated from its environment, so that energy is being exchanged between the system and some external field sources. Naturally, as mentioned in the footnote on p. 292, by including those sources as part of the system we arrive at a new, enlarged system whose Hamiltonian is not time-dependent, but the resulting mathematical model might be either too difficult to handle exactly, or an appropiate mathematical description might not even be available within the nonrelativistic context studied in this volume. A typical instance of the second kind is provided by a charged particle (electron, proton, ion, etc.) moving in a laser cavity and therefore continuously bombarded by photons. The exact mathematical description of the swarm of photons filling the cavity requires the quantization of the electromagnetic field within the cavity and therefore comes under the heading of (relativistic) quantum field theory. However, if the particle moves at nonrelativistic speeds in relation to the cavity, the time-dependent (nonrelativistic) Hamiltonian

$$H(t) \supseteq (2m)^{-1} [-i\hbar \nabla - (e/c) \mathbf{A}(\mathbf{r}, t)]^2 + eA_0(\mathbf{r}, t) \tag{7.46}$$

for a particle of charge e in an external four-potential $(A_0, \mathbf{A})$ related to the electromagnetic field in the cavity might represent a reasonable approximation. Rellich's theorem (Lemma 7.2) provides a suitable tool for establishing the self-adjointness of (7.46) under reasonable conditions on the functions A_0, $A^{(x)}$, $A^{(y)}$, and $A^{(z)}$. However, we are still faced with the problem of defining the time-evolution operator (see p. 292)

$$U(t, t_0) = T \exp \left[-(i/\hbar) \int_{t_0}^{t} H(\tau)\, d\tau\right] \tag{7.47}$$

by giving a suitable meaning to the above integration symbol, which in physical literature bears the name of *time-ordered integral*. Indeed, since $H(t_1)$ does not in general commute with $H(t_2)$ for $t_1 \neq t_2$, we have to exercise caution in preserving the appropriate time-ordering as we (heuristically) combine "infinitesimal" contributions in the following manner,

$$U(t_1 + dt, t_0) = \exp[-(i/\hbar)\, H(t_1)\, dt]\, U(t_1, t_0), \tag{7.48}$$

to arrive at a sensible definition of $U(t, t_0)$ which satisfies (3.16)–(3.18).

The clue for a general definition of (7.47) is provided by (3.17) which, if valid, by mathematical induction leads to the conclusion that

$$U(t, t_0) = U(t, t_n)\, U(t_n, t_{n-1}) \cdots U(t_2, t_1)\, U(t_1, t_0) \tag{7.49}$$

regardless of how small

$$\delta = \max\{|\, t - t_n \,|, |\, t_j - t_{j-1} \,| : j = 1,..., n\}$$

is chosen to be. Hence we set, by definition,

$$U(t, t_0) = \operatorname*{s-lim}_{\substack{n\to+\infty \\ \delta\to+0}} \prod_{j=n}^{1} \exp[-(i/\hbar)\, H(t_{j-1})(t_j - t_{j-1})] \tag{7.50}$$

where the order $j = n, n - 1,..., 1$ in the above operator product is very essential. Under suitable conditions on $H(t)$ [such as the existence of a common domain of definition for all $H(t)$], it can be shown not only that the strong limit (7.50) does exist, but also that Axiom S.3 on p. 292 is indeed satisfied.

In Chapter V, we apply the framework of this chapter to quantum scattering problems and we shall concentrate on the case of time-independent Hamiltonians, for which it is easily seen that (7.50) indeed equals (3.2). Hence we direct the reader interested in details of the general theory of time-ordered integrals (7.47) to Chapter XIV, Section 4, of Yosida [1974], and for details on the general scattering theory for time-dependent Hamiltonians extrapolating that of Chapter V to the articles by Prugovečki and Tip [1974].

Exercises

7.1. Let $\mathscr{H}_1{}',..., \mathscr{H}_m{}'$ and $\mathscr{H}_1{}'',..., \mathscr{H}_n{}''$ be Hilbert spaces. Prove that the mapping of

$$\mathscr{H}_1 = \Big(\bigoplus_{i=1}^{m} \mathscr{H}_i'\Big) \otimes \Big(\bigoplus_{j=1}^{n} \mathscr{H}_j''\Big)$$

into

$$\mathscr{H}_2 = \bigoplus_{i=1}^{m} \bigoplus_{j=1}^{n} (\mathscr{H}_i' \otimes \mathscr{H}_j''),$$

which takes $(\bigoplus_{i=1}^{m} f_i') \otimes (\bigoplus_{j=1}^{n} f_j'')$ into $\bigoplus_{i=1}^{m} \bigoplus_{j=1}^{n} (f_i' \otimes f_j'')$, is a unitary transformation.

7.2. Generalize and prove the statement contained in Exercise 7.1 to the cases when (a) $m = +\infty$, $n < +\infty$; (b) $m < +\infty$, $n = +\infty$; (c) $m = n = +\infty$.

7.3. Let U_k be a unitary transformation of the Hilbert space $\mathscr{H}_k'$ onto the Hilbert space $\mathscr{H}_k''$ for $k = 1, 2, \ldots$. Show that the linear transformation $U = U_1 \oplus U_2 \oplus \cdots$ of $\mathscr{H}_1 = \bigoplus_{k=1}^{\infty} \mathscr{H}_k'$ into $\mathscr{H}_2 = \bigoplus_{k=1}^{\infty} \mathscr{H}_k''$ is unitary, where U is defined by

$$U(f_1 \oplus f_2 \oplus \cdots) = U_1 f_1 \oplus U_2 f_2 \oplus \cdots, \qquad f_1 \in \mathscr{H}_1, f_2 \in \mathscr{H}_2, \ldots .$$

7.4. Prove that the unitary transformation U of the Hilbert space (7.5) onto the Hilbert space (7.9), defined in the text, maps the subspace of all symmetric (antisymmetric) functions from $\mathscr{H}_1$ in (7.5) *onto* the subspace of all symmetric (antisymmetric) functions from $\mathscr{H}$ in (7.9); more generally, it maps any subspace of "appropriately" symmetrized functions in (7.5) onto the subspace of functions in (7.9) "appropriately" symmetrized in the same way (see Definition 5.1).

7.5. Derive from Axioms W1–W3 the result that $\{P^{(x)}, P^{(y)}, P^{(z)}, S^{\mathbf{n}}\}$ is a complete set of observables.

7.6. Show that if A^{**} exists, then $A^{***} = A^*$.

7.7. Show that the Fourier transform of a function

$$P(\mathbf{r}_1, \ldots, \mathbf{r}_n) \exp\left[-\frac{1}{2\hbar}(\mathbf{r}_1^2 + \cdots + \mathbf{r}_n^2)\right]$$

is a function of the form $P_1(\mathbf{p}_1, \ldots, \mathbf{p}_n) \exp[-1/2\hbar(\mathbf{p}_1^2 + \cdots + \mathbf{p}_n^2]$.

7.8. Show that if $V(\mathbf{r})$ is locally square integrable and bounded at infinity, then

$$V(\mathbf{r}_1 - \mathbf{r}_2) f(\mathbf{r}_1, \mathbf{r}_2), \qquad f(\mathbf{r}_1, \mathbf{r}_2) = \mathbf{P}(\mathbf{r}_1, \mathbf{r}_2) \exp\left[-\frac{1}{2\hbar}(\mathbf{r}_1^2 + \mathbf{r}_2^2)\right]$$

is square integrable on $\mathbb{R}^6$.

7.9. Prove that the range of $1 + A$, where $\| A \| < 1$, is the entire Hilbert space $\mathscr{H}$, and that $(1 + A)^{-1} = \text{u-lim}_{n\to\infty} \sum_{k=0}^{n} (-1)^k A^k$.

*8. Completely Continuous Operators and Statistical Operators

8.1. Completely Continuous Operators

A compact set in a complete metric space—in particular, in a Hilbert space—is a set having the property that any infinite sequence of elements from that set has a convergent subsequence. Linear operators on Banach or Hilbert spaces which map bounded sets into compact sets are called completely continuous or compact. The following equivalent definition of completely continuous operators emphasizes their main feature which is of importance to us.

Definition 8.1. The linear operator A on the Hilbert space $\mathscr{H}$ is said to be *completely continuous* (or *compact*) if for every infinite sequence $f_1, f_2, \ldots \in \mathscr{H}$, such that $\|f_n\| \leqslant C$ for all $n = 1, 2, \ldots$, the sequence $Af_1, Af_2, \ldots$ contains a strongly convergent subsequence.

We shall see later that the density operator ρ describing states in statistical quantum mechanics, as well as the product ρA_0, where A_0 is any bounded observable, are completely continuous operators. The theory of completely continuous operators has many applications, some of which we shall encounter in Chapter V. For this reason, the foundations of this theory deserve special attention.

Theorem 8.1. Completely continuous operators have the following properties:

(a) They are bounded.

(b) The products AB and BA of a completely continuous operator A and any bounded operator B on $\mathscr{H}$ are completely continuous operators.

(c) The linear combination $a_1A_1 + a_2A_2$ of two completely continuous operators is a completely continuous operator.

Proof. (a) Suppose that A is not bounded, i.e., that there is no constant C such that $\|Af\| \leqslant C\|f\|$ for all $f \in \mathscr{H}$. In that case we can find a sequence $f_1, f_2, \ldots$ of normalized vectors, $\|f_n\| = 1$, such that $\|Af_1\|, Af_2\|, \ldots$ diverges to infinity. But then $Af_1, Af_2, \ldots$ contains no convergent subsequence, which is impossible if A is completely continuous.

(b) If $f_1, f_2, \ldots$ is any bounded sequence, then $Af_1, Af_2, \ldots$ contains a convergent subsequence $Af_{j_1}, Af_{j_2}, \ldots$. Since B is bounded and therefore continuous (see Chapter III, Theorem 1.3), $BAf_{j_1}, BAf_{j_2}, \ldots$ converges, i.e., BA is completely continuous. Further, $Bf_1, Bf_2, \ldots$ is bounded. Therefore, there must be a subsequence $Bf_{k_1}, Bf_{k_2}, \ldots$ such that

$A(Bf_{k_1}), A(Bf_{k_2}), \ldots$ is convergent. Consequently, AB is also completely continuous.

(c) If $g_1, g_2, \ldots$ is any bounded sequence, there is a subsequence $g_{i_1}, g_{i_2}, \ldots$ such that $A_1 g_{i_1}, \ldots$ is convergent. Moreover, since $g_{i_1}, g_{i_2}, \ldots$ is also bounded, it must contain a subsequence $g_{k_1}, g_{k_2}, \ldots$ such that $A_2 g_{k_1}, A_2 g_{k_2}, \ldots$ is convergent. Consequently, the sequence $a_1 A_1 g_{k_1} + a_2 A_2 g_{k_1}, \ldots$ is convergent, which shows that $a_1 A_1 + a_2 A_2$ is completely continuous. Q.E.D.

Theorem 8.2. If the linear operator A on $\mathscr{H}$ is the uniform limit of a sequence $A_1, A_2, \ldots$ of completely continuous operators, then A is completely continuous.

Proof. Let $f_1, f_2, \ldots$ be any bounded sequence, $\| f_n \| \leqslant C$, of vectors from $\mathscr{H}$. Construct a family of subsequences

$$\{f_n^{(0)} = f_n\} \supset \{f_n^{(1)}\} \supset \cdots \supset \{f_n^{(i)}\} \supset \cdots$$

of $f_1, f_2, \ldots$ by the recursive procedure in which, for every $i = 0, 1, 2, \ldots$, $\{f_n^{(i+1)}\}$ is a subsequence of $\{f_n^{(i)}\}$ such that $\{A_{i+1} f_n^{(i+1)}\}$ is convergent. Since $\{f_n^{(i)}\}$ is a subsequence of $\{f_n\}$ and therefore bounded, $\{f_n^{(i+1)}\}$ exists, due to the complete continuity of A_{i+1}.

From the inequality

$$\begin{aligned}
&\| Af_i^{(i)} - Af_k^{(k)} \| \\
&\quad \leqslant \| Af_i^{(i)} - A_n f_i^{(i)} \| + \| A_n f_i^{(i)} - A_n f_k^{(k)} \| + \| A_n f_k^{(k)} - Af_k^{(k)} \| \\
&\quad \leqslant \| A - A_n \| (\| f_i^{(i)} \| + \| f_k^{(k)} \|) + \| A_n (f_i^{(i)} - f_k^{(k)}) \| \\
&\quad \leqslant 2C \| A - A_n \| + \| A_n (f_i^{(i)} - f_k^{(k)}) \|,
\end{aligned}$$

we infer that $\{Af_i^{(i)}\}$ is convergent; namely, for any given $\epsilon > 0$ we have $2C \| A - A_n \| < \epsilon/2$ for sufficiently large n, while

$$\| A_n (f_i^{(i)} - f_k^{(k)}) \| < \epsilon/2$$

if sufficiently large $i, k \geqslant n$ are chosen. Q.E.D.

There are many continuous linear operators on an infinite-dimensional Hilbert space $\mathscr{H}$ which are not completely continuous. A conspicuous example is the identity operator on $\mathscr{H}$ (see Exercise 8.1).

As a generalization of the operator $| f \rangle \langle f |$, let us define for any two vectors $f, g \in \mathscr{H}$ the operator $| f \rangle \langle g |$ in the following way:

$$| f \rangle \langle g | f' = \langle g | f' \rangle f, \qquad f' \in \mathscr{H}. \tag{8.1}$$

Lemma 8.1. Every operator of the form $A = |f\rangle\langle g|$ is completely continuous.

Proof. In fact, if $f_1, f_2, \ldots$ is a bounded sequence, $\langle g \mid f_1\rangle, \langle g \mid f_2\rangle, \ldots$ is obviously a bounded sequence of numbers. Hence, the set $\{\langle g \mid f_n\rangle\}$ has at least one accumulation point α. Thus, we can choose a subsequence $\{\langle g \mid f_{k_n}\rangle\}$ converging to α. Consequently

$$\text{s-}\lim_{n\to\infty} Af_{k_n} = f \lim_{n\to\infty} \langle g \mid f_{k_n}\rangle = \alpha f$$

exists. Hence, $A = |f\rangle\langle g|$ is completely continuous. Q.E.D.

It follows from the above result and Theorem 8.1(c) that any *operator of finite rank*, i.e., an operator that can be reduced to the form

$$\sum_{k=1}^{n} \lambda_k \,|\, e_k'\rangle \langle e_k \,| = \sum_{k=1}^{n} |\, e_k'\rangle \,\lambda_k \,\langle e_k \,| \,, \tag{8.2}$$

is a completely continuous operator. Moreover, by Theorem 8.2, any operator which is the uniform limit of operators of the form (8.2) is completely continuous. We shall prove that, conversely, any completely continuous operator is such a uniform limit. To prove this result we need two additional results which will be derived as part of Theorem 8.3 and in Lemma 8.2.

It should be evident by now that all linear operators on a finite-dimensional Hilbert space $\mathscr{H}$ are completely continuous. In fact, if A is a linear operator on $\mathscr{H}$ and $\{e_1, \ldots, e_n\}$ is an orthornormal basis in $\mathscr{H}$, then it is easily seen that

$$A = \sum_{k=1}^{n} |\, h_k\rangle \langle e_k \,|,$$

where $h_k = Ae_k$, $k = 1, \ldots, n$. In view of Lemma 8.1, we have established that A is completely continuous.

Theorem 8.3. Every symmetric and completely continuous operator A has a pure point spectrum $S^A = S_p^A$, which has no accumulation points, with the possible exception of zero. The characteristic subspace of every nonzero eigenvalue is finite dimensional.

Proof. Assume that $\alpha \neq 0$ is an accumulation point of the point spectrum of A. This would imply that there is an infinite sequence $\lambda_1, \lambda_2, \ldots$ of distinct eigenvalues of A such that $|\lambda_1|, |\lambda_2|, \ldots > \frac{1}{2}|\alpha|$. If $f_1, f_2, \ldots$ are the corresponding unit eigenvectors

$$Af_n = \lambda_n f_n \,, \qquad n = 1, 2, \ldots,$$

the sequence $Af_1, Af_2, \ldots$ has no convergent subsequence since for any $m \neq n$ we have $\lambda_m \neq \lambda_n$, which implies that $\langle f_m \mid f_n \rangle = 0$ and

$$\| Af_m - Af_n \|^2 = | \lambda_m |^2 + | \lambda_n |^2 > \alpha^2/2.$$

Thus, A would not be completely continuous. Hence, S_p^A has no nonzero accumulation points.

Assume now that the continuous spectrum S_c^A of A is not empty, so that S_c^A contains infinitely many points (see Exercise 8.2). Choose an infinite sequence of nonzero $\lambda_1', \lambda_2', \ldots \in \mathsf{S}_\mathrm{c}^A$ such that $| \lambda_1' | < | \lambda_2' | < \cdots$, and positive numbers $r_1, r_2, \ldots$ such that the intervals I_n, $n = 1, 2, \ldots$ between the points $\lambda_n' - r_n$ and $\lambda_n' + r_n$ are disjoint and do not contain the origin. Since $\lambda_1', \lambda_2', \ldots$ are points of the spectrum of A, the projectors $E^A(I_n)$ are nonzero. Therefore, there are vectors $g_1, g_2, \ldots$ satisfying

$$E^A(I_n)\, g_n = g_n, \qquad \| g_n \| = 1, \qquad n = 1, 2, \ldots .$$

Due to the disjointness of the intervals I_n, $E^A(I_m)\, E^A(I_n) = 0$ when $m \neq n$. Using the spectral theorem, it is easy to compute that

$$\begin{aligned} \langle Ag_m \mid Ag_n \rangle &= \langle g_m \mid A^2 g_n \rangle \\ &= \int_{I_n} \lambda^2 d \langle g_m \mid E_\lambda^A g_n \rangle \\ &\begin{cases} = 0 & \text{for} \quad m \neq n \\ > (| \lambda_n' | - r_n)^2 & \text{for} \quad m = n. \end{cases} \end{aligned}$$

Thus, we have for $m \neq n$

$$\| Ag_m - Ag_n \|^2 = \| Ag_m \|^2 + \| Ag_n \|^2 > 2(| \lambda_1' | - r_1)^2,$$

i.e., $Ag_1, Ag_2, \ldots$ contains no convergent subsequence. Hence, if A is completely continuous, its continuous spectrum must be empty.

If M is a characteristic space of A corresponding to the eigenvalue λ, then $Af = \lambda f$ for all $f \in \mathsf{M}$. Consequently, if $\lambda \neq 0$, A is a multiple of the identity operator when restricted to M. Therefore, M has to be finite dimensional if A is completely continuous (see Exercise 8.1). Q.E.D.

Lemma 8.2. Every completely continuous operator A on $\mathscr{H}$ can be written in the form

$$A = VK, \qquad K = \sqrt{A^*A},$$

where K is a positive-definite completely continuous operator, and V is an isometric operator defined on the range of K.

Proof. Since A is completely continuous and A^* is continuous (see Chapter III, Theorem 2.7), A^*A is completely continuous. Hence, A^*A, and consequently also $K = \sqrt{A^*A}$, have a pure point spectrum, which has no accumulation points, with the possible exception of zero (see Theorem 8.3). This implies that K is a completely continuous operator (see Exercise 8.3).

Let us define now the operator V on the range $\mathscr{R}_K$ of K by setting for every $g \in \mathscr{R}_K$

$$Vg = Af, \qquad g = Kf \in \mathscr{R}_K .$$

The operator V is uniquely defined for every $g \in \mathscr{R}_K$. In fact, if $g = Kf = Kf_1$ for $f \neq f_1$, we have

$$\| Af - Af_1 \|^2 = \langle f - f_1 \mid A^*A(f - f_1)\rangle = \langle f - f_1 \mid K^2(f - f_1)\rangle = 0,$$

which implies that $Af = Af_1$.

It is easy to check that V is linear. From the relations

$$\begin{aligned} \| Vg \|^2 &= \langle Af \mid Af \rangle = \langle f \mid A^*Af \rangle = \langle f \mid K^2 f \rangle \\ &= \langle Kf \mid Kf \rangle = \| g \|^2, \qquad g \in \mathscr{R}_K , \end{aligned}$$

we infer that V is isometric. Q.E.D.

We are now ready to investigate the relation of completely continuous operators in general to operators of the form $| f \rangle \langle g |$.

Theorem 8.4. An operator A on any infinitely dimensional Hilbert space $\mathscr{H}$ is completely continuous if and only if there are orthonormal systems $\{e_1, e_2, \ldots\}$ and $\{e_1', e_2', \ldots\}$ in $\mathscr{H}$ and positive numbers $\lambda_1, \lambda_2, \ldots \to 0$, such that A can be reduced to the *canonical form*

$$A = \sum_k | e_k' \rangle \lambda_k \langle e_k | , \tag{8.3}$$

where, in case that the sum is infinite, we define

$$\sum_{k=1}^{\infty} | e_k' \rangle \lambda_k \langle e_k | = \underset{n \to +\infty}{\text{u-lim}} \sum_{k=1}^{n} | e_k' \rangle \lambda_k \langle e_k | , \tag{8.4}$$

and the set $\{\lambda_k\}$ consists of all the eigenvalues of $\sqrt{A^*A}$.

Proof. According to Lemma 8.1 and Theorem 8.1(c), an operator of the form

$$\sum_{k=1}^{n} | e_k' \rangle \lambda_k \langle e_k | \tag{8.5}$$

is completely continuous. Hence, by Theorem 8.2, any operator which is the uniform limit of operators (8.5) is also completely continuous.

Conversely, assume that A is a completely continuous operator. According to Lemma 8.2, $A = VK$, $K = \sqrt{A^*A}$, where K is positive definite and completely continuous. Hence, in view of Theorem 8.3 and Theorem 6.4 in Chapter III the spectral resolution of K is

$$K = \int_{\mathbb{R}^1} \lambda \, dE_\lambda^K = \sum_j \lambda_j' E_{\mathsf{M}_j}, \tag{8.6}$$

where $\lambda_1', \lambda_2',...$ are the eigenvalues of K; in addition, if K has infinitely many eigenvalues, the convergence of the series in (8.6) is uniform. By selecting an orthonormal basis in each of the characteristic subspaces M_j (which are finite dimensional by Theorem 8.3), we arrive at an orthonormal system $\{e_1, e_2, ...\}$ in $\mathscr{H}$.

Let us set $\lambda_k = \lambda_j'$ if $e_k \in \mathsf{M}_j$. Assume that we are dealing with the nontrivial case when K has infinitely many nonzero eigenvalues

$$K = \operatorname*{u-lim}_{n\to\infty} \sum_{k=1}^{n} \lambda_k \mid e_k\rangle \langle e_k \mid = \sum_{k=1}^{\infty} \mid e_k\rangle \lambda_k \langle e_k \mid .$$

Since V is isometric,

$$A = VK = \operatorname*{u-lim}_{n\to\infty} \sum_{k=1}^{n} \lambda_k V \mid e_k\rangle \langle e_k \mid . \tag{8.7}$$

Setting $e_k' = Ve_k$, we can easily check that

$$(V \mid e_k\rangle \langle e_k \mid) f = V(\langle e_k \mid f\rangle e_k) = \langle e_k \mid f\rangle e_k',$$

i.e., $V \mid e_k\rangle\langle e_k \mid = \mid e_k'\rangle\langle e_k \mid$. Inserting this result in (8.7), we arrive at (8.3). Moreover, the isometry of V implies that $\{e_1', e_2', ...\}$ is an orthonormal system since $\{e_1, e_2, ...\}$ is orthonormal. Q.E.D.

8.2. The Trace of a Linear Operator

In order to be able to study systematically different classes of completely continuous operators, we need the concept of trace.

Definition 8.2. A linear operator A defined on the separable Hilbert space $\mathscr{H}$ is said to be of the *trace class* if the series $\sum_k \langle e_k \mid Ae_k\rangle$ converges and has the same value in any orthonormal basis $\{e_k\}$ of $\mathscr{H}$. The sum

$$\operatorname{Tr} A = \sum_k \langle e_k \mid Ae_k\rangle \tag{8.8}$$

is called the *trace* of A.

We shall investigate the existence of the above trace in the context of future theorems. But first we would like to remark that every linear operator A on an n-dimensional Hilbert space $\mathscr{H}$ has a trace; namely, if $\{e_1, ..., e_n\}$ and $\{e_1', ..., e_n'\}$ are any two orthonormal basis, the sum in (8.8) is finite and the same in both bases:

$$\sum_{k=1}^{n} \langle e_k \mid Ae_k \rangle = \sum_{i,j,k=1}^{n} \langle e_k \mid e_i' \rangle \langle e_i' \mid Ae_j' \rangle \langle e_j' \mid e_k \rangle$$

$$= \sum_{i,j=1}^{n} \langle e_i' \mid Ae_j' \rangle \left(\sum_{k=1}^{n} \langle e_j' \mid e_k \rangle \langle e_k \mid e_i' \rangle \right)$$

$$= \sum_{i=1}^{n} \langle e_i' \mid Ae_i' \rangle.$$

We note that in the infinite-dimensional case we cannot immediately establish, using the same method, that the sum in (8.5) does not depend on the basis employed. The reason is that in dealing with multiple infinite sums the interchange of the order of summations is legitimate only if we know that the convergence of the multiple sum is unconditional. We shall see that this is the case when A is *positive definite*, i.e., when $\langle f \mid Af \rangle \geqslant 0$ for all $f \in \mathscr{H}$.

Lemma 8.3. If $A = K^*K$, where K is a linear operator defined on the infinite-dimensional separable Hilbert space $\mathscr{H}$, then the series

$$\sum_{k=1}^{\infty} \langle e_k \mid Ae_k \rangle = \sum_{k=1}^{\infty} \| Ke_k \|^2 \tag{8.9}$$

has the same sum in all orthonormal bases $\{e_1, e_2, ...\}$ on $\mathscr{H}$; thus, if that sum is finite, Tr A exists. Moreover, A is bounded and $\| A \| \leqslant \text{Tr } A$.

Proof. Let $\{e_1, e_2, ...\}$ and $\{e_1', e_2', ...\}$ be any two orthonormal bases in $\mathscr{H}$. The double series

$$\sum_{i,k=1}^{\infty} |\langle e_i' \mid Ke_k \rangle|^2 \tag{8.10}$$

is a series with positive terms; it converges if and only if the three sums in (8.11) are equal (see, e.g., Randolph [1968, p. 162, Theorem 3]),

$$\text{(8.11)} \qquad \sum_{i=1}^{\infty}\Big(\sum_{k=1}^{\infty} |\langle e_i' \mid Ke_k\rangle|^2\Big) = \sum_{k=1}^{\infty}\Big(\sum_{i=1}^{\infty} |\langle e_i' \mid Ke_k\rangle|^2\Big)$$

$$= \sum_{i,k=1}^{\infty} |\langle e_i' \mid Ke_k\rangle|^2.$$

According to Parseval's relation [(4.15) of Chapter I],

$$\sum_{k=1}^{\infty} |\langle e_i' \mid Ke_k\rangle|^2 = \sum_{k=1}^{\infty} \langle K^*e_i' \mid e_k\rangle\langle e_k \mid K^*e_i'\rangle = \| K^*e_i' \|^2,$$

$$\sum_{i=1}^{\infty} |\langle e_i' \mid Ke_k\rangle|^2 = \sum_{i=1}^{\infty} \langle Ke_k \mid e_i'\rangle\langle e_i' \mid Ke_k\rangle = \| Ke_k \|^2.$$

Hence, from (8.11), (8.9) converges if and only if (8.10) converges, and

$$\sum_{k=1}^{\infty} \| Ke_k \|^2 = \sum_{i=1}^{\infty} \| K^*e_i' \|^2 = \sum_{i,j=1}^{\infty} |\langle e_j' \mid K^*e_i'\rangle|^2$$

$$= \sum_{i,j=1}^{\infty} |\langle e_i' \mid Ke_j'\rangle|^2 = \sum_{j=1}^{\infty} \| Ke_j' \|^2,$$

thus proving that Tr A exists.

By using the Schwarz–Cauchy inequality, which yields $|\langle Ke_i \mid Ke_k\rangle| \leqslant \| Ke_i \| \, \| Ke_k \|$, we get

$$(\mathrm{Tr}\, A)^2 = \Big(\sum_{k=1}^{\infty} \| Ke_k \|^2\Big)^2 = \sum_{i,k=1}^{\infty} \| Ke_i \|^2 \, \| Ke_k \|^2$$

$$\geqslant \sum_{i,k=1}^{\infty} |\langle Ke_i \mid Ke_k\rangle|^2 = \sum_{i,k=1}^{\infty} |\langle e_i \mid Ae_k\rangle|^2$$

$$= \sum_{k=1}^{\infty}\Big(\sum_{i=1}^{\infty} \langle Ae_k \mid e_i\rangle\langle e_i \mid Ae_k\rangle\Big) = \sum_{k=1}^{\infty} \| Ae_k \|^2,$$

since the interchanges of the order of summation are permissible on account of the unconditional convergence of a series with positive terms. From the above we obtain

$$\text{(8.12)} \qquad \| Ae_1 \|^2 \leqslant (\mathrm{Tr}\, A)^2.$$

Given any nonzero vector $f \in \mathscr{H}$, we can adopt $f/\| f \|$ to be the first element e_1 of an orthonormal series, and, therefore, we have by (8.12)

$$\| Af \| \leqslant (\text{Tr } A)\| f \|.$$

The above inequality shows that A is bounded and $\| A \| \leqslant \text{Tr } A$.

Q.E.D.

8.3. Hilbert–Schmidt Operators

In Chapter V we shall encounter integral operators which belong to a special class of completely continuous operators called the Hilbert–Schmidt class.

Definition 8.3. A completely continuous operator, having the canonical decomposition (8.3), is called a *Hilbert–Schmidt operator* if

$$\sum_k \lambda_k^2 < +\infty \tag{8.13}$$

Theorem 8.5. A bounded linear operator A defined on the (separable*) Hilbert space $\mathscr{H}$ is a Hilbert–Schmidt operator if and only if A^*A is of trace class.

Proof. If A is a Hilbert–Schmidt operator, then by (8.3)

$$\sum_k \langle e_k \mid A^*Ae_k \rangle = \sum_k \| Ae_k \|^2 = \sum_k \lambda_k^2. \tag{8.14}$$

In the case where the orthonormal system $\{e_k\}$ is not already a basis, we can complete it to an orthonormal basis in $\mathscr{H}$ by adding to it an orthonormal basis $\{e_i''\}$ in the orthogonal complement of the closed linear subspace $[e_1, e_2, \ldots]$ spanned by $\{e_k\}$. Since $Ae_i'' = \mathbf{0}$, and therefore also $\langle e_i'' \mid A^*Ae_i'' \rangle = 0$, we have by virtue of (8.14) and Lemma 8.3 that the trace of A^*A exists and

$$\text{Tr}[A^*A] = \sum_k \lambda_k^2. \tag{8.15}$$

Consider now any bounded linear operator K for which $\text{Tr}(K^*K)$ exists. Since any normalized vector $f \in \mathscr{H}$ can be chosen as the first element of an orthonormal basis in $\mathscr{H}$, we have

$$\| Kf \|^2 = \langle f \mid K^*Kf \rangle \leqslant \text{Tr}[K^*K].$$

* We give the proof only for the separable case, but the result is true in general (see Schatten [1960]).

Thus

$$\| K \|^2 = \sup_{\|f\|=1} \| Kf \|^2 \leqslant \mathrm{Tr}[K^*K]. \tag{8.16}$$

Assume that A is a bounded linear operator with $\mathrm{Tr}[A^*A] < +\infty$, and let us prove that A is a Hilbert–Schmidt operator. Since the proof is trivial when $\mathscr{H}$ is finite dimensional, we shall consider only the case of operators A on infinite-dimensional separable Hilbert spaces.

Choose an orthonormal basis $\{e_1, e_2, \ldots\}$ in $\mathscr{H}$, and define the linear operators

$$A_n = \sum_{k=1}^{n} | Ae_k\rangle\langle e_k |, \qquad n = 1, 2, \ldots .$$

Using (8.16) and noting that $A_n e_j = Ae_j$ if $j \leqslant n$, we get

$$\begin{aligned} \| A - A_n \|^2 &\leqslant \mathrm{Tr}[(A - A_n)^* (A - A_n)] \\ &= \sum_{k=1}^{\infty} \|(A - A_n) e_k \|^2 = \sum_{k=n+1}^{\infty} \| Ae_k \|^2. \end{aligned}$$

Thus, $\| A - A_n \| \to 0$ when $n \to \infty$. Since $A_1, A_2, \ldots$ are completely continuous operators, it follows from Theorem 8.2 that A is also completely continuous. It is easy to reestablish (8.15) by using the earlier procedure, and thus prove that A is a Hilbert–Schmidt operator.

Q.E.D.

For any Hilbert–Schmidt operator A, the quantity

$$\| A \|_2 = \sqrt{\mathrm{Tr}[A^*A]} \tag{8.17}$$

exists, and is called the *Hilbert–Schmidt norm* of A. As a matter of fact, using first the triangle inequality in $\mathscr{H}$ and then the triangle inequality in l^2 spaces we obtain (see also Theorem 8.10)

$$\begin{aligned} \| a_1 A_1 + a_2 A_2 \|_2 &= \Big(\sum_k \|(a_1 A_1 + a_2 A_2) e_k \|^2\Big)^{1/2} \\ &\leqslant \Big(\sum_k [\| a_1 A_1 e_k \| + \| a_2 A_2 e_k \|]^2\Big)^{1/2} \\ &\leqslant \Big(\sum_k \| a_1 A_1 e_k \|^2\Big)^{1/2} + \Big(\sum_k \| a_2 A_2 e_k \|^2\Big)^{1/2} \\ &= | a_1 | \, \| A_1 \|_2 + | a_2 | \, \| A_2 \|_2 . \end{aligned}$$

Moreover, according to (8.16),

$$\| A \| \leqslant (\mathrm{Tr}[A^*A])^{1/2} = \| A \|_2 , \tag{8.18}$$

which shows that $\| A \|_2 = 0$ implies $A = 0$.

8.4. The Trace Norm and the Trace Class

Let us consider now a completely continuous operator A, having the decomposition (8.3), for which

$$\mathrm{Tr}[\sqrt{A^*A}] = \mathrm{Tr}\left[\sum_k \mid e_k\rangle\, \lambda_k \,\langle e_k \mid\right] = \sum_k \lambda_k < +\infty. \tag{8.19}$$

If the above series is infinite, its convergence implies that $\lambda_k \to 0$ when $k \to +\infty$. Consequently, $\lambda_k^2 \leqslant \lambda_k$ for sufficiently large values of k. Hence, $\sum_k \lambda_k^2$ converges when $\sum_k \lambda_k$ converges. This shows that any completely continuous operator A satisfying (8.19) is a Hilbert–Schmidt operator.

Lemma 8.4. If A is a completely continuous operator on the Hilbert space $\mathscr{H}$ having the canonical decomposition (8.3), then

$$\sum_k \lambda_k = \sup \sum_i |\langle f_i \mid Ag_i\rangle|, \tag{8.20}$$

where the supremum is taken over all choices of countable orthonormal systems $\{f_i\}$ and $\{g_i\}$.

Proof. Using (8.3) we get

$$\begin{aligned} |\langle f_i \mid Ag_i\rangle| &= \left|\sum_k \langle f_i \mid e_k'\rangle\, \lambda_k \,\langle e_k \mid g_i\rangle\right| \\ &\leqslant \sum_k \lambda_k \,|\langle f_i \mid e_k'\rangle\langle e_k \mid g_i\rangle| \\ &\leqslant \tfrac{1}{2} \sum_k \lambda_k(|\langle f_i \mid e_k'\rangle|^2 + |\langle g_i \mid e_k\rangle|^2). \end{aligned}$$

Applying Bessel's inequality (see Chapter I, Lemma 4.1) to the orthonormal systems $\{f_i\}$ and $\{g_i\}$ we obtain

$$\begin{aligned} &\sum_i |\langle f_i \mid e_k'\rangle|^2 \leqslant \| e_k' \|^2 = 1, \\ &\sum_i |\langle g_i \mid e_k\rangle|^2 \leqslant \| e_k \|^2 = 1. \end{aligned} \tag{8.21}$$

In proving Lemma 8.3, we used theorems on the convergence of double series. In view of (8.21), we can apply those theorems to justify the reversal of the order of summation in the following computation:

$$\begin{aligned}\sum_i |\langle f_i \mid Ag_i\rangle| &\leqslant \tfrac{1}{2} \sum_i \sum_k \lambda_k(|\langle f_i \mid e_k'\rangle|^2 + |\langle g_i \mid e_k\rangle|^2)\\ &= \tfrac{1}{2} \sum_k \lambda_k \left[\sum_i (|\langle f_i \mid e_k'\rangle|^2 + |\langle g_i \mid e_k\rangle|^2\right]\\ &\leqslant \sum_k \lambda_k .\end{aligned}$$

Since we have

$$\sum_k |\langle e_k' \mid Ae_k\rangle| = \sum_k \lambda_k ,$$

we conclude that (8.20) is true. Q.E.D.

We shall now show that the functional

$$\| A \|_1 = \mathrm{Tr}[\sqrt{A^*A}] = \sum_k \lambda_k \tag{8.22}$$

on the class of Hilbert–Schmidt operators satisfying (8.19) is a norm, called the *trace norm.*

Since for every series with positive terms

$$\sum_k \lambda_k^2 \leqslant \Big(\sum_k \lambda_k\Big)^2,$$

we obtain

$$\| A \|_2 \leqslant \| A \|_1 . \tag{8.23}$$

Hence, $\| A \|_1 = 0$ implies that $\| A \|_2 = 0$, and consequently $A = 0$.

For any complex number a,

$$\| aA \|_1 = \mathrm{Tr}[(| a |^2 A^*A)^{1/2}] = | a | \,\mathrm{Tr}[\sqrt{A^*A}] = | a | \, \| A \|_1 .$$

Finally, by making use of Lemma 8.4 we obtain

$$\begin{aligned}\| A_1 + A_2 \|_1 &= \sup \sum_i |\langle f_i \,|(A_1 + A_2)\, g_i\rangle|\\ &\leqslant \sup \sum_i |\langle f_i \mid A_1 g_i\rangle| + \sup \sum_i |\langle f_i \mid A_2 g_i\rangle|\\ &= \| A_1 \|_1 + \| A_2 \|_1 .\end{aligned}$$

Thus, $\|\cdot\|_1$ satisfies all the conditions of a norm on the class of Hilbert–Schmidt operator for which (8.19) is true. Theorem 8.7 states that this class is identical to the trace class of Definition 8.2 when $\mathscr{H}$ is a separable Hilbert space. To prove it, we need the following result.

Theorem 8.6. Any self-adjoint operator of trace class has a pure point spectrum.

Proof. Any self-adjoint operator A of trace class can be written as a difference of two positive-definite self-adjoint operators A_1 and A_2,

$$A = \int_{\mathbb{R}^1} \lambda \, dE_\lambda = A^{(+)} - A^{(-)}$$

(8.24)

$$A^{(+)} = \int_{[0,+\infty)} \lambda \, dE_\lambda, \qquad A^{(-)} = -\int_{(-\infty,0)} \lambda \, dE_\lambda,$$

both having a finite trace (see Exercise 8.6). Consequently, it is sufficient to prove the theorem for positive-definite operators.

Assume that A is positive definite, and choose any $\lambda_0 > 0$. Let $\lambda_1 < \lambda_2 < \cdots < \lambda_n$ be n numbers greater than λ_0 such that

$$(8.25) \qquad E_{\lambda_1} \neq E_{\lambda_0}, \qquad E_{\lambda_2} \neq E_{\lambda_1}, \qquad \ldots, \qquad E_{\lambda_n} \neq E_{\lambda_{n-1}}.$$

Since $E_{\lambda_k} - E_{\lambda_{k-1}} \neq 0$, there are n normalized vectors $f_1, \ldots, f_n$ such that

$$(E_{\lambda_k} - E_{\lambda_{k-1}}) f_k = f_k, \qquad k = 1, \ldots, n.$$

According to the spectral theorem we have

$$(8.26) \qquad \begin{aligned} \langle f_k \mid Af_k \rangle &= \int_{\mathbb{R}^1} \lambda \, d\langle f_k \mid E_\lambda f_k \rangle \\ &= \int_{(\lambda_{k-1},\lambda_k]} \lambda \, d\langle f_k \mid E_\lambda f_k \rangle \\ &\geqslant \lambda_0 \int_{(\lambda_{k-1},\lambda_k]} d\langle f_k \mid E_\lambda f_k \rangle = \lambda_0 . \end{aligned}$$

By completing $f_1, \ldots, f_n$ to an orthonormal basis $f_1, f_2, \ldots,$ in $\mathscr{H}$ we get, using the positivity of A and (8.26),

$$\operatorname{Tr} A = \sum_k \langle f_k \mid Af_k \rangle \geqslant \sum_{k=1}^{n} \langle f_k \mid Af_k \rangle \geqslant n\lambda_0 .$$

The above inequality shows that the number of points $\lambda_1, \ldots, \lambda_n$ for which

(8.25) is true is smaller than Tr A/λ_0, i.e., the spectral function E_λ can change only at a finite number of points in $(\lambda_0, +\infty)$. Thus, A has exclusively a point spectrum in $(\lambda_0, +\infty)$, and since $\lambda_0 > 0$ can be chosen arbitrarily small, A has a pure point spectrum. Q.E.D.

Theorem 8.7. A completely continuous operator A on the Hilbert space $\mathscr{H}$ is of trace class if and only if it is a Hilbert–Schmidt operator for which

$$\mathrm{Tr}[\sqrt{A^*A}] = \sum_k \lambda_k < +\infty,$$

where λ_k are the numbers appearing in the decomposition (8.3) of A.

Proof. The positive numbers λ_k occurring in the decomposition (8.3) of a Hilbert–Schmidt operator A are the eigenvalues of $\sqrt{A^*A}$. Hence, if $A^* = A$, we must have $e_k' = e_k$ when $\lambda_k \in \mathsf{S}^A$, and $e_k' = -e_k$ when $-\lambda_k \in \mathsf{S}^A$. In the decomposition

$$A = A^{(+)} - A^{(-)}, \qquad A^{(\pm)} = \sum_{\pm\lambda_k \in \mathsf{S}^A} |\, e_k\rangle\, \lambda_k\, \langle e_k \,|,$$

the operators $A^{(\pm)}$ are positive definite. Hence, we can write

$$A^{(\pm)} = \sqrt{A^{(\pm)}}\,\sqrt{A^{(\pm)}},$$

and apply Lemma 8.3 to infer that

$$\mathrm{Tr}\, A^{(\pm)} = \sum_{\pm\lambda_k \in \mathsf{S}^A} \lambda_k \leqslant \sum_k \lambda_k$$

exists if $\sum_k \lambda_k < +\infty$. Consequently

$$\mathrm{Tr}\, A = \mathrm{Tr}\, A^{(+)} - \mathrm{Tr}\, A^{(-)}$$

also exists.

The more general case of nonsymmetric A can be reduced to the above case by setting

$$A = A_1 + iA_2, \qquad A_1 = \tfrac{1}{2}(A + A^*), \qquad A_2 = -(i/2)(A - A^*).$$

In fact, the operators A_1 and A_2 are obviously symmetric Hilbert–Schmidt operators, for which we get

$$\mathrm{Tr}(\sqrt{A_1{}^*A_1}) = \| A_1 \|_1 \leqslant \| A \|_1 + \| A^* \|_1 < +\infty,$$

$$\mathrm{Tr}(\sqrt{A_2{}^*A_2}) = \| A_2 \|_1 \leqslant \| -iA \|_1 + \| iA^* \|_1 < +\infty,$$

using the properties of the trace norm $\| \cdot \|_1$.

Consider now a self-adjoint operator A of trace class. According to Theorem 8.6, A has a pure point spectrum. Hence,

$$A = \int_{\mathbb{R}^1} \lambda \, dE_\lambda^A = \sum_{\lambda \in \mathsf{S}^A} \lambda E^A(\{\lambda\}), \tag{8.27}$$

where the convergence of the above series is uniform if the series is infinite. Each characteristic subspace M_λ corresponding to a nonzero eigenvalue $\lambda \in \mathsf{S}^A$ has to be finite dimensional; in fact, if that were not true for some $\lambda \neq 0$, we would have $\mathrm{Tr}[E^A(\{\lambda\})] = +\infty$, which would imply that the trace of A does not exist. Choosing an orthonormal basis in each characteristic subspace M_λ, $\lambda \in \mathsf{S}^A$, we obtain, by taking the union of all such bases, an orthonormal basis $\{e_1, e_2, \ldots\}$ in $\mathscr{H}$. Since each M_λ is finite dimensional, (8.27) yields

$$A = \sum_{k=1}^{\infty} | e_k\rangle \lambda_k' \langle e_k |, \tag{8.28}$$

where λ_k' is the eigenvalue of A for the eigenvector e_k.

In taking the trace of A,

$$\mathrm{Tr}\, A = \sum_{k=1}^{\infty} \langle e_k | A e_k\rangle = \sum_{k=1}^{\infty} \lambda_k',$$

we can use any orthonormal basis, and therefore any order of $e_1, e_2, \ldots$ in the above series. Hence, the series $\sum_{k=1}^{\infty} \lambda_k'$ converges unconditionally, and consequently it converges absolutely (see Randolph [1968, Section 3.9]):

$$\sum_{k=1}^{\infty} \lambda_k < +\infty, \qquad \lambda_k = | \lambda_k' |.$$

It follows from (8.28) that A is a completely continuous operator. Since the convergence of $\sum_{k=1}^{\infty} \lambda_k$ implies the convergence of $\sum_{k=1}^{\infty} \lambda_k^2$ ($\lambda_k^2 \leqslant \lambda_k$ for sufficiently large k), we easily deduce that A is a Hilbert–Schmidt operator.

If A is any operator of trace class, then obviously A^* is also of trace class:

$$\mathrm{Tr}\, A^* = \sum_{k=1}^{\infty} \langle e_k | A^* e_k\rangle = \sum_{k=1}^{\infty} \langle e_k | A e_k\rangle^* = (\mathrm{Tr}\, A)^*.$$

Hence, $A_1 = \frac{1}{2}(A + A^*)$ and $A_2 = -(i/2)(A - A^*)$ are self-adjoint operators of trace class, and consequently they are Hilbert–Schmidt operators for which $\| A_1 \|_1 < +\infty$ and $\| A_2 \|_1 < +\infty$. This implies that

$A = A_1 + iA_2$ is also a Hilbert–Schmidt operator. Moreover, since $\| \cdot \|_1$ as defined in (8.22) is a norm,

$$\operatorname{Tr} \sqrt{A^*A} = \| A \|_1 \leqslant \| A_1 \|_1 + \| A_2 \|_1 < +\infty,$$

i.e., $\sum_k \lambda_k$ converges. Q.E.D.

8.5. Statistical Ensembles and the Process of Measurement

When we were discussing in §1 the concept of preparatory measurement in quantum mechanics, we saw that for n compatible observables represented in the Schroedinger picture by the operators $A_1, \ldots, A_n$ it is usually not the case that we can prepare at an instant t_0 n absolutely accurate simultaneous values $\lambda_1^{(0)}, \ldots, \lambda_n^{(0)}$ for $A_1, \ldots, A_n$. Instead, in general, a preparatory measurement yields the weaker result that the simultaneously prepared values of $A_1, \ldots, A_n$ are within an n-dimensional Borel set B. Therefore, after a preparatory measurement we have to assign to the system the family of all states $\Psi(t)$ which satisfy (1.26). This family contains, in general, more than one element. However, in practice, instead of dealing with such a family of state vectors, one prefers to make some reasonable statistical assumption based on some additional knowledge or insight which one might have regarding the ensemble of systems used in the experiment.

In order to simplify the discussion from the mathematical point of view, let us assume that $A_1, \ldots, A_n$ represent a *complete* set of observables with pure point spectra, and that for every $\lambda = (\lambda_1, \ldots, \lambda_n) \in S^{A_1} \times \cdots \times S^{A_n}$ we can choose a normalized vector e_λ such that $A_k e_\lambda = \lambda_k e_\lambda$, $k = 1, \ldots, n$. Consequently, the family of all vectors e_λ, $\lambda \in S^{A_1} \times \cdots \times S^{A_n}$, constitutes an orthonormal basis in $\mathscr{H}$.

Assume that the preparatory measurement which has been carried out on the system at t_0 provides us with the information that the prepared simultaneous values of $A_1, \ldots, A_n$ are at t_0 within the Borel set B, where $B \subset S^{A_1} \times \cdots \times S^{A_n}$. Let us further simplify the considerations by assuming that B is finite. Then the prepared family of states consists of all the states $\Psi(t)$ satisfying

$$\Psi_\lambda(t_0) = e_\lambda, \qquad \lambda \in B,$$

and any linear combination of such states.

Suppose that we submit a large ensemble of identical systems to the described preparatory procedure, and that we know that in this ensemble $(100 w_\lambda)\%$ of the systems have at time t_0 the simultaneous values $\lambda_1, \ldots, \lambda_n$, $(\lambda_1, \ldots, \lambda_n) \in B$, for $A_1, \ldots, A_n$. As we already know, the theory predicts that the mean value of the outcome of a determinative measurement at time $t > t_0$ of an observable represented by a bounded operator A_0 is

$$\langle A_0 \rangle_{\Psi_\lambda(t)} = \langle \Psi_\lambda(t) | A_0 \Psi_\lambda(t) \rangle \tag{8.29}$$

if the measurement is carried out only on that part of the ensemble consisting of particles which at t_0 had the simultaneous values $\lambda_1, \ldots, \lambda_n$ for $A_1, \ldots, A_n$, respectively. Hence, the mean value of A_0 for the entire ensemble is

$$\langle A_0 \rangle_{\text{ens}}(t) = \sum_{\lambda \in B} w_\lambda \langle \Psi_\lambda(t) | A_0 \Psi_\lambda(t) \rangle. \tag{8.30}$$

Let us introduce for each instant t the operator

$$\rho(t) = \sum_{\lambda \in B} w_\lambda E_{\Psi_\lambda(t)}, \qquad E_{\Psi_\lambda(t)} = | \Psi_\lambda(t) \rangle \langle \Psi_\lambda(t) |, \tag{8.31}$$

which is called the *density operator* or the *statistical operator* of the considered ensemble.

From the orthonormality property of the set of all vectors $\Psi_\lambda(t)$, $\lambda \in \mathsf{S}^{A_1} \times \cdots \times \mathsf{S}^{A_n}$, at any fixed t, it follows immediately that

$$\langle A_0 \rangle_{\text{ens}}(t) = \sum_{\lambda \in \mathsf{S}^{A_1} \times \cdots \times \mathsf{S}^{A_n}} \langle \Psi_\lambda(t) | A_0 \, \rho(t) \, \Psi_\lambda(t) \rangle. \tag{8.32}$$

The above relation can be written in the form

$$\langle A_0 \rangle_{\text{ens}}(t) = \operatorname{Tr}[A_0 \rho(t)] = \operatorname{Tr}[\rho(t) A_0], \tag{8.33}$$

provided that $A_0 \, \rho(t)$ is of trace class. Theorem 8.8 shows that this is indeed the case.

Theorem 8.8. Suppose $\{e_1, e_2, \ldots\}$ is an orthonormal basis in $\mathscr{H}$. If

$$A = \sum_k | e_k \rangle \, \lambda_k \, \langle e_k |, \tag{8.34}$$

where $\lambda_k \geqslant 0$ and

$$\operatorname{Tr}[A] = \sum_k \lambda_k < +\infty, \tag{8.35}$$

and if A_0 is a bounded operator, then $A_0 A$ is of trace class.

The above theorem can be easily derived from Theorem 8.7 by showing that the operators $(A_0 A)^* A_0 A$ and $(A A_0{}^* A_0 A)^{1/2}$ are of trace class. As a matter of fact, it is sufficient to show that $(A A_0{}^* A_0 A)^{1/2}$ is of trace class, since it is easy to see that in that case $AA_0{}^*A_0A$ is also of trace class.

To establish that $(AA_0{}^*A_0A)^{1/2}$ is of trace class, note that

$$(AA_0{}^*A_0A)^{1/2} = (AA_0{}^*A_0A)^{1/4}\,(AA_0{}^*A_0A)^{1/4}.$$

Then it follows from Lemma 8.3 that the sum

$$\sum_k \langle e_k \,|(AA_0{}^*A_0A)^{1/2}\, e_k\rangle$$

has the same value in any orthonormal basis $\{e_1, e_2, \ldots\}$. Let $e_1, e_2, \ldots$ be the vectors appearing in (8.34). We have

$$\begin{aligned}\langle e_k \,|(AA_0{}^*A_0A)^{1/2}\, e_k\rangle &\leqslant \|(AA_0{}^*A_0A)^{1/2}\, e_k\,\| \\ &= (\langle e_k \mid AA_0{}^*A_0Ae_k\rangle)^{1/2} \\ &= \lambda_k\langle e_k \mid A_0{}^*A_0e_k\rangle^{1/2} \leqslant \lambda_k \,\|\, A_0\,\|,\end{aligned}$$

and consequently

$$\sum_k \langle e_k \,|(AA_0{}^*A_0A)^{1/2}\, e_k\rangle \leqslant \|\, A_0\,\| \sum_k \lambda_k < +\infty.$$

Thus, $(AA_0{}^*A_0A)^{1/2}$ is of trace class and Theorem 8.8 is established.

8.6. The Quantum Mechanical State of an Ensemble

Theorem 8.8 establishes that (8.32) can be written in the form (8.33) when A_0 is bounded, since $\rho(t)$ is a positive-definite operator with a pure point spectrum and of finite trace:

$$\mathrm{Tr}[\rho(t)] = \sum_\lambda \langle \Psi_\lambda(t)|\, \rho(t)\, \Psi_\lambda(t)\rangle = 1.$$

In the experiment described at the beginning of Section 8.5, in which we were concerned with a statistical description of an ensemble of identical independent systems, the relevant information about the outcome of any measurement can be obtained from the operator-valued function $\rho(t)$; thus, it is justifiable to state that $\rho(t)$ represents the state of the ensemble in the Schroedinger picture.

In the light of the above discussion the following postulate is a natural generalization of (8.31).

Axiom E. Consider a statistical ensemble of identical systems, and let $\mathscr{H}$ be the Hilbert space associated with each system in the ensemble (Axiom O, §2). Any *state of the ensemble* can be represented in the Schroedinger picture by an operator-valued function $\rho(t)$

$$\rho(t) = \sum_{k=1}^{\infty} w_k E_{\Psi_k(t)} = \text{u-}\lim_{n\to\infty} \sum_{k=1}^{n} w_k E_{\Psi_k(t)} , \tag{8.36}$$

where $E_{\Psi_k}(t) = | \Psi_k(t)\rangle \langle\Psi_k(t) |$,

$$\sum_{k=1}^{\infty} w_k = 1, \qquad w_k \geqslant 0, \tag{8.37}$$

and $\{\Psi_1(t), \Psi_2(t),...\}$ is an orthonormal system of state vectors whose time dependence is governed by the relation

$$\Psi_k(t) = U(t, t_0)\, \Psi_k(t_0), \qquad t, t_0 \in \mathbb{R}^1, \tag{8.38}$$

where $U(t, t_0)$ is the evolution operator of the systems in the ensemble.

We note that the uniform limit in (8.36) exists on account of (8.37). This can be seen easily if we introduce

$$\rho = \text{s-}\lim_{n\to\infty} \sum_{k=1}^{n} w_k E_{\Psi_k} = \sum_{k=1}^{\infty} | \Psi_k\rangle\, w_k \langle\Psi_k |$$

by means of the above strong limit, which obviously defines a linear operator. Since

$$\left| \rho - \sum_{k=1}^{n} w_k E_{\Psi_k} \right| \leqslant \sum_{k=n+1}^{\infty} w_k \| E_{\Psi_k} \| = \sum_{k=n+1}^{\infty} w_k$$

converges to zero when $n \to +\infty$, ρ is the uniform limit of the partial sums in (8.36).

An ensemble in the state (8.36) is viewed as consisting of systems in the states $\Psi_1(t), \Psi_2(t),...$ in the respective proportions $w_1, w_2, ...$. If $w_k = 1$ for $k = k_0$ and therefore $w_k = 0$ for $k \neq k_0$, the described state of the ensemble is one in which each of the systems is in the state $\Psi_{k_0}(t)$. In that case $\rho(t)$ is called a *pure state* of the ensemble. If a few of the w_k are nonzero, $\rho(t)$ is called a *mixed state* or a *mixture* of all the single-system states $\Psi_k(t)$ for which $w_k > 0$.

It is easy to derive from (8.36) and (8.38)

$$\rho(t) = U(t, t_0)\, \rho(t_0)\, U^{-1}(t, t_0). \tag{8.39}$$

To see that, take any $f \in \mathscr{H}$ and note that for any $k = 1, 2,\ldots$

$$\begin{aligned} E_{\Psi_k(t)} f &= \langle \Psi_k(t) | f \rangle \, \Psi_k(t) \\ &= \langle U(t, t_0) \, \Psi_k(t_0) | f \rangle \, U(t, t_0) \, \Psi_k(t_0) \\ &= U(t, t_0) \langle \Psi_k(t_0) | \, U^{-1}(t, t_0) f \rangle \, \Psi_k(t_0) \\ &= U(t, t_0) \, E_{\Psi_k(t_0)} U^{-1}(t, t_0) f, \end{aligned}$$

which shows that

$$E_{\Psi_k(t)} = U(t, t_0) \, E_{\Psi_k(t_0)} \, U^{-1}(t, t_0).$$

From the above relation we get

$$\sum_{k=1}^{n} w_k E_{\Psi_k(t)} = U(t, t_0) \Big(\sum_{k=1}^{n} w_k E_{\Psi_k(t_0)} \Big) \, U^{-1}(t, t_0).$$

In the limit $n \to +\infty$, we obtain (8.39) by using the basic rules for dealing with uniform limits (see Exercise 8.7).

In view of (8.33) and of Born's correspondence rule in Definition 1.3 it is natural to assume that if a determinative measurement of a bounded observable A_0 is carried out at time t on each system in an ensemble in a state $\rho(t)$, then the mean value of A_0 will be $\mathrm{Tr}[\rho(t) \, A_0]$. Thus, as a generalization of (1.27), we can state that the ratio of the number $\nu_{A_1,\ldots,A_n}(B)$ of systems having at t simultaneous values of $A_1 ,\ldots, A_n$ within the set B, to the total number N of systems in the ensemble, should be equal to

$$\text{(8.40)} \qquad \frac{\nu_{A_1,\ldots,A_n}(B)}{N} \approx \mathrm{Tr}[\rho(t) \, E^{A_1,\ldots,A_n}(B)] = P^{A_1,\ldots,A_n}_{\rho(t)}(B).$$

If a preparatory procedure on the systems of an ensemble does not leave the ensemble in a pure state, i.e., if there are more than one linearly independent state vectors satisfying (1.26) in the preparatory measurement of a general type described in Definition 1.4, then usually the *principle of equal a priori probabilities* is invoked in order to assign to the ensemble a state. According to this principle, the proposed state (8.36) of the ensemble is the one in which equal nonzero weights are assigned to the single-system states $\Psi_k(t)$ satisfying (1.26), and zero weights to those states not satisfying (1.26). Thus, if $\Psi_1(t),\ldots, \Psi_n(t)$ are all the mutually orthogonal states satisfying (1.26), i.e., those states for which

$$\langle \Psi_k(t_0) | \, E^{A_1,\ldots,A_n}(B) \, \Psi_k(t_0) \rangle = 1,$$

or equivalently

$$E^{A_1,\ldots,A_n}(B)\,\Psi_k(t_0) = \Psi_k(t_0),$$

then obviously

$$\rho(t_0) = \frac{1}{n}\sum_{k=1}^{n} E_{\Psi_k(t_0)} = \frac{1}{n} E^{A_1,\ldots,A_n}(B).$$

Theorem 8.9 provides a simple algebraic criterion for distinguishing between pure states and mixed states of a statistical ensemble.

Theorem 8.9. The state $\rho(t)$ of an ensemble is a pure state if and only if $\rho(t) = \rho^2(t)$.

Proof. If the ensemble is in a pure state in which the states of each system in the ensemble is $\Psi(t)$, the statistical operator $\rho(t)$ is at each instant t the projector onto $\Psi(t)$, and therefore

$$\rho^2(t) = \rho(t). \tag{8.41}$$

Conversely, assume that (8.41) is true. If we write $\rho(t)$ in the form (8.36) and compare it with the following expression for $\rho^2(t)$ (see Chapter III, Exercise 5.10)

$$\rho^2(t) = \underset{n\to\infty}{\text{u-lim}} \left(\sum_{k=1}^{n} w_k E_{\Psi_k(t)}\right)^2$$

$$= \underset{n\to\infty}{\text{u-lim}} \sum_{k=1}^{n} w_k^2 E_{\Psi_k(t)} = \sum_{k=1}^{\infty} w_k^2 E_{\Psi_k(t)},$$

we obtain $w_k^2 = w_k$. Thus, w_k can assume only the values 0 and 1. This, in conjuction with the condition $\sum_{k=1}^{\infty} w_k = 1$, shows that $w_k = 0$ except for one value k_0 of k at which $w_{k_0} = 1$. Q.E.D.

We have arrived at the concept of "state vector in the Schroedinger picture" by considering the possible values of all states $\Psi(t)$ of the system at a given instant. In this same manner we arrive at the concept of statistical operators by considering the possible values of a state $\rho(t)$ of an ensemble at instant t.

Definition 8.4. Let $\mathscr{H}$ be a separable Hilbert space associated with a quantum mechanical system. A linear operator ρ on $\mathscr{H}$ is a *statistical* *operator* (*density operator*) if it can be written in the form

$$\rho = \underset{n\to\infty}{\text{u-lim}} \sum_{k=1}^{n} w_k E_{\Psi_k} = \sum_{k=1}^{\infty} w_k E_{\Psi_k}, \tag{8.42}$$

where the real numbers $w_1, w_2, \ldots$ satisfy (8.37) and $\{\Psi_1, \Psi_2, \ldots\}$ is an orthonormal system of state vectors in $\mathscr{H}$.

Any statistical operator ρ is obviously a positive definite self-adjoint operator. Hence, according to Lemma 8.3, $\operatorname{Tr}\rho$ exists if the sum $\sum_i \langle e_i \mid \rho e_i\rangle$ is finite in some orthonormal basis $\{e_1, e_2, \ldots\}$. We can obtain such a basis by completing $\{\Psi_1, \Psi_2, \ldots\}$ to an orthonormal basis in $\mathscr{H}$. Thus, we have

$$\operatorname{Tr}\rho = \sum_{k=1}^{\infty} \langle \Psi_k \mid \rho\Psi_k\rangle = \sum_{k=1}^{\infty} w_k = 1. \tag{8.43}$$

Conversely, Theorem 8.7 tells us that in the absence of superselection rules any positive self-adjoint operator of unit trace is a density operator.

8.7. The von Neumann Equation in Liouville Space

Let us concentrate now on the case of time-independent Hamiltonians H; i.e., in physical terms, on the case where each element of the ensemble, whose density operator is described (in accordance with Axiom E) by $\rho(t)$, is isolated from its environment so that its total energy is conserved. In that case $U(t, t_0)$ in (8.39) is given by (3.2). Differentiating the right-hand side of (8.39) formally, we arrive at the *von Neumann equation*:

$$i\hbar \frac{d\rho(t)}{dt} = i\hbar \underset{\Delta t\to 0}{\text{h-lim}} \frac{\rho(t+\Delta t) - \rho(t)}{\Delta t} = [H, \rho(t)]. \tag{8.44}$$

This equation bears a remarkable formal resemblance to the Heisenberg equation (3.25), but there are both mathematical and physical distinctions: in (3.25) $\hat{A}(t)$ is in general an unbounded self-adjoint operator representing an observable in the Heisenberg picture, whereas in (8.44) $\rho(t)$ is a (self-adjoint) trace-class operator of unit trace and of bound

$$\|\rho(t)\| = \sup_{k=1,2,\ldots} w_k \leqslant 1,$$

and this $\rho(t)$ represents a (mixed or pure) state in the Schroedinger picture. Hence we have to study the precise meaning of the limit on the left-hand side and of the commutator on the right-hand side of (8.44), respectively, on their own rather than as corollaries to the results of §3.4.

Such a study is greatly facilitated by the fact that the family $\mathscr{L}(\mathscr{H})$ of all Hilbert–Schmidt operators on $\mathscr{H}$ (called in mathematical literature the Hilbert–Schmidt class* and in physical literature the Liouville space) is a Hilbert space, and that as a trace-class operator $\rho(t)$ belongs to $\mathscr{L}(\mathscr{H})$ by Theorem 8.7.

Theorem 8.10. The family $\mathscr{L}(\mathscr{H})$ of all Hilbert–Schmidt operators on a given (separable) Hilbert space $\mathscr{H}$ is itself a (separable) Hilbert space (called the *Liouville space* over $\mathscr{H}$) with respect to the inner product

$$\langle A_1 \mid A_2 \rangle_2 = \mathrm{Tr}(A_1{}^* A_2), \qquad A_1, A_2 \in \mathscr{L}(\mathscr{H}). \tag{8.45}$$

Proof. The argument at the end of §8.3, which establishes the fact that (8.17) is a norm, simultaneously establishes that $\mathscr{L}(\mathscr{H})$ is a linear space. Hence, if $A_1, A_2 \in \mathscr{L}(\mathscr{H})$ then $A_1 + A_2$, $A_1 + iA_2 \in \mathscr{L}(\mathscr{H})$, and therefore

$$\begin{aligned} A_1{}^* A_2 = {} & \tfrac{1}{2}(A_1 + A_2)^*(A_1 + A_2) + (2i)^{-1}(A_1 + iA_2)^*(A_1 + iA_2) \\ & - \tfrac{1}{2}(1 - i)(A_1{}^* A_1 + A_2{}^* A_2) \end{aligned} \tag{8.46}$$

is of trace-class by Theorem 8.5, so that (8.45) exists. The linearity features of the trace operation,

$$\mathrm{Tr}(a_1 A^{(1)} + a_2 A^{(2)}) = a_1 \,\mathrm{Tr}\, A^{(1)} + a_2 \,\mathrm{Tr}\, A^{(2)} \tag{8.47}$$

that is evident from† (8.8), immediately leads to the conclusion that (8.45) has the properties (2)–(4) in Definition 2.1 of Chapter I, whereas property (1) follows from (8.18).

To establish that the Euclidean space $\mathscr{L}(\mathscr{H})$ is complete and separable if $\mathscr{H}$ is a separable infinite dimensional Hilbert space,‡ we introduce

* In Definition 8.2 we had restricted ourselves to separable Hilbert spaces, which are of exclusive interest in quantum mechanics, but that definition can be extended to the nonseparable case (see Schatten [1960]).

† The trace-class is often denoted by $\mathscr{B}_1(\mathscr{H})$, the Hilbert–Schmidt class by $\mathscr{B}_2(\mathscr{H})$, the class of compact operators by $\mathscr{B}_\infty(\mathscr{H})$, the C^*-algebra $\mathfrak{A}(\mathscr{H})$ of Theorem 1.1 in Chapter III by $\mathscr{B}(\mathscr{H})$. We then have $\mathscr{B}_1(\mathscr{H}) \subset \mathscr{B}_2(\mathscr{H}) \subset \cdots \subset \mathscr{B}_\infty(\mathscr{H}) \subset \mathscr{B}(\mathscr{H})$ where $\mathscr{B}_1(\mathscr{H})$, $\mathscr{B}_2(\mathscr{H})$, and $\mathscr{B}_\infty(\mathscr{H})$ are Banach spaces (and, moreover, so-called minimal norm ideals in $\mathscr{B}(\mathscr{H}) \equiv \mathfrak{A}(\mathscr{H})$) with respect to the trace-norm $\|\cdot\|_1$, Hilbert–Schmidt norm $\|\cdot\|_2$, and operator bound $\|\cdot\|$, respectively). However, among all $\mathscr{B}_i(\mathscr{H})$, $i = 1, 2, \ldots, \infty$ (where appropriate classes and norms are defined for $i = 3, 4, \ldots$), only $\mathscr{B}_2(\mathscr{H}) = \mathscr{L}(\mathscr{H})$ is a Hilbert space (see Schatten [1960] for details).

‡ See Schatten [1960] for the case when $\mathscr{H}$ is nonseparable.

an orthonormal basis $\{e_1, e_2, ...\}$ in $\mathscr{H}$, and note that for any $A_1, A_2 \in \mathscr{L}(\mathscr{H})$

$$(8.48)\quad \langle A_1 \mid A_2\rangle_2 = \sum_{k=1}^{\infty} \langle e_k \mid A_1{}^* A_2 e_k\rangle = \sum_{k=1}^{\infty}\sum_{i=1}^{\infty} \langle e_i \mid A_1 e_k\rangle^* \langle e_i \mid A_2 e_k\rangle.$$

Consequently, the mapping assigning to $A \in \mathscr{L}(\mathscr{H})$ the element $\{\{\langle e_i \mid Ae_k\rangle\}_{i,k}$ in the separable Hilbert space $l^2(\infty^2) = l^2(\infty) \otimes l^2(\infty)$ (see Theorem 4.7 in Chapter I and Theorems 6.9 and 6.10 in Chapter II) of matrices $\{a_{ik}\}$ with inner product

$$\langle \{a_{ik}^{(1)}\} \mid \{a_{ik}^{(2)}\}\rangle = \sum_{i,k=1}^{\infty} a_{ik}^{(1)*} a_{ik}^{(2)}$$

is an isometric transformation of $\mathscr{L}(\mathscr{H})$ into $l^2(\infty^2)$. To see that this transformation is actually unitary note that by Theorem 8.5 $A \in \mathscr{L}(\mathscr{H})$ if and only if

$$(8.49)\qquad \sum_{i,k=1}^{\infty} |\langle e_i \mid Ae_k\rangle|^2 = \operatorname{Tr}(A^*A) < \infty.$$

Thus $\mathscr{L}(\mathscr{H})$ is isomorphic to $l^2(\infty^2)$, and therefore it is a separable Hilbert space. Q.E.D.

In physical literature operators acting on a Liouville space $\mathscr{L}(\mathscr{H})$ are called *superoperators* (whereas in mathematical literature either the term operator is retained, since after all $\mathscr{L}(\mathscr{H})$ is a Hilbert space itself, or the term *transformer* might be applied). As examples of superoperators acting on $A \in \mathscr{L}(\mathscr{H})$ consider

$$(8.50)\qquad \mathbf{U}A = UAU^*, \qquad U^* = U^{-1},$$

$$(8.51)\qquad \mathbf{H}A = \hbar^{-1}(HA - AH), \qquad H = H^*.$$

The superoperator $\mathbf{U}$ has domain $\mathscr{D}_{\mathbf{U}} = \mathscr{L}(\mathscr{H})$ and is unitary. Indeed by (8.46) and (8.11) the infinite double series in (8.48) converges unconditionally, so that by setting $A = A_1^*$ and $B = A_2$ we conclude that

$$(8.52)\quad \operatorname{Tr}(AB) = \sum_{k=1}^{\infty} \langle e_k \mid ABe_k\rangle = \sum_{k=1}^{\infty}\sum_{i=1}^{\infty} \langle e_k \mid Ae_i\rangle\langle e_i \mid Be_k\rangle$$

$$= \sum_{i=1}^{\infty}\sum_{k=1}^{\infty} \langle e_i \mid Be_k\rangle\langle e_k \mid Ae_i\rangle = \sum_{k=1}^{\infty} \langle e_i \mid BAe_i\rangle = \operatorname{Tr}(BA)$$

for arbitrary A, $B \in \mathscr{L}(\mathscr{H})$. Therefore,

$$\langle \mathbf{U}A \mid \mathbf{U}A \rangle_2 = \mathrm{Tr}[U(A^*AU^*)] = \mathrm{Tr}[(A^*AU^*)U] = \langle A \mid A \rangle_2 ,$$

and since obviously $\mathbf{U}^{-1}A = U^*AU$, we also have that $\mathscr{R}_{\mathbf{U}} = \mathscr{L}(\mathscr{H})$, which establishes the unitarity of $\mathbf{U}$.

Similarly, assuming that in (8.51) H is some bouunded self-adjoint operator with $\mathscr{D}_H = \mathscr{H}$, (8.52) yields

$$\begin{aligned} \hbar\langle A_1 \mid \mathbf{H}A_2 \rangle_2 &= \mathrm{Tr}[A_1^*(HA_2 - A_2H)] \\ &= \mathrm{Tr}[(HA_1)^*A_2 - HA_1^*A_2] = \hbar\langle \mathbf{H}A_1 \mid A_2 \rangle_2 , \end{aligned}$$

so that $\mathbf{H}$ is a self-adjoint superoperator on $\mathscr{L}(\mathscr{H})$. On the other hand, if H is unbounded we run into technical problems with domains of definition, so that we have to be more precise as to the meaning of $\mathbf{H}$ in (8.51). The following result illucidates, however, the relationship between the domains, spectra, and spectral measures of the operator H on $\mathscr{H}$ and the superoperator $\mathbf{H}$ on $\mathscr{L}(\mathscr{H})$.

Lemma 8.5. If H is any self-adjoint operator in $\mathscr{H}$, then the superoperator $\mathbf{H}$ acting in $\mathscr{L}(\mathscr{H})$ as follows

$$\mathbf{H}A = \hbar^{-1}[H, A]^{**} = \hbar^{-1}(\overline{HA} - \overline{AH}) \tag{8.53}$$

and with domain $\mathscr{D}_{\mathbf{H}}$ consisting of all $A \in \mathscr{L}(\mathscr{H})$ for which $\overline{HA}, \overline{AH} \in \mathscr{L}(\mathscr{H})$, is self-adjoint in the Liouville space $\mathscr{L}(\mathscr{H})$; furthermore,

$$\mathsf{S}^{\mathbf{H}} = \left\{ \frac{\lambda' - \lambda''}{\hbar} : \lambda', \lambda'' \in \mathsf{S}^H \right\}, \qquad \mathsf{S}_{\mathrm{p}}^{\mathbf{H}} = \left\{ \frac{\lambda' - \lambda''}{\hbar} : \lambda', \lambda'' \in \mathsf{S}_{\mathrm{p}}^H \right\},$$

and the action of the spectral function $\mathbf{E}_\lambda^{\mathbf{H}}$ of the superoperator $\mathbf{H}$ on any $A \in \mathscr{D}_{\mathbf{H}}$ can be expressed in terms of the spectral function E_λ^H of the operator H by means of the following spectral integral*

$$\mathbf{E}_{\hbar\lambda}^{\mathbf{H}} A = \int_{-\infty}^{+\infty} E_{\lambda-\mu}^H A \, d_\mu E_\mu^H . \tag{8.54}$$

If we denote by h-lim the *limit in the Hilbert–Schmidt norm* on $\mathscr{L}(\mathscr{H})$ (so that $A = \text{h-lim}\, A_n$ means that $\| A - A_n \|_2 \to 0$ as $n \to +\infty$), then we easily deduce from Theorems 3.1 and 8.4, Lemma 8.5, and the properties of (8.50) that the following is true (see Prugovečki and Tip [1975]).

* See §3.7 in Chapter V for a definition of spectral integrals. The proof of (8.54) is a corollary of Theorem A.2 by Prugovečki [1972b], whereas the proof of the remainder of Lemma 8.5 follows from the results in Section 3 of Prugovečki and Tip [1975].

Theorem 8.11. If H is any self-adjoint operator in $\mathscr{H}$ and $\mathbf{U}_t$ is the superoperator that acts on all $A \in \mathscr{L}(\mathscr{H})$ as follows,

$$\mathbf{U}_t A = e^{(i/\hbar)Ht} A e^{-(i/\hbar)Ht}, \qquad t \in \mathbb{R}^1, \tag{8.55}$$

then $\mathbf{U}_t$ is unitary on $\mathscr{L}(\mathscr{H})$ and

$$\mathbf{U}_{t_1+t_2} = \mathbf{U}_{t_1}\mathbf{U}_{t_2}, \qquad t_1, t_2 \in \mathbb{R}^1, \tag{8.56}$$

$$\mathbf{U}_0 A = A = \operatorname*{h-lim}_{t\to 0} \mathbf{U}_t A, \qquad A \in \mathscr{L}(\mathscr{H}), \tag{8.57}$$

$$i\mathbf{H}A = \operatorname*{h-lim}_{t\to 0} (\mathbf{U}_t A - A)/t. \tag{8.58}$$

We recall how Theorem 3.1 was used in clarifying the meaning of the Schroedinger equation (3.1) and we observe that the same arguments lead to the conclusion that

$$\mathbf{U}(t, t_0) = \mathbf{U}_{t-t_0} = \mathbf{U}_t \mathbf{U}_{t_0}^{-1}$$

leaves $\mathscr{D}_\mathbf{H}$ invariant, and that

$$i\frac{dA(t)}{dt} = \operatorname*{h-lim}_{\Delta t\to 0} \frac{A(t+\Delta t) - A(t)}{\Delta t} = \mathbf{H}A(t) \tag{8.59}$$

whenever $A(t_0) \in \mathscr{D}_\mathbf{H}$ and $A(t)$ evolves in accordance with (8.39):

$$A(t) = \mathbf{U}(t, t_0)\, A(t_0) = U(t, t_0)\, A(t_0)\, U^*(t, t_0). \tag{8.60}$$

When we now compare (8.44) with (8.53) and (8.59) we immediately uncover the mathematical meaning as well as the physical significance of von Neumann's equation: (8.44) is in fact a Schroedinger equation on Liouville space, which consequently governs not the time evolution of state vectors (i.e., merely of pure states), but that of density operators (i.e., of either pure or mixed states)!

8.8. Density Matrices on Spectral Representation Spaces

In quantum mechanics actual computations of quantities describing the behavior of a given system are performed in some spectral representation space $L^2(\mathbb{R}^n, \mu)$ of a complete set of observables $\{A_1, ..., A_n\}$ for that system. Such spaces were introduced in the abstract in Definition 5.2, and concrete realizations of particular importance were encountered in §7 in the form of configuration, momentum, and angular momentum representation spaces.

The same observation stays true in quantum statistical mechanics by virtue of the identification of the Liouville space $\mathscr{L}(\mathscr{H})$ over $\mathscr{H} = L^2(\mathbb{R}^n, \mu)$ with the Hilbert space $L^2(\mathbb{R}^{2n}, \mu \times \mu)$.

Theorem 8.12. To each Hilbert–Schmidt operator A in the Liouville space $\mathscr{L}(\mathscr{H})$ over $\mathscr{H} = L^2(\mathbb{R}^n, \mu)$ there corresponds a unique $h_A \in L^2(\mathbb{R}^{2n}, \mu \times \mu)$ such that

$$\langle f \mid Ag \rangle = \int_{\mathbb{R}^{2n}} f^*(x')\, h_A(x', x'')\, g(x'')\, d\mu(x')\, d\mu(x'') \tag{8.61}$$

for all f, $g \in L^2(\mathbb{R}^n, \mu)$, and the mapping $A \mapsto h_A$ is a unitary transformation of $\mathscr{L}(L^2(\mathbb{R}^n, \mu))$ onto $L^2(\mathbb{R}^{2n}, \mu \times \mu)$.

Proof. If $e_k(x)$, $e_k'(x) \in L_{(2)}(\mathbb{R}^n, \mu)$ are functions representing the elements e_k, $e_k' \in L^2(\mathbb{R}^n, \mu)$ in the canonical representation (8.3) of $A \in \mathscr{L}(\mathscr{H})$, then we observe that by (8.14)

$$\| A_m - A_n \|_2^2 = \sum_{k=n+1}^{m} \lambda_k^2, \qquad A_n = \sum_{k=1}^{n} | e_k' \rangle \lambda_k \langle e_k |,$$

and if we set

$$h_{A_n}(x', x'') = \sum_{k=1}^{n} e_k'(x')\, \lambda_k e_k^*(x''),$$

then

$$\begin{aligned} \langle f \mid A_n g \rangle &= \int_{\mathbb{R}^{2n}} f^*(x')\, h_{A_n}(x', x'')\, g(x'')\, d\mu(x')\, d\mu(x''), \\ \| A_m - A_n \|_2^2 &= \int_{\mathbb{R}^{2n}} | h_{A_m}(x', x'') - h_{A_n}(x', x'')|^2\, d\mu(x')\, d\mu(x''). \end{aligned} \tag{8.62}$$

Consequently, by the Riesz–Fischer theorem (Theorem 4.4 in Chapter II),

$$h_A(x', x'') = \operatorname*{l.i.m.}_{n \to \infty} h_{A_n}(x', x'') \tag{8.63}$$

exists and belongs to $L_{(2)}(\mathbb{R}^{2n}, \mu \times \mu)$. On the other hand, by (8.3) and (8.4),

$$\langle f \mid Ag \rangle = \lim_{n \to \infty} \langle f \mid A_n g \rangle, \qquad f, g \in L^2(\mathbb{R}^n, \mu),$$

and since $f(x')\, g^*(x'') \in L_{(2)}(\mathbb{R}^{2n}, \mu \times \mu)$, we deduce from (8.62) that (8.61) is satisfied.

The function $h_A(x', x'')$ is not uniquely determined by A since, in general, the canonical decomposition (8.3) is not unique [e.g., that is the case if some λ_k is a degenerate eigenvalue of some $A = A^* \in \mathscr{L}(\mathscr{H})$]. However, if $h_A'(x', x'')$ also satisfies (8.61), then we have

$$\int_{\mathbb{R}^{2n}} [f(x')\, g^*(x'')]^* [h_A(x', x'') - h_A'(x', x'')]\, d\mu(x')\, d\mu(x'') = 0$$

for all $f, g \in \mathscr{H} = L^2(\mathbb{R}^n, \mu)$, and since by Theorem 6.9 in Chapter II the linear manifold spanned by all functions $f(x')\, g^*(x'')$ is dense in $L^2(\mathbb{R}^{2n}, \mu \times \mu)$, we conclude that $h_A(x', x'')$ and $h_A'(x', x'')$ have to be almost everywhere equal with respect to $\mu \times \mu$, and therefore they represent one and the same element of $L^2(\mathbb{R}^{2n}, \mu \times \mu)$.

Thus $A \mapsto h_A$ is defined for all $A \in \mathscr{L}(\mathscr{H})$, and it obviously represents a linear mapping. To prove that this mapping is an isometry we note that by (8.14)

$$\begin{aligned} \operatorname{Tr}(A^*A) &= \lim_{n\to\infty} \sum_{k=1}^{n} \lambda_k^2 = \lim_{n\to\infty} \operatorname{Tr}(A_n^* A_n) \\ &= \lim_{n\to\infty} \int_{\mathbb{R}^{2n}} | h_{A_n}(x', x'')|^2\, d\mu(x')\, d\mu(x''), \end{aligned}$$

and consequently, by (8.17),

$$\| A \|_2 = \left[\int_{\mathbb{R}^{2n}} | h_A(x', x'')|^2\, d\mu(x')\, d\mu(x'') \right]^{1/2} = \| h_A \|_{L^2}. \tag{8.64}$$

The unitarity of this same mapping is finally established by noting that to any element $h(x', x'')$ of $L_{(2)}(\mathbb{R}^{2n}, \mu \times \mu)$ there corresponds a unique operator A such that when h_A is replaced by h in (8.61) the outcome is an equality valid for all $f, g \in L^2(\mathbb{R}^n, \mu)$ (see Exercise 5.3 in Chapter V).

Q.E.D.

If $\rho \in \mathscr{L}(L^2(\mathbb{R}^{2n}))$ is a density operator, then (8.3) assumes the form (8.42), and we shall write

$$\langle x' \mid \rho \mid x'' \rangle = h_\rho(x', x'') = \underset{n\to\infty}{\text{l.i.m.}} \sum_{k=1}^{n} \Psi_k(x')\, w_k \Psi_k^*(x''). \tag{8.65}$$

Following a long-established tradition in physics literature, we shall call the function $\langle x' \mid \rho \mid x'' \rangle$ a *density matrix* on the spectral representation space $L^2(\mathbb{R}^n, \mu)$—albeit this function can be identified with a matrix only in the case where the spectra of $A_1, \ldots, A_n$ in Definition 5.2 are all point spectra so that the support S of the measure μ is a discrete set.

This last case, however, almost never occurs in practice. For example, in the important case of the configuration to the representation (6.8) of the canonical commutation relations (concretely, if in the considerations of §7.2 we have $N = n/3$ spinless particles), the Hilbert space $\mathscr{H}$ equals $L^2(\mathbb{R}^n)$, and therefore $\langle x' \mid \rho \mid x'' \rangle$ is an element of $L_{(2)}(\mathbb{R}^{2n})$.

To understand the physical significance of the density matrix, we note that the probability appearing in (8.40) can be expressed as follows:

$$P_\rho^{A_1,\ldots,A_n}(B) = \lim_{n\to\infty} \sum_{k=1}^{n} w_k \int_B |\Psi_k(x)|^2 \, d\mu(x). \tag{8.66}$$

Indeed, choosing an orthonormal basis $e_1, e_2, \ldots$ in the subspace of $L^2(\mathbb{R}^n, \mu)$ onto which $E^{A_1,\ldots,A_n}(B)$ projects, and then completing it to an orthonormal basis in $L^2(\mathbb{R}^n, \mu)$ by adding to it $e_1', e_2', \ldots$ with $\operatorname{supp} e_i'(x) \subset B' = \mathbb{R}^n - B$, we obtain

$$\begin{aligned} \operatorname{Tr}[\rho E^{A_1,\ldots,A_n}(B)] &= \sum_i \langle e_i \mid \rho e_i \rangle \\ &= \lim_{n\to\infty} \sum_{k=1}^{n} w_k \sum_i \langle e_i \mid \Psi_k \rangle \langle \Psi_k \mid e_i \rangle \end{aligned} \tag{8.67}$$

through the use of theorems on the unconditional convergence of double series with positive sums [already used in (8.11)]. The right-hand sides of (8.66) and (8.67), however, coincide.

We now set by definition

$$\langle x \mid \rho \mid x \rangle = \sum_{k=1}^{\infty} w_k \, |\Psi_k(x)|^2 \tag{8.68}$$

at all $x \in \mathbb{R}^n$ where the above series converges, and by Lebesgue's monotone convergence theorem we obtain in accordance with (8.66) the basic result

$$P_\rho^{A_1,\ldots,A_n}(B) = \operatorname{Tr}[\rho E^{A_1,\ldots,A_n}(B)] = \int_B \langle x \mid \rho \mid x \rangle \, d\mu(x), \tag{8.69}$$

which represents the most general integral-form expression for the probability that a simultaneous determinative measurement of the complete set $\{A_1, \ldots, A_n\}$ of observables of a system in a (pure or mixed) quantum state ρ would yield values in the Borel set $B \in \mathscr{B}^n$. For the special case $B = \mathbb{R}^n$ (8.69) assumes the form of a trace formula:

$$\operatorname{Tr} \rho = \int_{\mathbb{R}^n} \langle x \mid \rho \mid x \rangle \, d\mu(x). \tag{8.70}$$

It should be emphasized that in general the defining formula (8.68) of the so-called *diagonal elements* $\langle x \mid \rho \mid x \rangle$ of the density matrix $\langle x' \mid \rho \mid x'' \rangle$ is not derivable merely by setting $x' = x'' = x$ in (8.65). Indeed, (8.65) involves a limit-in-the-mean, which in general implies pointwise convergence only almost everywhere on $\mathbb{R}^{2n}$ for some suitable subsequence of the partial sums of the given infinite series (see Chapter II, Theorem 4.5), whereas the set $\{(x', x'') \mid x' = x''\}$ might be of μ-measure zero in $\mathbb{R}^{2n}$ (e.g., that is the case when μ is a Lebesgue–Stieltjes measure). On the other hand, all the terms of the series in (8.68) are nonnegative and

$$\int_B \sum_{k=1}^{n} w_k \mid \Psi_k(x)|^2 \, d\mu(x) \leqslant \sum_{k=1}^{n} w_k \leqslant 1$$

so that (8.69) is true by Theorem 3.10 in Chapter II (cf. also Fatou's lemma). Furthermore, if the equivalence class of all functions $h_\rho(x', x'') \in L_{(2)}(\mathbb{R}^{2n})$ which satisfy (8.65) contains a continuous function, then we can set by definition

$$\langle x' \mid \rho \mid x'' \rangle = \sum_{k=1}^{\infty} \Psi_k(x') \, w_k \Psi_k{}^*(x'')$$

for appropriate choices of $\Psi_k(x)$, $k = 1, 2,\ldots$. Indeed, in that case by Mercer's theorem (see Riesz and Sz. Nagy [1955], Section 98), the functions $\Psi_k(x) \in L_{(2)}(\mathbb{R}^n)$, $k = 1, 2,\ldots$, appearing in (8.65) can be chosen to be continuous, and then the above series converges pointwise (and, in fact, also uniformly on compact sets). Hence, in the practically important case of continuous density matrices $\langle x' \mid \rho \mid x'' \rangle \in L_{(2)}(\mathbb{R}^{2n})$, we can obtain the "diagonal" values defined in (8.68) simply by setting $x' = x'' = x$ in the chosen continuous density matrix $\langle x' \mid \rho \mid x'' \rangle$.

The most frequently used representations of density matrices are those in configuration and in momentum space. For example, for a system of N spinless particles possessing a total of $n = 3N$ degrees of freedom, the frequent choice of spectral representation space is $L^2(\mathbb{R}^n)$, and therefore the Liouville space can be identified with $L^2(\mathbb{R}^{2n})$, where the measure μ is in this case the Lebesgue measure on the measurable space $(\mathbb{R}^n, \mathscr{B}^n)$.

**8.9. Appendix: Classical and Quantum Statistical Mechanics in Master Liouville Space

Classical and quantum statistical mechanics are both statistical theories describing the same body of phenomena, and it is commonly held that when both classical and quantum theory are applied to one and the same system (e.g., a Brownian particle moving in some gas, such as air) then the classical description should provide in some sense an approximation to the quantum one. Yet, the conventional mathematical formalisms for these two cases are so very dissimilar that it is not at all evident that this observation is of universal validity. Thus, whereas, as we have seen in §§8.6–8.8, the states of an ensemble are represented in the quantum case by density operators ρ in a Liouville space, their evolution in time (in the Schroedinger picture) being governed by the von Neumann equation (8.44), in the classical case the states of the same ensemble are represented by distribution functions $f(q, p; t)$ on phase space Γ (where $\Gamma = \mathbb{R}^{2n}$ in case of n degrees of freedom) which obey the *Liouville equation*

$$\frac{\partial f}{\partial t} = \sum_{i=1}^{n} \left(\frac{\partial H^{\text{cl}}}{\partial q_i} \frac{\partial f}{\partial p_i} - \frac{\partial H^{\text{cl}}}{\partial p_i} \frac{\partial f}{\partial q_i} \right) \tag{8.71}$$

at almost all $(q, p) \in \Gamma$, where H^{cl} is the classical Hamiltonian; e.g., for time-independent potentials $V(q)$ and N particles moving in three dimensions

$$H^{\text{cl}}(q, p; t) = \sum_{i=1}^{n} \frac{p_i^2}{2m_i} + V(q), \tag{8.72}$$

where $n = 3N$, $q = (q_1, \ldots, q_n)$, $p = (p_1, \ldots, p_n)$, and $m_{3j+1} = m_{3j+2} = m_{3j+3}$, $j = 0, \ldots, N-1$.

In this appendix we shall indicate how one can take advantage of the very general approach to quantum mechanics developed in this chapter to treat classical and quantum statistical mechanics in a unified manner by working in both instances on the same Hilbert space. In the aforementioned case of n degrees of freedom this Hilbert space will be $L^2(\Gamma)$. We shall refer to the Liouville space $\mathscr{L}_\Gamma$ over $L^2(\Gamma)$ as the *master Liouville space* for both theories.*

We represent in $\mathscr{L}_\Gamma$ the classical state of an ensemble by the equivalence class of all density operators $\rho^{\text{cl}} \in \mathscr{L}_\Gamma$ for which

$$f(q, p) = \langle q, p \mid \rho^{\text{cl}} \mid q, p \rangle = \lim_{n\to\infty} \sum_{k=1}^{n} w_k \, |\Psi_k(q, p)|^2, \tag{8.73}$$

* It is also possible to recast quantum statistical mechanics in terms of a formalism based on phase space distribution functions. For these distribution functions one can adopt the Wigner transforms of density matrices (see, e.g., Balescu [1975]). However, this transform is not positive definite, so it cannot be interpreted as a probability density on phase space (see, e.g., Srinivas and Wolf [1975]). The alternative is to use the stochastic phase space probability densities briefly described later in this section (see the review articles by Prugovečki [1979, 1981 a, b] and Ali and Prugovečki [1980] for details and further references).

almost everywhere in $L^2(\Gamma)$, where $f(q, p)$ is the Γ distribution function describing that state. Introducing in $\mathscr{L}_\Gamma$ the superoperator

(8.74)

$$\mathbf{H}^{\mathrm{cl}} \supseteq$$

$$-i \sum_{k=1}^{n} \left[m_k^{-1} \left(p_k' \frac{\partial}{\partial q_k'} - p_k'' \frac{\partial}{\partial q_k''} \right) - \left(\frac{\partial V(q')}{\partial q_k'} \frac{\partial}{\partial p_k'} - \frac{\partial V(q'')}{\partial q_k''} \frac{\partial}{\partial p_k''} \right) \right],$$

it is easily verified* that if we restrict ourselves to diagonal elements, by setting $q' = q''$, $p' = p''$ in the (classical) von Neumann equation

$$i \left\langle q', p' \left| \frac{\partial \rho_t^{\mathrm{cl}}}{\partial t} \right| q'', p'' \right\rangle = \langle q', p' \mid \mathbf{H}^{\mathrm{cl}} \rho_t^{\mathrm{cl}} \mid q'', p'' \rangle \tag{8.75}$$

for any density matrix representing an evolving state in accordance with (8.73), then (8.75) coincides with the Liouville equation (8.71).

To formulate quantum statistical mechanics on $\mathscr{L}_\Gamma$, we adopt in $L^2(\Gamma)$ the following representation of the canonical commutation relations

$$\hat{Q}_k \supseteq i\hbar \frac{\partial}{\partial p_k}, \qquad \hat{P}_k \supseteq p_k - i\hbar \frac{\partial}{\partial q_k}, \qquad k = 1, \ldots, n. \tag{8.76}$$

Of course, this representation is highly reducible and according to Theorem 6.4 it can be decomposed into a (countable) direct sum of irreducible representations. Theorem 6.4, however, does not claim that the decomposition is unique, and indeed an uncountable infinity of such decompositions exists.

To uncover those closed subspaces of $L^2(\Gamma)$ carrying irreducible representations of physical interest, we consider *positive operator-valued measures* of the form

$$\mathbb{P}_\gamma(B) = \int_B \mid \gamma_{q,p} \rangle \langle \gamma_{q,p} \mid d^n q \, d^n p, \qquad B \in \mathscr{B}^{2n}, \tag{8.77}$$

where for each fixed $(q, p) \in \Gamma$ the function

$$\gamma_{q,p}(q', p') = \int_{\mathbb{R}^n} \exp[(i/\hbar)(p - p') \cdot x] \, \gamma^*(x - q') \, \gamma(x - q) \, d^n x \tag{8.78}$$

* Further details and proofs of all the statements in this section can be found in Ali and Prugovečki [1977a, b] and Prugovečki [1978a–d, 1981b].

represents an element of $L^2(\Gamma)$ if $\gamma \in L^2(\mathbb{R}^n)$. It turns out that $\mathbb{P}_\gamma(\Gamma)$ is the projector onto a closed subspace of $L^2(\Gamma)$ when $(2\pi\hbar)^{n/2}\gamma(x)$ is normalized, and that this subspace $L^2(\Gamma_\gamma)$ is left invariant by the operators (8.76) that induce in $L^2(\Gamma_\gamma)$ an irreducible representation of the canonical commutation relations. The last fact is established by proving that the mapping

$$\psi(x) \mapsto \psi(q, p) = \int_{\mathbb{R}^n} \exp[-(i/\hbar)\, p \cdot x]\, \gamma^*(x - q)\, \psi(x)\, d^n x \tag{8.79}$$

supplies a unitary transformation U_γ of $L^2(\mathbb{R}^n)$ onto $L^2(\Gamma_\gamma)$, and that the unitary transforms $U_\gamma Q_k U_\gamma^{-1}$ and $U_\gamma P_k U_\gamma^{-1}$ of the Schroedinger representation (6.8) of the canonical commutation relation for n degrees of freedom coincide with the restrictions of the operators $\hat{Q}_k$ and $\hat{P}_k$ in (8.76) to the closed subspace $L^2(\Gamma_\gamma)$ of $L^2(\Gamma)$ [note that $L^2(\Gamma_\gamma)$ is itself a Hilbert space!]. In general, distinct choices of $\gamma \in L^2(\mathbb{R}^n)$ will lead to distinct spaces $L^2(\Gamma_\gamma)$ (that have only the zero vector in common) as illustrated by the choice

$$\gamma^{(s)}(x) = (\pi\hbar^3 s^2)^{-n/4} \exp(-x^2/2\hbar s^2), \qquad s > 0, \tag{8.80}$$

providing a continuum of different subspaces $L^2(\Gamma_{\gamma^{(s)}})$ as the parameter s varies over $(0, \infty)$.

The fact that the mapping U_γ defined in (8.79) is unitary for each $\gamma \in L^2(\mathbb{R}^n)$ indicates that one and the same quantum state of an ensemble will be represented in $\mathscr{L}_\Gamma$ by an equivalence class of density operators, albeit the equivalence relation has to be defined in a way that is different from the method used in the classical case, namely to each $\rho \in \mathscr{L}(L^2(\Gamma)) = \mathscr{L}_\Gamma$, given in the configuration representation by the density matrix $\langle x' \mid \rho \mid x'' \rangle$, we assign a unique $\rho^\gamma = U_\gamma \rho U_\gamma^{-1}$ in each of the Liouville spaces $\mathscr{L}_{\Gamma_\gamma}$ over $L^2(\Gamma_\gamma)$. The essential point, however, is that a (quantum) von Neumann equation

$$i\langle q', p' \mid \partial\rho_t/\partial t \mid q'', p'' \rangle = \langle q', p' \mid \hat{\mathbf{H}}\rho_t \mid q'', p'' \rangle, \tag{8.81}$$

representing a counterpart of (8.75) can be introduced globally on $L^2(\Gamma)$ by setting

$$\hat{\mathbf{H}} = \hbar^{-1}[\hat{H}_0 + \hat{H}_\mathrm{I}, \cdot], \qquad \hat{H}_0 = \sum_{i=1}^{n} \hat{P}_i^2/2m_i, \quad \hat{H}_\mathrm{I} = V(\hat{Q}), \tag{8.82}$$

and that due to the unitarity of U_γ the restriction of this global equation to each $\mathscr{L}_{\Gamma_\gamma}$ supplies a time evolution for each

$$\rho_t^\gamma = \exp[i(\hat{\mathbf{H}}_0 + \hat{\mathbf{H}}_\mathrm{I})t]\rho^\gamma, \qquad t \in \mathbb{R}^1, \tag{8.83}$$

that is unitarily equivalent to that supplied by the conventional von Neumann equation in the configuration representation (naturally, the self-adjointness of $\hat{Q}_k$ and $\hat{P}_k$ guarantees that of $\hat{H}_0$ and $\hat{H}_{\mathrm{I}}$ defined as operator-valued functions in accordance with Theorem 2.5, and the question of the self-adjointness of $\hat{H}_0 + \hat{H}_{\mathrm{I}}$ can be handled with the methods of §7); furthermore, when the restriction to diagonal elements is executed in (8.75) and (8.81), a detailed comparison shows that in the zeroth order in $\hbar$ the two equations coincide (cf. Prugovečki [1978b]).

This last observation supplies the first indication as to the exact sense in which the classical theory approximates the quantum theory when in the later case terms of first and higher orders in $\hbar$ are neglected. However, the physical validity of this conclusion depends on the existence of a physical interpretation for the diagonal elements of density matrices in $\mathscr{L}_\Gamma$. In the classical case such an interpretation has been implicitly imposed from the very beginning by the relation (8.73) between Γ distribution functions and density matrices on $\mathscr{L}_\Gamma$.

To discover the physical meaning of $\langle q, p \mid \rho^\gamma \mid q, p\rangle$ in the quantum case, we use the mapping

(8.84) $$\rho \mapsto \rho^\gamma = U_\gamma \rho U_\gamma^{-1}, \qquad \rho \in \mathscr{L}(L^2(\mathbb{R}^n)), \qquad \rho^\gamma \in \mathscr{L}_{\Gamma_\gamma},$$

to establish that as a consequence of the definition (8.79) of U_γ,

(8.85) $$\int_{\mathbb{R}^n} \langle q, p \mid \rho^\gamma \mid q, p\rangle \, d^n p$$
$$= \int_{\mathbb{R}^n} \chi_q^\gamma(x) \langle x \mid \rho \mid x\rangle \, d^n x, \qquad \chi_q^\gamma(x) = (2\pi\hbar)^n \mid \gamma(x-q)\mid^2,$$

(8.86) $$\int_{\mathbb{R}^n} \langle q, p \mid \rho^\gamma \mid q, p\rangle \, d^n q$$
$$= \int_{\mathbb{R}^n} \hat{\chi}_p^\gamma(k) \langle k \mid \hat{\rho} \mid k\rangle \, d^n k, \qquad \hat{\chi}_p^\gamma(k) = (2\pi\hbar)^n \mid \tilde{\gamma}(k-p)\mid^2,$$

where $\langle k \mid \hat{\rho} \mid k\rangle$ denotes the density matrix in the momentum representation, i.e. (cf. §8.8),

$$\langle x \mid \rho \mid x\rangle = \sum_{j=1}^{\infty} w_j \mid \psi_j(x)\mid^2, \qquad \langle k \mid \hat{\rho} \mid k\rangle = \sum_{j=1}^{\infty} w_j \mid \tilde{\psi}_j(k)\mid^2.$$

The physical meaning of the integrals in (8.85) and (8.86) emerges as soon as the resulting measures

$$(8.87)\quad P_{\rho^\gamma}^{Q_1,\ldots,Q_n}(B) = \mathrm{Tr}[\rho^\gamma \mathbb{P}_\gamma(B \times \mathbb{R}^n)] = \int_B d^n q \int_{\mathbb{R}^n} \chi_q{}^\gamma(x)\langle x \mid \rho \mid x\rangle\, d^n x,$$

$$(8.88)\quad P_{\rho^\gamma}^{P_1,\ldots,P_n}(B) = \mathrm{Tr}[\rho^\gamma \mathbb{P}_\gamma(\mathbb{R}^n \times B)] = \int_B d^n p \int_{\mathbb{R}^n} \hat{\chi}^\gamma(k)\langle k \mid \hat{\rho} \mid k\rangle\, d^n k,$$

on $(\mathbb{R}^n, \mathscr{B}^n)$ are compared with their conventional counterparts obtained by specializing (8.69) to the configuration and momentum representations, respectively:

$$(8.89)\quad P_\rho^{Q_1,\ldots,Q_n}(B) = \mathrm{Tr}[\rho E^{Q_1,\ldots,Q_n}(B)] = \int_B \langle x \mid \rho \mid x\rangle\, d^n x,$$

$$(8.90)\quad P_\rho^{P_1,\ldots,P_n}(B) = \mathrm{Tr}[\rho E^{P_1,\ldots,P_n}(B)] = \int_B \langle k \mid \hat{\rho} \mid k\rangle\, d^n k.$$

Indeed, if we consider (8.78) and (8.88) for the case of γ given in (8.80), we easily establish that

$$(8.91)\quad P_\rho^{Q_1,\ldots,Q_n}(B) = \lim_{s\to+0} P_{\rho^{\gamma(s)}}^{Q_1,\ldots,Q_n}(B),$$

$$(8.92)\quad P_\rho^{P_1,\ldots,P_n}(B) = \lim_{s\to+\infty} P_{\rho^{\gamma(s)}}^{P_1,\ldots,P_n}(B).$$

For $s \in (0, +\infty)$, or for the case of a general γ, (8.87) and (8.88) also have an obvious interpretation: whereas (8.89) and (8.90) are the probabilities that would be measured if in each random sample of values of position and momentum, respectively, each value $x \in \mathbb{R}^n$ and $k \in \mathbb{R}^n$ would be determined with absolute precision, (8.87) and (8.88) are probabilities obtainable with realistic measurements performed with actual instruments whose readings necessarily involve a margin of uncertainty (cf. Dietrich [1973]). In other words, by executing the calibration of such instruments we always find out that, say, a *reading* q in reality means that for any interval I containing q the confidence probability of the *actual value* x being within I equals $\mu^\gamma(I)$, and that the measure $\mu^\gamma(B)$, $B \in \mathscr{B}^n$, is never a δ-measure centered at x. Thus, in our case the respective confidence measures would be

$$(8.93)\quad \mu_q{}^\gamma(B) = \int_B \chi_q{}^\gamma(x)\, d^n x, \qquad \hat{\mu}_p{}^\gamma(B) = \int_B \hat{\chi}_p{}^\gamma(k)\, d^n k,$$

and we can regard (8.78) and (8.88) as the probabilities of obtaining *stochastic* values (q, μ_q^γ) and $(p, \hat{\mu}_p^\gamma)$, respectively, with $q \in B$ and $p \in B$. Naturally, under normal circumstances the difference between (8.87) and (8.89), or (8.88) and (8.90), might be negligible by comparison with the statistical fluctuation displayed by different samples.

Yet, there is one obvious advantage of dealing with stochastic values in quantum mechanics: if we do that we can then consider the possibility of simultaneous stochastic values $((q, \mu_q^\gamma), (p, \hat{\mu}_p^\gamma))$ for position and momentum without violating the uncertainty principle, since in view of the definitions in (8.85) and (8.86) of the *confidence functions* $\chi_q^\gamma(x)$ and $\hat{\chi}_p^\gamma(k)$, the spreads of these functions automatically obey that principle. Denoting by Γ_γ the resulting space* of all such stochastic values obtained as (q, p) varies over Γ with γ fixed, we can consistently extend the standard interpretation of quantum mechanics based on Definition 1.3 by interpreting

$$P_{\rho^\gamma}^{Q_1,\ldots,Q_n,P_1,\ldots,P_n}(B) = \mathrm{Tr}[\rho^\gamma \mathbb{P}_\gamma(B)], \qquad B \in \mathscr{B}^{2n}, \tag{8.94}$$

as the probability that a determinative measurement of stochastic values in Γ_γ on an ensemble in state ρ would yield outcomes $((q, \mu_q^\gamma), (p, \hat{\mu}_p^\gamma))$ with $(q, p) \in B$. Indeed, by (8.87) and (8.88),

$$P_{\rho^\gamma}^{Q_1,\ldots,Q_n,P_1,\ldots,P_n}(B \times \mathbb{R}^n) = P_{\rho^\gamma}^{Q_1,\ldots,Q_n}(B), \tag{8.95}$$

$$P_{\rho^\gamma}^{Q_1,\ldots,Q_n,P_1,\ldots,P_n}(\mathbb{R}^n \times B) = P_{\rho^\gamma}^{P_1,\ldots,P_n}(B), \tag{8.96}$$

so that this interpretation is simply an extrapolation of the conventional one.

One of the many attractive features of such an interpretation is that it naturally leads to a concept of probability current in quantum mechanics without any need for extraneous new postulates. Consider for the sake of simplicity the case of a single particle, i.e., of three degrees of freedom and of $\Gamma = \mathbb{R}^6$. Setting in the classical case

$$\rho_t^{\mathrm{cl}}(\mathbf{r}) = \int_{\mathbb{R}^3} \langle \mathbf{r}, \mathbf{p} \mid \rho_t^{\mathrm{cl}} \mid \mathbf{r}, \mathbf{p} \rangle \, d\mathbf{p} = \int_{\mathbb{R}^3} f_t(\mathbf{r}, \mathbf{p}) \, d\mathbf{p} \tag{8.97}$$

$$\mathbf{j}_t^{\mathrm{cl}}(\mathbf{r}) = \int_{\mathbb{R}^3} \mathbf{v} f_t(\mathbf{r}, \mathbf{p}) \, d\mathbf{p}, \qquad \mathbf{v} = \mathbf{p}/m, \tag{8.98}$$

we see that as the average value of velocity, $\mathbf{j}_t^{\mathrm{cl}}(\mathbf{r})$ represents a *probability current* at $\mathbf{r} \in \mathbb{R}^3$. Specializing (8.71) and (8.72) to $n = 3$ and integrating

* This is called *stochastic phase space*; cf. Prugovečki [1981a].

both sides of the Liouville equation (8.71) in $\mathbf{p}$ over $\mathbb{R}^3$, we obtain the well-known *continuity equation*

$$\partial_t \rho_t^{\mathrm{cl}}(\mathbf{r}) + \nabla \cdot \mathbf{j}_t^{\mathrm{cl}}(\mathbf{r}) = 0 \tag{8.99}$$

expressing probability conservation.

Let us similarly define in the quantum case

$$\rho_t^{(s)}(\mathbf{r}) = \int_{\mathbb{R}^3} \langle \mathbf{r}, \mathbf{p} \mid \rho_t^{\gamma^{(s)}} \mid \mathbf{r}, \mathbf{p} \rangle \, d\mathbf{p}, \tag{8.100}$$

$$\mathbf{j}_t^{(s)}(\mathbf{r}) = \int (\mathbf{p}/m) \langle \mathbf{r}, \mathbf{p} \mid \rho_t^{\gamma^{(s)}} \mid \mathbf{r}, \mathbf{p} \rangle \, d\mathbf{p}. \tag{8.101}$$

We can then easily establish that (cf. Prugovečki [1978a])

$$\partial_t \rho_t^{(s)}(\mathbf{r}) + \nabla \cdot \mathbf{j}_t^{(s)}(\mathbf{r}) = 0, \tag{8.102}$$

so that $\mathbf{j}_t^{(s)}(\mathbf{r})$ possesses all the natural features of a probability current. Moreover,

$$\lim_{s \to +0} \rho^{(s)}(\mathbf{r}) = \sum_{k=1}^{\infty} w_k \mid \psi_k(\mathbf{r}) \mid^2 = \langle \mathbf{r} \mid \rho \mid \mathbf{r} \rangle \tag{8.103}$$

$$\lim_{s \to +0} \mathbf{j}^{(s)}(\mathbf{r}) = \frac{\hbar}{2im} \sum_{k=1}^{\infty} w_k [\psi_k^*(\mathbf{r}) \nabla \psi_k(\mathbf{r}) - \psi_k(\mathbf{r}) \nabla \psi_k^*(\mathbf{r})] \tag{8.104}$$

whenever $\nabla \psi_k(\mathbf{r})$, $k = 1, 2, \ldots$ are continuous in a neighborhood of $\mathbf{r}$ and the two series can be suitably majorized. The expression on the right-hand side of (8.104) coincides with the probability current found in all text-books on quantum mechanics (see, e.g., Messiah [1962]).

Exercises

8.1. Show that the projector E_{M} onto an infinite-dimensional closed subspace M of $\mathscr{H}$ is not completely continuous.

8.2. Prove that if the continuous spectrum $\mathsf{S}_{\mathrm{c}}^{A}$ of a self-adjoint operator A is not empty and contains points which are not accumulation points of the point spectrum, then $\mathsf{S}_{\mathrm{c}}^{A}$ contains infinitely many points.

8.3. Show that a bounded self-adjoint operator with a pure point spectrum, containing only eigenvalues of finite degeneracy which have no accumulation points with the possible exception of zero, is a completely continuous operator.

8.4. Show that the trace of a projector is equal to the dimension of the space onto which it projects.

8.5. Show that if a self-adjoint operator A is positive definite, i.e., $\langle f \mid Af \rangle \geqslant 0$ for all $f \in \mathscr{D}_A$, then its spectrum does not contain negative values.

8.6. Prove that if Tr A is finite and $A = A^*$, then the traces of the operators $A^{(+)}$ and $A^{(-)}$ defined in (8.24) are also finite.

8.7. Show that if the linear operators A, B, C are the uniform limits of the sequences $\{A_n\}$, $\{B_n\}$, $\{C_n\}$, then ABC is the uniform limit of the sequence $\{A_n B_n C_n\}$.

References for Further Study

The theory of quantum measurement has many ramifications that traditionally are surveyed primarily in monographs on the philosophy of quantum mechanics, such as d'Espagnat [1976] or Jammer [1974], or in specialized collections of review articles, such as those by Price and Chissick [1977] or Marlow [1978]; for basic aspects see, however, Messiah [1962], Gottfried [1966], and Ballentine [1970].

A great deal of functional analysis pertinent to the mathematical background of §§2–7 can be found in Riesz and Sz. Nagy [1955], Akhiezer and Glazman [1961], Stone [1964], Schechter [1971], Reed and Simon [1972, 1975], and Yosida [1974].

The operator–theoretical treatment of complete sets of observables of §5 is taken from Prugovečki [1969a], whereas other rigorous treatments are based on a W^*-algebra approach (Jauch and Misra [1975]). The case of infinitely many observables has been given a careful mathematical treatment by de Dormal and Gautrin [1975], and relevant results are also reported by Berezanskiĭ [1978] and Ascoli *et al.* [1978].

The subject of canonical commutation relations for a finite number of degrees of freedom treated in §6 is thoroughly surveyed by Putnam [1967], who also provides an extensive bibliography on the subject. An introduction to the case of infinitely many degrees of freedom that contains further references can be found in Bogolubov *et al.* [1975].

Many of the results on essential self-adjointness of Schroedinger operators that were only briefly dealt with in §7 are reviewed by Kato [1967] and by Reed and Simon [1975]. The subject of group theory in quantum mechanics (barely touched upon in §7.8 since it lies beyond the scope of the present book) has been given careful mathematical treatment in many monographs, such as those by Miller [1972] and by Barut and Rączka [1977].

The concept of density operator was first introduced by von Neumann [1955] in the original German edition of his classic book, and by now it forms the basis of all standard formulations of the concept of state in quantum statistical mechanics (see, e.g., Balescu [1975]). However, other formulations exist, such as those based on the C^*-algebra approach (see Emch [1972]). Many questions pertinent to the related subject of completely continuous, Hilbert–Schmidt and trace-class operators are systematically treated by Schatten [1960].

It should be mentioned that there have been many attempts to generalize the formalism of quantum mechanics by employing mathematical tools other than Hilbert space theory. Perhaps the best known is that employing lattice theory, and a very readable account can be found in Jauch [1968], whereas Varadarajan [1968] covers exhaustively the more technical aspects of this subject. In addition, other mathematical disciplines, ranging from W^*- and C^*-algebras (von Neumann [1936], Segal [1947]) to measure theory (Prugovečki [1966]) have been used. Yet, Hilbert space theory has remained the one and only practically viable alternative that enjoys universal acceptance in its role of providing the mathematical background for nonrelativistic quantum mechanics.

CHAPTER V

Quantum Mechanical Scattering Theory

In the preceding four chapters we have given the basic mathematical tools needed for a mathematically rigorous study of conventional nonrelativistic quantum mechanics. The actual applications have been limited either to obtaining very general, albeit fundamental, results of quantum mechanics or to computing the bound states of specific systems (e.g., the hydrogen atom).

Within the context of nonrelativistic quantum mechanics, scattering theory deals with collision phenomena of particles moving at velocities small by comparison with the speed of light. Since scattering theory is concerned with completely or partially free particles, it is of utmost interest from the experimental point of view. From the theoretical point of view, scattering theory is still the object of intensive research for the physicist as well as the mathematician.

We shall give in this chapter the most basic ideas of nonrelativistic quantum scattering theory. In illustrating these ideas, we shall limit ourselves to two-particle scattering, which is the only case which has received a thorough treatment until now. The scattering of n particles, usually referred to as n-body or multichannel scattering theory, is a field of active current research, which deserves at least one entire volume for a minimally adequate treatment. A very short introduction to this general case will be given in §8 of the present chapter.

1. Basic Concepts in Scattering Theory of Two Particles

1.1. Scattering Theory and the Initial-Value Problem

In studying in Chapter II n-particle systems in quantum mechanics, we were concerned exclusively with problems involving bound states.

Such states are related to the point spectrum of the internal energy operator of the system, as described in §5, Chapter II. We shall now turn our attention to two-particle states which are not bound states.

The solution of the general initial-value problem for an n-particle system is in essence contained in Axiom S3 of Chapter IV, §3—namely, if we could prepare by some means a state Ψ_0 of the system at the instant t_0, and if the Hamiltonian H were given, then, in principle, we could compute the state*

$$\Psi(t) = \exp[-iH(t - t_0)]\, \Psi_0$$

at all later instants t. However, such a computational job, besides being in most practical cases unmanageable due to its extremely great computational complexity, is not warranted by the needs of the experimentalist who carries out experiments at the present level of technical sophistication.

The experiments carried out at present in molecular, atomic, nuclear, and elementary particle physics which involve, from the theoretical point of view, an initial value problem are exclusively *scattering* experiments. Such experiments are characterized by the fact that the system undergoes a preparatory measurement at some instant t_0, a determinative measurement at a later instant t_1, and that for most of the time span $t_1 - t_0$ of the experiment there is no interaction (or practically negligeable interaction) between the different constituent parts of the system. As a matter of fact, these parts interact with one another only for a very small part of the duration $t_1 - t_0$ of the experiment. For the rest of the time they are spatially too far apart to have the very short-range forces make themselves felt. This and some other features of scattering experiments, which will be mentioned later, make it unnecessary to solve completely the initial-value problem in order to relate the theoretical predictions to the *available* experimental data.

It is clear from the above remarks that for the greater part of a scattering experiment, the different constituent parts of the physical system under observation move independently from one another, i.e., they are "almost free." In order to elucidate this intuitive concept of "almost free" motion in a mathematically clear and concise form, it is desirable to consider this concept first in the framework of classical mechanics, which readily appeals to intuition.

* In this chapter we adopt a system of units in which $\hbar = 1$.

1.2. Asymptotic States in Classical Mechanics

Let us consider a two-particle classical system. In the Newtonian formalism of classical mechanics, the state of such a system is given by a vector-valued function

$$x(t) = (x_1(t), y_1(t), z_1(t), x_2(t), y_2(t), z_2(t)) \in \mathbb{R}^6. \tag{1.1}$$

The trajectory of the kth particle, $k = 1, 2$, can be described by a three-dimensional vector-valued function $\mathbf{r}_k(t)$, which geometrically is a curve in three dimensions. If the two particles of the system undergo the kind of motion which was depicted earlier as typical in a scattering experiment, then the trajectories $\mathbf{r}_k(t)$, $k = 1, 2$, should have as characteristic the main qualitative features of the curves in Fig. 4: at the beginning $t \approx t_0$

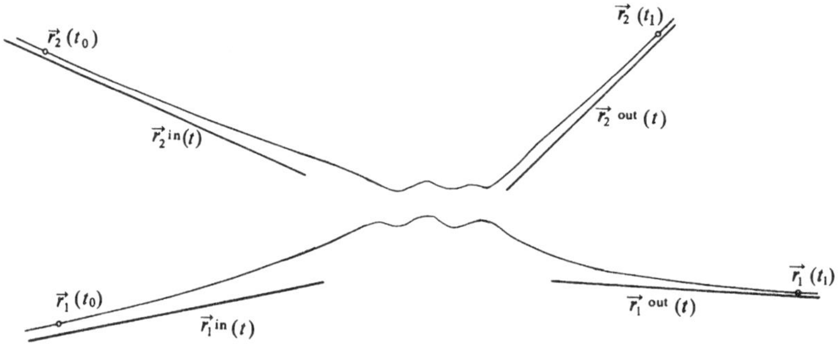

Fig. 4. Asymptotic states in classical mechanics.

of the experiment and for a relatively long time afterwards the motion of each particle should be "almost free," i.e., "almost" along a straight line $\mathbf{r}_k^{\text{in}}(t)$ at a uniform speed $\mathbf{v}_k^{\text{in}}$, $k = 1, 2$; then, during a relatively short period, while the particles are interacting with one another, the motion is, in general, very complex; finally, at the end, $t \approx t_1$, of the experiment and for some time before that, the motion of the particles should be again along straight lines $\mathbf{r}_k^{\text{out}}(t)$ and at "uniform speed" $\mathbf{v}_k^{\text{out}}$.

Now, in classical mechanics a particle is said to be in free motion if its state is of the form

$$\mathbf{r}(t) = \mathbf{r}_0 + \mathbf{v}t,$$

so that $\ddot{\mathbf{r}}(t) = 0$. Hence, at times $t \approx t_0$, we expect that

$$\mathbf{r}_k(t) \approx \mathbf{r}_k^{\text{in}} + \mathbf{v}_k^{\text{in}} t, \qquad k = 1, 2, \tag{1.2}$$

and similarly for $t \approx t_1$

$$\mathbf{r}_k(t) \approx \mathbf{r}_k^{\text{out}} + \mathbf{v}_k^{\text{out}} t, \qquad k = 1, 2. \tag{1.3}$$

The vectors $\mathbf{r}_k^{\text{in}}$ and $\mathbf{v}_k^{\text{in}}$ characterize a free state of the kth particle, $k = 1, 2$, which is called the *incoming asymptotic state*. Similarly, the free state characterized by $\mathbf{r}_k^{\text{out}}$ and $\mathbf{v}_k^{\text{out}}$ is called the *outgoing asymptotic state* of the kth particle.

It is not *a priori* clear in what precise sense the approximation signs in (1.2) and (1.3) should be understood. One obvious choice is to take them to mean that

$$\begin{aligned} |\,\mathbf{r}_k(t) - (\mathbf{r}_k^{\text{in}} + \mathbf{v}_k^{\text{in}} t)| &\approx 0, \qquad t \approx t_0\,, \\ |\,\dot{\mathbf{r}}_k(t) - \mathbf{v}_k^{\text{in}}\,| &\approx 0, \qquad t \approx t_0\,, \end{aligned} \tag{1.4}$$

where $|\cdot|$ above denotes the length of three-dimensional vectors. Since one can reasonably expect that the situation does not intrinsically change if we increase $t_1 - t_0$, and, moreover, the above approximations might hold with increasing accuracy when $t_0 \to -\infty$, it seems natural to give to (1.4) the following precise meaning:

$$\begin{aligned} \lim_{t\to-\infty} |\,\mathbf{r}_k(t) - (\mathbf{r}_k^{\text{in}} + \mathbf{v}_k^{\text{in}} t)| &= 0, \\ \lim_{t\to-\infty} |\,\dot{\mathbf{r}}_k(t) - \mathbf{v}_k^{\text{in}}\,| &= 0. \end{aligned} \tag{1.5}$$

Similar considerations lead to the following corresponding requirements:

$$\begin{aligned} \lim_{t\to+\infty} |\,\mathbf{r}_k(t) - (\mathbf{r}_k^{\text{out}} + \mathbf{v}_k^{\text{out}} t)| &= 0, \\ \lim_{t\to+\infty} |\,\dot{\mathbf{r}}_k(t) - \mathbf{v}_k^{\text{out}}\,| &= 0. \end{aligned} \tag{1.6}$$

The conditions (1.5) and (1.6) define in a precise way the asymptotic states of a single particle. Naturally, the asymptotic states $x^{\text{in}}(t)$ and $x^{\text{out}}(t)$ of the entire system is the aggregate of the asymptotic states of its two constituent particles.

Incoming asymptotic states defined by (1.5) can be thought of as being states in which the particles of the system would be at all times if there were no interaction between the two particles. Similarly, the outgoing states would correspond to the fictitious case of particles which have interacted in the "infinite past," and afterwards move independently.

We note that according to the definitions in (1.5) and (1.6), if asymptotic states exist at all for a given classical physical process, then they are unique for each state, since $\mathbf{v}_k^{\text{in}}$ and $\mathbf{v}_k^{\text{out}}$ are uniquely determined by the second of the respective relations in (1.5) and in (1.6), and then $\mathbf{r}_k^{\text{in}}$ and $\mathbf{r}_k^{\text{out}}$ are uniquely determined by the first relations in (1.5) and (1.6), respectively. It is, however, natural to ask whether asymptotic states in the above sense exist for every classical physical process of importance.

The answer to the above question is negative. In the case of long-range forces, i.e., forces which are of significant intensity even at arbitrarily large separations between particles, asymptotic states in the above sense do not exist. A notable example of such a force is the Coulomb force (see Exercises 1.1 and 1.2). The intuitive physical explanation of this phenomenon is that long-range forces never sufficiently loosen their grip on the particles to allow them to travel in a free manner when they are at large distances from one another. A generalization of the above concept of asymptotic state is obviously necessary to deal with this situation. However, the already introduced concept is of sufficient generality to cope with most of the practically important cases. Even the Coulomb potential is in practice most often "screened," i.e., instead of having to deal with the Coulomb potential $V(r) \sim 1/r$, we are faced with a potential which at large distances decreases much faster than $1/r$, and for which asymptotic states, in the above sense, exist.

1.3. Asymptotic States and Scattering States in the Schroedinger Picture

It is straightforward to adapt the concept of asymptotic states defined in the classical two-particle case by (1.5) and (1.6) to the case of quantum mechanics. We can achieve this by stating that in the Schroedinger picture, the "free states" $\Psi^{\text{in}}(t)$ and $\Psi^{\text{out}}(t)$ are, respectively, the *strong* (or *Møller*) *asymptotic incoming and outgoing states* of a two-particle system in state $\Psi(t)$ if and only if*

$$\lim_{t\to-\infty} \| \Psi(t) - \Psi^{\text{in}}(t)\| = \lim_{t\to+\infty} \| \Psi(t) - \Psi^{\text{out}}(t)\| = 0. \tag{1.7}$$

In order to define precisely the meaning of the concept of a "free state" we must assume that in addition to the "total" Hamiltonian $H^{(2)}$ of the two-particle quantum mechanical system we have at our disposal a "free" Hamiltonian $H_0^{(2)}$, which would describe a system consisting of two particles with the same physical characteristics (mass, intrinsic

* See §2.7 for a more general concept of quantum asymptotic state.

spin, etc.), but which do not interact among themselves. For instance, in the case where $H^{(2)}$ is given essentially* by

$$(1.8) \qquad -\frac{1}{2m_1}\Delta_1 - \frac{1}{2m_2}\Delta_2 + V(\mathbf{r}), \qquad \mathbf{r} = \mathbf{r}_2 - \mathbf{r}_1,$$

then $H_0^{(2)}$ is taken to be essentially

$$(1.9) \qquad -\frac{1}{2m_1}\Delta_1 - \frac{1}{2m_2}\Delta_2,$$

where the domains of definition of the above differential operators are adequately defined,† as was done in §7 of Chapter IV. However, if both particles move in an external force field, given by a potential $V_{\text{ext}}(\mathbf{r}_1, \mathbf{r}_2)$, then instead of (1.8) we have

$$-\frac{1}{2m_1}\Delta_1 - \frac{1}{2m_2}\Delta_2 + V(\mathbf{r}) + V_{\text{ext}}(\mathbf{r}_1, \mathbf{r}_2),$$

and instead of (1.9) we have

$$-\frac{1}{2m_1}\Delta_1 - \frac{1}{2m_2}\Delta_2 + V_{\text{ext}}(\mathbf{r}_1, \mathbf{r}_2).$$

This will be the case when, for example, the system under consideration consists of two spinless charged particles moving in the force field generated by a much heavier particle (such as the heavy nucleus of some atom). Since the heavier particle is practically unaffected by the motion of the lighter particles, such a problem can be treated fairly accurately as a two-body problem in an external field rather than a three-body problem.

Consider now the general case when the "free" Hamiltonian $H_0^{(2)}$ and the total Hamiltonian $H^{(2)}$ are given by the self-adjoint operators acting in the Hilbert space $\mathscr{H}^{(2)}$ associated with the system. We say that a nonzero vector-valued function $\Psi^{(f)}(t)$, $t \in \mathbb{R}^1$, is a *free state* in the Schroedinger picture if and only if it satisfies the relation

$$\Psi^{(f)}(t) = \exp(-iH_0^{(2)}t)\,\Psi^{(f)}(0), \qquad t \in \mathbb{R}^1,$$

and if $\Psi^{(f)}(0)$ is orthogonal to the closed linear subspace $\mathscr{H}_{0b}^{(2)}$ of bound states in $\mathscr{H}^{(2)}$ of the system with Hamiltonian $H_0^{(2)}$. In the absence of

* See the discussion in §7 of Chapter IV on the essential self-adjointness of the Schroedinger operator.

† We note that in the preceding two operators, as well as throughout this chapter, we have adopted a system of units in which $\hbar = 1$.

external forces both $H_0^{(2)}$ and its internal energy part $H_{0i}^{(2)}$, obtained after subtraction from $H_0^{(2)}$ of the center of mass motion, are operators with a pure continuous spectrum so that $\mathscr{H}_{0b}^{(2)} = \{0\}$. For example, this is the case when $H_0^{(2)}$ is given by (1.9), and therefore $H_{0i}^{(2)}$ is essentially $-(1/2m_0)\Delta$ acting on wave functions $\psi(\mathbf{R}, \mathbf{r})$, where Δ is the Laplacian containing derivatives with respect to the relative position coordinates of $\mathbf{r}$ and $m_0 = m_1 m_2(m_1 + m_2)^{-1}$.

In reality, the particles interact, and any Schroedinger picture state $\Psi(t)$ in which they can be found, which will be called occasionally an *interacting state*, has to satisfy the relation

$$\Psi(t) = \exp(-iH^{(2)}t)\,\Psi(0), \qquad t \in \mathbb{R}^1.$$

Due to the unitarity of the operator $\exp(-iH^{(2)}t)$ we have

$$(1.10)\quad \begin{aligned} \|\Psi(t) - \Psi^{\text{ex}}(t)\| &= \|\exp(-iH^{(2)}t)\,\Psi(0) - \exp(-iH_0^{(2)}t)\,\Psi^{\text{ex}}(0)\| \\ &= \|\Psi(0) - \exp(iH^{(2)}t)\exp(-iH_0^{(2)}t)\,\Psi^{\text{ex}}(0)\| \\ &= \|\exp(iH_0^{(2)}t)\exp(-iH^{(2)}t)\,\Psi(0) - \Psi^{\text{ex}}(0)\|, \end{aligned}$$

where "ex" is a symbol which will be frequently used in the future, and which stands for "in" and "out."

In view of the preceding notation, we see that (1.7) implies that

$$(1.11)\quad \lim_{t\to\mp\infty} \|\Psi(0) - \exp(iH^{(2)}t)\exp(-iH_0^{(2)}t)\,\Psi^{\text{ex}}(0)\| = 0,$$

or, equivalently,

$$(1.12)\quad \lim_{t\to\mp\infty} \|\Psi^{\text{ex}}(0) - \exp(iH_0^{(2)}t)\exp(-iH^{(2)}t)\,\Psi(0)\| = 0,$$

where the interpretation of the above notation is that one should let $t \to -\infty$ when "ex" stands for "in," and $t \to +\infty$ when "ex" stands for "out." Since, conversely, (1.11) or (1.12) implies that (1.7) is true, we can state the following theorem.

Theorem 1.1. The state $\Psi(t) = \exp(-iH^{(2)}t)\,\Psi(0)$ possesses strong incoming and outgoing asymptotic states $\Psi^{\text{ex}}(t)$:

(a) if and only if the strong limits

$$(1.13)\quad \Psi^{\text{ex}} = \operatorname*{s-lim}_{t\to\mp\infty} \exp(iH_0^{(2)}t)\exp(-iH^{(2)}t)\,\Psi(0),$$

exist and belong to $\mathscr{H}^{(2)} \ominus \mathscr{H}_{0b}^{(2)}$;

(b) if and only if there are vectors $\Psi^{\text{ex}}(0) \in \mathscr{H} \ominus \mathscr{H}_{0b}^{(2)}$, ex $=$ in, out, such that

$$\Psi(0) = \operatorname*{s-lim}_{t\to\mp\infty} \exp(iH^{(2)}t) \exp(-iH_0^{(2)}t)\, \Psi^{\text{ex}}(0). \tag{1.14}$$

An interacting state $\Psi(t)$ which has an incoming as well as an outgoing asymptotic state is called a *scattering state.*

We recall from relation (3.43) of Chapter IV that

$$\tilde{\Psi}(t) = \exp(iH_0^{(2)}t)\exp(-iH^{(2)}t)\, \Psi(0)$$

represents in the interaction picture the same state which is represented in the Schroedinger picture by $\Psi(t) = \exp(-iH^{(2)}t)\, \Psi(0)$. Thus, $\tilde{\Psi}^{\text{ex}}$, ex $=$ in, out, can be thought of as being the asymptotic states of $\tilde{\Psi}(t)$ in the interaction picture.

1.4. Møller Wave Operators

Denote by $\mathsf{R}_+^{(2)}$ and $\mathsf{R}_-^{(2)}$ the set of all vectors which are the limits (1.14) for some $\Psi^{\text{ex}}(0) \in \mathscr{H}^{(2)} \ominus \mathscr{H}_{0b}^{(2)}$ (where $\mathscr{H}_{0b}^{(2)}$ is spanned by the bound states of $H_0^{(2)}$) when $t \to -\infty$ and $t \to +\infty$, respectively, i.e.,

$$\mathsf{R}_+^{(2)} = \{f_+ : f_+ = \operatorname*{s-lim}_{t\to-\infty} \Omega^{(2)}(t)f, f \in \mathscr{H} \ominus \mathscr{H}_{0b}^{(2)}\},$$

$$\mathsf{R}_-^{(2)} = \{f_- : f_- = \operatorname*{s-lim}_{t\to+\infty} \Omega^{(2)}(t)f, f \in \mathscr{H} \ominus \mathscr{H}_{0b}^{(2)}\}, \tag{1.15}$$

$$\Omega^{(2)}(t) = \exp(iH^{(2)}t) \exp(-iH_0^{(2)}t).$$

We see from (1.14) that if a nonzero vector g belongs to either $\mathsf{R}_+^{(2)}$ or $\mathsf{R}_-^{(2)}$, then $\exp(-iH^{(2)}t)g$ does not represent a bound state, since it is intuitively obvious that a bound state cannot be asymptotically free. Hence, it seems natural to assume that in any realistic quantum mechanical two-particle theory

$$\mathsf{R}_+^{(2)} = \mathsf{R}_-^{(2)} = \mathscr{H}_b^{(2)\perp},$$

where $\mathscr{H}_b^{(2)}$ is the space of bound states of $H^{(2)}$. We shall see later that the conclusion of this heuristic argument is confirmed when dealing with potentials which usually occur in practice.

Lemma 1.1. If $A(t)$ is a uniformly bounded family of operators,

$$\| A(t)\| \leqslant C, \qquad -\infty < t < +\infty,$$

then the set M of all vectors $f \in \mathscr{H}$ for which

$$\operatorname*{s-lim}_{t\to t_0} A(t)f, \qquad -\infty \leqslant t_0 \leqslant +\infty,$$

exists is a closed linear subspace of $\mathscr{H}$.

Proof. Since the existence of the strong limits of $A(t)f$ and $A(t)g$ for $t \to t_0$ implies the existence of the strong limits of $A(t)(af + bg) = aA(t)f + bA(t)g$, it follows that M is a linear subspace of $\mathscr{H}$.

Let f be a vector in $\mathscr{H}$ which is the strong limit of a sequence $f_1, f_2, \ldots$ of vectors from M. If

$$g_n = \operatorname*{s-lim}_{t\to t_0} A(t) f_n,$$

then we can write

$$\begin{aligned}
(1.16)\quad & \|[A(t_1) - A(t_2)] f\| \\
&= \|[A(t_1) - A(t_2)](f - f_n) + [A(t_1) - A(t_2)] f_n\| \\
&\leqslant \|[A(t_1) - A(t_2)](f - f_n)\| + \| A(t_1) f_n - g_n - [A(t_2) f_n - g_n]\| \\
&\leqslant \| A(t_1)(f - f_n)\| + \| A(t_2)(f - f_n)\| \\
&\quad + \| A(t_1) f_n - g_n \| + \| A(t_2) f_n - g_n \| \\
&\leqslant 2C \|f - f_n\| + \| A(t_1) f_n - g_n \| + \| A(t_2) f_n - g_n \|.
\end{aligned}$$

By choosing some sufficiently large $n = n_0$, we can obtain

$$2C \|f - f_{n_0}\| < \epsilon/3.$$

Since, for this fixed n_0, we have

$$\| A(t) f_{n_0} - g_{n_0} \| < \epsilon/3$$

for all t sufficiently close* to t_0, we see by glancing at the right-hand side of (1.16) that

$$\| A(t_1) f - A(t_2) f \| < \epsilon$$

for all such t_1, t_2 which are sufficiently close to t_0. Hence, the strong limit of $A(t)f$ for $t \to t_0$ exists, and therefore $f \in$ M. Q.E.D.

The operator function $\Omega^{(2)}(t)$ defined in (1.15) assumes as value a unitary operator for every value of $t \in \mathbb{R}^1$, and consequently it is uniformly bounded:

$$\| \Omega^{(2)}(t)\| = 1, \qquad -\infty < t < +\infty.$$

* If $t_0 = \pm\infty$, sufficiently "close" to t_0 means sufficiently large.

Thus, we can apply Lemma 1.1 to conclude that the families $\mathsf{M}_{\pm}^{(2)}$ of all vectors $f \in \mathscr{H}^{(2)} \ominus \mathscr{H}_{0b}^{(2)}$ for which the strong limits $\Omega^{(2)}(t) f$ for $t \to -\infty$ and $t \to +\infty$, respectively, exist are closed linear subspaces of $\mathscr{H}^{(2)} \ominus \mathscr{H}_{0b}^{(2)}$. Hence, the family

$$\mathsf{M}_0^{(2)} = \mathsf{M}_+^{(2)} \cap \mathsf{M}_-^{(2)}$$

of all vectors f for which both limits,

$$\underset{t \to \mp\infty}{\text{s-lim}}\, \Omega^{(2)}(t) f = f_\pm ,$$

exist is a closed linear subspace of $\mathscr{H}^{(2)}$.

Let $E_{\mathsf{M}_{\pm}^{(2)}}$ be the projectors onto $\mathsf{M}_{\pm}^{(2)}$,

$$E_{\mathsf{M}_{\pm}^{(2)}} \mathscr{H}^{(2)} \equiv \mathsf{M}_{\pm}^{(2)}. \tag{1.17}$$

Then, according to the above definition of $\mathsf{M}_{\pm}^{(2)}$, the limits

$$\underset{t \to \mp\infty}{\text{s-lim}}\, \Omega^{(2)}(t)\, E_{\mathsf{M}_{\pm}^{(2)}} f$$

exist for *all* vectors $f \in \mathscr{H}$, because $E_{\mathsf{M}_{\pm}^{(2)}} f \in \mathsf{M}_{\pm}^{(2)}$. Then the operators

$$\Omega_{\pm}^{(2)} = \underset{t \to \mp\infty}{\text{s-lim}}\, \Omega^{(2)}(t)\, E_{\mathsf{M}_{\pm}^{(2)}} \tag{1.18}$$

exist and are defined on the entire Hilbert space $\mathscr{H}^{(2)}$. The operator $\Omega_{+}^{(2)}$ is called the *incoming Møller wave operator* and $\Omega_{-}^{(2)}$ is called the *outgoing Møller wave operator*. We shall refer to these two operators as the "in" and "out" wave operators.

1.5. The Scattering Operator

The *scattering operator* (or *S operator*) $S^{(2)}$ for the problem at hand is defined in terms of the wave operators $\Omega_{\pm}^{(2)}$,

$$S^{(2)} = \Omega_{-}^{(2)*} \Omega_{+}^{(2)}. \tag{1.19}$$

Since the wave operators $\Omega_{\pm}^{(2)}$ are strong limits of uniformly bounded operators $\Omega^{(2)}(t)$, $\| \Omega^{(2)}(t) \| \leqslant 1$, they are bounded. Furthermore, we have seen that $\Omega_{\pm}^{(2)}$ are defined everywhere in $\mathscr{H}^{(2)}$. Hence, the adjoints $\Omega_{\pm}^{(2)*}$ exist and are bounded (see Chapter III, Theorem 2.7). Thus, $S^{(2)}$ is a bounded linear operator, defined on the entire Hilbert space $\mathscr{H}^{(2)}$.

The importance of the scattering operator lies in the fact that many quantities which are readily measurable by scattering experiments are easy to compute if we know $S^{(2)}$.

To provide an illustration, assume that a scattering state $\Psi_+(t)$ of the system has been prepared at the beginning of the scattering experiment, and that at the instant t_0 of the preparatory measurement $\Psi_+(t)$ had been essentially a free state $\Psi_+^{\text{in}}(t)$,

$$\lim_{t\to-\infty} \| \Psi_+(t) - \Psi_+^{\text{in}}(t)\| = 0, \qquad \Psi_+^{\text{in}}(t) = \exp(-iH_0^{(2)}t)\, \Psi_+^{\text{in}}(0).$$

Furthermore, assume that a determinative measurement at an instant t_1, long after the two particles of the system have interacted, can determine the value of the observable E_{t_1} of the form

$$E_{t_1} = |\, \Psi_-(t_1)\rangle\langle\Psi_-(t_1)|, \qquad \| \Psi_-(t)\| = 1,$$

which is represented by the projector on the scattering state $\Psi_-(t)$ at $t = t_1$. Then, if $\Psi_+(t)$ is normalized, the mean value of E_{t_1} is given by the expression

$$(1.20) \qquad P_{\Psi_+\to\Psi_-} = \langle\Psi_+(t_1) \mid E_{t_1} \Psi_+(t_1)\rangle = |\langle\Psi_+(t_1) \mid \Psi_-(t_1)\rangle|^2.$$

We recall from the discussion at the end of §1 in Chapter IV that the above expression is usually referred to as the transition probability from $\Psi_+(t)$ to $\Psi_-(t)$, due to the usual assumption that if the outcome of the determinative measurement of E_{t_1} is $\lambda = 1$, the system is left afterwards in the state $\Psi_-(t)$.

We note that the expression in (1.20) is actually independent of t_1,

$$(1.21) \qquad P_{\Psi_+\to\Psi_-} = |\langle\Psi_+(0) \mid \Psi_-(0)\rangle|^2,$$

due to the fact that the operator $\exp(-iH^{(2)}t)$ is unitary. Let $\Psi_-^{\text{out}}(t)$ be the strong outgoing asymptotic state of $\Psi_-(t)$:

$$\lim_{t\to+\infty} \| \Psi_-(t) - \Psi_-^{\text{out}}(t)\| = 0, \qquad \Psi_-^{\text{out}}(t) = \exp(-iH_0^{(2)}t)\, \Psi_-^{\text{out}}(0).$$

Since according to (1.14)

$$\Psi_+(0) = \operatorname*{s-lim}_{t\to-\infty} \exp(iH^{(2)}t) \exp(-iH_0^{(2)}t)\, \Psi_+^{\text{in}}(0) = \Omega_+^{(2)}\Psi_+^{\text{in}}(0),$$

$$\Psi_-(0) = \operatorname*{s-lim}_{t\to+\infty} \exp(iH^{(2)}t) \exp(-iH_0^{(2)}t)\, \Psi_-^{\text{out}}(0) = \Omega_-^{(2)}\Psi_-^{\text{out}}(0),$$

we arrive at the following expression for the transition probability:

$$(1.22)\qquad \begin{aligned} P_{\Psi_+ \to \Psi_-} &= |\langle \Omega_-^{(2)} \Psi_-^{\text{out}}(0) \mid \Omega_+^{(2)} \Psi_+^{\text{in}}(0) \rangle|^2 \\ &= |\langle \Psi_-^{\text{out}}(0) \mid \Omega_-^{(2)*} \Omega_+^{(2)} \Psi_+^{\text{in}}(0) \rangle|^2 \\ &= |\langle \Psi_-^{\text{out}}(0) \mid S^{(2)} \Psi_+^{\text{in}}(0) \rangle|^2. \end{aligned}$$

This expression is very convenient and helpful in practice, since it involves free states which in the typical scattering experiment are computationally easy to derive from the available experimental data.

1.6. The Differential Scattering Cross Section

In practice, the typical two-particle scattering experiment is of the kind depicted in Figure 5: a beam of particles $\mathfrak{S}_1$ impinges on a target

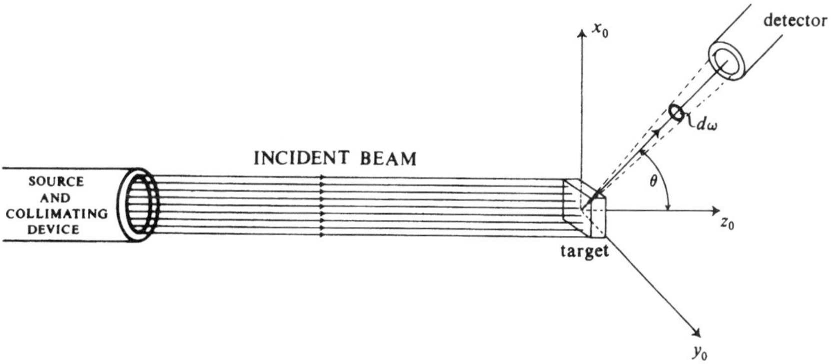

Fig. 5. Scattering of a beam of particles off a finite planar target.

consisting of particles $\mathfrak{S}_2$. Assuming that the particles in the beam do not interact with one another, and that each particle $\mathfrak{S}_1$ in the beam interacts with only one particle $\mathfrak{S}_2$ in the target, the experiment can be viewed as consisting of a large ensemble of independent scattering experiments of two-particle systems* $\mathfrak{S} = \{\mathfrak{S}_1, \mathfrak{S}_2\}$.

* The two mentioned conditions can be satisfied by taking a "weak" beam of particles (i.e., a beam with few particles per unit volume), and taking for the target a slab of material which is sufficiently thin to eliminate the possibility of a particle $\mathfrak{S}_1$ scattering from more than one particle in the target.

The above experimental setup is ideally suited to determine the percentage of the particles in the beam which can be found in a given volume of space, occupied by a detector, after they have been scattered by the target. In any scattering experiment the detector is placed as depicted in Fig. 5, i.e., sufficiently far from the target so that it never interferes with the scattering process. Hence, the experiment obviously provides information about the probability that a particle of the beam, coming towards the target from the direction

$$\omega_0 = (\phi_0, \theta_0), \qquad 0 \leqslant \phi_0 \leqslant 2\pi, \qquad 0 \leqslant \theta_0 \leqslant \pi,$$

will be scattered within the solid angle $d\omega$ around the direction

$$(1.23) \qquad \omega = (\phi, \theta), \qquad 0 \leqslant \phi \leqslant 2\pi, \qquad 0 \leqslant \theta \leqslant \pi.$$

If N denotes the number of two-particle scatterings per unit time at the energy of relative motion E, then the number of particles scattered within the solid angle $d\omega$ can be written in the form $N\pi(E, \omega_0, \omega)\, d\omega$, since it is obviously proportional to N as long as the two earlier mentioned conditions on beam and target are satisfied. It can be expected that $\pi(E, \omega_0, \omega)$ is dependent on the energies E and E' for the relative motion of the particles of the system $\mathfrak{S} = \{\mathfrak{S}_1, \mathfrak{S}_2\}$ before and after collision. However, since energy is conserved, we must have $E = E'$, so that the dependence on E' does not have to be displayed.

It has become customary to orient the frame of reference so that $\phi_0 = \theta_0 = 0$, and to call the quantities

$$(1.24) \qquad \sigma(E, \omega) = (N/J_0)\, \pi(E, \omega_0, \omega), \qquad \sigma(E) = \int_{\Omega_S} \sigma(E, \omega)\, d\omega$$

(where J_0 is the *incident flux*, i.e., the number of incident particles per unit time and unit beam cross section) the *differential cross section* and the *total cross section*, respectively. The notation $d\sigma/d\omega$ instead of $\sigma(E, \omega)$ is often encountered in literature.

1.7. The Transition Operator

In potential scattering $H^{(2)}$ and $H_0^{(2)}$ are given by (1.8) and (1.9), and one can easily relate $\sigma(E, \omega)$ to the S operator. To establish this relation, it is computationally advantageous to express $S^{(2)}$ in terms of a new operator,

$$(1.25) \qquad S^{(2)} = 1 - 2\pi i T^{(2)}.$$

We shall refer to $T^{(2)}$ as a *transition operator* [not to be confused, however, with the $\mathscr{T}(\zeta)$ operators introduced in §4.8] since in the case when there is no interaction present, the system stays in the same free state (i.e., there are no "transitions"), and the operator $T^{(2)}$ is zero. We easily establish that $T^{(2)} = 0$ when there is no interaction by noting that the "absence of interaction" means that $H^{(2)} = H_0^{(2)}$ and therefore $\Omega^{(2)}(t) \equiv 1$, i.e.,

$$\Omega_+^{(2)} = \Omega_-^{(2)} = S^{(2)} = 1.$$

We have seen already in §7 of Chapter II that in the absence of external forces the Schroedinger operator $H_S^{(2)}$ of a two-particle system can be written in the form

$$H_S^{(2)} = H_c^{(2)} + H_i^{(2)}$$

where $H_c^{(2)}$ is the center-of-mass energy operator and $H_i^{(2)}$ is the internal energy operator. The Hilbert space $\mathscr{H}^{(2)} \equiv L^2(\mathbb{R}^6)$ of wave functions $\psi(\mathbf{R}, \mathbf{r})$ is obviously the tensor product

$$\mathscr{H}^{(2)} = \mathscr{H}_c^{(1)} \otimes \mathscr{H}^{(1)} \tag{1.26}$$

of the Hilbert spaces $\mathscr{H}_c^{(1)} = L^2(\mathbb{R}^3)$ and $\mathscr{H}^{(1)} = L^2(\mathbb{R}^3)$ of wavefunctions $\psi_c(\mathbf{R})$ and $\psi(\mathbf{r})$, respectively, where $\mathbf{R}$ and $\mathbf{r}$ are the center of mass variables given in (7.2) of Chapter II. It becomes evident from looking at the differential operators

$$-\frac{1}{2M_0}\Delta_R \subseteq H_c^{(2)}, \qquad -\frac{1}{2m_0}\Delta + V(\mathbf{r}) \subseteq H_i^{(2)} \tag{1.27}$$

and (7.3) in Chapter II that $H_c^{(2)} = H_c^{(1)} \otimes 1$ and $H_i^{(2)} = 1 \otimes H_i^{(1)}$, where $H_c^{(1)}$ and $H_i^{(1)}$ act in $\mathscr{H}_c^{(1)}$ and $\mathscr{H}^{(1)}$, respectively.

It is natural to require that, in general and when no external forces are present, the free and total Hamiltonians can be written in the form

$$\begin{aligned} H_0^{(2)} &= H_c^{(2)} + H_{0i}^{(2)}, & H_{0i}^{(2)} &= 1 \otimes H_{0i}^{(1)}, \\ H^{(2)} &= H_c^{(2)} + H_i^{(2)}, & H_i^{(2)} &= 1 \otimes H_i^{(1)}. \\ H_c^{(2)} &= H_c^{(1)} \otimes 1 \end{aligned} \tag{1.28}$$

Since the operators $H_c^{(2)}$ and $H_{0i}^{(2)}$, as well as $H_c^{(2)}$ and $H_i^{(2)}$, obviously commute, we can use Theorem 2.4 of Chapter IV to infer that

$$\exp(-iH_0^{(2)}t) = \exp(-iH_c^{(2)}t)\exp(-iH_{0i}^{(2)}t),$$

$$\exp(iH^{(2)}t) = \exp(iH_c^{(2)}t)\exp(iH_i^{(2)}t).$$

The above two relations are derivable directly from (2.13) in Chapter IV by noting that in the above case all operators are bounded and defined on the entire Hilbert space $\mathscr{H}^{(2)}$.

From (1.28) we get

$$(1.29)\quad \begin{aligned} \Omega^{(2)}(t) &= \exp(iH^{(2)}t)\exp(-iH_0^{(2)}t) \\ &= \exp(iH_1^{(2)}t)\exp(iH_c^{(2)}t)\exp(-iH_c^{(2)}t)\exp(-iH_{01}^{(2)}t) \\ &= \exp(iH_1^{(2)}t)\exp(-iH_{01}^{(2)}t) \\ &= 1\otimes\exp(iH_1^{(1)}t)\exp(-iH_{01}^{(1)}t). \end{aligned}$$

Inserting the above expression for $\Omega^{(2)}(t)$ in (1.18) and (1.19), we easily arrive at the following result.

Theorem 1.2. If the free and total Hamiltonians $H_0^{(2)}$ and $H^{(2)}$ of a two-body system can be decomposed into sums (1.28) of commuting Hamiltonians for center of mass and internal motion, then

$$\Omega_\pm^{(2)} = 1\otimes\Omega_\pm^{(1)},\quad S^{(2)} = 1\otimes S^{(1)},\quad T^{(2)} = 1\otimes T^{(1)},\quad S^{(1)} = \Omega_-^{(1)*}\Omega_+^{(1)},$$

where $\Omega_\pm^{(1)}$ and $S^{(1)}$ are operators in the internal-motion Hilbert space $\mathscr{H}^{(1)}$ of the tensor product $\mathscr{H}^{(2)} = \mathscr{H}_c^{(1)}\otimes\mathscr{H}^{(1)}$.

The above theorem shows that in case of two-body potential scattering in which no external forces are involved, we can work in the Hilbert space $\mathscr{H}^{(1)}$ which is related to the internal motion of the system. If we work in momentum space, we can introduce the relative momentum variables $\mathbf{p}$, related to the relative position vector $\mathbf{r}$. We might hope that the $S^{(1)}$ operator could be written as an integral operator in $\mathscr{H}^{(1)}$, which, in case of spinless particles, would be of the form

$$(1.30)\qquad (S^{(1)}\psi)^{\tilde{}}(\mathbf{p}) = \int_{\mathbb{R}^3} S^{(1)}(\mathbf{p},\mathbf{p}')\tilde{\psi}(\mathbf{p}')\,d\mathbf{p}'.$$

However, we shall see in the §2 that $S^{(1)}$ commutes with the free Hamiltonian $H_{01}^{(1)}$. Lemma 1.2 shows that, on account of this fact, the representation (1.30) is not feasible.

Lemma 1.2. Let K be a bounded integral operator on $L^2(\mathbb{R}^3)$

$$(Kf)^{\tilde{}}(\mathbf{p}) = \int_{\mathbb{R}^3} K(\mathbf{p},\mathbf{p}')\tilde{f}(\mathbf{p}')\,d\mathbf{p}',$$

with a kernel $K(\mathbf{p}, \mathbf{p}')$ which is Borel measurable in $\mathbb{R}^6$. If K commutes with the self-adjoint operator $H_{0i}^{(1)}$,

$$(H_{0i}^{(1)}f)^{\sim}(\mathbf{p}) = \frac{\mathbf{p}^2}{2m_0}\tilde{f}(\mathbf{p}),$$

then K is the zero operator.

Proof. If $\tilde{f}(\mathbf{p})$ is a function of compact support Δ, then $E^{\mathbf{p}}(\Delta)f = f \in \mathscr{D}_{H_{0i}^{(1)}}$, where $E^{\mathbf{p}}(\Delta)$ is the spectral measure obtained as the product (1.13) in Chapter IV of the spectral measures of the three momentum component observables:

$$(E^{\mathbf{p}}(\Delta)g)^{\sim}(\mathbf{p}) = \chi_{\Delta}(\mathbf{p})\,\tilde{g}(\mathbf{p})$$

The operator $H_{0i}^{(1)}E^{\mathbf{p}}(\Delta)$ is obviously bounded, and commutes with K,

$$\begin{aligned}([K, H_{0i}^{(1)}E^{\mathbf{p}}(\Delta)]f)^{\sim}(\mathbf{p}) &= \int_{\mathbb{R}^3} K(\mathbf{p}, \mathbf{p}')\left(\frac{\mathbf{p}'^2}{2m_0} - \frac{\mathbf{p}^2}{2m_0}\chi_{\Delta}(\mathbf{p})\right)\tilde{f}(\mathbf{p}')\,d\mathbf{p}' \\ &= 0 \qquad \text{(almost everywhere).}\end{aligned}$$

Since the above expression has to vanish almost everywhere for all $\tilde{f}(\mathbf{p})$ of compact support, we have

$$K(\mathbf{p}, \mathbf{p}')\left(\frac{\mathbf{p}'^2}{2m_0} - \frac{\mathbf{p}^2}{2m_0}\right) = 0 \qquad \text{(almost everywhere in } \mathbb{R}^3 \times \mathbb{R}^3 = \mathbb{R}^6\text{).}$$

Now, the set of points in $\mathbb{R}^6$ for which $|\mathbf{p}| = |\mathbf{p}'|$ (called the *energy shell*) is of Lebesgue measure zero. Hence, we deduce that $K(\mathbf{p}, \mathbf{p}')$ vanishes almost everywhere in $\mathbb{R}^6$.

If $\tilde{f}(\mathbf{p})$ and $\tilde{g}(\mathbf{p})$, f, $g \in L^2(\mathbb{R}^3)$, are bounded functions of compact support, then $\tilde{g}^*(\mathbf{p})\,K(\mathbf{p}, \mathbf{p}')\tilde{f}(\mathbf{p}')$ is certainly integrable on $\mathbb{R}^6$ since $K(\mathbf{p}, \mathbf{p}')$ is Borel measurable and vanishes almost everywhere. From the relation

$$\int_{\mathbb{R}^6} \tilde{g}^*(\mathbf{p})\,K(\mathbf{p}, \mathbf{p}')\tilde{f}(\mathbf{p}')\,d\mathbf{p}\,d\mathbf{p}' = 0,$$

we get by applying Fubini's theorem and Theorem 4.6 in Chapter III,

$$\langle g \mid Kf\rangle = \int_{\mathbb{R}^3} d\mathbf{p}\tilde{g}^*(\mathbf{p}) \int_{\mathbb{R}^3} d\mathbf{p}'K(\mathbf{p}, \mathbf{p}')\tilde{f}(\mathbf{p}') = 0,$$

for all $\tilde{f}$, $\tilde{g} \in \mathscr{C}_{\mathrm{b}}^0(\mathbb{R}^3)$. Since $\mathscr{C}_{\mathrm{b}}^0(\mathbb{R}^3)$ is dense in $L^2(\mathbb{R}^3)$, we get $Kf = 0$ for all $\tilde{f} \in \mathscr{C}_{\mathrm{b}}^0(\mathbb{R}^3)$. Thus, K is a bounded linear operator which is zero on a dense subset $\mathscr{C}_{\mathrm{b}}^0(\mathbb{R}^3)$ of $L^2(\mathbb{R}^3)$, and therefore $K = 0$. Q.E.D.

Let us introduce spherical coordinates $(p, \omega) = (p, \theta, \phi)$, $0 \leqslant p < +\infty$, $0 \leqslant \theta \leqslant \pi$, $0 \leqslant \phi \leqslant 2\pi$, in the space $\mathbb{R}^3$ of the variables $\mathbf{p}$. We shall see later in this chapter [cf. (4.96) and (4.97)] that the $T^{(1)}$ operator can be written as an integral operator with respect to the variable ω,

$$(T^{(1)}\psi)^{\sim}(p, \omega) = \int_{\Omega_s} T^{(1)}(p; \omega, \omega') \tilde{\psi}(p, \omega') \, d\omega', \quad d\omega' = \sin\theta' \, d\theta' \, d\phi', \tag{1.31}$$

over the set Ω_s, defined in §7.2 of Chapter II, which is essentially isomorphic to the unit sphere in three dimensions. The function $T^{(1)}(p; \omega, \omega')$ is uniquely determined by the operator $T^{(1)}$ (see Exercise 1.6) if we restrict ourselves to continuous kernels for the integral operator in (1.31).

*1.8. The T-Matrix Formula for the Differential Cross Section

The differential cross section $\sigma(E, \omega)$ can be computed from the function $T^{(1)}(p; \omega, \omega')$ which occurs in the integral representation (1.31) of the T operator. In order to arrive at an expression relating $\sigma(E, \omega)$ and $T^{(1)}(p; \omega, \omega')$, we have to express in theoretical terms those essential features of a scattering experiment which enter in the definition of the differential cross section. In any such scattering experiment, a beam of particles scatters from the particles of a target. The experimental setup is such that the relative momentum $\mathbf{p}$ of the incoming particles lies within a very narrow range of values.

Let us introduce spherical coordinates (p, θ, ϕ) in the $\mathbf{p}$ space, and suppose that the incoming particles are prepared in a scattering state $\Psi(t)$, which is such that $\mathbf{p}$ lies within the set

$$\tilde{C}^{(0)} = \{\mathbf{p}: p \in I_{p_0}, (\theta, \phi) \in I_{\omega_0}\},$$
$$I_{p_0} = [p_0, p_0 + \Delta p_0], \qquad I_{\omega_0} = [\theta_0, \theta_0 + \Delta\theta_0] \times [\phi_0, \phi_0 + \Delta\phi_0]. \tag{1.32}$$

Denote by $\Psi^{\text{in}}(t)$ the incoming asymptotic state of $\Psi(t)$. Suppose that $\Psi^{\text{in}}(0)$ is represented by the function (see Exercise 1.5)

$$\tilde{\psi}^{\text{in}}(\mathbf{P}, \mathbf{p}) = \tilde{\psi}_c(\mathbf{P}) \tilde{\psi}_0(\mathbf{p}), \tag{1.33}$$

of the center of mass momentum $\mathbf{P}$ and relative momentum $\mathbf{p}$.

Let us denote by $E^{(\mathbf{p})}(B)$ the spectral measure

$$(E^{(\mathbf{p})}(B)\psi)^{\sim}(\mathbf{P}, \mathbf{p}) = \chi_B(\mathbf{p}) \tilde{\psi}(\mathbf{P}, \mathbf{p}). \tag{1.34}$$

It can be shown (see Exercise 1.8) that for any Borel set $B \in \mathscr{B}^3$

(1.35)
$$\lim_{t\to-\infty} \langle \Psi(t) \mid E^{(\mathbf{p})}(B)\, \Psi(t)\rangle = \langle \Psi^{\text{in}}(0) \mid E^{(\mathbf{p})}(B)\, \Psi^{\text{in}}(0)\rangle,$$
$$\int_{\mathbb{R}^3\times B} |\tilde{\psi}^{\text{in}}(\mathbf{P}, \mathbf{p})|^2\, d\mathbf{P}\, d\mathbf{p} = \int_B |\tilde{\psi}_0(\mathbf{p})|^2\, d\mathbf{p}.$$

Since the support of the function $\tilde{\psi}(\mathbf{P}, \mathbf{p}; t)$ representing $\Psi(t)$ is within $\mathbb{R}^3 \times \tilde{C}^{(0)}$, we deduce from the above relation that

$$\int_{\mathbb{R}^3-\tilde{C}^{(0)}} |\tilde{\psi}_0(\mathbf{p})|^2\, d\mathbf{p} = 0.$$

Hence, $\tilde{\psi}_0(\mathbf{p})$ vanishes almost everywhere outside $\tilde{C}^{(0)}$.

Denote by $\Psi^{\text{out}}(t)$ the outgoing asymptotic state of $\Psi(t)$. It is easy to show (see Exercise 1.4) that

(1.36)
$$\Psi^{\text{out}}(0) = S^{(2)}\Psi^{\text{in}}(0) = (1 \times S^{(1)})\, \Psi^{\text{in}}(0).$$

Hence, by applying the same type of argument which led to (1.34) to $\Psi^{\text{out}}(t)$, we obtain

$$\begin{aligned}\lim_{t\to+\infty} \langle \Psi(t) \mid E^{(\mathbf{p})}(B)\, \Psi(t)\rangle &= \langle \Psi^{\text{out}}(0) \mid E^{(\mathbf{p})}(B)\, \Psi^{\text{out}}(0)\rangle \\ &= \int_{\mathbb{R}^3\times B} |\tilde{\psi}^{\text{out}}(\mathbf{P}, \mathbf{p})|^2\, d\mathbf{P}\, d\mathbf{p} \\ &= \int_B |(S^{(1)}\psi_0)^{\sim}(\mathbf{p})|^2\, d\mathbf{p}.\end{aligned}$$

Consequently, the probability $P(I_{\omega_0} \to I_{\omega_0'})$ of finding that the relative momentum $\mathbf{p}$ of the system, a long time t after the scattering had taken place, is within the cone

$$C^{(+)} = \{\mathbf{p}: (\theta, \phi) \in I_{\omega_0'}\}, \qquad I_{\omega_0'} = [\theta_0', \theta_0' + \Delta\theta_0'] \times [\phi_0', \phi_0' + \Delta\phi_0]$$

is given by (see also Exercise 1.12)

(1.37)
$$P(I_{\omega_0} \to I_{\omega_0'}) = \int_{C^{(+)}} |(S^{(1)}\psi_0)^{\sim}(\mathbf{p})|^2\, d\mathbf{p}.$$

We shall see in the next section that $S^{(1)}$ commutes with $E^{H_{01}^{(1)}}(B)$ (see Theorem 2.3), where

$$(E^{H_{01}^{(1)}}(B)\psi)^{\sim}(\mathbf{p}) = \chi_{\{\mathbf{p}^2/2m_0 \in B\}}(\mathbf{p})\, \tilde{\psi}(\mathbf{p}).$$

It follows that for the complement $B_1 \in \mathscr{B}^1$ of the set $\{p^2/2m: p \in I_{p_0}\}$, we have

$$E^{H_{01}^{(1)}}(B_1)\, S^{(1)}\psi_0 = S^{(1)} E^{H_{01}^{(1)}}(B_1)\, \psi_0 = 0$$

since $E^{H_{01}^{(1)}}(B_1)\, \psi_0 = 0$ because $\tilde{\psi}_0(\mathbf{p})$ vanishes outside $\tilde{C}^{(0)}$. This means that $(S^{(1)}\psi_0)\tilde{}(\mathbf{p})$ vanishes when $p^2/2m \in B_1$, i.e., when $|\mathbf{p}| \notin I_{p_0}$. Therefore, if the integral in (1.37) is expressed in spherical coordinates, the integration in p extends in effect only over the set I_{p_0}. Thus, we can write

$$\text{(1.38)} \qquad P(I_{\omega_0} \to I_{\omega_0'}) = \int_{\mathbb{R}^3} d^3\mathbf{P} \int_{I_{p_0}} p^2\, dp \int_{I_{\omega_0'}} d\omega'\, |\tilde{\psi}^{\text{out}}(\mathbf{P}, \mathbf{p})|^2$$

$$= \int_{I_{p_0}} p^2\, dp \int_{I_{\omega_0'}} d\omega'\, |(S^{(1)}\psi_0)\tilde{}(\mathbf{p})|^2.$$

Let us consider the case when the direction of relative motion of the two particles in the system has changed as a result of interaction, i.e., when $\omega_0 \neq \omega_0'$ and $I_{\omega_0} \cap I_{\omega_0'} = \varnothing$. If we use the definition (1.25) of the transition operator to write

$$S^{(1)}\psi_0 = \psi_0 - 2\pi i T^{(1)}\psi_0$$

and recall that $\tilde{\psi}_0'(p, \omega) = \tilde{\psi}_0(\mathbf{p})$ vanishes when ω is outside I_{ω_0}, we arrive at the result

$$\text{(1.39)} \qquad P(I_{\omega_0} \to I_{\omega_0'}) = 4\pi^2 \int_{I_{p_0}} p^2\, dp \int_{I_{\omega_0'}} d\omega'\, |(T^{(1)}\psi_0)\tilde{}(\mathbf{p})|^2.$$

Hence, we get for the probability density $P(\omega_0')$ (per unit solid angle) of scattering in the direction ω_0'

$$\text{(1.40)} \quad P(\omega_0') = \lim_{\Delta\omega' \to 0} \frac{P(I_{\omega_0} \to I_{\omega_0'})}{\Delta\omega'} = 4\pi^2 \int_{I_{p_0}} |(T^{(1)}\psi_0')\tilde{}\ (p, \omega_0')|^2\, p^2\, dp,$$

provided that $(T^{(1)}\psi_0')\tilde{}(p, \omega')$ is continuous in ω' in some neighborhood of ω_0'. Substituting in the above relation for the operator $T^{(1)}$ its integral representation (1.31), we obtain

$$\text{(1.41)} \qquad P(\omega_0') = 4\pi^2 \int_{I_{p_0}} p^2\, dp \iint d\omega_1\, d\omega_2\, T^{(1)*}(p; \omega_0', \omega_1)$$

$$\times\, T^{(1)}(p; \omega_0', \omega_2)\, \tilde{\psi}_0'^*(p, \omega_1)\, \tilde{\psi}_0'(p, \omega_2),$$

where the integrations in ω_1 and ω_2 extend over the unit sphere in three dimensions. The request that $(T^{(1)}\psi_0')^{\sim}(p, \omega')$ be continuous can be met by demanding that $T^{(1)}(p; \omega', \omega)$ be continuous in ω'.

At this stage one must remember that the expression (1.41) is the probability density of scattering of two particles in the relative direction ω_0' with respect to one another. We have already indicated, however, that in a scattering experiment a beam of particles is being scattered on a target consisting of a large number of particles. Thus, in a scattering experiment, a great many two-body scatterings are observed simultaneously. These scatterings between particles in the beam and particles in the target are taking place at different locations with respect to the laboratory frame of reference. Hence, in order to arrive at an expression for a planar differential cross section, an averaging of $P(\omega_0')$ over all these two-body scatterings at different locations has to be carried out.

If a particle $\mathfrak{S}_2''$ in the target is at a location $\mathbf{r}_0$ in relation to another particle $\mathfrak{S}_2'$, while $\mathfrak{S}_2'$ is in position $\mathbf{r}$ with respect to a particle $\mathfrak{S}_1$ in the beam, then $\mathfrak{S}_2''$ is at the location $\mathbf{r} + \mathbf{r}_0$ in relation to the incident particle $\mathfrak{S}_1$. Since, with the exception of location, all the other variables of $\mathfrak{S}_2'$ and $\mathfrak{S}_2''$ are the same, we conclude that if

$$\psi_c(\mathbf{R})\,\psi_0(\mathbf{r}), \qquad \mathbf{R} = \frac{m_1\mathbf{r}_1 + m_2\mathbf{r}_2}{m_1 + m_2}, \qquad \mathbf{r} = \mathbf{r}_2 - \mathbf{r}_1,$$

represents $\Psi^{\text{in}}(0)$ for $\{\mathfrak{S}_1, \mathfrak{S}_2'\}$, the wave function

$$\psi_{c\mathbf{r}_0}(\mathbf{R})\,\psi_{\mathbf{r}_0}(\mathbf{r}) = \psi_c\Big(\mathbf{R} + \frac{m_2}{m_1 + m_2}\,\mathbf{r}_0\Big)\,\psi_0(\mathbf{r} + \mathbf{r}_0)$$

represents $\Psi^{\text{in}}(0)$ for $\{\mathfrak{S}_1, \mathfrak{S}_2''\}$. By taking Fourier transforms, for $\{\mathfrak{S}_1, \mathfrak{S}_2''\}$ we easily get

$$\tilde{\psi}_{\mathbf{r}_0}(\mathbf{p}) = \exp(i\mathbf{p}\mathbf{r}_0)\,\tilde{\psi}_0(\mathbf{p}),$$

where $\tilde{\psi}_0(\mathbf{p})$ refers to $\{\mathfrak{S}_1, \mathfrak{S}_2'\}$, and $\tilde{\psi}_{\mathbf{r}_0}(\mathbf{p})$ refers to $\{\mathfrak{S}_1, \mathfrak{S}_2''\}$. Hence, the probability $P_{\mathbf{r}_0}(\omega_0')$ for $\{\mathfrak{S}_1, \mathfrak{S}_2''\}$ is

$$(1.42)\quad P_{\mathbf{r}_0}(\omega_0') = 4\pi^2 \int_{I_{p_0}} dp\, p^2 \int d\omega_1\, d\omega_2\, T^*(p; \omega_0', \omega_1) \times T(p; \omega_0', \omega_2) \exp[i\mathbf{r}_0(\mathbf{p}_2 - \mathbf{p}_1)]\, \tilde{\psi}_0^*(\mathbf{p}_1)\tilde{\psi}_0(\mathbf{p}_2)|_{p_1=p_2}$$

instead of (1.41).

We shall now make the assumption that the scatterers in the target are uniformly distributed, so that N_0 scatterers interact with the incident particles per unit target volume and unit time. Then the total number of particles scattered* per unit time and solid angle in the direction ω_0' is on the average equal to

(1.43)

$$\nu(\omega_0') = N_0 \int_V P_{\mathbf{r}_0}(\omega_0')\, d\mathbf{r}_0$$

$$= 4\pi^2 N_0 \int_V d\mathbf{r}_0 \int dp\, p^2 \int d\omega_1\, d\omega_2 \exp[i\mathbf{r}_0(\mathbf{p}_2 - \mathbf{p}_1)]\, T^{(1)*}(p; \omega'_0, \omega_1)$$

$$\times\, T^{(1)}(p; \omega_0', \omega_2)\, \tilde{\psi}_0^*(\mathbf{p}_1)\, \tilde{\psi}_0(\mathbf{p}_2)|_{p_1=p_2=p}\,,$$

assuming that the scatterers in the target are sufficiently dense that we may, with sufficient accuracy, replace the sum over the individual scatterers with an integral over the volume V of the target.

We shall evaluate the integral (1.43) under the assumption that the target is a slab of thickness δ, which extends to infinity in the directions x_0 and y_0 (see Fig. 5). Setting $\mathbf{k} = \mathbf{p}_2 - \mathbf{p}_1$ and reversing the orders of integration in (1.43), we obtain

(1.44) $$\nu(\omega_0') = 4\pi^2 N_0 \int d\omega_1 \int d\mathbf{r}_0 \int dp\, d\omega_2\, p^2 e^{i\mathbf{k}\mathbf{r}_0} F(\omega_0'; p, \omega_1, \omega_2),$$

where we have introduced the function

(1.45) $$F(\omega_0'; p, \omega_1, \omega_2)$$

$$= T^{(1)*}(p; \omega_0', \omega_1)\, T^{(1)}(p; \omega_0', \omega_2)\, \tilde{\psi}_0'^*(p, \omega_1)\, \tilde{\psi}_0'(p, \omega_2).$$

Since $|\mathbf{p}_1| = |\mathbf{p}_2| = p$, only two of the components of $\mathbf{k}$ are independent. We can substitute for ω_2 the variable ω which represents the spherical coordinates θ and ϕ of $\mathbf{p}_2$ in a coordinate system with its z axis in the direction of $\mathbf{p}_1$. Hence, $p_{1x} = p_{2x} = 0$ and

$$k_x = p_{2x} = p \sin\theta \cos\phi,$$

$$k_y = p_{2y} = p \sin\theta \sin\phi.$$

By substituting for $\omega = (\theta, \phi)$ the variables k_x and k_y, we obtain

$$p^2\, dp\, d\omega = p^2 \sin\theta\, dp\, d\theta\, d\phi = p^2 \sin\theta \frac{\partial(\theta, \phi)}{\partial(k_x, k_y)}\, dp\, dk_x\, dk_y = \frac{1}{\cos\theta}\, dp\, dk_x\, dk_y\,.$$

* In practice, the scatterers in the target are approximately at rest with respect to the laboratory frame, so that the relative scattering direction ω_0' computed in the center-of-mass frame is very often almost equal to the scattering angle measured in the laboratory reference frame.

Using the properties of Fourier transforms applied to the variables k_x, k_y and their conjugates x_0 and y_0, we obtain

$$\int_{\mathbb{R}^2} dx_0\, dy_0 \int dk_x\, dk_y\, e^{i\mathbf{r}_0\mathbf{k}} \int dp\, \frac{1}{\cos\theta} F(\omega_0'; p, \omega_1, \omega_2)$$

$$= 4\pi^2 \int_{I_{p_0}} e^{iz_0k_z} \frac{F(\omega_0'; p, \omega_1, \omega_2)}{\cos\theta}\bigg|_{k_x=k_y=0} dp.$$

Since obviously $dx_0\, dy_0\, dz_0 = d\mathbf{r}_0$, the above relation can be used in computing the integral in (1.44). When $k_x = k_y = 0$, we have $\phi_1 = \phi_2$ and $\theta_1 = \theta_2$ or $\phi_1 = \phi_2$ and $\theta_1 = \pi - \theta_2$. Since $F(\omega_0'; p, \omega_1, \omega_2)$ vanishes when ω_1 or ω_2 are outside I_{ω_0}, it follows that when I_{ω_0} is sufficiently small only the first alternative $\phi_1 = \phi_2$ and $\theta_1 = \theta_2$ yields a nonzero result. This means that $k_z = 0$, $\theta = 0$, and the integration in z_0 in (1.44) can be immediately performed:

$$\nu(\omega_0') = (2\pi)^4 N_0\delta \int_{I_{\omega_0}} d\omega_1 \int_{I_{p_0}} dp\, F(\omega_0'; p, \omega_1, \omega_1) \tag{1.46}$$

$$= (2\pi)^4 N_0\delta \int_{I_{\omega_0}} d\omega_1 \int_{I_{p_0}} \frac{|T^{(1)}(p, \omega_0', \omega_1)|^2}{p^2} |\tilde{\psi}_0(p, \omega_1)|^2\; p^2\, dp.$$

We note that $N_0\delta = N_1$ is the total number of scatterings per unit time and target area, i.e., the incident flux J_0. Let us choose Δp_0, $\Delta\theta_0$, and $\Delta\phi_0$ very small, and the direction of the incident beam along the z axis, so that $\theta_0 = \phi_0 = 0$. Then, in the limit

$$\Delta p_0, \Delta\theta_0, \Delta\phi_0 \to +0, \qquad |\tilde{\psi}_0(\mathbf{p})|^2 \to \delta^3(\mathbf{p} - \mathbf{p}_0),$$

we get by the mean-value theorem of the integral calculus

$$\lim_{|\tilde{\psi}_0|^2 \to \delta^3} \nu(\omega_0') = \frac{(2\pi)^4}{p^2} J_0 |T^{(1)}(p; \omega_0', \omega_0)|^2, \qquad \omega_0 = (0, 0)$$

We note that the application of the mean-value theorem is warranted on account of the implict assumption that $T^{(1)}(p; \omega', \omega)$ is continuous in some neighborhood of $\omega' = (0, 0)$.

Since $\nu(\omega_0')$ is the number of scatterings off an infinite planar* target (cf. Fig. 5) in the direction ω_0' per unit time, we have by (1.24)

$$\sigma(E, \omega_0') = \frac{1}{J_0} \lim_{|\tilde{\psi}_0|^2 \to \delta^3} \nu(\omega_0').$$

* By qualifying the above differential cross section as "planar" we depart from conventional terminology, which in the quantum context recognizes only the cross section given by Eq. (1.47) or, equivalently, by its more standard counterparts (4.98) and (4.99) (see §4.10 for a discussion of this point).

Thus, we have obtained the expression

$$(1.47)\qquad \sigma(E,\omega)=\frac{(2\pi)^4}{p_0^2}\,|\,T^{(1)}(p_0;\omega,\omega_0)|^2,\qquad E=\frac{(m_1+m_2)p_0^2}{2m_1m_2},$$

for the planar differential scattering cross section under the assumption that $T^{(1)}(p;\omega_1',\omega_1)$ is continuous in ω_1' and ω_1 when ω_1' is in a neighborhood of ω' and ω_1 in a neighborhood of $\omega_0=(0,0)$ for a given $p\in[0,\infty)$.

Exercises

1.1. Let $\mathbf{r}_k^{\text{ex}}+\mathbf{v}_k^{\text{ex}}t$, $k=1,2$, be the asymptotic states for the state $\mathbf{r}_k(t)$, $k=1,2$, of a classical two-particle system. In the center of mass coordinates (7.2) of Chapter II the state of the system is given by $\mathbf{R}(t)$ and $\mathbf{r}(t)$, where $\mathbf{R}(t)=\mathbf{R}_c+\mathbf{V}_ct$ describes the uniform motion of the center of mass of the system. Show that if $\mathbf{r}^{\text{ex}}=\mathbf{r}_2^{\text{ex}}-\mathbf{r}_1^{\text{ex}}$, $\mathbf{v}^{\text{ex}}=\mathbf{v}_2^{\text{ex}}-\mathbf{v}_1^{\text{ex}}$, then $|\,\mathbf{r}(t)-(\mathbf{r}^{\text{ex}}+\mathbf{v}^{\text{ex}}\,t)|\to 0$, $(1/t)\,\mathbf{r}(t)\to\mathbf{v}^{\text{ex}}$ and $|\,\mathbf{r}(t)|-t\,|\,\mathbf{v}^{\text{ex}}\,|\to\mathbf{r}^{\text{ex}}\cdot\mathbf{v}^{\text{ex}}/|\,\mathbf{v}^{\text{ex}}\,|$ when $t\to\mp\infty$, respectively.

1.2. Prove that a system of two particles interacting via a potential $V(r)=1/r$ (gravitational or Coulomb potential) does not possess any asymptotic states satisfying (1.5) and (1.6).

1.3. Show that the adjoint K^* of a bounded integral operator K,

$$(Kf)(x)=\int_{\mathbb{R}^n}K(x,x')f(x')\,d^nx',$$

defined on $L^2(\mathbb{R}^n)$, is an integral operator of the form

$$(K^*g)(x)=\int_{\mathbb{R}^n}(K(x',x))^*\,g(x')\,d^nx'.$$

1.4. Prove that if $\Psi^{\text{ex}}(t)\in\mathsf{M}_0(\text{ex}=\text{in, out})$ are the asymptotic states of $\Psi(t)$, and S is the scattering operator, then $\Psi^{\text{out}}(0)=S\Psi^{\text{in}}(0)$.

1.5. Show that the formula (1.47) for the scattering cross section stays valid in the general case when $\Psi^{\text{in}}(0)$ is represented by an arbitrary wave function $\tilde{\psi}^{\text{in}}(\mathbf{P},\mathbf{p})$, which is not necessarily in the form of the product occurring in (1.33).

1.6. Prove that if $T^{(1)}(p;\omega,\omega')$ and $T_0^{(1)}(p;\omega,\omega')$ are two kernels satisfying (1.31) for a given operator $T^{(1)}$, then $T_0^{(1)}(p;\omega,\omega')=T^{(1)}(p;\omega,\omega')$ almost everywhere with respect to the measure $\mu_l^{(1)}\times\mu_{\mathrm{p}}\times\mu_{\mathrm{p}}$ on $[0,+\infty)\times\Omega_{\mathrm{p}}\times\Omega_{\mathrm{p}}$.

1.7. Suppose $\Psi(t)$ and $\Psi_0(t)$, $t \in \mathbb{R}^1$, are vector-valued functions assuming values in the Hilbert space $\mathscr{H}$, and that there is a constant C such that $\| \Psi(t)\| \leqslant C$ and $\| \Psi_0(t)\| \leqslant C$ for all $t \in \mathbb{R}^1$. Let A be a bounded operator on $\mathscr{H}$. Show that $\lim_{t\to\mp\infty} \| \Psi(t) - \Psi_0(t)\| = 0$ implies that

$$\lim_{t\to\mp\infty} [\langle \Psi(t) \mid A\Psi(t)\rangle - \langle \Psi_0(t) \mid A\Psi_0(t)\rangle] = 0.$$

1.8. Use the result of Exercise 1.7 to prove that

$$\lim_{t\to\mp\infty} \langle \Psi(t) \mid E^{(\mathbf{p})}(B)\, \Psi(t)\rangle = \langle \Psi^{\mathrm{ex}}(0) \mid E^{(\mathbf{p})}(B)\, \Psi^{\mathrm{ex}}(0)\rangle,$$

where $(H_0^{(2)}\, \psi)^{\sim}(\mathbf{P}, \mathbf{p}) = (\mathbf{P}^2/2M_0 + \mathbf{p}^2/2m_0)\, \tilde{\psi}(\mathbf{P}, \mathbf{p})$, with M_0 and m_0 denoting, respectively, the total and the reduced mass of the system.

1.9. Consider the case of one-particle wave mechanics, when $H_0 \supseteq (1/2m)\Delta$. Using the well-known relation

$$(\exp(-iH_0t)\psi)^{\sim}(\mathbf{k}) = \exp[-i(\mathbf{k}^2/2m)t]\, \tilde{\psi}(\mathbf{k})$$

derive that, for functions $\psi(\mathbf{r})$ which are Lebesgue integrable as well as square integrable,

$$(\exp(-iH_0t)\psi)(\mathbf{r}) = (m/2\pi it)^{3/2} \int_{\mathbb{R}^3} \exp[i(m/2t)(\mathbf{r} - \mathbf{r}')^2]\, \psi(\mathbf{r}')\, d\mathbf{r}'$$

almost everywhere in $\mathbb{R}^3$.

1.10. Verify that the operators $U_t^{(1)}$ and $U_t^{(2)}$, defined for real $t \neq 0$ by

$$(U_t^{(1)}\psi)(\mathbf{r}) = (m/it)^{3/2} \exp[im(\mathbf{r}^2/2t)]\, \tilde{\psi}(m\mathbf{r}/t),$$

$$(U_t^{(2)}\psi)(\mathbf{r}) = \exp[\, im(\mathbf{r}^2/2t)]\, \psi(\mathbf{r}),$$

are unitary. Use the result of the preceding exercise to show that for $H_0 \supseteq -(1/2m)\Delta$

$$e^{-iH_0t} = U_t^{(1)} U_t^{(2)}.$$

1.11. Use the result of Exercise 1.10 to prove that the operator families e^{-iH_0t} and $U_t^{(1)}$ defined there satisfy

$$\lim_{t\to\mp\infty} \|(e^{-iH_0t} - U_t^{(1)})\psi\| = 0$$

for all $\psi \in L^2(\mathbb{R}^3)$.

1.12. Consider the cones

$$C^{(+)} = \{(r, \theta, \phi): r \geqslant 0, \theta_0 \leqslant \theta \leqslant \theta_0 + \Delta\theta, \phi_0 \leqslant \phi \leqslant \phi_0 + \Delta\phi\}$$

and $C^{(-)} = \{\mathbf{r}: -\mathbf{r} \in C^{+}\}$ in $\mathbb{R}^3$. Use the results of Exercises 1.7 and 1.11 to prove that for $H_0 \supseteq -(1/2m)\Delta$

$$\lim_{t\to\mp\infty} \int_{C^{(+)}} |(e^{-iH_0t}\psi)(\mathbf{r})|^2 \, d\mathbf{r} = \int_{C^{(\mp)}} |\tilde{\psi}(\mathbf{k})|^2 \, d\mathbf{k}.$$

Remark. This result indicates that asymptotically in the future the probability of finding the particle spatially located in the cone $C^{(+)}$ is equal to the probability of finding its momentum within $C^{(+)}$.

2. General Time-Dependent Two-Body Scattering Theory

2.1. The Intertwining Property of Wave Operators

In §1 we defined the *Møller wave operators* $\Omega_\pm$ by means of strong limits of $\Omega(t)$,

$$\Omega_\pm = \operatorname*{s-lim}_{t\to\mp\infty} \Omega(t)\, E_{\mathsf{M}_\pm},$$

(2.1)

$$\Omega(t) = e^{iHt}\, e^{-iH_0t}, \qquad H = H^*, \qquad H_0 = H_0{}^*,$$

in the time parameter t. In this definition, the closed linear subspaces $\mathsf{M}_\pm$ of $\mathscr{H}$ are spanned by all vectors f which are such that $e^{-iH_0t} f$ is a free state and the respective strong limits $\text{s-lim}_{t\to\mp\infty}\, \Omega(t) f$ exist.* The version of scattering theory based on this procedure of computing the wave operators, and eventually the scattering operator

$$S = \Omega_-{}^*\Omega_+, \tag{2.2}$$

is called *time-dependent* scattering theory.

In this section we shall derive some of the basic properties of wave and scattering operators, which follow from the general definitions (2.1) and (2.2), and which do not require the specification of H_0 and H within

* A mathematically common (but physically unfounded) way of defining $\Omega_\pm$ features $E^H(\mathsf{S}^H_{\text{ac}})$ instead of $E_{\mathsf{M}_\pm}$ in (2.1), where S^H_{ac} is the absolutely continuous spectrum of H (in practice, $\mathsf{S}^H_{\text{c}} = \mathsf{S}^H_{\text{ac}}$, though in general $\mathsf{S}^H_{\text{ac}} \subset \mathsf{S}^H_{\text{c}}$). Since, in general, the strong limit of $\Omega(t)E^H(\mathsf{S}^H_{\text{ac}})$ for $t \to \mp\infty$ might not exist, this alternative definition gives rise to the question of *existence* of wave operators. In our version, the wave operators always exist, but one must check in each particular case on the size of $\mathsf{M}_\pm$ and $\mathsf{R}_\pm$ (see §§7.1–7.2).

the confines of some particular model. It can be easily seen by comparing (2.1), (2.2) with (1.18), (1.28), and (1.29) that the derived theorems apply to the operators $\Omega_\pm^{(2)}$ and $S^{(2)}$ as well as to their counterparts $\Omega_\mp^{(1)}$ and $S^{(1)}$ in the "internal motion" Hilbert space $\mathscr{H}^{(1)}$. Naturally, in the first case one should take $H = H^{(2)}$ and $H_0 = H_0^{(2)}$, while in the second case one should take $H_0 = H_0^{(1)}$ and $H = H^{(1)}$.

Theorem 2.1. The wave operators $\Omega_\pm$ have the *intertwining properties*

$$\Omega_\pm e^{itH_0} = e^{itH}\Omega_\pm, \qquad t \in \mathbb{R}^1, \tag{2.3}$$

$$\Omega_\pm E^{H_0}(B) = E^H(B)\,\Omega_\pm, \qquad B \in \mathscr{B}^1. \tag{2.4}$$

If $f \in \mathscr{D}_{H_0}$, then

$$\Omega_\pm H_0 f = H\Omega_\pm f. \tag{2.5}$$

Proof. If f_1 belongs to the respective subspaces $\mathsf{M}_\pm$, then according to the definition of $\mathsf{M}_\pm$, the respective strong limits of $\Omega(t_1)f_1$ exist when $t_1 \to \mp\infty$. Thus, we can write

$$\begin{aligned}
&\| \exp(iHt)\,\Omega_\pm f_1 - \operatorname*{s-lim}_{t_1\to\mp\infty}[\exp(iHt_1)\exp(-iH_0t_1)]\exp(iH_0t)f_1 \| \\
&= \lim_{t_1\to\mp\infty} \| \exp(iHt)\,\Omega_\pm f_1 - (\exp(iHt_1)\exp(-iH_0t_1))\exp(iH_0t)f_1 \| \\
&= \lim_{t_1\to\mp\infty} \| \exp(iHt)\,\Omega_\pm f_1 - \exp(iHt)\exp[iH(t_1-t)]\exp[-iH_0(t_1-t)]f_1 \| \\
&= \lim_{t_1\to\mp\infty} \| \Omega_\pm f_1 - \exp[iH(t_1-t)]\exp[-iH_0(t_1-t)]f_1 \| = 0.
\end{aligned}$$

Hence, $\exp(iH_0t)\,f_1$ also belongs to $\mathsf{M}_\pm$, and since

$$\operatorname*{s-lim}_{t_1\to\mp\infty} \exp(iHt_1)\exp(-iH_0t_1)\exp(iH_0t)f_1 = \Omega_\pm \exp(iH_0t)f_1,$$

we have proved that

$$\Omega_\pm \exp(iH_0t)f_1 = \exp(iHt)\,\Omega_\pm f_1, \qquad f_1 \in \mathsf{M}_\pm,$$

for all $t \in \mathbb{R}^1$.

Now, if $f_2 \in \mathsf{M}_\pm^\perp = \mathscr{H} \ominus \mathsf{M}_\pm$, $\exp(iH_0t)\,f_2$ also belongs to $\mathsf{M}_\pm^\perp$. In fact, for all $f_1 \in \mathsf{M}_\pm$

$$\langle \exp(iH_0t)f_2 \mid f_1\rangle = \langle f_2 \mid \exp(-iH_0t)f_1\rangle = 0,$$

since $\exp(iH_0t) f_1 \in \mathsf{M}_\pm$ according to the preceding discussion. Thus, due to definition (2.1) of $\Omega_\pm$, we have $\Omega_\pm f_2 = \mathbf{0}$ and $\Omega_\pm \exp(iH_0t) f_2 = \mathbf{0}$. Consequently

$$\Omega_\pm \exp(iH_0t)f_2 = \exp(iHt)\,\Omega_\pm f_2 = \mathbf{0}, \qquad f_2 \in \mathsf{M}_\pm^\perp. \tag{2.6}$$

Since every $f \in \mathscr{H}$ can be decomposed in the form

$$f = f_1 + f_2, \qquad f_1 = E_{\mathsf{M}_\pm} f \in \mathsf{M}_\pm, \qquad f_2 = E_{\mathsf{M}_\pm^\perp} f \in \mathsf{M}_\pm^\perp,$$

we immediately obtain (2.3) from (2.5) and (2.6).

The relation (2.4) can be easily derived from (2.3) by methods frequently employed in Chapter IV (see Exercise 2.1).

Finally, by making use of the result (3.6) in Theorem 3.1, Chapter IV, and of the fact that $\Omega_\pm$ are bounded and therefore continuous, we easily derive that for $f \in \mathscr{D}_{H_0}$

$$\begin{aligned}\Omega_\pm H_0 f &= \Omega_\pm \operatorname*{s-lim}_{t\to 0} \frac{1}{it}(\exp(iH_0t) - 1)f \\ &= \operatorname*{s-lim}_{t\to 0} \frac{1}{it}(\Omega_\pm \exp(iH_0t) - \Omega_\pm)f \\ &= \operatorname*{s-lim}_{t\to 0} \frac{1}{it}(\exp(iHt)\,\Omega_\pm - \Omega_\pm)f \\ &= \operatorname*{s-lim}_{t\to 0} \frac{1}{it}(\exp(iHt) - 1)\,\Omega_\pm f = H\Omega_\pm f. \quad \text{Q.E.D.}\end{aligned}$$

2.2. The Partial Isometry of Wave Operators

We recall that in the last section $\mathsf{M}_\pm$ were defined to be the sets of all vectors $f \in \mathscr{H} \ominus \mathscr{H}_{0\mathrm{b}}$ for which the respective limits

$$f_\pm = \operatorname*{s-lim}_{t\to\mp\infty} \Omega(t) f$$

exist, while $\mathsf{R}_\pm$ are the sets of all vectors $f_\pm$. According to Lemma 1.1, the sets $\mathsf{M}_\pm$ are a closed linear subspace of $\mathscr{H}$. We shall prove the same statement for $\mathsf{R}_\pm$, by showing that $\Omega_\pm$ are partially isometric operators.

Definition 2.1. A bounded linear operator U defined on the Hilbert space $\mathscr{H}$ is said to be *partially isometric* with initial domain M and final domain N if U maps M isometrically onto N and maps $\mathsf{M}^\perp = \mathscr{H} \ominus \mathsf{M}$ into the zero vector, i.e., if $\| Uf \| = \| f \|$ for $f \in \mathsf{M}$ and $\| Uf \| = 0$ for $f \in \mathscr{H} \ominus \mathsf{M}$.

Theorem 2.2. The ranges of $\Omega_\pm$ are closed linear subspaces of $\mathscr{H}$; the wave operators $\Omega_\pm$ are partially isometric with initial domains $\mathsf{M}_\pm$ and final domains $\mathsf{R}_\pm$, respectively.

Proof. According to the definition of R_+, any vector $f_+ \in \mathsf{R}_+$ is the strong limit

$$f_+ = \operatorname*{s-lim}_{t\to-\infty} \Omega(t) f = \Omega_+ f$$

for some vector $f \in \mathsf{M}_+$. Moreover, since $\Omega(t)$ is unitary,

$$\| \Omega_+ f \| = \lim_{t\to-\infty} \| \Omega(t) f \| = \| f \| \tag{2.7}$$

if $f \in \mathsf{M}_+$. Hence, the restriction $\hat{\Omega}_+$ of Ω_+ to M_+ has an inverse $\hat{\Omega}_+^{-1}$, which is bounded on account of (2.7).

To prove that R_+ is closed, let $h \in \mathscr{H}$ be the strong limit of a sequence $h_1, h_2, \ldots \in \mathsf{R}_+$. By (2.7), $f_1 = \hat{\Omega}_+^{-1} h_1$, $f_2 = \hat{\Omega}_+^{-1} h_2, \ldots$ is a Cauchy sequence of elements from M_+. Since M_+ is closed, $f_1, f_2, \ldots$ has a limit $f \in \mathsf{M}_+$. Therefore, $h = \hat{\Omega}_+ f \in \mathsf{R}_+$, which proves that R_+ is a closed set.

The linearity of R_+ is an obvious consequence of the linearity of $\hat{\Omega}_+$. Thus, R_+ is a closed linear subspace of $\mathscr{H}$, and Ω_+ maps M_+ isometrically onto R_+. Moreover, if $g \perp \mathsf{M}_+$, then

$$\Omega_+ g = \operatorname*{s-lim}_{t\to-\infty} \Omega(t)\, E_{\mathsf{M}_+} g = \mathbf{0}.$$

Thus, Ω_+ is partially isometric with initial domain M_+ and final domain R_+.

The corresponding result for R_- is established in a similar manner. Q.E.D.

Lemma 2.1. If U is a partially isometric operator on $\mathscr{H}$ with initial domain M and final domain N, then U^* is a partial isometric operator with initial domain N and final domain M, and the following relations are true:

$$U^* U = E_{\mathsf{M}}, \tag{2.8}$$

$$U U^* = E_{\mathsf{N}}. \tag{2.9}$$

Proof. If $g \perp \mathsf{M}$, then $Ug = \mathbf{0}$ and consequently $U^* U g = \mathbf{0}$. Hence, if $f \in \mathsf{M}$, then

$$\langle g \mid U^* U f \rangle = \langle U^* U g \mid f \rangle = 0$$

for all $g \in \mathsf{M}^\perp$, and therefore $U^* U f \in \mathsf{M}$. Since M is linear, we also have $f - U^* U f \in \mathsf{M}$ for any $f \in \mathsf{M}$. On the other hand,

$$\langle h \mid f - U^* U f \rangle = \langle h \mid f \rangle - \langle Uh \mid Uf \rangle = 0$$

for all $h, f \in \mathsf{M}$ (see Exercise 2.2), which implies that $f = U^*Uf$. Thus, $U^*Uf = f$ for all $f \in \mathsf{M}$ and $U^*Ug = \mathbf{0}$ for all $g \perp \mathsf{M}$, i.e., $U^*U = E_{\mathsf{M}}$ and (2.8) is established.

We shall prove now that U^* is partially isometric. If $h \in \mathsf{N}$, there is a unique $f \in \mathsf{M}$ such that $h = Uf$, $\|f\| = \|h\|$. Consequently, $U^*h = U^*Uf = f$ by (2.8). On the other hand, if $h_1 \perp \mathsf{N}$, then $\langle U^*h_1 \mid f_1\rangle = \langle h_1 \mid Uf_1\rangle = 0$ for all $f_1 \in \mathscr{H}$, since $Uf_1 \in \mathsf{N}$ always. Consequently, $U^*h_1 = \mathbf{0}$ whenever $h_1 \perp \mathsf{N}$. Thus, U^* is partially isometric with initial domain N and final domain M.

By reversing the roles of U and U^* in (2.8) we show that (2.9) is true. Q.E.D.

If we apply this lemma to the operators $\Omega_\pm$, we arrive at the following relations:

$$\Omega_\pm{}^*\Omega_\pm = E_{\mathsf{M}_\pm}, \tag{2.10}$$

$$\Omega_\pm\Omega_\pm{}^* = E_{\mathsf{R}_\pm}. \tag{2.11}$$

In particular, if $\mathsf{M}_+ = \mathsf{M}_- = \mathsf{M}_0 = \mathscr{H}$, we would have $\Omega_\pm{}^*\Omega_\pm = 1$. We shall see in later sections that in problems of physical interest $\mathsf{M}_0^{(2)}$ coincides with the subspace $\mathscr{H}^{(2)} \ominus \mathscr{H}_{0b}^{(2)}$ of all vectors which are orthogonal to the subspace $\mathscr{H}_{0b}^{(2)} = \mathscr{H}_c^{(1)} \otimes \mathscr{H}_{0b}^{(1)}$, where [see (1.26)] $\mathscr{H}_{0b}^{(1)} \subset \mathscr{H}^{(1)}$ is the space spanned by the eigenvectors of the "internal" free Hamiltonian $H_{0i}^{(1)}$ given in (1.28). Ordinarily, $H_{0i}^{(1)}$ has a pure continuous spectrum, in which case $\mathscr{H}_{0b}^{(2)} = \{0\}$ and $\mathsf{M}_0^{(2)} \equiv \mathscr{H}^{(2)}$.

2.3. Properties of the S Operator

We turn our attention now to the S operator, giving special attention to the practically very important property of its "unitarity."

Theorem 2.3. The S operator commutes with the free Hamiltonian H_0, and $\Psi^{\text{out}}(t) = S\Psi^{\text{in}}(t)$ if $\Psi^{\text{ex}}(t)$, ex = in, out, are the strong asymptotic states of some $\Psi(t) = e^{iHt}\Psi(0)$.

Proof. Let $E^{H_0}(B)$ be the spectral measure of H_0. Using the intertwining property (2.4) and its equivalent

$$\Omega_\pm{}^*E^H(B) = E^{H_0}(B)\,\Omega_\pm{}^*, \qquad B \in \mathscr{B}^1, \tag{2.12}$$

derived from (2.4) by taking adjoints, we obtain

$$E^{H_0}(B)S = E^{H_0}(B)\,\Omega_-{}^*\Omega_+ = \Omega_-{}^*E^H(B)\,\Omega_+ = \Omega_-{}^*\Omega_+E^{H_0}(B) = SE^{H_0}(B).$$

Thus, S commutes with H_0. Since by (2.1) [cf. (1.14)], $\Psi(0) = \Omega_{\pm}\Psi^{\text{ex}}(0)$ so that $\Psi(0) = \mathsf{R}_+ \cap \mathsf{R}_-$, we get $\Psi^{\text{out}}(t) = S\Psi^{\text{in}}(t)$ at $t = 0$ by using (2.11). By virtue of $[S, H_0] = 0$, this stays true for all $t \in \mathbb{R}^1$. Q.E.D.

We recall that Theorem 2.3 has been used in proving that (1.30) is not feasible. In Theorems 2.4 and 2.5 we show that $S^*S = E_{\mathsf{M}_0}$ if $\mathsf{M}_0 = \mathsf{M}_{\pm}$. Thus, the condition that $S \neq 0$ coincides with the condition that $\mathsf{M}_0 \neq \{0\}$, i.e., that there are asymptotic states and that the wave operators do not vanish.

Theorem 2.4. The S operator maps the subspace M_+ into M_-, and maps its orthogonal complement $\mathscr{H} \ominus \mathsf{M}_+$ into zero; moreover it maps M_+ isometrically into M_- (i.e., $\| Sf \| = \| f \|$ for all $f \in \mathsf{M}_+$), if and only if $\mathsf{R}_+ \subset \mathsf{R}_-$.

Proof. According to Lemma 2.1, the final domain of Ω_-^* is identical to the initial domain M_- of Ω_-. Therefore, $Sf = \Omega_-^*(\Omega_+ f) \in \mathsf{M}_-$ for all $f \in \mathscr{H}$. However, if $f \perp \mathsf{M}_+$, then $\Omega_+ f = \mathbf{0}$ and consequently $Sf = \mathbf{0}$, which establishes the first statement in the theorem.

According to Theorem 2.2, Ω_+ maps M_+ isometrically onto R_+, i.e., $\| \Omega_+ g \| = \| g \|$ for $g \in \mathsf{M}_+$. By Lemma 2.1, Ω_-^* maps R_- isometrically onto M_-. Thus, we have $\| \Omega_-^*\Omega_+ g \| = \| \Omega_+ g \| = \| g \|$ for all $g \in \mathsf{M}_+$ if $\mathsf{R}_+ \subset \mathsf{R}_-$. However, if $\mathsf{R}_+ \ominus (\mathsf{R}_- \cap \mathsf{R}_+) \neq \{0\}$, then the elements of $\mathsf{R}_+ \ominus (\mathsf{R}_- \cap \mathsf{R}_+)$ are mapped by Ω_-^* into zero, and $S = \Omega_-^*\Omega_+$ is not isometric on M_+. Q.E.D.

The above theorem indicates that the S operator might be unitary if $\mathsf{M}_+ = \mathsf{M}_- = \mathsf{M}_0$ and S is restricted to M_0. Theorem 2.5 gives an important necessary and sufficient condition for this case to occur.

Theorem 2.5. The S operator is partially isometric with initial domain M_+ and final domain M_- and satisfies the relations

$$S^*S = E_{\mathsf{M}_+}, \qquad SS^* = E_{\mathsf{M}_-} \tag{2.13}$$

if and only if $\mathsf{R}_+ \equiv \mathsf{R}_-$.

Proof. Assume that $\mathsf{R}_+ \equiv \mathsf{R}_-$. Using

$$E_{\mathsf{R}_\pm}\Omega_\pm = \Omega_\pm ,$$

and the identity (2.11), we obtain that (2.13) holds:

$$S^*S = \Omega_+^*\Omega_-\Omega_-^*\Omega_+ = \Omega_+^*E_{\mathsf{R}_-}\Omega_+ = \Omega_+^*E_{\mathsf{R}_+}\Omega_+ = \Omega_+^*\Omega_+ = E_{\mathsf{M}_+},$$

$$SS^* = \Omega_-^*\Omega_+\Omega_+^*\Omega_- = \Omega_-^*E_{\mathsf{R}_+}\Omega_- = \Omega_-^*E_{\mathsf{R}_-}\Omega_- = \Omega_-^*\Omega_- = E_{\mathsf{M}_-}.$$

Now, by Theorem 2.4, when $\mathsf{R}_- = \mathsf{R}_+$, the S operator is partially isometric, with initial domain M_+ and a final domain equal to its range $\mathscr{R}_S$. Since, according to the same theorem, $\mathscr{R}_S \subset \mathsf{M}_-$, we obtain by using (2.13) and Lemma 2.1 that $\mathscr{R}_S \equiv \mathsf{M}_-$.

Conversely, assume that the relations (2.13) are true. We get by expressing in (2.13) S^*, S, and $E_{\mathsf{M}_\pm}$ in terms of the wave operators $\Omega_\pm$:

$$\Omega_+{}^*\Omega_-\Omega_-{}^*\Omega_+ = \Omega_+{}^*\Omega_+, \qquad \Omega_-{}^*\Omega_+\Omega_+{}^*\Omega_- = \Omega_-{}^*\Omega_-.$$

Multiplying the above two relations from the left by $\Omega_\pm$ and from the right by $\Omega_\pm{}^*$, respectively, we obtain, after using (2.10) and (2.11),

$$E_{\mathsf{R}_+}E_{\mathsf{R}_-}E_{\mathsf{R}_+} = E^2_{\mathsf{R}_+} = E_{\mathsf{R}_+}, \tag{2.14}$$

$$E_{\mathsf{R}_-}E_{\mathsf{R}_+}E_{\mathsf{R}_-} = E^2_{\mathsf{R}_-} = E_{\mathsf{R}_-}. \tag{2.15}$$

Thus, we have for any $f \in \mathscr{H}$

$$\| E_{\mathsf{R}_+} f \|^2 = \langle f \mid E_{\mathsf{R}_+} f \rangle = \langle f \mid E_{\mathsf{R}_+}E_{\mathsf{R}_-}E_{\mathsf{R}_+} f \rangle$$
$$= \langle E_{\mathsf{R}_+} f \mid E_{\mathsf{R}_-}E_{\mathsf{R}_+} f \rangle = \| E_{\mathsf{R}_-}E_{\mathsf{R}_+} f \|^2.$$

The above equality implies (see Exercise 2.3) that $E_{\mathsf{R}_+} f = E_{\mathsf{R}_-}E_{\mathsf{R}_+} f$ for all $f \in \mathscr{H}$, i.e.,

$$E_{\mathsf{R}_-}E_{\mathsf{R}_+} = E_{\mathsf{R}_+} = E^*_{\mathsf{R}_+} = E_{\mathsf{R}_+}E_{\mathsf{R}_-}.$$

Inserting the above relation in (2.15), we get

$$E_{\mathsf{R}_-} = E_{\mathsf{R}_-}E_{\mathsf{R}_+}E_{\mathsf{R}_-} = E^2_{\mathsf{R}_-}E_{\mathsf{R}_+} = E_{\mathsf{R}_-}E_{\mathsf{R}_+} = E_{\mathsf{R}_+}. \quad \text{Q.E.D.}$$

In the case of two-body potential scattering, i.e., scattering theory in which H and H_0 are essentially of the form (1.8) and (1.9), we will be able to prove that indeed $\mathsf{M}_+ = \mathsf{M}_- = \mathsf{M}_0$ and $\mathsf{R}_+ = \mathsf{R}_-$ when the potential $V(\mathbf{r})$ fulfills certain conditions. However, it is still interesting to investigate the physical consequences of a theory in which $\mathsf{R}_+ \neq \mathsf{R}_-$, i.e., when the S operator is not a partial isometry from M_+ to M_-.

Let us assume that g_0 is a vector which belongs to R_+ but does not belong to R_-. According to the definition of R_+, the fact that $g_0 \in \mathsf{R}_+$ implies that there is a vector $g_{\text{in}} \in \mathsf{M}_+$ such that $g_0 = \Omega_+ g_{\text{in}}$, i.e.,

$$\lim_{t\to-\infty} \| e^{-iHt} g_0 - e^{-iH_0 t} g_{\text{in}} \| = \lim_{t\to-\infty} \| g_0 - e^{iHt} e^{-iH_0 t} g_{\text{in}} \| = 0.$$

Hence, the interacting state $e^{-iHt} g_0$ has an incoming asymptotic free state. This means that $e^{-iHt}g_0$ is not a bound state. However, since $g_0 \notin \mathsf{R}_-$,

$$\operatorname*{s-lim}_{t\to+\infty} e^{iHt}\, e^{-iH_0t}\, g_0$$

does not exist. Hence, by Theorem 1.1 $e^{-iHt}g_0$ has no outgoing asymptotic state. Thus, $e^{-iHt}g_0$ is a "quasi-bound" state, which is approximately a free state in the distant past, but which does not become a free state again at any time in the future (see Pearson [1975] for examples).

On the other hand, if there is a vector $g_1 \in \mathsf{R}_-$ which does not belong to R_+, then the theory possesses an interacting state $e^{-iHt}g_1$ which becomes a free state in the distant future, but wich has never been a free state in the past. Thus, the condition of "unitarity" of the S operator, which is usually imposed on scattering theories, is equivalent to the requirement of not having any quasi-bound states in the theory.

2.4. Initial and Final Domains of Wave Operators

Thus far in this section we have never used the basic assumption that a state $e^{-iH_0t}g$ is called a free state only if it is not a bound state of the Hamiltonian H_0. As a matter of fact, all the theorems of this section stay valid if we define $\mathsf{M}_\pm$ to be the sets of all vectors $f \in \mathscr{H}$ for which the respective strong limits of $\Omega(t)f$, $t \to \mp\infty$, exist. If $\mathsf{M}_\pm$ are defined this way and we choose H identical to H_0, we would have $\Omega(t) \equiv 1$ and $\mathsf{M}_\pm = \mathsf{M}_0 \equiv \mathscr{H}$, always. However, if H_0 has bound states, then not all $e^{-iH_0t}f$, $f \in \mathscr{H}$, are free states, and M_0 defined in §1 is a nontrivial subspace of $\mathscr{H}$.

Let $\mathsf{M}_1 = E^H(\mathsf{S}_c^H)\mathscr{H}$ denote the closed subspace of $\mathscr{H}$ onto which the spectral measure $E^H(\mathsf{S}_c^H)$ belonging to the continuous spectrum of H projects.

Theorem 2.6. The closed subspaces $\mathsf{R}_\pm$ of $\mathscr{H}$ are also subspaces of $\mathsf{M}_1 = E^H(\mathsf{S}_c^H)\mathscr{H}$, and are left invariant by $E^H(B)$, i.e., $E^H(B)f \in \mathsf{R}_\pm$ for all $B \in \mathscr{B}^1$ whenever $f \in \mathsf{R}_\pm$.

Proof. Since $\mathsf{R}_\pm$ are the ranges of $\Omega_\pm$, we can find for any $g_\pm \in \mathsf{R}_\pm$ vectors $f_\pm$, such that $g_\pm = \Omega_\pm f_\pm$. Hence, using (2.4), we get

$$E^H(B)\,g_\pm = \Omega_\pm E^{H_0}(B)\,f_\pm \in \mathsf{R}_\pm, \qquad B \in \mathscr{B}^1,$$

which establishes that $\mathsf{R}_\pm$ are left invariant by H.

We shall prove that $\mathsf{R}_\pm \subset \mathsf{M}_1$ by showing that $\Omega_\pm f_\pm$ is orthogonal to all the eigenvectors of H.

Let g be an eigenvector of H with the eigenvalue λ_0, so that $E^H(\{\lambda_0\})g = g$. Given any $f_1 \in \mathsf{M}_\pm$, we can write

$$\langle g \mid \Omega_\pm f_1\rangle = \langle E^H(\{\lambda_0\})g \mid \Omega_\pm f_1\rangle = \langle \Omega_\pm{}^* E^H(\{\lambda_0\})g \mid f_1\rangle$$
$$= \langle E^{H_0}(\{\lambda_0\})\,\Omega_\pm{}^* g \mid f_1\rangle = \langle g \mid \Omega_\pm E^{H_0}(\{\lambda_0\})f_1\rangle = 0,$$

where we have made use of (2.12) to derive the third equality; the last inner product vanishes because $E^{H_0}(\{\lambda_0\})f_1 = \mathbf{0}$, since no $f_1 \in \mathsf{M}_\pm$ is an eigenvector of H_0.

Since $\Omega_\pm f_2 = \mathbf{0}$ whenever $f_2 \perp \mathsf{M}_\pm$, we derive, by setting $f = f_1 + f_2$, $f_1 \in \mathsf{M}_\pm$, $f_2 \perp \mathsf{M}_\pm$,

$$\langle g \mid \Omega_\pm f\rangle = \langle g \mid \Omega_\pm f_1\rangle + \langle g \mid \Omega_\pm f_2\rangle = 0$$

for any $f \in \mathscr{H}$. Hence, $\mathsf{R}_\pm$ are orthogonal to any eigenvectors g of H. Q.E.D.

We will be able to prove in later sections that in potential scattering with certain classes of potentials, we have $\mathsf{R}_+ = \mathsf{R}_- = \mathsf{M}_1$. In order to arrive at results of this kind, one usually starts by proving first that $\mathsf{R}_+ = \mathsf{R}_-$. This establishes immediately that the S operator is unitary. Now we shall prove a theorem which states a sufficient condition for the identity of R_+ and R_-. This theorem applies to operators with a simple continuous spectrum.

The concept of one operator A with a simple spectrum was introduced in §5 of Chapter IV: the self-adjoint operator A has a simple spectrum if the one-element set $\{A\}$ is a complete set of operators. According to the theorems of §5 in Chapter IV it is equivalent to say that A has a simple spectrum if there is a vector f_0 which is cyclic with respect to A.

If S_c^A and S_p^A denote the continuous and point spectrum of A, then the closed subspaces $\mathscr{H}_\mathrm{c}^A = E^A(\mathsf{S}_\mathrm{c}^A)\mathscr{H}$ and $\mathscr{H}_\mathrm{p}^A = E^A(\mathsf{S}_\mathrm{p}^A)\mathscr{H}$ onto which $E^A(\mathsf{S}_\mathrm{c}^A)$ and $E^A(\mathsf{S}_\mathrm{p}^A)$ project, respectively, are left invariant by A. To see that, note that $\mathscr{H} = \mathscr{H}_\mathrm{c}^A \oplus \mathscr{H}_\mathrm{p}^A$ and that

$$\langle f \mid Ag\rangle = \int_{-\infty}^{+\infty} \lambda\, d\langle f \mid E_\lambda^A g\rangle = 0$$

if $f \in \mathscr{H}_\mathrm{c}^A$ and $g \in \mathscr{H}_\mathrm{p}^A \cap \mathscr{D}_A$, or if $f \in \mathscr{H}_\mathrm{p}^A$ and $g \in \mathscr{H}_\mathrm{c}^A \cap \mathscr{D}_A$.

The restrictions A_c and A_p of A to $\mathscr{H}_\mathrm{c}$ and $\mathscr{H}_\mathrm{p}$, respectively, are obviously self-adjoint operators with spectral resolutions

$$A_\mathrm{c} = \int \lambda\, dE_\lambda^{A_\mathrm{c}}, \qquad E^{A_\mathrm{c}}(B) = E^A(B \cap \mathsf{S}_\mathrm{c}^A), \tag{2.16}$$

$$A_\mathrm{p} = \int \lambda\, dE_\lambda^{A_\mathrm{p}}, \qquad E^{A_\mathrm{p}}(B) = E^A(B \cap \mathsf{S}_\mathrm{p}^A), \tag{2.17}$$

respectively. When A_c or A_p are operators with a simple spectrum, we say that A has a *simple continuous spectrum* or a *simple point spectrum*, respectively. It is easy to establish that A has a simple spectrum if and only if both its continuous spectrum and its point spectrum are simple (see Exercise 2.4).

***Lemma 2.2.** Let A be a self-adjoint operator with a simple spectrum, and let M_k, $k = 1, 2$, be invariant subspaces of A, i.e., for any $f \in \mathsf{M}_k \cap \mathscr{D}_A$ we have $Af \in \mathsf{M}_k$, $k = 1, 2$. If the restrictions A_k, $k = 1, 2$, of A to M_k have identical spectra, then $\mathsf{M}_1 \equiv \mathsf{M}_2$.

Proof. Since M_k is left invariant by A, the operators A and E_{M_k} commute (see Exercise 2.5). Hence, by Theorem 5.6 of Chapter IV, the projector E_{M_k} is a function of A, i.e., $E_{\mathsf{M}_k} = F_k(A)$. From the properties $E_{\mathsf{M}_k} = E^*_{\mathsf{M}_k}$ and $E^2_{\mathsf{M}_k} = E_{\mathsf{M}_k}$, by using (2.26) of Chapter IV we prove that $F_k(\lambda)$ is a real function satisfying

$$\int_{\mathbb{R}^1} F_k^2(\lambda)\, d\,\|E_\lambda^A f\|^2 = \int_{\mathbb{R}^1} F_k(\lambda)\, d\,\|E_\lambda^A f\|^2, \qquad f \in \mathscr{H},$$

and therefore $F_k(\lambda)$ can assume only the values 0 and 1 for $\lambda \in \mathsf{S}^A$. Thus, if we set $F_k(\lambda) = 0$ for $\lambda \notin \mathsf{S}^A$, we can state that $F_k(\lambda)$ is the characteristic function of a Borel subset B_k of S^A.

We obviously have

$$A_k = AE_{\mathsf{M}_k} = AF_k(A).$$

Hence, the spectrum S^{A_k} of A_k coincides with B_k. Consequently, $\mathsf{S}^{A_1} = \mathsf{S}^{A_2}$ implies that $F_1(A) \equiv F_2(A)$, i.e., $E_{\mathsf{M}_1} = E_{\mathsf{M}_2}$. Q.E.D.

The following theorem can be very useful in establishing that S is unitary.

Theorem 2.7. If the continuous spectrum of H is simple and $\mathsf{M}_+ = \mathsf{M}_-$, then $\mathsf{R}_+ \equiv \mathsf{R}_-$.

Proof. Let H_1 be the restriction of H to the closed subspace $\mathsf{M}_1 = E^H(\mathsf{S}_c^H)\mathscr{H}$. The operator H_1 is obviously a self-adjoint operator with a pure continuous spectrum which is identical to the continuous spectrum of H. Hence, according to the assumption, the spectrum S^{H_1} of H_1 is simple.

According to Theorem 2.6, $\mathsf{R}_\pm$ are closed subspaces of M_1 which H_1 leaves invariant. Since the spectrum of H_1 is simple, by (2.4) and (2.11) we can apply Lemma 2.2 and infer that $\mathsf{R}_+ \equiv \mathsf{R}_-$. Q.E.D.

2.5. Dyson's Perturbation Expansion

We saw in §1 that the knowledge of the S operator is sufficient for the calculation of the differential cross section and of transition probabilities. This shows that a complete solution of the dynamical problem, which would consist in computing the evolution operator $U(t, t_0)$ introduced in §3 of Chapter IV, is not necessary for the theoretical description of a scattering experiment. Since the computation of $\Psi(t) = U(t, t_0)\,\Psi_0$ for given Ψ_0 can be a difficult and intricate computational task, it is desirable to develop methods for computing the S operator in a direct manner. Dyson's perturbation theory provides one such method.

It follows from (2.1) that for any $f, g \in \mathscr{H}$

$$\langle f \mid \Omega_-^* g\rangle = \langle \Omega_- f \mid g\rangle = \lim_{t\to+\infty} \langle \Omega(t)\, E_{\mathsf{M}_-} f \mid g\rangle = \lim_{t\to+\infty} \langle f \mid E_{\mathsf{M}_-} \Omega^*(t)\, g\rangle. \tag{2.18}$$

Since $\Omega^*(t)$ is unitary, we have $\| \Omega^*(t) g \| = \| g \|$. On the other hand, if $g \in \mathsf{R}_-$, it follows from the fact that Ω_-^* is a partially isometric operator with initial domain R_- (see Theorem 2.2 and Lemma 2.1) that $\| \Omega_-^* g \| = \| g \|$, so that $\| \Omega^*(t) g \| = \| \Omega_-^* g \|$. According to Lemma 6.2 of Chapter IV, this result in conjunction with (2.18) implies that

$$\Omega_-^* g = \operatorname*{s-lim}_{t\to+\infty} E_{\mathsf{M}_-} \Omega^*(t)\, g, \qquad g \in \mathsf{R}_- .$$

Hence, if S is partially isometric, so that $\mathsf{R}_+ \equiv \mathsf{R}_-$, we have

$$\begin{aligned} S &= E_{\mathsf{M}_-} \operatorname*{s-lim}_{t\to+\infty} \Omega^*(t) (\operatorname*{s-lim}_{t_0\to-\infty} \Omega(t_0)\, E_{\mathsf{M}_+}) \\ &= E_{\mathsf{M}_-} \operatorname*{s-lim}_{t\to+\infty} (\operatorname*{s-lim}_{t_0\to-\infty} \Omega^*(t)\, \Omega(t_0))\, E_{\mathsf{M}_+}, \end{aligned} \tag{2.19}$$

where the operator

$$V(t, t_0) = \Omega^*(t)\, \Omega(t_0) = e^{iH_0 t}\, e^{-iH(t-t_0)}\, e^{-iH_0 t_0} \tag{2.20}$$

is obviously unitary for any $t, t_0 \in \mathbb{R}^1$.

Comparison of

$$\Omega^*(t) = e^{iH_0 t}\, e^{-iHt}$$

with (3.43) of Chapter IV shows that $\Omega^*(t)$ is the operator which determines the time dependence of the state $\tilde{\Psi}(t)$ in the interaction picture. Hence, we have (see Chapter IV, Exercise 3.4)

$$\frac{d\Omega^*(t)}{dt} f = \operatorname*{s-lim}_{\Delta t\to 0} \frac{\Omega^*(t + \Delta t) - \Omega^*(t)}{\Delta t} f = -iH_{\mathrm{I}}(t) \Omega^*(t)\, f, \tag{2.21}$$

$$H_{\mathrm{I}}(t) = e^{iH_0 t}(H - H_0)\, e^{-iH_0 t} \tag{2.22}$$

for any $f \in \mathscr{D}_H \cap \mathscr{D}_{H_0}$.

We have seen in §7 of Chapter IV that in potential scattering, under very reasonable assumptions on the potential, $\mathscr{D}_H$ coincides with $\mathscr{D}_{H_0}$. When this is the case, and the vector f belongs to $\mathscr{D}_H \equiv \mathscr{D}_{H_0}$, we have $\exp(-iH_0t_0)f \in \mathscr{D}_{H_0}$ and $\exp(iHt_0)\exp(-iH_0t_0)f \in \mathscr{D}_H$ (see Exercise 2.6), so that $\Omega(t_0)f \in \mathscr{D}_H \equiv \mathscr{D}_H \cap \mathscr{D}_{H_0}$ and

$$\frac{\partial V(t, t_0)}{\partial t} f = \operatorname*{s-lim}_{\Delta t \to 0} \frac{V(t + \Delta t, t_0) - V(t, t_0)}{\Delta t} f = -iH_{\mathrm{I}}(t)\, V(t, t_0)f. \tag{2.23}$$

Thus, for any vector $h \in \mathscr{H}$

$$(\partial/\partial t)\langle h \mid V(t, t_0)f\rangle = -i\langle h \mid H_{\mathrm{I}}(t)\, V(t, t_0)f\rangle. \tag{2.24}$$

If the function $\langle h \mid H_{\mathrm{I}}(t)\, V(t, t_0)f\rangle$ is integrable in t, we obtain from (2.24), after having observed that $V(t_0, t_0) = 1$, the following relation:

$$\langle h \mid V(t, t_0)f\rangle = \langle h \mid f\rangle - i \int_{t_0}^{t} \langle h \mid H_{\mathrm{I}}(t_1)\, V(t_1, t_0)f\rangle\, dt_1 . \tag{2.25}$$

Since $\langle h \mid V(t, t_0)f\rangle$ and $\langle h \mid f\rangle$ are bounded functionals,

$$-i \int_{t_0}^{t} \langle h \mid H_{\mathrm{I}}(t_1)\, V(t_1, t_0)f\rangle\, dt_1$$

is defined for any $h \in \mathscr{H}$, it determines for each fixed $f \in \mathscr{D}_{H_0}$ a continuous antilinear functional in h, and therefore by Riesz's Theorem (Theorem 2.3, Chapter III), it defines a mapping $f \mapsto W_1(t, t_0)f$. It is easy to see that $W_1(t, t_0)$ is a linear operator defined on $\mathscr{D}_{H_0}$. We write, symbolically,

$$W_1(t, t_0) = -i \int_{t_0}^{t} H_{\mathrm{I}}(t_1)\, V(t_1, t_0)\, dt_1 . \tag{2.26}$$

In this notation (2.25) is equivalent to the statement that

$$V(t, t_0)f = \left(1 - i \int_{t_0}^{t} H_{\mathrm{I}}(t_1)\, V(t_1, t_0)\, dt_1\right) f, \qquad f \in \mathscr{D}_{H_0} . \tag{2.27}$$

If we treat the above integral equation recursively in $V(t, t_0)$, we obtain after n steps

$$V(t, t_0)f = [1 + V_1(t, t_0) + \cdots + V_{n-1}(t, t_0) + W_n(t, t_0)]f, \qquad f \in \mathscr{D}_{H_0},$$

where $V_k(t, t_0)$ and $W_n(t, t_0)$ are defined by the relations

$$V_k(t, t_0) = (-i)^k \int_{t_0}^{t} dt_1 \int_{t_0}^{t_1} dt_2 \cdots \int_{t_0}^{t_{k-1}} dt_k\, H_{\mathrm{I}}(t_1) \cdots H_{\mathrm{I}}(t_k), \tag{2.28}$$

$$W_n(t, t_0) = (-i)^n \int_{t_0}^{t} dt_1 \int_{t_0}^{t_1} dt_2 \cdots \int_{t_0}^{t_{n-1}} dt_n\, H_{\mathrm{I}}(t_1) \cdots H_{\mathrm{I}}(t_n)\, V(t_n, t_0), \tag{2.29}$$

whose precise meanings are established in the same manner as in the case of $W_1(t, t_0)$.

It has become customary to introduce a chronological ordering operation T by setting

$$\mathsf{T}[H_I(t_1)\, H_I(t_2) \cdots H_I(t_k)] = H_I(t_{i_1})\, H_I(t_{i_2}) \cdots H_I(t_{i_k}), \qquad t_{i_1} \geqslant t_{i_2} \geqslant \cdots \geqslant t_{i_k}. \tag{2.30}$$

This operation orders the time-dependent operators in the product $H_I(t_1) \cdots H_I(t_k)$ in a chronological sequence of increasing values of the time variable from right to left, so that the integrand in (2.28) becomes symmetric under permutations of $t_1, \ldots, t_n$ when T is set in front of it. Since there are $n!$ such permutations, it is easy to check that

$$V_k(t, t_0) = \frac{(-i)^k}{k!} \int_{t_0}^{t} dt_1 \int_{t_0}^{t} dt_2 \cdots \int_{t_0}^{t} dt_k \mathsf{T}[H_I(t_1)\, H_I(t_2) \cdots H_I(t_k)]. \tag{2.31}$$

Suppose that

$$\begin{aligned} V_k f &= \frac{(-i)^k}{k!} \operatorname*{w-lim}_{t \to +\infty} \left(\operatorname*{w-lim}_{t_0 \to -\infty} \int_{t_0}^{t} dt_1 \cdots \int_{t_0}^{t} dt_k \mathsf{T}[H_I(t_1) \cdots H_I(t_k)] f \right) \\ &= \frac{(-i)^k}{k!} \int_{-\infty}^{+\infty} dt_1 \cdots \int_{-\infty}^{+\infty} dt_k \mathsf{T}[H_I(t_1) \cdots H_I(t_k)] f \end{aligned} \tag{2.32}$$

exists for every $k = 1, 2, \ldots$, and that

$$W_n f = \operatorname*{w-lim}_{t \to +\infty} (\operatorname*{w-lim}_{t_0 \to -\infty} W_n(t, t_0) f) \tag{2.33}$$

also exists and converges (at least in a weak sense) to the zero vector when $n \to +\infty$. Under these assumptions we obtain from (2.19), (2.20), (2.30), and (2.31) the following expansion:

$$S f = E_{\mathsf{M}_-}(1 + V_1 + V_2 + \cdots)\, E_{\mathsf{M}_+} f.$$

The above expansion is known as a perturbation expansion of the S operator. In the zero order for the case $\mathsf{M}_\pm = \mathsf{M}_0$, it becomes $S = E^2_{\mathsf{M}_0} = E_{\mathsf{M}_0}$, i.e., it describes the case when there is no scattering. The computation to higher orders is facilitated in practice by an explicit knowledge of the commutator $[H_I(t_1), H_I(t_2)] = D(t_1, t_2)$. This makes possible the computation of the integrand of V_k by successive use of the relation

$$\mathsf{T}[H_I(t_1)\, H_I(t_2)] = H_I(t_1)\, H_I(t_2) - \theta(t_2 - t_1)\, D(t_1, t_2).$$

Computations of this nature occur very frequently in quantum field theory, and techniques have been developed for coping with them in a systematic manner.

2.6. Criteria for Existence of Strong Asymptotic States

There are cases even in potential scattering in which $\mathsf{M}_\pm = \{0\}$, i.e., when there are no asymptotic states in the sense of (1.7). One notable example is the case of Coulomb scattering, when we are dealing with the Schroedinger operator H with a Coulomb potential. Namely, it can be shown that quantum Coulomb scattering shares with the classical Coulomb scattering the feature of not having asymptotic states in the conventional sense (1.7), though asymptotic states can still be introduced in a somewhat different sense of the word.*

This particular example indicates that it is desirable to produce some criteria for the existence of strong asymptotic states. The rest of this section deals with the derivation of Theorem 2.8, which will be found very useful in establishing the existence of strong asymptotic states in potential scattering (see Exercise 2.8 and §7.1).

Theorem 2.8. Suppose $\mathscr{D}_1$, $\mathscr{D}_1 \subset \mathscr{D}_H \cap \mathscr{D}_{H_0}$, is a linear submanifold of $\mathscr{H}$ which is such that $e^{-iH_0 t} f \in \mathscr{D}_H \cap \mathscr{D}_{H_0}$ for all $f \in \mathscr{D}_1$. Then the closure $\bar{\mathscr{D}}_1$ of $\mathscr{D}_1$ belongs to the initial domains $\mathsf{M}_\pm$ of the wave operators $\Omega_\pm$ (Møller, i.e. strong, operators) if and only if when $t_1 \to -\infty$ and $t_2 \to +\infty$ the respective strong limits of

$$I_1(t_1, t_2)f = \int_{t_1}^{t_2} e^{iHt} H_1 e^{-iH_0 t} f \, dt, \qquad H_1 = H - H_0, \tag{2.34}$$

exist for every $f \in \mathscr{D}_1$. A sufficient condition for the existence of these limits is that for some $t_1, t_2 \in \mathbb{R}^1$

$$\int_{-\infty}^{t_2} \| H_1 \exp(-iH_0 t) f \| \, dt < +\infty, \qquad \int_{t_1}^{+\infty} \| H_1 \exp(-iH_0 t) f \| \, dt < +\infty,$$

respectively.

To understand the above theorem, we must comprehend the meaning of the integral appearing in (2.34) (see also §3.7).

This integral is defined by the requirement that for fixed $f \in \mathscr{D}_1$

$$\phi_f(g; t_1, t_2) = \int_{t_1}^{t_2} \langle \exp(iHt) H_1 \exp(-iH_0 t) f \mid g \rangle \, dt \tag{2.35}$$

* For details, see §4.9.

determines a continuous linear functional in g. In fact, if the integral in (2.35) exists, then, by Riesz' theorem (Chapter III, Theorem 2.3), there is a unique vector $f_1 = I_1(t_1, t_2) f$ such that $\langle f_1 \mid g\rangle = \phi_f(g; t_1, t_2)$. Thus, in order to know that $I_1(t_1, t_2)$ is well defined we only have to establish that the Riemann integral in (2.35) really exists. For this task we need Lemma 2.3.

Lemma 2.3. For any two vectors $f \in \mathscr{D}_{H_0}$ and $g \in \mathscr{D}_H$ the function

(2.36)

$$F(t) = \langle \exp(-iH_0t)f \mid H\exp(-iHt)\,g\rangle - \langle H_0\exp(-iH_0t)f \mid \exp(-iHt)\,g\rangle$$

is continuous in $t \in \mathbb{R}^1$, and

(2.37)

$$\int_{t_1}^{t_2} (\langle \exp(-iH_0t)f \mid H\exp(-iHt)\,g\rangle - \langle H_0\exp(-iH_0t)f \mid \exp(-iHt)\,g\rangle)\,dt = i\,\langle[\Omega(t_2) - \Omega(t_1)]f \mid g\rangle.$$

Proof. To establish the continuity of the first term in (2.36), we note that for any $\tau \in \mathbb{R}^1$

(2.38)

$$\begin{aligned} &|\langle \exp[-iH_0(t+\tau)]f \mid H\exp[-iH(t+\tau)]\,g\rangle - \langle \exp(-iH_0t)f \mid H\exp(-iHt)g\rangle| \\ &\quad \leqslant |\langle \exp[-iH_0(t+\tau)]f \mid H(\exp[-iH(t+\tau)] - \exp(-iHt))\,g\rangle| \\ &\qquad + |\langle(\exp[-iH_0(t+\tau)] - \exp(-iH_0t))f \mid H\exp(-iHt)\,g\rangle| \\ &\quad \leqslant \|f\|\,\|H(\exp[-iH(t+\tau)] - \exp(-iHt))\,g\| \\ &\qquad + \|(\exp[-iH_0(t+\tau)] - \exp(-iH_0t))f\|\,\|H\exp(-iHt)\,g\|. \end{aligned}$$

We recall now that by Theorem 2.7 of Chapter IV, the operator $H(\exp[-iH(t+\tau)] - \exp(-iHt)$ is a function of the self-adjoint operator H, and consequently

$$H(\exp[-iH(t+\tau)] - \exp(-iHt))\,g = (\exp[-iH(t+\tau)] - \exp(-iHt))\,Hg.$$

Since $\exp(-iHt)$ and $\exp(-iH_0t)$ are strongly continuous operator-valued functions in t (see Chapter IV, §6.3), the expression on the right-hand side of (2.38) converges to zero when $\tau \to 0$.

The continuity in t of the second term in (2.36) is even easier to prove by following the same line of reasoning, since we have

$$\begin{aligned} &\langle H_0\exp[-iH_0(t+\tau)]f \mid \exp[-iH(t+\tau)]\,g\rangle - \langle H_0\exp(-iH_0t)f \mid \exp(-iHt)\,g\rangle| \\ &\quad \leqslant \|H_0(\exp[-iH_0(t+\tau)] - \exp(-iH_0t))f\|\,\|g\| \\ &\qquad + \|H_0\exp(-iH_0t)f\|\,\|(\exp[-iH(t+\tau)] - \exp(-iHt))\,g\|. \end{aligned}$$

The continuity of $F(t)$ in the variable t implies the existence of the integral in (2.37) for any given t_1, $t_2 \in \mathbb{R}^1$. Thus, the first part of the lemma is established.

To establish the second part, we recall from Theorem 3.1 of Chapter IV that whenever $g \in \mathscr{D}_A$

$$\operatorname*{s-lim}_{\tau\to 0} \frac{i}{\tau} (\exp[-iA(t+\tau)] - \exp(-iAt))\, g = A \exp(-iAt)\, g$$

for any $A = A^*$ in general, and consequently for $A = H$ and $A = H_0$ in particular. In view of this fact, a glimpse at (2.36) suffices to show that

$$F(t) = i\,(d/dt)\langle e^{-iH_0t}f \mid e^{-iHt}g\rangle.$$

Hence, (2.37) immediately follows. Q.E.D.

Let us return now to the proof of Theorem 2.8. When

$$f \in \mathscr{D}_1 \subset \mathscr{D}_H \cap \mathscr{D}_{H_0},$$

then $\exp(-iH_0t) f \in \mathscr{D}_{H_0} \cap \mathscr{D}_H$, so that

$$\langle \exp(-iH_0t)f \mid H \exp(-iHt)\, g\rangle = \langle \exp(iHt)\, H \exp(-iH_0t)f \mid g\rangle,$$

and $F(t)$ can be written in the form

$$F(t) = \langle \exp(iHt)\, H_1 \exp(-iH_0t)f \mid g\rangle.$$

Thus, according to (2.37), we have

$$\langle [\Omega(t_2) - \Omega(t_1)]f \mid g\rangle = i \int_{t_1}^{t_2} \langle \exp(iHt)\, H_1 \exp(-iH_0t)f \mid g\rangle\, dt \tag{2.39}$$

for any $f \in \mathscr{D}_1$ and $g \in \mathscr{D}_H$. Since $\mathscr{D}_H$ is dense in $\mathscr{H}$, the integral on the right-hand side of the above relation can be said to determine a unique linear functional in $g \in \mathscr{H}$. Thus, we can write symbolically

$$[\Omega(t_2) - \Omega(t_1)]f = -i \int_{t_1}^{t_2} \exp(iHt)\, H_1 \exp(-iH_0t)f\, dt. \tag{2.40}$$

It follows from the above relation that in the case s-lim $\Omega(t) f$ exists for $t \to -\infty$ or $t \to +\infty$, the respective strong limits of the integral in (2.34) for $t_1 \to -\infty$ or $t_2 \to +\infty$ also exist.

Conversely, let us assume that

$$\operatorname*{s-lim}_{t_1\to-\infty} \int_{t_1}^{t_2} \exp(iHt)\, H_1 \exp(-iH_0t)f\, dt$$

exist for all $f \in \mathscr{D}_1$. Then, it follows from (2.40) that s-lim $\Omega(t_1) f$ also exists when $t_1 \to -\infty$ and $f \in \mathscr{D}_1$. However, by Lemma 1.1, the set of all vectors f for which this last limit exists is closed, and consequently the closure $\bar{\mathscr{D}}_1$ of $\mathscr{D}_1$ must be contained in it. Thus, the theorem is established.

For all practical purposes, the investigation of the existence of the strong limits in t_1 and t_2 of the integral in (2.34) reduces to the investigation of the existence of integrals

$$\int_{t_1}^{+\infty} \| H_1 \exp(-iH_0 t) f \| \, dt \tag{2.41}$$

for some $t_1 \in \mathbb{R}^1$, and of

$$\int_{-\infty}^{t_2} \| H_1 \exp(-iH_0 t) f \| \, dt \tag{2.42}$$

for some $t_2 \in \mathbb{R}^1$. In fact, by using the Schwarz–Cauchy inequality we get

(2.43)

$$\left| \int_{t_1}^{t_2} \langle \exp(iHt) \, H_1 \exp(-iH_0 t) f \mid g \rangle \, dt \right| \leqslant \| g \| \int_{t_1}^{t_2} \| \exp(iHt) \, H_1 \exp(-iH_0 t) f \| \, dt.$$

Hence, the existence of (2.42) implies, in conjunction with (2.39), that $f_+ =$ w-lim $\Omega(t_1) f$ exists when $t_1 \to -\infty$. Since $\| \Omega(t_1) f \| \leqslant \| f \| = \| f_+ \|$, we infer by using Lemma 6.2 of Chapter IV that s-lim $\Omega(t_1) f$ also exists for $t_1 \to -\infty$.

The argument for $t_2 \to +\infty$ is completely analogous.

2.7. The Physical Asymptotic Condition

The condition which the Schroedinger-picture strong asymptotic states $\Psi^{\text{ex}}(t)$ (ex = in, out) of the interacting state $\Psi(t)$ have to satisfy,

$$\lim_{t \to \pm\infty} \| \Psi(t) - \Psi^{\text{ex}}(t) \| = 0, \tag{2.44}$$

has no *direct* physical meaning, since the above vector norm is not a measurable quantity. We recall from §§1 and 2 of Chapter IV that the directly measurable quantities are the probabilities

$$P_{\Psi(t)}^{A_1,\dots,A_n}(B) = \langle \Psi(t) \mid E^{A_1,\dots,A_n}(B) \, \Psi(t) \rangle, \qquad B \in \mathscr{B}^n, \quad \| \Psi(t) \| = 1, \tag{2.45}$$

for any set of compatible fundamental* observables. Hence, we expect asymptotic states to satisfy some kind of physical asymptotic condition which would specify that the probabilities (2.45) for the interacting state $\Psi(t)$ approximate arbitrarily closely the corresponding probabilities for the respective asymptotic states $\Psi^{\text{in,out}}(t)$, when $t \to \pm\infty$. That this is actually the case is stated in Theorem 2.9.

***Theorem 2.9.** Let the Hamiltonian H and a set $\mathcal{O}_0$ of additional fundamental observables be such that any other observables of the interacting system or the free system is a function of compatible observables from $\mathcal{O}_{\mathrm{f}} = \{H\} \cup \mathcal{O}_0$ or $\mathcal{O}_{\mathrm{f}}^{\mathrm{ex}} = \{H_0\} \cup \mathcal{O}_0$, respectively. Furthermore, assume that whenever $\{H, A_1',..., A_n'\}$, $A_1',..., A_n' \in \mathcal{O}_0$, is a set of compatible observables, then $H_0, A_1',..., A_n'$ are also compatible. In that case the strong asymptotic states $\Psi^{\mathrm{ex}}(t)$ (ex = in, out) of the scattering state $\Psi(t)$ satisfy the *physical asymptotic conditions*

$$\lim_{t\to\pm\infty} | P_{\Psi(t)}^{A_1,\ldots,A_m}(B) - P_{\Psi^{\mathrm{ex}}(t)}^{A_1,\ldots,A_m}(B)| = 0, \qquad B \in \mathscr{B}^m, \tag{2.46}$$

$$\lim_{t\to\pm\infty} | P_{\Psi(t)}^{H,A_1',\ldots,A_n'}(B) - P_{\Psi^{\mathrm{ex}}(t)}^{H_0,A_1',\ldots,A_n'}(B)| = 0, \qquad B \in \mathscr{B}^{n+1}, \tag{2.47}$$

for any sets $\{A_1, ..., A_m\} \subset \mathcal{O}_0$, $\{H, A_1',..., A_n'\} \subset \mathcal{O}_{\mathrm{f}}$ of compatible observables.

Proof. It is straightforward (see Exercise 1.7) to prove that (2.46) holds, i.e.,

$$\lim_{t\to\pm\infty} |\langle\Psi(t) \mid E^{A_1,\ldots,A_m}(B)\, \Psi(t)\rangle - \langle\Psi^{\mathrm{ex}}(t) \mid E^{A_1,\ldots,A_m}(B)\, \Psi^{\mathrm{ex}}(t)\rangle| = 0$$

when (2.44) is satisfied.

Since $A_1',..., A_n'$ commute with H as well as with H_0, we infer that for arbitrary $B_2 \in \mathscr{B}^n$

$$e^{iHt}\, e^{-iH_0t} E^{A_1',\ldots,A_n'}(B_2) = E^{A_1',\ldots,A_n'}(B_2)\, e^{iHt}\, e^{-iH_0t}.$$

By applying the operators on both sides of the above equation to vectors $f \in \mathsf{M}_\pm$ and letting $t \to \pm\infty$, we obtain immediately $E^{A_1',\ldots,A_n'}(B_2) f \in \mathsf{M}_\pm$ and

$$\Omega_\pm E^{A_1',\ldots,A_n'}(B_2) f = E^{A_1',\ldots,A_n'}(B_2)\, \Omega_\pm f. \tag{2.48}$$

* See the discussion in the beginning of §2 in Chapter IV concerning fundamental observables.

Furthermore, if $g \perp \mathsf{M}_{\pm}$, then

$$\langle E^{A_1',\ldots,A_n'}(B_2)\, g \mid f\rangle = \langle g \mid E^{A_1',\ldots,A_n'}(B_2) f\rangle = 0$$

for all $f \in \mathsf{M}_{\pm}$, and therefore $E^{A_1',\ldots,A_n'}(B_2)g \perp \mathsf{M}_{\pm}$. Consequently, (2.48) holds for arbitrary $f \in \mathscr{H}$. Combining this relation with (2.4), we conclude that

$$\Omega_{\pm} E^{H_0}(B_1)\, E^{A_1',\ldots,A_n'}(B_2) = E^{H}(B_1)\, E^{A_1',\ldots,A_n'}(B_2)\, \Omega_{\pm} \tag{2.49}$$

for any $B_1 \in \mathscr{B}^1$ and $B_2 \in \mathscr{B}^n$.

As a consequence of the basic properties of spectral measures (see Chapter III, the proof of Theorem 5.5), we can write for any $B \in \mathscr{B}^{n+1}$

$$E^{H_0,A_1',\ldots,A_n'}(B) = \inf \Big\{ \sum_k E^{H_0}(B_1^{(k)})\, E^{A_1',\ldots,A_n'}(B_2^{(k)}) : B \subset \bigcup_k B_1^{(k)} \times B_2^{(k)} \Big\},$$

where the projector infimum is taken only over finite unions of sets $B_1^{(k)} \times B_2^{(k)}$, with $B_1^{(k)} \in \mathscr{B}^1$, $B_2^{(k)} \in \mathscr{B}^n$. Hence, $E^{H_0,A_1,\ldots,A_n}(B)$ is the strong limit of some sequence of finite sums

$$\sum_k E^{H_0}(B_1^{(k)})\, E^{A_1',\ldots,A_n'}(B_2^{(k)})$$

corresponding to a sequence of sets $\bigcup_k (B_1^{(k)} \times B_2^{(k)})$, with B as an intersection. This result in conjunction with (2.49) implies that

$$\Omega_{\pm} E^{H_0,A_1',\ldots,A_n'}(B) = E^{H,A_1',\ldots,A_n'}(B)\, \Omega_{\pm} \tag{2.50}$$

for any $B \in \mathscr{B}^{n+1}$.

We recall that

$$\Psi(0) = \Omega_{\pm} \Psi^{\text{ex}}(0), \tag{2.51}$$

where ex = in for Ω_+ and ex = out for Ω_-. Hence, after noting that

$$[e^{-iHt}, E^{H,A_1',\ldots,A_n'}(B)] = [e^{-iH_0t}, E^{H_0,A_1',\ldots,A_n'}(B)] = 0,$$

by using (2.50) we easily obtain

$$\begin{aligned} P_{\Psi(t)}^{H,A_1',\ldots,A_n'}(B) &= \langle e^{-iHt}\Psi(0) \mid E^{H,A_1',\ldots,A_n'}(B)\, e^{-iHt}\Psi(0)\rangle \\ &= \langle \Omega_{\pm}\Psi^{\text{ex}}(0) \mid E^{H,A_1',\ldots,A_n'}(B)\, \Omega_{\pm}\Psi^{\text{ex}}(0)\rangle \\ &= \langle \Omega_{\pm}\Psi^{\text{ex}}(0) \mid \Omega_{\pm} E^{H_0,A_1',\ldots,A_n'}(B)\, \Psi^{\text{ex}}(0)\rangle \\ &= P_{\Psi^{\text{ex}}(t)}^{H_0,A_1',\ldots,A_n'}(B), \end{aligned} \tag{2.52}$$

where the last step follows from the fact that $\Omega_\pm{}^*\Omega_\pm\Psi^{\text{ex}}(0) = \Psi^{\text{ex}}(0)$, since $\Psi^{\text{ex}}(0) \in \mathsf{M}_\pm$. Hence, (2.47) is established. Q.E.D.

All nonrelativistic quantum mechanical theories satisfy the conditions of the above theorem. For example, in wave mechanics for a single spinless particle interacting with a potential $V(\mathbf{r})$, one can take

$$H \supseteq -(1/2m)\Delta + V(\mathbf{r})$$

and $\mathcal{O}_0$ consisting of $\mathbf{L}^2$, $L_\mathbf{n}(\mathbf{n} \in \mathbb{R}^3, |\mathbf{n}| = 1)$, Q_k and $P_k(k = 1, 2, 3)$; here, $\mathbf{L}^2$ denotes the total angular momentum operator (encountered in §7 of Chapter IV), $L_\mathbf{n} = \mathbf{n} \cdot \mathbf{L}$ is the angular momentum projection in the direction $\mathbf{n}$, and Q_k, P_k are the position and momentum components with respect to the kth axis of a Cartesian inertial reference frame.

From the physical point of view, it is sufficient if asymptotic states satisfy the physical asymptotic conditions (2.46) and (2.47), rather than (2.44). As a matter of fact, albeit that for long-range potentials (such as the Coulomb potential or $V(r) \sim r^{-\alpha}$, $0 < \alpha < 1$) there are no free states satisfying (2.44), there are free states satisfying (2.46) and (2.47). One could call such states *physical asymptotic states*, and develop* a scattering theory based on the existence of these states.

As a matter of fact, in this case one could *define* the wave operator Ω_+ to be the operator which maps an incoming physical asymptotic state into the corresponding interacting state, i.e., as the operator satisfying (2.51); Ω_- could be defined in a similar manner. As soon as the wave operators are defined, one can proceed in the usual manner and introduce the scattering operator by means of (2.2) and the transition operator by the old formula

$$T = (1 - S)/2\pi i. \tag{2.53}$$

For the rest of the quantities of importance in scattering theory (transition probabilities, differential scattering cross section, etc.), one can also retain the old definitions (see §4.9).

EXERCISES

2.1. Assume that C is a bounded operator in $\mathscr{H}$, and A_1, A_2 are self-adjoint operators in $\mathscr{H}$ with spectral measures $E^{A_1}(B)$ and $E^{A_2}(B)$, respectively. Prove that the relation $e^{iA_1t}C = Ce^{iA_2t}$ is satisfied for all $t \in \mathbb{R}^1$ if and only if $E^{A_1}(B)C = CE^{A_2}(B)$ for all $B \in \mathscr{B}^1$.

* For details see the articles by Prugovečki [1971a, 1978b], which also contain further references.

2.2. Show that if U is a partially isometric operator with initial domain M, then $\langle Uf \mid Ug\rangle = \langle f \mid g\rangle$ for all $f, g \in \mathsf{M}$.

2.3. Show that if E is a projector and $\| Ef \| = \| f \|$ for some $f \in \mathscr{H}$, then $f = Ef$.

2.4. Show that f_0 is a cyclic vector with respect to A if and only if $E^A(\mathsf{S}_c^A) f_0$ is a cyclic vector of A_c and $E^A(\mathsf{S}_p^A) f_0$ is a cyclic vector of A_p.

2.5. Prove that if a self-adjoint operator A leaves the closed linear subspace M invariant, then $[E^A(B), E_{\mathsf{M}}] = 0$ for all $B \in \mathscr{B}^1$.

2.6. Show that if A is self-adjoint and $f \in \mathscr{D}_A$, then $e^{iAt} f \in \mathscr{D}_A$ for all $t \in \mathbb{R}^1$.

2.7. Suppose that the potential $V(\mathbf{r})$ is Lebesgue square integrable on $\mathbb{R}^3$ so that $\| V \| < +\infty$. Apply the result of Exercise 1.9 to prove that if $\psi(\mathbf{r})$ is Lebesgue integrable as well as square integrable on $\mathbb{R}^3$, then

$$|(e^{-iH_0t}\psi)(\mathbf{r})| \leqslant \left|\frac{m}{2\pi t}\right|^{3/2} \|\psi\|_1, \qquad \|\psi\|_1 = \int_{\mathbb{R}^3} |\psi(\mathbf{r})|\, d\mathbf{r},$$

almost everywhere in $\mathbb{R}^3$, and consequently

$$\| Ve^{-iH_0t}\psi \| \leqslant \left|\frac{m}{2\pi t}\right|^{3/2} \| V \| \, \|\psi\|_1,$$

where $(Ve^{-iH_0t}\psi)(\mathbf{r}) = V(\mathbf{r})(e^{-iH_0t}\psi)(\mathbf{r})$.

2.8. If the potential $V(\mathbf{r})$ is Lebesgue square integrable on $\mathbb{R}^3$, then $\mathsf{M}_+ = \mathsf{M}_- = L^2(\mathbb{R}^3)$ in potential scattering for $V(\mathbf{r})$. Explain how this statement follows from Theorem 2.8 and Exercise 2.7.

3. General Time-Independent Two-Body Scattering Theory

3.1. The Relation of the Time-Independent to the Time-Dependent Approach

It is possible to construct the Møller wave operators $\Omega_\pm$ by taking limits with respect to parameters other than the time parameter t. Scattering theory based on such time-independent procedures for computing $\Omega_\pm$, S, and T is called *time independent* or *stationary*.

The relation of the time-dependent to the time-independent approach is best exhibited by showing that $\Omega_\pm$ can be obtained as strong limits of operator functions $\Omega_{\pm\epsilon}$ for $\epsilon \to +0$, where the parameter ϵ,

$0 < \epsilon < +\infty$, is not related to the time parameter t and has no physical meaning, and the operators $\Omega_{\pm\epsilon}$ satisfy the equations

$$(3.1)\qquad \begin{aligned} \langle g \mid \Omega_{-\epsilon} f\rangle &= \epsilon \int_0^\infty e^{-\epsilon t}\langle g \mid \Omega(t) f\rangle \, dt, \\ \langle g \mid \Omega_{+\epsilon} f\rangle &= \epsilon \int_{-\infty}^0 e^{\epsilon t}\langle g \mid \Omega(t) f\rangle \, dt \end{aligned}$$

for all $f, g \in \mathscr{H}$ and $0 < \epsilon < +\infty$.

Let us show that the relations (3.1) can play the role of definitions of linear operators $\Omega_{\pm\epsilon}$.

Since e^{iHt} and e^{-iH_0t} are strongly continuous functions of t (see §6 of Chapter IV), $\Omega(t) = e^{iHt}\, e^{-iH_0t}$ is also strongly continuous (see Chapter III, Exercise 5.6). Hence, for any $f, g \in \mathscr{H}$, the function

$$\langle g \mid \Omega(t) f\rangle = \langle g \mid e^{iHt}\, e^{-iH_0t} f\rangle, \qquad t \in \mathbb{R}^1,$$

is continuous, as well as bounded in t:

$$|\langle g \mid e^{iHt}\, e^{-iH_0t} f\rangle| \leqslant \| g \| \, \| e^{iHt}\, e^{-iH_0t} f \| = \| f \| \, \| g \|.$$

Consequently, the first improper Riemann integral in (3.1),

$$(3.2)\qquad \epsilon \int_0^{+\infty} e^{-\epsilon t}\langle g \mid e^{iHt}\, e^{-iH_0t} f\rangle \, dt = (g \mid f),$$

converges absolutely for any $f, g \in \mathscr{H}$:

$$(3.3)\qquad \epsilon \int_0^\infty e^{-\epsilon t}\, |\langle g \mid e^{iHt}\, e^{-iH_0t} f\rangle| \, dt \leqslant \| f \| \, \| g \| \, \epsilon \int_0^\infty e^{-\epsilon t} \, dt = \| f \| \, \| g \|.$$

Thus, the functional $(g \mid f)$ is defined on $\mathscr{H} \times \mathscr{H}$.

It is straightforward to verify that $(g \mid f)$, $f, g \in \mathscr{H}$, is a bilinear form. Furthermore, we see from (3.3) that $(\cdot \mid \cdot)$ is bounded. Hence, there is a unique bounded linear operator (see Chapter III, Exercise 2.5), which we denote by $\Omega_{-\epsilon}$, such that

$$(3.4)\qquad \langle g \mid \Omega_{-\epsilon} f\rangle = (g \mid f) = \epsilon \int_0^\infty e^{-\epsilon t}\langle g \mid e^{iHt}\, e^{-iH_0t} f\rangle \, dt$$

for all $f, g \in \mathscr{H}$. We write the above relation in the symbolic form

$$(3.5)\qquad \Omega_{-\epsilon} = \epsilon \int_0^\infty e^{-\epsilon t}\, e^{iHt}\, e^{-iH_0t} \, dt, \qquad \epsilon > 0,$$

under the agreement that its real meaning is given by (3.4). In fact, (3.5) is an example of a Bochner integral (see Theorem 3.4 in Appendix 3.7 to this section).

It can be shown in a completely analogous manner that for any $\epsilon > 0$ there is a unique bounded linear operator $\Omega_{+\epsilon}$ which satisfies

$$(3.6) \qquad \langle g \mid \Omega_\epsilon f\rangle = \epsilon \int_{-\infty}^{0} e^{\epsilon t}\langle g \mid e^{iHt}\, e^{-iH_0 t} f\rangle \, dt$$

for all $f, g \in \mathscr{H}$. The above relation is written in the symbolic form

$$\Omega_\epsilon = \epsilon \int_{-\infty}^{0} e^{\epsilon t}\, e^{iHt}\, e^{-iH_0 t}\, dt, \qquad \epsilon > 0.$$

We are interested in the existence of the strong limits of $\Omega_{\pm\epsilon} f$ as $\epsilon \to +0$.

From (3.3) we immediately infer that

$$(3.7) \qquad \| \Omega_{\pm\epsilon} \| \leqslant 1.$$

By applying Lemma 1.1 we deduce that the sets N_+ and N_- of all vectors $f \in \mathscr{H}$ for which the respective limits of $\Omega_{\pm\epsilon} f$ exist,

$$(3.8) \qquad \mathsf{N}_\pm = \{f\colon \operatorname*{s-lim}_{\epsilon\to+0} \Omega_{\pm\epsilon} f \in \mathscr{H}\}$$

are closed linear subspaces of $\mathscr{H}$. We shall prove now that $\mathsf{M}_+ \subset \mathsf{N}_+$ and $\mathsf{M}_- \subset \mathsf{N}_-$ (the converse not being generally true!).

Theorem 3.1. If for some $f \in \mathscr{H}$, $\operatorname{s-lim}_{t\to+\infty} \Omega(t) f$ exists, then $\operatorname{s-lim}_{\epsilon\to+0} \Omega_{-\epsilon} f$ also exists, and

$$(3.9) \qquad \operatorname*{s-lim}_{t\to+\infty} \Omega(t) f = \operatorname*{s-lim}_{\epsilon\to+0} \Omega_{-\epsilon} f.$$

Similarly, if for some $g \in \mathscr{H}$, $\operatorname{s-lim}_{t\to-\infty} \Omega(t)\, g$ exists, then $\operatorname{s-lim}_{\epsilon\to+0} \Omega_{+\epsilon} g$ also exists, and

$$(3.10) \qquad \operatorname*{s-lim}_{t\to-\infty} \Omega(t) g = \operatorname*{s-lim}_{\epsilon\to+0} \Omega_{+\epsilon} g.$$

Proof. Let us write

$$f_- = \operatorname*{s-lim}_{t\to+\infty} \Omega(t) f$$

when the above limit exists. Then, for any $g \in \mathscr{H}$

$$
\begin{aligned}
&|\langle g \mid \Omega_{-\epsilon} f - f_- \rangle| \\
&\quad = \left| \epsilon \int_0^\infty e^{-\epsilon t} \langle g \mid \Omega(t) f \rangle \, dt - \epsilon \int_0^\infty e^{-\epsilon t} \langle g \mid f_- \rangle \, dt \right| \\
&\quad \leqslant \epsilon \int_0^\infty e^{-\epsilon t} \, |\langle g \mid \Omega(t) f - f_- \rangle| \, dt \leqslant \epsilon \, \| g \| \int_0^\infty e^{-\epsilon t} \, \| \Omega(t) f - f_- \| \, dt \\
&\quad = \| g \| \int_0^\infty e^{-u} \, \| \Omega(u/\epsilon) f - f_- \| \, du.
\end{aligned}
$$

We set out to prove that the above integral is smaller than any given $\eta > 0$ for all sufficiently small $\epsilon > 0$.

Choose some $\alpha > 0$ and split the interval of integration:

$$\int_0^\infty = \int_0^\alpha + \int_\alpha^\infty .$$

For the first integral we obtain the following upper bound:

$$\int_0^\alpha e^{-u} \, \| \Omega(u/\epsilon) f - f_- \| \, du \leqslant \int_0^\alpha e^{-u} (\| \Omega(u/\epsilon) f \| + \| f_- \|) \, du \leqslant \alpha (\| f \| + \| f_- \|).$$

Now, take some α satisfying

$$0 < \alpha \leqslant \frac{1}{2} \frac{\eta}{\| f \| + \| f_- \|},$$

and then choose $\epsilon > 0$ so small that

$$\| \Omega(t) f - f_- \| < \eta/2 \tag{3.11}$$

for all $t \geqslant \alpha/\epsilon$. For any such ϵ, we compute that

$$
\begin{aligned}
&\int_0^\infty e^{-u} \, \| \Omega(u/\epsilon) f - f_- \| \, du \\
&\quad < \frac{1}{2} \frac{\eta}{\| f \| + \| f_- \|} (\| f \| + \| f_- \|) + \int_\alpha^\infty e^{-u} \tfrac{1}{2} \eta \, du \leqslant \tfrac{1}{2} \eta + \tfrac{1}{2} \eta \int_0^\infty e^{-u} \, du = \eta.
\end{aligned}
$$

Consequently, for any given $\eta > 0$ we have

$$|\langle g \mid \Omega_{-\epsilon} f - f_- \rangle| < \eta \, \| g \|$$

for all $\epsilon > 0$ which are such that (3.11) is true whenever $t \geqslant \alpha/\epsilon$. Hence, we have established that

$$f_- = \operatorname*{w-lim}_{\epsilon \to +0} \Omega_{-\epsilon} f. \tag{3.12}$$

Since $\| f_- \| = \lim_{t\to+\infty} \| \Omega(t) f \| = \| f \| \geqslant \| \Omega_{-\epsilon} f \|$, we have shown that the weak limit (3.12) is also a strong limit (see Chapter IV, Lemma 6.2),

$$f_- = \operatorname*{s-lim}_{\epsilon\to+0} \Omega_{-\epsilon} f,$$

i.e., (3.9) is true.

The proof of (3.10) proceeds along identical lines. Q.E.D.

As an immediate consequence of the above theorem we obtain the important result

$$\Omega_\pm = \operatorname*{s-lim}_{\epsilon\to+0} \Omega_{\pm\epsilon} E_{\mathsf{M}_\pm}, \tag{3.13}$$

which can be considered as a time-independent alternative definition of the wave operators. It should be noted, however, that Theorem 3.1 establishes only that $\mathsf{M}_\pm \subset \mathsf{N}_\pm$, and not that $\mathsf{M}_\pm \equiv \mathsf{N}_\pm$. Since the con- converse of Theorem 3.1 [which would state that the existence of $\operatorname{s-lim}_{\epsilon\to+0} \Omega_{\pm\epsilon} f$ implies the existence of $\operatorname{s-lim}_{t\to\pm\infty} \Omega(t) f$] is not generally true (see Howland [1967]), it has been suggested (see Jordan [1962a, Section 6]) that $\Omega_\pm$ could be defined in a time-independent manner, by taking limits of $\Omega_{\pm\epsilon} f$, even in cases when the time-dependent definition (2.1) is not adequate because $\mathsf{M}_\pm = \{0\}$.

In potential scattering, for classes of potentials which have been considered in the last section, the above problem does not arise, since in those cases $\mathsf{M}_\pm \equiv \mathscr{H}$, and consequently

$$\Omega_\pm = \operatorname*{s-lim}_{t\to\mp\infty} \Omega(t) = \operatorname*{s-lim}_{\epsilon\to+0} \Omega_{\pm\epsilon} .$$

From (3.13) we get for any $f, g \in \mathscr{H}$

$$\langle \Omega_\pm{}^* f \mid g \rangle = \lim_{\epsilon\to+0} \langle E_{\mathsf{M}_\pm} \Omega_{\pm\epsilon}^* f \mid g \rangle.$$

Since $\Omega_\pm{}^*$ are partially isometric with initial domains $\mathsf{R}_\pm$, we easily verify that

$$\langle \Omega_\pm{}^* E_{\mathsf{R}_\pm} f \mid g \rangle = \lim_{\epsilon\to+0} \langle E_{\mathsf{M}_\pm} \Omega_{\pm\epsilon}^* E_{\mathsf{R}_\pm} f \mid g \rangle$$

by noting that for $f \in \mathsf{R}_\pm$ the above relation is equivalent to the relation preceding it, while for $f \perp \mathsf{R}_\pm$, the expressions on both sides of the above relation vanish. In view of the fact that

$$\| E_{\mathsf{M}_\pm} \Omega_{\pm\epsilon}^* E_{\mathsf{R}_\pm} f \| \leqslant \| E_{\mathsf{R}_\pm} f \| = \| \Omega_\pm{}^* E_{\mathsf{R}_\pm} f \|,$$

we infer from this result (by using Lemma 6.2 of Chapter IV) that

$$\Omega_{\pm}{}^* = \operatorname*{s-lim}_{\epsilon \to +0} E_{M_\pm} \Omega^*_{\pm\epsilon} E_{R_\pm} . \tag{3.14}$$

3.2. Lippmann–Schwinger Equations in Hilbert Space

One of the advantages of the time-independent limits (3.13) for determining $\Omega_\pm$ is that they lead to integral equations which can be used to compute $\Omega_\pm$ without having to compute $\Omega(t)$.

Using the spectral decomposition $H = \int \lambda \, dE^H_\lambda$ and recalling that e^{iHt} is a function of H, by employing Theorem 2.1 of Chapter IV we obtain

$$\begin{aligned} \langle g \mid \Omega_{-\epsilon} f \rangle &= \epsilon \int_0^\infty e^{-\epsilon t} \langle g \mid e^{iHt} \, e^{-iH_0 t} f \rangle \, dt \\ &= \epsilon \int_0^\infty dt \, e^{-\epsilon t} \int_{\mathbb{R}^1} e^{i\lambda t} \, d_\lambda \langle g \mid E^H_\lambda \, e^{-iH_0 t} f \rangle. \end{aligned} \tag{3.15}$$

Now, we would like to interchange in the above relation the order of the integrations in λ and t, and write

$$\langle g \mid \Omega_{-\epsilon} f \rangle = \epsilon \int_{-\infty}^{+\infty} d_\lambda \int_0^\infty \exp[-(\epsilon - i\lambda)t] \langle g \mid E^H_\lambda \exp(-iH_0 t) f \rangle \, dt. \tag{3.16}$$

However, before we ask whether such an interchange of the order of integration is correct, we must make clear the meaning of the integration in λ occurring in (3.16). We will show that the interchange of the order of integration can be easily justified if the integral in λ is taken to be a Riemann–Stieltjes integral.*

Definition 3.1. The *Riemann–Stieltjes integral*

$$\int_a^b F(\lambda) \, d_\lambda \, \sigma(\lambda), \qquad a < b,$$

of the complex function $F(\lambda)$ with respect to the complex function $\sigma(\lambda)$, $a \leqslant \lambda \leqslant b$, is said to exist if for any sequence of subdivisions $a = \lambda^{(0)} < \lambda^{(1)} < \cdots < \lambda^{(n)} = b$ of $[a, b]$ which is such that the norm $\delta = \max_{k=1,\ldots,n}(\lambda^{(k)} - \lambda^{(k-1)})$ of the subdivisions in the sequence

* Measure theory (i.e., Fubini's theorem) cannot be applied since in the resulting formula (3.30) the Riemann–Stieltjes integral cannot be interpreted as an integral with respect to a measure on $(\mathbb{R}^2, \mathscr{B}^2)$. Indeed, if such a measure existed, then, by the Riemann–Lebesgue lemma we would obtain that $\langle g \mid \Omega_\pm f \rangle \equiv 0$ if S^H_c and $\mathsf{S}^{H_0}_c$ are absolutely continuous.

tends to zero, and for any choice of $\lambda'^{(k)} \in [\lambda^{(k-1)}, \lambda^{(k)}]$, $k = 1,\ldots, n$, the limit of *Riemann–Stieltjes sums*

$$\lim_{\delta\to 0} \sum_{k=1}^{n} F(\lambda'^{(k)})(\sigma(\lambda^{(k)}) - \sigma(\lambda^{(k-1)})) \tag{3.17}$$

exists, and is independent of the selected sequence of subdivisions; if the integral (3.17) exists, then its value is taken to be

$$\int_a^b F(\lambda)\, d_\lambda\, \sigma(\lambda) = \lim_{\delta\to 0} \sum_{k=1}^{n} F(\lambda'^{(k)})[\sigma(\lambda^{(k)}) - \sigma(\lambda^{(k-1)})]. \tag{3.18}$$

If $F(\lambda)$ and $\sigma(\lambda)$ are defined on $\mathbb{R}^1$, then the *improper Riemann–Stieltjes* integral on $\mathbb{R}^1$ is defined as

$$\int_{-\infty}^{+\infty} F(\lambda)\, d_\lambda \sigma(\lambda) = \lim_{\substack{b\to+\infty \\ a\to-\infty}} \int_a^b F(\lambda)\, d_\lambda \sigma(\lambda), \tag{3.19}$$

provided that the two limits in a and b exist. For functions $F(\lambda, \lambda_0)$ and $\sigma(\lambda, \lambda_0)$ of two variables $\lambda, \lambda_0 \in \mathbb{R}^1$, we define the *cross-iterated Riemann–Stieltjes integral* on $[a, b] \times I$, where I is a nondegenerate interval in $\mathbb{R}^1$, by requiring that for any of the above sequences of subdivisions of $[a, b]$,

$$\begin{aligned} &\int_a^b d_\lambda \int_I F(\lambda, \lambda_0)\, d_{\lambda_0}\sigma(\lambda, \lambda_0) \\ &\quad = \lim_{\delta\to 0} \sum_{k=1}^{n} \int_I F(\lambda'^{(k)}, \lambda_0)\, d_{\lambda_0}[\sigma(\lambda^{(k)}, \lambda_0) - \sigma(\lambda^{(k-1)}, \lambda_0)], \end{aligned} \tag{3.20}$$

where the above limit should be finite and of a value which is independent of the chosen sequence of subdivisions.

It is easy to relate Riemann–Stieltjes integrals to integrals with respect to measures when $F(\lambda)$ and $\sigma(\lambda)$ are Borel measurable functions (see Exercise 3.3). It is equally easy to prove in a direct manner that most of the usual properties of Riemann integrals (which are obviously special cases of Riemann–Stieltjes integrals in which $\sigma(\lambda) = \lambda$) are also properties of Riemann–Stieltjes integrals (see Exercises 3.4–3.8).

We find the concept of Riemann–Stieltjes integration of importance because it enables us to give a meaning to the integral in (3.16), namely, the integral in (3.16) can be written as a cross-iterated Riemann–Stieltjes integral with respect to the function (see also Definition 3.4)

$$\sigma(\lambda, \lambda_0) = \int_0^{\lambda_0} \langle g \mid E_\lambda^H\, e^{-iH_0 t} f\rangle\, dt. \tag{3.21}$$

Let us introduce the following convenient notation:

$$\sigma((\lambda, \lambda'] \times (\lambda_0, \lambda_0']) = \sigma(\lambda', \lambda_0') - \sigma(\lambda', \lambda_0) - \sigma(\lambda, \lambda_0') + \sigma(\lambda, \lambda_0). \tag{3.22}$$

We can establish that (3.16) is true by means of Lemma 3.1.

***Lemma 3.1.** Suppose the real function $\sigma(\lambda, \lambda_0)$ is of *bounded variation* on $[a, b] \times [c, d]$, i.e., that for any $\lambda \in [a, b]$ and $\lambda_0 \in [c, d]$ we have

$$v(\lambda, \lambda_0) = \sup \sum_{i=1}^{m} \sum_{k=1}^{n} \left| \sigma((\lambda^{(i-1)}, \lambda^{(i)}] \times (\lambda_0^{(k-1)}, \lambda_0^{(k)}]) \right| < +\infty, \tag{3.23}$$

where the supremum is taken over all subdivisions

$$a = \lambda^{(0)} < \lambda^{(1)} < \cdots < \lambda^{(m)} = \lambda, \qquad c = \lambda_0^{(0)} < \lambda_0^{(1)} < \cdots < \lambda_0^{(n)} = \lambda_0 \tag{3.24}$$

of $[a, \lambda] \times [c, \lambda_0]$. If the complex function $F(\lambda, \lambda_0)$ is continuous on $[a, b] \times [c, d]$, then both its cross-iterated Riemann–Stieltjes integrals with respect to $\sigma(\lambda, \lambda_0)$ exist and

$$\int_a^b d_\lambda \int_c^d F(\lambda, \lambda_0)\, d_{\lambda_0}\sigma(\lambda, \lambda_0) = \int_c^d d_{\lambda_0} \int_a^b F(\lambda, \lambda_0)\, d_\lambda \sigma(\lambda, \lambda_0). \tag{3.25}$$

Proof. It is easy to see that the functions

$$\sigma_\pm(\lambda, \lambda_0) = \tfrac{1}{2}[v(\lambda, \lambda_0) + \sigma(a, c) \pm \sigma(\lambda, \lambda_0)]$$

are nondecreasing in $[a, b] \times [c, d]$ in the sense that $\sigma_\pm((\lambda, \lambda'] \times (\lambda_0, \lambda_0']) \geqslant 0$ whenever $\lambda \leqslant \lambda'$ and $\lambda_0 \leqslant \lambda_0'$. Since

$$\sigma(\lambda, \lambda_0) = \sigma_+(\lambda, \lambda_0) - \sigma_-(\lambda, \lambda_0),$$

we can exploit the linearity properties (see Exercises 6.4 and 6.7) of cross-iterated integrals to reduce the integrals in (3.25) to a sum of integrals of real functions with respect to $\sigma_+(\lambda, \lambda_0)$ and $\sigma_-(\lambda, \lambda_0)$. Hence, it is sufficient to prove the lemma for the case of a nondecreasing $\sigma(\lambda, \lambda_0)$ of bounded variation and for real $F(\lambda, \lambda_0)$.

By using the mean-value theorem for Riemann–Stieltjes integrals (see Exercise 3.8), we obtain

$$\int_{\lambda_0^{(k-1)}}^{\lambda_0^{(k)}} F(\lambda'^{(i)}, \lambda_0)\, d_{\lambda_0}[\sigma(\lambda^{(i)}, \lambda_0) - \sigma(\lambda^{(i-1)}, \lambda_0)]$$
$$= F(\lambda'^{(i)}, \lambda_0'^{(k)})\, \sigma((\lambda^{(i-1)}, \lambda^{(i)}] \times (\lambda_0^{(k-1)}, \lambda_0^{(k)}])$$

for some $\lambda_0^{(k-1)} \leqslant \lambda_0'^{(k)} \leqslant \lambda_0^{(k)}$. Combining this result with the defining relation (3.20), we get

$$\int_a^b d_\lambda \int_c^d F(\lambda, \lambda_0)\, d_{\lambda_0}\sigma(\lambda, \lambda_0)$$

$$= \lim \sum_{i=1}^{m} \sum_{k=1}^{n} F(\lambda'^{(i)}, \lambda_0'^{(k)})\, \sigma((\lambda^{(i-1)}, \lambda^{(i)}] \times (\lambda_0^{(k-1)}, \lambda_0^{(k)}])$$

in the limit of finer and finer subdivisions (3.24) of $[a, b] \times [c, d]$. The existence of this limit and its independence of the chosen sequence of subdivisions $\{(\lambda^{(i)}, \lambda_0^{(k)})\}$ is obviously a necessary and sufficient condition for the existence of the integral. To establish the existence of this limit and its independence on the sequence, it is sufficient to note that if $\{(\lambda^{(i,j)}, \lambda_0^{(k,l)})\}$ is a subdivision of $[a, b] \times [c, d]$, which contains and is finer than the subdivision $\{(\lambda^{(i)}, \lambda_0^{(k)})\}$, and if $(\lambda'^{(i,j)}, \lambda_0'^{(k,l)})$ are points within the meshes determined by $\{(\lambda^{(i,j)}, \lambda_0^{(k,l)})\}$, then due to the uniform continuity of $F(\lambda, \lambda_0)$ we can achieve for any given $\epsilon > 0$

$$| F(\lambda'^{(i,j)}, \lambda_0'^{(k,l)}) - F(\lambda'^{(i)}; \lambda_0'^{(k)})| < \epsilon$$

for all sufficiently fine subdivisions $\{(\lambda^{(i)}, \lambda_0^{(k)})\}$. Consequently, if $\hat{\lambda}^{(i,j)}$ and $\hat{\lambda}_0^{(k,l)}$ denote the points preceding $\lambda^{(i,j)}$ and $\lambda_0^{(k,l)}$, respectively, in the subdivision, then

$$\Bigg| \sum_{i,j} \sum_{k,l} F(\lambda'^{(i,j)}, \lambda_0'^{(k,l)})\, \sigma((\hat{\lambda}^{(i,j)}, \lambda^{(i,j)}] \times (\hat{\lambda}_0^{(k,l)}, \lambda_0^{(k,l)}])$$

$$- \sum_{i} \sum_{k} F(\lambda'^{(i)}, \lambda_0'^{(k)})\, \sigma((\lambda^{(i-1)}\ \lambda^{(i)}] \times (\lambda_0^{(k-1)}, \lambda_0^{(k)}]) \Bigg|$$

$$< \epsilon \sum_{i,j} \sum_{k,l} \sigma((\hat{\lambda}^{(i,j)}, \lambda^{(i,j)}] \times (\hat{\lambda}_0^{(k,l)}, \lambda_0^{(k,l)}])$$

$$= \epsilon\sigma((a, b] \times (c, d]).$$

This establishes the existence of the considered limit, as well as its independence of the sequence of subdivisions.

By the same token

$$\int_c^d d_{\lambda_0} \int_a^b F(\lambda, \lambda_0)\, d_\lambda\sigma(\lambda, \lambda_0)$$

$$= \lim \sum_{i=1}^{m} \sum_{k=1}^{n} F(\lambda''^{(i)}, \lambda_0''^{(k)})\, \sigma((\lambda^{(i-1)}, \lambda^{(i)}] \times (\lambda_0^{(k-1)}, \lambda_0^{(k)}])$$

also exists. By exploiting again the uniform continuity of $F(\lambda, \lambda_0)$ in $[a, b] \times [c, d]$, we conclude that for any $\epsilon > 0$

$$|F(\lambda''^{(i)}, \lambda_0''^{(k)}) - F(\lambda'^{(i)}, \lambda_0'^{(k)})| < \epsilon$$

for all sufficiently fine subdivisions (3.24) of $[a, b] \times [c, d]$. Hence, the absolute value of the difference of the two integrals in (3.25) is smaller than

$$\sum_{i=1}^{m} \sum_{k=1}^{n} |F(\lambda''^{(i)}, \lambda_0''^{(k)} - F(\lambda'^{(i)}, \lambda_0'^{(k)})| \, |\sigma((\lambda^{(i-1)}, \lambda^{(i)}] \times (\lambda_0^{(k-1)}, \lambda_0^{(k)}])|$$

$$< \epsilon \sum_{i=1}^{m} \sum_{k=1}^{n} \sigma((\lambda^{(i-1)}, \lambda^{(i)}] \times (\lambda_0^{(k-1)}, \lambda_0^{(k)}])$$

$$= \epsilon\sigma((a, b] \times (c, d]).$$

Since this is true for arbitrarily small values of $\epsilon > 0$, we conclude that (3.25) holds. Q.E.D.

Let us compute the *total variation* function $v(\lambda, \lambda_0)$, defined by (3.23), for the function $\sigma(\lambda, \lambda_0)$ given in (3.21). By using in the process a decomposition of the type (5.13) in Chapter III, we easily derive

(3.26)

$$v(\lambda, \lambda_0) \leqslant \sup \sum_{i=1}^{m} \sum_{k=1}^{n} \int_{\lambda_0^{(k-1)}}^{\lambda_0^{(k)}} |\langle E^H((\lambda^{(i-1)}, \lambda^{(i)}])g \mid e^{-iH_0t}f\rangle| \, dt$$

$$= \sup \int_c^{\lambda_0} \sum_{i=1}^{m} |\langle e^{-iH_0t}f \mid E^H((\lambda^{(i-1)}, \lambda^{(i)}])g\rangle| \, dt$$

$$\leqslant \tfrac{1}{2} \int_c^{\lambda_0} \{\langle e^{-iH_0t}f + g \mid E^H((a, \lambda])(e^{-iH_0t}f + g)\rangle$$

$$+ \langle e^{-iH_0t} f + ig \mid E^H((a, \lambda])(e^{-iH_0t}f + ig)\rangle$$

$$+ \sqrt{2}\langle e^{-iH_0t}f \mid E^H((a, \lambda])\, e^{-iH_0t} f\rangle + \sqrt{2}\langle g \mid E^H((a, \lambda])g\rangle\} \, dt.$$

Since the integrand in the above integral is continuous and therefore integrable, it follows that $\sigma(\lambda, \lambda_0)$ is of bounded variation. Consequently, Lemma 3.1 can be applied to infer that

$$\int_0^T dt \int_a^b e^{-(\epsilon - i\lambda)t} \, d_\lambda\langle g \mid E_\lambda^H \, e^{-iH_0t}f\rangle = \int_a^b d_\lambda \int_0^T e^{-(\epsilon - i\lambda)t}\langle g \mid E_\lambda^H \, e^{-iH_0t}f\rangle \, dt.$$

To establish the validity of (3.16), we have to prove that the above relation remains true in the limit when $a \to -\infty$ and $b, \tau \to +\infty$.

In view of the relations (see Exercise 3.5)

$$\int_0^{+\infty}\int_{-\infty}^{+\infty} - \int_0^{\tau}\int_a^{b} = \int_{\tau}^{+\infty}\int_{-\infty}^{+\infty} + \int_0^{\tau}\int_{-\infty}^{a} + \int_0^{\tau}\int_b^{+\infty},$$
$$\int_{-\infty}^{+\infty}\int_0^{+\infty} - \int_a^{b}\int_0^{\tau} = \int_{-\infty}^{+\infty}\int_{\tau}^{+\infty} + \int_{-\infty}^{a}\int_0^{\tau} + \int_b^{+\infty}\int_0^{\tau}, \tag{3.27}$$

and (3.26), we see that this will be the case if and only if the integrals on the right-hand side of the equalities (3.27) exist and vanish when $a \to -\infty$ and $b, \tau \to +\infty$.

Employing the procedure used in deriving (3.26) we arrive at the following estimate:

$$\begin{aligned}
&\left|\int_{c_0}^{d_0} dt e^{-\epsilon t}\int_{a_0}^{b_0} e^{i\lambda t}\, d_\lambda\langle g \mid E_\lambda^H\, e^{-iH_0 t}f\rangle\right| \\
&\quad \leqslant \int_{c_0}^{d_0} e^{-\epsilon t}\left|\int_{a_0}^{b_0} e^{i\lambda t}\, d_\lambda\langle E_\lambda^H g \mid e^{-iH_0 t}f\rangle\right| dt \\
&\quad \leqslant \tfrac{1}{2}\int_{c_0}^{d_0} e^{-\epsilon t}\{\langle e^{-iH_0 t}f + g \mid E^H((a_0\,,b_0])(e^{-iH_0 t}f + g)\rangle \\
&\qquad + \langle e^{-iH_0 t} + ig \mid E^H((a_0\,,b_0])(e^{-iH_0 t}f + ig)\rangle \\
&\qquad + \sqrt{2}\langle e^{-iH_0 t}f \mid E^H((a_0\,,b_0])\, e^{-iH_0 t}f\rangle + \sqrt{2}\langle g \mid E^H((a_0\,,b_0])g\rangle\}\, dt.
\end{aligned}$$

By Lemma 3.1 in Chapter IV, the integral on the right-hand side of the above inequality can be made arbitrarily small, e.g., for sufficiently large values of either c_0 (for arbitrary values of $d_0 > c_0$), or of a_0 (for arbitrary values of $b_0 > a_0$). Since, according to Lemma 3.1, the order of integration in λ and t can be reversed without changing the value of this integral, we conclude that all integrals on the right-hand sides of the relations (3.27) vanish in the limit when $a \to -\infty$ or $b, \tau \to +\infty$; for example, for the representative case of the first of these integrals, we can choose for given $\eta > 0$ a $\tau(\eta)$ and a sequence $\tau(\eta) = \tau_0 < \tau_1 < \tau_2 < \cdots$ diverging to infinity which is such that

$$\left|\int_{\tau_{k-1}}^{\tau_k}\int_{-\infty}^{+\infty}\right| < \eta^k,$$

and consequently

$$\left|\int_{\tau(\eta)}^{+\infty}\int_{-\infty}^{+\infty}\right| \leqslant \sum_{k=1}^{\infty}\left|\int_{\tau_{k-1}}^{\tau_k}\int_{-\infty}^{+\infty}\right| \leqslant \sum_{k=1}^{\infty}\eta^k = \frac{\eta}{1-\eta}\xrightarrow[\eta\to 0]{} 0.$$

Thus, we conclude that (3.26) stays valid in the limit $a \to -\infty$ and $b, \tau \to +\infty$, and therefore (3.16) is true.

Using the fact that e^{-iH_0t} is a function of H_0, and applying Lemma 3.1 in the same manner as before to interchange orders of integration, we arrive at the result

$$\begin{aligned}\int_0^{\infty} &\exp[-(\epsilon - i\lambda')t]\langle g \mid E_\lambda^H \exp[-iH_0t]f\rangle\, dt \\ &= \int_0^{\infty} dt \exp[-(\epsilon - i\lambda')t] \int_{-\infty}^{+\infty} \exp[-i\lambda_0 t]\, d_{\lambda_0}\langle E_\lambda^H g \mid E_{\lambda_0}^{H_0} f\rangle \\ &= \int_{-\infty}^{+\infty} d_{\lambda_0} \int_0^{\infty} \exp[-(\epsilon - i\lambda' + i\lambda_0)t]\langle E_\lambda^H g \mid E_{\lambda_0}^{H_0} f\rangle\, dt \\ &= \int_{-\infty}^{+\infty} \frac{1}{\epsilon - i\lambda' + i\lambda_0}\, d_{\lambda_0}\langle g \mid E_\lambda^H E_{\lambda_0}^{H_0} f\rangle.\end{aligned}$$

Inserting this result in (3.16), we obtain

$$\text{(3.28)} \qquad \langle g \mid \Omega_{-\epsilon} f\rangle = \int_{-\infty}^{+\infty} d_\lambda \int_{-\infty}^{+\infty} \frac{i\epsilon}{\lambda - \lambda_0 + i\epsilon}\, d_{\lambda_0}\langle g \mid E_\lambda^H E_{\lambda_0}^{H_0} f\rangle.$$

In a completely analogous manner we can derive a corresponding relation for $\Omega_{+\epsilon}$:

$$\text{(3.29)} \qquad \langle g \mid \Omega_{+\epsilon} f\rangle = \int_{-\infty}^{+\infty} d_\lambda \int_{-\infty}^{+\infty} \frac{-i\epsilon}{\lambda - \lambda_0 - i\epsilon}\, d_{\lambda_0}\langle g \mid E_\lambda^H E_{\lambda_0}^{H_0} f\rangle.$$

The two relations (3.28) and (3.29) are equivalent to the single relation

$$\text{(3.30)} \quad \langle g \mid \Omega_\eta f\rangle = \int_{-\infty}^{+\infty} d_\lambda \int_{-\infty}^{+\infty} \frac{-i\eta}{\lambda - \lambda_0 - i\eta}\, d_{\lambda_0}\langle E_\lambda^H g \mid E_{\lambda_0}^{H_0} f\rangle, \qquad \eta \neq 0,$$

if the parameter $\epsilon > 0$ is replaced by the new parameter $\eta \neq 0$, which can assume also negative values.

Using the defining formula (3.20) for cross-iterated integrals, we obtain from (3.30)

$$\begin{aligned}\langle \Omega_\eta{}^* g \mid f\rangle &= \langle g \mid \Omega_\eta f\rangle \\ &= \lim_{\substack{b \to +\infty \\ a \to -\infty}} \lim_{\delta \to 0} \sum_{k=1}^{n} \int_{-\infty}^{+\infty} \frac{-i\eta}{\lambda'^{(k)} - \lambda_0 - i\eta}\, d_{\lambda_0}\langle (E_{\lambda^{(k)}}^H - E_{\lambda^{(k-1)}}^H)\, g \mid E_{\lambda_0}^{H_0} f\rangle \\ &= \lim_{\substack{b \to +\infty \\ a \to -\infty}} \lim_{\delta \to 0} \sum_{k=1}^{n} \left\langle g \,\middle|\, E^H((\lambda^{(k-1)}, \lambda^{(k)}])\, \frac{-i\eta}{\lambda'^{(k)} - H_0 - i\eta} f\right\rangle \\ &= \left\langle g \,\middle|\, \int_{-\infty}^{+\infty} d_\lambda E_\lambda^H \frac{-i\eta}{\lambda - H_0 - i\eta} f\right\rangle,\end{aligned}$$

where the last integral is *defined* by the limit preceding it.*

* See also Definition 3.4 for integrals of this type in Appendix 3.7 to this section.

According to Theorem 2.2 of Chapter IV,

$$\left(\frac{-i\eta}{\lambda - H_0 - i\eta}\right)^* = \frac{i\eta}{\lambda - H_0 + i\eta}, \qquad \eta = \eta^* \neq 0,$$

and consequently we have

$$\langle \Omega_\eta{}^* g \mid f \rangle = \left\langle \int_{-\infty}^{+\infty} \frac{i\eta}{\lambda - H_0 + i\eta} d_\lambda\, E_\lambda^H g \mid f \right\rangle. \tag{3.31}$$

Since $i\eta(\lambda - H_0 + i\eta)^{-1}$ and $(\lambda - H_0)(\lambda - H_0 + i\eta)^{-1}$ are bounded functions of H_0, they are defined everywhere in $\mathscr{H}$. According to Theorem 2.4 of Chapter IV, we can add and multiply these operator functions as we would ordinary functions, and therefore we have the identity

$$\frac{i\eta}{\lambda - H_0 + i\eta} = 1 - \frac{\lambda - H_0}{\lambda - H_0 + i\eta}. \tag{3.32}$$

Substituting the right-hand side of the above equation in (3.31), and using symbolic notation (see also Appendix 3.7 to this section), we arrive at the following expression for $\Omega_\eta{}^*$:

$$\Omega_\eta{}^* = 1 - \int_{-\infty}^{+\infty} \frac{\lambda - H_0}{\lambda - H_0 + i\eta} d_\lambda E_\lambda^H. \tag{3.33}$$

Consider a free state $e^{-iH_0 t}\Psi$, $\Psi \in \mathsf{M}_\pm$, which is the incoming asymptotic state of $e^{-iHt}\Psi^+$ or the outgoing asymptotic state of $e^{-iHt}\Psi^-$. Since obviously

$$\Psi^\pm = \operatorname*{s-lim}_{t\to\mp\infty} e^{iHt}\, e^{-iH_0 t}\Psi = \Omega_\pm \Psi,$$

we deduce from (3.14) and (2.10)

$$\operatorname*{s-lim}_{\eta\to\pm 0} E_{\mathsf{M}_\pm} \Omega_\eta{}^* \Psi^\pm = \Omega_\pm{}^* \Omega_\pm \Psi = \Psi.$$

By Theorem 3.5 in Appendix 3.7 this result, combined for $\mathsf{M}_\pm = \mathscr{H}$ with (3.33) leads, after a rearrangement of terms, to the equations

$$\Psi^\pm = \Psi + \operatorname*{s-lim}_{\eta\to\pm 0} \int_{-\infty}^{+\infty} \frac{1}{\lambda - H_0 + i\eta} H_1\, d_\lambda E_\lambda^H \Psi^\pm, \qquad H_1 = H - H_0. \tag{3.34}$$

The above two relations can be considered to be equations for determining Ψ^- and Ψ^+, respectively. We shall refer to them as *Lippmann–Schwinger equations in Hilbert space*. Indeed, in the next sections we shall see that in the special case of potential scattering these equations lead to a commonly encountered form of integral equations for distorted plane waves, usually called Lippmann–Schwinger equations.

3.3. Spectral Integral Representations of Wave Operators

The main drawback of (3.34), viewed as equations for finding $\Psi^{\pm}$ when Ψ is given, is that they require a knowledge of the spectral function E_λ^H of the total Hamiltonian H. Now, in practice, E_λ^H is an unknown quantity which is very difficult to compute. In fact, if E_λ^H were known, then one could easily compute the time evolution operator $U(t, t_0) = \exp[-iH(t - t_0)]$, and thus solve the dynamics of the problem generally, without even having to resort to scattering theory techniques.

A practically more suitable set of equations relating Ψ and $\Psi^{\pm}$, which yield themselves to perturbation techniques, can be derived from (3.1) by applying the spectral theorem first to H_0 and then to H, and otherwise proceeding in the same manner as in the derivation of (3.16):

$$(3.35)\quad \begin{aligned}\langle g \mid \Omega_{-\epsilon} f\rangle &= \epsilon \int_0^{+\infty} dt e^{-\epsilon t} \int_{-\infty}^{+\infty} e^{-i\lambda_0 t}\, d_{\lambda_0}\langle e^{-iHt} g \mid E_{\lambda_0}^{H_0} f\rangle \\ &= \epsilon \int_{-\infty}^{+\infty} d_{\lambda_0} \int_0^{+\infty} e^{-(\epsilon+i\lambda_0)t} \langle g \mid e^{iHt} E_{\lambda_0}^{H_0} f\rangle\, dt \\ &= \epsilon \int_{-\infty}^{+\infty} d_{\lambda_0} \int_0^{+\infty} dt e^{-(\epsilon+i\lambda_0)t} \int_{-\infty}^{+\infty} e^{i\lambda t}\, d_\lambda \langle g \mid E_\lambda^H E_{\lambda_0}^{H_0} f\rangle \\ &= \epsilon \int_{-\infty}^{+\infty} d_{\lambda_0} \int_{-\infty}^{+\infty} \frac{1}{\epsilon + i\lambda_0 - i\lambda}\, d_\lambda \langle g \mid E_\lambda^H E_{\lambda_0}^{H_0} f\rangle.\end{aligned}$$

Hence, we end up with the relation

$$(3.36)\quad \begin{aligned}\langle g \mid \Omega_{-\epsilon} f\rangle &= \int_{-\infty}^{+\infty} d_{\lambda_0} \int_{-\infty}^{+\infty} \frac{i\epsilon}{\lambda - \lambda_0 + i\epsilon}\, d_\lambda \langle g \mid E_\lambda^H E_{\lambda_0}^{H_0} f\rangle \\ &= \left\langle g \,\middle|\, \int_{-\infty}^{+\infty} \frac{i\epsilon}{H - \lambda_0 + i\epsilon}\, d_{\lambda_0} E_{\lambda_0}^{H_0} f \right\rangle.\end{aligned}$$

An analogous relation can be derived for $\Omega_{+\epsilon}$. After reintroducing the parameter $\eta \neq 0$, we can write (3.36) and the corresponding relation for $\Omega_{+\epsilon}$ jointly in the symbolic form

$$(3.37)\quad \Omega_\eta = \int_{-\infty}^{+\infty} \frac{-i\eta}{H - \lambda_0 - i\eta}\, d_{\lambda_0} E_{\lambda_0}^{H_0}.$$

Naturally, if we exploit the relation

$$(3.38)\quad \frac{-i\eta}{H - \lambda_0 - i\eta} = 1 - \frac{H - \lambda_0}{H - \lambda_0 - i\eta},$$

(3.37) can be recast as

$$\Omega_\eta = 1 + \int_{-\infty}^{+\infty} \frac{H - \lambda_0}{\lambda_0 - H + i\eta}\, d_{\lambda_0} E_{\lambda_0}^{H_0}. \tag{3.39}$$

Hence, if we again write

$$\Psi^{\pm} = \Omega_{\pm}\Psi = \operatorname*{s-lim}_{\eta\to\pm 0} \Omega_\eta \Psi, \qquad \Psi \in \mathsf{M}_{\pm}, \tag{3.40}$$

then in view of (3.39) and of Theorem 3.5,

$$\Psi^{\pm} = \Psi + \operatorname*{s-lim}_{\eta\to\pm 0} \int_{-\infty}^{+\infty} \frac{1}{\lambda_0 - H + i\eta} H_1\, d_{\lambda_0} E_{\lambda_0}^{H_0} \Psi, \qquad H_1 = H - H_0. \tag{3.41}$$

The above two equations are the *solution-type Lippmann–Schwinger equations in Hilbert space*, since they can be viewed as solutions of (3.34). However, they actually are not explicit solutions of the equations (3.34) since they contain the operator function $(\lambda_0 - H + i\eta)^{-1}(H - H_0)$, which is unknown in practice. On the other hand, we shall see later in this section [see (3.67)] that (3.41) can be solved by iterative procedures.

Let us write (3.39) in the form

$$\langle f \mid \Omega_\eta g\rangle = \langle f \mid g\rangle + \left\langle f \,\middle|\, \int_{-\infty}^{+\infty} \frac{H - \lambda_0}{\lambda_0 - H + i\eta}\, d_{\lambda_0} E_{\lambda_0}^{H_0} g \right\rangle. \tag{3.42}$$

Taking adjoints on both sides, we obtain

$$\langle g \mid \Omega_\eta^* f\rangle = \langle g \mid f\rangle + \left\langle g \,\middle|\, \int_{-\infty}^{+\infty} d_{\lambda_0} E_{\lambda_0}^{H_0} \frac{H - \lambda_0}{\lambda_0 - H - i\eta} f \right\rangle.$$

In symbolic notation the above relation assumes the form

$$\Omega_\eta^* = 1 + \int_{-\infty}^{+\infty} d_{\lambda_0} E_{\lambda_0}^{H_0} \frac{H - \lambda_0}{\lambda_0 - H - i\eta}, \tag{3.43}$$

which will be very useful further on.

3.4. The Transition Amplitude

It was mentioned in §1 of Chapter IV that if $\Psi_+(t)$ and $\Psi_-(t)$ are states in the Schroedinger picture, the $|\langle \Psi_+(t) \mid \Psi_-(t)\rangle|^2$ is called the *transition probability* from $\Psi_-(t)$ to $\Psi_+(t)$; furthermore, the complex number $\langle \Psi_+(t) \mid \Psi_-(t)\rangle$ is called the *transition* (*probability*) *amplitude* from $\Psi_-(t)$ to $\Psi_+(t)$.

Let us assume that $\Psi_+(t)$ and $\Psi_-(t)$ are scattering states, with asymptotic states $\Psi_+^{\text{ex}}(t)$ and $\Psi_-^{\text{ex}}(t)$, respectively. Since

$$\Psi_\pm(0) = \Omega_+ \Psi_\pm^{\text{in}}(0) = \Omega_- \Psi_\pm^{\text{out}}(0),$$

we easily obtain

$$\langle \Psi_+(t) \mid \Psi_-(t) \rangle = \langle \Psi_+(0) \mid \Psi_-(0) \rangle = \langle \Psi_+^{\text{out}}(0) \mid S\Psi_-^{\text{in}}(0) \rangle.$$

Let us simplify the notation by writing

$$\Psi_-^{\text{in}}(0) = \Psi_{\text{i}}, \qquad \Psi_+^{\text{out}}(0) = \Psi_{\text{f}},$$

where the indices i and f stand for "initial" and "final," respectively. Using the properties of strong limits, we obtain

$$\begin{aligned} \text{(3.44)} \qquad \langle \Psi_{\text{f}} \mid S\Psi_{\text{i}} \rangle &= \langle \Psi_{\text{f}} \mid \Omega_-{}^* \Omega_+ \Psi_{\text{i}} \rangle = \langle \Omega_- \Psi_{\text{f}} \mid \Omega_+ \Psi_{\text{i}} \rangle \\ &= \langle \operatorname*{s-lim}_{\epsilon \to +0} \Omega_{-\epsilon} \Psi_{\text{f}} \mid \Omega_+ \Psi_{\text{i}} \rangle \\ &= \lim_{\epsilon \to +0} \langle \Omega_{-\epsilon} \Psi_{\text{f}} \mid \Omega_+ \Psi_{\text{i}} \rangle. \end{aligned}$$

Substituting the expression (3.39) for $\Omega_{-\epsilon}$ in $\langle \Omega_{-\epsilon} \Psi_{\text{f}} \mid \Omega_+ \Psi_{\text{i}} \rangle$, we get

(3.45)
$$\langle \Omega_{-\epsilon} \Psi_{\text{f}} \mid \Omega_+ \Psi_{\text{i}} \rangle = \langle \Psi_{\text{f}} \mid \Omega_+ \Psi_{\text{i}} \rangle + \left\langle \int_{-\infty}^{+\infty} \frac{H - \lambda_0}{\lambda_0 - H - i\epsilon} \, d_{\lambda_0} \, E_{\lambda_0}^{H_0} \Psi_{\text{f}} \mid \Omega_+ \Psi_{\text{i}} \right\rangle.$$

On the other hand, after using (3.43) to express the identity operator in terms of the other two operators occuring in that equation and then applying it to the vector $\Omega_+ \Psi_{\text{i}}$, we arrive at the relation

$$\text{(3.46)} \qquad \Omega_+ \Psi_{\text{i}} = \Omega_{+\epsilon}^* \Omega_+ \Psi_{\text{i}} - \int_{-\infty}^{+\infty} d_{\lambda_0} E_{\lambda_0}^{H_0} \frac{H - \lambda_0}{\lambda_0 - H - i\epsilon} \, \Omega_+ \Psi_{\text{i}} \, .$$

After substituting the above expression for $\Omega_+ \Psi_{\text{i}}$ in the first term on the right-hand side of (3.45) and carrying out an obvious summation, we obtain

$$\begin{aligned} \text{(3.47)} \quad & \langle \Omega_{-\epsilon} \Psi_{\text{f}} \mid \Omega_+ \Psi_{\text{i}} \rangle \\ &= \langle \Psi_{\text{f}} \mid \Omega_{+\epsilon}^* \Omega_+ \Psi_{\text{i}} \rangle - \left\langle \Psi_{\text{f}} \,\middle|\, \int_{-\infty}^{+\infty} d_{\lambda_0} \, E_{\lambda_0}^{H_0} \frac{2i\epsilon(H - \lambda_0)}{(\lambda_0 - H)^2 + \epsilon^2} \, \Omega_+ \Psi_{\text{i}} \right\rangle. \end{aligned}$$

In view of (3.44) and the fact that if $\mathsf{M}_+ = \mathsf{M}_-$,

$$\text{(3.48)} \qquad \lim_{\epsilon \to +0} \langle \Psi_{\text{f}} \mid \Omega_{+\epsilon}^* \Omega_+ \Psi_{\text{i}} \rangle = \langle \Psi_{\text{f}} \mid E_{\mathsf{M}_+} \Psi_{\text{i}} \rangle = \langle \Psi_{\text{f}} \mid \Psi_{\text{i}} \rangle,$$

we deduce from (3.47) that

(3.49)
$$\langle \Psi_f \mid S\Psi_i \rangle = \langle \Psi_f \mid \Psi_i \rangle - \lim_{\epsilon \to +0} \left\langle \Psi_f \,\middle|\, \int_{-\infty}^{+\infty} d_{\lambda_0} E_{\lambda_0}^{H_0} \frac{2i\epsilon(H - \lambda_0)}{(H - \lambda_0)^2 + \epsilon^2} \Omega_+ \Psi_i \right\rangle .$$

In terms of the transition operator

(3.50)
$$T = \frac{1 - S}{2\pi i},$$

and provided that $\mathsf{M}_+ = \mathsf{M}_-$, we can write (3.49) in the form

(3.51)
$$\langle \Psi_f \mid T\Psi_i \rangle$$
$$= \frac{1}{\pi} \lim_{\epsilon \to +0} \left\langle \Psi_f \,\middle|\, \int_{-\infty}^{+\infty} d_{\lambda_0} E_{\lambda_0}^{H_0} \frac{\epsilon(H - \lambda_0)}{(H - \lambda_0)^2 + \epsilon^2} \Omega_+ \Psi_i \right\rangle$$
$$= \frac{1}{\pi} \lim_{\epsilon \to +0} \left\langle \Psi_f \,\middle|\, \int_{-\infty}^{+\infty} d_{\lambda_0} E_{\lambda_0}^{H_0} (H - \lambda_0) \Omega_+ \frac{\epsilon}{(H_0 - \lambda_0)^2 + \epsilon^2} \Psi_i \right\rangle$$

by taking advantage of the intertwining property (2.4). Using Lemma 3.3, we immediately obtain from (3.51)

(3.52)
$$\langle \Psi_f \mid T\Psi_i \rangle = \frac{1}{\pi} \lim_{\epsilon \to +0} \left\langle \Psi_f \,\middle|\, \int_{-\infty}^{+\infty} d_{\lambda_0} E_{\lambda_0}^{H_0} H_1 \Omega_+ \frac{\epsilon}{(H_0 - \lambda_0)^2 + \epsilon^2} \Psi_i \right\rangle .$$

It is interesting to note that on account of the relations

(3.53)
$$\Psi_i = \Psi_-^{\text{in}}(0) = \operatorname*{s-lim}_{t \to -\infty} e^{iH_0 t} e^{-iHt} \Psi_-(0),$$

(3.54)
$$\Psi_f = \Psi_+^{\text{out}}(0) = \operatorname*{s-lim}_{t \to +\infty} e^{iH_0 t} e^{-iHt} \Psi_+(0),$$

the vectors Ψ_f and Ψ_i can be considered to be the asymptotic states of the interaction-picture states [see (3.43) of Chapter IV]

(3.55)
$$\tilde{\Psi}_\pm(t) = e^{iH_0 t} e^{-iHt} \Psi_\pm(0),$$

respectively. Thus, it can be said that (3.52) provides an expression for the transition amplitude $\langle \Psi_f \mid T\Psi_i \rangle$ from the incoming interaction-picture asymptotic state Ψ_i to the outgoing interaction-picture asymptotic state Ψ_f .

3.5. The Resolvent of an Operator

In the Lippmann–Schwinger equations (3.34) and (3.41) we encountered operators of the form $(A - \lambda - i\eta)^{-1}$, where in the first case A stands for H_0 and in the second case A stands for H . Operators of the form $(A - \zeta)^{-1}$, where ζ is in general a complex number and A is a linear operator in $\mathscr{H}$, play a very important role in functional analysis. Since they have been systematically studied, it is advisable to acquaint oneself with the results of such studies, and then apply these results to the special problems at hand.

Definition 3.2. If A is any linear operator in $\mathscr{H}$, the *resolvent* $R_A(\zeta)$ of A is the operator-valued function

$$R_A(\zeta) = (A - \zeta)^{-1}, \qquad \zeta \in \mathbb{C}^1, \tag{3.56}$$

defined at all complex values of ζ at which $(A - \zeta)^{-1}$ exists.

We note that, according to Theorem 4.2 of Chapter III, $(A - \zeta)^{-1}$ exists if and only if the equation $(A - \zeta)f = 0$ has the unique solution $f = \mathbf{0}$, i.e., if and only if ζ is not an eigenvalue of A.

We can classify the points ζ in the complex plane $\mathbb{C}^1$ in relation to a densely defined linear operator A as follows.

Definition 3.3. The set of complex numbers ζ consisting of all points $\zeta \in \mathbb{C}^1$ at which the resolvent $R_A(\zeta)$ is a bounded operator defined densely in $\mathscr{H}$ is called the *resolvent set of* A. The complement of the resolvent set is called the *spectrum* S^A of A. The spectrum S^A is the disjoint union of the *point spectrum* S_p^A, *continuous spectrum* S_c^A, and *residual spectrum* S_r^A, which consist of all points which have the following respective properties: S_p^A of the points $\zeta \in \mathrm{S}^A$ at which $R_A(\zeta)$ does not exist; S_c^A of all $\zeta \in \mathrm{S}^A$ for which $R_A(\zeta)$ exists, is unbounded and defined on a dense subset of $\mathscr{H}$; S_r^A of all $\zeta \in \mathrm{S}^A$ at which $R_A(\zeta)$ exists but it is not defined in a dense subset of $\mathscr{H}$.

In the case where A is a self-adjoint operator, the above definition of the spectrum coincides with Definition 6.2 of Chapter III. To establish this equivalence, we need to show that

$$\mathscr{D}_{R_A(\zeta)} = \left\{ f: \int_{\mathbb{R}^1} \frac{1}{|\lambda - \zeta|^2} \, d\, \| E_\lambda^A f \|^2 < +\infty \right\}. \tag{3.57}$$

If f belongs to the domain of $(A - \zeta)^{-1}$, there is a vector $g \in \mathscr{D}_A$ such that $f = (A - \zeta)g$. Consequently, we have

$$\begin{aligned} \int_{\mathbb{R}^1} \frac{1}{|\lambda - \zeta|^2} \, d\, \| E_\lambda^A f \|^2 &= \int_{\mathbb{R}^1} \frac{1}{|\lambda - \zeta|^2} \, d_\lambda \left(\int_{-\infty}^{\lambda} |\lambda' - \zeta|^2 \, d_{\lambda'} \| E_{\lambda'}^A g \|^2 \right) \\ &= \int \frac{1}{|\lambda - \zeta|^2} \, |\lambda - \zeta|^2 \, d\, \| E_\lambda^A g \|^2 = \| g \|^2 < +\infty \end{aligned}$$

Conversely, if

$$\int_{\mathbb{R}^1} \frac{1}{|\lambda - \zeta|^2} \, d\, \| E_\lambda^A f \|^2 < +\infty,$$

then, according to Theorem 2.5 of Chapter IV

$$\frac{1}{A-\zeta}f = \int_{\mathbb{R}^1} \frac{1}{\lambda-\zeta}\, dE_\lambda^A f \tag{3.58}$$

is defined. Moreover, $g = [1/(A-\zeta)]f$ belongs to $\mathscr{D}_A$ because

$$\int_{\mathbb{R}^1} \lambda^2\, d\left\| E_\lambda^A \frac{1}{A-\zeta} f\right\|^2 = \int_{\mathbb{R}^1} \frac{\lambda^2}{|\lambda-\zeta|^2}\, d\| E_\lambda^A f\|^2 < +\infty.$$

Hence, by (2.25) of Chapter IV,

$$(A-\zeta)g = (A-\zeta)\frac{1}{A-\zeta}f = \int_{\mathbb{R}^1} (\lambda-\zeta)\frac{1}{\lambda-\zeta}\, dE_\lambda^A f = f,$$

i.e., f is in the domain of $(A-\zeta)^{-1}$. Thus, (3.57) is established.

By comparing (3.58) here and (2.19) of Chapter IV, we see that the above argument also establishes that when A is self-adjoint,

$$R_A(\zeta) = \int_{\mathbb{R}^1} \frac{1}{\lambda-\zeta}\, dE_\lambda^A = \frac{1}{A-\zeta}\cdot \tag{3.59}$$

Hence, for self-adjoint operators, Definition 6.3 of Chapter III and Definition 3.3 of the point spectrum S_p^A of A coincide, since $E^A(\{\zeta\}) \neq 0$, $\zeta \in \mathbb{R}^1$, if and only if ζ is an eigenvalue of A, in which case $(A-\zeta)^{-1}$ does not exist.

We shall prove now the equivalence of the two definitions (6.3 of Chapter III and 3.3) for the continuous spectrum S_c^A of a self-adjoint operator.

When, for a fixed real ζ, $E^A([\zeta-\epsilon, \zeta+\epsilon]) \neq 0$ for all $\epsilon > 0$ and $E^A(\{\zeta\}) = 0$, we must have either $E^A([\zeta-\epsilon, \zeta)) \neq 0$ or $E^A((\zeta, \zeta+\epsilon]) \neq 0$ for all $\epsilon > 0$. Let us say that the first alternative is the case. Then the equations

$$E^A([\zeta-\epsilon, \zeta))f_\epsilon = f_\epsilon, \qquad \|f_\epsilon\| = 1,$$

can be satisfied for all $\epsilon > 0$. Inasmuch as

$$\| R_A(\zeta) f_\epsilon\|^2 = \int_{[\zeta-\epsilon,\zeta)} \frac{1}{(\lambda-\zeta)^2}\, d\| E_\lambda^A f_\epsilon\| \geqslant \frac{1}{\epsilon^2}\int_{[\zeta-\epsilon,\zeta)} d\| E_\lambda^A f_\epsilon\|^2 = \frac{\|f_\epsilon\|^2}{\epsilon^2},$$

we conclude that $R_A(\zeta)$ is an unbounded operator. On the other hand, $R_A(\zeta)$ is defined on the dense set of all vectors g satisfying $E^A([\zeta-\epsilon, \zeta+\epsilon])g = \mathbf{0}$ for some $\epsilon > 0$. Thus, if ζ is an element of the continuous spectrum in the sense of Definition 6.3 in Chapter III, then it belongs to the continuous spectrum in the sense of Definition 3.3. The complete equivalence of these two definitions of S_c^A for a self-adjoint operator A follows from Lemma 3.2.

Lemma 3.2. The resolvent $R_A(\zeta)$ of a self-adjoint operator A is a bounded linear operator if and only if either Im $\zeta \neq 0$ or ζ is real but $E^A([\zeta - \epsilon, \zeta + \epsilon]) = 0$ for some $\epsilon > 0$; in either case $\mathscr{D}_{R_A(\zeta)} \equiv \mathscr{H}$.

Proof. We have seen earlier that if ζ is real and $E^A([\zeta - \epsilon, \zeta + \epsilon)] \neq 0$ for all $\epsilon > 0$, then $R_A(\zeta)$ either does not exist or, if it exists, it is an unbounded operator.

Assume now that ζ is real and $E^A((\zeta - \epsilon_0, \zeta + \epsilon_0)) = 0$ for some $\epsilon_0 > 0$. Since the function $1/(\zeta - \lambda)$ is bounded on

$$(-\infty, \zeta - \epsilon_0] \cup [\zeta + \epsilon_0, +\infty),$$

we conclude that

$$\int_{\mathbb{R}^1} \frac{1}{(\lambda - \zeta)^2} d\| E_\lambda^A f \|^2 = \int_{-\infty}^{\xi-\epsilon_0} \frac{1}{(\lambda - \zeta)^2} d\| E_\lambda^A f \|^2 + \int_{\zeta+\epsilon_0}^{\infty} \frac{1}{(\lambda - \zeta)^2} d\| E_\lambda^A f \|^2$$
$$\leqslant \frac{1}{\epsilon_0^2} \int_{-\infty}^{\zeta-\epsilon_0} d\| E_\lambda^A f \|^2 + \frac{1}{\epsilon_0^2} \int_{\zeta+\epsilon_0}^{\infty} d\| E_\lambda^A f \|^2 = \frac{2}{\epsilon_0^2} \| f \|^2.$$

Hence, it follows from (3.57) that $R_A(\zeta)$ is bounded and defined for all $f \in \mathscr{H}$.

In the case where Im $\zeta \neq 0$, we have

$$\int_{\mathbb{R}^1} \frac{1}{|\lambda - \zeta|^2} d\| E_\lambda^A f \|^2 \leqslant \frac{1}{|\operatorname{Im} \zeta|^2} \int_{\mathbb{R}^1} d\| E_\lambda^A f \|^2 = \frac{\| f \|^2}{|\operatorname{Im} \zeta|^2}.$$

The above inequality establishes that when Im $\zeta \neq 0$,

$$\| R_A(\zeta) \| \leqslant \frac{1}{|\operatorname{Im} \zeta|}, \tag{3.60}$$

and $R_A(\zeta)$ is defined for all $f \in \mathscr{H}$. Q.E.D.

It follows from the above lemma that a self-adjoint operator has no residual spectrum.

3.6. The Resolvent Method in Scattering Theory

The two resolvents we encounter in time-independent scattering theory are the resolvent $R_0(\zeta)$ of the free Hamiltonian H_0 and the resolvent $R(\zeta)$ of the total Hamiltonian H. However, while $R_0(\zeta)$ is easy to compute in practice, $R(\zeta)$ is an unknown quantity. Thus, Theorem 3.2, which relates $R(\zeta)$ and $R_0(\zeta)$, proves to be very useful.

Theorem 3.2. Let H and H_0 be two closed (not necessarily self-adjoint) operators in the Hilbert space $\mathscr{H}$, having dense domains of definition $\mathscr{D}_H \equiv \mathscr{D}_{H_0}$. The following *second resolvent equations* are satisfied by the resolvents $R(\zeta)$ and $R_0(\zeta)$ of H and H_0, respectively,

$$(3.61)\quad R(\zeta) - R_0(\zeta) = -R_0(\zeta)(H - H_0)R(\zeta) = -R(\zeta)(H - H_0)\, R_0(\zeta)$$

for any ζ which belongs to the resolvent sets of both H and H_0, i.e., for $\zeta \notin \mathsf{S}^H \cup \mathsf{S}^{H_0}$.

Proof. Since H and H_0 are closed, $R(\zeta)$ and $R_0(\zeta)$ are defined on the entire Hilbert space $\mathscr{H}$ whenever ζ does not belong to the spectra of H and H_0. Hence, for every $f \in \mathscr{H}$ there are vectors $g \in \mathscr{D}_H$ and $h \in \mathscr{D}_{H_0}$ such that $f = (H - \zeta)\, g = (H_0 - \zeta)h$. Since $\mathscr{D}_H = \mathscr{D}_{H_0}$, g also belongs to $\mathscr{D}_{H_0}$, and therefore

$$R_0(\zeta)(H - H_0)g = R_0(\zeta)(H - H_0)\, R(\zeta)f$$

is defined. Setting above $(H - H_0)g = [H - \zeta - (H_0 - \zeta)]g$, we get the relation

$$R_0(\zeta)(H - H_0)g = R_0(\zeta)(H - \zeta)g - g = R_0(\zeta)f - R(\zeta)f,$$

which is equivalent to the first of the identities (3.61).

The second identity in (3.61) can be established by a similar procedure. Q.E.D.

Before proceeding with the applications of the second resolvent equations, we shall derive, for the sake of completeness, the *first* resolvent equations.

Theorem 3.3. If ζ_1 and ζ_2 belong to the resolvent set of the closed and densely defined linear operator A, then the *first resolvent equations*

$$(3.62)\qquad R_A(\zeta_1) - R_A(\zeta_2) = (\zeta_1 - \zeta_2)\, R_A(\zeta_1)\, R_A(\zeta_2)$$

are satisfied.

Proof. The range of $A - \zeta$ is identical to the domain of definition of $R_A(\zeta) = (A - \zeta)^{-1}$, which coincides with $\mathscr{H}$ when ζ is in the resolvent set of A. Hence, for any $f \in \mathscr{H}$, a vector g can be found so that $f = (A - \zeta_1)(A - \zeta_2)g$, and therefore

$$(3.63)\qquad (\zeta_1 - \zeta_2)\, R_A(\zeta_1)\, R_A(\zeta_2)f = (\zeta_1 - \zeta_2)g.$$

On the other hand, we have

$$(3.64)\quad [R_A(\zeta_1) - R_A(\zeta_2)]f = (A - \zeta_2)g - (A - \zeta_1)g = (\zeta_1 - \zeta_2)g.$$

Since (3.63) and (3.64) hold for any $f \in \mathscr{H}$, (3.62) follows. Q.E.D.

Let us return now to our main task of giving a perturbational method for computing transition probabilities.

Using Theorem 3.2 we infer that, if $H_0 = H_0{}^*$, $H = H^*$, and $\mathscr{D}_H = \mathscr{D}_{H_0}$,

$$R(\zeta) = R_0(\zeta) - R_0(\zeta)\, H_1 R(\zeta), \qquad H_1 = H - H_0, \tag{3.65}$$

whenever Im $\zeta \neq 0$. After n successive iterations of the above relation, we obtain

$$R(\zeta) = R_0(\zeta) \sum_{k=0}^{n} (-1)^k \,(H_1 R_0(\zeta))^k + R_n, \tag{3.66}$$

where the remainder R_n is

$$R_n = (-1)^{n+1} (R_0(\zeta)\, H_1)^{n+1}\, R(\zeta). \tag{3.67}$$

If the above remainder converges to zero, in some sense,* when $n \to +\infty$, then (3.66) provides us with a means of computing $R(\zeta)$ when $R_0(\zeta)$ is known:

$$R(\zeta) = R_0(\zeta) - R_0(\zeta)\, H_1 R_0(\zeta) + R_0(\zeta)\, H_1 R_0(\zeta)\, H_1 R_0(\zeta) \pm \cdots. \tag{3.68}$$

This result combined with (3.67) leads eventually to a "perturbation" solution of the Lippmann–Schwinger equations. This solution is obtained by substituting for $R(\lambda_0 + i\eta) = (H - \lambda_0 - i\eta)^{-1}$ in (3.41) the series on the right-hand side of (3.68). If the series so obtained for the integral in (3.41) converges, and the order of summation and integration can be interchanged, then $\Psi^{\pm}$ can be computed to any order.

3.7. Appendix: Integration of Vector- and Operator-Valued Functions

In the last two sections we encountered integrals, such as (3.5), which are special cases of Bochner integrals. The theory of Bochner integration is based on the following theorem.

Theorem 3.4. Suppose μ is a measure in the measurable space $\mathscr{X}$ and that $f(\xi)$, $\xi \in R$, is a vector-valued function on the measurable set R, assuming values in a Hilbert space $\mathscr{H}$. If $\langle g \mid f(\xi)\rangle$ is measurable on R for any $g \in \mathscr{H}$ and if

$$\int_R \|f(\xi)\|\, d\mu(\xi) < +\infty, \tag{3.69}$$

* See also Lemma 5.2 and Theorem 5.9 of this chapter.

then:

(a) there is a unique vector $h \in \mathscr{H}$, called the *Bochner integral* of $f(\xi)$ on R and denoted by

$$\int_R f(\xi)\, d\mu(\xi),$$

which is such that for all $g \in \mathscr{H}$

$$\langle g \mid h\rangle = \int_R \langle g \mid f(\xi)\rangle\, d\mu(\xi); \tag{3.70}$$

(b) if A is a bounded operator on $\mathscr{H}$ and $\| Af(\xi)\|$ is measurable, then

$$A \int_R f(\xi)\, d\mu(\xi) = \int_R Af(\xi)\, d\mu(\xi). \tag{3.71}$$

Proof. (a) The linear functional

$$\phi(g) = \int_R \langle f(\xi) \mid g\rangle\, d\mu(\xi)$$

is bounded on account of (3.69):

$$|\phi(g)| \leqslant \int_R |\langle f(\xi) \mid g\rangle|\, d\mu(\xi) \leqslant \| g \| \int_R \| f(\xi)\|\, d\mu(\xi).$$

Hence, by Riesz' theorem there is a unique vector h which is such that $\phi(g) = \langle h \mid g\rangle$ for all $g \in \mathscr{H}$.

(b) The Bochner integral of $Af(\xi)$ exists since $\langle g \mid Af(\xi)\rangle = \langle A^*g \mid f(\xi)\rangle$ is measurable for any $g \in \mathscr{H}$, and

$$\int_R \| Af(\xi)\|\, d\mu(\xi) \leqslant \| A \| \int_R \| f(\xi)\|\, d\mu(\xi) < +\infty.$$

In order to prove (3.71) it is necessary and sufficient to show that

$$\langle g \mid Ah\rangle = \int_R \langle g \mid Af(\xi)\rangle\, d\mu(\xi) \tag{3.72}$$

for all $g \in \mathscr{H}$. Using the definition of the Bochner integral, we get

$$\langle g \mid Ah\rangle = \langle A^*g \mid h\rangle = \int_R \langle A^*g \mid f(\xi)\rangle\, d\mu(\xi),$$

thus establishing that (3.72) is true. Q.E.D.

The Bochner integral for vector-valued functions can be used to define Bochner integrals for operator-valued functions $A(\xi)$. As a matter of fact, if the Bochner integral

$$A_R[f] = \int_R [A(\xi)f] \, d\mu(\xi)$$

exists for all $f \in \mathscr{H}$, then the mapping $f \to A_R[f]$ is linear since

$$\begin{aligned}\langle g \mid A_R(a_1 f_1 + a_2 f_2)\rangle &= \int_R \langle g \mid A(\xi)(a_1 f_1 + a_2 f_2)\rangle \, d\mu(\xi) \\ &= a_1 \int_R \langle g \mid A(\xi) f_1\rangle \, d\mu(\xi) + a_2 \int_R \langle g \mid A(\xi) f_2\rangle \, d\mu(\xi) \\ &= a_1\langle g \mid A_R f_1\rangle + a_2\langle g \mid A_R f_2\rangle = \langle g \mid a_1 A_R f_1 + a_2 A_R f_2\rangle\end{aligned}$$

for all $g \in \mathscr{H}$. Hence,

$$A_R = \int_R A(\xi) \, d\mu(\xi) \tag{3.73}$$

can be defined to be the Bochner integral of $A(\xi)$ on R.

It is useful to note that the integral in (3.73) has the property (see Exercise 3.9)

$$A_R{}^* = \int_R A^*(\xi) \, d\mu(\xi). \tag{3.74}$$

Naturally, all of the above results can be immediately generalized to complex measures μ by decomposing such measures into a sum of real measures.

A certain kind of generalization of Bochner integrals has been encountered in (3.33), (3.34), (3.41), and (3.43). We define such integrals in a more general context by means of the following concept of integration.

Definition 3.4. Let f_λ and $A(\lambda)$ be functions in λ which assign to every $\lambda \in [a, b]$ a vector f_λ in $\mathscr{H}$ and a bounded operator $A(\lambda)$ on $\mathscr{H}$, respectively. Suppose that for any sequence of subdivisions $a = \lambda^{(0)} < \lambda^{(1)} < \cdots < \lambda^{(n)} = b$ of $[a, b]$ with norm

$$\delta = \max\{| \lambda^{(k)} - \lambda^{(k-1)} | : k = 1,\ldots, n\}$$

converging to zero, and that for any $\lambda'^{(k)} \in (\lambda^{(k-1)}, \lambda^{(k)}]$

$$\lim_{\delta \to 0} \sum_{k=1}^{n} \langle g \mid A(\lambda'^{(k)})(f_{\lambda^{(k)}} - f_{\lambda^{(k-1)}})\rangle \tag{3.75}$$

exists for some fixed $g \in \mathcal{H}$. If this limit is the same for any choice of sequences of subdivisions with shrinking norm $\delta \to 0$ and for any choice of $\lambda'^{(k)} \in (\lambda^{(k-1)}, \lambda^{(k)}]$, then it is denoted by the integral symbol

$$\left\langle g \,\middle|\, \int_a^b A(\lambda)\, d_\lambda f_\lambda \right\rangle . \tag{3.76}$$

A corresponding improper integral is defined as

$$\left\langle g \,\middle|\, \int_a^{+\infty} A(\lambda)\, d_\lambda f_\lambda \right\rangle = \lim_{b \to +\infty} \left\langle g \,\middle|\, \int_a^b A(\lambda)\, d_\lambda f_\lambda \right\rangle , \tag{3.77}$$

with a similar definition holding when $a \to -\infty$.

We encountered integrals of the type (3.76) in (3.31) and (3.39). In those cases, f_λ is of the form $E_\lambda f$, where E_λ is a spectral function and f is a fixed vector in $\mathcal{H}$.

For any given interval $I \subseteq \mathbb{R}^1$, the functional

$$\phi_I(g) = \left\langle \int_I A(\lambda)\, d_\lambda f_\lambda \mid g \right\rangle$$

is obviously linear. If $\phi_I(g)$ is defined for all $g \in \mathcal{H}$ and if it is bounded, then by Riesz' theorem there is a unique vector h_I which satisfies the equality $\langle h_I \mid g \rangle = \phi_I(g)$ for all $g \in \mathcal{H}$. We denote this vector by the integration symbol

$$\int_I A(\lambda)\, d_\lambda f_\lambda . \tag{3.78}$$

Suppose K_λ is a operator-valued function. It is easy to see that

$$\begin{aligned} \int_I A(\lambda)\, d_\lambda[K_\lambda(a_1 f_1 + a_2 f_2)] &= \int_I A(\lambda)\, d_\lambda[a_1 K_\lambda f_1 + a_2 K_\lambda f_2] \\ &= a_1 \int_I A(\lambda)\, d_\lambda K_\lambda f_1 + a_2 \int_I A(\lambda)\, d_\lambda K_\lambda f_2 \end{aligned}$$

and that the existence of the integrals on the right-hand side of the above relation implies the existence of the integral on the left-hand side. Hence, the mapping which takes f into

$$A_I(f) = \int_I A(\lambda)\, d_\lambda K_\lambda f$$

is linear. We denote the linear operator A_I by the integration symbol

$$A_I = \int_I A(\lambda)\, d_\lambda K_\lambda , \tag{3.79}$$

and refer to it as the *weak* Riemann–Stieltjes integral* over I of $A(\lambda)$ with respect to K_λ. If K_λ is a spectral function then (3.79) is called a (weak*) *spectral integral.*

Lemma 3.3. Suppose that E_λ^A is the spectral function of a self-adjoint operator A and that $F(\lambda)$, $\lambda \in \mathbb{R}^1$, is a family of bounded operators such† that $F^*(\lambda)f \in \mathscr{D}_A$ for all $f \in \mathscr{H}$. If $\| F(\lambda) \| \leqslant C$ for all $\lambda \in \mathbb{R}^1$ then

$$\int_{-\infty}^{+\infty} F(\lambda)\lambda \, d_\lambda E_\lambda^A g = \int_{-\infty}^{+\infty} F(\lambda) A \, d_\lambda E_\lambda^A g, \qquad g \in \mathscr{D}_A, \tag{3.80}$$

and the existence of one of these weak spectral integrals implies the existence of the other.

Proof. We shall prove first that for any $a < b$,

$$\left\langle f \,\middle|\, \int_a^b F(\lambda)(\lambda - A) \, d_\lambda E_\lambda^A g \right\rangle = 0, \qquad f \in \mathscr{H}, \quad g \in \mathscr{D}_A, \tag{3.81}$$

by establishing that the corresponding Riemann–Stieltjes sum converges weakly to zero in limit $\delta \to +0$ described in Definition 3.4. Indeed,

$$\begin{aligned} &\left\langle f \,\middle|\, \sum_{k=1}^{n} F(\lambda'^{(k)})(\lambda'^{(k)} - A)(E^A_{\lambda(k)} - E^A_{\lambda(k-1)})g \right\rangle \\ &\quad = \sum_{k=1}^{n} \langle (E^A_{\lambda(k)} - E^A_{\lambda(k-1)})(\lambda'^{(k)} - A) F^*(\lambda'^{(k)}) f \mid (E^A_{\lambda(k)} - E^A_{\lambda(k-1)})g \rangle \end{aligned} \tag{3.82}$$

since $E^A_{\lambda(k)} - E^A_{\lambda(k-1)}$ is a projector and therefore self-adjoint and equal to its own square and $F^*(\lambda'^{(k)})f \in \mathscr{D}_A$. The absolute value of (3.82) is obviously smaller than

$$\sum_{k=1}^{n} \| (E^A_{\lambda(k)} - E^A_{\lambda(k-1)})(\lambda'^{(k)} - A) F^*(\lambda'^{(k)}) f \| \, \| (E^A_{\lambda(k)} - E^A_{\lambda(k-1)}) g \|$$

(equation continues)

* In defining (3.79) we used weak limits of Riemann–Stieltjes sums such as (3.75), but the limit might exist and yield the same operator also in other topologies, such as the strong topology (see Prugovečki [1972a] for a systematic treatment).

† See Lemma 2.1 by Prugovečki [1973a] for less restrictive conditions under which (3.80) is still true.

$$\leqslant \left(\sum_{k=1}^{n} \| (E^A_{\lambda^{(k)}} - E^A_{\lambda^{(k-1)}})(\lambda'^{(k)} - A) F^*(\lambda'^{(k)}) f \|^2 \right)^{1/2}$$

$$\times \left(\sum_{k=1}^{n} \| (E^A_{\lambda^{(k)}} - E^A_{\lambda^{(k-1)}}) g \|^2 \right)^{1/2}$$

$$\leqslant \left(\sum_{k=1}^{n} (\lambda^{(k)} - \lambda^{(k-1)})^2 \| F^*(\lambda'^{(k)}) f \|^2 \right)^{1/2} \left(\left\langle g \,\middle|\, \sum_{k=1}^{n} (E^A_{\lambda^{(k)}} - E^A_{\lambda^{(k-1)}}) g \right\rangle \right)^{1/2}$$

$$\leqslant \left(\delta \sum_{k=1}^{n} (\lambda^{(k)} - \lambda^{(k-1)})\, C^2 \| f \|^2 \right)^{1/2} (\langle g \mid (E^A_b - E^A_a) g \rangle)^{1/2}$$

$$\leqslant \delta^{1/2} C \| f \| (b-a)^{1/2} \| (E^A_b - E^A_a) g \| \leqslant C(b-a)^{1/2} \| f \| \| g \| \delta^{1/2},$$

where the second inequality follows by the Schwarz–Cauchy inequality on $l^2(n)$ and the third inequality follows from the following relations:

$$\| (\lambda'^{(k)} - A)(E^A_{\lambda^{(k)}} - E^A_{\lambda^{(k-1)}}) h \|$$

$$\leqslant \left\| \int_{\lambda^{(k-1)}}^{\lambda^{(k)}} (\lambda'^{(k)} - \lambda)\, dE^A_\lambda h \right\|$$

$$\leqslant (\lambda^{(k)} - \lambda^{(k-1)}) \| (E^A_{\lambda^{(k)}} - E^A_{\lambda^{(k-1)}}) h \|,$$

$$(E^A_{\lambda^{(k)}} - E^A_{\lambda^{(k-1)}})(E^A_{\lambda^{(l)}} - E^A_{\lambda^{(l-1)}}) = \delta_{kl} (E^A_{\lambda^{(k)}} - E^A_{\lambda^{(k-1)}}).$$

Thus, we see that in the limit $\delta \to +0$, (3.82) approaches zero, and therefore (3.81) is true. In view of the obvious fact that

$$\int_{-\infty}^{+\infty} F(\lambda)(\lambda - A)\, d_\lambda E^A_\lambda g = \underset{N\to\infty}{\text{w-lim}} \sum_{n=-N}^{+N} \int_n^{n+1} F(\lambda)(\lambda - A)\, d_\lambda E^A_\lambda g,$$

we can conclude not only that (3.80) is true, but also that the existence of one of the integrals in (3.80) indeed implies the existence of the other. Q.E.D.

By using the Lemma 3.3 we can deduce the following general result, which in view of Theorem 3.1 immediately leads to the Hilbert space Lippmann–Schwinger equations (3.34) and (3.41).

Theorem 3.5. If H and H_0 are any two self-adjoint operators having a common domain of definition in a Hilbert space $\mathscr{H}$, then for any $\epsilon > 0$

(3.83)
$$\Omega_{\pm\epsilon} = \mp\epsilon \int_0^{\mp\infty} e^{\pm\epsilon t}\, e^{iHt}\, e^{-iH_0 t}\, dt$$
$$= 1 - \int_{-\infty}^{+\infty} (H - \lambda_0 \mp i\epsilon)^{-1}(H - H_0)\, d_{\lambda_0} E_{\lambda_0}^{H_0}$$
$$= 1 - \int_{-\infty}^{+\infty} d_\lambda E_\lambda^H (H - H_0)(\lambda - H_0 \mp i\epsilon)^{-1}.$$

Proof. By reviewing the derivation of (3.33) and (3.39) we easily establish that the only prerequisite has been the self-adjointness of H and H_0. If in addition $\mathscr{D}_H \subset \mathscr{D}_{H_0}$ then $(H - \lambda_0 \pm i\epsilon)^{-1}$ in the role of $F(\lambda_0)$ satisfies both conditions of Lemma 3.3 since by (3.60)

$$\|(H - \lambda_0 \pm i\epsilon)^{-1}\| \leqslant \epsilon^{-1}, \qquad \lambda_0 \in \mathbb{R}^1,$$

and obviously $F(\lambda_0)f \in \mathscr{D}_H$. Hence the equality of the expressions in the first and second line of (3.83) follows by applying this lemma to (3.39). The last of the equalities in (3.83) can be derived in a similar manner by taking adjoints in both sides of (3.33). Q.E.D.

The most remarkable feature of the Hilbert space versions of the Lippmann–Schwinger equations (3.34) and (3.41) is their great generality due to very simple and modest assumptions required in Theorem 3.5—from which they immediately follow. Consequently, not only do these equations lead to Lippmann–Schwinger equations for distorted waves* in two-body scattering (see §8.7), but their applicability extends beyond the realm of quantum mechanics proper (see §4). It should be noted, however, that these equations become trivial identities if $\mathsf{M}_\pm$ consist of only the zero vector. This is exactly what happens even in two-body potential scattering if the potential is of long range, and in particular for the practically important case of Coulomb-like potentials. One can then impose, however, space cutoffs on the potential, thus recovering nontrivial Lippmann–Schwinger equations, and then study the behavior of these equations as the cutoff is removed (see §4.9 as well as Prugovečki and Zorbas [1973] for details and further references). Alternatively,

* These were the equations actually derived by Lippmann and Schwinger [1950]. Hilbert space equations equivalent to (3.34) and (3.41) made their first appearance in a paper by Tixaire [1959] (and were independently derived and generalized by Prugovečki [1969, 1971b, 1973a]), whereas the Riemann–Stieltjes spectral integrals, which form their basis, were first introduced by Daletskiĭ and Krein [1956]. In studying and applying these Hilbert space equations some authors (e.g., Amrein, Georgescu, and Jauch [1971], Chandler and Gibson [1973], Thomas [1975]) prefer using strong instead of weak spectral integrals, but then additional restrictions have to be imposed on H and H_0.

one can develop modified Lippmann–Schwinger equations (Prugovečki [1973a]) that can be reduced to a form that contains a new effective potential and can be then solved by iterative techniques (see Masson and Prugovečki [1976]).

**3.8. Appendix: Scattering Theory in Liouville Space

Scattering theory for density operators plays an important role in nonequilibrium statistical mechanics.* Through the use of the Liouville space formalism of §8 in Chapter IV we can immediately adapt the results of §§2–3 to quantum statistical mechanics.

Indeed, a Liouville space $\mathscr{L}(\mathscr{H})$ is itself a Hilbert space. Therefore, all the general theorems that we have derived in §§2–3 are as applicable to $\mathscr{L}(\mathscr{H})$ as they were to $\mathscr{H}$ itself once it is established that their presuppositions are satisfied. Thus, the basic difference between quantum scattering theory in $\mathscr{H}$ and statistical-mechanics scattering theory on $\mathscr{L}(\mathscr{H})$ stems from somewhat different physical asymptotic conditions in the two cases. The difference in these conditions reflects the difference between equations (1.14) and (8.66) in Chapter IV for state vectors and density operators, respectively. This leads to appropriate changes in the detailed form of (2.46) and (2.47), which in the case of density operators, requires trace operations rather than inner products. For example, if we work with the density matrices of a complete set of operators, the density operator counterpart of the asymptotic condition (2.46) states that, by (8.69) in Chapter IV,

$$
\begin{aligned}
(3.84)\qquad & P^{A_1,\ldots,A_m}_{\rho(t)}(B) - P^{A_1,\ldots,A_m}_{\rho^{\mathrm{ex}}(t)}(B) \\
&\quad = \int_B [\langle x \mid \rho(t) \mid x\rangle - \langle x \mid \rho^{\mathrm{ex}}(t) \mid x\rangle]\, d\mu(x)
\end{aligned}
$$

should converge to zero as $t \to \pm\infty$.

It is quite easy to establish that a condition such as (3.84) converging to zero for $t \to \pm\infty$ is satisfied if

$$
\text{(3.85)}\qquad \operatorname*{h-lim}_{t\to\mp\infty}[\rho(t) - \rho^{\mathrm{ex}}(t)] = 0, \qquad \mathrm{ex} = \mathrm{in, out},
$$

* See Prigogine [1962], Kirczenov and Marro [1974], and Balescu [1975]. A sample of specific applications can be found in the articles by Zwanzig [1963], Fano [1963], Snider and Sanctuary [1971], Tip [1971], Eu [1975], and Coombe *et al.* [1975]. The proofs for statements made in this section can be found in Jauch *et al.* [1968] and Prugovečki [1972, 1978b].

although the converse is by no means true. On the other hand, since the limit in Hilbert–Schmidt norm appearing in (3.85) is in fact a strong limit in $\mathscr{L}(\mathscr{H})$, and $\mathscr{L}(\mathscr{H})$ is itself a Hilbert space with an inner product that gives rise to that norm, the general part of Theorem 1.1 is applicable in the present case so that we can state that (3.85) is satisfied if and only if

(3.86) $$\rho(0) = \operatorname*{h-lim}_{t\to\mp\infty} \mathbf{\Omega}(t)\, \rho^{\text{ex}}(0), \qquad \mathbf{\Omega}(t) = e^{i\mathbf{H}t}\, e^{-i\mathbf{H}_0 t}.$$

If we take into consideration Theorem 8.11 in Chapter IV and recall (2.1), we see that for any $A \in \mathscr{L}(\mathscr{H})$,

(3.87) $$\mathbf{\Omega}(t)A = \Omega(t)\, A\Omega^*(t)$$

(3.88) $$\Omega_\pm A \Omega_\pm{}^* = \operatorname*{h-lim}_{t\to\mp\infty} \mathbf{\Omega}(t)\, \mathbf{E}_\pm A,$$

where the superoperators $\mathbf{E}_\pm$,

(3.89) $$\mathbf{E}_\pm A = E_{\mathsf{M}_\pm} A E_{\mathsf{M}_\pm},$$

are obviously idempotent, i.e., $\mathbf{E}_\pm = \mathbf{E}_\pm{}^2$, and also self-adjoint in $\mathscr{L}(\mathscr{H})$,

$$\langle A_1 \mid \mathbf{E}_\pm A_2\rangle_2 = \operatorname{Tr}[A_1{}^* E_{\mathsf{M}_\pm} A_2 E_{\mathsf{M}_\pm}] = \operatorname{Tr}[E_{\mathsf{M}_\pm} A_1{}^* E_{\mathsf{M}_\pm} A_2] = \langle \mathbf{E}_\pm A_1 \mid A_2\rangle_2 .$$

Therefore $\mathbf{E}_\pm$ are superprojectors in $\mathscr{L}(\mathscr{H})$. Consequently, (3.86) suggests the introduction of the *wave superoperators*

(3.90) $$\mathbf{\Omega}_\pm = \operatorname*{h-lim}_{t\to\mp\infty} \mathbf{\Omega}(t)\mathbf{E}_\pm ,$$

which by (3.87) we related to wave operators as follows:

(3.91) $$\mathbf{\Omega}_\pm A = \Omega_\pm A \Omega_\pm{}^*, \qquad A \in \mathscr{L}(\mathscr{H}).$$

Starting with the definition (3.89) we can restate in a most striaght-forward manner every single theorem in §§2–3 in terms of $\mathbf{\Omega}_\pm$ and the *scattering superoperator* $\mathbf{S}$, where

(3.92) $$\mathbf{S}\rho = \mathbf{\Omega}_-^* \mathbf{\Omega}_+ \rho = S\rho S^*.$$

In particular, we shall have by Theorem 3.1

(3.93) $$\mathbf{\Omega}_\pm = \operatorname*{h-lim}_{\epsilon\to+0} \mathbf{\Omega}_{\pm\epsilon} \mathbf{E}_\pm ,$$

(3.94) $$\mathbf{\Omega}_{\mp\epsilon} = \pm\epsilon \int_0^{\pm\infty} e^{\mp\epsilon t}\, e^{i\mathbf{H}t}\, e^{-i\mathbf{H}_0 t}\, dt,$$

and by the results of §§3.2–3.4, for $\mathbf{H}_1 = \mathbf{H} - \mathbf{H}_0$ and $\mathbf{E}_\pm = \mathbf{1}$,

$$\boldsymbol{\Omega}_{\pm\epsilon} = \mathbf{1} + \int_{-\infty}^{+\infty} (\lambda - \mathbf{H} \pm i\epsilon)^{-1}\mathbf{H}_1 \, d_\lambda E_\lambda^{\mathbf{H}_0} \tag{3.95}$$

$$= \mathbf{1} - \int_{-\infty}^{+\infty} d_\lambda E_\lambda^{\mathbf{H}}\mathbf{H}_1(\lambda - \mathbf{H}_0 \pm i\epsilon)^{-1},$$

thus obtaining a spectral integral formula for the *transition super-operator*

$$\mathbf{T} = (2\pi i)^{-1}(\mathbf{1} - \mathbf{S}) = (2\pi i)^{-1} \operatorname*{h-lim}_{\epsilon \to +0}(\boldsymbol{\Omega}_{+\epsilon}^* - \boldsymbol{\Omega}_{-\epsilon}^*)\boldsymbol{\Omega}_+ \tag{3.96}$$

$$= \operatorname*{h-lim}_{\epsilon \to +0} \int_{-\infty}^{+\infty} d_\lambda E_\lambda^{\mathbf{H}_0}\mathbf{H}_1\{\epsilon/[(\mathbf{H} - \lambda)^2 + \epsilon^2]\}\boldsymbol{\Omega}_+ .$$

One of the advantages of the Liouville space approach to quantum scattering theory is greater generality. Indeed, if we define $\mathbf{E}_\pm$ as the superprojector onto the closed subspace of $\mathscr{L}(\mathscr{H})$ spanned by all $\rho^{(\mathrm{ex})}(0)$ for which (3.86) exists [rather than in terms of $\mathsf{M}_\pm$ by means of (3.89)], then we might have $\mathbf{E}_\pm = \mathbf{1}$ even when $E_{\mathsf{M}_\pm} = 0$. For example, assume that $\mathsf{M}_\pm = \mathscr{H}$ for some choice of H_0 and H in (2.1). If we then shift the energy scale by adding a constant $c \neq 0$ to H, thus arriving at $H' = H + c$, we shall have that

$$\operatorname*{s-lim}_{t \to \mp\infty} e^{iH't}\, e^{-iH_0t}f = e^{ict} \operatorname*{s-lim}_{t \to \mp\infty} e^{iHt}\, e^{-iH_0t}f$$

obviously does not exist except if $f = \mathbf{0}$. Thus, although nonrelativistic theories should not be sensitive to shifts in the energy scale, since (nonrelativistically) only relative and not absolute energy values are measurable, the standard scattering theory in $\mathscr{H}$ espoused in §§2–3 displays a definite (and totally unphysical) dependence on the choice of origin of the energy scale of H in relation to H_0. This weakness is, however, removed in the Liouville space approach, since then obviously

$$e^{i\mathbf{H}'t}\, e^{-i\mathbf{H}_0t}\rho = e^{i\mathbf{H}t}\, e^{-i\mathbf{H}_0t}\rho$$

for any density operator ρ, so that $\mathbf{E}_\pm$ are unaffected by such shifts in energy scale.

An even more significant advantage of the Liouville space approach to scattering theory is its immediate applicability to classical statistical mechanics. This enables not only the application of techniques developed in the quantum context to classical theory (see, e.g., Resibois [1959], Miles and Dahler [1970], St. Pierre [1973], Leaf [1975]), but through the use of the master Liouville space $\mathscr{L}_\Gamma$ introduced in §8.9 of Chapter IV, it makes possible the step by step comparison of classical and quantum effects when classical and quantum models for the same system are formulated in a common $\mathscr{L}_\Gamma$ (see Ali and Prugovečki [1977b], Prugovečki [1978b, c], as well as §4.10).

Exercises

3.1. Show that there are unique operators $\Omega_{-\epsilon}(s)$ and $\Omega_{+\epsilon}(s)$, $0 < \epsilon \leqslant s$, which satisfy the relations

$$\langle g \mid \Omega_{-\epsilon}(s) f\rangle = \epsilon \int_0^s e^{-\epsilon t} \langle g \mid \Omega(t) f\rangle \, dt,$$

$$\langle g \mid \Omega_{+\epsilon}(s) f\rangle = \epsilon \int_{-s}^0 e^{\epsilon t} \langle g \mid \Omega(t) f\rangle \, dt$$

for all $f, g \in \mathscr{H}$, and that $\| \Omega_{\pm\epsilon}(s) \| \leqslant 1$.

3.2. Prove that

$$\Omega_{\pm\epsilon} = \operatorname*{s-lim}_{s \to +\infty} \Omega_{\pm\epsilon}(s).$$

3.3. Suppose that the Riemann–Stieltjes integral of $F(\lambda)$ with respect to $\sigma(\lambda)$ exists in $[a, b]$, that $F(\lambda)$ is Borel measurable, and that $\sigma(\lambda)$ is nondecreasing. Show that $F(\lambda)$ is integrable with respect to a measure $\mu(B)$ which satisfies the relation $\mu((-\infty, \lambda]) = \sigma(\lambda)$, and that

$$\int_a^b F(\lambda) \, d_\lambda \sigma(\lambda) = \int_{(a,b]} F(\lambda) \, d\mu(\lambda).$$

3.4. Derive from the basic Definition 3.1 that the integrals (3.18)–(3.20) represent linear operations; for example, that

$$\int_a^b d_\lambda \int_I [a_1 F_1(\lambda, \lambda_0) + a_2 F_2(\lambda, \lambda_0)] \, d_{\lambda_0} \sigma(\lambda, \lambda_0)$$

$$= a_1 \int_a^b d_\lambda \int_I F_1(\lambda, \lambda_0) \, d_{\lambda_0} \sigma(\lambda, \lambda_0) + a_2 \int_a^b d_\lambda \int_I F_2(\lambda, \lambda_0) \, d_{\lambda_0} \sigma(\lambda, \lambda_0).$$

3.5. Prove that if the Riemann–Stieltjes integral on $[a, b]$, $[a, c]$ and $[c, b]$ exists ($a < c < b$), then

$$\int_a^c F(\lambda) \, d\sigma(\lambda) = \int_a^c F(\lambda) \, d\sigma(\lambda) + \int_c^b F(\lambda) \, d\sigma(\lambda), \qquad a < c < b.$$

Extend the proof to show that for cross-iterated integrals

$$\int_a^b \int_I = \int_a^c \int_I + \int_c^b \int_I, \quad \int_I \int_a^c \quad \int_I \int_c^b .$$

Remark. If $\sigma(\lambda)$ as well as $F(\lambda)$ are discontinuous at c, the integral on $[a, b]$ might not exist, although the integrals on $[a, c]$ and $[c, d]$ exist.

3.6. Show that if $\sigma(\lambda)$ is nondecreasing and $F_1(\lambda) \leqslant F_2(\lambda)$, then

$$\int F_1(\lambda)\, d\sigma(\lambda) \leqslant \int F_2(\lambda)\, d\sigma(\lambda).$$

Using this result, show that if $\sigma(\lambda, \lambda_0)$ is nondecreasing [i.e., $\sigma(\lambda, \lambda_0) \leqslant \sigma(\lambda', \lambda_0')$ whenever $\lambda \leqslant \lambda'$ and $\lambda_0 \leqslant \lambda_0'$] and $F_1(\lambda, \lambda_0) \leqslant F_2(\lambda, \lambda_0)$, then

$$\int d_\lambda \int F_1(\lambda, \lambda_0)\, d_{\lambda_0}\sigma(\lambda, \lambda_0) \leqslant \int d_\lambda \int F_2(\lambda, \lambda_0)\, d_{\lambda_0}\sigma(\lambda, \lambda_0).$$

3.7. Assuming that the respective Riemann–Stieltjes integrals for σ_1 and σ_2 exist, prove that the corresponding integrals for the linear combinations $a_1\sigma_1 + a_2\sigma_2$, $a_1, a_2 \in \mathbb{C}^1$, exist and

$$\int_I F(\lambda)\, d_\lambda[a_1\sigma_1(\lambda) + a_2\sigma_2(\lambda)] = a_1 \int_I F(\lambda)\, d_\lambda\sigma_1(\lambda) + a_2 \int_I F(\lambda)\, d_\lambda\sigma_2(\lambda),$$

$$\int_I d_\lambda \int_{I_0} F(\lambda)\, d_{\lambda_0}[a_1\sigma_1(\lambda, \lambda_0) + a_2\sigma_2(\lambda, \lambda_0)] = a_1 \int_I d_\lambda \int_{I_0} F(\lambda, \lambda_0)\, d_{\lambda_0}\sigma_1(\lambda, \lambda_0) + a_2 \int_I d_\lambda \int_{I_0} F(\lambda, \lambda_0)\, d_{\lambda_0}\sigma_2(\lambda, \lambda_0),$$

where I and I_0 are arbitrary nondegenerate intervals in $\mathbb{R}^1$.

3.8. Using the result of Exercise 3.6, prove the *mean-value theorem* for Riemann–Stieltjes integrals:

If $F(\lambda)$ is real and continuous, and if $\sigma(\lambda)$ is nondecreasing and bounded on $[a, b]$, then there is a point $\lambda' \in [a, b]$ for which

$$\int_a^b F(\lambda)\, d_\lambda\sigma(\lambda) = F(\lambda') \int_a^b d_\lambda\sigma(\lambda) = F(\lambda')\, \sigma((a, b]).$$

Show that the above integral exists if $F(\lambda)$ and $\sigma(\lambda)$ satisfy the assumptions made earlier.

3.9. Prove that if (3.73) exists, then the relation (3.74) holds.

3.10. Prove that the resolvent $R_A(\zeta)$ of a self-adjoint operator A can be represented for $\operatorname{Im} \zeta < 0$ by a Bochner integral:

$$R_A(\zeta) = -i \int_0^{+\infty} e^{i(A-\zeta)t}\, dt.$$

4. Eigenfunction Expansions in Two-Body Potential Scattering Theory

4.1. Free Plane Waves in Three Dimensions

In §3 we presented a very general framework for time-independent scattering theory. However, in the practically important case of potential scattering one encounters much more specialized versions of that framework, which are, on the other hand, more convenient to deal with from a computational point of view. To derive these formulations rigorously from the general framework of §3, we need the existence of eigenfunction expansions for the Hamiltonians H_0 and H.

Before formulating the concept of eigenfunction expansions on a general level, we shall elucidate this concept by studying it in the special case of the kinetic energy (or "free") Schroedinger operator H_0 on $L^2(\mathbb{R}^3)$, defined by means of the differential operator form

$$H_0^{\mathbf{r}} = -\frac{1}{2m}\Delta, \qquad \Delta = \frac{\partial^2}{\partial x^2} + \frac{\partial^2}{\partial y^2} + \frac{\partial^2}{\partial z^2}. \tag{4.1}$$

If we consider $H_0^{\mathbf{r}}$ to be an operator acting on the space of all everywhere twice-differentiable complex functions $f(\mathbf{r})$, then $H_0^{\mathbf{r}}$ has *eigenfunctions* $\Phi_{\mathbf{k}}(\mathbf{r})$

$$(H_0^{\mathbf{r}}\Phi_{\mathbf{k}})(\mathbf{r}) = \frac{\mathbf{k}^2}{2m}\Phi_{\mathbf{k}}(\mathbf{r}), \qquad \Phi_{\mathbf{k}}(\mathbf{r}) = \frac{1}{(2\pi)^{3/2}} e^{i\mathbf{k}\mathbf{r}}, \tag{4.2}$$

associated with the *eigennumbers*

$$\Lambda(\mathbf{k}) = \mathbf{k}^2/2m. \tag{4.3}$$

In physical literature the functions $\Phi_{\mathbf{k}}(\mathbf{r})$ are called *free plane waves.* However, these eigenfunctions are not square integrable on $\mathbb{R}^3$ in the Lebesgue measure, and therefore they do not represent eigenvectors of H_0, considered as a self-adjoint operator in $L^2(\mathbb{R}^3)$.

The family $\{\Phi_{\mathbf{k}}(\mathbf{r}): \mathbf{k} \in \mathbb{R}^3\}$ of eigenfunctions of $H_0^{\mathbf{r}}$ has some remarkable properties, which follow from Theorems 4.5 and 4.6 of Chapter III.

According to Theorem 4.5 of Chapter III, if $\psi \in L^2(\mathbb{R}^2)$, and if $B_1, B_2, \ldots$ are any bounded measurable sets, the functions $(\chi_{B_n}\psi)(\mathbf{r})$ have Fourier transforms $U_{\mathrm{F}}(\chi_{B_n}\psi)$, since for every $n = 1, 2, \ldots$ the function $\chi_{B_n}(\mathbf{r})\, \psi(\mathbf{r})$ is integrable as well as square integrable with respect to the Lebesgue measure. On the other hand, if $\bigcup_{n=1}^{\infty} B_n = \mathbb{R}^3$, then $\| \chi_{B_n}\psi - \psi \| \to 0$ when $n \to +\infty$, and therefore, in view of Theorem 4.6 in Chapter III,

$$U_{\mathrm{F}}\psi = \operatorname*{s-lim}_{n\to+\infty} U_{\mathrm{F}}(\chi_{B_n}\psi). \tag{4.4}$$

Let us introduce some convenient notation in Definition 4.1.

Definition 4.1. Suppose $F(x, y)$ is the kernel of an integral operator F from $L^2(\mathscr{Y}, \nu)$ into $L^2(\mathscr{X}, \mu)$

$$(Fg)(x) = \int_{\mathscr{Y}} F(x, y)\, g(y)\, d\nu(y), \tag{4.5}$$

and that $\mathscr{F}$ is the family of all functions $g(x)$ for which the integral in (4.5) exists. Then we shall write for some $h \in L^2(\mathscr{Y}, \nu)$

$$(Fh)(x) = \text{l.i.m.} \int_{\mathscr{Y}} F(x, y)\, h(y)\, d\nu(y) \tag{4.6}$$

if the linear operator defined by (4.5) has an unique extension $f = Fh$ to h, i.e., if for any sequence $g_1, g_2, \ldots \in \mathscr{F}$ converging strongly to h there is an unique $f \in L^2(\mathscr{X}, \mu)$ such that

$$\lim_{n\to\infty} \int_{\mathscr{X}} | f(x) - (Fg_n)(x)|^2\, d\mu(x) = 0. \tag{4.7}$$

We note that if F happens to be a bounded operator on $\mathscr{F}$, then by the extension principle of bounded operators (Chapter III, Theorem 2.6), l.i.m. $\int F(x, y)\, h(y)\, d\nu(y)$ is defined for all $h \in L^2(\mathscr{Y}, \nu)$ and represents the extension of the operator F to h. In particular, if F is the Fourier–Plancherel transform U_{F}, then

$$\tilde{\psi}(\mathbf{k}) = (U_{\mathrm{F}}\psi)(\mathbf{k}) = (2\pi)^{-3/2}\, \text{l.i.m.} \int_{\mathbb{R}^3} e^{-i\mathbf{k}\mathbf{r}}\psi(\mathbf{r})\, d\mathbf{r} \tag{4.8}$$

is well defined. Moreover, by Theorem 4.6 in Chapter III we also have

$$(E^{\mathrm{P}}(B_n)\psi)(\mathbf{r}) = (2\pi)^{-3/2} \int_{B_n} e^{i\mathbf{r}\mathbf{k}}\tilde{\psi}(\mathbf{k})\, d\mathbf{k}, \tag{4.9}$$

and consequently

$$\psi(\mathbf{r}) = (2\pi)^{-3/2}\ \text{l.i.m.} \int_{\mathbb{R}^3} e^{i\mathbf{r}\mathbf{k}}\, \tilde{\psi}(\mathbf{k})\, d\mathbf{k}. \tag{4.10}$$

The formal analogy of (4.10) and (4.8) with the expansion formula of a vector $\psi \in \mathscr{H}$,

$$\psi = \sum_n \langle e_n \mid \psi\rangle\, e_n,$$

in an orthonormal basis e_k becomes obvious if we rewrite these two formulas in the form

$$\psi(\mathbf{r}) = \text{l.i.m.} \int_{\mathbb{R}^3} \tilde{\psi}(\mathbf{k})\, \Phi_{\mathbf{k}}(\mathbf{r})\, d\mathbf{k}, \qquad \tilde{\psi}(\mathbf{k}) = \text{l.i.m.} \int \Phi_{\mathbf{k}}^*(\mathbf{r})\, \psi(\mathbf{r})\, d\mathbf{r} \tag{4.11}$$

and introduce the convenient notation

$$(\Phi_{\mathbf{k}} \mid \psi\rangle = \text{l.i.m.} \int \Phi_{\mathbf{k}}^*(\mathbf{r})\, \psi(\mathbf{r})\, d\mathbf{r} = \tilde{\psi}(\mathbf{k}). \tag{4.12}$$

The fact that $(\Phi_{\mathbf{k}} \mid f\rangle$ is *not* related to the inner product in $L^2(\mathbb{R}^3)$ is emphasized in this notation by the round bracket in $(\cdot \mid \cdot\rangle$.

Due to such analogies, the relations (4.11) are referred to as eigenfunction expansions of $\psi(\mathbf{r})$ by means of free plane waves $\Phi_{\mathbf{k}}(\mathbf{r})$.

It is important to realize that we can easily express the spectral measure $E^{H_0}(B)$ of H_0 with the help of $\Phi_{\mathbf{k}}(\mathbf{r})$. Indeed, we recall that the momentum operators are essentially multiplication operators when they act on $\tilde{\psi}(\mathbf{k})$,

$$(P^{(x)}\psi)^{\sim}(\mathbf{k}) = k_x\tilde{\psi}(\mathbf{k}), \quad (P^{(y)}\psi)^{\sim}(\mathbf{k}) = k_y\tilde{\psi}(\mathbf{k}), \quad (P^{(z)}\psi)^{\sim}(\mathbf{k}) = k_z\tilde{\psi}(\mathbf{k}), \tag{4.13}$$

so that the $\mathbf{k}$ variables are in this case identical to the momentum variables $\mathbf{p}$. Since H_0 is a function of $\mathbf{P}$,

$$H_0 = \frac{\mathbf{P}^2}{2m} = \Lambda(\mathbf{P}), \tag{4.14}$$

we have, in the momentum space,

$$(E^{H_0}(B)\psi)^{\sim}(\mathbf{k}) = \chi_{\Lambda^{-1}(B)}(\mathbf{k})\, \tilde{\psi}(\mathbf{k}),$$

or equivalently, in the configuration space,

$$(E^{H_0}(B)\psi)(\mathbf{r}) = \text{l.i.m.} \int_{\Lambda^{-1}(B)} \Phi_{\mathbf{k}}(\mathbf{r})\, \tilde{\psi}(\mathbf{k})\, d\mathbf{k}. \tag{4.15}$$

4.2. Distorted Plane Waves

Let us now consider a total Schroedinger operator H defined by means of the differential operator form

$$H^{\mathrm{r}} = -(1/2m)\Delta + V(\mathbf{r}). \tag{4.16}$$

In §§6–7 we shall extend to H^{r} the above results on H_0^{r} by proving that, under certain assumptions on the potential $V(\mathbf{r})$, the following statements are true.

For each vector $\mathbf{k} \in \mathbb{R}^3 - \mathscr{S}_V^3$, there is a unique solution*

$$\Phi_{\mathbf{k}}^{(+)}(\mathbf{r}) = (2\pi)^{-3/2}\,(e^{i\mathbf{kr}} + v_{\mathbf{k}}(\mathbf{r}))$$

of the differential equation

$$-\frac{1}{2m}\,\Delta\Phi_{\mathbf{k}}^{(+)}(\mathbf{r}) + V(\mathbf{r})\,\Phi_{\mathbf{k}}^{(+)}(\mathbf{r}) = \frac{\mathbf{k}^2}{2m}\,\Phi_{\mathbf{k}}^{(+)}(\mathbf{r}) \tag{4.17}$$

for which, in spherical coordinates r, θ, and ϕ, $v_{\mathbf{k}}(\mathbf{r})$ behaves asymptotically as follows:

$$v_{\mathbf{k}}(\mathbf{r}) \sim \frac{f_{\mathbf{k}}(\theta, \phi)}{r}\,e^{ikr}, \qquad r = |\,\mathbf{r}\,| \to +\infty, \qquad k = |\,\mathbf{k}\,|, \tag{4.18}$$

where $f_{\mathbf{k}}(\theta, \phi)$ is a function uniquely determined by $v_{\mathbf{k}}(\mathbf{r})$. The family $\{\Phi_{\mathbf{k}}^{(+)}(\mathbf{r})\colon \mathbf{k} \in \mathbb{R}^3 - \mathscr{S}_V^3\}$ provides an eigenfunction expansion for any element $\psi_+ \in L^2_{H_c}(\mathbb{R}^3)$ in the sense that

$$\psi_+(\mathbf{r}) = \mathrm{l.i.m.} \int_{\mathbb{R}^3} \Phi_{\mathbf{k}}^{(+)}(\mathbf{r})\,\tilde{\psi}(\mathbf{k})\,d\mathbf{k}, \tag{4.19}$$

where $L^2_{H_c}(\mathbb{R}^3) = E^H(\mathsf{S}_c^H)\,L^2(\mathbb{R}^3)$ denotes the closed linear subspace of $L^2(\mathbb{R}^3)$ corresponding to the continuous spectrum S_c^H of H.

The function $\tilde{\psi}(\mathbf{k})$ appearing in the eigenfunction expansion (4.19) is *not* the Fourier–Plancherel transform of $\psi_+(\mathbf{r})$, but rather the Fourier–Plancherel transform of another function related to ψ_+ by the equation $\psi_+ = \Omega_+\psi$. In fact, one of the most important results (which will be derived in §7) of time-independent potential scattering is that Ω_+ is in the present case a partial isometry with initial domain $\mathsf{M}_0 \equiv L^2(\mathbb{R}^3)$ and final domain $\mathsf{R}_+ \equiv L^2_{H_c}(\mathbb{R}^3)$. Hence, to every $\psi_+ \in L^2_{H_c}(\mathbb{R}^3)$ corresponds a unique $\psi \in L^2(\mathbb{R}^3)$ such that $\psi_+ = \Omega_+\psi$.

* $\mathscr{S}_3^V$ is an exceptional set of Lebesgue measure zero on which Fredholm techniques do not yield a unique $\Phi_{\mathbf{k}}^{(\pm)}(\mathbf{r})$ (see Theorem 6.1).

In physical literature the functions $\Phi_{\mathbf{k}}^{(+)}(\mathbf{r})$ are referred to as the *outgoing* (or *retarded*) *distorted plane waves*, while $f_{\mathbf{k}}(\theta, \phi)$ is called the *scattering amplitude* since, as we shall see in §6, it is intimately related to the function $T^{(1)}(p; \omega, \omega')$ introduced in (1.31):

(4.20)

$$f_{\mathbf{k}}(\omega) = -(2\pi)^2 \,|\,\mathbf{k}\,|^{-1}\, T^{(1)}(|\,\mathbf{k}\,|; \omega, \omega_0), \qquad \omega = (\theta, \phi), \quad \omega_0 = (0, 0).$$

The justification of the term "distorted plane wave" is that in the special case when $V(\mathbf{r}) \equiv 0$ we have $v_{\mathbf{k}}(\mathbf{r}) \equiv 0$, and $\Phi_{\mathbf{k}}^{(+)}(\mathbf{r})$ becomes the free plane wave $\Phi_{\mathbf{k}}(\mathbf{r})$. In addition, the function

$$\Phi_{\mathbf{k}}^{(+)}(\mathbf{r}; t) = \Phi_{\mathbf{k}}^{(+)}(\mathbf{r}) \exp[-i(\mathbf{k}^2/2m)t] \tag{4.21}$$

provides a solution of the time-dependent Schroedinger equation

$$i\frac{\partial}{\partial t}\Phi_{\mathbf{k}}^{(+)}(\mathbf{r}; t) = [-(1/2m)\Delta + V(\mathbf{r})]\, \Phi_{\mathbf{k}}^{(+)}(\mathbf{r}; t). \tag{4.22}$$

The "distorted part," $v_{\mathbf{k}}(\mathbf{r}) \exp[-i(\mathbf{k}^2/2m)t]$ of this wave describes, heuristically speaking, a process which recedes away from the scattering center $\mathbf{r} = 0$.

The functions

$$\Phi_{\mathbf{k}}^{(-)}(\mathbf{r}) = \Phi_{-\mathbf{k}}^{(+)*}(\mathbf{r}) \tag{4.23}$$

are called *incoming* (or *advanced*) *distorted plane waves* for analogous reasons. They also satisfy (4.17), and for any $\psi_- \in L^2_{H_c}(\mathbb{R}^3)$, we have the expansion

$$\psi_-(\mathbf{r}) = \text{l.i.m.} \int_{\mathbb{R}^3} \Phi_{\mathbf{k}}^{(-)}(\mathbf{r})\, \tilde{\psi}(\mathbf{k})\, d\mathbf{k}, \tag{4.24}$$

where $\psi_- = \Omega_- \psi$, with ψ uniquely determined by ψ_- due to the fact that $\mathsf{M}_0 \equiv L^2(\mathbb{R}^3)$ and $\mathsf{R}_- \equiv L^2_{H_c}(\mathbb{R}^3)$.

A further significant result of §6, which in fact represents a generalization of (4.19) and (4.24), is that the spectral measure $E^H(B)$ can be computed on $L^2_{H_c}(\mathbb{R}^3)$ by means of $\Phi_{\mathbf{k}}^{(\pm)}(\mathbf{r})$:

$$(E^H(B)\psi_\pm)(\mathbf{r}) = \text{l.i.m.} \int_{\Lambda^{-1}(B)} \Phi_{\mathbf{k}}^{(\pm)}(\mathbf{r})\, \tilde{\psi}(\mathbf{k})\, d\mathbf{k}, \qquad \psi_\pm = \Omega_\pm \psi. \tag{4.25}$$

In particular, for $B = \mathbb{R}^1$, the relations (4.25) assume the form (4.19) or (4.24), respectively. It is important to realize that (4.15) is also a special case of (4.25), since when $V(\mathbf{r}) \equiv 0$, we have $\Omega_+ = \Omega_- = 1$ and $L^2_{H_0}(\mathbb{R}^3) = L^2(\mathbb{R}^3)$.

4.3. Free and Distorted Spherical Waves

The free plane waves $\Phi_{\mathbf{k}}(\mathbf{r})$ are not the only eigenfunctions of the differential operator $H_0^{\mathbf{r}}$. It is easy to verify that the functions

$$(4.26) \qquad \Phi_{klm}(r, \theta, \phi) = (2/\pi)^{1/2} i^l j_l(kr)\, Y_l^m(\theta, \phi)$$

are also eigenfunctions of $H_0^{\mathbf{r}}$:

$$(4.27) \qquad H_0^{\mathbf{r}} \Phi_{klm}(r, \theta, \phi) = \frac{1}{2m}\left[\frac{\partial^2}{\partial r^2} + \frac{2}{r}\frac{\partial}{\partial r} - \frac{1}{r^2}\mathbf{L}^2\right] \Phi_{klm}(r, \theta, \phi)$$
$$= \frac{k^2}{2m} \Phi_{klm}(r, \theta, \phi),$$

$$(4.28) \qquad -\mathbf{L}^2 = \frac{1}{\sin\theta}\left[\frac{\partial}{\partial\theta} \sin\theta \frac{\partial}{\partial\theta}\right] + \frac{1}{\sin^2\theta}\frac{\partial^2}{\partial\phi^2},$$

for any values

$$(4.29) \qquad 0 \leqslant k < +\infty, \qquad l = 0, 1, 2, \ldots, \qquad m = -l, -l+1, \ldots, +l.$$

The function $\Phi_{klm}(r, \theta, \phi)$ is called an *outgoing free spherical* or *partial wave*. It follows from (4.10) and the relation (7.70) in §7 that the family of all functions (4.26) provides us with an eigenfunction expansion:

$$(4.30) \quad \psi(\mathbf{r}) = \sum_{l=0}^{\infty} \sum_{m=-l}^{+l} \text{l.i.m.} \int_0^\infty \Phi'_{klm}(\mathbf{r})\, \hat{\psi}_{lm}(k)\, k^2\, dk, \quad \Phi'_{klm}(\mathbf{r}) = \Phi_{klm}(r, \theta, \phi),$$

where, using the notation Ω_{p} introduced in (7.10) of Chapter II, we have

$$(4.31) \qquad \hat{\psi}_{lm}(k) = \text{l.i.m.} \int_{\Omega_{\mathrm{p}}} \Phi'^{*}_{klm}(\mathbf{r})\, \psi(r, \theta, \phi)\, r^2 \sin\theta\, dr\, d\theta\, d\phi.$$

In the case that $V(\mathbf{r})$ is spherically symmetric, i.e., $V(\mathbf{r}) = V_0(r)$, the differential operator $H^{\mathbf{r}}$ has the eigenfunctions

$$(4.32) \qquad \Phi^{(\pm)}_{klm}(r, \theta, \phi) = (2/\pi)^{1/2} i^l R^{(\pm)}_{kl}(r)\, Y_l^m(\theta, \phi).$$

These eigenfunctions are called, respectively, outgoing and incoming *distorted spherical waves* and are generalizations of Φ_{klm}; i.e.,

$$(4.33) \quad H^{\mathbf{r}}\Phi^{(\pm)}_{klm}(r, \theta, \phi) = \left(-\frac{1}{2m}\Delta + V_0(r)\right)\Phi^{(\pm)}_{klm}(r, \theta, \phi) = \frac{k^2}{2m}\Phi^{(\pm)}_{klm}(r, \theta, \phi),$$

and for any $\psi_\pm \in L^2_{H_c}(\mathbb{R}^3)$

$$\psi_\pm(\mathbf{r}) = \sum_{l=0}^{\infty} \sum_{m=-l}^{+l} \text{l.i.m.} \int_0^\infty \Phi^{(\pm)'}_{klm}(\mathbf{r})\, \hat{\psi}_{lm}(k)\, k^2\, dk, \tag{4.34}$$

where $\psi_\pm = \Omega_\pm \psi$, and $\hat{\psi}_{lm}(k)$ is given by (4.31).

It is well known (see Butkov [1968, Section 9.10]) that the asymptotic behavior for large r of the spherical Bessel function $j_l(kr)$ is

$$j_l(kr) \sim (1/kr)\sin(kr - l\pi/2), \qquad r \to +\infty. \tag{4.35}$$

We shall see in §7 that the functions $R^{(+)}_{kl}(r)$ have a similar asymptotic behavior

$$R^{(\pm)}_{kl}(r) \sim (1/kr)\sin(kr - (l\pi/2) + \delta_l(k)). \tag{4.36}$$

In physical literature $\delta_l(k)$ are called *phase shifts*, since they represent the change of phase from the free spherical waves corresponding to the noninteraction case $V(r) = 0$, to the distorted spherical waves. We shall see in §7 that they play a crucial role in the computation of the S operator, which can be written in the form

$$(S\psi)\hat{}_{lm}(k) = \exp[2i\delta_l(k)]\, \hat{\psi}_{lm}(k). \tag{4.37}$$

The function $S_l(k) = \exp[2i\delta_l(k)]$ is called the *partial-wave S matrix*. The reason for this terminology is that if we use Dirac's bra and ket notation to write $\mid klm) = \Phi_{klm}$, then taking in (4.31) and (4.36) $\psi = \Phi_{klm}$ and working formally, without paying too much attention to the real mathematical meaning of these expressions, we get

$$(k'l'm' \mid S \mid klm) = (kk')^{-1}\, \delta(k' - k)\, \delta_{l'l}\, \delta_{m'm} \exp[2i\, \delta_l(k)].$$

In §7 we shall establish the following generalization of (4.34):

$$(E^H(B)\, \psi_\pm)(\mathbf{r}) = \sum_{l=0}^{\infty} \sum_{m=-l}^{+l} \text{l.i.m.} \int_{\Lambda_0^{-1}(B)} \Phi^{(\pm)'}_{klm}(\mathbf{r})\, \hat{\psi}_{lm}(k)\, k^2\, dk, \tag{4.38}$$

where

$$\Lambda_0(k) = k^2/2m. \tag{4.39}$$

These formulas can be recast in a more compact form if we introduce the measure $\mu_0 = \mu^{(k)} \times \mu^{(\omega)}$ on $\mathscr{B}^3$, where

$$\mu^{(k)}(B_1) = \int_{B_1} k^2\, dk, \qquad B_1 \in \mathscr{B}^1,$$

(4.40)

$$\mu^{(\omega)}(B_2) = \sum_{l=0}^{\infty} \sum_{m=-l}^{l} \chi_{B_2}(l, m), \qquad B_2 \in \mathscr{B}^2.$$

It is easy to verify (see Exercise 4.1) that, written in terms of the measure μ_0, (4.38) becomes

$$(4.41) \qquad (E^H(B)\,\psi_\pm)(\mathbf{r}) = \text{l.i.m.} \int_{\Lambda_0^{-1}(B)} \hat{\Phi}^{(\pm)}_{klm}(\mathbf{r})\, \hat{\psi}(k, l, m)\, d\mu_0(k, l, m),$$

where $\hat{\psi}(k, l, m)$ and $\Phi^{(\pm)}_{klm}(\mathbf{r})$ are arbitrary extensions of the functions $\hat{\psi}_{lm}(k)$ and $\Phi^{(\pm)\prime}_{klm}(\mathbf{r})$, respectively, outside the support of the measure μ_0 and to the whole of $\mathbb{R}^3$. In particular, for $H = H_0$, we have

$$(4.42) \qquad (E^{H_0}(B)\psi)(\mathbf{r}) = \text{l.i.m.} \int_{\Lambda_0^{-1}(B)} \hat{\Phi}_{klm}(\mathbf{r})\, \hat{\psi}(k, l, m)\, d\mu_0(k, l, m)$$

as a special case of (4.41).

4.4. Eigenfunction Expansions for Complete Sets of Operators

The free plane and spherical waves introduced in the preceding two sections set the pattern for both the mathematical and physical meaning, as well as the role, of free (or unperturbed) eigenfunction expansions in scattering theory*: in both cases we are dealing with integral kernels of unitary operators U appearing in Definition 5.2 of Chapter IV when executing the transition from the configuration representation space $L^2(\mathbb{R}^3)$ (namely the spectral representation space of position observables) to the spectral representation space $L^2(\mathbb{R}^n, \mu)$ of another complete set of observables, namely $\{P^{(x)}, P^{(y)}, P^{(z)}\}$ and $\{|\mathbf{P}|, |\mathbf{L}|, L^{(z)}\}$, respectively. Indeed, in the first case $L^2(\mathbb{R}^n, \mu)$ is the momentum representation space $L^2(\mathbb{R}^3)$ of wave functions $\tilde{\psi}(\mathbf{k})$ on which $P^{(x)}$, $P^{(y)}$,

* More abstract approaches based on rigged (Gel'fand and Vilenkin [1968]) or equipped (Berezanskiĭ [1968, 1978]) Hilbert spaces are feasible (see, e.g., Antoine [1969], Prugovečki [1973b], Ascoli *et al.* [1978]), but on one hand a definitive treatment is still pending, whereas on the other hand the more elementary approach adopted in this section suffices in potential scattering and is also close to computational techniques actually used in practice.

and $P^{(z)}$ act in accordance with (4.13), whereas in the second case $L^2(\mathbb{R}^n, \mu)$ equals $L^2(\mathbb{R}^3, \mu_0)$ since

$$(4.43) \qquad (|\mathbf{P}|\psi)\hat{}_{lm}(k) = k\hat{\psi}_{l,m}(k), \qquad (L^{(z)}\psi)\hat{}_{lm}(k) = m\hat{\psi}_{lm}(k),$$

and for L' in (7.44), Chapter IV,

$$(4.44) \qquad (L'\psi)\hat{}_{lm}(k) = l\hat{\psi}_{lm}(k), \qquad \mathbf{L}^2 = L'(L'+1).$$

We shall generalize now the above features of free plane and spherical waves into a concept of eigenfunction expansions that is applicable to all other choices of complete sets of observables in two-body as well as in multichannel scattering theory.

Definition 4.2. Let $\{X_1, ..., X_m\}$ and $\{Y_1, ..., Y_n\}$ be two complete sets of operators in $L^2(\mathbb{R}^m, \mu)$ (in the sense of Definition 5.2 in Chapter IV). We shall say that $\Phi_y(x)$ is an *eigenfunction expansion for* $\{Y_1, ..., Y_n\}$ *in the spectral representation space* $L^2(\mathbb{R}^m, \mu)$ *of* $\{X_1, ..., X_m\}$ if there is a spectral representation space $L^2(\mathbb{R}^n, \nu)$ for $\{Y_1, ..., Y_n\}$ and a unitary transformation U of $L^2(\mathbb{R}^m, \mu)$ onto $L^2(\mathbb{R}^n, \nu)$ such that

$$(4.45) \qquad \check{f}(y) = (Uf)(y) = \text{l.i.m.} \int \Phi_y^*(x) f(x)\, d\mu(x), \qquad f \in L^2(\mathbb{R}^m, \mu),$$

$$(4.46) \qquad f(x) = (U^{-1}\check{f})(x) = \text{l.i.m.} \int \Phi_y(x) \check{f}(y)\, d\nu(y), \qquad \check{f} \in L^2(\mathbb{R}^n, \nu),$$

$$(4.47) \qquad (UY_jU^{-1}\check{f})(y) = y_j\check{f}(y), \qquad y = (y_1, ..., y_n), \qquad j = 1, ..., n.$$

Clearly, the free Hamiltonian H_0 is in potential scattering a function of the complete set $\{P^{(x)}, P^{(y)}, P^{(z)}\}$ as well as of $\{|\mathbf{P}|, |\mathbf{L}|, L^{(z)}\}$, as seen from (4.14) and (4.39):

$$H_0 = |\mathbf{P}|^2/2m = \Lambda_0(|\mathbf{P}|).$$

Hence, in general we can expect that the free Hamiltonian (or, in the language of perturbation theory, the unperturbed operator) shall be a function

$$(4.48) \qquad H_0 = \Lambda(Y_1, ..., Y_n)$$

of the complete set $\{Y_1, ..., Y_n\}$, so that by Theorem 5.3 in Chapter IV,

$$(UH_0U^{-1}\check{f})(y) = \Lambda(y)\check{f}(y).$$

Consequently, by (4.46) we arrive at the following generalizations of

(4.15) and (4.42):

$$(E^{H_0}(B)f)(x) = \text{l.i.m.} \int_{\Lambda^{-1}(B)} \Phi_y(x) \check{f}(y)\, d\nu(y). \tag{4.49}$$

As partial isometries with final domains $\mathsf{R}_\pm$, the wave operators $\Omega_\pm$ can also be viewed as unitary transformations of $\mathsf{M}_\pm$ onto $\mathsf{R}_\pm$ treated as Hilbert spaces in their own right. If $Y_1, ..., Y_n$ leave $\mathsf{M}_\pm$ invariant, then so do

$$Y_j^{(\pm)} = \Omega_\pm Y_j \Omega_\pm^*, \qquad j = 1, ..., n, \tag{4.50}$$

leave $\mathsf{R}_\pm$ invariant, and from Theorem 5.1 in Chapter IV it can be easily inferred that $\{Y_1^{(\pm)}, ..., Y_n^{(\pm)}\}$ constitute complete sets of operators in $\mathsf{R}_\pm$. By (2.4), (2.11), and Theorem 2.5 in Chapter IV,

$$H^\pm = \Lambda(Y_1^{(\pm)}, ..., Y_n^{(\pm)}) = \Omega_\pm \Lambda(Y_1, ..., Y_n)\Omega_\pm^* \tag{4.51}$$

are the restrictions of H to $\mathsf{R}_\pm$. Thus, if eigenfunction expansions $\Phi_y^{(\pm)}(x)$ of $\{Y_y^{(\pm)}, ..., Y_n^{(\pm)}\}$ in the spectral representation space $L^2(\mathbb{R}^m, \mu)$ of $\{X_1, ..., X_m\}$ exist, so that in accordance with Definition 4.2,

(4.52)

$$\check{g}^{(\pm)}(y) = (U^{(\pm)}g)(y) = \text{l.i.m.} \int \Phi_y^{(\pm)*}(x)\, g(x)\, d\mu(x), \qquad g \in \mathsf{R}_\pm \subset L^2(\mathbb{R}^m, \mu)$$

(4.53)

$$g(x) = (U^{(\pm)-1}\check{g}^{(\pm)})(x) = \text{l.i.m.} \int \Phi_y^{(\pm)}(x)\, \check{g}^{(\pm)}(y)\, d\nu(y), \qquad \check{g}^{(\pm)} \in U\mathsf{M}_\pm,$$

$$(U^{(\pm)} Y_j^{(\pm)} U^{(\pm)-1} \check{g})(y) = y_j \check{g}^{(\pm)}(y), \qquad j = 1, ..., n, \tag{4.54}$$

where* $U^{(\pm)}g = U\Omega_\pm^* g$ for all $g \in \mathsf{R}_\pm$, then the argument leading to (4.49) in the present context yields

$$(E^{H^\pm}(B)g)(x) = \text{l.i.m.} \int_{\Lambda^{-1}(B)} \Phi_y^{(\pm)}(x)\, \check{g}^{(\pm)}(y)\, d\nu(y). \tag{4.55}$$

By setting above $g = f_\pm = \Omega_\pm f$, so that $\check{g}^{(\pm)} = \check{f}$, we arrive at the following generalization of (4.25) and (4.41):

$$(E^{H^\pm}(B) f_\pm)(x) = \text{l.i.m.} \int_{\Lambda^{-1}(B)} \Phi_y^{(\pm)}(x) \check{f}(y)\, d\nu(y). \tag{4.56}$$

* In the notation introduced in (4.12) and (4.81), this condition ensures that $(\Phi_y^{(\pm)} \mid g\rangle = (\Phi_y \mid \Omega_\pm^* g\rangle$, so that in a sense (cf. Prugovečki [1973b]) $\Phi_y^{(\pm)} = (\Omega_\pm^{*+})\Phi_y$, where $\Omega_\pm^{*+}$ is the extension of $\Omega_\pm$ to a larger space $\mathscr{H}^+ \supset \mathscr{H}$.

Although in perturbation theory terminology $\Phi_y(x)$ and $\Phi_y^{(\pm)}(x)$ represent the eigenfunction expansions for the unperturbed and perturbed problem, respectively, motivated by their physical interpretation, the common names in physical literature are as follows: $\Phi_y(x)$ is called a *free wave*, $\Phi_y^{(+)}(x)$ an *outgoing* (or *retarded*) *distorted wave*, and $\Phi_y^{(-)}(x)$ an *incoming* (or *advanced*) *distorted wave*.

4.5. Green's Operators and Green Functions

There is an intimate relationship between the spectral measure $E^A(B)$ of a self-adjoint operator A and what in physical literature is called the *Green's operator*

$$G_A(\zeta) = -R_A(\zeta) = (\zeta - A)^{-1}, \qquad \zeta \in \mathbb{C}^1 - \mathsf{S}^A,$$

i.e., the negative value of the resolvent in resolvent set of A. Part of that connection will be investigated in §5.8, whereas in the present section we shall concentrate on those aspects in which eigenfunction expansions play a crucial role.

To make our remarks immediately applicable to both the case of H_0 as well as H in potential scattering, let us be general and assume that A acts in the spectral representation space $\mathscr{H} = L^2(\mathbb{R}^m, \mu)$ of a complete set $\{X_1, \ldots, X_m\}$, which we can always decompose into the direct sum

$$(4.57) \qquad \mathscr{H} = \mathscr{H}_\mathrm{p}{}^A \oplus \mathscr{H}_\mathrm{c}{}^A, \qquad \mathscr{H}_\mathrm{p}{}^A = E^A(\mathsf{S}_\mathrm{p}^A)\mathscr{H}, \qquad \mathscr{H}_\mathrm{c}{}^A = E^A(\mathsf{S}_\mathrm{c}^A)\mathscr{H}.$$

The restriction A_p of A to $\mathscr{H}_\mathrm{p}{}^A$ is an operator with a pure point spectrum, and therefore the considerations of §5.1 in Chapter III apply to it, so that on $\mathscr{H}_\mathrm{p}{}^A$

$$(4.58) \qquad E^{A_\mathrm{p}}(B) = E^A(B)\, E^A(\mathsf{S}_\mathrm{p}^A) = E^A(B \cap \mathsf{S}_\mathrm{p}^A) = \sum_{\lambda \in B \cap \mathsf{S}_\mathrm{p}{}^A} \sum_{\alpha=1}^{n_\lambda} |f_{\lambda,\alpha}\rangle\langle f_{\lambda,\alpha}|,$$

where $\{f_{\lambda,\alpha} : \alpha = 1, \ldots, n_\lambda\}$ is an orthonormal basis in $E^A(\{\lambda\})\mathscr{H}$. Consequently, by (3.59)

$$(4.59) \qquad \langle f \mid G_A(\zeta)\, E^A(\mathsf{S}_\mathrm{p}^A) g\rangle = \sum_{\lambda \in \mathsf{S}_\mathrm{p}{}^A} \sum_{\alpha=1}^{n_\lambda} \frac{\langle f \mid f_{\lambda,\alpha}\rangle\langle f_{\lambda,\alpha} \mid g\rangle}{\zeta - \lambda}.$$

Expressing the $L^2(\mathbb{R}^m, \mu)$ inner products on the right-hand side of the above relation in terms of integrals and introducing

$$(4.60) \qquad G_{A_\mathrm{p}}(x, x'; \zeta) = \operatorname{l.i.m.} \sum_{\lambda \in \mathsf{S}_\mathrm{p}{}^A} \sum_{\alpha=1}^{n_\lambda} \frac{f_{\lambda,\alpha}(x)\, f_{\lambda,\alpha}^*(x')}{\zeta - \lambda},$$

we can rewrite (4.59) in the form

$$\langle f \mid G_A(\zeta)\, E^A(S_p^A)g\rangle = \int d\mu(x)\, f^*(x) \int d\mu(x')\, G_{A_p}(x, x'; \zeta)\, g(x') \tag{4.61}$$

if we assume that the orders of summation and integration can be interchanged.

To arrive at a similar formula for the continuous spectrum of A, we shall assume that the restriction A_c of A to $\mathscr{H}_c^A$ is a function $F(Y_1, \ldots, Y_n)$ of a complete set $\{Y_1, \ldots, Y_n\}$ that possesses an eigenfunction expansion $\Phi_y(x)$ on $L^2(\mathbb{R}^m, \mu)$. Thus, by Theorem 5.3 in Chapter IV

$$(U(\zeta - A_c)^{-1}U^{-1}\check{g})(y) = \check{g}(y)/[\zeta - F(y)],$$

and consequently according to (4.45) and (4.46) for f and g from a dense set $\mathscr{F}$ (cf. Definition 4.1) in $L^2(\mathbb{R}^m, \mu)$, we shall have

$$\begin{aligned} &\langle f \mid G_A(\zeta)\, E^A(S_c^A)g\rangle \\ &\quad = \int_{\mathbb{R}^n} d\nu(y)[\zeta - F(y)]^{-1} \int_{\mathbb{R}^m} d\mu(x)\, f^*(x)\, \Phi_y(x) \int_{\mathbb{R}^m} d\mu(x')\, \Phi_y^*(x')\, g(x'). \end{aligned} \tag{4.62}$$

Assuming that the function

$$G_{A_c}(x, x'; \zeta) = \text{l.i.m.} \int_{\mathbb{R}^n} \frac{\Phi_y(x)\, \Phi_y^*(x')}{\zeta - F(y)}\, d\nu(y) \tag{4.63}$$

exists and that orders of integration can be interchanged in (4.62), we finally obtain from (4.61)–(4.63):

$$\langle f \mid G_A(\zeta)g\rangle = \int d\mu(x)\, f^*(x) \int d\mu(x')\, G_A(x, x'; \zeta)\, g(x'), \tag{4.64}$$

$$G_A(x, x'; \zeta) = G_{A_p}(x, x'; \zeta) + G_{A_c}(x, x'; \zeta). \tag{4.65}$$

The $\mu \times \mu$-almost everywhere unique (see Exercise 4.2) function $G_A(x, x'; \zeta)$ that satisfies (4.64) for $f, g \in L^2(\mathbb{R}^m, \mu)$ is called the *Green function* of A on the spectral representation space $L^2(\mathbb{R}^m, \mu)$. In Theorem 5.1 we shall explicitly compute the Green function $G_0(\mathbf{r}, \mathbf{r}'; \zeta)$ of the free two-body relative motion Hamiltonian H_0 in the configuration representation space $L^2(\mathbb{R}^3)$. In physical literature $G_0(\mathbf{r}, \mathbf{r}'; \zeta)$ is called the *free Green function.* As we shall see in §5.3, the *full* (or *total*) *Green function* $G(\mathbf{r}, \mathbf{r}'; \zeta)$ corresponding to H can be computed from the second resolvent equation (3.61) rewritten in terms of Green functions [see (4.76)].

A closely related concept is that of *retarded* and *advanced Green function* $G_A^{(+)}(x, x'; \lambda)$ and $G_A^{(-)}(x, x'; \lambda)$, respectively, defined for $\lambda \in \mathsf{S}_c^A$. If they exist, these two functions by definition have to satisfy the respective conditions

$$(4.66)\qquad \int_{\mathbb{R}^m} d\mu(x) f^*(x) \int_{\mathbb{R}^m} d\mu(x')\, G_A^{(\pm)}(x, x'; \lambda)\, g(x')$$
$$= \lim_{\epsilon\to\pm 0} \int_{\mathbb{R}^m} d\mu(x) f^*(x) \int_{\mathbb{R}^m} d\mu(x')\, G_A(x, x'; \lambda \pm i\epsilon)\, g(x')$$

for at least a dense set of values $f, g \in L^2(\mathbb{R}^m, \mu)$.

4.6. Lippmann–Schwinger Equations for Eigenfunction Expansions

By taking advantage of the fact that, as we shall establish in Theorem 5.1, the free Green function $G_0(\mathbf{r}, \mathbf{r}'; \zeta)$, $\zeta \in \mathbb{C}^1 - [0, \infty)$, is a known quantity in two-body potential scattering, and that the corresponding advanced and retarded Green functions $G_0^{(\pm)}(\mathbf{r}, \mathbf{r}'; \lambda)$, $\lambda \in [0, \infty)$, exist, we derive from (3.34) the *Lippmann–Schwinger equations*

$$(4.67)\qquad \Phi_{\mathbf{k}}^{(\pm)}(\mathbf{r}) = \Phi_{\mathbf{k}}(\mathbf{r}) + \int_{\mathbb{R}^3} G_0^{(\pm)}(\mathbf{r}, \mathbf{r}'; \mathbf{k}^2/2m)\, V(\mathbf{r}')\, \Phi_{\mathbf{k}}^{(\pm)}(\mathbf{r}')\, d\mathbf{r}'$$

for the distorted plane waves $\Phi_{\mathbf{k}}^{(\pm)}(\mathbf{r})$ introduced in §4.2. As we shall see in §6.2, (4.67) provides a viable method of computing $\Phi_{\mathbf{k}}^{(\pm)}(\mathbf{r})$ for a wide class of short-range potentials.

First of all, we note that by Exercise 2.7 (and more generally by Theorem 7.1 in §7), $\mathsf{M}_{\pm} = L^2(\mathbb{R}^3)$ for the considered class of potentials, which includes all short-range potentials of practical interest (see Appendix 4.9 for long-range potentials). Hence, we can rewrite (3.34) in the form

$$(4.68)\qquad \langle g \mid f_\pm\rangle - \langle g \mid f\rangle$$
$$= \lim_{\epsilon\to+0} \left\langle g \,\middle|\, \int_{-\infty}^{+\infty} G_0(\lambda \pm i\epsilon) V\, d_\lambda E_\lambda{}^H f_\pm \right\rangle, \qquad f_\pm = \Omega_\pm f,$$

where by (4.11) and (4.19)

$$(4.69)\qquad \langle g \mid f_\pm - f\rangle = \int_{\mathbb{R}^3} d\mathbf{r} g^*(\mathbf{r}) \int_{\mathbb{R}^3} d\mathbf{k} \tilde{f}(\mathbf{k})[\Phi_{\mathbf{k}}^{(\pm)}(\mathbf{r}) - \Phi_{\mathbf{k}}(\mathbf{r})].$$

The key idea in the derivation of (4.67) is to show that

$$\left\langle g \middle| \int_{-\infty}^{+\infty} G_0(\lambda \pm i\epsilon) V \, d_\lambda E_\lambda{}^H f_\pm \right\rangle \tag{4.70}$$

$$= \int_{\mathbb{R}^3} d\mathbf{r} g^*(\mathbf{r}) \int_{\mathbb{R}^3} d\mathbf{k} \tilde{f}(\mathbf{k}) \int_{\mathbb{R}^3} d\mathbf{r}' G_0(\mathbf{r}, \mathbf{r}'; \mathbf{k}^2/2m \pm i\epsilon) \, V(\mathbf{r}') \, \Phi_{\mathbf{k}}^{(\pm)}(\mathbf{r}')$$

for all functions $g(\mathbf{r}) \in \mathscr{C}_{\mathrm{b}}{}^0(\mathbb{R}^3)$ and $\tilde{f}(\mathbf{k}) \in \mathscr{C}_V{}^0 \subset \mathscr{C}_{\mathrm{b}}{}^0(\mathbb{R}^3)$. $\mathscr{C}_V{}^0$ consists, by definition, of all continuous functions whose compact support is disjoint from an exceptional set $\mathscr{S}_V^3$ (assumed to be of Lebesgue measure zero—see §§6.2 and 6.5) of values $\mathbf{k} \in \mathbb{R}^3$ at which $\Phi_{\mathbf{k}}^{(\pm)}(\mathbf{r})$ might not exist, and is such that for $\tilde{\psi}(\mathbf{k}) \in \mathscr{C}_V{}^0$, (4.19) and (4.24) become

$$\psi_\pm(\mathbf{r}) = (\Omega_\pm \psi)(\mathbf{r}) = \int_{\mathbb{R}^3} \Phi_{\mathbf{k}}^{(\pm)}(\mathbf{r}) \, \tilde{\psi}(\mathbf{k}) \, d\mathbf{k}.$$

Then, upon taking in (4.70) the limit $\epsilon \to +0$ we can argue that the outcome can equal (4.69) if and only if the square bracket in (4.69) equals the last integral in (4.70).

Indeed, by (4.25) and Definition 3.4, for any a, $b \in \mathbb{R}^1$, $a < b$,

$$\left\langle g \middle| \int_a^b G_0(\lambda \pm i\epsilon) V \, d_\lambda E_\lambda{}^H f_\pm \right\rangle \tag{4.71}$$

$$= \lim_{\delta \to 0} \sum_{j=1}^{n} \int_{\mathbb{R}^3} d\mathbf{r} g^*(\mathbf{r}) \int_{\mathbb{R}^3} d\mathbf{r}' G_0(\mathbf{r}, \mathbf{r}'; \lambda_j' \pm i\epsilon) \, V(\mathbf{r}')$$

$$\times \int_{\lambda_{j-1} < \mathbf{k}^2/2m \leqslant \lambda_j} d\mathbf{k} \Phi_{\mathbf{k}}^{(\pm)}(\mathbf{r}') \tilde{f}(\mathbf{k}),$$

where the notable fact is that $\lambda_j' \in (\lambda_{j-1}, \lambda_j]$ so that, heuristically speaking, λ_j' and $\mathbf{k}^2/2m$ merge together in the limit $\delta \to 0$. On the rigorous level, we have to assume that $\Phi_{\mathbf{k}}^{(\pm)}(\mathbf{r})$ is uniformly bounded in $\mathbf{r}$ when $\mathbf{k}$ is restricted to supp $\tilde{f}$, which is compact. This condition is certainly satisfied by $\Phi_{\mathbf{k}}(\mathbf{r})$, and in §6 we shall establish that it is also satisfied by $\Phi_{\mathbf{k}}^{(\pm)}(\mathbf{r})$ for a large class of potentials.* The potentials in that class are locally square integrable and of faster than r^{-3} decrease as $r \to \infty$, so that considering the form (5.6) of the free Green function we see

* This will be achieved by solving the Lippmann–Schwinger equation (4.67), yet the reasoning is not circular. The procedure is rather analogous to, say, the establishing of power series solutions for some ordinary linear differential equations (see Ince [1956, Chapter XVI]), whereby one starts by assuming the existence of such solutions and upon computing the coefficients, one proves that the assumption was justified.

that the interchange in orders of integration in $\mathbf{r}'$ and $\mathbf{k}$ can be carried out in (4.71) due to Fubini's theorem (see Exercise 4.3). Furthermore, Lebesgue's dominated convergence theorem can be invoked to establish that the right-hand sides in (4.70) and (4.71) are indeed equal, since upon taking the sum under the integral sign in (4.71) the resulting integrand is majorized in $\mathbf{r}$, $\mathbf{r}'$, $\mathbf{k} \in \mathbb{R}^3$ by the integrable function

$$\text{(4.72)} \qquad \text{const.}\, |g(\mathbf{r})|\, |V(\mathbf{r}')|\, |\mathbf{r}-\mathbf{r}'|^{-1}\, |\tilde{f}(\mathbf{k})| \exp[-(2m\epsilon)^{1/2} |\mathbf{r}-\mathbf{r}'|].$$

Finally, upon inserting the outcome into the right-hand side of (4.68), the same theorem can be applied again (see Exercise 4.3).

Obviously, the main ideas behind the derivation of (4.67) are not confined to position and momentum observables, but are equally well applicable to arbitrary eigenfunction expansions $\Phi_y(x)$ (see Exercise 4.4), and in particular to partial waves, when the role of $\{Y_1,..., Y_n\}$ is played by $\{|\mathbf{P}|, |\mathbf{L}|, L^{(z)}\}$, and that of $\{X_1,..., X_m\}$ by the position observables $\{|\mathbf{Q}|, \Theta_\mathbf{Q}, \Phi_\mathbf{Q}\}$ corresponding to the spherical coordinates (r, θ, ϕ). If in this instance the potential is spherically symmetric, we easily arrive at the following *partial wave Lippmann–Schwinger equations*,

$$\text{(4.73)} \qquad \Phi^{(\pm)}_{klm}(r,\theta,\phi) = \Phi_{klm}(r,\theta,\phi) + \int_0^\infty G_0^{(\pm)l}(r,r';k^2/2m)\, V(r')\, \Phi^{(\pm)}_{klm}(r',\theta,\phi)\, r'^2\, dr'.$$

where $G_0^{(\pm)l}(r, r')$ is given in (5.22).

The solution-type Lippmann–Schwinger equations (3.41) in Hilbert space also have their counterparts for eigenfunction expansions. For example, duplicating the argument leading to (4.67) with (3.41) rather than (3.34) as a starting point, we obtain

$$\text{(4.74)} \qquad \Phi^{(\pm)}_\mathbf{k}(\mathbf{r}) = \Phi_\mathbf{k}(\mathbf{r}) + \int_{\mathbb{R}^3} G^{(\pm)}(\mathbf{r},\mathbf{r}';\mathbf{k}^2/2m)\, V(\mathbf{r}')\, \Phi_\mathbf{k}(\mathbf{r}')\, d\mathbf{r}'.$$

The only difference is that in the present derivation we have to make assumptions about the full retarded and advanced Green functions $G^{(\pm)}(\mathbf{r}, \mathbf{r}'; \lambda)$, rather than about $\Phi^{(\pm)}_\mathbf{k}(\mathbf{r})$. In §5.4 we shall see, however, that these functions can be computed by means of the second resolvent equation (3.65), which can be written in the form

$$\text{(4.75)} \qquad \langle g \mid G(\zeta) f\rangle = \langle g \mid G_0(\zeta) f\rangle + \langle g \mid G_0(\zeta)\, H_1 G(\zeta) f\rangle, \qquad f, g \in \mathscr{H}.$$

Indeed, upon expressing these inner products on the spectral representation space of, in general, a complete set $\{X_1,..., X_m\}$, and considering that (4.75) holds for all $f(x)$, $g(x) \in \mathscr{C}_b{}^0(\mathbb{R}^m)$, we infer that

$$G(x, x'; \zeta) = G_0(x, x'; \zeta) + \int G_0(x, x''; \zeta)[H_1^{x''} G(x'', x'; \zeta)]\, d\mu(x''). \tag{4.76}$$

As we shall see in Theorem 5.6, in two-body potential scattering the equation (4.76) can be solved by the Fredholm method. Combined with (4.74) this method provides a means for the computation of distorted waves that is totally equivalent to the method based on solving the Lippmann–Schwinger equations (4.67).

4.7. The On-Shell T-Matrix

Upon introducing the spectral integral

$$\begin{aligned} T_\epsilon &= \frac{1}{\pi}\int_{-\infty}^{+\infty} d_\lambda E_\lambda^{H_0} H_1 \Omega_+ \frac{\epsilon}{(H_0 - \lambda)^2 + \epsilon^2} \\ &= \frac{1}{\pi}\int_{-\infty}^{+\infty} d_\lambda E_\lambda^{H_0} H_1 \frac{\epsilon}{(H - \lambda)^2 + \epsilon^2}\, \Omega_+ , \end{aligned} \tag{4.77}$$

we see that (3.52) is equivalent to the simple statement that

$$T = \operatorname*{w-lim}_{\epsilon \to +0} T_\epsilon . \tag{4.78}$$

To recast (4.77) in terms of eigenfunction expansions, we note that by (4.48) and by Theorems 2.3 and 5.3 in Chapter IV,

$$\langle g \mid E^{H_0}(B) f\rangle = \int_{\Lambda^{-1}(B)} \check{g}^*(y) \check{f}(y)\, d\nu(y). \tag{4.79}$$

Upon generalizing the notation in (4.12), we write

$$\langle g \mid E^{H_0}(B) f\rangle = \int_{\Lambda^{-1}(B)} \langle g \mid \Phi_y)(\Phi_y \mid f\rangle\, d\nu(y), \tag{4.80}$$

$$(\Phi_y \mid f\rangle = \text{l.i.m.} \int_{\mathbb{R}^m} \Phi_y{}^*(x) f(x)\, d\mu(x) = \langle f \mid \Phi_y)^*, \tag{4.81}$$

or using a simpler, symbolic notation, we have

$$E^{H_0}(B) = \int_{\Lambda^{-1}(B)} \mid \Phi_y)\, d\nu(y) (\Phi_y \mid . \tag{4.82}$$

Thus, proceeding formally, we see that (4.77) can be written as

$$T_\epsilon = \frac{1}{\pi}\int \mid \Phi_y)\, d\nu(y) (\Phi_y \mid H_1 \Omega_+ \frac{\epsilon}{[H_0 - \Lambda(y)]^2 + \epsilon^2} . \tag{4.83}$$

The exact meaning of this expression follows from the next general result.

Lemma 4.1. Suppose that $F(\lambda)$ is a strongly continuous operator-valued function for which $\| F(\lambda) \| \leqslant C$, that $(H_1{}^x\Phi_y)(x)$ is μ-square integrable in $x \in \mathbb{R}^m$, and that $H_1{}^x\Phi_y \in L^2(\mathbb{R}^m, \mu)$ represents a strongly continuous vector-valued function in $y \in \mathrm{S} \subset \mathbb{R}^n$. If $\Lambda(y)$ is continuous and $\check{g}(y) \in \mathscr{C}_{\mathrm{b}}{}^0(\mathbb{R}^n)$ then

$$\left\langle g \,\middle|\, \int_{-\infty}^{+\infty} d_\lambda E_\lambda^{H_0} H_1 F(\lambda) f \right\rangle = \int_{\mathbb{R}^n} \check{g}^*(y) \langle H_1{}^x\Phi_y \mid F(\Lambda(y)) f \rangle \, d\nu(y). \tag{4.84}$$

Proof. In view of the strong continuity of $F(\Lambda(y))$ and $H_1{}^x\Phi_y$,

$$\langle H_1{}^x\Phi_y \mid F(\Lambda(y)) f \rangle = \int_{\mathbb{R}^m} [H_1{}^x\Phi_y{}^*(x)][F(\Lambda(y)) f](x) \, d\mu(x) \tag{4.85}$$

is continuous in y, so that the integral on the right-hand side of (4.84) certainly exists if $\check{g}(y) \in \mathscr{C}_{\mathrm{b}}{}^0(\mathbb{R}^n)$. On the other hand, by (4.82) and Definition 3.4,

$$\begin{aligned} &\left\langle g \,\middle|\, \int_a^b d_\lambda E_\lambda^{H_0} H_1 F(\lambda) f \right\rangle \\ &\quad = \lim_{\delta \to 0} \sum_{j=1}^n \int_{\lambda_{j-1} < \Lambda(y) \leqslant \lambda_j} \langle g \mid \Phi_y) \, d\nu(y) (\Phi_y \mid H_1 F(\lambda_j') f \rangle \end{aligned} \tag{4.86}$$

and this limit can be shown (see Exercise 4.5) to equal that same integral when we let $a \to -\infty$ and $b \to +\infty$. Q.E.D.

Under the restrictions imposed in §6 on $V(\mathbf{r})$ (see also Exercise 4.3), the conditions of Lemma 4.1 are satisfied, so we can certainly write

$$\langle g \mid T_\epsilon f \rangle = \frac{1}{\pi} \int_{\mathbb{R}^3} \langle g \mid \Phi_{\mathbf{k}}) \left\langle V\Phi_{\mathbf{k}} \,\middle|\, \Omega_+ \frac{\epsilon}{[H_0 - (\mathbf{k}^2/2m)]^2 + \epsilon^2} f \right\rangle d\mathbf{k} \tag{4.87}$$

whenever $\tilde{g}(\mathbf{k}) \in \mathscr{C}_{\mathrm{b}}{}^0(\mathbb{R}^3)$. By (4.19) for $\tilde{f}(\mathbf{k}') \in \mathscr{C}_{\mathrm{b}}{}^0(\mathbb{R}^3 - \mathscr{S}_V{}^3)$,

$$\begin{aligned} &\left(\Omega_+ \frac{\epsilon}{[H_0 - (\mathbf{k}^2/2m)]^2 + \epsilon^2} f\right)(\mathbf{r}) \\ &\quad = \int_{\mathbb{R}^3} \Phi_{\mathbf{k}'}^{(+)}(\mathbf{r}) \frac{\epsilon \tilde{f}(\mathbf{k}')}{[(\mathbf{k}'^2/2m) - (\mathbf{k}^2/2m)]^2 + \epsilon^2} \, d\mathbf{k}' \end{aligned}$$

provided that (as shown in §6) $\Phi_{\mathbf{k}}^{(+)}(\mathbf{r})$ has the same continuity and asymptotic properties as $\Phi_{\mathbf{k}}(\mathbf{r})$. Thus, under the same conditions,

$$\left\langle V\Phi_{\mathbf{k}} \mid \Omega_+ \frac{\epsilon}{[H_0 - (\mathbf{k}^2/2m)]^2 + \epsilon^2} f\right\rangle \tag{4.88}$$

$$= \int_{\mathbb{R}^3} d\mathbf{k}' \frac{\epsilon\langle\Phi_{\mathbf{k}} \mid f\rangle}{[(\mathbf{k}'^2/2m) - (\mathbf{k}^2/2m)]^2 + \epsilon^2} \int_{\mathbb{R}^3} d\mathbf{r}\, V(\mathbf{r})\, \Phi_{\mathbf{k}}{}^*(\mathbf{r})\, \Phi_{\mathbf{k}'}^{(+)}(\mathbf{r}),$$

since orders of integration can be interchanged by Fubini's theorem if we assume the integrability of $V(\mathbf{r})$ on $\mathbb{R}^3$, as that assures the existence of the following integral for $\mathbf{k} \in \mathbb{R}^3$ and $\mathbf{k}' \in \mathbb{R}^3 - \mathscr{S}_V{}^3$:

$$(\Phi_{\mathbf{k}} \mid V \mid \Phi_{\mathbf{k}'}^{(+)}) = \int_{\mathbb{R}^3} \Phi_{\mathbf{k}}{}^*(\mathbf{r})\, V(\mathbf{r})\, \Phi_{\mathbf{k}'}^{(+)}(\mathbf{r})\, d\mathbf{r}. \tag{4.89}$$

The matrix-type notation on the left-hand side of (4.89) is a variant of the Dirac-type of notation that is in very wide use in physical literature, and the function of $\mathbf{k}, \mathbf{k}' \in \mathbb{R}^3$ given by (4.89) is referred to as a *transition matrix* or *T-matrix*. By using this expression and (4.88), and inserting the outcome in (4.87), we get

(4.90)

$$\langle g \mid T_\epsilon f\rangle = \int_{\mathbb{R}^3} d\mathbf{k} \langle g \mid \Phi_{\mathbf{k}}) \int_{\mathbb{R}^3} d\mathbf{k}'\, \delta_\epsilon\left(\frac{\mathbf{k}^2}{2m} - \frac{\mathbf{k}'^2}{2m}\right) (\Phi_{\mathbf{k}} \mid V \mid \Phi_{\mathbf{k}'}^{(+)})(\Phi_{\mathbf{k}'} \mid f\rangle,$$

where we have introduced the abbreviation

$$\delta_\epsilon(\lambda - \lambda') = \pi^{-1}\{\epsilon/[(\lambda - \lambda')^2 + \epsilon^2]\}. \tag{4.91}$$

In (4.90) we can then explicitly take the limit $\epsilon \to +0$ whenever the T-matrix (4.89) is continuous in $\mathbf{k} \in \mathbb{R}^3$ and $\mathbf{k}' \in \mathbb{R}^3 - \mathscr{S}_V{}^3$ so that the following lemma is applicable.

Lemma 4.2. If the function $f(x)$ is bounded on any finite interval in $[0, +\infty)$ (or $(-\infty, 0]$) and such that its limit $f(+0)$ from the right [or $f(-0)$ from the left] exists at the origin, and if $\delta_\epsilon(x)f(x)$ is integrable on $[0, +\infty)$ (or $(-\infty, 0]$) for any $0 < \epsilon \leqslant \epsilon_0$, then

$$\lim_{\epsilon \to +0} (2/\pi) \int_0^{\pm\infty} [\epsilon/(x^2 + \epsilon^2)]\, f(x)\, dx = \pm f(\pm 0). \tag{4.92}$$

Proof. To prove the lemma for the first case of integrals over $[0, +\infty)$, choose for given $\eta > 0$ some $N(\eta) > 0$ so large that

$$(2/\pi) \int_{N(\eta)}^{+\infty} [\epsilon_0/(x^2 + \epsilon_0{}^2)]\, |f(x)|\, dx < \eta,$$

and also such that whenever $0 < \epsilon \leqslant \epsilon_0$, we have

$$\delta_\epsilon(x) = \epsilon/\pi(x^2 + \epsilon^2) \leqslant \epsilon_0/\pi(x^2 + \epsilon_0{}^2), \qquad x \geqslant N(\eta).$$

Consequently, for any δ such that $0 < \delta \leqslant N(\epsilon)$ we have

$$\left| \int_0^{N(\eta)} \delta_\epsilon(x) f(x)\, dx \right| \leqslant \epsilon \int_0^{N(\eta)} [| f(x)|/x^2]\, dx \leqslant N(\eta)\, M(\eta)(\epsilon/\delta^2),$$

where $M(\eta) = \sup\{| f(x) | : 0 \leqslant x \leqslant N(\eta)\}$. Since

$$\int_0^\delta [\epsilon/(x^2 + \epsilon^2)]\, dx = \arctan(\delta/\epsilon),$$

we obtain by using the generalized mean-value theorem of integral calculus

$$\left| \frac{f(+0)}{2} - \int_0^{N(\eta)} \delta_\epsilon(x) f(x)\, dx \right|$$
$$\leqslant \left| \frac{f(+0)}{2} - \frac{\gamma(\epsilon, \delta)}{\pi} \arctan \frac{\delta}{\epsilon} \right| + N(\eta)\, M(\eta) \frac{\epsilon}{\delta^2},$$

where $\inf\{f(x) : 0 \leqslant x \leqslant \delta\} \leqslant \gamma(\epsilon, \delta) \leqslant \sup\{f(x) : 0 \leqslant x \leqslant \delta\}$, so that $\gamma(\epsilon, \delta) \to f(+0)$ uniformly in ϵ as $\delta \to +0$. Thus, both terms on the right-hand side of the above inequality can be made smaller than η by choosing some appropriate $\delta = \delta(\eta)$, then making $\epsilon/\delta(\eta)$ sufficiently small, and taking advantage of $2 \arctan x$ approaching π as $x \to +\infty$.
The proof for the case of $f(-0)$ runs along similar lines. Q.E.D.

We note that if $f(x)$ is actually continuous at the origin, then

$$\lim_{\epsilon \to +0} \int_{-\infty}^{+\infty} \delta_\epsilon(x) f(x)\, dx = f(0) = \int_{-\infty}^{+\infty} \delta(x) f(x)\, dx, \tag{4.93}$$

where, following a widespread custom, we have employed Dirac's $\delta(x)$ symbol, albeit this so-called δ-"function" is interpreted in mathematical literature as a generalized function or distribution,* whereas (4.93) is valid even when $f(x)$ is not a so-called test-function. This notation has enabled us, however, to formally combine (4.89) and (4.90) by writing

$$\langle g \mid Tf \rangle = \int_{\mathbb{R}^6} \tilde{g}^*(\mathbf{k})\, \delta\Big(\frac{\mathbf{k}^2}{2m} - \frac{\mathbf{k}'^2}{2m}\Big) \langle \mathbf{k} \mid T \mid \mathbf{k}' \rangle \tilde{f}(\mathbf{k}')\, d\mathbf{k}\, d\mathbf{k}', \tag{4.94}$$

* See Dennery and Krzywicki [1967] for an elementary introduction, and Gel'fand and Vilenkin [1968] or Treves [1967] for a systematic treatment.

where, again according to widespread custom, we have introduced the following Dirac-type notation for the T-matrix (4.89):

$$\langle \mathbf{k} \mid T \mid \mathbf{k}' \rangle = (\Phi_{\mathbf{k}} \mid V \mid \Phi_{\mathbf{k}'}^{(+)}) = (\Phi_{\mathbf{k}} \mid V\Omega_+ \mid \Phi_{\mathbf{k}'}). \tag{4.95}$$

In a rigorous approach to deriving (4.94) we can make in (4.40) the transition to spherical coordinates for $\mathbf{k}'$, and afterwards apply (4.93) to the variable $k' = |\mathbf{k}'|$ when taking the limit $\epsilon \to +0$. The easily computable outcome for $\tilde{g}(\mathbf{k}) \in \mathscr{C}_b{}^0(\mathbb{R}^3)$ and $\tilde{f}(\mathbf{k}) \in \mathscr{C}_b{}^0(\mathbb{R}^3 - \mathscr{S}_V{}^3)$ is

$$\langle g \mid Tf \rangle = m \int_{\mathbb{R}^3} d\mathbf{k}\, \tilde{g}^*(\mathbf{k}) k \int_{\Omega_S} d\omega' \langle k\hat{\mathbf{k}} \mid T \mid k\hat{\mathbf{k}}' \rangle \tilde{f}(k\hat{\mathbf{k}}'), \qquad \hat{\mathbf{k}} = \mathbf{k}/k, \tag{4.96}$$

where $\hat{\mathbf{k}}'$ is the unit vector in direction of $\mathbf{k}'$, and it has the angular spherical coordinates $\omega' = (\theta', \phi')$. A comparison of (1.31) and (4.96) reveals that

$$T^{(1)}(k; \omega, \omega') = mk\langle k\hat{\mathbf{k}} \mid T \mid k\hat{\mathbf{k}}' \rangle \tag{4.97}$$

where ω and ω' are the angular spherical coordinates of $\hat{\mathbf{k}}$ and $\hat{\mathbf{k}}'$, respectively. Consequently, the formula (1.47) for the differential cross section $\sigma(\mathbf{p} \to \hat{\mathbf{p}}')$ of particles of (relative to the target) momentum $\mathbf{p}$ scattering off a planar target in the direction of the unit vector $\boldsymbol{\omega}'$ (with angular spherical coordinates ω') assumes the form

$$\sigma(\mathbf{p} \to \boldsymbol{\omega}') = \sigma(\mathbf{p}^2/2m, \omega') = (2\pi)^4 m^2 |\langle p\boldsymbol{\omega}' \mid T \mid \mathbf{p} \rangle|^2. \tag{4.98}$$

Due to the relation (6.67) (to be established as part of Theorem 6.3) between the T matrix and the scattering amplitude $f_{\mathbf{p}}(\omega')$ we arrive at the simple yet fundamental formula

$$\sigma(\mathbf{p} \to \boldsymbol{\omega}') = |f_{\mathbf{p}}(\omega')|^2, \qquad \omega' = (\theta', \phi'). \tag{4.99}$$

A notable feature of the differential cross-section formula (4.98) is that it involves only T-matrix values $\langle \mathbf{p}' \mid T \mid \mathbf{p} \rangle$ for which $|\mathbf{p}'| = |\mathbf{p}|$. From the physical point of view this fact is taken to reflect energy conservation. Upon introducing in $\mathbb{R}^6$ the surface of points $(\mathbf{p}', \mathbf{p})$ for which $\mathbf{p}'^2/2m = \mathbf{p}^2/2m$ (which in physical literature is called the *energy shell*), it can be said that (4.98) involves only on-shell values of the T matrix.

4.8. The Off-Shell T Matrix and $\mathscr{T}$ Operators

The obvious approach to computing the T matrix for a given potential $V(\mathbf{r})$ is to solve (e.g., by the methods discussed in §6) the Lippmann–

Schwinger equation (4.67) for the distorted plane wave $\Phi_{\mathbf{k}}^{(+)}(\mathbf{r})$ and then insert the result into (4.89). If, however, the full Green function $G(\mathbf{r}, \mathbf{r}'; \zeta)$ is known, then the solution-type Lippmann–Schwinger equation (4.74) can be used to obtain directly the desired solution from (4.89):

$$\text{(4.100)}\quad \langle \mathbf{k}' \mid T \mid \mathbf{k}\rangle = \lim_{\epsilon\to+0} (\Phi_{\mathbf{k}'} \mid V\{1 + G[(\mathbf{k}^2/2m) + i\epsilon]V\} \mid \Phi_{\mathbf{k}})$$

$$= \int_{\mathbb{R}^3} d\mathbf{r}\Phi_{\mathbf{k}'}^*(\mathbf{r})\, V(\mathbf{r})$$

$$\times \left[\Phi_{\mathbf{k}}(\mathbf{r}) + \int_{\mathbb{R}^3} G^{(+)}(\mathbf{r}, \mathbf{r}'; \mathbf{k}^2/2m)\, V(\mathbf{r}')\, \Phi_{\mathbf{k}}(\mathbf{r}')\, d\mathbf{r}'\right].$$

Equation (4.100) suggests the introduction of the following $\mathscr{T}(\zeta)$ *operators**

$$\text{(4.101)}\qquad \mathscr{T}(\zeta) = V + VG(\zeta)V, \qquad \zeta \in \mathbb{C}^1 - \mathsf{S}^H,$$

which consitute a family of in general unbounded operators in $L^2(\mathbb{R}^3)$ when ζ varies over the resolvent set of the total Hamiltonian H.

When the free Green's operator is applied to equation (4.101) from the left we obtain

$$\text{(4.102)}\qquad G_0(\zeta)\,\mathscr{T}(\zeta) = [G_0(\zeta) + G_0(\zeta)\, VG(\zeta)]V = G(\zeta)V,$$

where the last expression follows from the second resolvent equation (3.61) recast in terms of Green's operators. We substitute this expression for $G(\zeta)V$ into (4.101) and arrive at the relation

$$\text{(4.103)}\qquad \mathscr{T}(\zeta) = V + VG_0(\zeta)\,\mathscr{T}(\zeta), \qquad \zeta \in \mathbb{C}^1 - \mathsf{S}^H,$$

known as the *Lippmann–Schwinger equation for* $\mathscr{T}(\zeta)$.

Although the $\mathscr{T}(\zeta)$ operators are well defined in $L^2(\mathbb{R}^3)$, their primary importance stems from their close relation to the so-called *off-shell T matrix*

$$\text{(4.104)}\qquad \langle \mathbf{p}' \mid \mathscr{T}(\zeta) \mid \mathbf{p}\rangle = (\Phi_{\mathbf{p}'} \mid V + VG(\zeta)V \mid \Phi_{\mathbf{p}}),$$

which, generally speaking, is a function of the real variables $\mathbf{p}$, $\mathbf{p}' \in \mathbb{R}^3$, and the complex variable $\zeta \in \mathbb{C}^1 - \mathsf{S}^H$. The interest of this quantity

* It is the custom in physical literature to refer to (4.101) as "the T operator," albeit one is actually dealing with an operator-valued function. To avoid notational confusion, authors who follow this custom (see, e.g., Taylor [1972], Amrein *et al.* [1977]) instead of dealing with the transition operator (3.50) work with the "R-operator" $R = -2\pi i T$.

lies in the possibility of deriving from (4.103) the integral equation (see Exercise 4.6)

$$\langle \mathbf{p}' \mid \mathcal{T}(\zeta) \mid \mathbf{p} \rangle = \langle \mathbf{p}' \mid V \mid \mathbf{p} \rangle + \int_{\mathbb{R}^3} \frac{\langle \mathbf{p}' \mid V \mid \mathbf{p}'' \rangle}{\zeta - (\mathbf{p}''^2/2m)} \langle \mathbf{p}'' \mid \mathcal{T}(\zeta) \mid \mathbf{p} \rangle \, d\mathbf{p}'' \tag{4.105}$$

where $\langle \mathbf{p}' \mid V \mid \mathbf{p} \rangle$ is proportional to the Fourier–Plancherel transform $\tilde{V}(\mathbf{p} - \mathbf{p}')$ of $V(\mathbf{r})$:

(4.106)

$$\langle \mathbf{p}' \mid V \mid \mathbf{p} \rangle = (\Phi_{\mathbf{p}'} \mid V \mid \Phi_{\mathbf{p}}) = (2\pi)^{-3} \operatorname{l.i.m.} \int_{\mathbb{R}^3} V(\mathbf{r}) \exp[i(\mathbf{p} - \mathbf{p}')\mathbf{r}] \, d\mathbf{r}.$$

Thus, a strategy often followed in computing the on-shell T matrix $\langle \mathbf{p}' \mid T \mid \mathbf{p} \rangle$ is to solve (4.105) for the off-shell T-matrix (4.104), then set $|\mathbf{p}| = |\mathbf{p}'|$ and $\zeta = (\mathbf{p}^2/2m) + i\epsilon$ in that solution, and thus finally obtain the desired quantity by letting $\epsilon \to +0$. This procedure is symbolized by the standard formal expression for the S matrix,

(4.107)

$$\langle \mathbf{p}' \mid S \mid \mathbf{p} \rangle = \delta^3(\mathbf{p}' - \mathbf{p}) - 2\pi i \, \delta\Big(\frac{\mathbf{p}'^2}{2m} - \frac{\mathbf{p}^2}{2m}\Big) \Big\langle \mathbf{p}' \Big| \mathcal{T}\Big(\frac{\mathbf{p}^2}{2m} + i0\Big) \Big| \mathbf{p} \Big\rangle$$

which, in fact, simply reflects the aforementioned computation of the on-shell T matrix

$$\langle \mathbf{p}' \mid T \mid \mathbf{p} \rangle = \lim_{\epsilon \to +0} \Big\langle \mathbf{p}' \Big| \mathcal{T}\Big(\frac{\mathbf{p}^2}{2m} + i\epsilon\Big) \Big| \mathbf{p} \Big\rangle, \tag{4.108}$$

the insertion of the outcome into (4.94), and finally the computation of the transition amplitude

$$\langle g \mid Sf \rangle = \langle g \mid f \rangle - 2\pi i \langle g \mid Tf \rangle, \tag{4.109}$$

using in the process the notationally expedient device of writing

$$\langle g \mid f \rangle = \int_{\mathbb{R}^3} \tilde{g}^*(\mathbf{p}) \tilde{f}(\mathbf{p}) \, d\mathbf{p} = \int_{\mathbb{R}^3} d\mathbf{p}' \tilde{g}^*(\mathbf{p}') \int_{\mathbb{R}^3} d\mathbf{p} \, \delta^3(\mathbf{p}' - \mathbf{p}) \tilde{f}(\mathbf{p}).$$

Although easily avoidable as well as, strictly speaking, mathematically unjustified when used as before in conjunction with functions $\tilde{f}(\mathbf{p})$, $\tilde{g}(\mathbf{p}) \in L_{(2)}(\mathbb{R}^3)$ (which, in general, might display discontinuities or even singular behavior on sets of Lebesgue measure zero), the δ-"function" notation has achieved general acceptance in physical literature. One other of the many illustrations of its usage can be found in the *generalized optical theorem*, which states that for $|\mathbf{p}| = |\mathbf{p}'|$,

$$(4.110)\quad \langle \mathbf{p}' \mid T \mid \mathbf{p} \rangle^* - \langle \mathbf{p}' \mid T \mid \mathbf{p} \rangle = 2\pi i \int_{\mathbb{R}^3} \langle \mathbf{p}'' \mid T \mid \mathbf{p}' \rangle^* \,\delta\left(\frac{\mathbf{p}'^2}{2m} - \frac{\mathbf{p}''^2}{2m}\right) \langle \mathbf{p}'' \mid T \mid \mathbf{p} \rangle \, d\mathbf{p}''.$$

Without the use of δ-symbolism, (4.110) can be written in terms of the scattering amplitude (4.20) as follows,

$$(4.111)\quad f_{\mathbf{p}}(\omega') - f_{\mathbf{p}}^{\,*}(\omega') = (ip/2\pi) \int_{\Omega_S} f_{\mathbf{p}'}^{*}(\omega'') f_{\mathbf{p}}(\omega'')\, d\omega'', \qquad |\mathbf{p}'| = |\mathbf{p}|,$$

where $\omega' = (\theta', \phi')$ are angular spherical coordinates of $\mathbf{p}' = p\boldsymbol{\omega}'$. The relation (4.111) is a reflection of the unitarity of the S operator, as can be easily established (see Exercise 4.7) by observing that

$$S^*S = 1 - 2\pi i(T - T^*) + 4\pi^2 T^*T = (SS^*)^*,$$

and therefore that $S^*S = SS^* = 1$ if and only if

$$(4.112)\qquad T^*T = (T^* - T)/2\pi i = TT^*.$$

By setting $\mathbf{p} = \mathbf{p}'$ in (4.111) we obtain the *optical theorem*

$$(4.113)\qquad \operatorname{Im} f_{\mathbf{p}}(\omega) = (p/4\pi) \int_{\Omega_S} \sigma(\mathbf{p} \to \omega'')\, d\omega'' = (p/4\pi)\, \sigma(E)$$

that relates the scattering amplitude $f_{p\boldsymbol{\omega}_0}(\omega_0)$ in the forward direction $\boldsymbol{\omega}_0$ to the total cross-section $\sigma(E)$ at energy $E = p^2/2m$.

Finally, it is interesting to note that if we use the solution-type Lippmann–Schwinger equation (4.74) in (4.90) and take (4.78) and (4.108) into consideration, we can then directly relate the transition operator T to the $\mathscr{T}(\zeta)$ operators by means of a double spectral integral,

$$(4.114)\quad T = \operatorname*{w-lim}_{\epsilon \to +0} \operatorname*{w-lim}_{\eta \to +0} \int_{-\infty}^{+\infty} d_\lambda E_\lambda^{H_0} \int_{-\infty}^{+\infty} \delta_\epsilon(\lambda - \mu)\, \mathscr{T}(\lambda + i\eta)\, d_\mu E_\mu^{H_0},$$

by applying Lemma 4.1 to express the $\mathbf{k}$ and $\mathbf{k}'$ integrations of (4.90) in terms of integrations with respect to the spectral function of H_0.

**4.9. Appendix: Scattering Theory for Long-Range Potentials

We have mentioned in §2.7 that the standard time-dependent theory developed in §§1–2 does not apply to long-range interactions corresponding in general to potentials of the form*

$$V_\gamma(\mathbf{r}) = V_0(\mathbf{r}) + \gamma r^{-\alpha}, \qquad \tfrac{1}{2} < \alpha \leqslant 1, \tag{4.115}$$

where $V_0(\mathbf{r})$ is some short-range potential of the kind to which the standard theory is applicable. The same remark is, of course, valid for the stationary theory.

When we have $V_0 \equiv 0$, $\alpha = 1$, and $\gamma = e_1 e_2$, we are dealing with the important case of the Coulomb interaction between two charges e_1 and e_2. Hence it is of both physical and mathematical interest to learn how the standard theory should be modified in order to cope with long-range cases.†

The rigorous time-dependent scattering theory for long-range potential is based on the observation (first made by Dollard [1964] for the Coulomb case) that the *modified* (or *renormalized*) *wave operators*

$$\mathring{\Omega}_\pm = \underset{t \to \mp\infty}{\text{s-lim}} \exp(iHt) \exp\{-i[H_0 t + D_\alpha(t; H_0)]\} \tag{4.116}$$

exist for the following (asymptotically unique) choices of operator-valued functions of H_0 and $t > 0$

$$D_\alpha(\pm t, H_0) = \begin{cases} \pm\gamma(m/2H_0)^{1/2} \ln(4tH_0), & \alpha = 1, \\ \pm \dfrac{\gamma m^{\alpha/2} t^{1-\alpha}}{(1-\alpha)(2H_0)^{\alpha/2}}, & \tfrac{1}{2} < \alpha < 1. \end{cases} \tag{4.117}$$

The most mathematically interesting feature of these operators is that they intertwine H_0 in (4.14) and $H = H_0 + V_\gamma$ given in (4.115), so that (2.3)–(2.5) are satisfied if $\Omega_\pm$ is replaced by $\mathring{\Omega}_\pm$, and perturbation theory can then be pursued in the ordinary manner. Physically, however, this fact does not suffice, but it turns out‡ that expectation values for

* The case of $0 < \alpha \leqslant \frac{1}{2}$ requires iterative techniques involving derivatives of $V_\gamma(\mathbf{r})$—see Buslaev and Matveev [1970].

† See Amrein [1974] for a review of early rigorous results. For some more recent results see, e.g., Lavine [1973], Klein and Zinnes [1973], Gibson and Chandler [1974], Cattapan *et al.* [1975], Ikebe [1975], Narnhofer [1975], Semon and Taylor [1975, 1976], Gersten [1976], Hörmander [1976], Masson and Prugovečki [1976], Rejto [1976], Saenz and Zachary [1976], Zachary [1976], Zorbas [1976a,b, 1977], Alsholm [1977], Kitada [1977], Saito [1977], Alt *et al.* [1978], Lapicki and Losonsky [1979], and Rosenberg [1979].

‡ See Manoukian and Prugovečki [1971] and Prugovečki [1971, 1976] for proofs concerning position, momentum, and angular momentum observables, and Amrein *et al.* [1970], Corbett [1970], and Zachary [1972] for proofs based on the "algebra of observables" approach.

the modified asymptotic states

$$\mathring{\Psi}^{\text{ex}}(t) = \exp\{i[H_0 t + D_\alpha(t; H_0)]\}\, \mathring{\Psi}^{\text{ex}}(0) \tag{4.118}$$

approach in the limit $t \to \pm\infty$ those of truly free asymptotic states of $\Psi(t)$,

$$\Psi^{\text{ex}}(t) = e^{iH_0 t}\Psi^{\text{ex}}(0), \qquad \Psi^{\text{ex}}(0) = \mathring{\Psi}^{\text{ex}}(0) = \mathring{\Omega}_\pm{}^*\Psi(0),$$

so that the physical asymptotic conditions (2.46) and (2.47) are satisfied by these free states $\Psi^{\text{ex}}(t)$. This in turn implies that all basic physical relationships that were studied in §1 can be recovered in the context of long-range interactions.

The oscillatory behavior of $\exp[iD_\alpha(t; H_0)]$ as a function of t and the existence of the strong limit in (4.116) implies that in the long-range case

$$\operatorname*{w-lim}_{t\to\mp\infty} e^{iHt}\, e^{-iH_0 t} = 0, \tag{4.119}$$

so that the Møller wave operators (2.1) have trivial initial domains $\mathsf{M}_\pm = \{\mathbf{0}\}$. This in turn implies that (see Prugovečki and Zorbas [1973a, Section 2]),

$$\mathring{\Psi}^\pm = \operatorname*{w-lim}_{\eta\to\pm 0} \int_{-\infty}^{+\infty} (\lambda - H_0 + i\eta)^{-1}\, V_\gamma d_\lambda E_\lambda^H \mathring{\Psi}^\pm; \qquad \mathring{\Psi}^\pm = \mathring{\Omega}_\pm \Psi, \tag{4.120}$$

so that the Lippmann–Schwinger equations (3.34) are not satisfied. It is, of course, possible to reproduce the considerations of §§3.1–3.2 adopting (4.116) as a starting point (see Prugovečki [1971b], Chandler and Gibson [1974]), and thus arrive at a modified version of (3.34) corresponding to a replacement of the integrand of the spectral integral in (3.33) by

$$\eta \int_0^{+\infty} \exp[i(H_0 - \lambda \mp i \mid \eta \mid)t + iD_\alpha(t; H_0)]\, dt - 1,$$

but from a computational point of view one runs into difficulties with explicitly taking the limit $\eta \to \pm 0$, due to the presence of oscillatory terms generated by $\exp[iD_\alpha(t; H_0)]$.

One way of eliminating these oscillatory terms (characteristic of long-range problems) is to introduce an *asymptotically compensating operator* Z (see Matveev and Skriganov [1972], Prugovečki and Zorbas [1973a, b]) in terms of which we can rewrite (4.116) as follows:

$$\mathring{\Omega}_\pm \Psi = \operatorname*{s-lim}_{t\to\mp\infty} e^{iHt} Z e^{-iH_0 t}\Psi. \tag{4.121}$$

Formally, we are then faced with the same kind of expressions as in the two-Hilbert space formulations of multichannel scattering theory described in §8.6, the difference being that the embedding operator J has to be replaced by the asymptotically compensating operator Z. Indeed, one can show (Prugovečki and Zorbas [1973a, b]) that the equations (8.71) with Z replacing J hold, and that they can be solved by iterative methods (Masson and Prugovečki [1976]).

The preceding outline indicates that two-body scattering theory for long-range potentials can be brought in line with the short-range case. However, these results are actually of little practical interest, since due to screening effects (such as those for the Coulomb potential of a nucleus by the electron cloud around it) one can always replace γ in (4.115) by a *screening* (or *space cutoff*) function $\gamma(\mathbf{r})$ that equals γ on some neighborhood of the origin, but rapidly decreases to zero as $|\mathbf{r}| \to \infty$. Of course, we can incorporate those particles that give rise to screening into the problem, but then one is dealing with a many-body problem, so that the phenomenological approach based on a screening function presents great computational advantages. Furthermore, it can be shown rigorously (see Prugovečki and Zorbas [1973b, Section 7], Zorbas [1974]) that in its basic aspects the theory with a space cutoff merges into the exact theory as the cutoff is removed.

A fortuitous aspect of two-body scattering for Coulomb potentials is that its eigenfunctions can be computed explicitly by solving (4.17) for $V(\mathbf{r}) = \gamma \mathbf{r}^{-1}$ (see, e.g., Messiah [1962, Chapter IX, §§7–11]), and the addition in (4.115) of the short-range potential $V_0(\mathbf{r})$ can be then treated as a perturbation of this pure Coulomb problem. Consequently, the main thrust of most of the work on scattering with long-range forces quoted in this section has been in the direction of multichannel scattering, the two-body case providing mainly a testing ground of the proposed ideas.

**4.10. Appendix: Eigenfunctions and Transition Density Matrices in Statistical Mechanics

The general concepts and results of §§4.4–4.7 can be adapted with ease to Liouville space. In fact, a Liouville space $\mathscr{L}(\mathscr{H})$ is itself a Hilbert space, and therefore from a purely mathematical point of view one might simply consider such constructs as eigenfunction expansions and T matrices directly in $\mathscr{L}(\mathscr{H})$, without reference to $\mathscr{H}$. However, such an oversimplified approach would ignore the basic differences between the formulas (1.14) and (8.69) in Chapter IV, which reflect a difference in the physical interpretation in $\mathscr{L}(\mathscr{H})$ as opposed to $\mathscr{H}$.

Indeed, whereas for a wave packet $\psi(x)$ it is $|\psi(x)|^2$ that is interpreted as a probability density, for a density matrix $\rho(x', x'') = \langle x' | \rho | x'' \rangle$ it is its "diagonal" values $\rho(x, x)$, and *not* $|\rho(x', x'')|^2$, that play an analogous role. Consequently, in a physically correct approach to $\mathscr{L}(\mathscr{H})$ we cannot altogether ignore $\mathscr{H}$.

For example, in treating eigenfunctions related to $\mathbf{H}_0 = \hbar^{-1}[H_0, \cdot]$ [as defined by (8.51) in Chapter IV] we shall use as a starting point an eigenfunction expansion $\Phi_y(x)$ for a complete set of observables represented by operators $Y_1, \ldots, Y_n$ in $\mathscr{H}$ and satisfying (4.45)–(4.48). The functions

$$\boldsymbol{\Phi}_{y',y''}(x', x'') = \Phi_{y'}^{*}(x')\, \Phi_{y''}(x'') \tag{4.122}$$

then provide an eigenfunction expansion for the complete set of superoperators $[Y_j, \cdot]$ and $\{Y_j, \cdot\}$, $j = 1, \ldots, n$, but these superoperators do not always possess a physical interpretation, although $Y_1, \ldots, Y_n$ might. On the other hand, we have

$$\mathbf{H}_0^{x',x''} \boldsymbol{\Phi}_{y',y''}(x', x'') = \hbar^{-1}[\Lambda(y') - \Lambda(y'')]\, \boldsymbol{\Phi}_{y',y''}(x', x''), \tag{4.123}$$

so that the concepts and results of §§4.4 and 4.5 can be transferred with ease to $\mathbf{H}_0$ and $\boldsymbol{\Phi}_{y',y''}(x', x'')$.

As mentioned at the end of §3.8, the methods of quantum scattering theory can be adapted to classical statistical mechanics for the study of transport phenomena. In particular, it has proven advantageous to study classical differential cross sections within a superoperator formalism inspired by the quantum approach.* In fact, through the use of the master Liouville space $\mathscr{L}_\Gamma$ described in §8.9 of Chapter IV, it is possible to provide a unified treatment of both classical and quantum differential cross sections.†

The idea is to start with a density operator describing a state in which the incoming reduced particle is to be found within a finite cylinder of base B and length l (i.e., within a *finite* beam), its momentum probability distribution being a δ-like function $\delta_\epsilon^3(\mathbf{p} - \mathbf{p}_0)$ peaked around $\mathbf{p}_0$. Proceeding very much as in §1.8, one then uses the T superoperator formula (3.89) to express the probability of that particle scattering in a potential (centered at the origin) within a cone $C^{(+)}$, which is then treated as in (1.40) by letting $C^{(+)}$ become infinitely

* See, e.g., Miles and Dahler [1970], Coombe *et al.* [1975], and Eu [1975].

† This treatment has been espoused in Sections 3 and 4 of Prugovečki [1978c], where proofs of the subsequently outlined results can be found.

narrow. The key difference between the treatment of §1.8 and the present case emerges, however, as one makes the transition to an infinite beam [by letting in the present case B and l go to infinity, so that $\delta_\epsilon^3(\mathbf{p} - \mathbf{p}_0) \to \delta^3(\mathbf{p} - \mathbf{p}_0)$]. Indeed, whereas in §1.8 prior to taking this limit it had proven to be mandatory* to carry out an averaging over an infinite plane (reflecting a planar distribution of particles in the target), that is not at all the case in the present context. As a matter of fact, since the present derivation is applicable not only to the quantum but also the classical case, any such averaging would be mathematically ad hoc, since the well-known formula (see, e.g., Balescu [1975])

$$\sigma_{\mathrm{cl}}(\mathbf{p} \to \boldsymbol{\omega}') = (b\,db\,d\phi)/d\boldsymbol{\omega}' \tag{4.124}$$

for the classical differential cross section in terms of the impact parameter b can be derived without any kind of averaging. Furthermore, in the context of statistical mechanics, any such averaging would be also physically unjustified since in such realms of applicability as collisions within gases obviously no planar pattern can be claimed to exist.

The resulting general formula for the (point-target) differential cross section is

$$\sigma_s(\mathbf{p} \to \boldsymbol{\omega}') = \int_0^\infty dp p^2 \int d\mathbf{u} \hat{\chi}^{(s)}_{p\boldsymbol{\omega}'}(\mathbf{u})(m/u)\langle \mathbf{u} \mid \mathbf{T} \mid \mathbf{p}\rangle_s \tag{4.125}$$

where the transition density supermatrix

$$\langle \mathbf{u} \mid \mathbf{T} \mid \mathbf{p}\rangle_s = -i(2\pi)^3(\boldsymbol{\Phi}^{(s)}_{0,\mathbf{u}} \mid \mathbf{H}_{\mathrm{I}}^{(s)} \mid \boldsymbol{\Phi}^{(s)+}_{0,\mathbf{p}}) \tag{4.126}$$

makes its appearance. The classical cross section (notationally represented by $s = \mathrm{cl}$) formally corresponds to $\hat{\chi}^{(\mathrm{cl})}$ taken to be a δ-function, and it indeed turns out to equal (4.124) upon adopting for $\mathbf{H}_{\mathrm{I}}^{(\mathrm{cl})}$ the interaction term in equation (8.74) of Chapter IV specialized to the case of a single particle. In the quantum case $\hat{\chi}^{(s)}$ is given by (8.80) and (8.86) in Chapter IV, and for very precise momentum measurements (i.e., very large values of s) σ_s in (4.125) appears to be approximately equal to σ in (4.98) under realistic assumptions about relative orders of magnitude of the basic parameters appearing in a scattering experiment (see Prugovečki [1978c, Section 4]), but no exact equality has been established.

* This averaging procedure is common to all rigorous derivations of the formula $\sigma(\mathbf{p} \to \omega) = |f_{\mathbf{p}}(\omega)|^2$, and is not peculiar to the specific derivation in §1.8 (see, e.g., Messiah [1961, Chapter X, §§5 and 6], Taylor [1972, Section 3-e] and Newton [1979]).

EXERCISES

4.1. Show that the expressions on the right-hand sides of (4.38) and (4.41) are identical.

4.2. Show that if $G_A^{(1)}(x, x'; \zeta)$ and $G_A^{(2)}(x, x'; \zeta)$ are two Green functions of A and μ is σ finite, then $G_A^{(1)}(x, x'; \zeta) = G_A^{(2)}(x, x'; \zeta)$ almost everywhere in x and x' (with respect to the measure $\mu \times \mu$).

4.3. In potential scattering $G(\mathbf{r}, \mathbf{r}'; \lambda \pm i\epsilon)$ and $G^{(\pm)}(\mathbf{r}, \mathbf{r}'; \lambda)$ are given by (5.6) and (5.17), respectively. Prove that if $V(\mathbf{r})$ is square integrable and $| V(\mathbf{r})| = O(1/r^{2+\epsilon_0})$ for some $\epsilon_0 > 0$, and if $| \Phi_{\mathbf{k}}^{(\pm)}(\mathbf{r})| \leqslant M_B(\mathbf{r})$ for all $\mathbf{k} \in B$, where $M_B(\mathbf{r})$ is integrable on $\mathbb{R}^3$ for any compact set B disjoint from the exceptional set $\mathscr{S}_V^3$ mentioned in §4.2, then provided $\int | \mathbf{r} - \mathbf{r}' |^{-1} V(\mathbf{r}'^1)\, d\mathbf{r}' \leqslant$ const (cf. Excercise 6.2),

$$\lim_{\eta \to \pm 0} \int \psi_0^*(\mathbf{r})\, G_0(\mathbf{r}, \mathbf{r}'; \lambda + i\eta)\, V(\mathbf{r}')\, \Phi_{\mathbf{k}}^{(\pm)}(\mathbf{r}')\, \tilde{\psi}(\mathbf{k})\, d\mathbf{r}\, d\mathbf{r}'\, d\mathbf{k}$$
$$= \int \psi_0^*(\mathbf{r})\, G_0^{(\pm)}(\mathbf{r}, \mathbf{r}'; \lambda)\, V(\mathbf{r}')\, \Phi_{\mathbf{k}}^{(\pm)}(\mathbf{r})\, \tilde{\psi}(\mathbf{k})\, d\mathbf{r}\, d\mathbf{r}'\, d\mathbf{k}$$

for $\psi_0(\mathbf{r}) \in \mathscr{C}_{\mathrm{b}}^0(\mathbb{R}^3)$ and $\tilde{\psi}(\mathbf{k}) \in \mathscr{C}_{\mathrm{b}}^0(\mathbb{R}^3)$ with supp $\tilde{\psi} \cap \mathscr{S}_3^V = \varnothing$.

4.4. State conditions under which you can derive Lippmann–Schwinger equations from (3.34) for arbitrary eigenfunction expansions:

$$\Phi_y^{(\pm)}(x) = \Phi_y(x) + \int_{\mathbb{R}^m} G_0^{(\pm)}(x, x'; A(y))\, V(x')\, \Phi_y^{(\pm)}(x')\, d\mu(x').$$

4.5. Show that if the integration on the right-hand side of (4.84) is restricted to $A^{-1}((a, b])$, the outcome equals the limit in (4.86).

4.6. Prove that if $V(\mathbf{r})$ is square integrable on $\mathbb{R}^3$, then

$$\langle \mathbf{k} \mid \mathscr{T}(\zeta) | \mathbf{k}' \rangle = (\Phi_{\mathbf{k}} \mid [1 + VG(\zeta)]\, V\Phi_{\mathbf{k}'}\rangle$$

exists [in the sense of (4.81) and Definition 4.2] and that (4.105) is satisfied almost everywhere in $(\mathbf{p}', \mathbf{p}) \in \mathbb{R}^6$.

Note. If $V(\mathbf{r})$ is also integrable, then its Fourier transform exists and is continuous, and therefore (4.105) holds whenever $\langle \mathbf{k} \mid \mathscr{T}(\zeta) \mid \mathbf{k}' \rangle$ is continuous in $\mathbf{k}, \mathbf{k}' \in \mathbb{R}^3$.

4.7. Show that (4.112) implies (4.111) if (6.62) holds true.

5. Green Functions in Potential Scattering

5.1. The Free Green Function

Potential quantum scattering is the special case of quantum scattering theory in which the underlying formalism is that of wave mechanics, and the interaction is determined by a potential. Proceeding as in §§4.7–9, in applying the general methods of scattering theory to potential scattering of two particles we shall limit ourselves to systems of two spinless, nonidentical particles interacting via a potential $V(\mathbf{r})$, where $\mathbf{r} = \mathbf{r}_2 - \mathbf{r}_1$, and $\mathbf{r}_1$, $\mathbf{r}_2$ are the position vectors of the two particles. This case exhibits all the essential features of all other cases. For example, the case of nonzero-spin particles can be treated essentially by the same methods as the zero-spin case when group-theoretical methods are applied.

In the case of systems of two identical particles one must also take into account the presence of the Fermi–Dirac or Bose–Einstein statistics. From the practical point of view this means that in these last two cases the calculations have to be carried out in the respective subspaces of antisymmetric or symmetric functions of $L^2(\mathbb{R}^3)$, rather than in the space $L^2(\mathbb{R}^3)$ itself.* Upon removal of the center-of-mass motion (see §1.7), the resulting subspaces of $L^2(\mathbb{R}^3)$ containing the states of the reduced particle consist of wave functions of even or odd parity, respectively.

In the absence of external fields and mutual interaction, the relative motion Hamiltonian H_0 of a system of two spinless particles is the kinetic energy operator in the center of mass system that can be viewed also as the kinetic energy operator for a particle of reduced mass m [see (5.93) in Chapter II]. We recall from Theorem 7.3 of Chapter IV that $H_0 = T_S$ is a self-adjoint operator for which

$$(H_0\psi)(\mathbf{r}) = -\frac{1}{2m}\Delta\psi(\mathbf{r}), \qquad \psi \in \mathscr{D}_0, \tag{5.1}$$

and that H_0 is the only self-adjoint operator having this property. On wave functions $\tilde{\psi}(\mathbf{k}) \in U_F\mathscr{D}_{T_S} \subset L^2(\mathbb{R}^3)$ in the momentum space, H_0 acts in the following manner:

$$(H_0\psi)^\sim(\mathbf{k}) = \frac{\mathbf{k}^2}{2m}\tilde{\psi}(\mathbf{k}), \qquad \mathbf{k}^2\tilde{\psi}(\mathbf{k}) \in L_{(2)}(\mathbb{R}^3). \tag{5.2}$$

* See Rys [1965] and Taylor [1972, Chapter 22] for a general treatment of the scattering theory of two identical particles. Naturally, for particles of nonzero spin one has to work on direct sums of $L^2(\mathbb{R}^3)$ spaces [see (7.9) in Chapter IV].

The above operator is obviously self-adjoint and, since it is a function of the momentum operators $P^{(x)}$, $P^{(y)}$, $P^{(z)}$, it immediately follows that it has a pure continuous spectrum

$$\mathsf{S}^{H_0} = \mathsf{S}_{\mathrm{c}}^{H_0} = [0, +\infty). \tag{5.3}$$

Thus, its resolvent

$$R_0(\zeta) = \frac{1}{H_0 - \zeta} \tag{5.4}$$

is defined for all values of $\zeta \in \mathbb{C}^1$. Moreover, $R_0(\zeta)$ is a bounded linear operator defined on the entire Hilbert space $\mathscr{H} = L^2(\mathbb{R}^3)$ whenever $\operatorname{Re} \zeta < 0$ or $\operatorname{Im} \zeta \neq 0$, i.e., when the argument $\arg \zeta$ of ζ is within the open interval $(0, 2\pi)$ and for such values in scattering theory $-R_0(\zeta)$ is called the *free Green's operator*.

Theorem 5.1. The Green's operator $G_0(\zeta)$ of the free Hamiltonian (5.2) is an integral operator

$$(G_0(\zeta)\psi)(\mathbf{r}) = -(R_0(\zeta)\psi)(\mathbf{r}) = \int_{\mathbb{R}^3} G_0(\mathbf{r}, \mathbf{r}'; \zeta)\, \psi(\mathbf{r}')\, d\mathbf{r}' \tag{5.5}$$

for values of ζ restricted to the range $0 < \arg \zeta < 2\pi$. The kernel $G_0(\mathbf{r}, \mathbf{r}'; \zeta)$, called the free Green function, is

$$G_0(\mathbf{r}, \mathbf{r}'; \zeta) = \frac{-m}{2\pi \mid \mathbf{r} - \mathbf{r}' \mid} \exp[i \sqrt{2m\zeta} \mid \mathbf{r} - \mathbf{r}' \mid], \tag{5.6}$$

where $\sqrt{2m\zeta}$ is the square root for which $\operatorname{Im} \sqrt{2m\zeta} > 0$.

Proof. According to (5.2) and Theorems 2.5–2.7 in Chapter IV,

$$(R_0(\zeta)\psi)^{\sim}(\mathbf{k}) = \frac{1}{(\mathbf{k}^2/2m) - \zeta}\, \tilde{\psi}(\mathbf{k}). \tag{5.7}$$

Assume that the wave function $\psi(\mathbf{r})$ in the configuration space belongs to $\mathscr{C}_{\mathrm{b}}^{\infty}(\mathbb{R}^3)$. Then $\psi(\mathbf{r})$ and all its derivatives have Fourier transforms

$$(ik_x)^{n_1}(ik_y)^{n_2}(ik_z)^{n_3}\tilde{\psi}(\mathbf{k}) = (2\pi)^{-3/2} \int_{\mathbb{R}^3} e^{-i\mathbf{k}\mathbf{r}'} \frac{\partial^{n_1}}{\partial x^{n_1}} \frac{\partial^{n_2}}{\partial y^{n_2}} \frac{\partial^{n_3}}{\partial z^{n_3}} \psi(\mathbf{r}')\, d\mathbf{r}', \tag{5.8}$$

which by Theorem 4.5 of Chapter III are continuous and bounded, and therefore $\tilde{\psi}(\mathbf{k})\, e^{i\mathbf{k}\mathbf{r}}$ is integrable* on $\mathbb{R}^3$. Since $(\mathbf{k}^2/2\mathbf{m} - \zeta)^{-1}$ is bounded

* Throughout this section, whenever we deal with integration, we have in mind integration with respect to the Lebesgue measure, except if otherwise explicitly stated.

in $\mathbf{k}$ over $\mathbb{R}^3$ when $0 < \arg \zeta < 2\pi$, the inverse Fourier transform of

$$(\mathbf{k}^2/2m - \zeta)^{-1}\tilde{\psi}(\mathbf{k})$$

exists. By virtue of (5.7), (5.8), and Lebesgue's dominated convergence theorem,

$$(5.9)\quad (R_0(\zeta)\psi)(\mathbf{r}) = \frac{1}{(2\pi)^3} \lim_{R\to\infty} \int_{|\mathbf{k}|\leqslant R} d\mathbf{k} \left(\frac{\mathbf{k}^2}{2m} - \zeta\right)^{-1} e^{i\mathbf{r}\mathbf{k}} \int_{\mathbb{R}^3} d\mathbf{r}'\, e^{-i\mathbf{k}\mathbf{r}'}\, \psi(\mathbf{r}').$$

Due to the fact that $\psi(\mathbf{r}')$ is of compact support, the function

$$\left|\left(\frac{\mathbf{k}^2}{2m} - \zeta\right)^{-1} \exp[i\mathbf{k}(\mathbf{r} - \mathbf{r}')]\, \psi(\mathbf{r}')\right| = \left|\frac{\psi(\mathbf{r}')}{\mathbf{k}^2/2m - \zeta}\right|$$

is Lebesgue integrable on $\mathbb{R}^3 \times \{\mathbf{k}\colon |\,\mathbf{k}\,| \leqslant R\}$ with respect to the variables $\mathbf{r}$ and $\mathbf{k}$. Hence, we can apply Fubini's theorem to interchange the order of integration in (5.9):

$$(5.10)\quad (R_0(\zeta)\psi)(\mathbf{r}) = \frac{1}{(2\pi)^3} \int_{\mathbb{R}^3} d\mathbf{r}'\psi(\mathbf{r}') \lim_{R\to\infty} \int_{|\mathbf{k}|\leqslant R} d\mathbf{k} \left(\frac{\mathbf{k}^2}{2m} - \zeta\right)^{-1} \exp[i\mathbf{k}(\mathbf{r} - \mathbf{r}')].$$

To carry out the above integration in $\mathbf{k}$, let us write (for fixed $\boldsymbol{\rho} = \mathbf{r} - \mathbf{r}'$) the variable $\mathbf{k}$ in spherical coordinates introduced in such a manner that

$$\mathbf{k}(\mathbf{r} - \mathbf{r}') = \mathbf{k}\boldsymbol{\rho} = k\rho \cos\theta.$$

It is very easy to carry out immediately the integrations in $0 \leqslant \theta \leqslant \pi$ and $0 \leqslant \phi \leqslant 2\pi$:

$$(5.11)\quad \int_{|\mathbf{k}|\leqslant R} \left(\frac{\mathbf{k}^2}{2m} - \zeta\right)^{-1} e^{i\mathbf{k}\boldsymbol{\rho}}\, d\mathbf{k}$$
$$= \int_0^R dk \int_0^\pi d\theta \int_0^{2\pi} d\phi\, k^2 \sin\theta \left(\frac{k^2}{2m} - \zeta\right)^{-1} e^{ik\rho\cos\theta}$$
$$= -4\pi \int_0^R \frac{k \sin k\rho}{\rho(k^2/2m - \zeta)}\, dk = -4m\pi/\rho \int_{-R}^{+R} \frac{k \sin k\rho}{k^2 - 2m\zeta}\, dk.$$

In the last of the above integrals we have to integrate with respect to $k \in [-R, +R]$ the function

$$(5.12)\quad \frac{k \sin k\rho}{k^2 - 2m\zeta} = \frac{k\, e^{ik\rho}}{2i(k - \sqrt{2m\zeta})(k + \sqrt{2m\zeta})} - \frac{k\, e^{-ik\rho}}{2i(k - \sqrt{2m\zeta})(k + \sqrt{2m\zeta})}$$

and then let $R \to \infty$. If we work with the *complex* variable $k \in \mathbb{C}^1$, the above functions are analytic everywhere in the complex plane, except for two simple poles at $\pm\sqrt{2m}\,\sqrt{\zeta}$, $\operatorname{Im}\sqrt{\zeta} > 0$. Let us close the contour of integration by semicircles in the upper-half and lower-half complex plane, respectively, when integrating on the real line the first and second function on the right-hand side of (5.12). Since the contributions from the respective integrations on the semicircles vanish in the limit of their radii R becoming infinite, we easily determine, by using Cauchy's integral formula and computing the respective residua:

$$\lim_{R\to\infty}\int_{-R}^{+R}\frac{k\sin k\rho}{k^2-2m\zeta}\,dk = \frac{1}{2i}\int\frac{k\,e^{ik\rho}}{k^2-2m\zeta}\,dk - \frac{1}{2i}\int\frac{k\,e^{-ik\rho}}{k^2-2m\zeta}\,dk$$

$$= -\pi\,\frac{k\,e^{ik\rho}}{k+\sqrt{2m\zeta}}\bigg|_{k=\sqrt{2m\zeta}} + \pi\,\frac{k\,e^{-ik\rho}}{k-\sqrt{2m\zeta}}\bigg|_{k=-\sqrt{2m\zeta}}$$

$$= -\pi e^{i\rho\sqrt{2m\zeta}}$$

Inserting this result in (5.11), we get from (5.10)

$$(5.13)\quad (R_0(\zeta)\psi)(\mathbf{r}) = \frac{m}{2\pi}\int_{\mathbb{R}^3}\frac{1}{|\,\mathbf{r}'-\mathbf{r}\,|}\exp(i\,\sqrt{2m\zeta}|\,\mathbf{r}'-\mathbf{r}\,|)\,\psi(\mathbf{r}')\,d\mathbf{r}'$$

for all $\psi \in \mathscr{C}_b^\infty(\mathbb{R}^3)$. Since $\mathscr{C}_b^\infty(\mathbb{R}^3)$ is dense in $L^2(\mathbb{R}^3)$ (see Chapter II, Theorem 5.6), while both $R_0(\zeta)$ and the integral operator with the kernel $(m/2\pi)|\,r-r'\,|^{-1}\exp(i\sqrt{2m\zeta}\,|\,r-r'\,|)$ are bounded linear operators defined on the entire Hilbert space $L^2(\mathbb{R}^3)$, we infer with the help of the extension principle (Chapter III, Theorem 2.6) that (5.13) is true for all $\psi \in L^2(\mathbb{R}^3)$. Q.E.D.

It follows from (5.6) that the free Green function has the following symmetry properties:

$$(5.14)\qquad G_0(\mathbf{r},\mathbf{r}';\zeta) = G_0^*(\mathbf{r},\mathbf{r}';\zeta^*),$$

$$(5.15)\qquad G_0(\mathbf{r},\mathbf{r}';\zeta) = G_0(\mathbf{r}',\mathbf{r};\zeta)$$

for any $\zeta \in \mathbb{C}^1$ not on the positive real axis.

We note that considered as a function on $\mathbb{R}^6$, $G_0(\mathbf{r},\mathbf{r}';\zeta)$ is continuous and infinitely differentiable everywhere except on the set

$$(5.16)\qquad D_{G_0} = \{(\mathbf{r},\mathbf{r}') : |\,\mathbf{r}-\mathbf{r}'\,| = 0\},$$

which is of Lebesgue measure zero on $\mathbb{R}^6$ (see Exercise 5.1). Furthermore,

$$G_0^{(\pm)}(\mathbf{r}, \mathbf{r}'; k^2/2m) = \lim_{\epsilon \to +0} G_0[\mathbf{r}, \mathbf{r}'; (k^2/2m) \pm i\epsilon] \tag{5.17}$$

$$= [-m/(2\pi \mid \mathbf{r} - \mathbf{r}' \mid)] \exp(\pm ik \mid \mathbf{r} - \mathbf{r}' \mid)$$

also exists pointwise for any $k \geqslant 0$ and all $(\mathbf{r}, \mathbf{r}') \in \mathbb{R}^6 - D_{G_0}$, and by Lebesgue's dominated convergence theorem (4.66) is certainly satisfied in the present context for all $f(\mathbf{r})$, $g(\mathbf{r}) \in \mathscr{C}_b{}^0(\mathbb{R}^3)$. Consequently the *retarded and advanced free Green* functions $G_0^{(+)}(\mathbf{r}, \mathbf{r}'; \lambda)$ and $G_0^{(-)}(\mathbf{r}, \mathbf{r}'; \lambda)$ do exist for all $\lambda \in S_c^{H_0} = [0, +\infty)$. These functions inherit the symmetry property (5.14), and we have

$$G_0^{(\pm)}(\mathbf{r}, \mathbf{r}'; \lambda) = G_0^{(\pm)}(\mathbf{r}', \mathbf{r}; \lambda) = G_0^{(\mp)*}(\mathbf{r}, \mathbf{r}'; \lambda) \tag{5.18}$$

for all $\lambda \geqslant 0$, as is easily verified by the use of (5.17).

5.2. Partial Wave Free Green Functions

As we have shown in §4.6, the free Green function $G_0(\mathbf{r}, \mathbf{r}'; \zeta)$ is a key ingredient of the Lippmann–Schwinger equations (4.67) used in computing distorted plane waves $\Phi_{\mathbf{k}}^{(\pm)}(\mathbf{r})$. In such computations one implicitly works with the complete sets of observables $\mathbf{Q}$ and $\mathbf{P}$, as reflected by the fact that the plane waves $\Phi_{\mathbf{k}}(\mathbf{r})$ are functions of the position variables $\mathbf{r}$, and are parametrized in terms of the momentum variables $\mathbf{k}$.

We shall see in §7.4 that when the potential is central (i.e., spherically symmetric) it can be advantageous to work with the complete sets $\{\mid \mathbf{Q} \mid, \Theta_{\mathbf{Q}}, \Phi_{\mathbf{Q}}\}$ and $\{\mid \mathbf{P} \mid, \mid \mathbf{L} \mid, L^{(z)}\}$ instead. It is obvious that H_0 commutes with $\mid \mathbf{P} \mid$, $\mid \mathbf{L} \mid$, and $L^{(z)}$, and in fact, in accordance with (4.39),

$$H_0 = \Lambda_0(\mid \mathbf{P} \mid, \mid \mathbf{L} \mid, L^{(z)}) = \mid \mathbf{P} \mid^2/2m.$$

Since H_0 has no point spectrum, by (4.42) and (4.63),

$$G_0(\mathbf{r}, \mathbf{r}'; \zeta) = \text{l.i.m.} \int_{\mathbb{R}^3} \frac{\hat{\Phi}_{klm}(\mathbf{r})\, \hat{\Phi}^*_{klm}(\mathbf{r}')}{\zeta - (k^2/2m)}\, d\mu_0(k, l, m).$$

Using spherical coordinates (r, θ, ϕ) for $\mathbf{r}$, and taking into consideration (4.26) and (4.40), we arrive at the following result.

Theorem 5.2. The free Green function equals a sum (convergent in the mean),

$$G_0(\mathbf{r}, \mathbf{r}'; \zeta) = \sum_{l=0}^{\infty} \sum_{m=-l}^{+l} G_0{}^l(r, r'; \zeta)\, Y_l{}^m(\theta, \phi)\, Y_l{}^m(\theta', \phi')^*, \tag{5.20}$$

of the *partial wave free Green functions*

$$(5.21)\qquad G_0{}^l(r, r'; \zeta) = \frac{2}{\pi}\int_0^\infty \frac{j_l(kr)\, j_l(kr')}{\zeta - (k^2/2m)}\, k^2\, dk$$
$$= \begin{cases} -2mi\kappa j_l(\kappa r)\, h_l^{(1)}(\kappa r'), & r < r' \\ -2mi\kappa j_l(\kappa r')\, h_l^{(1)}(\kappa r), & r > r', \end{cases}$$

where $\kappa = (2m\zeta)^{1/2}$, $\operatorname{Im} \kappa > 0$.

Upon noting that the integrand in (5.21) is an even function, the computation of the previous explicit expression for $G_0{}^l(r, r'; \zeta)$ proceeds by contour integration and the use of the spherical Hankel functions $h_l^{(1,2)}(kr)$. Indeed (see, e.g., Butkov [1968]), as $|k| \to \infty$ in the complex plane,

$$j_l(kr') = \tfrac{1}{2}[h_l^{(1)}(kr') + h_l^{(2)}(kr')]$$
$$\sim (2ikr')^{-1}\{\exp\{i[kr' - \tfrac{1}{2}(l\pi)]\} - \exp\{-i[kr' - \tfrac{1}{2}(l\pi)]\},$$

so that upon inserting this result in the integral in (5.21) we see that for $r < r'$, $j_l(kr)h_l^{(1)}(kr')$ and $h_l^{(2)}(kr)j_l(kr')$ converge to zero as $|k| \to \infty$ in the upper-half and lower-half complex plane, respectively. Hence, proceeding as we did from (5.12) to (5.13), we close the contour of integration appropriately, and using the residuum theorem arrive at the first of the explicit expressions in (5.21) upon noting that $j_l(-kr) = (-1)^l j_l(kr)$ and $h_l^{(2)}(-kr) = (-1)^l h_l^{(1)}(kr)$. For $r > r'$ the roles of r and r' in the above procedure are obviously reversed, and the second expression immediately follows.

In (5.21) taking the limits $\operatorname{Im} \zeta \to \pm 0$ for fixed $\operatorname{Re} \zeta \geqslant 0$, we arrive at

$$(5.22)\qquad G_0^{(\pm)l}(r, r'; k^2/2m) = \begin{cases} \mp 2mikj_l(\pm kr)\, h_l^{(1)}(\pm kr'), & r < r', \\ \mp 2mikj_l(\pm kr')\, h_l^{(1)}(\pm kr), & r > r'. \end{cases}$$

The two functions in (5.22) are obviously the *advanced and retarded partial wave free Green functions.*

5.3. Fredholm Integral Equations with Hilbert–Schmidt Kernels

One of the main significant features of the free Green function $G_0(\mathbf{r}, \mathbf{r}'; \zeta)$ is that it provides an integral equation for the full Green function $G(\mathbf{r}, \mathbf{r}'; \zeta)$, corresponding to the Schroedinger operator H, for which

$$(5.23)\qquad (H\psi)(\mathbf{r}) = -(1/2m)\, \Delta\psi(\mathbf{r}) + V(\mathbf{r})\, \psi(\mathbf{r}), \qquad \psi \in \mathscr{D}_0 .$$

This integral equation is the second resolvent integral equation (4.76), which in the present case becomes

$$G(\mathbf{r}, \mathbf{r}'; \zeta) = G_0(\mathbf{r}, \mathbf{r}'; \zeta) + \int_{\mathbb{R}^3} G_0(\mathbf{r}, \mathbf{r}''; \zeta)\, V(\mathbf{r}'')\, G(\mathbf{r}'', \mathbf{r}'; \zeta)\, d\mathbf{r}''. \tag{5.24}$$

This equation can be solved by the Fredholm method if its kernel

$$K(\mathbf{r}, \mathbf{r}''; \zeta) = G_0(\mathbf{r}, \mathbf{r}''; \zeta)\, V(\mathbf{r}'') \tag{5.25}$$

is of Hilbert–Schmidt type.*

Before attacking the problem of solving (5.24), let us first review some basic facts about integral operators with Hilbert–Schmidt kernels.

Definition 5.1. The kernel $K(\alpha, \alpha')$, $\alpha, \alpha' \in \mathbb{R}^m$, of the integral operator K on $L^2(\mathbb{R}^m)$,

$$(K\psi)(\alpha) = \int_{\mathbb{R}^m} K(\alpha, \alpha')\, \psi(\alpha')\, d^m\alpha', \qquad \psi \in L^2(\mathbb{R}^m), \tag{5.26}$$

is a *Hilbert–Schmidt kernel* (or *L^2 kernel*) if $|K(\alpha, \alpha')|^2$ is integrable on $\mathbb{R}^{2m}$:

$$\langle K \mid K \rangle_2 = \int_{\mathbb{R}^{2m}} |K(\alpha, \alpha')|^2\, d^m\alpha\, d^m\alpha' < +\infty. \tag{5.27}$$

A Hilbert–Schmidt kernel which is continuous and such that $K(\alpha, \alpha)$ is integrable on $\mathbb{R}^m$ is called a *trace-class kernel.*

The terminology for kernels of integral operators is related to the terminology for the corresponding operators in a straightforward fashion: integral operators with trace-class kernels are of trace class* with $\operatorname{Tr} K = \int K(\alpha, \alpha)\, d^m\alpha$, and integral operators with Hilbert–Schmidt kernels are Hilbert–Schmidt operators (see Exercise 5.3).

The *Fredholm integral equation* is

$$\psi(\alpha) = f(\alpha) + \omega \int_{\mathbb{R}^m} K(\alpha, \alpha')\, \psi(\alpha')\, d^m\alpha', \tag{5.28}$$

where ω is, in general, some complex number. If we search for solutions $\psi(\alpha)$ which are Lebesgue square integrable, so that $K(\alpha, \alpha')$ is the kernel of an integral operator on $L^2(\mathbb{R}^m)$, and if $f \in L^2(\mathbb{R}^m)$, then the integral equation (5.28) can be written in operator form:

$$\psi = f + \omega K \psi. \tag{5.29}$$

* A condensed and very lucid presentation of the Fredholm theory for Hilbert–Schmidt kernels (L^2 kernels) is given by Smithies [1965] in Sections 6.5–6.7.

Hence, (5.28) has a unique solution in $L^2(\mathbb{R}^m)$ if and only if $(1 - \omega K)^{-1}$ exists and if f is in the domain of definition of $(1 - \omega K)^{-1}$.

For $\omega \neq 0$, $(1 - \omega K)^{-1}$ is related in a straightforward manner to the resolvent $R_K(\zeta)$ of the operator K:

$$(1 - \omega K)^{-1} = \frac{1}{\omega}\left(\frac{1}{\omega} - K\right)^{-1} = -\frac{1}{\omega} R_K\left(\frac{1}{\omega}\right).$$

If $K(\alpha, \alpha')$ is a Hilbert–Schmidt kernel, then K is a Hilbert–Schmidt operator, and as such, K has a pure point spectrum. Hence, $(1 - \omega K)^{-1}$ will exist if and only if $1/\omega \notin S^K$. When this is the case, the domain of definition of $(1 - \omega K)^{-1}$ is the entire space $L^2(\mathbb{R}^m)$.

The Fredholm theory essentially provides a means for the computation of $(1 - \omega K)^{-1}$ from the kernel $K(\alpha, \alpha')$. The main results of this theory for L^2 kernels are contained in Theorems 5.3 and 5.4. The reader interested in the proofs of these theorems is advised to consult Smithies [1965], Theorems 6.5.1–6.7.1.

Theorem 5.3. Let $K(\alpha, \alpha')$ be a Hilbert–Schmidt kernel and let us introduce the *modified Fredholm determinant of* $K(\alpha, \alpha')$

$$(5.30)\quad d_n = \frac{(-1)^n}{n!} \int_{\mathbb{R}^{mn}} \begin{vmatrix} 0 & K(\alpha_1, \alpha_2) & \cdots & K(\alpha_1, \alpha_n) \\ K(\alpha_2, \alpha_1) & 0 & \cdots & K(\alpha_2, \alpha_n) \\ \vdots & \vdots & & \vdots \\ K(\alpha_n, \alpha_1) & K(\alpha_n, \alpha_2) & \cdots & 0 \end{vmatrix} d^m\alpha_1 \cdots d^m\alpha_n,$$

and the *modified first Fredholm minor of* $K(\alpha, \alpha')$

$$(5.31)\quad D_n(\alpha, \alpha') = \frac{(-1)^n}{n!} \int_{\mathbb{R}^{mn}} \begin{vmatrix} K(\alpha, \alpha') & K(\alpha, \alpha_1) & \cdots & K(\alpha, \alpha_n) \\ K(\alpha_1, \alpha') & 0 & \cdots & K(\alpha_1, \alpha_n) \\ \vdots & \vdots & & \vdots \\ K(\alpha_n, \alpha') & K(\alpha_n, \alpha_1) & \cdots & 0 \end{vmatrix} d^m\alpha_1 \cdots d^m\alpha_n,$$

for $n = 1, 2, \ldots$; for $n = 0$, $d_0 = 1$ and $D_0(\alpha, \alpha') = K(\alpha, \alpha')$. In that case the series

$$(5.32)\quad d(\omega) = \sum_{n=0}^{\infty} d_n \omega^n$$

is convergent for all complex values of ω. Likewise, for any fixed

* A proof of this statement can be based on Mercer's theorem (see Riesz and Sz. Nagy [1955, Section 98]), or it can be deduced from (8.10) in Chapter IV.

$\omega \in \mathbb{C}^1$, the series

$$D(\alpha, \alpha'; \omega) = \sum_{n=0}^{\infty} D_n(\alpha, \alpha')\omega^n \tag{5.33}$$

is convergent in the mean [in $L^2(\mathbb{R}^{2m})$] for all ω. Moreover, the function $D(\alpha, \alpha'; \omega)$ is a Hilbert–Schmidt kernel for any fixed complex ω.

Theorem 5.4. If $d(\omega) \neq 0$ and $f \in L^2(\mathbb{R}^m)$, the integral equation (5.28) has the unique Lebesgue square-integrable solution

$$\psi(\alpha) = f(\alpha) + \frac{\omega}{d(\omega)} \int_{\mathbb{R}^m} D(\alpha, \alpha'; \omega) f(\alpha')\, d^m\alpha'. \tag{5.34}$$

Theorem 5.5. The function $d(\omega)$ vanishes if and only if ω^{-1} is an eigenvalue of the integral operator K.

5.4. The Full Green Function

We can apply the Fredholm theory to the integral equation (5.24) when its kernel (5.25) is of the Hilbert–Schmidt type. Under these circumstances, (5.24) is a special case of the integral equation (5.28), in which the variables $\alpha, \alpha' \in \mathbb{R}^m$ become $\mathbf{r}, \mathbf{r}'' \in \mathbb{R}^3$, and $\omega = 1$. Moreover, we note that since in (5.6) ζ is not on the positive real axis,

$$| G_0(\mathbf{r}, \mathbf{r}'; \zeta)| = \frac{m}{2\pi \mid \mathbf{r} - \mathbf{r}' \mid} \exp(-\operatorname{Im} \sqrt{2m\zeta} \mid \mathbf{r} - \mathbf{r}' \mid) \tag{5.35}$$

is square integrable in $\mathbf{r}$ or $\mathbf{r}'$ because according to Theorem 5.1 we have to choose the square root of ζ for which $\operatorname{Im} \sqrt{\zeta} > 0$. Hence, for any fixed $\mathbf{r}''$, (5.24) has a unique solution, provided that $d(1) \neq 0$ for the given kernel (5.25).

To establish that $d(1)$ is not zero for the kernel $K(\mathbf{r}, \mathbf{r}'; \zeta)$ in (5.25), note that $K(\mathbf{r}, \mathbf{r}'; \zeta)$ is the kernel of the closed operator

$$\overline{G_0(\zeta)H_1} = \overline{(\zeta - H_0)^{-1}(H - H_0)}. \tag{5.36}$$

According to Theorem 5.5, $d(1) = 0$ if and only if $\omega = 1$ is an eigenvalue of this operator, i.e., if there is a nonzero ψ_1 for which

$$\langle(\zeta - H_0)\psi \mid \psi_1\rangle = \langle(\zeta - H_0)\psi \mid \overline{G_0(\zeta)H_1}\psi_1\rangle = \langle H_1\psi \mid \psi\rangle$$

for all $\psi \in \mathscr{D}_H \equiv \mathscr{D}_{H_0} \subset \mathscr{D}_{H_1}$. However, the above relation implies that $H\psi_1 = \zeta\psi_1$. Thus, as long as ζ does not belong to the spectrum of H, the Green function $G(\mathbf{r}, \mathbf{r}'; \zeta)$ of H exists, and can be computed by means of (5.34) with $\omega = 1$ if K is Hilbert–Schmidt.

Let us find out now the restrictions imposed on the potential $V(\mathbf{r})$ by the demand that $K(\mathbf{r}, \mathbf{r}'; \zeta)$ be square integrable. Applying Fubini's Theorem, we find that

$$\text{(5.37)} \quad \int_{\mathbb{R}^6} |K(\mathbf{r}, \mathbf{r}'; \zeta)|^2 \, d\mathbf{r} \, d\mathbf{r}' = \int_{\mathbb{R}^3} d\mathbf{r}' |V(\mathbf{r}')|^2 \int_{\mathbb{R}^3} d\mathbf{r} |G_0(\mathbf{r}, \mathbf{r}'; \zeta)|^2.$$

Using the expression (5.6) for $G_0(\mathbf{r}, \mathbf{r}'; \zeta)$ we easily compute that

$$\text{(5.38)} \qquad \int_{\mathbb{R}^3} |G_0(\mathbf{r}, \mathbf{r}'; \zeta)|^2 \, d\mathbf{r} = \frac{m}{2\pi} \sqrt{\frac{m}{2}} \frac{1}{\operatorname{Im} \sqrt{\zeta}}, \qquad \operatorname{Im} \sqrt{\zeta} > 0.$$

Thus, we get the relation

$$\text{(5.39)} \qquad \int_{\mathbb{R}^6} |K(\mathbf{r}, \mathbf{r}'; \zeta)|^2 \, d\mathbf{r} \, d\mathbf{r}' = \frac{m}{2\pi} \sqrt{\frac{m}{2}} \frac{1}{\operatorname{Im} \sqrt{\zeta}} \int_{\mathbb{R}^3} |V(\mathbf{r}')|^2 \, d\mathbf{r}',$$

which shows that $G_0(\mathbf{r}, \mathbf{r}'; \zeta) \, V(\mathbf{r}')$ is for $\operatorname{Im} \zeta \neq 0$ a kernel of Hilbert–Schmidt type if and only if $V(\mathbf{r})$ is square integrable on $\mathbb{R}^3$.

The main conclusions of the above discussion are contained in Theorem 5.6.

Theorem 5.6. If the potential $V(\mathbf{r})$ is square integrable on $\mathbb{R}^3$, then the full Green function $G(\mathbf{r}, \mathbf{r}'; \zeta)$ of H [where $H \supseteq -(1/2m)\Delta + V(\mathbf{r})$] exists for $\zeta \notin S^H$ and can be written in the form

$$\text{(5.40)} \quad G(\mathbf{r}, \mathbf{r}'; \zeta) = G_0(\mathbf{r}, \mathbf{r}'; \zeta) + \frac{1}{d(1; \zeta)} \int_{\mathbb{R}^3} D(\mathbf{r}, \mathbf{r}''; 1) \, G_0(\mathbf{r}'', \mathbf{r}'; \zeta) \, d\mathbf{r}'',$$

where $d(1; \zeta)$ and $D(\mathbf{r}, \mathbf{r}''; 1)$ are obtained by taking

$$\text{(5.41)} \qquad K(\mathbf{r}, \mathbf{r}'; \zeta) = G_0(\mathbf{r}, \mathbf{r}'; \zeta) V(\mathbf{r}')$$

as a kernel in (5.30)–(5.33).

5.5. Fredholm Expansion of the Full Green Function

The kernel $D(\mathbf{r}, \mathbf{r}''; 1)$ of the integral operator in (5.40) is given by the infinite series

$$\text{(5.42)} \qquad D(\mathbf{r}, \mathbf{r}''; \zeta) = K(\mathbf{r}, \mathbf{r}''; \zeta) + \sum_{n=1}^{\infty} D_n(\mathbf{r}, \mathbf{r}''; \zeta),$$

in which by (5.31)

$$D_n(\mathbf{r}, \mathbf{r}''; \zeta) = \frac{(-1)^n}{n!} \int_{\mathbb{R}^{3n}} \begin{vmatrix} K(\mathbf{r}, \mathbf{r}''; \zeta) & K(\mathbf{r}, \mathbf{r}_1; \zeta) & \cdots & K(\mathbf{r}, \mathbf{r}_n; \zeta) \\ K(\mathbf{r}_1, \mathbf{r}''; \zeta) & 0 & \cdots & K(\mathbf{r}_1, \mathbf{r}_n; \zeta) \\ \vdots & \vdots & & \vdots \\ K(\mathbf{r}_n, \mathbf{r}''; \zeta) & K(\mathbf{r}_n, \mathbf{r}_1; \zeta) & \cdots & 0 \end{vmatrix} d\mathbf{r}_1 \cdots d\mathbf{r}_n . \tag{5.43}$$

This would imply that the full Green function can be expanded in a series

$$G(\mathbf{r}, \mathbf{r}'; \zeta) = \sum_{n=0}^{\infty} G_n(\mathbf{r}, \mathbf{r}'; \zeta), \tag{5.44}$$

where $G_0(\mathbf{r}, \mathbf{r}'; \zeta)$ is the free Green function, and

$$G_n(\mathbf{r}, \mathbf{r}'; \zeta) = \frac{1}{d(1; \zeta)} \int_{\mathbb{R}^3} D_{n-1}(\mathbf{r}, \mathbf{r}''; \zeta)\, G_0(\mathbf{r}'', \mathbf{r}'; \zeta)\, d\mathbf{r}'' \tag{5.45}$$

for $n = 1, 2, \ldots$, if the order of integration and summation can be interchanged. To prove that this is indeed the case, we need Lemma 5.1.

Lemma 5.1. If $K(\alpha, \alpha')$ is a kernel of Hilbert–Schmidt type, then the Hilbert–Schmidt norm

$$\| D_n \|_2 = \left\{ \iint_{\mathbb{R}^{2m}} | D_n(\alpha, \alpha')|^2 \, d^m\alpha \, d^m\alpha \right\}^{1/2} \tag{5.46}$$

of the modified first Fredholm minor $D_n(\alpha, \alpha')$ satisfies the inequality

$$\| D_n \|_2 \leqslant n^{-n/2} \{ e^{1/2} \| K \|_2 \}^{n+1}, \qquad n = 1, 2, \ldots, \tag{5.47}$$

and the modified Fredholm determinant d_n satisfies the inequality

$$| d_n | \leqslant \{ (e/n)^{1/2} \| K \|_2 \}^n, \qquad n = 1, 2, \ldots, \tag{5.48}$$

where $\| K \|_2$ is the Hilbert–Schmidt norm of the integral operator K, with the kernel $K(\alpha, \alpha')$:

$$\| K \|_2 = \left\{ \iint_{\mathbb{R}^{2m}} | K(\alpha, \alpha')|^2 \, d^m\alpha \, d^m\alpha' \right\}^{1/2}. \tag{5.49}$$

The proof of the above result can be found in the work of Smithies [1965, Section 6.5].

Applying in (5.45) the Schwarz–Cauchy inequality at fixed $\mathbf{r}$, $\mathbf{r}' \in \mathbb{R}^3$ and using afterwards (5.47), we get by (5.38)

$$(5.50)\quad |d(1;\zeta)|^2 \int_{\mathbb{R}^3} d\mathbf{r}\, |G_{n+1}(\mathbf{r},\mathbf{r}';\zeta)|^2$$

$$\leqslant \int_{\mathbb{R}^3} d\mathbf{r} \left[\int_{\mathbb{R}^3} |D_n(\mathbf{r},\mathbf{r}'';\zeta)|^2\, d\mathbf{r}''\right]\left[\int_{\mathbb{R}^3} |G_0(\mathbf{r}'',\mathbf{r}';\zeta)|^2\, d\mathbf{r}''\right]$$

$$\leqslant n^{-n}\sqrt{\frac{m}{2}}\,\frac{m\{e\langle K \mid K\rangle_2\}^{n+1}}{2\pi\,\mathrm{Im}\,\sqrt{\zeta}}\,.$$

Due to n^{-n} this expression converges very rapidly to zero as $n \to +\infty$, so that we can state the following theorem.

Theorem 5.7. If $V(\mathbf{r})$ is square integrable on $\mathbb{R}^3$, then the full Green function $G(\mathbf{r},\mathbf{r}';\zeta)$, $\zeta \notin S^H$, is square integrable in $\mathbf{r} \in \mathbb{R}^3$ for almost all $\mathbf{r}' \in \mathbb{R}^3$, and the series (5.44) converges in the mean with respect to the variable $\mathbf{r} \in \mathbb{R}^3$ to $G(\mathbf{r},\mathbf{r}';\zeta)$ for all such $\mathbf{r}' \in \mathbb{R}^3$.

5.6. Symmetry Properties of the Full Green Function

The two symmetry properties (5.14) and (5.15) of the free Green function are inherited by the full Green function, i.e., for $\zeta \notin S^H$

$$(5.51)\qquad G(\mathbf{r},\mathbf{r}';\zeta) = G^*(\mathbf{r},\mathbf{r}';\zeta^*),$$

$$(5.52)\qquad G(\mathbf{r},\mathbf{r}';\zeta) = G(\mathbf{r}',\mathbf{r};\zeta)$$

holds almost everywhere in $\mathbb{R}^6$.

The first of these two relations is a straightforward consequence of the formulas (5.42) and (5.44) for $G(\mathbf{r},\mathbf{r}';\zeta)$, resulting from the fact that $V(\mathbf{r}')$ is real, and consequently

$$K(\mathbf{r},\mathbf{r}';\zeta) = G_0(\mathbf{r},\mathbf{r}';\zeta)\,V(\mathbf{r}') = G_0^*(\mathbf{r},\mathbf{r}';\zeta^*)\,V(\mathbf{r}') = K^*(\mathbf{r},\mathbf{r}';\zeta^*).$$

To prove (5.52) we have to resort to a lengthier argument. First of all, we have to note that the function

$$\Gamma_1(\mathbf{r},\mathbf{r}';\zeta) = \int_{\mathbb{R}^3} D(\mathbf{r},\mathbf{r}'';1)\,G_0(\mathbf{r}'',\mathbf{r}';\zeta)\,d\mathbf{r}''$$

is the kernel of a Hilbert–Schmidt operator. This follows from the fact that $G_0(\mathbf{r}'',\mathbf{r}';\zeta)$ is the kernel of the bounded operator $(\zeta - H_0)^{-1}$, while $D(\mathbf{r},\mathbf{r}'';1)$ is a kernel of the Hilbert–Schmidt type, since by

(5.47) and Theorem 5.3 it is the limit in the mean of the Hilbert–Schmidt kernels (see Exercise 5.4)

$$\sum_{n=0}^{N} D_n(\mathbf{r}, \mathbf{r}''; 1), \qquad N = 1, 2, \ldots$$

Consequently, $D(\mathbf{r}, \mathbf{r}''; 1)$ is the kernel of a Hilbert–Schmidt operator $D(1)$. Thus, $\mathbf{S}_1(\mathbf{r}, \mathbf{r}'; \zeta)$ is the kernel of the operator $D(1)(\zeta - H_0)^{-1}$, which is the product of a Hilbert–Schmidt operator and a bounded operator, and therefore is of Hilbert–Schmidt type (see Exercise 5.5).

The function $f^*(\mathbf{r})\, g(\mathbf{r}')$ is square integrable in $\mathbb{R}^6$ whenever $f, g \in L^2(\mathbb{R}^3)$. Since $\Gamma_1(\mathbf{r}, \mathbf{r}'; \zeta)$ is a kernel of Hilbert–Schmidt type, it is also square integrable in $\mathbb{R}^6$. Hence, the function $f^*(\mathbf{r})\, \Gamma_1(\mathbf{r}, \mathbf{r}'; \zeta)\, g(\mathbf{r}')$ is integrable in $\mathbb{R}^6$. Consequently, Fubini's theorem can be applied to infer that

$$\int_{\mathbb{R}^3} d\mathbf{r}\, f^*(\mathbf{r}) \int_{\mathbb{R}^3} d\mathbf{r}'\, \Gamma_1(\mathbf{r}, \mathbf{r}'; \zeta)\, g(\mathbf{r}') = \int_{\mathbb{R}^3} d\mathbf{r}'\, g(\mathbf{r}') \int_{\mathbb{R}^3} d\mathbf{r}\, \Gamma_1(\mathbf{r}, \mathbf{r}'; \zeta) f^*(\mathbf{r}).$$

It can be seen from (5.6) that a corresponding relation holds for $G_0(\mathbf{r}, \mathbf{r}'; \zeta)$. Consequently, in view of (5.40), we have

$$\int_{\mathbb{R}^3} d\mathbf{r}\, f^*(\mathbf{r}) \int_{\mathbb{R}^3} d\mathbf{r}'\, G(\mathbf{r}, \mathbf{r}'; \zeta)\, g(\mathbf{r}') = \int_{\mathbb{R}^3} d\mathbf{r}'\, g(\mathbf{r}') \int_{\mathbb{R}^3} d\mathbf{r}\, G(\mathbf{r}, \mathbf{r}'; \zeta) f^*(\mathbf{r}). \tag{5.53}$$

On the other hand, the relation

$$\left\langle f \,\middle|\, \frac{1}{H - \zeta}\, g \right\rangle = \left\langle \frac{1}{H - \zeta^*}\, f \,\middle|\, g \right\rangle, \qquad \zeta \notin S^H,$$

in conjunction with (5.51) implies that

$$\begin{aligned} \int_{\mathbb{R}^3} d\mathbf{r}\, f^*(\mathbf{r}) \int_{\mathbb{R}^3} d\mathbf{r}'\, G(\mathbf{r}, \mathbf{r}'; \zeta)\, g(\mathbf{r}') &= \int_{\mathbb{R}^3} d\mathbf{r}'\, g(\mathbf{r}') \int_{\mathbb{R}^3} d\mathbf{r}\, G^*(\mathbf{r}', \mathbf{r}; \zeta^*) f^*(\mathbf{r}) \\ &= \int_{\mathbb{R}^3} d\mathbf{r}'\, g(\mathbf{r}') \int_{\mathbb{R}^3} d\mathbf{r}\, G(\mathbf{r}', \mathbf{r}; \zeta) f^*(\mathbf{r}). \end{aligned} \tag{5.54}$$

Comparing (5.53) and (5.54), we see that

$$\int_{\mathbb{R}^3} d\mathbf{r}'\, g(\mathbf{r}') \int_{\mathbb{R}^3} d\mathbf{r}\, G(\mathbf{r}, \mathbf{r}'; \zeta) f^*(\mathbf{r}')$$
$$= \int_{\mathbb{R}^3} d\mathbf{r}'\, g(\mathbf{r}') \int_{\mathbb{R}^3} d\mathbf{r}\, G(\mathbf{r}', \mathbf{r}; \zeta) f^*(\mathbf{r})$$

for all square-integrable functions $g(\mathbf{r}')$. This is possible if and only if

$$\int_{\mathbb{R}^3} G(\mathbf{r}, \mathbf{r}'; \zeta) f^*(\mathbf{r})\, d\mathbf{r} = \int_{\mathbb{R}^3} G(\mathbf{r}', \mathbf{r}; \zeta) f^*(\mathbf{r})\, d\mathbf{r}$$

for almost all $\mathbf{r}' \in \mathbb{R}^3$. Since the above equality holds for all square-integrable functions $f(\mathbf{r})$, we infer that (5.52) holds almost everywhere in $\mathbb{R}^3 \times \mathbb{R}^3 = \mathbb{R}^6$.

This completes the proof of (5.52).

*5.7. Appendix: The Spectrum of the Schroedinger Operator

We have seen already that if $K(\mathbf{r}, \mathbf{r}'; \zeta)$ is a Hilbert–Schmidt kernel the full Green function exists as long as $\zeta \notin S^H$. This is certainly the case if $\operatorname{Im} \zeta \neq 0$. The following theorem shows that this will be also the case if ζ is real but negative, with the possible exception of a finite or at most countably infinite number of values.

Theorem 5.8. Suppose the potential $V(\mathbf{r})$ is locally square integrable and $V(\mathbf{r}) \to 0$ as $r \to \infty$. Then the continuous spectrum S_c^H of the Schroedinger operator $H \supseteq -(1/2m)\Delta + V(\mathbf{r})$ contains no negative values; moreover, there are only at most countably many negative eigenvalues, all of finite multiplicity,* and having no negative number as an accumulation point.

In order to prove the above theorem, we need a number of auxiliary results. Since these results are of interest in themselves, we shall present them as lemmas and theorems.

We recall from §3 that if H and H_0 are any two operators (not necessarily self-adjoint) with identical domains of definition and ζ belongs to their resolvent sets, then the resolvent $R(\zeta)$ of H can be expressed in terms of $H_1 = H - H_0$ and of the resolvent $R_0(\zeta)$ of H_0 by making use of the infinite series in (3.68), provided that this series converges in some sense. Lemma 5.2 represents, to some extent, a converse to this result, valid under the additional assumption that

* Recall that the *multiplicity* of an eigenvalue is the number of linearly independent eigenvectors corresponding to that eigenvalue. *Local (square) integrability* means (square) integrability over arbitrary bounded Borel sets.

$H_1R_0(\zeta)$ has a bound less than one; in this context, it should be noted that when $\mathscr{D}_H \equiv \mathscr{D}_{H_0}$, the operator $H_1R_0(\zeta)$ is defined everywhere on $\mathscr{H}$ since $R_0(\zeta)$ is defined everywhere on $\mathscr{H}$ and it maps any $f \in \mathscr{H}$ into $R_0(\zeta)f \in \mathscr{D}_{H_0} \equiv \mathscr{D}_{H_1}$, so that $H_1R_0(\zeta)f$ is defined.

Lemma 5.2. Suppose that H and H_0 are two operators (not necessarily self-adjoint) with identical domains of definition $\mathscr{D}_H \equiv \mathscr{D}_{H_0}$, and that the complex number ζ belongs to the resolvent set of H_0. If $\| H_1R_0(\zeta)\| < 1$, $H_1 = H - H_0$, then ζ belongs also to the resolvent set of H, and

$$R(\zeta) = R_0(\zeta) \operatorname*{u-lim}_{n\to+\infty} \sum_{k=0}^{n} (-1)^k (H_1R_0(\zeta))^k. \tag{5.55}$$

Proof. Since $\| H_1R_0(\zeta)\| < 1$, the uniform limit in (5.55) exists (see Exercise 5.7). For any $f \in \mathscr{D}_H$ we have

$$\begin{aligned}
&\left[R_0(\zeta) \sum_{k=0}^{n} (-1)^k (H_1R_0(\zeta))^k\right] (H - \zeta)f \\
&\quad = R_0(\zeta)(H_0 - \zeta)f - R_0(\zeta)\, H_1R_0(\zeta)(H_0 - \zeta)f + \cdots \\
&\quad\quad + (-1)^n R_0(\zeta)\, H_1 \cdots R_0(\zeta)(H_0 - \zeta)f \\
&\quad\quad + R_0(\zeta)\, H_1 f - R_0(\zeta)\, H_1R_0(\zeta)\, H_1 f + \cdots \\
&\quad\quad + (-1)^n R_0(\zeta)\, H_1 \cdots R_0(\zeta)\, H_1 f \\
&\quad = f + (-1)^n R_0(\zeta)\, H_1 \cdots R_0(\zeta)\, H_1 f \xrightarrow[n\to+\infty]{} f.
\end{aligned}$$

This shows that if $A(\zeta)$ denotes the operator on the right-hand side of the equality (5.55), then we have $A(\zeta)(H - \zeta)f = f$ for all $f \in \mathscr{D}_H$.

Since $\mathscr{D}_H \equiv \mathscr{D}_{H_0}$ and $R_0(\zeta)$ has the range $\mathscr{D}_{H_0}$, the operator $(H - \zeta)\, R_0(\zeta)$ is defined everywhere on $\mathscr{H}$. Consequently, we can write

$$\begin{aligned}
&(H - \zeta)\, R_0(\zeta) \sum_{k=0}^{n} (-1)^k (H_1R_0(\zeta))^k \\
&\quad = (H_0 - \zeta)\, R_0(\zeta)[1 - H_1R_0(\zeta) + \cdots + (-1)^n (H_1R_0(\zeta))^n] \\
&\quad\quad + H_1R_0(\zeta) - H_1R_0(\zeta)\, H_1R_0(\zeta) + \cdots + (-1)^n (H_1R_0(\zeta))^{n+1} \\
&\quad = 1 + (-1)^n (H_1R_0(\zeta))^{n+1} \underset{n\to+\infty}{\Longrightarrow} 1.
\end{aligned}$$

Thus, we also have $(H - \zeta)\, A(\zeta) = 1$, i.e., $A(\zeta)$ is identical to $(H - \zeta)^{-1} = R(\zeta)$. Q.E.D.

If A is an unbounded self-adjoint operator, then its spectrum S^A is an unbounded set. However, the set $S^A \subset \mathbb{R}^1$ could be still bounded from above or from below, in which case we shall say that A is *bounded from above* or *from below*, respectively. In the case where A is bounded from below, we denote by m_A the greatest lower bound of the set S^A:

$$m_A = \inf_{\lambda \in S^A} \lambda.$$

We note in passing that A is bounded from below if and only if the set

$$\{\|f\|^{-2}\langle f \mid Af\rangle : f \in \mathscr{D}_A, f \neq 0\}$$

is bounded from below, with the same greatest lower bound (see Exercise 5.8).

Theorem 5.9. Suppose H and H_0 are two self-adjoint operators with identical domains of definition, that H_0 is bounded from below, and that for some $0 \leqslant a < 1$, $b \geqslant 0$,

$$\| H_1 f \| \leqslant a \| H_0 f \| + b \| f \|, \qquad H_1 = H - H_0, \tag{5.56}$$

for all $f \in \mathscr{D}_{H_0} \equiv \mathscr{D}_H$. Then H is also bounded from below, and

$$m_H \geqslant m_{H_0} - \max\left\{\frac{b}{1-a}, b + a| m_{H_0} |\right\}. \tag{5.57}$$

Proof. We shall prove the main statement of the theorem by showing that for any real

$$\zeta < m_{H_0} - \max\left\{\frac{b}{1-a}, b + a| m_{H_0} |\right\}, \tag{5.58}$$

we have $\| H_1 R_0(\zeta)\| < 1$, and consequently, according to Lemma 5.2, such a $\zeta \in \mathbb{R}^1$ belongs to the resolvent set of H. This means that the spectrum of H contains only such points which do not satisfy (5.58); i.e., it is bounded from below and its greatest lower bound m_H satisfies (5.57).

To prove that $\| H_1 R_0(\zeta)\| < 1$, we use (5.56) to derive the inequality

$$\| H_1 R_0(\zeta)\| \leqslant a \| H_0 R_0(\zeta)\| + b \| R_0(\zeta)\|. \tag{5.59}$$

Since $S^{H_0} \subset [m_{H_0}, +\infty)$, and therefore for any $f \in \mathscr{H}$ and real $\zeta < m_{H_0}$,

$$\begin{aligned} \| R_0(\zeta) f \|^2 &= \int_{m_{H_0}}^{+\infty} \frac{1}{|\lambda - \zeta|^2}\, d\| E_\lambda f \|^2 \\ &\leqslant \frac{1}{| m_{H_0} - \zeta |^2} \int_{m_{H_0}}^{+\infty} d\| E_\lambda f \|^2 = \frac{\| f \|^2}{| m_{H_0} - \zeta |^2}, \end{aligned}$$

(*equation continues*)

$$\| H_0R_0(\zeta)f \|^2 = \int_{S^{H_0}} \frac{\lambda^2}{|\lambda - \zeta|^2} d\| E_\lambda f \|^2$$
$$\leqslant \sup_{\lambda \in S^{H_0}} \frac{\lambda^2}{|\lambda - \zeta|^2} \int_{S^{H_0}} d\| E_\lambda f \|^2 = \sup_{m_{H_0} \leqslant \lambda < +\infty} \frac{\lambda^2 \| f \|^2}{|\lambda - \zeta|^2},$$

we conclude that

$$\| R_0(\zeta) \| \leqslant \frac{1}{m_{H_0} - \zeta},$$

$$\| H_0R_0(\zeta) \| \leqslant \sup_{\lambda \in S^{H_0}} \left| \frac{\lambda}{\lambda - \zeta} \right| \leqslant \max \left\{ 1, \frac{|m_{H_0}|}{m_{H_0} - \zeta} \right\}.$$

Combining the above two inequalities with (5.59), we obtain

$$\| H_1R_0(\zeta) \| \leqslant \max \left\{ a, \frac{a|m_{H_0}|}{m_{H_0} - \zeta} \right\} + \frac{b}{m_{H_0} - \zeta}. \tag{5.60}$$

Since it is easily seen that the expression on the right-hand side of the inequality (5.60) is smaller than one when ζ satisfies (5.58), it follows from Lemma 5.2 that such ζ is in the resolvent set of H. Consequently, H is bounded from below by the expression on the right-hand side of (5.57). Q.E.D.

We recall from §7 in Chapter IV that when $H = T_S + V_S$ is the Schroedinger operator with a potential satisfying the conditions of Theorem 7.4 in Chapter IV, then $\mathscr{D}_H \equiv \mathscr{D}_{H_0}$ (Chapter IV, Theorem 7.5) if $H_0 = T_S$ is defined by (7.15) in Chapter IV and (5.56) is satisfied by any $0 < a < 1$. Hence, we can state Theorem 5.10.

Theorem 5.10. The n-body Schroedinger operator $H = H_0 + V$,

$$(H_0\psi)^\sim (\mathbf{p}_1, \ldots, \mathbf{p}_n) = \sum_{k=1}^{n} \frac{\mathbf{p}_k^{\,2}}{2m_k} \tilde{\psi}(\mathbf{p}_1, \ldots, \mathbf{p}_n),$$
$$(V\psi)(\mathbf{r}, \ldots, \mathbf{r}_n) = V(\mathbf{r}_1, \ldots, \mathbf{r}_n)\, \psi(\mathbf{r}_1, \ldots, \mathbf{r}_n), \tag{5.61}$$

corresponding to a potential $V(\mathbf{r}_1, \ldots, \mathbf{r}_n)$ satisfying the conditions of Theorem 7.4 in Chapter IV, is bounded from below.

As a by-product of the above theorem and (5.57), we can obtain estimates on the lower bound of an n-body Schroedinger operator H by computing the constants a and b, as was done in the proof of Theorem 7.4 in Chapter IV.

A single-particle Schroedinger operator H with a potential satisfying the requirements of Theorem 5.8 also satisfies the conditions of Theorem 7.4 in Chapter IV. Hence, such a Hamiltonian H is bounded from below.

The second resolvent equation

$$(H - \lambda)^{-1} = (H_0 - \lambda)^{-1} - (H_0 - \lambda)^{-1}V(H - \lambda)^{-1} \tag{5.62}$$

holds for any value of λ in the resolvent set, and therefore, in particular, for any real $\lambda < m_H$. The desired result about the spectrum of H, stated in Theorem 5.8, can be then obtained from Lemma 5.3, which will be derived with the help of the next theorem, which implicitly introduces the following classification of the spectrum of a self-adjoint operator.

Definition 5.2. A point λ_0 in the spectrum S^A of a self-adjoint operator A is a *limit point* of S^A if it is either an eigenvalue of infinite multiplicity or an accumulation point of S^A. The set S_l^A of all limit points of S^A is called the *essential spectrum* of A, and its complement $S_d^A = S^A - S_l^A$, i.e., the set of all isolated eigenvalues of finite multiplicity, is called the *discrete spectrum* of A.

***Theorem 5.11** (*Weyl's criterion for limit points*). A real number λ_0 is a limit point of the spectrum of a self-adjoint operator A if and only if there is a sequence $f_1, f_2, \ldots \in \mathscr{D}_A$ which converges weakly to the zero vector, and is such that $\| f_n \| = 1$ and

$$\lim_{n\to+\infty} \|(A - \lambda_0) f_n \| = 0. \tag{5.63}$$

Proof. Suppose that $\| f_n \| = 1$, that $\text{w-lim}_{n\to+\infty} f_n = 0$, and that (5.63) is true. For an arbitrary open interval (a_0, b_0) containing λ_0 we can write

$$\begin{aligned}
\|(A - \lambda_0)f_n\|^2 &= \int_{\mathbb{R}^1} (\lambda - \lambda_0)^2 \, d\| E_\lambda^A f_n \|^2 \\
&\geqslant (b_0 - \lambda_0)^2 \int_{b_0}^{\infty} d\| E_\lambda^A f_n \|^2 + (a_0 - \lambda_0)^2 \int_{-\infty}^{a_0} d\| E_\lambda^A f_n \|^2 \\
&\geqslant (b_0 - \lambda_0)^2 \|(1 - E_{b_0}^A) f_n \|^2 + (a_0 - \lambda_0)^2 \| E_{a_0}^A f_n \|^2.
\end{aligned}$$

Now, the expression on the left-hand side of the above inequality approaches zero in the limit $n \to +\infty$. Hence, we have

$$\lim_{n\to+\infty} \| E_{a_0}^A f_n \| = \lim_{n\to+\infty} \|(1 - E_{b_0}^A) f_n \| = 0,$$

and consequently

$$\begin{aligned}
\lim_{n\to+\infty} \| E^A((a_0, b_0]) f_n \|^2 &= \lim_{n\to+\infty} \|(E_{b_0}^A - E_{a_0}^A) f_n \|^2 \\
&= \lim_{n\to\infty} \| E_{b_0}^A f_n \|^2 - \lim_{n\to+\infty} \| E_{a_0}^A f_n \|^2 = 1.
\end{aligned}$$

This implies that the subspace $E^A((a_0, b_0])\mathscr{H}$ onto which $E^A((a_0, b_0])$ projects cannot be finite dimensional; in fact, if this subspace were of finite dimension N, we could select an orthonormal basis $\{e_1, ..., e_N\}$ in it, and write

$$\| E^A((a_0, b_0])f_n \|^2 = |\langle e_1 | f_n \rangle|^2 + \cdots + |\langle e_N | f_n \rangle|^2 \to 0,$$

since $\lim_{n\to+\infty} \langle e_1 | f_n \rangle = \cdots = \lim_{n\to+\infty} \langle e_N | f_n \rangle = 0$ due to the weak convergence to zero of $f_1, f_2, ...$ Thus, we conclude that λ_0 is either an accumulation point of S^A or an eigenvalue with an infinite-dimensional characteristic subspace, i.e., that λ_0 is a limit point of S^A.

Conversely, if λ_0 is a limit point of the spectrum of A and $I_1 \supset I_2 \supset \cdots$ are open intervals containing λ_0, then each $E^A(I_n)$ is the projector on an infinite-dimensional subspace of $\mathscr{H}$. Hence, we can choose an infinite orthonormal system $\{e_1, e_2, ...\}$ such that $E^A(I_n)\, e_n = e_n$, $n = 1, 2, ...$ Since we have

$$\|(A - \lambda_0)e_n\|^2 = \int_{I_n} (\lambda - \lambda_0)^2\, d\| E_\lambda^A e_n \|^2 \leqslant | I_n |^2 \| e_n \|^2 \to 0$$

when these intervals were so chosen that their lengths $| I_n |$ shrink to zero, and since $\lim_{n\to+\infty} \langle e_n | f \rangle = 0$ for any $f \in \mathscr{H}$ because of Bessel's inequality, we see that the three conditions of the theorem are satisfied by this sequence $e_1, e_2, ... \in \mathscr{D}_A$. Q.E.D.

***Lemma 5.3.** If A and K are self-adjoint bounded operators on $\mathscr{H}$ and K is also completely continuous, then essential spectra of A and $A + K$ coincide.

Proof. If λ_0 is a limit point of the spectrum of A, according to Theorem 5.11 we can find a sequence $f_1, f_2, ... \in \mathscr{H}$ such that $\| f_n \| = 1$, $\text{w-lim}_{n\to+\infty} f_n = \mathbf{0}$ and $\lim_{n\to+\infty} \|(A - \lambda_0) f_n \| = 0$. Since $f_1, f_2, ...$ is a bounded sequence and K is completely continuous, it must contain a subsequence $g_1, g_2, ...$ such that $Kg_1, Kg_2, ...$ converges strongly to some vector $h \in \mathscr{H}$. However, since $\text{w-lim}_{n\to+\infty} g_n = \mathbf{0}$, we infer that $\| h \|^2 = \lim_{n\to+\infty} \langle h | Kg_n \rangle = \lim_{n\to+\infty} \langle K^*h | g_n \rangle = 0$. Hence, $\lim_{n\to+\infty} \|(A + K - \lambda_0) g_n \| = 0$, which in conjunction with $\| g_n \| = 1$ and $\text{w-lim}_{n\to+\infty} g_n = \mathbf{0}$ implies, by virtue of Theorem 5.11, that λ_0 is a limit point of the spectrum of $A_1 = A + K$.

Conversely, if λ_1 is a limit point of the spectrum of A_1, then, according to the same argument applied to $(-K)$, we deduce that λ_1 is also a limit point of the spectrum of $A = A_1 - K$. Q.E.D.

We can return now to the proof of Theorem 5.8, and observe first that since the spectrum of H_0 is $[0, +\infty)$, the spectrum of $(H_0 - \lambda)^{-1}$,

$\lambda < 0$, is $[0, -1/\lambda]$, as seen from the relation

$$E^{R_0(\lambda)}(B) = E^{H_0}(F^{-1}(B)), \qquad F(\lambda') = (\lambda' - \lambda)^{-1}.$$

The operator $(\lambda - H_0)^{-1} V$ is an integral operator with the kernel $K(\mathbf{r}, \mathbf{r}'; \lambda) = G_0(\mathbf{r}, \mathbf{r}'; \lambda)V(\mathbf{r}')$. To prove that the operator $K(\lambda)$ is completely continuous, we set $V_n(\mathbf{r}) = V(\mathbf{r})$ for $r \leqslant n$ and $V_n(\mathbf{r}) = 0$ for $r > n$, and note that due to the local square integrability of $V(\mathbf{r})$ the operator $K_n(\lambda)$ with kernel $G_0(\mathbf{r}, \mathbf{r}'; \lambda)V_n(\mathbf{r}')$ is Hilbert–Schmidt. Since obviously

$$\| K(\lambda) - K_n(\lambda)\| \leqslant \|(\lambda - H_0)^{-1}\| \sup_{r>n} | V(\mathbf{r})| \to 0, \quad n \to \infty,$$

we deduce from Theorem 8.3 in Chapter IV that $K(\lambda)$ is indeed compact. For $\lambda < m_H$, the operator $(H - \lambda)^{-1}$ is bounded, and consequently, by Theorem 8.1(b) of Chapter IV, $(H_0 - \lambda)^{-1}V(H - \lambda)^{-1}$ is also a completely continuous operator. Hence, according to (5.62) and Lemma 5.3, the limit points of the spectrum of $(H - \lambda)^{-1}$ are the same as the limit points of the spectrum of $(H_0 - \lambda)^{-1}$, i.e., they constitute the interval $[0, -1/\lambda]$. This implies that the limit points of the spectrum of H constitute the interval $[0, +\infty)$, i.e., H has no negative limit points. This means that the continuous spectrum of H is contained in $[0, +\infty)$, that any negative eigenvalues of H must be of finite multiplicity, and that these eigenvalues have no accumulation point on the negative real axis. Thus, Theorem 5.8 is established.

Theorem 5.8 deals with the point spectrum of the Schroedinger operator on the negative real axis. For the more restricted (but still very large) class of potentials $V(\mathbf{r})$ for which there is an $R_0 > 0$ such that for all $r \geqslant R_0$ the potential $V(\mathbf{r})$ is continuous and $| r\, V(\mathbf{r})| \leqslant$ const, it can be shown (see Kato [1959, Theorem 1]) that the Schroedinger operator has no positive point spectrum, i.e., no positive eigenvalues.* The reader should now recall that these results on the point spectrum of the Schroedinger operator are well illustrated by the case of the hydrogen atom treated in §7 of Chapter II. In that case we were dealing with a point spectrum of the form $S_p^H = \{-a/n^2 : n = 1, 2, \ldots\}$, where a is a constant characteristic of the hydrogen atom.

*5.8. Appendix: Relations between Resolvents and Spectral Functions

We know already that if the spectral function E_λ of any self-adjoint operator H is given, then its resolvent $R(\zeta) = (H - \zeta)^{-1}$ can be

* See Iorio [1978] for recent results. General surveys can be found in Jörgens and Weidmann [1973], Reed and Simon [1978].

computed by means of the formula

$$\langle f \mid R(\zeta)g\rangle = \int_{\mathbb{R}^1} \frac{1}{\lambda - \zeta}\, d\langle f \mid E_\lambda g\rangle. \tag{5.64}$$

It is not so obvious, however, that the converse of this statement is also true—namely, that the spectral function E_λ of H can be computed from the resolvent $R(\zeta)$ (i.e., from the Green function of H, if it exists). To show that E_λ can always be computed from $R(\zeta)$, we need the following lemma.

Lemma 5.4. If the function $F(\lambda)$ has a continuous derivative $F'(\lambda)$ on $[a, b]$, and is Riemann–Stieltjes integrable on $[a, b]$ with respect to $\sigma(\lambda)$, then

$$\int_a^b F(\lambda)\, d_\lambda \sigma(\lambda) = F(b)\,\sigma(b) - F(a)\,\sigma(a) - \int_a^b F'(\lambda)\,\sigma(\lambda)\, d\lambda, \tag{5.65}$$

where the integral on the right-hand side of the above relation is a Riemann integral.

Proof. Taking $\lambda_k' = \lambda_k$ in the defining formula (3.18) and recalling that $\lambda_0 = a$ and $\lambda_n = b$, we get

$$\int_a^b F(\lambda)\, d_\lambda \sigma(\lambda) = \lim_{\substack{n\to\infty \\ \delta\to 0}} \left\{ \sum_{k=1}^n F(\lambda_k)\,\sigma(\lambda_k) - \sum_{k=1}^n F(\lambda_k)\,\sigma(\lambda_{k-1}) \right\}.$$

According to the mean value theorem of differential calculus,

$$F(\lambda_k) = F(\lambda_{k-1}) + F'(\bar\lambda_{k-1})(\lambda_k - \lambda_{k-1}), \qquad \lambda_{k-1} \leqslant \bar\lambda_{k-1} \leqslant \lambda_k .$$

Consequently, we have

$$\sum_{k=1}^n F(\lambda_k)\,\sigma(\lambda_k) - \sum_{k=1}^n F(\lambda_k)\,\sigma(\lambda_{k-1}) \tag{5.66}$$
$$= F(b)\,\sigma(b) - F(\lambda_1)\,\sigma(a) - \sum_{k=2}^n F'(\bar\lambda_{k-1})\,\sigma(\lambda_{k-1})(\lambda_k - \lambda_{k-1}).$$

Since $F'(\lambda)$ is uniformly continuous on $[a, b]$,

$$\lim_{\substack{n\to+\infty \\ \delta\to+0}} \sum_{k=2}^n F'(\bar\lambda_{k-1})\,\sigma(\lambda_{k-1})(\lambda_k - \lambda_{k-1}) = \int_a^b F'(\lambda)\,\sigma(\lambda)\, d\lambda,$$

and the expression in (5.66) converges to the right-hand side of (5.65).
Q.E.D.

Let us apply the result of the above lemma to the integral in (5.64). For Im $\zeta \neq 0$ this integral can be considered to be a Stieltjes integral, and since $(\lambda - \zeta)^{-1}$ is continuously differentiable, we get

$$\int_a^b \frac{1}{\lambda - \zeta}\, d\langle f \mid E_\lambda g\rangle = \frac{\langle f \mid E_b g\rangle}{b - \zeta} - \frac{\langle f \mid E_a g\rangle}{a - \zeta} + \int_a^b \frac{\langle f \mid E_\lambda g\rangle}{(\lambda - \zeta)^2}\, d\lambda.$$

Letting $a \to -\infty$ and $b \to +\infty$, we arrive at the result

$$(5.67) \qquad \int_{-\infty}^{+\infty} \frac{1}{\lambda - \zeta}\, d\langle f \mid E_\lambda g\rangle = \int_{-\infty}^{+\infty} \frac{\langle f \mid E_\lambda g\rangle}{(\lambda - \zeta)^2}\, d\lambda, \qquad \operatorname{Im} \zeta \neq 0,$$

from which we can derive Theorem 5.12.

Theorem 5.12. For any fixed $f, g \in \mathscr{H}$, the function $\langle f \mid R(\zeta)\, g\rangle$ is analytic at any complex ζ with Im $\zeta \neq 0$, and for $\lambda_1 \leqslant \lambda_2$

$$(5.68) \qquad \langle f \mid (E_{\lambda_2} + E_{\lambda_2 - 0}) g\rangle - \langle f \mid (E_{\lambda_1} + E_{\lambda_1 - 0}) g\rangle$$
$$= \frac{1}{\pi i} \lim_{\epsilon \to 0} \left\{ \int_{\lambda_1 + i\epsilon}^{\lambda_2 + i\epsilon} \langle f \mid R(\zeta) g\rangle\, d\zeta - \int_{\lambda_1 - i\epsilon}^{\lambda_2 - i\epsilon} \langle f \mid R(\zeta) g\rangle\, d\zeta \right\},$$

where the above integrations should be carried out along any piecewise smooth curves which do not intersect the real axis.

Proof. According to (5.67) the complex derivative

$$\frac{d}{d\zeta} \langle f \mid R(\zeta) g\rangle = \int_{-\infty}^{+\infty} \frac{2\langle f \mid E_\lambda g\rangle}{(\lambda - \zeta)^3}\, d\lambda$$

exists at any point $\zeta \in \mathbb{C}^1$ with Im $\zeta \neq 0$ since the above improper Riemann integral converges absolutely and uniformly in any closed neighborhood of ζ which is disjoint from the real axis. Hence, $\langle f \mid R(\zeta)\, g\rangle$ is analytic in the upper and in the lower complex plane.

Due to the analyticity of $\langle f \mid R(\zeta)\, g\rangle$ the integrals in (5.68) are independent of path. If we carry out the integration along straight lines and use (5.64) and (5.67), we get

$$I_\epsilon = \frac{1}{\pi i} \left\{ \int_{\lambda_1 + i\epsilon}^{\lambda_2 + i\epsilon} \langle f \mid R(\zeta) g\rangle\, d\zeta - \int_{\lambda_1 - i\epsilon}^{\lambda_2 - i\epsilon} \langle f \mid R(\zeta) g\rangle\, d\zeta \right\}$$
$$= \frac{1}{\pi i} \int_{\lambda_1}^{\lambda_2} \{\langle f \mid R(s + i\epsilon) g\rangle - \langle f \mid R(s - i\epsilon) g\rangle\}\, ds$$

(equation continues)

$$= \frac{1}{\pi i} \int_{\lambda_1}^{\lambda_2} ds \int_{-\infty}^{+\infty} d\lambda \langle f \mid E_\lambda g \rangle \left\{ \frac{1}{(s - \lambda + i\epsilon)^2} - \frac{1}{(s - \lambda - i\epsilon)^2} \right\}.$$

The integrand in the last of the above integrals is Lebesgue integrable in s and λ on the set $[\lambda_1, \lambda_2] \times \mathbb{R}^1$, since it is majorized by a function integrable on that set:

$$\left| \frac{\langle f \mid E_\lambda g \rangle}{(s - \lambda \pm i\epsilon)^2} \right| \leqslant \frac{\| f \| \, \| g \|}{(s - \lambda)^2 + \epsilon^2}.$$

Hence, the order of integration in s and λ can be reversed and, after carrying out the integration in s, we obtain

$$I_\epsilon = \frac{2}{\pi} \int_{-\infty}^{+\infty} \left\{ \frac{\epsilon}{(\lambda - \lambda_2)^2 + \epsilon^2} - \frac{\epsilon}{(\lambda - \lambda_1)^2 + \epsilon^2} \right\} \langle f \mid E_\lambda g \rangle \, d\lambda. \tag{5.69}$$

Upon splitting the interval of integration into $(-\infty, \lambda_i)$ and $[\lambda_i, +\infty)$, $i = 1, 2$, whereupon Lemma 4.2 can be applied separately to the first and second term in the above braces, we arrive at the relation

$$\lim_{\epsilon \to +0} I_\epsilon - \langle f \mid E_{\lambda_2 - 0} g \rangle + \langle f \mid E_{\lambda_1 - 0} g \rangle = \langle f \mid E_{\lambda_2} g \rangle - \langle f \mid E_{\lambda_1} g \rangle,$$

which is identical to (5.68). Q.E.D.

That the above theorem provides an important means of computing the spectral measure of a self-adjoint operator from its Green function.

Exercises

5.1. Prove that the set D_{G_0} defined in (5.16) is a Borel set of Lebesgue measure zero in $\mathbb{R}^6$.

5.2. Assuming that $\int_{\mathbb{R}^{2n}} | K(\alpha, \alpha')|^2 \, d\mu(\alpha) \, d\mu(\alpha') < +\infty$, show that the adjoint K^* of the operator $(Kf)(\alpha) = \int_{\mathbb{R}^n} K(\alpha, \alpha') f(\alpha') \, d\mu(\alpha')$ on $L^2(\mathbb{R}^n, \mu)$ is also an integral operator with the kernel $(K^*)(\alpha, \alpha') = K^*(\alpha', \alpha)$, and that the kernel $(K^*K)(\alpha, \alpha')$ of K^*K is equal to $\int_{\mathbb{R}^3} K^*(\alpha'', \alpha) \, K(\alpha'', \alpha') \, d\mu(\alpha'')$.

5.3. Prove that any integral operator K on $L^2(\mathbb{R}^n, \mu)$, which has an L^2 kernel $K(\alpha, \alpha')$, is a Hilbert–Schmidt operator, and that

$$\mathrm{Tr}(K^*K) = \int_{\mathbb{R}^{2n}} | K(\alpha, \alpha')|^2 \, d\mu(\alpha) \, d\mu(\alpha').$$

5.4. Give the reason why the limit in the mean $K(\alpha, \alpha')$ of a sequence of Hilbert–Schmidt kernels $K_n(\alpha, \alpha')$,

$$\lim_{n\to+\infty} \int_{\mathbb{R}^{2m}} | K(\alpha, \alpha') - K_n(\alpha, \alpha')|^2 \, d\mu(\alpha) \, d\mu(\alpha') = 0$$

is a kernel of Hilbert–Schmidt type.

5.5. Show that if A is Hilbert–Schmidt operator in a separable Hilbert space and B is a bounded operator then BA is a Hilbert–Schmidt operator.

5.6. Let $A_1, A_2, \ldots$ be a sequence of bounded operators for which $\sum_{k=1}^{\infty} \| A_k \| < +\infty$. Prove that the uniform limit

$$A = \underset{n\to+\infty}{\text{u-lim}} \sum_{k=1}^{n} A_k$$

exists and is a bounded operator with

$$\| A \| \leqslant \sum_{k=1}^{\infty} \| A_k \|.$$

5.7. Use the results of Exercise 5.6 to show that

$$\underset{n\to+\infty}{\text{u-lim}} \sum_{k=1}^{n} c_k A^k$$

exists if $\| A \| < 1$ and $| c_k | \leqslant M$ for all $k = 1, 2, \ldots$.

5.8. Prove that the self-adjoint operator A is bounded from below if and only if $\langle f \mid Af \rangle \geqslant m_A \| f \|^2$ for all $f \in \mathscr{D}_A$.

6. Distorted Plane Waves in Potential Scattering

6.1. Potentials of Rollnik Class

In §5.5 we have learned that for square integrable potentials the equation (5.24) could be solved by the Fredholm method, and on first sight it might appear that exactly the same method is applicable to the full advanced and retarded Green functions $G^{(\pm)}(\mathbf{r}, \mathbf{r}'; \lambda)$, $\lambda \in \mathsf{S}_c^H$, that are required in the solution-type Lippmann–Schwinger equations (4.74). Indeed, as discussed in §5.8, under reasonable conditions on $V(\mathbf{r})$ (which from now on we shall take for granted) we have

$$\mathsf{S}_c^H \subset \mathsf{S}_c^{H_0} = [0, \infty), \qquad \mathsf{S}_p^H \cap [0, \infty) = \varnothing, \tag{6.1}$$

and formally taking in (5.24) and (5.25) the limit $\operatorname{Im} \zeta \to \pm 0$ at fixed

$\lambda = \mathrm{Re}\,\zeta \in \mathrm{S}_c{}^H$ we obtain

$$G^{(\pm)}(\mathbf{r}, \mathbf{r}'; \lambda) = G_0^{(\pm)}(\mathbf{r}, \mathbf{r}'; \lambda) + \int_{\mathbb{R}^3} K^{(\pm)}(\mathbf{r}, \mathbf{r}''; \lambda)\, G^{(\pm)}(\mathbf{r}'', \mathbf{r}'; \lambda)\, d\mathbf{r}'', \tag{6.2}$$

$$K^{(\pm)}(\mathbf{r}, \mathbf{r}'; \lambda) = G_0^{(\pm)}(\mathbf{r}, \mathbf{r}'; \lambda)\, V(\mathbf{r}'). \tag{6.3}$$

Unfortunately, the kernels in (6.3) are not Hilbert–Schmidt since

$$| K^{(\pm)}(\mathbf{r}, \mathbf{r}'; \lambda)|^2 = \frac{m^2}{4\pi^2} \frac{| V(\mathbf{r}')|^2}{| \mathbf{r} - \mathbf{r}' |^2} \tag{6.4}$$

obviously is not integrable in $\mathbf{r}$ over $\mathbb{R}^3$.

The Lippmann–Schwinger equation (4.67), which shares the same kernel with (6.2),

$$\Phi_{\mathbf{k}}^{(\pm)}(\mathbf{r}) = \Phi_{\mathbf{k}}(\mathbf{r}) + \int_{\mathbb{R}^3} K^{(\pm)}(\mathbf{r}, \mathbf{r}'; \mathbf{k}^2/2m)\, \Phi_{\mathbf{k}}^{(\pm)}(\mathbf{r}')\, d\mathbf{r}' \tag{6.5}$$

runs into exactly the same problem.

The most frequently employed* way out of this impasse consists of factorizing $V(\mathbf{r})$ as $| V(\mathbf{r}) |^{1/2} V^{1/2}(\mathbf{r})$, where

$$V^{1/2}(\mathbf{r}) = \begin{cases} | V(\mathbf{r})|^{1/2} \operatorname{sgn} V(\mathbf{r}) & \text{if } V(\mathbf{r}) \neq 0, \\ \eta \exp(-\mathbf{r}^2),\ \eta > 0, & \text{if } V(\mathbf{r}) = 0, \end{cases} \tag{6.6}$$

and then multiplying (6.2) or (6.5) from the left by $V^{1/2}(\mathbf{r})$. For example, in case of (6.5), upon setting

$$\tilde{\Phi}_{\mathbf{k}}(\mathbf{r}) = V^{1/2}(\mathbf{r})\, \Phi_{\mathbf{k}}(\mathbf{r}), \qquad \tilde{\Phi}_{\mathbf{k}}^{(\pm)}(\mathbf{r}) = V^{1/2}(\mathbf{r})\, \Phi_{\mathbf{k}}^{(\pm)}(\mathbf{r}), \tag{6.7}$$

the resulting equations can be written in the form

$$\tilde{\Phi}_{\mathbf{k}}^{(\pm)}(\mathbf{r}) = \tilde{\Phi}_{\mathbf{k}}(\mathbf{r}) + \int_{\mathbb{R}^3} \tilde{K}^{(\pm)}(\mathbf{r}, \mathbf{r}'; \mathbf{k}^2/2m)\, \tilde{\Phi}_{\mathbf{k}}^{(\pm)}(\mathbf{r}')\, d\mathbf{r}' \tag{6.8}$$

where the new kernels

$$\tilde{K}^{(\pm)}(\mathbf{r}, \mathbf{r}'; \lambda) = V^{1/2}(\mathbf{r})\, G_0^{(\pm)}(\mathbf{r}, \mathbf{r}'; \lambda)\, | V(\mathbf{r}')|^{1/2} \tag{6.9}$$

are Hilbert–Schmidt for the class of potentials specified in the following definition.

* Apparently, this factorization method was first discovered by Rollnik [1956], and then independently rediscovered by many others (Schwartz [1960], Grossmann and Wu [1962], Scadron *et al.* [1964], etc.). For the sake of simplicity, in (6.11) and (6.12) we shall assume that $V(\mathbf{r}) \neq 0$ almost everywhere in $\mathbb{R}^3$.

Definition 6.1. A real function $V(\mathbf{r})$ is of *Rollnik class* if it is Borel measurable on $\mathbb{R}^3$ and if its Rollnik norm $\| V \|_{\mathrm{R}}$ exists, where

$$\| V \|_{\mathrm{R}}^2 = \int_{\mathbb{R}^6} \frac{| V(\mathbf{r}) V(\mathbf{r}')|}{| \mathbf{r} - \mathbf{r}' |^2} \, d\mathbf{r} \, d\mathbf{r}'. \tag{6.10}$$

We easily see that if $V(\mathbf{r})$ in (6.9) is of Rollnik class then $\tilde{K}^{(\pm)}(\lambda)$ are Hilbert–Schmidt:

$$\langle \tilde{K}^{(\pm)}(\lambda) \mid \tilde{K}^{(\pm)}(\lambda) \rangle_2 = \int_{\mathbb{R}^6} | \tilde{K}^{(\pm)}(\mathbf{r}, \mathbf{r}'; \lambda)|^2 \, d\mathbf{r} \, d\mathbf{r}' = (m^2/4\pi^2) \, \| V \|_{\mathrm{R}}^2 . \tag{6.11}$$

A frequently encountered type of potential $V(\mathbf{r})$ of Rollnik class is one that is continuous, except possibly for a divergent behavior $V(\mathbf{r}) \sim r^{-\alpha}$, $0 < \alpha < 1$, as $r \to +0$, and bounded by $r^{-2-\epsilon_0}$, $\epsilon_0 > 0$, as $r \to \infty$ (see Exercise 6.6).

6.2. Fredholm Series Expressions for Distorted Plane Waves

Another great advantage of (6.8) over (6.5) is that the inhomogeneous term $\tilde{\Phi}_{\mathbf{k}}(\mathbf{r})$ in (6.8) is square integrable if $V(\mathbf{r})$ is integrable on $\mathbb{R}^3$:

$$\| \tilde{\Phi}_{\mathbf{k}} \|^2 = \int_{\mathbb{R}^3} | \tilde{\Phi}_{\mathbf{k}}(\mathbf{r})|^2 \, d\mathbf{r} = (2\pi)^{-3} \int_{\mathbb{R}^3} | V(\mathbf{r})| \, d\mathbf{r} = (2\pi)^{-3} \, \| V \|_1 . \tag{6.12}$$

Hence the Fredholm theory becomes immediately applicable, and we can state the following theorem.

Theorem 6.1. Suppose that $V(\mathbf{r})$ is of Rollnik class and that* for some $\epsilon_0 > 0$

$$V(\mathbf{r}) = O(r^{-3-\epsilon_0}), \qquad r \to \infty. \tag{6.13}$$

Then the modified Fredholm determinants $\tilde{d}_n^{(\pm)}(\lambda)$ and minors $\tilde{D}_n^{(\pm)}(\mathbf{r}, \mathbf{r}'; \lambda)$ of $\tilde{K}^{(\pm)}(\mathbf{r}, \mathbf{r}'; \lambda)$ in (6.9) [as given by (5.30) and (5.31), respectively] exist, and the infinite series

$$\tilde{d}^{(\pm)}(\lambda) = \sum_{n=0}^{\infty} \tilde{d}_n^{(\pm)}(\lambda) \tag{6.14}$$

converge and are different from zero for all $\lambda \in (0, \infty)$, with the possible

* Recall that $f(\mathbf{r}) = O(g(\mathbf{r}))$ when $| \mathbf{r} | \to \infty$ if there are constants C and R such that $| f(\mathbf{r}) \leqslant C \, | g(\mathbf{r})|$ for all $| \mathbf{r} | \geqslant R$.

exception of a compact set $\mathscr{S}_V^1$ of Lebesgue measure zero for which $[0, +\infty) - \mathscr{S}_V^1 \subset S_c^H$. Furthermore, for all $\mathbf{k}$ outside the (compact) exceptional set

$$\mathscr{S}_V^3 = \{\mathbf{k} : (\mathbf{k}^2/2m) \in \mathscr{S}_V^1\} \tag{6.15}$$

of Lebesgue measure zero in $\mathbb{R}^3$, the Lippmann–Schwinger equations (6.5) have the unique solution

$$\Phi_{\mathbf{k}}^{(\pm)}(\mathbf{r}) = \Phi_{\mathbf{k}}(\mathbf{r}) + \frac{V^{-1/2}(\mathbf{r})}{\tilde{d}^{(\pm)}(\mathbf{k}^2/2m)} \times \int_{\mathbb{R}^3} \tilde{D}^{(\pm)}\left(\mathbf{r}, \mathbf{r}'; \frac{\mathbf{k}^2}{2m}\right) V^{1/2}(\mathbf{r}')\, \Phi_{\mathbf{k}}(\mathbf{r}')\, d\mathbf{r}', \tag{6.16}$$

$$\tilde{D}^{(\pm)}(\mathbf{r}, \mathbf{r}'; \lambda) = \sum_{n=0}^{\infty} \tilde{D}_n^{(\pm)}(\mathbf{r}, \mathbf{r}'; \lambda). \tag{6.17}$$

The fact that $\mathscr{S}_V^1$ (and therefore also $\mathscr{S}_V^3$) is compact and of measure zero is established in Appendix 6.7, whereas the rest of the statements are an immediate consequence of Theorems 5.3 and 5.4 since $V(\mathbf{r})$ is integrable on $\mathbb{R}^3$ so that by (6.12) $\| \tilde{\Phi}_{\mathbf{k}} \| < \infty$. Indeed, due to (6.13) $V(\mathbf{r})$ is integrable outside some sphere of radius R around the origin, whereas inside that sphere its integrability follows by Theorems 3.9 and 3.13 in Chapter II from the existence of $\| V \|_{\mathbf{R}}$.

Theorem 6.1 establishes that the Lippmann–Schwinger equations (4.67) possess for almost all $\mathbf{k} \in \mathbb{R}^3$ unique solutions, but it does not establish that these solutions are the distorted plane waves for which (4.25) is true. Indeed, (4.25) has been assumed in the course of deriving (4.67), and now we have to prove that this assumption was justified. The following lemma provides the first of many steps in that direction.

Lemma 6.1. If $V(\mathbf{r}) = O(r^{-3-\epsilon_0})$ is locally square integrable and of Rollnik class then

$$\hat{\psi}_{\pm}(\mathbf{k}) = \langle \Phi_{\mathbf{k}}^{(\pm)} \mid \psi \rangle = \text{l.i.m.} \int_{\mathbb{R}^3} \Phi_{\mathbf{k}}^{(\pm)*}(\mathbf{r})\, \psi(\mathbf{r})\, d\mathbf{r} \tag{6.18}$$

exist for any $\mathbf{k} \in \mathbb{R}^3 - \mathscr{S}_V^3$ provided $\Phi_{\mathbf{k}}^{(\pm)}(\mathbf{r})$ are the solutions (6.16) of (6.5), and if $B \subset (0, +\infty)$ is a compact set disjoint from $\mathscr{S}_V^1$ then

$$\| E^H(B)\psi \|^2 = \int_{\mathbf{k}^2/2m \in B} | \hat{\psi}_{\pm}(\mathbf{k})|^2\, d\mathbf{k}. \tag{6.19}$$

Proof. To prove the lemma, we first have to make a few remarks which are also of independent interest.

Comparing (5.5) with (5.10) we see that for fixed $\mathbf{r} \in \mathbb{R}^3$,

$$G_0(\mathbf{r}, \mathbf{r}'; \zeta) = (2\pi)^{-3/2} \operatorname{l.i.m.} \int_{\mathbb{R}^3} \frac{\Phi_{\mathbf{k}}(\mathbf{r})}{\zeta - (\mathbf{k}^2/2m)} e^{-i\mathbf{k}\mathbf{r}'} \, d\mathbf{k}. \tag{6.20}$$

Consequently, by Theorem 4.6 in Chapter III, for any given $\mathbf{r} \in \mathbb{R}^3$,

$$\frac{\Phi_{\mathbf{k}}(\mathbf{r})}{\zeta - (\mathbf{k}^2/2m)} = (2\pi)^{-3/2} \int_{\mathbb{R}^3} G_0(\mathbf{r}, \mathbf{r}'; \zeta) \, e^{i\mathbf{k}\mathbf{r}'} \, d\mathbf{r}' \tag{6.21}$$

where l.i.m. could be dropped, since by (5.6) $G_0(\mathbf{r}, \mathbf{r}'; \zeta)$ is obviously integrable in $\mathbf{r}'$ over $\mathbb{R}^3$ at fixed $\mathbf{r} \in \mathbb{R}^3$. Since by (5.52) and Theorem 5.7, $G(\mathbf{r}, \cdot; \zeta) \in L^2(\mathbb{R}^3)$ for almost all $\mathbf{r} \in \mathbb{R}^3$, by analogy with (6.21) we can define

$$\frac{\Phi_{\mathbf{k}}(\mathbf{r}; \zeta)}{\zeta - (\mathbf{k}^2/2m)} = (2\pi)^{-3/2} \operatorname{l.i.m.} \int_{\mathbb{R}^3} G(\mathbf{r}, \mathbf{r}'; \zeta) \, e^{i\mathbf{k}\mathbf{r}'} \, d\mathbf{r}' \tag{6.22}$$

for $\zeta \in \mathbb{C}^1 - S^H$. Hence by (5.24) and (5.25),

$$\Phi_{\mathbf{k}}(\mathbf{r}; \zeta) = \Phi_{\mathbf{k}}(\mathbf{r}) + \int_{\mathbb{R}^3} K(\mathbf{r}, \mathbf{r}''; \zeta) \, \Phi_{\mathbf{k}}(\mathbf{r}''; \zeta) \, d\mathbf{r}'', \tag{6.23}$$

and upon introducing by analogy with (6.7)

$$\tilde{\Phi}_{\mathbf{k}}(\mathbf{r}; \zeta) = V^{1/2}(\mathbf{r}) \, \Phi_{\mathbf{k}}(\mathbf{r}; \zeta), \tag{6.24}$$

we see that (6.23) is equivalent to

$$\tilde{\Phi}_{\mathbf{k}}(\mathbf{r}; \zeta) = \tilde{\Phi}_{\mathbf{k}}(\mathbf{r}) + \int_{\mathbb{R}^3} \tilde{K}(\mathbf{r}, \mathbf{r}''; \zeta) \, \tilde{\Phi}_{\mathbf{k}}(\mathbf{r}''; \zeta) \, d\mathbf{r}'', \tag{6.25}$$

$$\tilde{K}(\mathbf{r}, \mathbf{r}'; \zeta) = V^{1/2}(\mathbf{r}) \, G_0(\mathbf{r}, \mathbf{r}'; \zeta) \, | \, V(\mathbf{r}')|^{1/2}. \tag{6.26}$$

Since $K(\mathbf{r}, \mathbf{r}'; \zeta)$ is obviously Hilbert–Schmidt, we get in complete analogy with (6.16):

$$\Phi_{\mathbf{k}}(\mathbf{r}; \zeta) = \Phi_{\mathbf{k}}(\mathbf{r}) + \frac{V^{-1/2}(\mathbf{r})}{\tilde{d}(\zeta)} \int_{\mathbb{R}^3} \tilde{D}(\mathbf{r}, \mathbf{r}'; \zeta) \, \tilde{\Phi}_{\mathbf{k}}(\mathbf{r}') \, d\mathbf{r}', \tag{6.27}$$

$$\tilde{D}(\mathbf{r}, \mathbf{r}'; \zeta) = \sum_{n=0}^{\infty} \tilde{D}_n(\mathbf{r}, \mathbf{r}'; \zeta). \tag{6.28}$$

If we now define by analogy with (6.18)

$$\hat{\psi}(\mathbf{k}; \zeta) = \text{l.i.m.} \int_{\mathbb{R}^3} \Phi_{\mathbf{k}}^*(\mathbf{r}; \zeta^*)\, \psi(\mathbf{r})\, d\mathbf{r}, \tag{6.29}$$

we see from (6.22), (5.51), and (5.52) that

$$(G(\zeta)\psi)^{\sim}(\mathbf{k}) = (\zeta - (\mathbf{k}^2/2m)^{-1}\hat{\psi}(\mathbf{k}; \zeta). \tag{6.30}$$

Consequently, we obtain from the first resolvent equation (3.62)

$$\begin{aligned} \langle \psi \mid [G(\zeta) - G(\zeta^*)]\psi\rangle &= (\zeta^* - \zeta)\langle G(\zeta)\, \psi \mid G(\zeta)\psi\rangle \\ &= 2i \operatorname{Im} \zeta \int_{\mathbb{R}^3} \left| \frac{\hat{\psi}(\mathbf{k}; \zeta)}{\zeta - (\mathbf{k}^2/2m)} \right|^2 d\mathbf{k}. \end{aligned} \tag{6.31}$$

Thus the connection of $\hat{\psi}_{\pm}(\mathbf{k})$ in (6.18) with the spectral function of H finally becomes evident as we consider that by Theorem 5.12 we have

$$\langle \psi \mid (E_b{}^H - E_a{}^H)\psi\rangle = \frac{1}{\pi} \lim_{\epsilon\to+0} \int_a^b d\lambda \int_{\mathbb{R}^3} \frac{\epsilon \mid \hat{\psi}(\mathbf{k}; \lambda + i\epsilon)|^2}{(\lambda - (\mathbf{k}^2/2m))^2 + \epsilon^2}\, d\mathbf{k} \tag{6.32}$$

when we take into account that $E_\lambda{}^H = E^H_{\lambda-0}$ for $\lambda \notin \mathscr{S}_V{}^1$ since no point in $[0, \infty) - \mathscr{S}_V{}^1$ belongs to $S_p{}^H$. Indeed, if $H\psi_0 = \lambda_0\psi_0$ for some $\lambda_0 \geqslant 0$ then $\psi_0 \in \mathscr{D}_H \subset \mathscr{D}_V$, so that $V(\mathbf{r}) \mid \psi_0(\mathbf{r}) \mid^2$ is integrable, and therefore its absolute value is integrable (see Chapter II, Definition 3.5); consequently, $V^{1/2}\psi \in L^2(\mathbb{R}^3)$ and

$$\tilde{K}^{(+)}(\lambda_0)(V^{1/2}\psi) = V^{1/2}(\lambda_0 - H_0)^{-1}V\psi = V^{1/2}(\lambda_0 - H_0)^{-1}(\lambda_0 - H_0)\psi = V^{1/2}\psi,$$

which implies that $(1 - \tilde{K}^{(+)}(\lambda_0))^{-1}$ does not exist, and therefore by Theorem 5.4, $\tilde{d}(\lambda_0) = 0$, i.e., $\lambda_0 \in \mathscr{S}_V{}^1$.

On the formal level, (6.19) with $B = [a, b] \subset (0, \infty)$ follows from (6.32) and Lemma 4.2 by noting that

$$\Phi_{\mathbf{k}}^{(+)}(\mathbf{r}) = \Phi_{-\mathbf{k}}^{(-)*}(\mathbf{r}) = \lim_{\epsilon\to+0} \Phi_{\mathbf{k}}(\mathbf{r}; (\mathbf{k}^2/2m) + i\epsilon). \tag{6.33}$$

and consequently for $\mathbf{k} \notin \mathscr{S}_V{}^3$

$$\hat{\psi}_{\pm}(\mathbf{k}) = \lim_{\epsilon\to+0} \hat{\psi}(\mathbf{k}; (\mathbf{k}^2/2m) \pm i\epsilon). \tag{6.34}$$

The rigorous proof is, however, considerably lengthier since one has to justify the interchange of orders of integration in λ and $\mathbf{k}$ in (6.32) and the application of Lemma 4.2 under the double integral sign with an integrand $|\hat{\psi}(\mathbf{k}; \lambda + i\epsilon)|^2$ that is ϵ-dependent. As we shall see, this can be achieved by proving that $|\hat{\psi}(\mathbf{k}; \zeta)|^2$ is uniformly bounded in $\mathbf{k}$ and ζ as well as continuous in ζ in the complex half-planes Im $\zeta \geqslant 0$ or Im $\zeta \leqslant 0$, respectively, wherever $\mathscr{d}(\zeta) \neq 0$, so that Fubini's theorem and Lebesgue's dominated convergence theorems can be applied in a routine manner.

Expanding the determinant in (5.31) along the first row or first column we get, respectively,

$$D_n(\alpha, \alpha') = d_n K(\alpha, \alpha') + \int_{\mathbb{R}^m} D_{n-1}(\alpha, \alpha_1) K(\alpha_1, \alpha') \, d^m\alpha_1$$
$$= d_n K(\alpha, \alpha') + \int_{\mathbb{R}^m} K(\alpha, \alpha_1) D_{n-1}(\alpha_1, \alpha') \, d^m\alpha_1 . \tag{6.35}$$

Combining these two relations we obtain

$$|D_n(\alpha, \alpha')| = \Big| d_n K(\alpha, \alpha') + d_{n-1} \int_{\mathbb{R}^m} K(\alpha, \alpha_1) K(\alpha_1, \alpha') \, d^m\alpha_1$$
$$+ \int_{\mathbb{R}^m} K(\alpha, \alpha_1) K(\alpha_2, \alpha') D_{n-2}(\alpha_1, \alpha_2) \, d^m\alpha_1 \, d^m\alpha_2 \Big|$$
$$\leqslant |d_n| \, |K(\alpha, \alpha')|$$
$$+ (|d_{n-1}| + \| D_{n-2} \|_2) \, \| K(\alpha, \cdot) \| \, \| K(\cdot, \alpha') \|. \tag{6.36}$$

In the case of (6.26)–(6.28) this inequality yields

$$|\tilde{D}_n(\mathbf{r}, \mathbf{r}'; \zeta)| \leqslant C_n \, |V^{1/2}(\mathbf{r})| \left[\frac{1}{|\mathbf{r} - \mathbf{r}'|} + \left(\int_{\mathbb{R}^3} \frac{|V(\mathbf{r}_1)| \, d\mathbf{r}_1}{|\mathbf{r} + \mathbf{r}_1|^2} \right)^{1/2} \right.$$
$$\left. \times \left(\int_{\mathbb{R}^3} \frac{|V(\mathbf{r}_2)| \, d\mathbf{r}_2}{|\mathbf{r}_2 - \mathbf{r}'|^2} \right)^{1/2} \right] |V(\mathbf{r}')|^{1/2}, \tag{6.37}$$

where due to (5.47) and (5.48) we can choose the positive constants C_n to be independent of $\mathbf{r}$, $\mathbf{r}'$, and ζ, and such that $C = \sum_{n=0}^{\infty} C_n < \infty$. Consequently, by (6.27),

$$|\Phi_{\mathbf{k}}^{(+)}(\mathbf{r}; \zeta)| \leqslant (2\pi)^{-3/2} [1 + F_V(\mathbf{r}; \zeta)], \qquad \zeta \in \{\zeta \mid \zeta \in \mathbb{C}^1, \mathscr{d}(\zeta) \neq 0\}, \tag{6.38}$$

$$F_V(\mathbf{r}; \zeta) = \frac{C}{|\mathscr{d}(\zeta)|} \left[\int_{\mathbb{R}^3} \frac{|V(\mathbf{r}')| \, d\mathbf{r}'}{|\mathbf{r} - \mathbf{r}'|} + \left(\int_{\mathbb{R}^3} \frac{|V(\mathbf{r}_1)| \, d\mathbf{r}_1}{|\mathbf{r} - \mathbf{r}_1|^2} \right)^{1/2} \right.$$
$$\left. \times \int_{\mathbb{R}^3} d\mathbf{r}' \, |V(\mathbf{r}')| \left(\int_{\mathbb{R}^3} \frac{V(\mathbf{r}_2) \, d\mathbf{r}_2}{|\mathbf{r}' - \mathbf{r}_2|^2} \right)^{1/2} \right]. \tag{6.39}$$

Due to the conditions imposed on the potential, $F_V(\mathbf{r}; \zeta)$ is locally square integrable and bounded at infinity (see Exercise 6.5). Since (5.47) yields the uniform convergence and therefore, due to the continuity of $\tilde{d}(\zeta)$ for $\operatorname{Im} \zeta \geqslant 0$ or $\operatorname{Im} \zeta \leqslant 0$, also the continuity of $\tilde{d}(\zeta)$, this boundedness is uniform when ζ varies over any compact set not containing zeros of $\tilde{d}(\zeta)$. Hence, the procedure described earlier leading to (6.19) for $B = [a, b]$ can be applied if $\psi(\mathbf{r}) \in \mathscr{C}_b^\infty(\mathbb{R}^3)$. The outcome can be then extended to arbitrary $\psi \in L^2(\mathbb{R}^3)$ by taking the limits in the mean appearing in (6.18).

The general case of compact $B \subset (0, \infty)$ disjoint from $\mathscr{S}_V^1$ can be reduced to the above special case by using Theorems 1.3 and 1.7 in Chapter II. Q.E.D.

6.3. Asymptotic Completeness and the Generalized Parseval's Equality

We recall from Theorem 5.8 that locally square integrable potentials which satisfy (6.13) give rise to a Hamiltonian H whose continuous spectrum contains no negative values, so that by Theorem 6.1

$$[0, \infty) - \mathscr{S}_V^1 \subset \mathsf{S}_c^H \subset [0, \infty). \tag{6.40}$$

Let us introduce for each eigenvalue $\lambda \in \mathsf{S}_p^H$ an orthonormal basis $\{\Phi_{\lambda,\gamma}\}$ in the proper subspace of that eigenvalue (which by Theorem 5.8 is of finite multiplicity if $\lambda < 0$). Then for any $f, g \in L^2(\mathbb{R}^3)$,

$$\langle f \mid E^H(\mathsf{S}_p^H) g\rangle = \sum_{\lambda \in \mathsf{S}_p^H} \sum_{\gamma} \hat{f}^*(\lambda, \gamma)\, \hat{g}(\lambda, \gamma), \tag{6.41}$$

$$\hat{f}^*(\lambda, \gamma) = \langle f \mid \Phi_{\lambda,\gamma}\rangle, \qquad \hat{g}(\lambda, \gamma) = \langle \Phi_{\lambda,\gamma} \mid g\rangle. \tag{6.42}$$

Lemma 6.1 suggests that for the continuous spectrum we might expect a similar relation, so that the following *generalized Parseval's equality* [cf. (4.15) in Chapter I] might hold:

$$\begin{aligned}\langle f \mid g\rangle &= \langle f \mid E^H(\mathsf{S}_p^H) g\rangle + \langle f \mid E^H(\mathsf{S}_c^H) g\rangle \\ &= \sum_{\lambda \in \mathsf{S}_p^H} \sum_{\gamma} \hat{f}^*(\lambda, \gamma)\, \hat{g}(\lambda, \gamma) + \int_{\mathbb{R}^3} \hat{f}_\pm^*(\mathbf{k})\, \hat{g}_\pm(\mathbf{k})\, d\mathbf{k}.\end{aligned} \tag{6.43}$$

The nontrivial part of equation (6.43) is the last expression, which follows from the next theorem.

Theorem 6.2. Suppose that $V(\mathbf{r})$ is locally square integrable, of Rollnik class, and $O(r^{-3-\epsilon_0})$ at infinity. Then the continuous spectrum of H satisfies (6.40), and for any Borel set $B \subset S_c^H$,

$$(E^H(B)\psi)(\mathbf{r}) = \text{l.i.m.} \int_{\mathbf{k}^2/2m \in B} \Phi_{\mathbf{k}}^{(\pm)}(\mathbf{r})\, \hat{\psi}_\pm(\mathbf{k})\, d\mathbf{k}, \tag{6.44}$$

$$\hat{\psi}_\pm(\mathbf{k}) = \langle \Phi_{\mathbf{k}}^{(\pm)} \mid \psi \rangle = \text{l.i.m.} \int_{\mathbb{R}^3} \Phi_{\mathbf{k}}^{(\pm)*}(\mathbf{r})\, \psi(\mathbf{r})\, d\mathbf{r}, \tag{6.45}$$

for all $\psi \in L^2(\mathbb{R}^3)$. Furthermore, for any f, $g \in L^2(\mathbb{R}^3)$,

$$\langle f \mid E^H(B) g \rangle = \int_{\mathbf{k}^2/2m \in B} \hat{f}_\pm^*(\mathbf{k})\, \hat{g}_\pm(\mathbf{k})\, d\mathbf{k}. \tag{6.46}$$

Proof. When B is compact and disjoint from $\mathscr{S}_V^1$, (6.46) is obtained from (6.19) by using formula (5.13) in Chapter III.

To arrive at (6.44) for such B we choose $f(\mathbf{r}) \in \mathscr{C}_b^\infty(\mathbb{R}^3)$, so that by (6.38)

$$(2\pi)^{3/2} \mid \Phi_{\mathbf{k}}^{(\pm)}(\mathbf{r}) f(\mathbf{r}) \mid \leqslant [1 + F_V(\mathbf{r}; \mathbf{k}^2/2m)] \mid f(\mathbf{r}) \mid \tag{6.47}$$

is integrable in $(\mathbf{r}, \mathbf{k})$ over $\mathbb{R}^3 \times \Lambda_0^{-1}(B)$. Hence, for arbitrary $g \in L^2(\mathbb{R}^3)$, the reversal in order of integrations in

$$\begin{aligned} \langle f \mid E^H(B) g \rangle &= \int_{\Lambda_0^{-1}(B)} d\mathbf{k}\, \tilde{g}_\pm(\mathbf{k}) \int_{\mathbb{R}^3} d\mathbf{r}\, \Phi_{\mathbf{k}}^{(\pm)}(\mathbf{r}) f^*(\mathbf{r}) \\ &= \int_{\mathbb{R}^3} d\mathbf{r}\, f^*(\mathbf{r}) \int_{\Lambda_0^{-1}(B)} d\mathbf{k}\, \Phi_{\mathbf{k}}^{(\pm)}(\mathbf{r})\, \tilde{g}_\pm(\mathbf{k}) \end{aligned} \tag{6.48}$$

is justified by Fubini's theorem, and we conclude that

$$(E^H(B) g)(\mathbf{r}) = \int_{\Lambda_0^{-1}(B)} \Phi_{\mathbf{k}}^{(\pm)}(\mathbf{r})\, \tilde{g}_\pm(\mathbf{k})\, d\mathbf{k}. \tag{6.49}$$

If we interpret "l.i.m." in (6.44) for any Borel set $B \subset S_c^H$ as a procedure in which the integration is first performed over sets

$$\Lambda_0^{-1}(B \cap B_n) = \{\mathbf{k} : \Lambda_0(\mathbf{k}) = (\mathbf{k}^2/2m) \in B \cap B_n\}, \tag{6.50}$$

with B_n compact and disjoint from $\mathscr{S}_V^1$, and then the limit in the mean is taken for $B_1 \subset B_2 \subset \cdots \to [0, \infty) - \mathscr{S}_V^1$, we finally arrive at the general result (6.46). Q.E.D.

In view of (6.45), the formula (6.46) can be written in the symbolic notation introduced in (4.82) as follows,

$$E^H(B) = \int_{\Lambda_0^{-1}(B)} |\Phi_{\mathbf{k}}^{(\pm)}) \, d\mathbf{k} (\Phi_{\mathbf{k}}^{(\pm)}|, \tag{6.51}$$

and therefore (6.41)–(6.43) implies that

$$\sum_{\lambda \in \mathsf{S}_p{}^H} \sum_{\gamma} |\Phi_{\lambda,\gamma}\rangle\langle\Phi_{\lambda,\gamma}| + \int_{(\mathbf{k}^2/2m)\in \mathsf{S}_c{}^H} |\Phi_{\mathbf{k}}^{(\pm)}) \, d\mathbf{k} (\Phi_{\mathbf{k}}^{(\pm)}| = 1. \tag{6.52}$$

Thus, Equation (6.52) is intimately related to the following notion of asymptotic completeness.

Definition 6.2. A quantum mechanical theory on $\mathscr{H}$ with given free and total Hamiltonians H_0 and H is said to be *asymptotically complete* if any interacting state $\Psi(t)$ orthogonal to the all bound states of the theory is a scattering state, i.e., it has incoming or outgoing (physical) asymptotic states $\Psi^{\text{ex}}(t)$, ex $=$ in, out. The wave operators $\Omega_\pm$ of a theory are said to be *complete* if and only if the theory is asymptotically complete.

In §1.4 we denoted by $\mathscr{H}_{\text{b}}^{(2)}$ the closed subspace spanned by all bound states in the Hilbert space $\mathscr{H}^{(2)}$ of a two-particle system. In §1.7, upon eliminating from the problem the center of mass motion, we were left with the Hilbert space $\mathscr{H}^{(1)}$ of the reduced particle whose bound states spanned the closed subspace $\mathscr{H}_{\text{b}}^{(1)}$. In the present context of wave mechanics for spinless particles we have $\mathscr{H}^{(1)} = L^2(\mathbb{R}^3)$, so that

$$\mathscr{H}_{\text{b}}^{(1)} = E^H(\mathsf{S}_p{}^H) L^2(\mathbb{R}^3) = L^2(\mathbb{R}^3) \ominus L^2_{H_c}(\mathbb{R}^3). \tag{6.53}$$

Hence the theory is asymptotically complete (and $\Omega_\pm$ are complete) if and only if*

$$\mathsf{R}_\pm = L^2_{H_c}(\mathbb{R}^3) = E^H(\mathsf{S}_c{}^H) L^2(\mathbb{R}^3). \tag{6.54}$$

On first sight it might appear that by establishing (6.43) or, equivalently, (6.52), we have already established the completeness of $\Omega_\pm$ for

* Motivated by primarily mathematical considerations, many authors follow the custom of calling $\Omega_\pm$ complete if and only if the projector onto $\mathsf{R}_\pm$ coincides with $E^H(\mathsf{S}^H_{\text{ac}})$ rather than $E^H(\mathsf{S}_c{}^H)$, where S^H_{ac} is the absolutely continuous spectrum of H. As a rule they then proceed to prove (see, e.g., Amrein *et al.* [1977]) that the *singularly continuous spectrum* $\mathsf{S}^H_{\text{sc}} = \mathsf{S}_c{}^H - \mathsf{S}^H_{\text{ac}}$ is empty, so that in practical terms their definition coincides with Definition 6.2, which is dictated by physical rather than mathematical circumstances (see, e.g., Schechter [1978]).

the class of potentials specified in Theorem 6.2. The trouble is, however, that $\Phi_{\mathbf{k}}^{(\pm)}(\mathbf{r})$ in (6.45) and (6.52) appear only as solutions of the Lippmann–Schwinger equations (6.5), rather than as eigenfunction expansions that are related to $\Omega_{\pm}$ via (4.25). Having already proven in §4.6 that (4.25) implies (6.5), we still have to complete the circle by proving that (6.5) in turn implies (4.25) under the conditions of Theorem 6.2 (and thus, incidently, establish asymptotic completeness in two-body potential scattering). This final task will be carried out in the context of proving Theorem 7.2 in §7 by first establishing in Lemma 7.2 the existence of operators $W_{\pm}$ for which (4.25) holds with $\psi_{\pm} = W_{\pm}\psi$ if we adopt the solutions (6.16) of (6.5) in the role of $\Phi_{\mathbf{k}}^{(\pm)}(\mathbf{r})$, and then, at long last, settling the issue in §7.3 by proving that $W_{\pm} = \Omega_{\pm}$. In the meantime, we shall derive some additional properties of $\Phi_{\mathbf{k}}^{(\pm)}(\mathbf{r})$ that are of great interest physically as well as mathematically.

6.4. The Scattering Amplitude

There are two distinct methods of defining the scattering amplitude in potential scattering: by means of the T matrix [i.e., by adopting formula (6.67), which we shall derive later, as a definition], or in terms of the asymptotic behavior of $\Phi_{\mathbf{k}}^{(\pm)}(\mathbf{r})$ as $|\mathbf{r}| \to \infty$. We shall adopt the second alternative since it is in that context that the scattering amplitude makes its first appearance in most textbooks on quantum mechanics.

Definition 6.3. The function $f_{\mathbf{k}}(\theta, \phi)$ is called the *scattering amplitude* of the incoming distorted plane wave $\Phi_{\mathbf{k}}^{(+)}(\mathbf{r})$ in the direction (θ, ϕ) if for a given $\mathbf{k} \in \mathbb{R}^3 - \mathscr{S}_V{}^3$ and $(\theta, \phi) \in \Omega_s$ there is an $\epsilon > 0$ such that

$$(6.55) \qquad v_{\mathbf{k}}(\mathbf{r}) = (2\pi)^{3/2}[\Phi_{\mathbf{k}}^{(+)}(\mathbf{r}) - \Phi_{\mathbf{k}}(\mathbf{r})] = [f_{\mathbf{k}}(\theta, \phi)/r]\, e^{ikr} + O(r^{-1-\epsilon})$$

as $r \to \infty$, where (r, θ, ϕ) are the spherical coordinates of $\mathbf{r} \in \mathbb{R}^3$.

Theorem 6.3. If $V(\mathbf{r}) = O(r^{-3-\epsilon_0})$, $r \to \infty$, is square integrable and of Rollnik class then $v_{\mathbf{k}}(\mathbf{r})$ is locally square integrable at each fixed $\mathbf{k} \in \mathbb{R}^3 - \mathscr{S}_V{}^3$. Furthermore, for each compact set B disjoint from $\mathscr{S}_V{}^3$ and for a certain $R_0 > 0$ there are constants $C_1(B, R)$ and $C_2(B, R)$ such that for all $\mathbf{k} \in B$,

$$(6.56) \qquad \int_{r \leqslant R_0} |v_{\mathbf{k}}(\mathbf{r})|^2 \, d\mathbf{r} \leqslant C_1(B, R_0),$$

$$(6.57) \qquad \left| v_{\mathbf{k}}(\mathbf{r}) - \frac{f_{\mathbf{k}}(\theta, \phi)}{r} e^{ikr} \right| \leqslant \frac{C_2(B, R_0)}{r^{1+\epsilon}}, \qquad r \geqslant R_0,$$

where $\epsilon = \epsilon_0(1 + \epsilon_0)^{-1}$ and

$$(6.58) \quad f_{\mathbf{k}}(\theta, \phi) = - m(2\pi)^{1/2} \int_{\mathbb{R}^3} \exp\left[-ik\left(\mathbf{r}' \cdot \frac{\mathbf{r}}{r}\right)\right] V(\mathbf{r}')\, \Phi_{\mathbf{k}}^{(+)}(\mathbf{r}')\, d\mathbf{r}'.$$

Proof. The conditions imposed on $V(\mathbf{r})$ in the present theorem are the same ones as in Lemma 6.1. Consequently, $v_{\mathbf{k}}(\mathbf{r})$ satisfies (6.56) on account of (6.38).

Let us introduce the function

$$(6.59) \quad w_{\mathbf{k}}(\mathbf{r}) = -\frac{m}{2\pi}(e^{i\mathbf{kr}} + v_{\mathbf{k}}(\mathbf{r}))\, V(\mathbf{r}) = -m(2\pi)^{1/2} V(\mathbf{r})\, \Phi_k^{(+)}(\mathbf{r})$$

in terms of which by (6.3), (6.5), and (5.17),

$$(6.60) \quad v_{\mathbf{k}}(\mathbf{r}) = \int_{\mathbb{R}^3} \frac{\exp(ik|\mathbf{r} - \mathbf{r}_1|)}{|\mathbf{r} - \mathbf{r}_1|} w_{\mathbf{k}}(\mathbf{r}_1)\, d\mathbf{r}_1 .$$

We note that $v_{\mathbf{k}}(\mathbf{r})$ is certainly uniformly bounded at infinity for $\mathbf{k} \in B$ (see Exercise 6.7), so that there is an $R_B > 0$ independent of $\mathbf{k} \in B$ such that

$$(6.61) \quad |w_{\mathbf{k}}(\mathbf{r})| \leqslant \text{const}\, |V(\mathbf{r})|, \; |\mathbf{r}| \geqslant R_B .$$

Let us write now (6.60) in the form

$$(6.62) \quad v_{\mathbf{k}}(\mathbf{r}) = \int_{r_1 \leqslant R} \frac{\exp(ik|\mathbf{r} - \mathbf{r}_1|)}{|\mathbf{r} - \mathbf{r}_1|} w_{\mathbf{k}}(\mathbf{r}_1)\, d\mathbf{r}_1 + \int_{r_1 > R} \frac{\exp(ik|\mathbf{r} - \mathbf{r}_1|)}{|\mathbf{r} - \mathbf{r}_1|} w_{\mathbf{k}}(\mathbf{r}_1)\, d\mathbf{r}_1 = I_R' + I_R''.$$

According to (6.61) there is a constant M such that for suitably large R

$$(6.63) \quad |w_{\mathbf{k}}(\mathbf{r})| \leqslant M \frac{1}{r^{3+\epsilon_0}}, \qquad r > R, \quad \mathbf{k} \in B.$$

Consequently, for such values of R

$$|I_R''| \leqslant M \int_{r_1 > R} \frac{d\mathbf{r}_1}{|\mathbf{r} - \mathbf{r}_1|\,|\mathbf{r}_1|^{3+\epsilon_0}} .$$

By selecting, for fixed $\mathbf{r}$, the z_1 coordinate axis in the direction of the vector $\mathbf{r}$, and working in spherical coordinates, we arrive at the following estimate of the above integral:

$$(6.64)\qquad |I_R''| \leqslant 2\pi M \int_R^\infty dr_1 \int_0^\pi d\theta_1 \frac{r_1^2 \sin\theta_1}{r_1^{3+\epsilon_0}(r^2 + r_1^2 - 2rr_1\cos\theta_1)^{1/2}}$$

$$= 4\pi M \int_R^r \frac{dr_1}{r r_1^{1+\epsilon_0}} + 4\pi M \int_r^\infty \frac{dr_1}{r_1^{2+\epsilon_0}}$$

$$\leqslant 4\pi M \left[\frac{1}{\epsilon_0 r R^{\epsilon_0}} + \frac{1}{(1+\epsilon_0) r^{1+\epsilon_0}}\right].$$

Furthermore, using Taylor's formula, we get for $r_1/r \to +0$,

$$\exp(ik|\mathbf{r} - \mathbf{r}_1|) = \exp\left(ikr\left[1 - \frac{\mathbf{r}\cdot\mathbf{r}_1}{r^2} + O\left(\frac{r_1^2}{r^2}\right)\right]\right),$$

$$\frac{1}{|\mathbf{r} - \mathbf{r}_1|} = \frac{1}{r}\left[1 + O\left(\frac{r_1}{r}\right)\right].$$

Consequently, we can write as $r_1/r \to +0$

$$(6.65)\qquad \frac{\exp(ik|\mathbf{r} - \mathbf{r}_1|)}{|\mathbf{r} - \mathbf{r}_1|}$$

$$= \frac{\exp(ikr)}{r} \exp\left(-ik\frac{\mathbf{r}\cdot\mathbf{r}_1}{r}\right) \exp\left[ikr\, O\left(\frac{r_1^2}{r^2}\right)\right]\left[1 + O\left(\frac{r_1}{r}\right)\right].$$

Let us take in (6.62) $R = r^{1-\delta}$, with some fixed $\delta < 1$, and then choose r so large that (6.64) becomes valid. Inserting (6.65) in I_R', we obtain, after noting that for the selected R we have $R/r = r^{-\delta}$,

$$I_R' = \frac{\exp(ikr)}{r} \int_{r_1 \leqslant R} \exp\left(-ik\frac{\mathbf{r}\cdot\mathbf{r}_1}{r}\right) w_{\mathbf{k}}(\mathbf{r}_1)\, d\mathbf{r}_1 + O\left(\frac{1}{r^{1+\delta}}\right).$$

Now, by virtue of (6.63), we also have for all $\mathbf{k} \in B$

$$\left|\int_{r_1 > R} \exp\left(-ik\frac{\mathbf{r}\cdot\mathbf{r}_1}{r}\right) w_{\mathbf{k}}(\mathbf{r}_1)\, d\mathbf{r}_1\right| \leqslant \int_{r_1 > R} |w_{\mathbf{k}}(\mathbf{r}_1)|\, d\mathbf{r}_1 = O\left(\frac{1}{r^{(1-\delta)\epsilon_0}}\right),$$

and thus, uniformly in $\mathbf{k} \in B$,

$$I_R' = \frac{\exp(ikr)}{r} \int_{\mathbb{R}^3} \exp\left(-ik\frac{\mathbf{r}\cdot\mathbf{r}_1}{r}\right) w_{\mathbf{k}}(\mathbf{r}_1)\, d\mathbf{r}_1$$

$$+ O\left(\frac{1}{r^{1+\delta}}\right) + O\left(\frac{1}{r^{1+(1-\delta)\epsilon_0}}\right).$$

On the other hand, for the chosen value of R, (6.64) yields

$$I''_R = O\left(\frac{1}{r^{1+(1-\delta)\epsilon_0}}\right) + O\left(\frac{1}{r^{1+\epsilon_0}}\right),$$

so that (with asymptotic contants independent of $\mathbf{k} \in B$),

$$v_{\mathbf{k}}(\mathbf{r}) = I_R' + I''_R = \frac{\exp(ikr)}{r} \int_{\mathbb{R}^3} \exp\left(-ik\frac{\mathbf{r}\cdot\mathbf{r}_1}{r}\right) w_{\mathbf{k}}(\mathbf{r}_1)\, d\mathbf{r}_1$$

$$+ O\left(\frac{1}{r^{1+\delta}}\right) + O\left(\frac{1}{r^{1+(1-\delta)\epsilon_0}}\right) + O\left(\frac{1}{r^{1+\epsilon_0}}\right).$$

Setting in the above estimate $\delta = \epsilon_0/(1+\epsilon_0) < \epsilon_0$, we finally obtain (6.57). Q.E.D.

The main significance of the above theorem lies in the fact that it provides us with the explicit formula (6.58) for the scattering amplitude $f_{\mathbf{k}}(\theta, \phi)$. This formula requires, however, a full knowledge of $\Phi_{\mathbf{k}}^{(+)}(\mathbf{r})$. On the other hand, even in the case where $\Phi_{\mathbf{k}}^{(+)}(\mathbf{r})$ is not known, we can still obtain an approximation of $f_{\mathbf{k}}(\theta, \phi)$ by using (6.16) and neglecting the second term. Then we obtain the *first Born approximation* of the scattering amplitude:

$$\text{(6.66)} \qquad f_{\mathbf{k}}(\theta, \phi) \approx -\frac{m}{2\pi}\int_{\mathbb{R}^3} \exp\left(-ik\frac{\mathbf{r}\cdot\mathbf{r}'}{r}\right) V(\mathbf{r}')\exp(i\mathbf{k}\mathbf{r}')\, d\mathbf{r}'$$

$$= -m(2\pi)^{1/2}\tilde{V}(k\hat{\mathbf{r}} - \mathbf{k}).$$

Higher Born approximations can be obtained by an iteration method presented in the next section

By denoting with $\boldsymbol{\omega}$ the unit vector with angular spherical coordinates (θ, ϕ) and considering $f_{\mathbf{k}}(\theta, \phi)$ as a function $f_{\mathbf{k}}(\boldsymbol{\omega})$ of $\boldsymbol{\omega}$ rather than (θ, ϕ), we immediately note the intimate relationship between the scattering amplitude (6.58) and the T matrix (4.95):

$$\text{(6.67)} \qquad f_{\mathbf{k}}(\boldsymbol{\omega}) = -(2\pi)^2 m \int_{\mathbb{R}^3} \Phi^*_{k\boldsymbol{\omega}}(\mathbf{r}')\, V(\mathbf{r}')\, \Phi_{\mathbf{k}}^{(+)}(\mathbf{r}')\, d\mathbf{r}'$$

$$= -(2\pi)^2 m(\Phi_{k\boldsymbol{\omega}} \mid V \mid \Phi_{\mathbf{k}}^{(+)})$$

$$= -(2\pi)^2 m\langle k\boldsymbol{\omega} \mid T \mid \mathbf{k}\rangle.$$

Written in this form, $f_{\mathbf{k}}(\boldsymbol{\omega})$ exhibits a remarkable symmetry (intimately related to the time invariance of the Schroedinger equation):

$$\text{(6.68)} \qquad f_{k\hat{\mathbf{k}}}(\boldsymbol{\omega}) = f_{-k\boldsymbol{\omega}}(-\hat{\mathbf{k}}).$$

Indeed, using (6.5) to eliminate the plane wave $\Phi_{\mathbf{k}}{}^*(\mathbf{r}) = \Phi_{-\mathbf{k}}(\mathbf{r})$ from (6.67) we obtain

$$f_{\mathbf{k}}(\omega) = -(2\pi)^2 m \int_{\mathbb{R}^3} \Phi_{k\hat{\mathbf{k}}}^{(+)}(\mathbf{r}')\, V(\mathbf{r}')\, \Phi_{-k\omega}^{(+)}(\mathbf{r}')\, d\mathbf{r}' \tag{6.69}$$

$$+ 2\pi m^2 \int_{\mathbb{R}^3} d\mathbf{r}' \Phi_{k\hat{\mathbf{k}}}^{(+)}(\mathbf{r}')\, V(\mathbf{r}')$$

$$\times \int_{\mathbb{R}^3} d\mathbf{r} \frac{\exp(-ik \mid \mathbf{r}' - \mathbf{r} \mid)}{\mid \mathbf{r}' - \mathbf{r} \mid} V(\mathbf{r})\, \Phi_{-k\omega}^{(+)}(\mathbf{r}).$$

Under the conditions of Theorem 6.2 the integrand of the previous double integral is integrable in $(\mathbf{r}, \mathbf{r}')$ on $\mathbb{R}^6$ (see Exercise 6.2, and note that $V(\mathbf{r})$ is integrable, whereas $\Phi_{\mathbf{k}}^{(+)}(\mathbf{r})$ is bounded in $\mathbf{r} \in \mathbb{R}^3$) so that Fubini's theorem can be applied. Hence the right-hand side of (6.69) stays unchanged if we replace $\hat{\mathbf{k}}$ and $\boldsymbol{\omega}$ with $-\boldsymbol{\omega}$ and $-\hat{\mathbf{k}}$, respectively.

The relations (3.50), (4.96), and (6.67) between S operator, transition operator, T matrix, and scattering amplitude show that the action of the scattering operator on a wave packet $\tilde{\psi}(\mathbf{k}) = \tilde{\psi}(k\boldsymbol{\omega})$ can be expressed directly in terms of the scattering amplitude:

$$(S\psi)^{\sim}(\mathbf{k}) = \tilde{\psi}(\mathbf{k}) + (ik/2\pi) \int_{\Omega_S} f_{k\omega'}(\hat{\mathbf{k}})\, \tilde{\psi}(k\omega')\, d\omega'. \tag{6.70}$$

This represents yet another rigorous version of the formal relation (4.107).

6.5. The Born Series

The simplest method of solving an operator equation such as (5.29) is by repeated iteration, which then yields

$$\psi = (1 + \omega K + \omega^2 K^2 + \cdots + \omega^n K^n + \cdots) f, \tag{6.71}$$

provided that this infinite series converges in some sense. Clearly, the expression in between parentheses is nothing else than the series for $(1 - \omega K)^{-1}$ (cf. Exercise 7.9 in Chapter IV), and we shall obtain convergence in operator bound if $| \omega | \| K \| < 1$ (see Exercise 5.7). Furthermore, if K is Hilbert–Schmidt and $| \omega | \| K \|_2 < 1$, then this convergence in operator bound and is also convergent in the Hilbert–Schmidt norm (recall that $\| K \| \leqslant \| K \|_2$). Since for operators with Hilbert–Schmidt kernels $K(\alpha, \alpha')$ such as those appearing in (5.28) we have

$$\| K \|_2^2 = \langle K \mid K \rangle_2 = \mathrm{Tr}(K^*K) = \int_{\mathbb{R}^{2m}} | K(\alpha, \alpha')|^2\, d^m\alpha\, d^m\alpha', \tag{6.72}$$

this convergence becomes (see Exercise 5.4) convergence in the mean on $L^2(\mathbb{R}^{2m})$:

$$(6.73)\qquad \left(\sum_{n=0}^{\infty} \omega^n K^n\right)(\alpha, \alpha') = \underset{N\to\infty}{\text{l.i.m.}} \sum_{n=0}^{N} \omega^n K^n(\alpha, \alpha'),$$

$$(6.74)\qquad K^n(\alpha, \alpha') = \int_{\mathbb{R}^m} d^m\alpha_{n-1} K(\alpha, \alpha_{n-1}) \int_{\mathbb{R}^m} d^m\alpha_{n-2} K(\alpha_{n-1}, \alpha_{n-2}) \cdots \int_{\mathbb{R}^m} d^m\alpha_1 K(\alpha_2, \alpha_1)\, K(\alpha_1, \alpha').$$

These remarks can be applied in particular to (6.8), so that

$$(6.75)\qquad \tilde{\Phi}_{\mathbf{k}}^{(\pm)}(\mathbf{r}) = \tilde{\Phi}_{\mathbf{k}}(\mathbf{r}) + \sum_{n=1}^{\infty} \int_{\mathbb{R}^3} (\tilde{K}^{(\pm)n})(\mathbf{r}, \mathbf{r}'; \mathbf{k}^2/2m)\, \tilde{\Phi}_{\mathbf{k}}(\mathbf{r}')\, d\mathbf{r}'.$$

Upon eliminating from both sides of (6.75) the common factor $V^{1/2}(\mathbf{r})$, we get a series solution of the Lippmann–Schwinger equations (6.5):

$$(6.76)\qquad \Phi_{\mathbf{k}}^{(\pm)}(\mathbf{r}) = \Phi_{\mathbf{k}}(\mathbf{r}) + \int_{\mathbb{R}^3} G_0^{(\pm)}(\mathbf{r}, \mathbf{r}'; \mathbf{k}^2/2m)\, V(\mathbf{r}')\, \Phi_{\mathbf{k}}(\mathbf{r}')\, d\mathbf{r}' + \int_{\mathbb{R}^3} d\mathbf{r}_1 G_0^{(\pm)}(\mathbf{r}, \mathbf{r}_1; \mathbf{k}^2/2m)\, V(\mathbf{r}_1) \times \int d\mathbf{r}' G_0^{(\pm)}(\mathbf{r}_1, \mathbf{r}'; \mathbf{k}^2/2m)\, V(\mathbf{r}')\, \Phi_{\mathbf{k}}(\mathbf{r}') + \cdots.$$

This is the *Born series* for distorted plane waves, which inserted in (6.58) provides the Born series for the scattering amplitude, whose first term we had encountered in (6.66). Working with the on-shell T matrix (4.95) instead of the scattering amplitude, we can write this Born series in the symbolic form

$$(6.77)\qquad \langle \mathbf{k} \mid T \mid \mathbf{k}' \rangle = \sum_{n=0}^{\infty} \langle \mathbf{k} \mid V(G_0^{(+)}(\lambda)V)^n \mid \mathbf{k}' \rangle, \qquad \lambda = \mathbf{k}^2/2m = \mathbf{k}'^2/2m,$$

in which the nth term equals

$$(6.78)\qquad (\Phi_{\mathbf{k}} \mid V(G_0^{(+)}(\mathbf{k}^2/2m)V)^n \mid \Phi_{\mathbf{k}'}) = \int_{\mathbb{R}^3} d\mathbf{r} \Phi_{\mathbf{k}}^*(\mathbf{r})\, V(\mathbf{r}) \int_{\mathbb{R}^3} d\mathbf{r}' (G_0^{(+)}(\mathbf{k}^2/2m)V)^n(\mathbf{r}, \mathbf{r}')\, \Phi_{\mathbf{k}'}(\mathbf{r}').$$

Theorem 6.4. If $V(\mathbf{r}) = O(r^{-3-\epsilon_0})$, $r \to \infty$, is locally square integrable and $\| V \|_{\mathbf{R}} < \infty$, then there is an energy value λ_V such that the Born series (6.77) for the on-shell T matrix converges uniformly in $\mathbf{k}$ and $\mathbf{k}'$ at all higher energies $\lambda = \mathbf{k}^2/2m \geqslant \lambda_V$. Furthermore, if $\| V \|_{\mathbf{R}} < 2\pi/m$ then the series (6.77) converges uniformly in $\mathbf{k}, \mathbf{k}' \in \mathbb{R}^3$ at all energies, and the exceptional set $\mathscr{S}_V{}^1$ is empty.

Proof. According to (6.78), (6.8), and (6.9),

$$\begin{aligned}
\text{(6.79)} \qquad & \langle \mathbf{k} \mid V(G_0^{(+)}(\lambda)V)^n \mid \mathbf{k}' \rangle \\
&= (\Phi_{\mathbf{k}} \mid V(G_0^{(+)}(\lambda)V)^{n-1}G_0^{(+)}(\lambda) \mid V \mid^{1/2}\tilde{\Phi}_{\mathbf{k}'}\rangle \\
&= (\Phi_{\mathbf{k}} \mid \mid V \mid^{1/2}(V^{1/2}G_0^{(+)}(\lambda) \mid V \mid^{1/2})^n\tilde{\Phi}_{\mathbf{k}'}\rangle \\
&= \langle \mid V \mid^{1/2}\Phi_{\mathbf{k}} \mid (\tilde{K}^{(+)}(\lambda))^n\tilde{\Phi}_{\mathbf{k}'}\rangle
\end{aligned}$$

is a bona fide inner product in $L^2(\mathbb{R}^3)$. Since by (6.12),

$$\begin{aligned}
\text{(6.80)} \qquad |\langle \mid V \mid^{1/2}\Phi_{\mathbf{k}} \mid (\tilde{K}^{(+)}(\lambda))^n\tilde{\Phi}_{\mathbf{k}'}\rangle| &\leqslant \|(\tilde{K}^{(+)}(\lambda))^n\| \, \| \mid V \mid^{1/2}\Phi_{\mathbf{k}} \| \, \| \tilde{\Phi}_{\mathbf{k}'} \| \\
&\leqslant (2\pi)^{-3} \| \tilde{K}^{(+)}(\lambda)\|^n \int_{\mathbb{R}^3} \mid V(\mathbf{r})| \, d\mathbf{r},
\end{aligned}$$

we see that (6.77) converges if

$$\begin{aligned}
\text{(6.81)} \qquad \| \tilde{K}^{(+)}(\lambda)\|^4 &= \|[(\tilde{K}^{(+)}(\lambda))^*\tilde{K}^{(+)}(\lambda)]\|^2 \\
&\leqslant \|(\tilde{K}^{(+)}(\lambda))^*\tilde{K}^{(+)}(\lambda)\|_2^2 < 1.
\end{aligned}$$

However, by specializing (6.72) we get from (6.9),

$$\begin{aligned}
\text{(6.82)} \qquad & \| \tilde{K}^{(+)}(\lambda))^*\tilde{K}^{(+)}(\lambda)\|_2^2 \\
&= \mathrm{Tr}\{[(\tilde{K}^{(+)}(\lambda))^*\tilde{K}^{(+)}(\lambda)]^2\} \\
&= \int_{\mathbb{R}^6} d\mathbf{r}' \, d\mathbf{r}'' \mid V(\mathbf{r}') \, V(\mathbf{r}'')| \\
&\quad \times \left\{\int G_0^{(+)*}(\mathbf{r}, \mathbf{r}'; \lambda) \, G_0^{(+)}(\mathbf{r}, \mathbf{r}''; \lambda) \mid V(\mathbf{r})| \, d\mathbf{r}\right\}^2.
\end{aligned}$$

By (5.17) the term in braces equals

$$\text{(6.83)} \qquad \frac{m^2}{4\pi^2} \int_{\mathbb{R}^3} \frac{\exp[ik(\mid \mathbf{r} - \mathbf{r}'' \mid - \mid \mathbf{r} - \mathbf{r}' \mid)]}{\mid \mathbf{r} - \mathbf{r}' \mid \mid \mathbf{r} - \mathbf{r}'' \mid} \mid V(\mathbf{r})| \, d\mathbf{r}, \qquad k = (2m\lambda)^{1/2}.$$

In view of the Riemann–Lebesgue lemma* we conclude that upon introducing spherical coordinates, the radial integral in $r \in [0, \infty)$ approaches zero as $k \to \infty$ for any fixed $\mathbf{k}'$, $\mathbf{k}''$, and (θ, ϕ). Since for all $k \in [0, \infty)$, the integrand in (6.83) is majorized by an integrable and k-independent function that is square integrable in $\mathbf{k}'$, $\mathbf{k}''$, and (θ, ϕ) (see Exercise 6.4), we deduce from Lebesgue's dominated convergence theorem that (6.82) approaches zero as $\lambda \to \infty$, so that there is a λ_V for which (6.81) holds true whenever $\lambda \geqslant \lambda_V$. Taking into account that by (6.79) and (6.80) for any such λ,

$$(6.84)\qquad \sum_{n=0}^{\infty} |\langle \mathbf{k} \mid V(G_0^{(+)}(\lambda)V)^n \mid \mathbf{k}'\rangle| \leqslant (2\pi)^{-3} \| V \|_1 \sum_{n=0}^{\infty} \| \tilde{K}^{(+)}(\lambda_V)\|^n,$$

we see that the first part of the theorem is true.

According to (6.11), (6.79), and (6.80), we also have

$$(6.85)\qquad \begin{aligned} |\langle \mathbf{k} \mid V(G_0^{(+)}(\lambda)V)^n \mid \mathbf{k}'\rangle| &\leqslant (2\pi)^{-3} \| V \|_1 \| \tilde{K}^{(+)}(\lambda)\|_2^n \\ &\leqslant (2\pi)^{-3} \| V \|_1 (m \| V \|_{\mathrm{R}}/2\pi)^n, \end{aligned}$$

so that (6.77) converges uniformly and absolutely for all $\mathbf{k}$ and $\mathbf{k}'$ on the energy shell if $\| V \|_{\mathrm{R}} < 2\pi/m$. The same inequality also implies that

$$(6.86)\qquad (1 - \tilde{K}^{(+)}(\lambda))^{-1} = \underset{N\to\infty}{\text{u-lim}} \sum_{n=0}^{N} (\tilde{K}^{(+)}(\lambda))^n$$

exists for all $\lambda \in [0, +\infty)$, and therefore by Theorem 5.5, $\tilde{d}(\lambda) \neq 0$ at all energies, so that $\mathscr{S}_V^1$ is empty. Q.E.D.

6.6. Distorted Plane Waves as Solutions of the Schroedinger Equation

For the complete identification of $\Phi_{\mathbf{k}}^{(\pm)}(\mathbf{r})$ as distorted plane waves, we have to establish that $\Phi_{\mathbf{k}}^{(+)}(\mathbf{r})$ is a solution of the Schroedinger equation with the eigennumber $\mathbf{k}^2/2m$. Theorem 6.6 provides a set of precise criteria under which this is true.

Theorem 6.5. **Suppose that the potential $V(\mathbf{r})$ is continuous in the neighborhood of a point $\mathbf{r} \in \mathbb{R}^3$, and that it satisfies the conditions**

* See Lemma 4.1 in Chapter III, and note that its conclusion remains valid if we change variables of integration by setting $x = h(y)$, where $h'(y)$ exists and is positive (or negative) almost everywhere.

imposed on it in Theorem 6.2. Then the distorted plane waves $\Phi_{\mathbf{k}}^{(\pm)}(\mathbf{r})$ satisfy the time-independent Schroedinger differential equation

$$\left(-\frac{1}{2m}\Delta + V(\mathbf{r})\right)\Phi_{\mathbf{k}}^{(\pm)}(\mathbf{r}) = \frac{\mathbf{k}^2}{2m}\Phi_{\mathbf{k}}^{(\pm)}(\mathbf{r}) \tag{6.87}$$

at that point $\mathbf{r} \in \mathbb{R}^3$, and for any $\mathbf{k} \in \mathbb{R}^3 - \mathscr{S}_V^3$.

Before proceeding with the proof of the above theorem, let us make a few remarks.

For the proof of this theorem we need, essentially, only to know that $V(\mathbf{r})$ and $\Phi_{\mathbf{k}}^{(\pm)}(\mathbf{r})$ are continuous in some neighborhood of $\mathbf{r}$, and that

$$|V(\mathbf{r})| = O\left(\frac{1}{r^{2+\epsilon_0}}\right), \qquad |\Phi_{\mathbf{k}}^{(\pm)}(\mathbf{r})| = O(1).$$

The restrictions on $\Phi_{\mathbf{k}}^{(\pm)}(\mathbf{r})$ are fulfilled automatically if $V(\mathbf{r})$ satisfies the conditions of Theorem 6.2 and the restrictions imposed on it in the present theorem. It can happen, however, that $\Phi_{\mathbf{k}}^{(\pm)}(\mathbf{r})$ satisfies these restrictions even when $V(\mathbf{r})$ does not obey all the conditions of Theorem 6.2 (see §7.2). Hence, the validity of Theorem 6.5 is somewhat independent of the validity of Theorem 6.2.

The proof of Theorem 6.5 is straightforward, and it essentially consists in applying the Laplacian Δ to the right-hand side of equation (6.5) and computing the result. This computation is easily carried out with the help of Lemma 6.2.

Lemma 6.2. Suppose $f(\mathbf{r}) = O(r^{-2-\epsilon_0})$ for some $\epsilon_0 > 0$, and that there is a neighborhood $\mathscr{N}$ of the point $\mathbf{r} \in \mathbb{R}^3$ in which $f(\mathbf{r})$ is continuous and the function

$$g(\mathbf{r}) = \int_{\mathbb{R}^3} \frac{\exp(ik|\mathbf{r} - \mathbf{r}'|)}{|\mathbf{r} - \mathbf{r}'|} f(\mathbf{r}')\, d\mathbf{r}' \tag{6.88}$$

is well defined by the above integral. Then $g(\mathbf{r})$ is twice differentiable at that point in each of the variables x, y, and z, and

$$(-\Delta g)(\mathbf{r}) = k^2 g(\mathbf{r}) + 4\pi f(\mathbf{r}). \tag{6.89}$$

Proof. Let S be a neighborhood of $\mathbf{r}$ and write

$$(\Delta g)(\mathbf{r}) = \Delta \int_{\mathbb{R}^3 - S} \frac{\exp(ik|\mathbf{r} - \mathbf{r}'|)}{|\mathbf{r} - \mathbf{r}'|} f(\mathbf{r}')\, d\mathbf{r}' + \Delta \int_S \frac{\exp(ik|\mathbf{r} - \mathbf{r}'|)}{|\mathbf{r} - \mathbf{r}'|} f(\mathbf{r}')\, d\mathbf{r}'.$$

We shall compute separately the two terms on the right-hand side of the above identity.

Straightforward differentiation shows that

$$\text{(6.90)}\quad \frac{\partial}{\partial x}\frac{\exp(ik|\mathbf{r}-\mathbf{r}'|)}{|\mathbf{r}-\mathbf{r}'|} = \frac{(x-x')\exp(ik|\mathbf{r}-\mathbf{r}'|)}{|\mathbf{r}-\mathbf{r}'|^2}\left(ik - \frac{1}{|\mathbf{r}-\mathbf{r}'|}\right),$$

$$\text{(6.91)}\quad \frac{\partial^2}{\partial x^2}\frac{\exp(ik|\mathbf{r}-\mathbf{r}'|)}{|\mathbf{r}-\mathbf{r}'|}$$

$$= \frac{\exp(ik|\mathbf{r}-\mathbf{r}'|)}{|\mathbf{r}-\mathbf{r}'|^2}$$

$$\times\left(ik - \frac{1+k^2(x-x')^2}{|\mathbf{r}-\mathbf{r}'|} - \frac{3ik(x-x')^2}{|\mathbf{r}-\mathbf{r}'|^2} + \frac{3(x-x')^2}{|\mathbf{r}-\mathbf{r}'|^3}\right).$$

On any bounded domain which does not contain some closed neighborhood of $\mathbf{r}$, we have

$$|f(\mathbf{r}')| \leqslant \text{const}\,\frac{|f(\mathbf{r}')|}{|\mathbf{r}-\mathbf{r}'|}.$$

Now, the function on the right-hand side of the above inequality is integrable on account of the existence of the integral in (6.88) and the fact that if a function is integrable then it is also absolutely integrable. Hence, we conclude that $f(\mathbf{r})$ is locally integrable.

It is easy to infer from (6.90) and (6.91) that for sufficiently large $R_0 > 0$ and all $\mathbf{r}_1$ from some open neighborhood $S \subset \mathcal{N}$ of $\mathbf{r}$

$$\left|\frac{\partial^n}{\partial x^n}\frac{\exp(ik|\mathbf{r}-\mathbf{r}'|)}{|\mathbf{r}-\mathbf{r}'|}f(\mathbf{r}')\right| \leqslant \begin{cases} \dfrac{\text{const}}{|\mathbf{r}'|^{3+\epsilon_0}}, & r' \geqslant R_0 \\ \text{const}\, f(\mathbf{r}'), & \mathbf{r}' \in \mathbb{R}^3 - S \\ & \text{and } |\mathbf{r}'| \leqslant R_0 \end{cases}$$

for $n = 1, 2$. Thus, we deduce that

$$\int_{\mathbb{R}^3-S}\frac{\partial^n}{\partial x^n}\frac{\exp(ik|\mathbf{r}-\mathbf{r}'|)}{|\mathbf{r}-\mathbf{r}'|}f(\mathbf{r}')\,d\mathbf{r}', \qquad n = 1, 2,$$

exist for $\mathbf{r} \in S$. Consequently, $\partial/\partial x$ and $\partial^2/\partial x^2$ can be carried under the integral sign when applied to the right-hand side of (6.88) (see Exercise 6.8). Since the same obviously holds for the variables y and z, and since by (6.91),

$$\text{(6.92)}\quad \left(\frac{\partial^2}{\partial x^2} + \frac{\partial^2}{\partial y^2} + \frac{\partial^2}{\partial z^2}\right)\frac{\exp(ik|\mathbf{r}-\mathbf{r}'|)}{|\mathbf{r}-\mathbf{r}'|} = -k^2\frac{\exp(ik|\mathbf{r}-\mathbf{r}'|)}{|\mathbf{r}-\mathbf{r}'|},$$

we obtain

(6.93)
$$-\Delta \int_{\mathbb{R}^3 - S} \frac{\exp(ik|\mathbf{r}-\mathbf{r}'|)}{|\mathbf{r}-\mathbf{r}'|} f(\mathbf{r}')\, d\mathbf{r}' = k^2 \int_{\mathbb{R}^3 - S} \frac{\exp(ik|\mathbf{r}-\mathbf{r}'|)}{|\mathbf{r}-\mathbf{r}'|} f(\mathbf{r}')\, d\mathbf{r}'.$$

Let us choose

(6.94) $$S = \{\mathbf{r}_1 : |x_1 - x| < a, |y_1 - y| < a, |z_1 - z| < a\}.$$

If $a > 0$ is sufficiently small, then the generalized mean-value theorem of the integral calculus can be applied to obtain

(6.95)
$$\frac{\partial^2}{\partial x^2} \int_S \frac{\exp(ik|\mathbf{r}-\mathbf{r}'|)}{|\mathbf{r}-\mathbf{r}'|} f(\mathbf{r}')\, d\mathbf{r}'$$
$$= \lim_{\Delta x \to 0} \frac{1}{\Delta x} \int_{x-x_1-\Delta x}^{x-x_1} dx' \int_{y-a}^{y+a} dy' \int_{z-a}^{z+a} dz' \frac{\partial}{\partial x} \frac{\exp(ik|\mathbf{r}-\mathbf{r}'|)}{|\mathbf{r}-\mathbf{r}'|} f(\mathbf{r}') \Big|_{x_1=+a}^{x_1=-a}$$
$$= -\int_{y-a}^{y+a} \int_{z-a}^{z+a} \frac{\partial}{\partial x} \frac{\exp(ik|\mathbf{r}-\mathbf{r}'|)}{|\mathbf{r}-\mathbf{r}'|} f(\mathbf{r}') \Big|_{x'=x-a}^{x'=x+a} dy'\, dz'$$
$$= -f(x', y_0, z_0) \int_{y-a}^{y+a} \int_{z-a}^{z+a} \frac{\partial}{\partial x} \frac{\exp(ik|\mathbf{r}-\mathbf{r}'|)}{|\mathbf{r}-\mathbf{r}'|} dy'\, dz' \Big|_{x'=x-a}^{x'=x+a},$$

where $|y_0 - y| < a$ and $|z_0 - z| < a$. Using (6.90) and the mean value theorem of integral calculus, we get

$$\int_{y-a}^{y+a} \int_{z-a}^{z+a} \frac{\partial}{\partial x} \frac{\exp(ik|\mathbf{r}-\mathbf{r}'|)}{|\mathbf{r}-\mathbf{r}'|} \Big|_{x'=-a+x}^{x'=a+x} dy'\, dz'$$
$$= -\frac{\exp[ik(a^2+\bar{y}^2+\bar{z}^2)^{1/2}]}{a^2+\bar{y}^2+\bar{z}^2} \left(ik - \frac{1}{(a^2+\bar{y}^2+\bar{z}^2)^{1/2}} \right) (2a)^3,$$

where $|\bar{y}| < a$ and $|\bar{z}| < a$. This shows that, due to the continuity of $f(\mathbf{r})$ in $\mathcal{N}$, by taking the limit $a \to 0$ and using (6.90) we shall get in (6.95) a contribution of the form

(6.96) $$f(\mathbf{r}) \lim_{a \to +0} \int_{y-a}^{y+a} \int_{z-a}^{z+a} \frac{-2a \exp(ik[a^2 + (y-y')^2 + (z-z')^2]^{1/2})}{(a^2 + (y-y')^2 + (z-z')^2)^{3/2}} dy'\, dz'.$$

To compute the above limit, we must compute the following integral explicitly:

$$\int_{-a}^{+a} \int_{-a}^{+a} \frac{\exp[ik(a^2+y_1^2+z_1^2)^{1/2}]}{(a^2+y_1^2+z_1^2)^{3/2}} dy_1\, dz_1$$
$$= 8 \exp[ik(a^2 + r_{\text{av}}^2)^{1/2}] \int_0^{\pi/4} d\phi \int_0^{a/\cos\phi} \frac{r\, dr}{(a^2+r^2)^{3/2}}$$
$$= \frac{2\pi}{3} \frac{1}{a} \exp[ik(a^2 + r_{\text{av}}^2)^{1/2}],$$

where r_{av} is obtained by applying again the generalized mean-value theorem to the first of the above two integrals, and it satisfies the inequality $| r_{av} | < a\sqrt{2}$. Hence, the limit (6.96) is equal to $-(4\pi/3)f(\mathbf{r})$.

Due to the symmetry of the considered expressions under the exchange of x, y, and z, when S is the box in (6.94), we have

$$\text{(6.97)} \qquad \lim_{a\to 0} \Delta \int_S \frac{\exp(ik| \mathbf{r} - \mathbf{r}' |)}{| \mathbf{r} - \mathbf{r}' |} f(\mathbf{r}')\, d\mathbf{r}$$
$$= 3 \lim_{a\to 0} \frac{\partial^2}{\partial x^2} \int_S \frac{\exp(ik| \mathbf{r} - \mathbf{r}' |)}{| \mathbf{r} - \mathbf{r}' |} f(\mathbf{r}')\, d\mathbf{r}' = 4\pi f(\mathbf{r}).$$

Hence, the desired result in (6.89) is obtained by combining (6.97) with the result (6.93) in which we let $a \to +0$. Q.E.D.

In order to be able to apply the above lemma to derive (6.87) from (6.5), we have to establish that

$$\text{(6.98)} \qquad f(\mathbf{r}) = -\frac{m}{2\pi} V(\mathbf{r})\, \Phi_{\mathbf{k}}^{(\pm)}(\mathbf{r}) = O\left(\frac{1}{r^{2+\epsilon_0}}\right), \qquad \epsilon_0 > 0,$$

and that the above function is continuous in some neighborhood of the considered point $\mathbf{r} \in \mathbb{R}^3$.

Since $| \Phi_{\mathbf{k}}^{(\pm)}(\mathbf{r})| = O(1)$ for $\mathbf{k} \notin \mathscr{S}_V^3$ (see Exercise 6.7), it immediately follows that (6.98) is true. The continuity of $\Phi_{\mathbf{k}}^{(\pm)}(\mathbf{r})$ is an easy consequence of (6.5) (see Exercise 6.9). Hence, in computing the effect of applying the Laplacian Δ to the right-hand side of (6.5), we can use Lemma 6.2; then (6.87) easily follows.

*6.7. Appendix: Analytic Operator-Valued Functions

We have seen in §5.8 that $\langle f \mid R_A(\zeta)g\rangle$ is an analytic function on the resolvent set $\mathbb{C}^1 - S^A$ of a self-adjoint operator for all $f, g \in \mathscr{H}$. This categorizes $R_A(\zeta)$ as an analytic operator-valued function. The general theory of such functions is based on the following definition.

Definition 6.4. The vector-valued function $f(\zeta) \in \mathscr{H}$ defined on an open domain D of the complex plane $\mathbb{C}^1$ is said to be analytic at $\zeta_0 \in D$ if

$$\text{(6.99)} \qquad f'(\zeta) = \operatorname*{w-lim}_{\Delta\zeta\to 0}\{[f(\zeta + \Delta\zeta) - f(\zeta)]/\Delta\zeta\}$$

exists in some neighborhood of ζ_0.

Lemma 6.3. If $f(\zeta)$ is analytic at ζ_0 then

$$\text{(6.100)} \qquad f'(\zeta_0) = \operatorname*{s-lim}_{\Delta\zeta\to 0}\{[f(\zeta_0 + \Delta\zeta) - f(\zeta_0)]/\Delta\zeta\}.$$

Proof. Let $\mathscr{C}$ be a circle centered at ζ_0 that lies within an open neighborhood where (6.99) exists. Since obviously $\langle g \mid f(\zeta) \rangle$ is for any fixed $g \in \mathscr{H}$ an analytic complex-valued function in that neighborhood, we can apply Cauchy's integral formula from complex analysis and write

$$(6.101) \quad \langle g \mid \{[f(\zeta_0 + \Delta\zeta) - f(\zeta_0)]/\Delta\zeta\} - f'(\zeta_0)\rangle$$
$$= (2\pi i)^{-1} \int_{\mathscr{C}} [(\Delta\zeta)^{-1}([\zeta - (\zeta_0 + \Delta\zeta)]^{-1} - (\zeta - \zeta_0)^{-1})$$
$$- [(\zeta - \zeta_0)^{-2}]\langle g \mid f(\zeta)\rangle \, d\zeta$$

if $\zeta_0 + \Delta\zeta$ lies within the interior of the disk enclosed by $\mathscr{C}$. Furthermore, $|\langle g \mid f(\zeta)\rangle|$ is continuous and therefore bounded on the closed and bounded (and therefore compact) set $\mathscr{C}$ for each $g \in \mathscr{H}$. By Lemma 2.2 in Chapter III (generalized from sequences to arbitrary sets) $|\langle f(\zeta) \mid g\rangle|$ is therefore bounded uniformly in $g \in \mathscr{H}$, i.e.,

$$|\langle g \mid f(\zeta)\rangle| \leqslant C \| g \|, \qquad \zeta \in \mathscr{C}.$$

Hence the absolute value of the right-hand side of (6.101) is majorized by

$$(6.102) \quad (C/2\pi) \| g \| \int_{\mathscr{C}} |(\zeta - \zeta_0 - \Delta\zeta)^{-1} - (\zeta - \zeta_0)^{-1}| | \zeta - \zeta_0 |^{-1} \, d | \zeta |,$$

where the integral can be as small as desired by choosing $|\Delta\zeta|$ sufficiently small. Consequently, the same is true of the absolute value of the left-hand side of (6.101), since it is majorized by (6.102) for all $g \in \mathscr{H}$. Hence, (6.100) follows. Q.E.D.

Definition 6.5. Let $A(\zeta)$, $\zeta \in D$, be bounded operators defined everywhere in $\mathscr{H}$. The operator-valued function $A(\zeta)$ is said to be analytic on the open domain $D \subset \mathbb{C}^1$ if $A(\zeta)f$ is an analytic vector-valued function at all $\zeta \in D$ for any given $f \in \mathscr{H}$.

Theorem 6.6. If $A(\zeta)$ is an analytic operator-valued function on the open connected domain D in $\mathbb{C}^1$ and $A(\zeta)$ is a completely continuous operator for each $\zeta \in D$, then $[1 - A(\zeta)]^{-1}$ either does not exist at any $\zeta \in D$, or it exists at all $\zeta \in D$ with the possible exception of a subset $\mathscr{S}_0$, which has no accumulation points in D, in which case $[1 - A(\zeta)]^{-1}$ is analytic on $D - \mathscr{S}_0$.

Proof. Since $A(\zeta_0)$ is completely continuous for any given $\zeta_0 \in D$, according to Theorem 8.4 in Chapter IV, there is a finite-rank operator

$$A_N = \sum_{k=1}^{N} | e_k' \rangle \lambda_k \langle e_k |$$

such that $\| A_N - A(\zeta_0) \| < \frac{1}{2}$. On the other hand, the analyticity of $A(\zeta)$ at ζ_0 implies* the existence of a neighborhood D_{ζ_0} of ζ_0 in which $\| A(\zeta) - A(\zeta_0) \| < \frac{1}{2}$, so that $\| A(\zeta) - A_N \| < 1$ on D_{ζ_0}. Consequently (see Chapter IV, Exercise 7.9 and Theorem 8.2),

$$(6.103) \qquad (1 - A(\zeta) + A_N)^{-1} = \underset{n\to\infty}{\text{u-lim}} \sum_{k=0}^{n} (A(\zeta) - A_N)^k$$

exists as a compact operator for all $\zeta \in D_{\zeta_0}$, and it obviously defines an analytic operator-valued function. Furthermore, since upon introducing the compact operators

$$(6.104) \qquad \begin{aligned} A_N(\zeta) &= A_N(1 - A(\zeta) + A_N)^{-1} \\ &= \sum_{k=1}^{N} | e_k' \rangle \lambda_k \langle e_k | (1 - A(\zeta) + A_N)^{-1}, \end{aligned}$$

we can obviously claim the validity of the operator identity

$$1 - A(\zeta) = [1 - A_N(\zeta)](1 - A(\zeta) + A_N), \qquad \zeta \in D_{\zeta_0},$$

we conclude that $1 - A(\zeta)$ has an inverse for $\zeta \in D_{\zeta_0}$ if and only if $1 - A_N(\zeta)$ has an inverse.

By (6.104) the operator $A_N(\zeta)$ is of finite rank and therefore $(A_N(\zeta) - 1)^{-1}$ does not exist if and only if there is a nonzero $f \in \mathscr{H}$ for which

$$f = A_N(\zeta) f = \sum_{i=1}^{N} \lambda_i \langle e_i | (1 - A(\zeta) + A_N)^{-1} f \rangle \, e_i'.$$

Since $e_1', \ldots, e_N'$ is an orthonormal system, the previous equation is equivalent to the following system of linear algebraic equations:

(6.105)

$$\langle e_i' | f \rangle - \sum_{k=1}^{N} \lambda_i \langle e_i | (1 - A(\zeta) + A_N)^{-1} e_k' \rangle \langle e_k' | f \rangle = 0, \qquad i = 1, \ldots, N.$$

* The detailed argument requires the generalization of Lemma 3.3 to Banach spaces, and then the treatment of $A(\zeta)$ as a function in the C^*-algebra $\mathfrak{A}(\mathscr{H})$ viewed as a Banach space (see Dunford and Schwartz [1957]).

In turn, this system has a nontrivial solution for $\langle e_i' \mid f\rangle$, $i = 1,\ldots, N$, if and only if the corresponding determinant

$$\delta(\zeta) = \det\{\delta_{ik} - \lambda_i\langle e_i \mid (1 - A(\zeta) + A_N)^{-1} e_k'\rangle\} \tag{6.106}$$

vanishes. Since, however, (6.106) is analytic on D_{ζ_0} due to the analyticity of (6.103), $\delta(\zeta)$ cannot vanish on any set in D_{ζ_0} that has accumulation points in D_{ζ_0} without vanishing identically on D_{ζ_0}. Thus the theorem is established in D_{ζ_0}, and by standard methods used in analytic continuations, its validity can be therefore ascertained on all of D. Q.E.D.

The family $\tilde{K}(\zeta)$ of Hilbert–Schmidt operators with kernels (6.26) is obviously analytic in the upper as well as lower open complex half-planes, so that Theorem 6.6 is applicable to it. However, in solving (6.5), we were interested in the existence of $(1 - \tilde{K}(\zeta))^{-1}$ when $\operatorname{Im} \zeta \to \pm 0$, rather than for $\operatorname{Im} \zeta \neq 0$. Hence we need the following variant of Theorem 6.6.

Theorem 6.7. If $A(\zeta)$ is a compact operator-valued function that is analytic whenever $\operatorname{Im} \zeta > 0$ (or $\operatorname{Im} \zeta < 0$) and uniformly continuous for $\operatorname{Im} \zeta \geqslant 0$ (or $\operatorname{Im} \zeta \leqslant 0$, respectively), then either $(1 - A(\zeta))^{-1}$ does not exist anywhere on the domain of definition of $A(\zeta)$, or it exists everywhere on the real axis with the possible exception of a closed set $\mathscr{S}_0$ of real values of $\mathscr{S}$, which has Lebesgue measure zero in $\mathbb{R}^1$.

The proof of the Theorem 6.7 proceeds in exactly the same manner as that of Theorem 6.6 until (6.106) is reached. Then, in ascertaining the properties of the set $\mathscr{S}_0$ on which $\delta(\zeta)$ in (6.106) vanishes, we can apply a theorem* from complex analysis to reach the desired conclusion.

By the argument used in the first part of the proof of Theorem 6.4 we know that there is a $\lambda_V > 0$ such that $(1 - \tilde{K}(\zeta))^{-1}$ exists if $|\zeta| > \lambda_V$. This tells us, first of all, that the set $\mathscr{S}_V^1$ of real points at which $(1 - \tilde{K}(\zeta))^{-1}$ does not exist is bounded, and second, that due to Theorem 6.7 it has to be closed and of Lebesgue measure zero.

Exercises

6.1. Prove that the integral in $\mathbf{r}'$ of the function $|\mathbf{r}'|^{-2-\epsilon_0} |\mathbf{r} - \mathbf{r}'|^{-1}$ over the region $\{\mathbf{r}': |\mathbf{r}'| \geqslant R_1\}$ exists when $R_1 > 0$ and $\epsilon_0 > 0$, and

* That theorem states that if $\delta(\zeta)$ is any function that is analytic in the open upper (or lower) complex half-plane and continuous in the closed upper (or lower) complex half-plane, then the set $\mathscr{S}_0$ of points on the real axis at which $\delta(\zeta)$ vanishes is closed and of Lebesgue measure zero if $\delta(\zeta)$ is not identically equal to zero (see, e.g., Lemma 8.21 in Amrein *et al.* [1977]).

that when $r \to +\infty$, it behaves as $O(1/r^{\epsilon_0})$ if $0 < \epsilon_0 \leqslant 1$, and as $O(1/r)$ if $\epsilon_0 > 1$.

6.2. Prove that if the potential $V(\mathbf{r})$ obeys the restriction of Lemma 6.1, then the following integral is continuous in $\mathbf{r} \in \mathbb{R}^3$ and

$$\int_{\mathbb{R}^3} \frac{|V(\mathbf{r}')|}{|\mathbf{r} - \mathbf{r}'|}\, d\mathbf{r}' \leqslant C, \qquad \mathbf{r} \in \mathbb{R}^3,$$

where the constant C is independent of $\mathbf{r}$.

6.3. Show that if the potential $V(\mathbf{r}) = O(r^{-2-\epsilon_0})$ satisfies the remaining conditions of Lemma 6.1, then as $r \to \infty$

$$\int_{\mathbb{R}^3} \frac{|V(\mathbf{r}_1)|\, d\mathbf{r}_1}{|\mathbf{r} - \mathbf{r}_1|^2} = \begin{cases} O\left(\dfrac{1}{r^{1+\epsilon_0}}\right) & \text{for } 0 < \epsilon_0 \leqslant 1 \\ O\left(\dfrac{1}{r^2}\right) & \text{for } \epsilon_0 > 1, \end{cases}$$

and that the above integral defines a locally integrable function of $\mathbf{r}$.

6.4. Show that if the potential $V(\mathbf{r}) = O(r^{-2-\epsilon_0})$ satisfies the conditions of Lemma 6.1, then as $r_1 \to \infty$ for fixed $\mathbf{r} \in \mathbb{R}^3$,

$$\int_{\mathbb{R}^3} \frac{|V(\mathbf{r}_2)|}{|\mathbf{r} - \mathbf{r}_2|\,|\mathbf{r}_2 - \mathbf{r}_1|}\, d\mathbf{r}_2 = \begin{cases} O\left(\dfrac{1}{r_1^{(1+\epsilon_0)/2}}\right) & \text{for } 0 < \epsilon_0 \leqslant 1 \\ O\left(\dfrac{1}{r_1}\right) & \text{for } \epsilon_0 > 1, \end{cases}$$

and the above function is locally square integrable in $\mathbf{r}_1$ and $\mathbf{r}$.

6.5. Prove that if the potential $V(\mathbf{r})$ satisfies the conditions of Lemma 6.1, then the integrals appearing on the right-hand side of (6.39) exist, and that $F_V(\mathbf{r}; \zeta)$ is locally square integrable and bounded at infinity in $\mathbf{r}$ whenever $\tilde{d}(\zeta) \neq 0$.

6.6. Show that if the potential $V(\mathbf{r})$ is locally square integrable, and $|V(\mathbf{r})| \leqslant C_0 r^{-2-\epsilon_0}$, $\epsilon_0 > 0$, for $r \geqslant R_0$, then the function $V(\mathbf{r})\,|\mathbf{r} - \mathbf{r}'|^{-2}\, V(\mathbf{r}')$ is integrable on $\mathbb{R}^6$ if it is integrable on the set $\{(\mathbf{r}, \mathbf{r}'): r \leqslant R_0,\ r' \leqslant R_0\}$.

6.7. Prove that when the potential $V(\mathbf{r})$ satisfies the conditions of Lemma 6.1 then the function $v_{\mathbf{k}}(\mathbf{r})$ defined by (6.55) is uniformly bounded (i.e., $|v_{\mathbf{k}}(\mathbf{r})| \leqslant \text{const}$) for $r \geqslant R_0$ and $\mathbf{k}$ assuming values from a compact set D_0 containing no points for which $\mathbf{k}^2/2m \in \mathcal{S}_V^1$.

6.8. Let $h(\mathbf{r}, u)$ be a function with a continuous (in $\mathbf{r}$ and u) partial derivative $h_u(\mathbf{r}, u) = \partial h(\mathbf{r}, u)/\partial u$, and assume that for $r \geqslant R_0$

$$| h_u(\mathbf{r}, u) f(\mathbf{r})| \leqslant \rho(\mathbf{r}),$$

where $\rho(\mathbf{r})$ is integrable on $\{\mathbf{r}: r \geqslant R_0\}$. Show that if $h_u(\mathbf{r}, u) f(\mathbf{r})$ and $f(\mathbf{r})$ are integrable over the sphere $\{\mathbf{r}: r \leqslant R_0\}$, then

$$\frac{d}{du} \int_{\mathbb{R}^3} h(\mathbf{r}, u) f(\mathbf{r})\, d\mathbf{r} = \int_{\mathbb{R}^3} h_u(\mathbf{r}, u) f(\mathbf{r})\, d\mathbf{r}.$$

6.9. Show that the function

$$g(\mathbf{r}) = \int \frac{\exp(ik| \mathbf{r} - \mathbf{r}' |)}{| \mathbf{r} - \mathbf{r}' |} f(\mathbf{r}')\, d\mathbf{r}'$$

is continuous in $\mathbf{r}$ if $f(\mathbf{r})$ is locally square integrable and integrable on $\mathbb{R}^3$.

6.10. Show that the family $\mathscr{C}_V^\infty$ of all functions in $\mathscr{C}_b^\infty(\mathbb{R}^2)$ with supports disjoint from the set $\mathscr{S}_V^3$ (mentioned in Theorem 6.1) is dense in $L^2(\mathbb{R}^3)$.

7. Wave and Scattering Operators in Potential Scattering

7.1. The Existence of Strong Asymptotic States

In the study of a particular scattering model it is crucial to determine from the start the magnitude of the family of strong asymptotic states. In other words, we have to establish the size of the initial domain M_0 of $\Omega_\pm$.

In potential scattering, under very mild restrictions imposed on the potential, this question is answered by Theorem 7.1.

Theorem 7.1. Suppose the potential $V(\mathbf{r})$ is locally square integrable and that for some fixed $\epsilon > 0$

$$V(\mathbf{r}) = O\left(\frac{1}{r^{1+\epsilon}}\right), \qquad r = | \mathbf{r} | \to \infty. \tag{7.1}$$

Then the initial domain M_0 of the wave operators $\Omega_\pm$ corresponding to $H_0 \supseteq -(1/2m)\Delta$ and $H \supseteq -(1/2m)\Delta + V(\mathbf{r})$ coincides with the entire Hilbert space $L^2(\mathbb{R}^3)$.

The proof of this theorem is based on Theorem 2.8, and it will be carried out in several stages. It is important, however, to verify before starting with the proof that in the present case H is indeed self-adjoint. This self-adjointness of H is a consequence of the restrictions imposed on $V(\mathbf{r})$, which has to be locally square integrable and bounded at infinity to satisfy these restrictions. Consequently, $-(1/2m)\Delta + V(\mathbf{r})$ determines a unique self-adjoint operator by virtue of Theorem 7.5 in Chapter IV.

The starting point of the proof of the above theorem consists in choosing an adequate domain $\mathscr{D}_1$ which satisfies the conditions of Theorem 2.8. We shall prove now that we can choose for $\mathscr{D}_1$ the linear manifold spanned by all functions

$$\tilde{\psi}_{\boldsymbol{\rho}}(\mathbf{k}) = \exp[-(1/2m)\,\mathbf{k}^2 - i\mathbf{k}\boldsymbol{\rho}], \qquad \boldsymbol{\rho} \in \mathbb{R}^3, \tag{7.2}$$

where $\boldsymbol{\rho}$ varies over all vectors in $\mathbb{R}^3$.

As a straightforward consequence of (5.2) we obtain

$$\begin{aligned}(\exp(-itH_0)\psi_{\boldsymbol{\rho}})^\sim(\mathbf{k}) &= \exp\left(-it\frac{\mathbf{k}^2}{2m}\right)\tilde{\psi}_{\boldsymbol{\rho}}(\mathbf{k}) \\ &= \exp\left(\frac{-1-it}{2m}\,\mathbf{k}^2 - i\mathbf{k}\boldsymbol{\rho}\right).\end{aligned} \tag{7.3}$$

Hence, by virtue of Theorems 7.3 and 7.5 in Chapter IV we have* $\exp(-itH_0)\,\psi_{\boldsymbol{\rho}} \in \mathscr{D}_{H_0} \cap \mathscr{D}_H$, and consequently $\exp(-itH_0)\,\mathscr{D}_1 \subset \mathscr{D}_{H_0} \cap \mathscr{D}_H$.

The second condition imposed on $\mathscr{D}_1$ in Theorem 2.8 is that when $t_1 \to -\infty$ and $t_2 \to +\infty$, the strong limits of $W_1(t_1, t_2)\,\psi_{\boldsymbol{\rho}}$ exist. We have already pointed out, at the end of §2, that the existence of the integrals (2.41) and (2.42) is sufficient for the fulfillment of this condition. In view of this fact, we shall settle the question by showing that

$$\int_{-\infty}^{+\infty} \| V\,e^{-itH_0}\,\psi_{\boldsymbol{\rho}} \|\,dt < +\infty \tag{7.4}$$

for all $\boldsymbol{\rho} \in \mathbb{R}^3$.

* Note that $T_S \equiv H_0$ and $T_S + V_S \equiv H$.

By taking the Fourier inverse transform of both sides of the relation (7.3), we obtain for any real δ

$$
\begin{aligned}
&|(\exp(-itH_0)\psi_\rho)(\mathbf{r})| \\
&= (2\pi)^{-3/2} \left| \int_{\mathbb{R}^3} \tilde{\psi}_\rho(\mathbf{k}) \exp\left[-it\frac{\mathbf{k}^2}{2m} + i\mathbf{k}\mathbf{r}\right] d\mathbf{k} \right| \\
&= (2\pi)^{-3/2} \prod_{w\in\{x,y,z\}} \left| \int_{-\infty}^{+\infty} \exp\left[-\frac{1+it}{2m} k_w{}^2 + ik_w(w-\rho_w)\right] dk_w \right| \\
&= \left| \frac{m^{3/2}}{(1+it)^{3/2}} \prod_{w\in\{x,y,z\}} \exp\left[-\frac{m}{2}\frac{(w-\rho_w)^2}{1+it}\right] \right| \\
&\leqslant \frac{m^{3/2}}{|1+it|^{3/2}} \exp\left[-\frac{m}{2}\frac{(\mathbf{r}-\boldsymbol{\rho})^2}{1+t^2}\right] \\
&= m^{3/2} \frac{|\mathbf{r}-\boldsymbol{\rho}|^{-1/2+\delta}}{|1+it|^{1+\delta}} \left| \frac{\mathbf{r}-\boldsymbol{\rho}}{1+it} \right|^{1/2-\delta} \exp\left[-\frac{m}{2}\frac{(\mathbf{r}-\boldsymbol{\rho})^2}{1+t^2}\right].
\end{aligned}
$$

If we now take $0 < \delta < 1/2$, so that $1/2 - \delta > 0$, we get, by multiplying both sides of the above inequality by $V(\mathbf{r})$,

(7.5)

$$
\begin{aligned}
&|(V\, e^{-itH_0}\, \psi_\rho)(\mathbf{r})| \\
&\leqslant m^{3/2} \frac{|V(\mathbf{r})|\,|\mathbf{r}-\boldsymbol{\rho}|^{\delta-1/2}}{(1+it)^{1+\delta}} \left| \frac{(\mathbf{r}-\boldsymbol{\rho})^2}{1+t^2} \right|^{1/4-\delta/2} \exp\left(-\frac{m}{2}\frac{(\mathbf{r}-\boldsymbol{\rho})^2}{1+t^2}\right) \\
&\leqslant C \frac{|\mathbf{r}-\boldsymbol{\rho}|^{\delta-1/2}}{|1+it|^{1+\delta}} |V(\mathbf{r})|,
\end{aligned}
$$

where C is a constant which majorizes the function

$$m^{3/2} \left| \frac{(\mathbf{r}-\boldsymbol{\rho})^2}{1+t^2} \right|^{1/4-\delta/2} \exp\left[-\frac{m}{2}\frac{(\mathbf{r}-\boldsymbol{\rho})^2}{1+t^2}\right]$$

for all values of $\mathbf{r}$, $\boldsymbol{\rho}$, and t. If we choose δ so that in addition $\delta < \epsilon$, then by virtue of (7.1) the function on the right-hand side of (7.5) is square integrable. Consequently, we can write

$$\| V\, e^{-itH_0}\, \psi_\rho \| \leqslant C_\rho \frac{1}{|1+it|^{1+\delta}},$$

where C_ρ is constant with respect to the variable $t \in \mathbb{R}^1$. Thus, (7.4) holds and the second condition of Theorem 2.8 is fulfilled.

This establishes that the closure $\bar{\mathscr{D}}_1$ of the linear manifold $\mathscr{D}_1$ spanned by all functions (7.2) is contained in M_0.

To conclude the proof of Theorem 7.1, we still have to show that $\bar{\mathscr{D}}_1 \equiv L^2(\mathbb{R}^3)$, and thus establish that $\mathsf{M}_0 \equiv L^2(\mathbb{R}^3)$. To arrive at this result we need the following lemma, which is a straightforward generalization, to arbitrarily many dimensions, of a theorem by Wiener.

***Lemma 7.1.** Let us denote by $\mathscr{N}_f$ the set on which the Lebesgue square-integrable function $f(x)$, $x \in \mathbb{R}^n$, vanishes. The Lebesgue measure of $\mathscr{N}_f$ is zero if and only if the linear manifold $(\mathscr{F}_f)$ spanned by the family $\mathscr{F}_f$ of all functions $\exp[ip \cdot x]f(x)$, obtained by varying p over all vectors $p \in \mathbb{R}^n$, is dense in $L^2(\mathbb{R}^n)$.

Proof. We set out to prove that, given any $g \in L^2(\mathbb{R}^n)$ and $\epsilon > 0$, we can find vectors $p_1, \ldots, p_s \in \mathbb{R}^n$ and complex numbers $a_1, \ldots, a_s$ such that

$$\int_{\mathbb{R}^n} \left| g(x) - f(x) \sum_{\nu=1}^{s} a_\nu e^{ix \cdot p_\nu} \right|^2 d^n x < 16\epsilon. \tag{7.6}$$

We note first that for any $\epsilon > 0$ we can satisfy the inequality

$$\int_{B'} |g(x)|^2 \, d^n x < \epsilon, \qquad B' = \mathbb{R}^n - B, \tag{7.7}$$

with a compact Borel set B, by choosing that set sufficiently large.

Let us define the functions

$$h_M(x) = \begin{cases} \dfrac{g(x)}{f(x)} & \text{if } \left| \dfrac{g(x)}{f(x)} \right| \leqslant M \\[2ex] M \dfrac{g(x)}{|g(x)|} \dfrac{|f(x)|}{f(x)} & \text{if } \left| \dfrac{g(x)}{f(x)} \right| > M \end{cases}$$

for any positive integer M; note that $h_M(x)$ is almost everywhere well defined, since $f(x)$ vanishes only on a set of measure zero.

It is easy to see that by virtue of Theorem 3.10 in Chapter II

$$\lim_{M \to \infty} \int_{\mathbb{R}^n} | h_M(x) f(x) - g(x)|^2 \, d^n x = 0.$$

Consequently, we can find an M for which

$$\int_{\mathbb{R}^n} | h_M(x) f(x) - g(x)|^2 \, d^n x < \epsilon. \tag{7.8}$$

Since (7.7) obviously implies that

$$\int_{B'} |h_M(x)f(x)|^2 \, d^n x < \epsilon,$$

we easily obtain from this inequality and (7.8), by using in the process the triangle inequality in $L^2(\mathbb{R}^n)$,

$$\int_{R^n} |g(x) - \chi_B(x)\, h_M(x) f(x)|^2 \, d^n x < 4\epsilon. \tag{7.9}$$

Let L be a box

$$L = \{x\colon -l \leqslant x_i \leqslant l, i = 1, \ldots, n\}$$

which contains B, and let $g_l(x)$ be a function equal to $\chi_B(x)\, h_M(x)$ in L and periodic in each of the variables $x_1, \ldots, x_n$, with the period $2l$. For sufficiently large l we have

$$\int_{\mathbb{R}^n} |\chi_B(x)\, h_M(x) f(x) - g_l(x) f(x)|^2 \, d^n x$$

$$= \sum_{k_1, \ldots, k_n = -\infty}^{+\infty\,\prime} \int_L |\chi_B(x)\, h_M(x) f(x_1 + 2lk_1, \ldots, x_n + 2lk_n)|^2 \, d^n x$$

$$\leqslant M^2 \int_{L'} |f(x)|^2 \, d^n x < \epsilon,$$

where $L' = \mathbb{R}^n - L$, and the prime on the summation sign indicates that the sum does not include the term with $k_1 = \cdots = k_n = 0$. Combining (7.9) and the above inequality we obtain

$$\int_{\mathbb{R}^n} |g(x) - g_l(x) f(x)|^2 \, dx < 9\epsilon \tag{7.10}$$

by making use of the triangle inequality.

Let us expand $g_l(x)$ in a Fourier series,

$$g_l(x) = \sum_{k_1, \ldots, k_n = -\infty}^{+\infty} c_k \exp\left(i\pi \frac{k \cdot x}{l}\right),$$

$$c_k = \frac{1}{(2l)^n} \int_L g_l(x) \exp\left(-i\pi \frac{k \cdot x}{l}\right) d^n x,$$

where $k = (k_1, \ldots, k_n)$ and the convergence of the above series is in the mean. A straightforward computation yields, for any positive integers N and $N_0 < N$,

$$\int_L \left| g_l(x) - \sum_{k_1,\ldots,k_n=-N}^{+N} \left(1 - \frac{|k_1|}{N}\right) \cdots \left(1 - \frac{|k_n|}{N}\right) c_k \exp\left(i\pi \frac{k \cdot x}{l}\right)\right|^2 d^n x$$

$$\leqslant (2l)^n \left\{ \sum_{k_1,\ldots,k_n=-N_0}^{N_0} \frac{N_0^2}{N^2} n^2 |c_k|^2 + \sum_k{}'' |c_k|^2 \right\},$$

where the summation $\sum''$ extends over all $k = (k_1, \ldots, k_n)$ in which $|k_j| > N_0$ for at least one value of $j = 1, \ldots, n$. Due to the convergence of $\sum |c_k|^2$, the second sum on the right-hand side of the above inequality can be made arbitrarily small by choosing N_0 large enough, while the first sum can be then made arbitrarily small by taking N sufficiently larger than N_0. Hence, the sequence of functions

$$f_N(x) = g_l(x) - \sum_{k_1,\ldots,k_n=-N}^{N} \left(1 - \frac{|k_1|}{N}\right) \cdots \left(1 - \frac{|k_n|}{N}\right) c_k \exp\left(i\pi \frac{k \cdot x}{l}\right),$$
$$N = 1, 2, \ldots,$$

converges in the mean to zero when $N \to +\infty$. Thus, it follows from Theorem 4.5 of Chapter II that there is a subsequence $f_{N_1}(x), f_{N_2}(x), \ldots$, which converges to zero for almost all values of $x \in \mathbb{R}^n$.

Now we can apply Lemma 3.1 of Chapter IV to infer that

$$\lim_{r \to +\infty} \int_{\mathbb{R}^n} |f_{N_r}(x) f(x)|^2 \, d^n x = 0, \tag{7.11}$$

after noting that the use of this lemma is justified by virtue of the inequality

$$|f_N(x) f(x)| \leqslant \{|g_l(x)| + M\} |f(x)|.$$

In fact, the above inequality follows from the estimate

$$\left| \sum_{k_1,\ldots,k_n=-N}^{N} \left(1 - \frac{|k_1|}{N}\right) \cdots \left(1 - \frac{|k_n|}{N}\right) c_k \exp\left(i\pi \frac{k \cdot x}{l}\right)\right|$$

$$= \left| \frac{1}{(2l)^n} \int_L g_l(x') \left\{ \sum_{k_1,\ldots,k_n=-N}^{N} \left(1 - \frac{|k_1|}{N}\right) \cdots \right.\right.$$

(equation continues)

$$\times \left(1 - \frac{|k_n|}{N}\right) \exp\left(i\pi \frac{k \cdot (x - x')}{l}\right)\Bigg\} d^n x' \Bigg|$$

$$\leqslant \frac{M}{(2l)^n} \left| \int_L \prod_{j=1}^{n} \sum_{k_j=-N}^{N} \left(1 - \frac{|k_j|}{N}\right) \exp\left[i\pi \frac{k_j(x_j - x_j')}{l}\right] d^n x' \right| = M;$$

the estimation of the last integral in the above relation could be carried out due to the formula

$$\sum_{k_j=-N}^{N} \left(1 - \frac{|k_j|}{N}\right) \exp\left[i \cdot \frac{\pi k_j}{l}(x_j - x_j')\right] = \frac{\sin^2[N\pi(x_j - x_j')/2l]}{\sin^2[\pi(x_j - x_j')/2l]},$$

which can be proven by mathematical induction in N.

Since (7.11) is true, we can choose one $N_r = N'$ for which

$$\text{(7.12)} \qquad \int_{\mathbb{R}^n} \Bigg| g_l(x) f(x) - f(x) \sum_{k_1,\ldots,k_n=-N'}^{N'} \left(1 - \frac{|k_1|}{N'}\right) \cdots$$

$$\times \left(1 - \frac{|k_n|}{N'}\right) c_k \exp\left(i\pi \frac{k \cdot x}{l}\right) \Bigg|^2 d^n x < \epsilon.$$

Combining (7.10) and (7.12), we immediately obtain, by means of the triangle inequality,

$$\int_{\mathbb{R}^n} \Bigg| g(x) - f(x) \sum_{k_1,\ldots,k_n=-N'}^{N'} \left(1 - \frac{|k_1|}{N'}\right)$$

$$\cdots \left(1 - \frac{|k_n|}{N'}\right) c_k \exp\left(i\pi \frac{k \cdot x}{l}\right) \Bigg|^2 d^n x < 16\epsilon.$$

Since the above inequality becomes identical to (7.6) after an appropriate change of symbols, we have proven the statement that the linear manifold $(\mathscr{F}_f)$ spanned by $\mathscr{F}_f$ is dense in $L^2(\mathbb{R}^3)$.

Conversely, if $(\mathscr{F}_f)$ is dense in $L^2(\mathbb{R}^3)$, then the measure of $\mathscr{N}_f$ is zero. In fact, if this were not so, then by choosing $g(x) = \chi_{\mathscr{N}_f}(x)$ we would have

$$\int_{\mathbb{R}^n} \left| g(x) - f(x) \sum_{\nu=1}^{s} a_\nu e^{ip_\nu \cdot x} \right|^2 d^n x \geqslant \int_{\mathscr{N}_f} |g(x)|^2 > 0,$$

so that (7.6) would not be satisfied for all $\epsilon > 0$ by any $a_1, \ldots, a_s \in \mathbb{C}^1$ and $p_1, \ldots, p_s \in \mathbb{R}^n$. Q.E.D.

If $\tilde{f}(p)$ is the Fourier–Plancherel transform of $f(x)$, then it is easy to see that the Fourier–Plancherel transform of $f(x)e^{ip_\nu \cdot x}$ is $\tilde{f}(p - p_\nu)$.

The function $\tilde{f}(p - p_\nu)$ is usually called the *translation* of $\tilde{f}(p)$ by the amount p_ν.

By taking the Fourier–Plancherel transform of the integrand in (7.6) we obtain

$$\int_{\mathbb{R}^n} \left| \tilde{g}(p) - \sum_{\nu=1}^{s} a_\nu \tilde{f}(p - p_\nu) \right|^2 d^n p < 16\epsilon.$$

Hence, we conclude that the linear manifold spanned by all translations of $\tilde{f}(p)$ is dense in $L^2(\mathbb{R}^n)$. This statement constitutes *Wiener's theorem* on the closure of translations.

Let us return now to the proof of Theorem 7.1. If we set

$$f(\mathbf{k}) = \exp\left(-\frac{\mathbf{k}^2}{2m}\right),$$

then (7.2) assumes the form

$$\tilde{\psi}_{\mathbf{p}}(\mathbf{k}) = f(\mathbf{k})\, e^{-i\mathbf{k}\mathbf{p}}.$$

Since $f(\mathbf{k})$ is different from zero everywhere on $\mathbb{R}^3$, we deduce from Lemma 7.1 that the linear manifold $\mathscr{D}_1$ spanned by all function (7.2) is dense in $L^2(\mathbb{R}^3)$. Consequently, $\bar{\mathscr{D}}_1 \equiv L^2(\mathbb{R}^3)$, and since we already know that $\mathscr{D}_1 \subset \mathsf{M}_0$, we finally obtain $\mathsf{M}_0 \equiv L^2(\mathbb{R}^3)$.

This concludes the proof of Theorem 7.1. We note that this theorem guarantees that $\mathsf{M}_0 \equiv L^2(\mathbb{R}^3)$ for all almost everywhere continuous potentials $V(\mathbf{r})$ which decrease at infinity faster than the Coulomb potential. However, at this stage we cannot say anything about the unitarity of the S operator in potential scattering. Only when additional restrictions on the potential are imposed, as required in the theorems of the next section, will we be able to prove that in potential scattering the scattering operator S is unitary.

7.2. The Completeness of the Møller Wave Operators

The next theorem will relate the Møller wave operators to distorted plane waves, providing at the same time the answers to all the basic questions regarding $\mathsf{R}_\pm$, M_0, and the unitarity of the scattering operator S in potential scattering.

Theorem 7.2. Suppose the potential $V(\mathbf{r})$ is locally square integrable and of Rollnik class, and that

$$|V(\mathbf{r})| = O\left(\frac{1}{r^{3+\epsilon_0}}\right), \qquad r \to \infty, \tag{7.13}$$

for some $\epsilon_0 > 0$. Then the initial domain M_0 of the Møller wave operators $\Omega_\pm$ is $L^2(\mathbb{R}^3)$, $\Omega_\pm$ are complete [i.e., the final domains $\mathsf{R}_\pm$ of $\Omega_\pm$ are both identical to $L^2_{H_c}(\mathbb{R}^3) = E^H(S^H_c)L^2(\mathbb{R}^3)$] and the S operator is unitary on $\mathsf{M}_0 \equiv L^2(\mathbb{R}^3)$; furthermore,

$$(\Omega_\pm\psi)(\mathbf{r}) = \text{l.i.m.} \int_{\mathbb{R}^3} \Phi_{\mathbf{k}}^{(\pm)}(\mathbf{r})\,\tilde{\psi}(\mathbf{k})\,d\mathbf{k}, \tag{7.14}$$

where $\tilde{\psi}(\mathbf{k})$ is the Fourier–Plancherel transform of $\psi(\mathbf{r})$,

$$\tilde{\psi}(\mathbf{k}) = (2\pi)^{-3/2}\,\text{l.i.m.} \int_{\mathbb{R}^3} \psi(\mathbf{r})\,e^{-i\mathbf{kr}}\,d\mathbf{r}, \tag{7.15}$$

and $\Phi_{\mathbf{k}}^{(+)}(\mathbf{r})$ are the outgoing distorted plane waves, while $\Phi_{\mathbf{k}}^{(-)}(\mathbf{r}) = \Phi_{-\mathbf{k}}^{(+)*}(\mathbf{r})$ are the incoming distorted plane waves given as solutions (6.16) of (6.5).

The validity of statements which are essentially equivalent to the assertions made in the above theorem about M_0, $\mathsf{R}_\pm$, S, and $\Omega_\pm$ has been taken for granted by physicists from the very beginning of quantum scattering theory. In proving the above theorem, we confirm that this faith was justified only to the extent that the potential satisfies the conditions imposed on it at the beginning of the theorem. Consequently, it is important to realize that these conditions are essential only to the extent of making applicable Theorems 6.1 and 6.2 on distorted plane waves. To avoid getting into a maze of technical details we have opted for (7.13) from the beginning of our systematic study of $\Phi_{\mathbf{k}}^{(\pm)}(\mathbf{r})$ in §6, but in fact (7.13) can be relaxed to

$$|\,V(\mathbf{r})| = O(r^{-2-\epsilon}), \qquad \epsilon > 0. \tag{7.16}$$

However, in that case one has to base (6.7)–(6.9) on factorizations $V^\beta\,|\,V\,|^{1-\beta}$ with $\beta \in (\frac{1}{4}, \frac{3}{4})$ that, in general, might* not equal $\frac{1}{2}$. Various other approaches exist (such as those of Ikebe [1960] based on Banach space techniques) that can achieve essentially the same result, but clearly, as demonstrated by a counterexample of Pearson [1975], the previous unitarity of the S operator does not hold for arbitrary short-range potentials.

In case the potential is spherically symmetric, i.e., $V(\mathbf{r}) = V_0(r)$, the above conditions can be further relaxed by replacing (7.16) with

$$|\,V_0(r)| = O\left(\frac{1}{r^{1+\epsilon}}\right), \qquad \epsilon > 0, \tag{7.17}$$

* See, e.g., Amrein *et al.* [1977], pp. 372–373, or Reed and Simon [1979], p. 353.

and making the additional requirement that $rV_0(r)$ is locally integrable on $[0, +\infty)$ (see Green and Lanford [1960]). Since the assertions of Theorem 7.2 are not true for the Coulomb potential, it is clear that for spherically symmetric potentials one cannot hope to do any better by further relaxing (7.17). However, in the general case, it is an open question whether one could relax (7.16) any further.

We shall prove Theorem 7.2 in a few stages. The first step consists in verifying that the integral operator in the right-hand side of the equation (7.14) is a well-defined mathematical entity.

***Lemma 7.2.** If $\Phi_{\mathbf{k}}^{(\pm)}(\mathbf{r})$ are the solutions (6.16) of (6.5) with a potential satisfying the conditions of Theorem 7.2, then there are unique operators W_+ and W_- defined by

(7.18) $$(W_\pm\psi)(\mathbf{r}) = \underset{\psi_n\to\psi}{\text{l.i.m.}} \int_{\mathbb{R}^3} \Phi_{\mathbf{k}}^{(\pm)}(\mathbf{r})\,\tilde{\psi}_n(\mathbf{k})\,d\mathbf{k}$$

for arbitrary sequences $\tilde{\psi}_1(k), \tilde{\psi}_2(k),..., \in \mathscr{C}_V^\infty$ converging in the mean to $\tilde{\psi} \in L^2(\mathbb{R}^3)$, and these operators satisfy the relations

(7.19) $$W_\pm W_\pm{}^* = E^H(\mathsf{S}_{\mathrm{c}}^H),$$

(7.20) $$W_\pm{}^* E^H(B) = E^{H_0}(B) W_\pm{}^*, \qquad B \in \mathscr{B}^1.$$

Proof. We shall prove the lemma only for W_+, since the corresponding results for W_- easily follow from the relation $\Phi_{\mathbf{k}}^{(-)}(\mathbf{r}) = \Phi_{-\mathbf{k}}^{(+)*}(\mathbf{r})$.

First of all, we note that under the present restrictions on the potential, Theorem 6.3 holds. It follows that for any given bounded closed set B which is disjoint from the exceptional set $\mathscr{S}_V^3$ defined in (6.15) we can find a locally square integrable and bounded at infinity function $C_B(\mathbf{r})$ such that

(7.21) $$|\Phi_{\mathbf{k}}^{(+)}(\mathbf{r})| = |(2\pi)^{-3/2}[e^{i\mathbf{kr}} + v_{\mathbf{k}}(\mathbf{r})]| \leqslant C_B(\mathbf{r})$$

as long as $\mathbf{k}$ assumes only values from B.

Let us consider the bilinear form

(7.22) $$(\psi \mid \psi') = \int_{\mathbb{R}^3} \tilde{\psi}^*(\mathbf{k})\,\hat{\psi}_+{}'(\mathbf{k})\,d\mathbf{k}$$

in ψ and ψ', where

(7.23) $$\tilde{\psi}(\mathbf{k}) = (2\pi)^{-3/2} \int_{\mathbb{R}^3} e^{-i\mathbf{kr}}\,\psi(\mathbf{r})\,d\mathbf{r},$$

$$\hat{\psi}_+'(\mathbf{k}) = \int_{\mathbb{R}^3} \Phi_{\mathbf{k}}^{(+)*}(\mathbf{r})\, \psi'(\mathbf{r})\, d\mathbf{r}. \tag{7.24}$$

With the help of Theorem 6.2 we show that this bilinear form is bounded:

$$|(\psi \mid \psi')| \leqslant \left\{\int_{\mathbb{R}^3} |\tilde{\psi}(\mathbf{k})|^2\, d\mathbf{k}\right\}^{1/2} \left\{\int_{\mathbb{R}^3} |\hat{\psi}_+'(\mathbf{k})|^2\, d\mathbf{k}\right\}^{1/2}$$
$$= \| \psi \| \| E^H(\mathsf{S}_{\mathrm{c}}^H)\psi' \| \leqslant \| \psi \| \| \psi' \|.$$

Consequently (see Chapter III, Exercise 2.5), there is a unique bounded linear operator W_+ satisfying the relation

$$\langle W_+\psi \mid \psi'\rangle = (\psi \mid \psi') \tag{7.25}$$

for all $\psi, \psi' \in L^2(\mathbb{R}^3)$.

Let us choose now $\tilde{\psi}(\mathbf{k}) \in \mathscr{C}_V^\infty$ and $\psi'(\mathbf{r}) \in \mathscr{C}_{\mathrm{b}}^\infty(\mathbb{R}^3)$, where $\mathscr{C}_V^\infty$ is the family of all functions in $\mathscr{C}_{\mathrm{b}}^\infty(\mathbb{R}^3)$ with supports disjoint from $\mathscr{S}_V{}^3$. For such $\tilde{\psi}(\mathbf{k})$, "l.i.m." in (7.18) can be dropped, and (7.22) can be written in the form

$$(\psi \mid \psi') = \int_{\mathbb{R}^3} d\mathbf{k}\, \tilde{\psi}^*(\mathbf{k}) \left\{\int_{\mathbb{R}^3} \Phi_{\mathbf{k}}^{(+)*}(\mathbf{r})\, \psi'(\mathbf{r})\, d\mathbf{r}\right\}. \tag{7.26}$$

Since (7.21) is satisfied by $\Phi_{\mathbf{k}}(r)$, we immediately find that for such choices of $\psi'(\mathbf{r})$ and $\tilde{\psi}(\mathbf{k})$ the function $\tilde{\psi}^*(\mathbf{k})\Phi_{\mathbf{k}}^{(+)*}(\mathbf{r})\psi'(\mathbf{r})$ is integrable in $\mathbb{R}^6$. Consequently, by Fubini's theorem, we can interchange the order of integration in (7.26), thus arriving at the result

$$\langle W_+\psi \mid \psi'\rangle = \int_{\mathbb{R}^3} d\mathbf{r}\, \psi'(\mathbf{r}) \left\{\int_{\mathbb{R}^3} \Phi_{\mathbf{k}}^{(+)}(\mathbf{r})\, \tilde{\psi}(\mathbf{k})\, d\mathbf{k}\right\}^*. \tag{7.27}$$

Since $\mathscr{C}_{\mathrm{b}}^\infty(\mathbb{R}^3)$ is dense in $L^2(\mathbb{R}^3)$, equality (7.27) holds for all $\psi'(\mathbf{r}) \in \mathscr{C}_{\mathrm{b}}^\infty(\mathbb{R}^3)$ if and only if

$$(W_+\psi)(\mathbf{r}) = \int_{\mathbb{R}^3} \Phi_{\mathbf{k}}^{(+)}(\mathbf{r})\, \tilde{\psi}(\mathbf{k})\, d\mathbf{k} \tag{7.28}$$

almost everywhere. Hence, (7.18) is established for $\tilde{\psi} \in \mathscr{C}_V^\infty$. Moreover, due to the fact that $\mathscr{C}_V^\infty$ is dense in $L^2(\mathbb{R}^3)$ (see Exercise 6.10) the general validity of (7.18) for W_+ and arbitrary $\psi \in L^2(\mathbb{R}^3)$ is an immediate consequence of (7.28).

In order to express $W_+{}^*$ as an integral operator, let us take in (7.22) $\psi(\mathbf{r}) \in \mathscr{C}_{\mathrm{b}}^\infty(\mathbb{R}^3)$ and write for compact $B_n \to \mathbb{R}^3 - \mathscr{S}_V^3$

$$\langle \psi \mid W_+{}^*\psi'\rangle = \langle W_+\psi \mid \psi'\rangle = (2\pi)^{-3/2} \lim_{n\to\infty} \int_{B_n} d\mathbf{k}\hat{\psi}_+'(\mathbf{k}) \left\{\int_{\mathbb{R}^3} e^{-i\mathbf{k}\mathbf{r}}\psi(\mathbf{r})\, d\mathbf{r}\right\}^*.$$

Since, under the present conditions, the function $\psi^*(\mathbf{r})\, e^{i\mathbf{kr}}\, \psi_+'(\mathbf{k})$ is integrable in $\mathbf{r}$ and $\mathbf{k}$ on $\mathbb{R}^3 \times B_n$, we can invoke Fubini's theorem to invert the order of integration in $\mathbf{k}$ and $\mathbf{r}$:

$$\langle \psi \mid W_+{}^*\psi' \rangle = (2\pi)^{-3/2} \lim_{n\to\infty} \int_{\mathbb{R}^3} d\mathbf{r}\, \psi^*(\mathbf{r}) \left\{ \int_{B_n} e^{i\mathbf{kr}}\, \hat{\psi}_+'(\mathbf{k})\, d\mathbf{k} \right\}.$$

Now, the type of reasoning earlier applied to (7.27) leads us to the conclusion that

$$(W_+{}^*\psi)(\mathbf{r}) = (2\pi)^{-3/2}\, \text{l.i.m.} \int_{\mathbb{R}^3} e^{i\mathbf{kr}}\, \hat{\psi}_+(\mathbf{k})\, d\mathbf{k} \tag{7.29}$$

for arbitrary $\hat{\psi}_+(k)$ given by (6.45).

Combining (6.48), (7.18) and (7.29), we obtain

$$\begin{aligned} (W_+W_+{}^*\psi)(\mathbf{r}) &= [W_+(W_+{}^*\psi)](\mathbf{r}) = \text{l.i.m.} \int_{\mathbb{R}^3} \Phi_{\mathbf{k}}^{(+)}(\mathbf{r})(W_+{}^*\psi)^{\sim}(\mathbf{k})\, d\mathbf{k} \\ &= \text{l.i.m.} \int_{\mathbb{R}^3} \Phi_{\mathbf{k}}^{(+)}(\mathbf{r})\, \hat{\psi}_+(\mathbf{k})\, d\mathbf{k} = (E^H(\mathsf{S}_{\mathrm{c}}^H)\psi)(\mathbf{r}), \end{aligned}$$

where the last step follows from Theorem 6.2, and in particular from (6.44). Hence (7.19) is established.

Finally, combining (6.44)–(6.46) with (7.29), we easily derive

$$\begin{aligned} (W_+{}^*E^H(B)\psi)(\mathbf{r}) &= (2\pi)^{-3/2}\, \text{l.i.m.} \int_{\mathbb{R}^3} e^{i\mathbf{kr}}(E^H(B)\psi)^{\wedge}_{+}(\mathbf{k})\, d\mathbf{k} \\ &= (2\pi)^{-3/2}\, \text{l.i.m.} \int_{\mathbf{k}^2/2m\in B} e^{i\mathbf{kr}}\, \hat{\psi}_+(\mathbf{k})\, d\mathbf{k}, \end{aligned}$$

$$\begin{aligned} (E^{H_0}(B)\, W_+{}^*\psi)(\mathbf{r}) &= (2\pi)^{-3/2}\, \text{l.i.m.} \int_{\mathbf{k}^2/2m\in B} e^{i\mathbf{kr}}(W_+{}^*\psi)^{\sim}(\mathbf{k})\, d\mathbf{k} \\ &= (2\pi)^{-3/2}\, \text{l.i.m.} \int_{\mathbf{k}^2/2m\in B} e^{i\mathbf{kr}}\, \hat{\psi}_+(\mathbf{k})\, d\mathbf{k}. \end{aligned}$$

This establishes (7.20). Q.E.D.

*7.3. Proof of Asymptotic Completeness

We are now ready to carry through the proof of Theorem 7.2 to its completion by showing that $W_\pm = \Omega_\pm$.

Let us adapt the relation (2.37) to the present case, setting $t_1 = 0$,

$t_2 = \tau$, and $f = \psi$ to obtain

$$(7.30)\quad \langle(\Omega(\tau) - 1)\psi \mid g\rangle = i \int_0^\tau (\langle e^{-iH_0t}\psi \mid He^{-iHt} g\rangle - \langle H_0 e^{-iH_0t}\psi \mid e^{-iHt} g\rangle)\, dt$$

for any $g \in \mathscr{D}_H$.

Since the restrictions imposed at present on the potential meet the demands of Theorem 7.5 in Chapter IV, we infer that if $\psi \in \mathscr{D}_{H_0}$, then $\psi \in \mathscr{D}_V$. Hence, for $\psi \in \mathscr{D}_{H_0}$, $(H - H_0)e^{-iH_0t}\psi = Ve^{-iH_0t}\psi$, and the relation (7.30) assumes the form

$$(7.31)\qquad \langle\Omega(\tau)\psi \mid g\rangle = \langle\psi \mid g\rangle + i \int_0^\tau \langle e^{iHt} V e^{-iH_0t}\psi \mid g\rangle\, dt.$$

The above relation can be extended to arbitrary $g \in L^2(\mathbb{R}^3)$. Setting $g = W_-\phi$, $\phi \in L^2(\mathbb{R}^3)$, and using the relation $W_-^* e^{iHt} = e^{iH_0t} W_-^*$, which is a direct consequence of (7.20) (see also Exercise 2.1), we obtain

$$\langle\phi \mid W_-^*\Omega(\tau)\psi\rangle = \langle\phi \mid W_-^*\psi\rangle - i \int_0^\tau \langle\phi \mid e^{iH_0t} W_-^* V e^{-iH_0t}\psi\rangle\, dt.$$

Since it can be readily shown (see Exercise 7.1) that

$$(7.32)\qquad \int_0^{+\infty} \langle\phi \mid e^{iH_0t} W_-^* V e^{-iH_0t}\psi\rangle\, dt = \lim_{\epsilon\to+0} \int_0^{+\infty} e^{-\epsilon t}\langle\phi \mid e^{iH_0t} W_-^* V e^{-iH_0t}\psi\rangle\, dt,$$

we easily arrive at the result

$$(7.33)\qquad \langle\phi \mid W_-^*\Omega_-\psi\rangle = \lim_{\tau\to+\infty} \langle W_-\phi \mid \Omega(\tau)\psi\rangle = \langle\phi \mid W_-^*\psi\rangle - \lim_{\epsilon\to+0} i \int_0^{+\infty} e^{-\epsilon t}\langle\phi \mid e^{iH_0t} W_-^* V e^{-iH_0t}\psi\rangle\, dt.$$

It should be emphasized that, in this context, all the integrals over the infinite interval $[0, +\infty)$ are improper Riemann integrals, and not Lebesgue integrals.

We can calculate the integrand in the above relation by using (7.29) and (4.8),

$$(7.34)\quad \langle\phi \mid e^{iH_0t} W_-^* V e^{-iH_0t}\psi\rangle = \int_{\mathbb{R}^3} \tilde{\phi}^*(\mathbf{k})\, e^{i(\mathbf{k}^2/2m)t} (W_-^* V e^{-iH_0t}\psi)^\sim(\mathbf{k})\, d\mathbf{k}$$

(equation continues)

$$= \int_{\mathbb{R}^3} \tilde{\phi}^*(\mathbf{k})\, e^{i(\mathbf{k}^2/2m)t} (Ve^{-iH_0t}\psi)^\wedge_-(\mathbf{k})\, d\mathbf{k}.$$

We resort now to the relation (6.45), which for $\psi(\mathbf{r}) \in \mathscr{C}_b^\infty(\mathbb{R}^3)$ yields

$$(7.35) \qquad (Ve^{-iH_0t}\psi)^\wedge_-(\mathbf{k}) = \int_{\mathbb{R}^3} \Phi_{\mathbf{k}}^{(-)*}(\mathbf{r})\, V(\mathbf{r})(e^{-iH_0t}\psi)(\mathbf{r})\, d\mathbf{r}.$$

We note that the limit in the mean in (6.45) has been dropped in the above relation, since the integrand in (7.35) is integrable on $\mathbb{R}^3$. To arrive at this conclusion, we recall (7.21) and note that $(e^{-iH_0t}\psi)(\mathbf{r})$ is uniformly bounded for all $t \in \mathbb{R}^1$ (see Exercise 7.2).

By combining (7.34) and (7.35) we obtain

$$(7.36) \qquad \langle\phi \mid \exp(iH_0t)\, W_-^*V \exp(-iH_0t)\psi\rangle$$
$$= \int_{\mathbb{R}^3} d\mathbf{k}\, \tilde{\phi}^*(\mathbf{k}) \int_{\mathbb{R}^3} d\mathbf{r}\, \Phi_{\mathbf{k}}^{(-)*}(\mathbf{r}) V(\mathbf{r})(\exp[-i(H_0 - \mathbf{k}^2/2m)t]\psi)(\mathbf{r}).$$

Let us show now that when $\psi(\mathbf{r}) \in \mathscr{C}_b^\infty(\mathbb{R}^3)$ and B is compact and disjoint from $\mathscr{S}_V^3$ we have

$$(7.37) \qquad \left| \int_{\mathbb{R}^3} \Phi_{\mathbf{k}}^{(-)*}(\mathbf{r}) V(\mathbf{r})(\exp[-i(H_0 - \mathbf{k}^2/2m)t]\psi)(\mathbf{r})\, d\mathbf{r} \right| \leqslant C_0,$$

where C_0 is a constant independent of $\mathbf{k} \in B$ and $t \in \mathbb{R}^1$. Indeed, since $(\exp[-i(H_0 - \mathbf{k}^2/2m)t]\psi)(\mathbf{r})$ is uniformly bounded in $\mathbf{r}$, $\mathbf{k} \in \mathbb{R}^3$ and $t \in \mathbb{R}^1$ (see Exercise 7.2), by using (7.21) we get

$$\int_{\mathbb{R}^3} | \Phi_{\mathbf{k}}^{(-)*}(\mathbf{r})\, V(\mathbf{r})(\exp[-i(H_0 - \mathbf{k}^2/2m)t]\psi)(\mathbf{r})|\, d\mathbf{r} \leqslant \text{const} \int_{\mathbb{R}^3} V(\mathbf{r})\, C_B(\mathbf{r})\, d\mathbf{r},$$

thus establishing the validity of (7.37).

Let us insert (7.36) in (7.33) and choose $\tilde{\phi}(\mathbf{k}) \in \mathscr{C}_b^\infty(\mathbb{R}^3)$, thus obtaining

$$(7.38) \qquad \langle\phi \mid W_-^*\Omega_-\psi\rangle = \langle\phi \mid W_-^*\psi\rangle - \lim_{\epsilon\to+0} I_\epsilon(\phi, \psi),$$

where

$$(7.39) \qquad I_\epsilon(\phi, \psi) = i \int_0^{+\infty} dt\, e^{-\epsilon t} \int_{\mathbb{R}^3} d\mathbf{k}\, \tilde{\phi}^*(\mathbf{k})$$
$$\times \int_{\mathbb{R}^3} d\mathbf{r}\, \Phi_{\mathbf{k}}^{(-)*}(\mathbf{r})\, V(\mathbf{r})(\exp[-i(H_0 - \mathbf{k}^2/2m)t]\psi)(\mathbf{r}).$$

Using (7.37), we easily infer that if $\operatorname{supp} \tilde{\phi} \subset B$ the above integrand is integrable in $t \in [0, +\infty)$ and $\mathbf{k} \in \mathbb{R}^3$. Consequently, the reversal of the order of integration in t and $\mathbf{k}$ is justified by Fubini's theorem. Hence, we have

(7.40)

$$I_\epsilon(\phi,\psi) = i \int_{\mathbb{R}^3} d\mathbf{k}\, \tilde{\phi}^*(\mathbf{k}) \int_0^{+\infty} dt \,\langle F_{\mathbf{k}}^B \mid \exp[-i(H_0 - \mathbf{k}^2/2m - i\epsilon)t]\psi\rangle,$$

(7.41) $$F_{\mathbf{k}}^B(\mathbf{r}) = \chi_B(\mathbf{k})\, \Phi_{\mathbf{k}}^{(-)}(\mathbf{r})\, V(\mathbf{r}).$$

Using Theorem 3.1 of Chapter IV, we infer that

$$\left\langle F_{\mathbf{k}}^B \middle| \exp\left[-i\left(H_0 - \frac{\mathbf{k}^2}{2m} - i\epsilon\right)t\right]\psi\right\rangle$$
$$= i\frac{d}{dt}\left\langle F_{\mathbf{k}}^B \middle| \exp\left[-i\left(H_0 - \frac{\mathbf{k}^2}{2m} - i\epsilon\right)t\right] R_0\left(\frac{\mathbf{k}^2}{2m} + i\epsilon\right)\psi\right\rangle.$$

Consequently, we easily compute

$$\int_0^{+\infty} \left\langle F_{\mathbf{k}}^B \middle| \exp\left[-i\left(H_0 - \frac{\mathbf{k}^2}{2m} - i\epsilon\right)t\right]\psi\right\rangle dt$$
$$= i \lim_{\tau\to+\infty} \left\langle F_{\mathbf{k}}^B \middle| \left(\exp\left[-i\left(H_0 - \frac{\mathbf{k}^2}{2m} - i\epsilon\right)\tau\right] - 1\right) R_0\left(\frac{\mathbf{k}^2}{2m} + i\epsilon\right)\psi\right\rangle$$
$$= -i\left\langle F_{\mathbf{k}}^B \middle| R_0\left(\frac{\mathbf{k}^2}{2m} + i\epsilon\right)\psi\right\rangle = -i\left\langle R_0\left(\frac{\mathbf{k}^2}{2m} - i\epsilon\right)F_{\mathbf{k}}^B \middle| \psi\right\rangle.$$

If we combine (5.5) and (5.6) with the above result, we get

$$I_\epsilon(\phi,\psi) = \frac{m}{2\pi}\int_{\mathbb{R}^6}\left\{\int_{\mathbb{R}^3}\frac{\exp[i(\mathbf{k}^2 - 2mi\epsilon)^{1/2}\,|\,\mathbf{r}-\mathbf{r}'\,|]}{|\,\mathbf{r}-\mathbf{r}'\,|}\right.$$
$$\left.\times\, V(\mathbf{r}')\,\Phi_{\mathbf{k}}^{(-)}(\mathbf{r}')\,d\mathbf{r}'\right\}^* \tilde{\phi}^*(\mathbf{k})\,\psi(\mathbf{r})\,d\mathbf{r}\,d\mathbf{k}.$$

Inserting this result in (7.38), and noting that

(7.42) $$\lim_{\epsilon\to+0}\int_{\mathbb{R}^3}\frac{\exp[i(\mathbf{k}^2 - 2mi\epsilon)^{1/2}\,|\,\mathbf{r}-\mathbf{r}'\,|]}{|\,\mathbf{r}-\mathbf{r}'\,|}V(\mathbf{r}')\,\Phi_{\mathbf{k}}^{(-)}(\mathbf{r}')\,d\mathbf{r}'$$
$$= \int_{\mathbb{R}^3}\frac{\exp(-ik\,|\,\mathbf{r}-\mathbf{r}'\,|)}{|\,\mathbf{r}-\mathbf{r}'\,|}V(\mathbf{r}')\,\Phi_{\mathbf{k}}^{(-)}(\mathbf{r}')\,d\mathbf{r}',$$

and that the absolute value of the integrand in the above limit is majorized by $|\mathbf{r}-\mathbf{r}'|^{-1}|V(\mathbf{r}')|\,C_B(\mathbf{r}')$ when $\mathbf{k}\in B$, we arrive at the following result:

$$(7.43)\quad \langle\phi\mid W_-^*\Omega_-\psi\rangle = \langle W_-\phi\mid\psi\rangle - \frac{m}{2\pi}\int_{\mathbb{R}^6}\left\{\int_{\mathbb{R}^3}\frac{\exp(-ik\,|\mathbf{r}-\mathbf{r}'|)}{|\mathbf{r}-\mathbf{r}'|}\right.$$
$$\left.\times\, V(\mathbf{r}')\,\Phi_{\mathbf{k}}^{(-)}(\mathbf{r}')\,d\mathbf{r}'\right\}^*\tilde{\phi}^*(\mathbf{k})\,\psi(\mathbf{r})\,d\mathbf{r}\,d\mathbf{k}.$$

From (7.18) we obtain

$$(7.44)\qquad \langle W_-\phi\mid\psi\rangle = \int_{\mathbb{R}^3}d\mathbf{r}\,\psi(\mathbf{r})\int_{\mathbb{R}^3}d\mathbf{k}\,\Phi_{\mathbf{k}}^{(-)*}(\mathbf{r})\tilde{\phi}^*(\mathbf{k})$$
$$= \int_{\mathbb{R}^6}\Phi_{\mathbf{k}}^{(-)*}(\mathbf{r})\tilde{\phi}^*(\mathbf{k})\,\psi(\mathbf{r})\,d\mathbf{r}\,d\mathbf{k},$$

where in view of (7.21) the second equality follows from Fubini's theorem.

Substituting the expression for $\langle W_-\phi\mid\psi\rangle$ from (7.44) in (7.43), we obtain

$$\langle\phi\mid W_-^*\Omega_-\psi\rangle = \int_{\mathbb{R}^3}d\mathbf{k}\,\tilde{\phi}^*(\mathbf{k})\int_{\mathbb{R}^3}d\mathbf{r}\,\psi(\mathbf{r})\left\{\Phi_{\mathbf{k}}^{(-)}(\mathbf{r}) - \frac{m}{2\pi}\right.$$
$$\left.\times\int_{\mathbb{R}^3}\frac{\exp(-ik\,|\mathbf{r}-\mathbf{r}'|)}{|\mathbf{r}-\mathbf{r}'|}V(\mathbf{r}')\,\Phi_{\mathbf{k}}^{(-)}(\mathbf{r}')\,d\mathbf{r}'\right\}^*$$
$$= (2\pi)^{-3/2}\int_{\mathbb{R}^3}d\mathbf{k}\,\tilde{\phi}^*(\mathbf{k})\int_{\mathbb{R}^3}d\mathbf{r}\,e^{-i\mathbf{k}\mathbf{r}}\,\psi(\mathbf{r}) = \int_{\mathbb{R}^3}d\mathbf{k}\,\tilde{\phi}^*(\mathbf{k})\,\tilde{\psi}(\mathbf{k})$$

by using in the process the Lippmann–Schwinger equation (6.5) for $\Phi_{\mathbf{k}}^{(-)}(\mathbf{r})$.

In the above relation $\psi(\mathbf{r})\in\mathscr{C}_{\mathrm{b}}^{\infty}(\mathbb{R}^3)$ and $\tilde{\phi}(\mathbf{k})\in\mathscr{C}_V^{\infty}$. Since $\mathscr{C}_{\mathrm{b}}^{\infty}(\mathbb{R}^3)$ is dense in $L^2(\mathbb{R}^3)$, it follows that $W_-^*\Omega_- = 1$. Consequently, by virtue of (7.19) and the fact that $E_{R_-}\leqslant E^H(\mathsf{S}_{\mathrm{c}}^H)$, we conclude that

$$W_- = W_-W_-^*\Omega_- = E^H(\mathsf{S}_{\mathrm{c}}^H)\,\Omega_- = \Omega_-\,.$$

Completely analogous considerations lead to the conclusion that $W_+ = \Omega_+$. Hence, we have thus established the representation (7.14) for the wave operators $\Omega_\pm$ in potential two-body scattering for arbitrary $\psi\in L^2(\mathbb{R}^3)$. This shows that in the present case $\mathsf{M}_0\equiv L^2(\mathbb{R}^3)$. Moreover, in view of (7.19), we have also established that $\mathsf{R}_+ = \mathsf{R}_-\equiv L^2_{H_{\mathrm{c}}}(\mathbb{R}^3)$. Consequently, the proof of Theorem 7.2 has been completed.

7.4. Phase Shifts for Scattering in Central Potentials

We had mentioned in §7.8 of Chapter IV that $\{|\mathbf{P}|, |\mathbf{L}|, L^{(z)}\}$ is a complete set of observables next in importance to the position and momentum complete sets of observables. In §4.3 we had briefly outlined how the stationary scattering theory can be formulated in terms of partial (or spherical) waves constituting an eigenfunction expansion for $\{|\mathbf{P}|, |\mathbf{L}|, L^{(z)}\}$. We now propose to prove by a very general argument that indeed for spherically symmetric potentials

$$V(\mathbf{r}) = V_0(r), \qquad r = |\mathbf{r}|, \tag{7.45}$$

the S operator can be expressed in terms of the phase shifts $\delta_l(k)$ mentioned in §4.3.

We already know that H_0 in (4.14) or (5.1) and (5.2) commutes with $\{|\mathbf{P}|, |\mathbf{L}|, L^{(z)}\}$. Obviously, the interaction Hamiltonian $H_1 = V$ does not commute in general with $|\mathbf{P}|$, but it does commute with $|\mathbf{L}|$ and $L^{(z)}$ (in the sense of Definition 1.2 in Chapter IV) when (7.45) is satisfied. Indeed, in that case $V = V_0(|\mathbf{Q}|)$, and by Theorems 2.5 and 2.6 in Chapter IV, all we have to prove is that the spectral measure $E^{|\mathbf{Q}|}(B)$ of $|\mathbf{Q}|$,

$$(E^{|\mathbf{Q}|}(B)\psi)(r, \theta, \phi) = \chi_B(r)\, \psi(r, \theta, \phi), \tag{7.46}$$

commutes with the spectral measures of $|\mathbf{L}|$ and $L^{(z)}$—a fact that is implicit in the relation (7.43) of Chapter IV.

Since H_0 and V do not commute, we cannot argue that $|\mathbf{L}|$ and $L^{(z)}$ commute with $H = H_0 + V$ (in the sense of their spectral measures commuting) simply because they commute with H_0 and V. This result, however, can be deduced from the following theorem by Trotter [1959], which finds many other applications in quantum mechanics.

***Lemma 7.3.** If A_1 and A_2 are self-adjoint operators and $A_1 + A_2$ is self-adjoint on $\mathscr{D} = \mathscr{D}_{A_1} \cap \mathscr{D}_{A_2}$, then *Trotter's product formula*

$$e^{i(A_1+A_2)t} = \operatorname*{s-lim}_{n\to\infty} (e^{iA_1t/n}\, e^{iA_2t/n})^n \tag{7.47}$$

holds true for any $t \in \mathbb{R}^1$.

Proof. By Theorem 3.1 in Chapter IV and Exercise 5.6 in Chapter III we have for any $f \in \mathscr{D}$,

$$\operatorname*{s-lim}_{\tau\to 0} \tau^{-1}(e^{iA_1\tau}\, e^{iA_2\tau} f - f) = \operatorname*{s-lim}_{\tau\to 0} e^{iA_1\tau}\tau^{-1}(e^{iA_2\tau} f - f)$$
$$+ \operatorname*{s-lim}_{\tau\to 0} \tau^{-1}(e^{iA_1\tau} f - f)$$

(equation continues)

$$= i(A_1 f + A_2 f)$$
$$= \operatorname{s-lim}_{\tau\to 0} \tau^{-1}(e^{i(A_1+A_2)\tau} f - f).$$

Consequently, if we set

$$M(\tau) = \tau^{-1}(e^{i(A_1+A_2)\tau} - e^{iA_1\tau} e^{iA_2\tau}),$$

then $\| M(\tau)f \| \to 0$ as $\tau \to 0$ for all $f \in \mathscr{D}$.

To prove that as $\tau \to 0$, $\| M(\tau)f \|$ converges to zero uniformly on a dense set of vectors $f \in \mathscr{D}$, we shall employ Lemma 2.2 from Chapter III. Hence we introduce in $\mathscr{D}$ the new inner product

$$\langle f \mid g\rangle_{\mathscr{D}} = \langle (A_1 + A_2)f \mid (A_1 + A_2)g\rangle + \langle f \mid g\rangle, \tag{7.48}$$

and we note that since $A_1 + A_2$ is self-adjoint (and therefore closed) on $\mathscr{D}$, the linear set $\mathscr{D}$ is a Hilbert space with respect to this inner product. Furthermore, the linear functionals

$$\phi_{\tau,g}(f) = \langle M(\tau)g \mid M(\tau)f\rangle, \qquad f \in \mathscr{D}, \quad \| g \|_{\mathscr{D}} = 1, \quad \tau \in [-1, +1],$$

are bounded in $f, g \in \mathscr{D}$ (with inner product $\langle \cdot \mid \cdot \rangle_{\mathscr{D}}$), so that by Lemma 2.2 in Chapter III (which can be easily extended from sequences to arbitrary sets of linear functionals), we conclude that

$$|\phi_{\tau,g}(f)| \leqslant C_1, \qquad \| f \|_{\mathscr{D}} = \| g \|_{\mathscr{D}} = 1, \quad \tau \in [-1, +1].$$

Since obviously a similar constant C_2 can be found if $\tau \in \mathbb{R}^1 - [-1, +1]$, we can state that

$$\| M(\tau)f \| \leqslant C \| f \|_{\mathscr{D}}, \qquad f \in \mathscr{D}, \quad \tau \in \mathbb{R}^1,$$

so that in accordance with (1.16) in Lemma 1.1, $\| M(\tau)f \| \to 0$ uniformly in $f \in \mathscr{D}$, $\| f \|_{\mathscr{D}} = 1$, as $\tau \to 0$.

To apply this result we write

$$\|\{\exp[i(A_1 + A_2)t/n]^n\} f - [\exp(iA_1 t/n) \exp(iA_2 t/n)]^n f \|$$
$$= \Big\| \sum_{j=0}^{n-1} [\exp(iA_1 t/n) \exp(iA_2 t/n)]^j$$
$$\times \{\exp[i(A_1 + A_2)t/n] - \exp(iA_1 t/n) \exp(iA_2 t/n)\}$$
$$\times \{\exp[i(A_1 + A_2)t/n]\}^{n-j-1} f \Big\|$$

(equation continues)

$$\leqslant |t| \sup_{|u|\leqslant|t|} \|(t/n)^{-1}\{\exp[i(A_1 + A_2)t/n] - \exp(iA_1t/n)\exp(iA_2t/n)\}$$
$$\times \exp[i(A_1 + A_2)u] f\|,$$

and note that as $n \to \infty$ (and therefore $t/n \to 0$), the last of the above norms converges to zero uniformly in $u \in [-|t|, +|t|]$ since $\|\exp[i(A_1 + A_2)u] f\|_{\mathscr{D}} = \|f\|_{\mathscr{D}}$ for any fixed $f \in \mathscr{D}$. Consequently, (7.47) is established when applied to $f \in \mathscr{D}$.

Since $\mathscr{D}$ is dense in $\mathscr{H}$ and the operator on the left-hand side of (7.47) is bounded and defined everywhere in $\mathscr{H}$, the extension principle can be invoked to immediately extend this result to all $f \in \mathscr{H}$. Q.E.D.

Since by Theorems 7.3–7.5 in Chapter IV we have that $\mathscr{D}_H = \mathscr{D}_{H_0} \subset \mathscr{D}_V$, Lemma 7.3 can be applied if $V(\mathbf{r})$ is locally square integrable and bounded at infinity, so that

$$\exp(iHt_1)\exp(iL^{(z)}t_2) = \operatorname*{s-lim}_{n\to\infty}[\exp(iH_0t_1/n)\exp(iVt_1/n)]^n \exp(iL^{(z)}t_2) \tag{7.49}$$
$$= \operatorname*{s-lim}_{n\to\infty} \exp(iL^{(z)}t_2)[\exp(iH_0t_1/n)\exp(iVt_1/n)]^n$$
$$= \exp(iL^{(z)}t_2)\exp(iHt_1),$$

and similarly for $|\mathbf{L}|$. Hence by Theorem 6.2 in Chapter IV, we conclude that $|\mathbf{L}|$ and $L^{(z)}$ commute with H.

Theorem 7.3. If $V(\mathbf{r}) = O(r^{-1-\epsilon})$, $r \to \infty$, is spherically symmetric and locally square integrable, then the S operator is a function of $\{\mathbf{P}, |\mathbf{L}|, L^{(z)}\}$,

$$S = F_S(|\mathbf{P}|, L', L^{(z)}), \qquad |\mathbf{L}| = [L'(L'+1)]^{1/2}, \tag{7.50}$$

$$F_S(k, l, m) = \exp[2i\delta_l(k)], \tag{7.51}$$

where $\delta_l(k)$ are unique modulo π real functions that do not depend on m.

Proof. Under the present conditions on $V(\mathbf{r})$, our earlier considerations leading to (7.50) are valid, so that $|\mathbf{L}|$ and $L^{(z)}$ commute with both H_0 and H. Consequently, $|\mathbf{L}|$ and $L^{(z)}$ commute also with the Møller wave operators, since, e.g.,

$$E^{L^{(z)}}(B)\Omega_\pm = E^{L^{(z)}}(B) \operatorname*{s-lim}_{t\to\mp\infty} e^{iHt} e^{-iH_0t} \tag{7.52}$$
$$= \operatorname*{s-lim}_{t\to\mp\infty} e^{iHt} e^{-iH_0t} E^{L^{(z)}}(B)$$
$$= \Omega_\pm E^{L^{(z)}}(B)$$

for any $B \in \mathscr{B}^1$. Since $S = \Omega_-^*\Omega_+$, we also have

(7.53) $$[E^{L^{(z)}}(B), S] = [E^{|\mathbf{L}|}(B), S] = 0, \qquad B \in \mathscr{B}^1.$$

On the other hand, since $|\mathbf{P}| = (2mH_0)^{1/2}$, we have by Theorem 2.3 in Chapter IV,

(7.54) $$E^{|\mathbf{P}|}(B) = E^{H_0}(F^{-1}(B)), \qquad F(\lambda) = (2m\lambda)^{1/2},$$

so that according to Theorem 2.3, $|\mathbf{P}|$ also commutes with S. Therefore, by Theorem 5.6 in Chapter IV the S operator must be a function of the complete set $\{|\mathbf{P}|, |\mathbf{L}|, L^{(z)}\}$, i.e., (7.51) is satisfied.

To show that this function is of the form

(7.55) $$F_S(k, l, m) = \exp[2i\, \delta(k, l, m)], \qquad \delta(k, l, m) = \delta^*(k, l, m),$$

we note that by Theorem 7.1, $\mathsf{M}_0 = L^2(\mathbb{R}^3)$, and that [as we had remarked when discussing (7.17) in §7.2], $\mathsf{R}_+ = \mathsf{R}_-$, so that by Theorem 2.5 S is unitary on $L^2(\mathbb{R}^3)$. Consequently, by Theorems 2.6 and 2.7 in Chapter IV,

(7.56) $$F_S^*(k, l, m)\, F_S(k, l, m) = 1$$

almost everywhere on $\mathbb{R}^3$ with respect to the measure μ_0 introduced in §4.3. Thus, (7.55) must be satisfied, the functions $\delta(k, l, m)$ being obviously μ_0-almost everywhere unique modulo π.

The fact that $\delta(k, l, m)$ does not depend on m, and that therefore (7.55) reduces to (7.51), follows from a general group-theoretical theorem*. Q.E.D.

We note that Theorem 7.3 establishes the existence of the phase shifts $\delta_l(k)$ without any reference to distorted spherical waves. Therefore, as yet we have not proven that the functions $\delta_l(k)$ appearing in (7.51) coincide with those in (4.36). However, this result follows from the further considerations of §7.6. In the meantime, we shall establish in §7.5 that (7.50) and (7.51) have a generalization of sorts valid even when $V(\mathbf{r})$ is not central.

7.5. The General Phase-Shift Formula for the Scattering Operator

Let us introduce in the space $L^2(\Omega_s, \mu_s)$ defined in §7 of Chapter II [see (7.16) of Chapter II] operators analogous to (3.50) [c.f. also (4.97)

* The Wigner–Eckart theorem (see, e.g., Miller [1972], Section 7.10).

and (6.67)]:

$$S(k) = 1 - 2\pi i T(k), \qquad k \in [0, \infty), \tag{7.57}$$

$$(T(k)u)(\omega) = \frac{-k}{(2\pi)^2} \int_{\Omega_S} f_{(k\omega')}(\omega)\, u(\omega')\, d\omega', \qquad u \in L^2(\Omega_S, \mu_S). \tag{7.58}$$

The unitarity of the S operator imposes the condition that $S(k)$ should be unitary in $L^2(\Omega_S, \mu_S)$ for almost all $k \geqslant 0$:

$$S^*(k)\, S(k) = S(k)\, S^*(k) = 1. \tag{7.59}$$

As a matter of fact, for arbitrary $u \in L^2(\Omega_S, \mu_S)$ and $\rho \in L^2(\Omega_r, \mu_r)$ [see (7.16) of Chapter II] it easily follows from (4.96), (6.67), and (7.58) that for $(\rho \cdot u)(k, \omega) = \rho(k)\, u(\omega)$

$$(S^*S\rho \cdot u)(k, \omega) = \rho(k)(S^*(k)\, S(k)u)(\omega). \tag{7.60}$$

Consequently, we have for arbitrary $u, v \in L^2(\Omega_S, \mu_S)$

$$\begin{aligned}
&\int_0^\infty |\rho(k)|^2\, k^2\, dk \int_{\Omega_S} v^*(\omega)(S^*(k)\, S(k)u)(\omega)\, d\omega \\
&= \int_{\Omega_p} (\rho \cdot v)^*(k, \omega)(S^*S\rho \cdot u)(\omega)\, k^2\, dk\, d\omega \\
&= \int_{\Omega_p} (\rho \cdot v)^*(k, \omega)(\rho \cdot u)(k, \omega)\, d\omega \\
&= \int_0^\infty |\rho(k)|^2 k^2\, dk \int_{\Omega_S} v^*(\omega)\, u(\omega)\, d\omega.
\end{aligned}$$

This means that $\langle S(k)u_i \mid S(k)u_j\rangle = \delta_{ij}$ at almost all $k \in [0, \infty)$ for any given $i, j = 1, 2, \ldots$, if $u_1, u_2, \ldots,$ is an orthonormal basis in $L^2(\Omega_S, \mu_S)$. Since the set $\{(i, j)\colon i, j = 1, 2, \ldots, \}$ is countable, this implies that $S^*(k)S(k) = 1$ for almost all $k \in [0, \infty)$. A similar argument establishes that $S(k)\, S^*(k) = 1$ also almost everywhere in $[0, \infty)$.

The unitarity of the operators $S(k)$, $0 \leqslant k < +\infty$, enables us to derive Theorem 7.4.

Theorem 7.4. Suppose the potential $V(\mathbf{r})$ satisfies all the conditions imposed on it in Theorem 7.2. Then for almost all $k \in [0, +\infty)$, there is an orthonormal basis $\{w_1(k), w_2(k), \ldots\}$ in $L^2(\Omega_S, \mu_S)$ such that

$$S(k) = \sum_{\nu=1}^{\infty} |\, w_\nu(k)\rangle\, e^{2i\delta_\nu(k)} \langle w_\nu(k)| = \operatorname*{s-lim}_{n\to+\infty} \sum_{\nu=1}^{n} |\, w_\nu(k)\rangle\, e^{2i\delta_\nu(k)} \langle w_\nu(k)|, \tag{7.61}$$

where the numbers $\delta_1(k)$, $\delta_2(k)$,... are real and such that

$$\text{(7.62)} \qquad \operatorname{Tr}[T^*(k)\, T(k)] = \sum_{\nu=1}^{\infty} |\sin \delta_\nu(k)|^2 < +\infty.$$

Proof. The kernel $T^{(1)}(k; \omega, \omega') = -(2\pi)^{-2}\, k f_{k\omega'}(\omega)$ of the operator $T(k)$ is square integrable in (ω, ω') on $\Omega_s \times \Omega_s$ with respect to the measure $\mu_s \times \mu_s$ (see Exercise 7.3). Hence, the operator $T(k)$ on $L^2(\Omega_s, \mu_s)$ is of the Hilbert–Schmidt type* and therefore completely continuous.

Since $S(k)$ is unitary for almost all $k \in [0, \infty)$, we can apply Theorem 6.1 of Chapter III and write

$$\text{(7.63)} \qquad S(k) = \int_0^{2\pi} e^{i\lambda}\, dE_\lambda(k),$$

where the above integral converges in the uniform sense of (6.1) in Chapter III. We shall prove now that the complete continuity of the operator $T(k)$ implies that the spectral function $E_\lambda(k)$ is constant in λ except for an at most countable number of discontinuities $\lambda_1^{(0)}(k)$, $\lambda_2^{(0)}(k)$,..., where $E_{\lambda_i^{(0)}}(k) \neq E_{\lambda_i^{(0)}-0}(k)$.

Suppose to the contrary that there is a point $\lambda_0 \in (0, 2\pi)$ such that $E_{\lambda_0}(k) = E_{\lambda_0 - 0}(k)$ but $E_a(k) \neq E_b(k)$ for any $a < \lambda_0 < b$ such that $a, b \in (0, 2\pi)$. Then we can choose a sequence $\lambda_1, \lambda_2, \ldots \to \lambda_0$ which is monotonically increasing or decreasing, and such that $E_{\lambda_n}(k) \neq E_{\lambda_{n+1}}(k)$. To be specific, let us say that $\lambda_1 < \lambda_2 < \cdots < \lambda_0$, and let us choose vectors $f_1, f_2, \ldots \in \mathscr{H}$ which are such that $(E_{\lambda_{n+1}}(k) - E_{\lambda_n}(k)) f_n = f_n$ and $\| f_n \| = 1$. We shall prove that no subsequence of $T(k) f_1$, $T(k) f_2$,... is convergent, thus contradicting the complete continuity of $T(k)$.

In fact, it is easily seen that $T(k) f_m \perp T(k) f_n$ for $m \neq n$ (see Exercise 7.4), and consequently

$$\| T(k) f_m - T(k) f_n \|^2 = \| T(k) f_m \|^2 + \| T(k) f_n \|^2.$$

On the other hand, $\| T(k) f_n \| \geqslant C > 0$ for all $n = 1, 2, \ldots$,

$$\begin{aligned} \| T(k) f_n \|^2 &= \frac{1}{4\pi^2} \langle (S(k) - 1) f_n \mid (S(k) - 1) f_n \rangle \\ &= \frac{1}{4\pi^2} \int_{[\lambda_n, \lambda_{n+1})} | e^{i\lambda} - 1 |^2\, d \| E_\lambda(k) f_n \|^2 \\ &\geqslant \frac{1}{4\pi^2} \int_{[\lambda_n, \lambda_{n+1})} | \sin \lambda |^2\, d \| E_\lambda(k) f_n \|^2 \\ &\geqslant \frac{1}{4\pi^2} \inf_{n=1,2,\ldots} | \sin \lambda_n |^2 > 0. \end{aligned}$$

* This can be easily inferred by setting $\mu_s(B) = 0$ for $B \subset \mathbb{R}^2 - \Omega_s$, thus extending μ_s to $\mathbb{R}^2$, and then applying the result of Exercise 5.3 and of Theorem 8.6 in Chapter IV.

Thus, no subsequence of $T(k)f_1, T(k)f_2, \ldots$ is a Cauchy sequence, which is impossible if $T(k)$ is of Hilbert–Schmidt type. Hence, $E_\lambda(k)$ can increase only in a discontinuous manner.

Since $L^2(\Omega_S, \mu_S)$ is separable, there can be at most countably many such points $\lambda_1^{(0)}(k), \lambda_2^{(0)}(k), \ldots$ of discontinuity. Thus, (7.63) becomes

$$S(k) = \sum_{n=1}^{\infty} e^{i\lambda_n^{(0)}(k)}(E_{\lambda_n^{(0)}(k)}(k) - E_{\lambda_n^{(0)}(k)-0}(k)),$$

where it is easily seen that the convergence of the series is in the strong sense, but not necessarily in the uniform sense (think of the case $S(k) = 1$, when the convergence is *not* uniform).

If we introduce an orthonormal basis in each one of the mutually orthogonal subspaces on which $E_{\lambda_n^{(0)}(k)}(k) - E_{\lambda_n^{(0)}(k)-0}(k)$, $n = 1, 2, \ldots$, project, we obtain an orthonormal basis in $L^2(\Omega_S, \mu_S)$ by taking the union of all these bases. After adequately labeling the elements of these bases, and replacing $\lambda_n^{(0)}(k)$ with the adequately labeled $\delta_\nu(k) = \frac{1}{2}\lambda_n^{(0)}(k)$, we obtain (7.61).

The corresponding expression for $T(k)$ is

$$(7.64)\qquad T(k) = \frac{1}{2\pi i}(S(k) - 1) = \frac{1}{\pi}\sum_{\nu=1}^{\infty} | w_\nu(k)\rangle\, e^{i\delta_\nu(k)} \sin \delta_\nu(k) \langle w_\nu(k)|.$$

Thus, we have

$$(7.65)\qquad T^*(k)\, T(k) = \frac{1}{\pi^2}\sum_{\nu=1}^{\infty} | w_\nu(k)\rangle\, | \sin \delta_\nu(k)|^2 \langle w_\nu(k)|,$$

and since $T(k)$ is a Hilbert–Schmidt operator, (7.62) follows. Q.E.D.

We have seen in the preceeding section that when the potential $V(\mathbf{r})$ is spherically symmetric, physicists call the numbers $\delta_\nu(k)$ *phase shifts*. We can apply this terminology (see also Ikebe [1965]) to the general case, when $V(\mathbf{r})$ is not necessarily spherically symmetric. Such a name for $\delta_\nu(k)$ seems appropriate, since the absence of interaction is characterized by vanishing phase shifts $\delta_\nu(k) \equiv 0$; in fact, when there is no interaction we have $S = 1$, and therefore

$$S(k) = 1 = \sum_{\nu=1}^{\infty} | w_\nu(k)\rangle\langle w_\nu(k)|.$$

Furthermore, if $w_\nu(k) = Y_l^m$ for all $k \in [0, \infty)$ then $\delta_\nu(k)$ equals $\delta_l(k)$ [see (7.76)–(7.78)].

If we expand $f_{(k\omega')}(\omega)$ in the orthonormal basis $\{w_1(k), w_2(k), \ldots\}$ and compare the Fourier coefficients of this expansion with (7.64), we get

$$f_{(k\boldsymbol{\omega}')}(\boldsymbol{\omega}) = -4\pi k^{-1} \sum_{\nu=1}^{\infty} w_\nu(k; \boldsymbol{\omega})\, e^{i\delta_\nu(k)} \sin \delta_\nu(k) w_\nu{}^*(k; \boldsymbol{\omega}'). \tag{7.66}$$

Hence, the total scattering cross section $\sigma(k^2/2m)$ at a given energy $k^2/2m$ can be expressed conveniently in terms of phase shifts:

$$\sigma\left(\frac{k^2}{2m}\right) = \int_{\Omega_s} \sigma\left(\frac{k^2}{2m}, \boldsymbol{\omega}\right) d\omega = \frac{16\pi^2}{k^2} \sum_{\nu=1}^{\infty} | w_\nu(k; \mathbf{e}_z) \sin \delta_\nu(k)|^2. \tag{7.67}$$

Naturally, both formulas (7.66) and (7.67) are unambiguous only if $T(k; \boldsymbol{\omega}', \boldsymbol{\omega})$ and $w_\nu(k; \boldsymbol{\omega}')$ are continuous functions of $\boldsymbol{\omega}'$, so that (1.47) is valid.

7.6. Partial-Wave Analysis for Spherically Symmetric Potentials

If the potential $V(\mathbf{r})$ is spherically symmetric, i.e., if (7.45) is satisfied, then the functions $w_\nu(k; \boldsymbol{\omega})$ can be easily computed.

Using the well-known Bauer's formula (see Exercise 7.5),

$$\exp(ikz) = \exp(ikr \cos \theta) = \sum_{l=0}^{\infty} i^l(2l+1)\, j_l(kr)\, P_l(\cos \theta), \tag{7.68}$$

and combining it with the addition theorem for spherical harmonics,*

$$P_l(\boldsymbol{\omega} \cdot \boldsymbol{\omega}_1) = \frac{4\pi}{2l+1} \sum_{m=-l}^{l} Y_l{}^m(\omega)(Y_l{}^m(\omega_1))^*, \tag{7.69}$$

we obtain upon setting $Y_{lm}(\boldsymbol{\omega}) = Y_l{}^m(\omega)$,

$$\exp[ikr(\boldsymbol{\omega} \cdot \boldsymbol{\omega}')] = 4\pi \sum_{l=0}^{\infty} \sum_{m=-l}^{l} i^l j_l(kr)\, Y_{lm}(\boldsymbol{\omega})\, Y_{lm}^*(\boldsymbol{\omega}'), \tag{7.70}$$

where $\omega = (\theta, \phi)$ and $\omega_1 = (\theta_1, \phi_1)$ are spherical coordinates which correspond to the unit vectors $\boldsymbol{\omega} \in \mathbb{R}^3$ and $\boldsymbol{\omega}_1 \in \mathbb{R}^3$, respectively. Let us insert the above series in the integral in (6.58) defining $f_{\mathbf{k}}(\boldsymbol{\omega})$. Due to the uniform convergence of the series for $\boldsymbol{\omega}, \boldsymbol{\omega}' \in \Omega_s$ (see Exercise 7.5)

* See, e.g., Messiah [1962], p. 496.

we can integrate term by term, thus arriving at the following relation:

$$(7.71)\qquad f_{\mathbf{k}}(\boldsymbol{\omega}) = \sqrt{2\pi}\, m \sum_{l=0}^{\infty} \sum_{m=-l}^{l} 4\pi i^l Y_{lm}(\boldsymbol{\omega}) \int_0^{\infty} dr\, r^2 V_0(r)\, j_l(kr)$$

$$\times \int_{\Omega_s} d\omega_1 Y_{lm}^*(-\omega_1)\, \Phi_{\mathbf{k}}^{(+)}(r\omega_1).$$

From the Lippmann–Schwinger equations (6.5) we immediately see that when $V(\mathbf{r})$ is spherically symmetric $\Phi_{\mathbf{k}}^{(+)}(\mathbf{r})$ is a function of only r, k, and $\cos\theta = \mathbf{r}\cdot\mathbf{k}/rk$:

$$(7.72)\qquad \Phi_{\mathbf{k}}^{(+)}(\mathbf{r}) = \Phi_k^{(+)}(r, \cos\theta).$$

Hence, by carrying out a change of variables in which $\boldsymbol{\omega}_1$ is replaced by spherical coordinates of $\mathbf{r}$ with respect to $\mathbf{k}$, we obtain by (7.18) in Chapter II

$$(7.73)\qquad \int_{\Omega_s} Y_{lm}^*(-\boldsymbol{\omega}_1)\, \Phi_{\mathbf{k}}^{(+)}(r\boldsymbol{\omega}_1)\, d\omega_1 = \begin{cases} 0 & \text{for } m \neq 0 \\ \sqrt{\pi(2l+1)} \displaystyle\int_{-1}^{+1} \Phi_k^{(+)}(r, u)\, P_l(u)\, du & \text{for } m = 0. \end{cases}$$

Consequently, (7.71) can be written in the form

$$(7.74)\qquad T^{(1)}(k; \boldsymbol{\omega}, \boldsymbol{\omega}') = \frac{-k}{(2\pi)^2} f_{k\boldsymbol{\omega}'}(\boldsymbol{\omega}) = \sum_{l=0}^{\infty} \frac{2l+1}{4\pi} a_l(k)\, P_l(\boldsymbol{\omega}\cdot\boldsymbol{\omega}')$$

$$= \sum_{l=0}^{\infty} \sum_{m=-l}^{l} a_l(k)\, Y_{lm}(\boldsymbol{\omega}) Y_{lm}^*(\boldsymbol{\omega}'),$$

where the convergence is in the mean on $\Omega_s \times \Omega_s$, and

$$(7.75)\qquad a_l(k) = \sqrt{8\pi}\, kmi^l \int_0^{\infty} dr\, r^2 V_0(r)\, j_l(kr) \int_{-1}^{+1} du\, \Phi_k^{(+)}(r, u)\, P_l(u).$$

Thus, we have arrived at the following expansion for $S(k)$:

$$(7.76)\qquad S(k) = 1 - 2\pi i\, T(k) = \sum_{l=0}^{\infty} \sum_{m=-l}^{l} |Y_l^m\rangle(1 - 2\pi i a_l(k))\langle Y_l^m|.$$

Comparison with (7.61) yields the following explicit expression for the phase shifts:

$$(7.77)\qquad e^{2i\delta_l(k)} = 1 - 2\pi i a_l(k).$$

The reader can compare the above results with the considerations in Section 7.4, and easily arrive at the conclusion that

$$(7.78)\qquad S_l(k) = e^{2i\delta_l(k)}, \qquad k \in [0, \infty), \quad l = 0, 1, 2, \ldots,$$

with $\delta_l(k)$ given by (7.75) and (7.77) is indeed the S matrix in partial wave representation, whose existence has been established in Theorem 7.3.

The great importance of the above formula for the S matrix lies in the fact that in conjunction with (7.75) and (7.77) it enables us to analytically continue $S_l(k)$ in the complex planes of the variables l and k. The properties of the resulting complex S matrix $S(k, l)$ provide the foundations of a very fruitful theory of the S matrix, which has become the fundamental stepping-stone of much of the theoretical physics in the 1950's and 1960's. Since this book is concerned exclusively with functional-analytic methods, this topic is beyond its scope, but the reader is urged to consult the References for literature dealing with this very important subject.*

Exercises

7.1. Show that if $f(t)$ is continuous and the improper Riemann integral

$$I_0 = \lim_{\alpha \to +\infty} \int_0^{\alpha} f(t)\, dt = \int_0^{+\infty} f(t)\, dt$$

exists, then the improper Riemann integral

$$I_\epsilon = \int_0^{+\infty} e^{-\epsilon t} f(t)\, dt$$

exists for every $\epsilon > 0$, and $\lim_{\epsilon \to +0} I_\epsilon = I_0$.

7.2. Show that for any given $\psi(\mathbf{r}) \in \mathscr{C}_b^\infty(\mathbb{R}^3)$ there is a constant C_0 such that $|(e^{-iH_0 t}\psi)(\mathbf{r})| \leqslant C_0$ for all $\mathbf{r} \in \mathbb{R}^3$ and all $t \in \mathbb{R}^1$.

* A mathematically rigorous treatment of analyticity properties of the S matrix in potential scattering is given in the work by de Alfaro and Regge [1965], but most other textbooks on scattering theory also treat this subject at an adequate level of mathematical rigor—whereas the same certainly cannot be claimed when it comes to the functional analysis aspects of scattering theory.

7.3. Prove that for $k \in [0, +\infty)$ with $k^2/2m \notin \mathscr{S}_V^1$ the integral

$$\int_{\Omega_8 \times \Omega_8} | T^{(1)}(k; \omega, \omega')|^2 \, d\omega \, d\omega'$$

exists, provided $V(\mathbf{r})$ satifies the prerequisites in Theorem 6.4 and $T^{(1)}(k; \omega, \omega') = -(2\pi)^{-2} k f_{k\omega'}(\omega)$.

7.4. Suppose $[E_{\lambda_i'}(k) - E_{\lambda_i}(k)] f_i = f_i \, , i = 1, 2$ for

$$(\lambda_1 \, , \lambda_1'] \cap (\lambda_2 \, , \lambda_2'] = \varnothing,$$

where $E_\lambda(k)$ is the spectral function of $S(k)$, and $\lambda_1 \, , \lambda_2 \, , \lambda_1', \lambda_2' \in (0, 2\pi)$. Prove that $T(k) f_1 \perp T(k) f_2$.

7.5. Prove that for any $\alpha \in \mathbb{R}^1$

$$e^{i\alpha u} = \sum_{l=0}^{\infty} i^l (2l + 1) j_l(\alpha) P_l(u), \qquad -1 \leqslant u \leqslant 1,$$

and that the convergence is uniform in $u \in [-1, +1]$.

7.6. Show that for any $\beta > 0$ and any $\mathbf{r}_0 \in \mathbb{R}^3$

$$(\exp(-iH_0 t) \, \phi_{\mathbf{r}_0})(\mathbf{r}) = \left(\frac{m}{\beta^2 + it} \right)^{3/2} \exp \left[-\frac{m}{2} \frac{(\mathbf{r} - \mathbf{r}_0)^2}{\beta^2 + it} \right]$$

if $\phi_{\mathbf{r}_0}(\mathbf{r})$ is a wave packet with a Fourier transform of the form

$$\tilde{\phi}_{\mathbf{r}_0}(\mathbf{k}) = \exp \left[-\frac{\beta^2}{2m} \mathbf{k}^2 - i\mathbf{k}\mathbf{r}_0 \right].$$

Use this result to show that for any bounded Borel set B

$$\| E^{\mathrm{Q}}(B) \, e^{-iH_0 t} \phi_{\mathbf{r}_0} \|^2$$
$$= \int_B |(e^{-iH_0 t} \phi_{\mathbf{r}_0})(\mathbf{r})|^2 \, d\mathbf{r} \leqslant \mu_l^{(3)}(B) \left(\beta + \frac{t^2}{m^2} \right)^{-3/2}.$$

7.7. Any Schroedinger-picture free state ψ_t^{f},

$$\psi_t^{\mathrm{f}}(\mathbf{r}) = (e^{-iH_0 t} \, \psi_0^{\mathrm{f}})(\mathbf{r}),$$

corresponding to $\psi_0^{\mathrm{f}} \in L^2(\mathbb{R}^3)$, displays a tendency of "spreading out" in space as $t \to -\infty$. This conclusion can be reached by establishing the *evanescence* of the wave packet $\psi_0^{\mathrm{f}}(\mathbf{r})$ from any bounded Borel set B:

$$\lim_{t\to\mp\infty} \| E^{Q}(B)\, \psi_t^{\mathrm{f}} \|^2 = \lim_{t\to\mp\infty} \int_B | \psi_t^{\mathrm{f}}(\mathbf{r})|^2\, d\mathbf{r} = 0.$$

Combine the result of Exercise 7.6 with Lemma 1.1 and Lemma 7.1 to prove that the above limit is indeed equal to zero.

7.8. Prove that any interacting state ψ_t represented by a wave packet,

$$\psi_t(\mathbf{r}) = (e^{-iHt}\,\psi_0)(\mathbf{r}), \qquad \psi_0 \in L^2(\mathbb{R}^3),$$

which has an incoming (outgoing) asymptotic state $\psi_t^{\mathrm{in}}(\psi_t^{\mathrm{out}})$ evanesces from any bounded Borel set B in the configuration space when $t \to -\infty$ $(t \to +\infty)$.

8. Fundamental Concepts in Multichannel Scattering Theory

8.1. The Concept of Channel

The outcome of the mutual interaction of three or more particles can lead to results which are qualitively different from the possibilities encountered in two-particle interactions. Indeed, a system of two isolated particles can either be found forever in a bound state, or the two particles are free before and after collision. However, in the case of three particles $\mathfrak{S}_1$, $\mathfrak{S}_2$, and $\mathfrak{S}_3$, there already are other possibilities in addition to the two alternatives of $\mathfrak{S}_1$, $\mathfrak{S}_2$, $\mathfrak{S}_3$ becoming eventually free or constituting a bound system $\{\mathfrak{S}_1\text{–}\mathfrak{S}_2\text{–}\mathfrak{S}_3\}$. For example, it is possible that $\mathfrak{S}_1$ and $\mathfrak{S}_2$ will stay bound forever, thus building a new system $\mathfrak{S}_1\text{–}\mathfrak{S}_2$, while $\mathfrak{S}_3$ becomes free, etc. Each one of these possibilities determines a particular outgoing arrangement channel of the scattering process between $\mathfrak{S}_1$, $\mathfrak{S}_2$, and $\mathfrak{S}_3$.

In general, an *arrangement channel* (or *clustering*) in the scattering process of n particles $\mathfrak{S}_1$, $\mathfrak{S}_2$,..., $\mathfrak{S}_n$ is a particular partitioning of the set $\{\mathfrak{S}_1, \mathfrak{S}_2,..., \mathfrak{S}_n\}$ into a number of nontrivial subsets, called *clusters*. When all the particles in each cluster are actually bounded together either before or after collision with particles in other clusters, then we say that the clusters constitute *fragments* of the collision process. Clearly, we can talk about *incoming* channels and *outgoing* channels, depending on whether the particular channel we are considering refers to the system before or after the scattering had taken place.

In the case where some of the particles in the system $\{\mathfrak{S}_1, \mathfrak{S}_2, ..., \mathfrak{S}_n\}$ are identical, not all partitions of the set $\{\mathfrak{S}_1, \mathfrak{S}_2, ..., \mathfrak{S}_n\}$ will represent distinct arrangement channels. This is due to the presence of Bose–Einstein or Fermi–Dirac statistics, respectively (see Chapter IV, §4) which rules that identical particles are indistinguishable. For example, if $\mathfrak{S}_1$, $\mathfrak{S}_2$, and $\mathfrak{S}_3$ are identical, then the channel $\{\mathfrak{S}_1\text{–}\mathfrak{S}_2, \mathfrak{S}_3\}$ in which $\mathfrak{S}_1$ and $\mathfrak{S}_2$ are bound, while $\mathfrak{S}_3$ is free, is identical to the channel $\{\mathfrak{S}_1, \mathfrak{S}_2\text{–}\mathfrak{S}_3\}$ in which $\mathfrak{S}_1$ is free and $\mathfrak{S}_2$ is bound to $\mathfrak{S}_3$. As a matter of fact, in this example both above arrangement channels are also identical to $\{\mathfrak{S}_2, \mathfrak{S}_1\text{–}\mathfrak{S}_3\}$ so that we have only three distinct arrangement channels: $\{\mathfrak{S}_1, \mathfrak{S}_2\text{–}\mathfrak{S}_3\}$, and $\{\mathfrak{S}_1, \mathfrak{S}_2, \mathfrak{S}_3\}$.

If each fragment $\mathscr{F}$ is in an eigenstate of a set $\mathscr{O}_{\mathscr{F}}$ of commuting observables which are constants of motion in that particular arrangement channel, then the system is said to be in the channel determined by eigenvalues of the observables in the family $\bigcup_{\mathscr{F}} \mathscr{O}_{\mathscr{F}}$.

In practice, it is often convenient to take an $\mathscr{O}_{\mathscr{F}}$ containing the internal (binding) energy of the fragment, its internal angular momentum (called the spin of the fragment), etc. If the internal energy of each fragment $\mathscr{F}$ has been prescribed, then we call a channel determined by the given internal energy eigenvalues an *energy channel.**

In the case of many-body (three or more particles) scattering we can distinguish between elastic and inelastic scattering† by looking at the distribution of energy among fragments before and after the scattering has taken place. *Elastic scattering* is, by definition, the scattering process in which all the fragments and their internal energies are preserved; for example, if the initial energy channel consists of a particle $\mathfrak{S}_1$ being scattered from a bound system $\mathfrak{S}_2\text{–}\mathfrak{S}_3$ of two particles, then the scattering is elastic if after the scattering has taken place the particles $\mathfrak{S}_2$ and $\mathfrak{S}_3$ are still in a bound state of the same energy as the initial bound state, i.e., if no kinetic energy has been transferred from $\mathfrak{S}_1$ to the internal

* We depart somewhat from conventional terminology, in which a channel with respect to some of the above-specified observables is simply called a channel. However, since there does not seem to be complete agreement among different authors on the precise contents of the set $\mathscr{O}_{\mathscr{F}}$ when a channel is defined, the above general approach seems quite desirable.

† The distinction can be made also in the case of two-body scattering in an external field. In fact, the external field describes the interaction of the two particles with one or more additional particles which are not explicitly included in the system, but are effectively represented in an approximative manner by the "external" field. In that case, each one of the particles in the system can be bound to the field. However, it is clear that two-body scattering in an external field is, in fact, many-body scattering.

energy of $\mathfrak{S}_2$–$\mathfrak{S}_3$. Naturally, any scattering which is not elastic is called *inelastic.*

Suppose all the fragments in an energy channel are in their internal energy ground states and that ΔE is the smallest energy gap between the internal energies of these fragments and the energetically nearest nonground eigenstate of their respective internal energies. Then, obviously, no inelastic scattering can take place as long as there is not sufficient kinetic energy to be transfered to the k_0th fragment to raise it to an "excited" state of internal energy $E_{k_0}^{\text{int}} + \Delta E$. This amount of energy is then called the *threshold* of the inelastic process in that particular energy channel.

If in an inelastic scattering the fragments themselves are decomposed and new fragments result as an outcome of the scattering process, then we talk of a *rearrangement collision*; for instance, in the above example this is the case when a bound state of $\mathfrak{S}_1$ and $\mathfrak{S}_2$ is formed, while $\mathfrak{S}_3$ goes free after the interaction has taken place.

It should be immediately realized that a rearrangement collision cannot take place at any distribution of energy between the different fragments. For example, in an energy channel of $\{\mathfrak{S}_1, \mathfrak{S}_2–\mathfrak{S}_3\}$ a minimal energy equal to the internal binding energy of $\mathfrak{S}_2$–$\mathfrak{S}_3$ is necessary for the decomposition of the fragment $\mathfrak{S}_2$–$\mathfrak{S}_3$ occurring in the initial state. This energy is then the threshold energy for a rearrangement collision.

8.2. Channel Hamiltonians and Wave Operators

We shall illustrate the mathematical counterparts of some of the physical concepts introduced above on the case of three-particle potential scattering. For this purpose we consider three distinct particles which interact with one another via two-body forces. Here, the statement that we are dealing with two-body force means the total potential can be written in the form

$$V(\mathbf{r}_1, \mathbf{r}_2, \mathbf{r}_3) = V_{12}(\mathbf{r}_1 - \mathbf{r}_2) + V_{23}(\mathbf{r}_2 - \mathbf{r}_3) + V_{13}(\mathbf{r}_1 - \mathbf{r}_3), \tag{8.1}$$

and therefore each one of the particles interacts with each other separately. In addition, we simplify our considerations by assuming that these forces are of finite ranges ρ_{ij}, i.e., $V_{ij}(\mathbf{r})$, $i < j = 1, 2, 3$, vanishes for $r \geqslant \rho_{ij}$.

The total Hamiltonian of the above system is given by the Schroedinger operator

$$H = H_0 + V_{12} + V_{23} + V_{13}, \tag{8.2}$$

where H_0 is the free Schroedinger operator

$$(8.3)\qquad (H_0\psi)^\sim(\mathbf{p}_1, \mathbf{p}_2, \mathbf{p}_3) = \left(\frac{\mathbf{p}_1^2}{2m_1} + \frac{\mathbf{p}_2^2}{2m_2} + \frac{\mathbf{p}_3^2}{2m_3}\right) \tilde{\psi}(\mathbf{p}_1, \mathbf{p}_2, \mathbf{p}_3),$$

and V_{ij}, $i < j$, is the potential energy operator between the ith and the jth particle,

$$(8.4)\qquad (V_{ij}\psi)(\mathbf{r}_1, \mathbf{r}_2, \mathbf{r}_3) = V_{ij}(\mathbf{r}_i - \mathbf{r}_j)\, \psi(\mathbf{r}_1, \mathbf{r}_2, \mathbf{r}_3).$$

Naturally, these operators act in the Hilbert space $L^2(\mathbb{R}^9)$.

The arrangement channel $\{\mathfrak{S}_1, \mathfrak{S}_2, \mathfrak{S}_3\}$ in which all three particles are free is characterized by the fact that all these particles are so distant from one another as to be outside the ranges of the forces with which they interact on one another. Hence, for such distances $V(\mathbf{r}_1, \mathbf{r}_2, \mathbf{r}_3)$ vanishes and the Hamiltonian $H_{(1,2,3)}$ of the system is given effectively by the operator H_0, which is said to be the Hamiltonian of that particular arrangement channel.

In the arrangement channel $\{\mathfrak{S}_1, \mathfrak{S}_2\text{–}\mathfrak{S}_3\}$, the particles $\mathfrak{S}_2$ and $\mathfrak{S}_3$ are within the range ρ_{23} of their mutual interaction forces, while $\mathfrak{S}_1$ is outside the range ρ_{12} from $\mathfrak{S}_2$, and outside the range ρ_{13} from $\mathfrak{S}_3$. Consequently, in this case $V(\mathbf{r}_1, \mathbf{r}_2, \mathbf{r}_3)$ becomes equal to $V_{23}(\mathbf{r}_2 - \mathbf{r}_3)$, and the Hamiltonian of the arrangement channel is $H_{(1,2-3)} = H_0 + V_{23}$.

We see that in the present situation we are dealing with a much more intricate framework then in two-body scattering: instead of a single "free" Hamiltonian, we have a different Hamiltonian for each one of the five arrangement channels.

We can write the Hamiltonian of each one of the arrangement channels as the sum of the Hamiltonians of each one of the fragments in that channel. Thus, for example,

$$(8.5)\qquad H_{(1,2-3)} = H_{(1)} + H_{(2-3)},$$

where, in general, $H_{(i)}$ denotes the kinetic energy operator of the ith particle, so that in the momentum representation,

$$(8.6)\qquad (H_{(i)}\psi)^\sim(\mathbf{p}_1, \mathbf{p}_2, \mathbf{p}_3) = \mathbf{p}_i^2/2m_i\, \tilde{\psi}(\mathbf{p}_1, \mathbf{p}_2, \mathbf{p}_3), \qquad i = 1, 2, 3,$$

and $H_{(2-3)}$ describes the system $\{\mathfrak{S}_2\text{–}\mathfrak{S}_3\}$; we can write $H_{(2-3)}$ in the form

$$(8.7)\qquad H_{(2-3)} = H_{(2,3)} + V_{23},$$

where $H_{(2,3)}$ is the kinetic energy of the fragment $\{\mathfrak{S}_2\text{–}\mathfrak{S}_3\}$,

$$(8.8)\qquad H_{(2,3)} = H_{(2)} + H_{(3)},$$

and V_{23} describes the interaction between the two particles in the fragment.

When the system in some arrangement α and there is no interaction or negligible interaction between the fragments, then the state $\Psi(t) = e^{-iHt}\Psi(0)$ of the system is described quite well by the states

$$\Psi_\alpha^{\text{ex}}(t) = e^{-iH_\alpha t}\,\Psi_\alpha^{\text{ex}}(0) \tag{8.9}$$

provided they are *α-channel strong asymptotic states* in the sense that*

$$\lim_{t\to\mp\infty} \| \Psi(t) - \Psi_\alpha^{\text{ex}}(t) \| = 0, \tag{8.10}$$

where "ex" stands for "in" ($t \to -\infty$) or "out" ($t \to +\infty$), depending on whether the channel is an incoming or outgoing channel, respectively.

In two-body scattering we could formulate a general consistent time-dependent scattering theory for any two self-adjoint operators H and H_0. However, in multichannel scattering one has to be careful and establish from the very beginning whether the channel description is indeed consistent, since it is not *a priori* clear that some interacting state might not have two or more distinct asymptotic states satisfying (8.10). In order to be able to find the conditions under which such ambiguities in the asymptotic description do not occur, we have to understand more fully the structure of the arrangement channel Hamiltonian H_α.

Let us consider the arrangement channel $\{\mathfrak{S}_1, \mathfrak{S}_2\text{–}\mathfrak{S}_3\}$ in the scattering of the three particles $\mathfrak{S}_1$, $\mathfrak{S}_2$, and $\mathfrak{S}_3$ interacting via the two-body potential (8.1). The fragment $\mathfrak{S}_2\text{–}\mathfrak{S}_3$ is asymptotically in a bound state

$$\Psi_{(2-3)}(t) = \exp(-iH_{(2-3)}t)\,\Psi_{(2-3)}(0) \in L^2(\mathbb{R}^6) \tag{8.11}$$

of $\mathfrak{S}_2$ and $\mathfrak{S}_3$.

Let us assume that $\Psi_{(2-3)}(0)$ is represented by the wave function

$$\psi(\mathbf{r}_2, \mathbf{r}_3) = \psi'\left(\frac{m_2\mathbf{r}_2 + m_3\mathbf{r}_3}{m_2 + m_3}\right)\psi''(\mathbf{r}_3 - \mathbf{r}_2), \tag{8.12}$$

where according to §5 of Chapter II, ψ' describes the state of the center of mass of the system $\mathfrak{S}_2\text{–}\mathfrak{S}_3$, while $\psi''(\mathbf{r}_3 - \mathbf{r}_2)$ describes the state of the two particles in relation to one another. The two particles are in a bound

* In fact, the remarks made in §2.6 still apply, i.e., the use of norm topology is too restrictive to apply when long-range forces are present—only expectation values of observables actually being of importance.

state when $\psi''(\mathbf{r}_3 - \mathbf{r}_2)$ is an eigenvector

$$H^{\text{int}}_{(2-3)}\psi'' = E_{(2-3)}\psi'', \qquad E_{(2-3)} < 0, \tag{8.13}$$

of the internal energy operator

$$H^{\text{int}}_{(2-3)} \supseteq -\frac{1}{2m_{23}}\Delta + V_{23}, \tag{8.14}$$

where Δ is the Laplacian in the variable $\mathbf{r} = \mathbf{r}_3 - \mathbf{r}_2$. Naturally, the most general $\Psi_{(2-3)}(0)$ has to be an element of the closed subspace of $L^2(\mathbb{R}^6)$ spanned by all such vectors (8.12), corresponding to all eigenvectors ψ'' of $H^{\text{int}}_{(2-3)}$.

When $\mathfrak{S}_2$–$\mathfrak{S}_3$ is asymptotically in the state $\Psi_{(2-3)}(t) \in L_2(\mathbb{R}^6)$ and $\mathfrak{S}_1$ is asymptotically in the state $\Psi_{(1)}(t)$, then the entire system is in the state

$$\Psi^{\text{ex}}(t) = \Psi_{(1)}(t) \otimes \Psi_{(2-3)}(t) \in L^2(\mathbb{R}^9). \tag{8.15}$$

If $\Psi_{(2-3)}(0)$ is of the form (8.12) and ψ'' is represented by the eigenvector in (8.13), then it is easily seen that

$$\begin{aligned} H_{(1,2-3)}\,\Psi^{\text{ex}}(0) &= (H_{(1)} + H^{\text{kin}}_{(2-3)} + H^{\text{int}}_{(2-3)})\,\Psi^{\text{ex}}(0) \\ &= (H_{(1)} + H^{\text{kin}}_{(2-3)} + E_{(2-3)})\,\Psi^{\text{ex}}(0), \end{aligned} \tag{8.16}$$

where $H^{\text{kin}}_{(2-2)}$ is the kinetic energy operator of the fragment $\mathfrak{S}_2$–$\mathfrak{S}_3$,

$$(H^{\text{kin}}_{(2-3)}\psi)^\sim(\mathbf{p}_2, \mathbf{p}_3) = \frac{\mathbf{P}_{23}^2}{2(m_2 + m_3)}\tilde{\psi}(\mathbf{p}_2, \mathbf{p}_3), \qquad \mathbf{P}_{23} = \mathbf{p}_2 + \mathbf{p}_3. \tag{8.17}$$

The quantity $|E_{(2-3)}|$ is called the *internal energy* of the fragment $\mathfrak{S}_2$–$\mathfrak{S}_3$.

Naturally, it is often desirable to consider energy channels with respect to such internal energy operators. This is possible since in general, as well as in the above particular case, these internal energy operators commute with the arrangement channel Hamiltonian, and therefore they are constants of motion for the clusters in the given arrangement channel.

Let us assume that $\{\mathfrak{S}_1, \mathfrak{S}_2\text{–}\mathfrak{S}_3\}$ is an incoming energy channel, and that the kinetic energy part of $\psi(\mathbf{r}_2, \mathbf{r}_3)$ has a sharp distribution $|\tilde{\psi}'(\mathbf{P}_{23})|^2$ of momentum, i.e., that the incoming momentum of $\mathfrak{S}_2$–$\mathfrak{S}_3$ is prepared very accurately around some value $\mathbf{P}^0_{23}$. We can choose the system of reference in such a manner that $\mathbf{P}^0_{23} \approx 0$, and consequently

$$(H_\alpha\psi^{\text{in}})^\sim(\mathbf{p}_1, \mathbf{p}_2, \mathbf{p}_3) \approx \left(\frac{\mathbf{p}_1{}^2}{2m_1} + E_{(2-3)}\right)\tilde{\psi}^{\text{in}}(\mathbf{p}_1, \mathbf{p}_2, \mathbf{p}_3). \tag{8.18}$$

Hence, we see that in this case the incoming particle $\mathfrak{S}_1$ must have an amount of kinetic energy at least equal to $|E_{(2-3)}|$ in order to be able to impart it to $\mathfrak{S}_2$–$\mathfrak{S}_3$ in the collision process and break up that fragment, thus giving rise to a rearrangement collision. Consequently, in this case $-E_{(2-3)}$ represents the threshold energy for a rearrangement collision of the system $\{\mathfrak{S}_1, \mathfrak{S}_2, \mathfrak{S}_3\}$ in the given incoming energy channel.

The reader can easily verify that the Hamiltonian H_α of any of the above arrangement channels of the system $\{\mathfrak{S}_1, \mathfrak{S}_2, \mathfrak{S}_3\}$ is the sum of a kinetic energy part H_α^{kin} of the center of mass of the fragment, and of an internal energy part H_α^{int}. For example, in the arrangement channel $\{\mathfrak{S}_1, \mathfrak{S}_2\text{–}\mathfrak{S}_3\}$,

$$(8.19) \qquad H^{\text{kin}}_{(1,2-3)} = H_{(1)} + H^{\text{kin}}_{(2-3)}, \qquad H^{\text{int}}_{(1,2-3)} = H^{\text{int}}_{(2-3)};$$

in the arrangement channel $\{\mathfrak{S}_1, \mathfrak{S}_2, \mathfrak{S}_3\}$,

$$(8.20) \qquad H^{\text{kin}}_{(1,2,3)} = H_{(1)} + H_{(2)} + H_{(3)} = H_0, \qquad H^{\text{kin}}_{(1,2,3)} = 0;$$

in the bound-state channel $\{\mathfrak{S}_1\text{–}\mathfrak{S}_2\text{–}\mathfrak{S}_3\}$, $H^{(1-2-3)}_{\text{kin}}$ is the kinetic energy operator of the center of mass of the entire system, and

$$(8.21) \qquad H^{\text{int}}_{(1-2-3)} = H^{\text{int}}_{(1-2-3)} + V_{12} + V_{23} + V_{13}$$

It is obvious that the kinetic energy operators of all channels commute with one another since they are all functions of the compatible momentum observables of the three particles. We also know from §5 of Chapter II that each center-of-mass kinetic energy operator commutes with the corresponding internal energy operator. However, in general, the Hamiltonians for different arrangement channels do not commute among themselves. This is easily seen to be so, for example, in the case of $\{\mathfrak{S}_1, \mathfrak{S}_2, \mathfrak{S}_3\}$ and $\{\mathfrak{S}_1\text{–}\mathfrak{S}_2, \mathfrak{S}_3\}$ when the channel Hamiltonians are H_0 and $H = H_0 + V_{12}$, respectively.

Following the guidelines set up by the above example, we postulate that in any multichannel scattering process a unique channel Hamiltonian H_α is attached to every arrangement channel α. Let us denote by $\mathsf{M}_\pm^{(\alpha)}$ the sets of all vectors f which are not eigenvectors of H_α and for which the respective strong limits

$$\underset{t \to \mp\infty}{\text{s-lim}}\, e^{iHt} e^{-iH_\alpha t} f$$

exist. By applying Lemma 1.2, we immediately conclude that $\mathsf{M}_\pm^{(\alpha)}$ are closed subspaces of $\mathscr{H}$ which might contain only the zero vector, as often happens in practice (see Exercises 8.1 and 8.2). Hence, we can introduce the projectors $E_{\mathsf{M}_{\pm}^{(\alpha)}}$ onto $\mathsf{M}_{\pm}^{(\alpha)}$ and define the *Møller wave operators* $\Omega_{\pm}^{(\alpha)}$ *for the channel* α by the formula

$$\Omega_{\pm}^{(\alpha)} = \operatorname*{s-lim}_{t\to\mp\infty} e^{iHt}\, e^{-iH_\alpha t}\, E_{\mathsf{M}_{\pm}^{(\alpha)}}\,. \tag{8.22}$$

The above formulas are obviously straightforward generalizations of (2.1). Hence, as mathematical objects, $\Omega_{\pm}^{(\alpha)}$ will have for each channel α the same properties as $\Omega_{\pm}$. In particular $\Omega_{\pm}^{(\alpha)}$ are partially isometric operators with initial domains $\mathsf{M}_{\pm}^{(\alpha)}$ and final domains $\mathsf{R}_{\pm}^{(\alpha)}$, which coincide with the ranges of these operators. The intertwining properties (2.3)–(2.5)

$$\Omega_{\pm}^{(\alpha)}\, e^{itH_\alpha} = e^{itH}\, \Omega_{\pm}^{(\alpha)}\,, \qquad t \in \mathbb{R}^1, \tag{8.23}$$

$$\Omega_{\pm}^{(\alpha)}\, E^{H_\alpha}(B) = E^{H}(B)\, \Omega_{\pm}^{(\alpha)}\,, \qquad B \in \mathscr{B}^1, \tag{8.24}$$

$$\Omega_{\pm}^{(\alpha)}\, H_\alpha f = H\Omega_{\pm}^{(\alpha)} f, \qquad f \in \mathscr{D}_{H_\alpha}\,, \tag{8.25}$$

and the relations (2.10) and (2.11)

$$\Omega_{\pm}^{(\alpha)*}\, \Omega_{\pm}^{(\alpha)} = E_{\mathsf{M}_{\pm}^{(\alpha)}}\,, \tag{8.26}$$

$$\Omega_{\pm}^{(\alpha)}\, \Omega_{\pm}^{(\alpha)*} = E_{\mathsf{R}_{\pm}^{(\alpha)}}\,, \tag{8.27}$$

will hold in the present case. Theorem 3.1 is also still valid, so that

$$\begin{aligned} \Omega_{+}^{(\alpha)} &= \operatorname*{s-lim}_{\epsilon\to+0} \epsilon \int_{-\infty}^{0} e^{\epsilon t}\, e^{iHt}\, e^{-iH_\alpha t}\, E_{\mathsf{M}_{\pm}^{(\alpha)}}\, dt, \\ \Omega_{-}^{(\alpha)} &= \operatorname*{s-lim}_{\epsilon\to+0} \epsilon \int_{0}^{\infty} e^{-\epsilon t}\, e^{iHt}\, e^{-iH_\alpha t}\, E_{\mathsf{M}_{\pm}^{(\alpha)}}\, dt. \end{aligned} \tag{8.28}$$

Moreover, the above two formulas can be taken to be the starting point of the time-independent approach to multichannel scattering (see Section 8.6).

8.3. The Uniqueness of Channel Strong Asymptotic States

Take now $f \in \mathsf{M}_0^{(\alpha)} = \mathsf{M}_{+}^{(\alpha)} \cap \mathsf{M}_{-}^{(\alpha)}$ and write $f_{\pm} = \Omega_{\pm}^{(\alpha)} f$. Since

$$\lim_{t\to\mp\infty} \| e^{-iHt} f_{\pm}^{(\alpha)} - e^{-iH_\alpha t} f \| = \lim_{t\to\mp\infty} \| f_{\pm}^{(\alpha)} - e^{iHt}\, e^{-iH_\alpha t} f \| = 0, \tag{8.29}$$

we see that $e^{-iH_\alpha t} f$ is the incoming asymptotic state of the interacting state $e^{-iHt} f_+^{(\alpha)}$, and the outgoing asymptotic state of the interacting state $e^{-iHt} f_-^{(\alpha)}$. The questions now arises, however, whether $e^{-iH_\alpha t} f$ is the *only* incoming (outgoing) asymptotic state of $e^{-iHt} f_+^{(\alpha)}$ $(e^{-iHt} f_-^{(\alpha)})$; namely, there certainly cannot be another incoming asymptotic state $e^{-iH_\alpha t} g$ of $e^{-iHt} f_+^{(\alpha)}$ in the same arrangement channel, but, in general, there could be some other incoming asymptotic state $e^{-iH_\beta t} g$ of $e^{-iHt} f_+^{(\alpha)}$ in some other arrangement channel $\beta \neq \alpha$.

It was pointed out earlier (in the three-body case) that we cannot expect that such a uniqueness of asymptotic states would hold for any arrangement channel Hamiltonians H_α picked up at random from the family of self-adjoint operators in $\mathscr{H}$. To prove such a uniqueness of asymptotic states we have to restrict the families of candidates for arrangement channel Hamiltonians by requiring that such operators obey certain conditions, which are dictated by the physical situation at hand.

Extrapolating from the example considered earlier of three-body scattering, it seems sensible to require that every channel Hamiltonian H_α on $\mathsf{M}_+^{(\alpha)}$ and on $\mathsf{M}_-^{(\alpha)}$ be the sum

$$H_\alpha = H_\alpha^{\text{kin}} + H_\alpha^{\text{int}} \tag{8.30}$$

of a center-of-mass kinetic energy part H_α^{kin} and an internal energy part H_α^{int}; here H_α^{kin} is the operator which represents the sum of the kinetic energies of the centers of mass of all clusters, while the operator H_α^{int} represents the sum of the internal energies of all clusters in the arrangement channel α. Judging from the wave mechanical three-body problem, it is also reasonable to postulate that H_α^{int} has a pure point spectrum and commutes with H_α^{kin}, while H_α^{kin} has only a continuous spectrum, and that all kinetic energy operators H_α^{kin} commute among themselves.

When the channel Hamiltonians have the above indicated structure we can prove that incoming and outgoing (strong) asymptotic states of any interacting state are uniquely determined by that interacting state, and consequently the description of the scattering experiment in terms of the asymptotic behavior of the system is completely unambiguous.

Let us first understand better the problem of uniqueness in mathematical terms. We have seen in (8.19) that if $\Psi(t) = e^{-iHt}\Psi_0$ is an interacting state, and if $\Psi_0 = \Omega_+^{(\alpha)}\Psi_0^{\text{in}}$, then $\Psi^{\text{in}}(t) = e^{-iH_\alpha t}\Psi_0^{\text{in}}$ is an incoming asymptotic state of $\Psi(t)$. If $\Psi_1^{\text{in}}(t) = \exp(-iH_{\alpha'}t)\Psi_1^{\text{in}}$ were another incoming asymptotic state of $\Psi(t)$, then we would have $\Psi_0 = \Omega_+^{(\alpha')}\Psi_1^{\text{in}}$. Thus, $\Psi^{\text{in}}(t)$ could be a state different from $\Psi_1^{\text{in}}(t)$ if and only if $\alpha \neq \alpha'$ and $\Psi_0 = \Omega_+^{(\alpha)}\Psi_0^{\text{in}} = \Omega_+^{(\alpha')}\Psi_1^{\text{in}}$, i.e., if Ψ_0 belonged to both ranges $\mathsf{R}_+^{(\alpha)}$ and $\mathsf{R}_+^{(\alpha')}$ of $\Omega_+^{(\alpha)}$ and $\Omega_+^{(\alpha')}$, respectively. Consequently, the necessary and sufficient condition which has to be fulfilled in order to have a unique asymptotic state corresponding to an interacting state is that the different ranges $\mathsf{R}_+^{(\alpha)}$ should have no state vectors in common. Naturally, a similar conclusion holds for the ranges $\mathsf{R}_-^{(\alpha)}$, which are also required to have no state vectors in common.

The two ranges $\mathsf{R}_+^{(\alpha)}$ and $\mathsf{R}_+^{(\alpha')}$, $\alpha \neq \alpha'$, will have no state vectors in common if, in particular, they are orthogonal to one another; in fact, in that case $\mathsf{R}_+^{(\alpha)}$ and $\mathsf{R}_+^{(\alpha')}$ have in common only the zero vector which is not a state vector.

The following theorem shows that if the channel Hamiltonians have the earlier mentioned structure, which is reflected in (8.30), then $\mathsf{R}_+^{(\alpha)} \perp \mathsf{R}_+^{(\alpha')}$ for all $\alpha \neq \alpha'$ and $\mathsf{R}_-^{(\beta)} \perp \mathsf{R}_-^{(\beta')}$ for all $\beta \neq \beta'$. Consequently, under those conditions every interacting state has at most one incoming and one outgoing (strong) asymptotic state.

Theorem 8.1. Suppose that for a given system the arrangement-channel Hamiltonians H_α, acting in the separable Hilbert space $\mathscr{H}$, are of the form (8.30), where all H_α^{kin} commute among themselves, while each H_α^{int} has a pure point spectrum when restricted to $\mathsf{M}_\pm^{(\alpha)}$ and it commutes with the corresponding H_α^{kin}. If α and β are two distinct arrangement channels and $H_\alpha^{\text{kin}} - H_\beta^{\text{kin}}$ has no point spectrum, then $\mathsf{R}_+^{(\alpha)} \perp \mathsf{R}_+^{(\beta)}$ and $\mathsf{R}_-^{(\alpha)} \perp \mathsf{R}_-^{(\beta)}$.

Proof. It follows from the basic definitions of wave operators that

$$(8.31)\qquad \Omega_\pm^{(\beta)*}\,\Omega_\pm^{(\alpha)} = (\operatorname*{s-lim}_{t\to\mp\infty} E_{\mathsf{M}_\pm^{(\beta)}}\,\Omega^{(\beta)*}(t))(\operatorname*{s-lim}_{t\to\mp\infty} \Omega^{(\alpha)}(t)\,E_{\mathsf{M}_\pm^{(\alpha)}})$$

$$= \operatorname*{s-lim}_{t\to\mp\infty} E_{\mathsf{M}_\pm^{(\beta)}}\,\Omega^{(\beta)*}(t)\,\Omega^{(\alpha)}(t)\,E_{\mathsf{M}_\pm^{(\alpha)}}$$

$$= \operatorname*{s-lim}_{t\to\mp\infty} E_{\mathsf{M}_\pm^{(\beta)}}\,e^{iH_\beta t}\,e^{-iH_\alpha t}\,E_{\mathsf{M}_\pm^{(\alpha)}}\,.$$

Since H_α^{int} has a pure point spectrum on $\mathsf{M}_\pm^{(\alpha)}$, the integral in the spectral decomposition of H_α^{int} can be written as a sum:

$$(8.32)\qquad H_\alpha^{\text{int}} f = \sum_{\lambda\in \mathsf{S}_{\text{int}}^{(\alpha)}} \lambda E^{(\alpha)}(\{\lambda\}) f, \qquad f \in \mathsf{M}_\pm^{(\alpha)} \cap \mathscr{D}_{H_\alpha^{\text{int}}}\,;$$

here the summation extends over all eigenvalues of H_α^{int} corresponding to eigenvectors in $M_{\pm}^{(\alpha)}$, and $E^{\alpha}(\{\lambda\})$ is the projector onto the eigenspace corresponding to the eigenvalue λ. It should be observed that in this context it is indeed possible to sum over all eigenvalues λ on account of the fact that there can be at most countably many eigenvalues of a self-adjoint operator in a separable Hilbert space.

We immediately get from (8.32), by using in the process the commutativity of H_α^{kin} and H_α^{int},

$$\exp(-iH_\alpha t)\, E_{M_\pm^{(\alpha)}} = \exp(-iH_\alpha^{\text{kin}}t)\exp(-iH_\alpha^{\text{int}}t)\sum_{\lambda\in S_{\text{int}}^{(\alpha)}} E^{(\alpha)}(\{\lambda\})\, E_{M_\pm^{(\alpha)}}$$

$$= \exp(-iH_\alpha^{\text{kin}}t)\sum_{\lambda\in S_{\text{int}}^{(\alpha)}} \exp(-i\lambda t)\, E^{(\alpha)}(\{\lambda\})\, E_{M_\pm^{(\alpha)}},$$

where the above sum, if infinite, converges in the strong sense. Since a similar relation can be derived for $E_{M_\pm^{(\beta)}}\, e^{iH_\beta t}$, and since H_α^{kin} and H_β^{kin} commute, we can write for any $f \in M_\pm^{(\beta)}$ and $g \in M_\pm^{(\alpha)}$

$$\text{(8.33)} \qquad \langle f \mid E_{M_\pm^{(\beta)}}\, e^{iH_\beta t}\, e^{-iH_\alpha t}\, E_{M_\pm^{(\alpha)}} g\rangle$$

$$= \sum_{\lambda'\in S_{\text{int}}^{(\beta)}}\Big(\sum_{\lambda\in S_{\text{int}}^{(\alpha)}} \langle f_{\lambda'} \mid \exp[i(H_\beta^{\text{kin}} - H_\alpha^{\text{kin}} + \lambda' - \lambda)t]\, g_\lambda\rangle\Big),$$

where $f_{\lambda'} = E^{(\beta)}(\{\lambda'\})\, f$ and $g_\lambda = E^{(\alpha)}(\{\lambda\})\, g$.

Let us take any two vectors $f \in M_-^{(\beta)}$ and $g \in M_-^{(\alpha)}$ which have only a finite number of nonzero components $f_{\lambda'}$ and g_λ, respectively. Then, in view of the fact that $\langle f \mid e^{iH_\beta t}e^{-iH_\alpha t}g\rangle$ is continuous and therefore integrable in t on any finite interval, we have

(8.34)

$$\int_0^T \langle f \mid e^{iH_\beta t}\, e^{-iH_\alpha t} g\rangle\, dt = \sum_{\lambda'}\sum_{\lambda}\int_0^T \langle f_{\lambda'} \mid \exp[i(H_\beta^{\text{kin}} + \lambda' - H_\alpha^{\text{kin}} - \lambda)t]\, g_\lambda\rangle\, dt.$$

Since $H_\beta^{\text{kin}} + \lambda' - H_\alpha^{\text{kin}} - \lambda$ is a self-adjoint operator, we can apply the mean ergodic theorem (see Theorem 8.3 in Appendix 8.9 to this section) to each term in the sum of (8.34). According to this theorem

$$\text{(8.35)} \qquad \lim_{T\to+\infty}\frac{1}{T}\int_0^T \langle f_{\lambda'} \mid \exp[i(H_\beta^{\text{kin}} - H_\alpha^{\text{kin}} + \lambda' - \lambda)t]\, g_\lambda\rangle\, dt$$

$$= \langle f_{\lambda'} \mid E^{H_\beta^{\text{kin}} - H_\alpha^{\text{kin}}}(\{\lambda - \lambda'\})\, g_\lambda\rangle = 0,$$

where the above inner product is zero because the projector onto the characteristic subspace of $H_\beta^{\text{kin}} - H_\alpha^{\text{kin}}$ corresponding to the eigenvalue $\lambda - \lambda'$ has to be zero on account of the assumption that $H_\beta^{\text{kin}} - H_\alpha^{\text{kin}}$ has no eigenvalues.

By using (8.31), we obtain (see also Exercise 8.3)

$$\begin{aligned}\langle f \mid \Omega_-^{(\beta)*} \Omega_-^{(\alpha)} g\rangle &= \lim_{t\to+\infty} \langle f \mid E_{\mathsf{M}_-^{(\beta)}} e^{iH_\beta t} e^{-iH_\alpha t} E_{\mathsf{M}_-^{(\alpha)}} g\rangle \\ &= \lim_{T\to+\infty} \frac{1}{T} \int_0^T \langle f \mid E_{\mathsf{M}_-^{(\beta)}} e^{iH_\beta t} e^{-iH_\alpha t} E_{\mathsf{M}_-^{(\alpha)}} g\rangle \, dt.\end{aligned}$$

If $f \in \mathsf{M}_-^{(\beta)}$ and $g \in \mathsf{M}_-^{(\alpha)}$ have only a finite number of nonvanishing components $f_{\lambda'}$ and g_λ, respectively, then in view of (8.34) and (8.35), we conclude that $\langle \Omega_-^{(\beta)} f \mid \Omega_-^{(\alpha)} g\rangle = 0$ for all such f and g. Since the set of all such vectors $f \in \mathsf{M}_-^{(\beta)}$ and $g \in \mathsf{M}_-^{(\alpha)}$ is dense in $\mathsf{M}_-^{(\beta)}$ and $\mathsf{M}_-^{(\alpha)}$, respectively, it follows that the same is true for all $f, g \in \mathscr{H}$. This establishes the assertion that $\mathsf{R}_-^{(\alpha)} \perp \mathsf{R}_-^{(\beta)}$ for $\alpha \neq \beta$.

A similar argument yields that $\mathsf{R}_+^{(\alpha)} \perp \mathsf{R}_+^{(\beta)}$ for $\alpha \neq \beta$. Q.E.D.

The above theorem is valid under the restriction that $H_\alpha^{\text{kin}} - H_\beta^{\text{kin}}$ should have no point spectrum when $\alpha \neq \beta$. It should be noted that this condition is certainly satisfied if H_α^{kin} and H_β^{kin} are two distinct functions of the momentum operators of the n particles partaking in the scattering process. The reader can easily convince himself that such is the case, for example, in potential scattering, so that Theorem 8.1 is always applicable in that case.

8.4. Interchannel Scattering Operators

In multichannel scattering we are faced with the possibility that a system prepared in some given arrangement channel does not have to end up in the same arrangement channel. Hence, obviously, no single scattering operator S could describe all the transitions which the system can undergo. Instead, to any pair (β, α) consisting of an incoming arrangement channel α and an outgoing arrangement channel β we have to attach a different scattering operator,

$$S_{\beta\alpha} = \Omega_-^{(\beta)*} \Omega_+^{(\alpha)}, \tag{8.36}$$

describing the transitions between these two channels. In fact, if $\Psi(t)$ has the incoming asymptotic state $\Psi_+^{\text{in}}(t) = e^{-iH_\alpha t}\, \Psi_-^{\text{in}}(0)$, then the transition amplitude to the outgoing state $\Psi_+^{\text{out}}(t) = e^{-iH_\alpha t}\,\Psi_+^{\text{out}}(0)$ of free fragments in channel β is

$$(8.37) \quad \lim_{t\to+\infty} \langle \Psi_+^{\text{out}}(t) \mid \Psi(t)\rangle = \lim_{t\to+\infty} \langle e^{-iH_\beta t}\,\Psi_+^{\text{out}}(0) \mid e^{-iHt}\,\Psi(0)\rangle$$

$$= \langle \Omega_-^{(\beta)}\,\Psi_+^{\text{out}}(0) \mid \Psi(0)\rangle = \langle \Omega_-^{(\beta)}\,\Psi_+^{\text{out}}(0) \mid \Omega_+^{(\alpha)}\,\Psi_-^{\text{in}}(0)\rangle$$

$$= \langle \Psi_+^{\text{out}}(0) \mid S_{\beta\alpha}\,\Psi_-^{\text{in}}(0)\rangle.$$

From (8.23) we easily obtain for all $t \in \mathbb{R}^1$

$$(8.38) \quad S_{\beta\alpha}\,e^{iH_\alpha t} = \Omega_-^{(\beta)*}\,\Omega_+^{(\alpha)}\,e^{iH_\alpha t} = \Omega_-^{(\beta)*}\,e^{iHt}\,\Omega_+^{(\alpha)}$$

$$= e^{iH_\beta t}\,\Omega_-^{(\beta)*}\,\Omega_+^{(\alpha)} = e^{iH_\beta t}\,S_{\beta\alpha}\,.$$

We can also derive in similar manner

$$(8.39) \quad S_{\beta\alpha}\,E^{H_\alpha}(B) = E^{H_\beta}(B)\,S_{\beta\alpha}\,, \qquad B \in \mathscr{B}^1.$$

One of the most important properties of the single-channel scattering operator is its unitarity. To establish the presence of some analogous property for multichannel scattering we have to consider simultaneously the entire array of all interchannel scattering operators. Then, using (8.27) we obtain

$$(8.40) \quad \sum_\gamma S_{\gamma\alpha}^* S_{\gamma\beta} = \sum_\gamma (\Omega_+^{(\alpha)*}\Omega_-^{(\gamma)})(\Omega_-^{(\gamma)*}\Omega_+^{(\beta)})$$

$$= \Omega_+^{(\alpha)*}\left(\sum_\gamma \Omega_-^{(\gamma)}\Omega_-^{(\gamma)*}\right)\Omega_+^{(\beta)} = \Omega_+^{(\alpha)*}\sum_\gamma E_{\mathsf{R}_+^{(\gamma)}}\Omega_+^{(\beta)}$$

In the case of single channel scattering the identity of R_+ and R_- is a necessary and sufficient condition for the unitarity of S on M_0 (see Theorem 2.5). A reasonable generalization of that condition to the multichannel case would be to require that

$$(8.41) \quad \bigoplus_\gamma \mathsf{R}_+^{(\gamma)} = \bigoplus_\gamma \mathsf{R}_-^{(\gamma)} = \mathsf{R},$$

where we tacitly assume that the conditions of Theorem 8.1 are satisfied, and therefore $\mathsf{R}_+^{(\alpha)} \perp \mathsf{R}_+^{(\beta)}$ and $\mathsf{R}_-^{(\alpha)} \perp \mathsf{R}_-^{(\beta)}$ for $\alpha \neq \beta$. The assumption (8.41) implies that

$$\sum_\gamma E_{\mathsf{R}_-^{(\gamma)}}\Omega_+^{(\beta)} = \sum_\gamma E_{\mathsf{R}_+^{(\gamma)}}\Omega_+^{(\beta)} = E_{\mathsf{R}_+^{(\beta)}}\Omega_+^{(\beta)} = \Omega_+^{(\beta)}\,,$$

since $\Omega_+^{(\beta)}$ is a partial isometry with final domain $\mathsf{R}_+^{(\beta)}$. Hence, (8.40) yields

$$\sum_\gamma S_{\gamma\alpha}^* S_{\gamma\beta} = \Omega_+^{(\alpha)*} \Omega_+^{(\beta)} = \delta_{\alpha\beta} E_{\mathsf{M}_+^{(\alpha)}} , \tag{8.42}$$

where $\delta_{\alpha\beta} = 1$ for $\alpha = \beta$ and $\delta_{\alpha\beta} = 0$ for $\alpha \neq \beta$.

A similar procedure leads to the conclusion that

$$\sum_\gamma S_{\alpha\gamma} S_{\beta\gamma}^* = \Omega_-^{(\alpha)*} \Omega_-^{(\beta)} = \delta_{\alpha\beta} E_{\mathsf{M}_-^{(\alpha)}} \tag{8.43}$$

when (8.41) is satisfied.

8.5. The Existence of Strong Asymptotic States in n-Particle Potential Scattering

Let us consider the wave mechanical case of n particles without spin interacting via two-body forces, so that

$$(H\psi)(\mathbf{r}_1, ..., \mathbf{r}_n) = (H_0\psi)(\mathbf{r}_1, ..., \mathbf{r}_n) + \sum_{\substack{i,j=1 \\ i<j}}^n V_{ij}(\mathbf{r}_i - \mathbf{r}_j)\, \psi(\mathbf{r}_1, ..., \mathbf{r}_n), \tag{8.44}$$

where, using the momentum representation,

$$(H_0\psi)^\sim(\mathbf{p}_1, ..., \mathbf{p}_n) = \sum_{i=1}^n \frac{\mathbf{p}_i^2}{2m_i} \tilde{\psi}(\mathbf{p}_1, ..., \mathbf{p}_n).$$

Let us choose an arrangement channel α containing the fragments $\mathscr{F}_\alpha^{(1)}, ..., \mathscr{F}_\alpha^{(n_\alpha)}$. The free motion of each one of these fragments $\mathscr{F}_\alpha^{(k)}$ is described by the Hamiltonian

$$H_\alpha^{(k)} = \sum_{i\in\mathscr{F}_\alpha^{(k)}} \frac{\mathbf{P}_i^2}{2m_i} + \sum_{\substack{i,j\in\mathscr{F}_\alpha^{(k)} \\ i<j}} V_{ij}(\mathbf{Q}_i - \mathbf{Q}_j), \tag{8.45}$$

where the summation is over the indices of the particles in that fragment. The channel Hamiltonian is

$$H_\alpha = H_\alpha^{(1)} + \cdots + H_\alpha^{(n_\alpha)}. \tag{8.46}$$

Take $\mathbf{R}_\alpha^{(k)}$ to be the center-of-mass position vector of the kth fragment, and let $\rho_\alpha^{(k)}$ stand for some choice of internal motion coordinates, so that

$(\mathbf{R}_\alpha^{(k)}, \rho_\alpha^{(k)})$ completely describes the positions of all the particles in the fragment. Then proceeding in the manner indicated in §5 of Chapter II, we can write

$$H_\alpha^{(k)} = -(1/2M_\alpha^{(k)})\, \Delta_{\mathbf{R}_\alpha^{(k)}} + H_\alpha^{(k)\,\mathrm{int}}, \qquad M_\alpha^{(k)} = \sum_{i \in \mathscr{F}_\alpha^{(k)}} m_i, \tag{8.47}$$

where $M_\alpha^{(k)}$ is the mass of the fragment $\mathscr{F}_\alpha^{(k)}$ and $H_\alpha^{(k)\,\mathrm{int}}$ is the Hamiltonian of the internal motion of the particles in that fragment. If $\psi_\alpha^{(k)}(\rho_\alpha^{(k)})$, $k = 1,\ldots, n_\alpha$, represent any eigenvectors of $H_\alpha^{(k)\,\mathrm{int}}$, then $e^{-iH_\alpha t}\,\Psi_\alpha(0)$, with $\Psi_\alpha(0)$ represented by

$$\psi_\alpha(\mathbf{r}_1, \ldots, \mathbf{r}_n) = \prod_{k=1}^{n_\alpha} \Psi_\alpha^{(k)}(\mathbf{R}_\alpha^{(k)})\, \psi_\alpha^{(k)}(\rho_\alpha^{(k)}), \qquad \Psi_\alpha^{(1)}, \ldots, \Psi_\alpha^{(n_\alpha)} \in L^2(\mathbb{R}^3), \tag{8.48}$$

is a state in which all fragments in the channel α are moving independently. Obviously, the closed linear manifold $\mathsf{N}^{(\alpha)}$ spanned by all the functions of the form (8.48) coincides with the set of all free state vectors in the channel α.

Theorem 8.2. (*Hack's Theorem*). Suppose the functions $V_{ij}(\mathbf{r})$, $i < j$, $i, j = 1,\ldots, n$, are Lebesgue square integrable on $\mathbb{R}^3$. Then the initial domains $\mathsf{M}_0^{(\alpha)} \equiv \mathsf{M}_\pm^{(\alpha)}$ of the Moller channel wave operators $\Omega_\pm^{(\alpha)}$ coincide with the closed linear subspace $\mathsf{N}^{(\alpha)}$ of $L^2(\mathbb{R}^{3n})$ spanned by all functions ψ_α in (8.48) obtained by letting $\psi_\alpha^{(k)}$ vary over all eigenvectors of $H_\alpha^{(k)\,\mathrm{int}}$ and $\Psi_\alpha^{(k)}$ over all of $L^2(\mathbb{R}^3)$ for $k = 1,\ldots, n_\alpha$.

In order to avoid very involved notation, we shall prove Theorem 8.2 for the special case of the arrangement channel $\{\mathfrak{S}_1, \mathfrak{S}_2\text{–}\mathfrak{S}_3\}$ in three-body scattering. In this case

$$H_\alpha \supseteq -\frac{1}{2m}\Delta_{\mathbf{r}_1} - \frac{1}{2(m_2+m_3)}\Delta_{\mathbf{R}} - \frac{1}{2m_{23}}\Delta_{\mathbf{r}_{23}} + V_{23}(\mathbf{r}_{23}), \tag{8.49}$$

where m_{23} is the reduced mass of the second fragment $\mathfrak{S}_2\text{–}\mathfrak{S}_3$ and

$$\mathbf{r}_{23} = \mathbf{r}_2 - \mathbf{r}_3, \qquad \mathbf{R} = \frac{m_2\mathbf{r}_2 + m_3\mathbf{r}_3}{m_2 + m_3}. \tag{8.50}$$

The space $\mathsf{N}^{(\alpha)}$ of free state vectors in this channel is the closed linear manifold spanned by all the functions

$$\psi_\alpha(\mathbf{r}_1, \mathbf{r}_2, \mathbf{r}_3) = \Psi_\alpha^{(1)}(\mathbf{r}_1)\, \Psi_\alpha^{(2)}(\mathbf{R})\, \psi_\alpha^{(2)}(\mathbf{r}_{23}) \tag{8.51}$$

corresponding to all $\Psi_\alpha^{(1)}$, $\Psi_\alpha^{(2)} \in L^2(\mathbb{R}^3)$ and all eigenvectors $\psi_\alpha^{(2)}$ of

$$H_\alpha^{(2-3)\text{int}} \supseteq -\frac{1}{2m_{23}} \Delta_{\mathbf{r}} + V_{23}(\mathbf{r}).$$

On account of Wiener's theorem (see Lemma 7.1) the linear set spanned by all functions (8.51) with $\Psi_\alpha^{(k)}$, $k = 1, 2$, represented by functions which are inverse Fourier transforms of Gaussian wave packets $\exp[-(\mathbf{p} - \mathbf{p}_k)^2]$, $k = 1, 2$, is dense in $\mathsf{N}^{(\alpha)}$ when $\mathbf{p}_1$ and $\mathbf{p}_2$ are allowed to vary over $\mathbb{R}^3$. For such functions we can easily compute

$$\left| \left(\exp\left(-i \frac{\mathbf{p}_k^{\,2}}{2M_\alpha^{(k)}} t \right) \Psi_\alpha^{(k)} \right)(\mathbf{r}) \right|^2 \tag{8.52}$$

$$= \left\{ 4 \left[1 + \left(\frac{t}{2M_\alpha^{(k)}} \right)^2 \right] \right\}^{-3/2} \exp[-a_k(\mathbf{r} - \mathbf{b}_k)^2],$$

where $M_\alpha^{(1)} = m_1$, $M_\alpha^{(2)} = m_2 + m_3$ and

$$\mathbf{b}_k = \frac{t}{M_\alpha^{(k)}} \mathbf{p}_k, \qquad a_k = \frac{1}{2} \left(1 + \left(\frac{t}{2M_k} \right)^2 \right]^{-1}.$$

Hence, according to Theorem 2.8, the strong limits

$$\operatorname*{s-lim}_{t \to \mp\infty} e^{iHt} e^{-iH_\alpha t} \psi, \qquad H = H_\alpha + V_{12} + V_{13}$$

exist for all $\Psi \in \mathsf{N}^{(\alpha)}$ if the improper Riemann integral

$$\int_{-\infty}^{+\infty} \|(V_{12} + V_{13})\, e^{-iH_\alpha t} \psi\|\, dt \tag{8.53}$$

is convergent for all ψ of the above indicated form. Now, using (8.52) we get the following kind of estimate:

$$\| V_{12}\, e^{-iH_\alpha t} \psi_\alpha \|^2$$

$$= \int_{\mathbb{R}^9} V_{12}^{\,2}(\mathbf{r}_1 - \mathbf{r}_2) \left[4 \left(1 + \left(\frac{t}{2m_1} \right)^2 \right) \right]^{-3/2} \exp[-a_1(\mathbf{r}_1 - \mathbf{b}_1)^2]$$

$$\times \left[4 \left(1 + \left(\frac{t}{2(m_2 + m_3)} \right)^2 \right) \right]^{-3/2}$$

$$\times \exp\left[-a_2 \left(\frac{m_2 \mathbf{r}_2 + m_3 \mathbf{r}_3}{m_2 + m_3} - \mathbf{b}_2 \right)^2 \right] |\psi_\alpha^{(2)}(\mathbf{r}_2 - \mathbf{r}_3)|^2\, d\mathbf{r}_1\, d\mathbf{r}_2\, d\mathbf{r}_3$$

$$\leqslant \int_{\mathbb{R}^9} V_{12}^2(\mathbf{r}_{12}) \left[16 \left(1 + \left(\frac{t}{2m_1} \right)^2 \right) \left(1 + \left(\frac{t}{2(m_2 + m_3)} \right)^2 \right) \right]^{-3/2}$$

(equation continues)

$$\times \exp[-a_2(\mathbf{R} - \mathbf{b}_2)^2]\, \psi_\alpha^{(2)}(\mathbf{r}_{23})|^2 \frac{\partial(\mathbf{r}_1, \mathbf{r}_2, \mathbf{r}_3)}{\partial(\mathbf{r}_{12}, \mathbf{r}_{23}, \mathbf{R})}\, d\mathbf{r}_{12}\, d\mathbf{r}_{23}\, d\mathbf{R}$$

$$\leqslant \text{const} \left[4\left(1 + \left(\frac{t}{2m_1}\right)^2\right]^{-3/2}.$$

Hence, we conclude that (8.53) converges, and consequently $\mathsf{N}^{(\alpha)} \equiv \mathsf{M}_+^{(\alpha)} = \mathsf{M}_-^{(\alpha)} = \mathsf{M}_0^{(\alpha)}$.

With the existence of asymptotic states established, the next question of general interest is asymptotic completeness. If $\mathscr{H}_b$ denotes the closed subspace of bounded states of a general n-particle system described in the Hilbert space $\mathscr{H}$, then in accordance with Definition 6.2, the theory is said to be asymptotically complete if (8.41) is satisfied and

$$\mathscr{H} = \mathsf{R} \oplus \mathscr{H}_n. \tag{8.54}$$

In multichannel potential scattering theory there are at the present time proofs of asymptotic completeness in the three-body case for large classes of short-range interactions (Faddeev [1965], Thomas [1975]), as well as for special instances of n-body ($n \geqslant 4$) scattering (Hagedorn [1978], Sigal [1978a, b]), but the general situation is as yet far from being as satisfactory as in the two-body case.

**8.6. Two-Hilbert Space Formulation of Multichannel Scattering Theory

We have seen in §8.4 that in multichannel scattering theory we cannot express all relevant physical information in terms of a single scattering operator acting in the Hilbert space $\mathscr{H}$ of the system, but that we have to deal rather with operators $S_{\beta\alpha}$ of the form (8.36) for each pair (β, α) consisting of an incoming and an outgoing arrangement channel α and β, respectively. We can, however, represent all these interchannel scattering operators in terms of a single operator S^J from a new Hilbert space of incoming asymptotic states $\mathscr{H}^{\text{in}}$ into another new Hilbert space $\mathscr{H}^{\text{out}}$ of outgoing asymptotic states if we build the direct sums (Ekstein [1956])

$$\mathscr{H}^{\text{in}} = \bigoplus_\alpha \mathsf{M}_+^{(\alpha)}, \qquad \mathscr{H}^{\text{out}} = \bigoplus_\beta \mathsf{M}_-^{(\beta)} \tag{8.55}$$

over all incoming and outgoing arrangement channels α and β, respectively. We shall refer to $\mathscr{H}^{\text{in}}$ and $\mathscr{H}^{\text{out}}$ as the *in-* and *out-asymptotic Hilbert space*, respectively.

Let us therefore denote by $J_\pm$ the *identification* (or *injection*) *operators* which, by definition, assign to the vectors

$$f = \bigoplus_\alpha f_\alpha \in \mathscr{H}^{\text{in}}, \qquad g = \bigoplus_\beta g_\beta \in \mathscr{H}^{\text{out}}, \tag{8.56}$$

the respective vectors $\sum_\alpha f_\alpha$ and $\sum_\beta g_\beta$ from $\mathscr{H}$. Naturally, the $J_\pm$ operators are not in general invertible, but they are bounded transformations (in the sense of Definition 1.5 in Chapter III) from $\mathscr{H}^{\text{ex}}$, ex = in, out, into $\mathscr{H}$. Therefore they have adjoints $J_\pm^\dagger$, which, by definition, satisfy*

$$\langle h \mid J_\pm f\rangle = \langle J_\pm^\dagger h \mid f\rangle_{\text{ex}}, \tag{8.57}$$

for all $f \in \mathscr{H}^{\text{ex}}$ and $h \in \mathscr{H}$, where $\langle\cdot \mid \cdot\rangle_{\text{ex}}$, ex = in, out, represent the inner products in the asymptotic Hilbert space $\mathscr{H}^{\text{ex}}$ constructed in accordance with Theorem 6.1 in Chapter II. It is a routine matter to extend Theorem 2.7 of Chapter III to bounded transformations from one Hilbert space into another and thus conclude that $J_\pm^\dagger$ are bounded transformations from $\mathscr{H}$ into $\mathscr{H}^{\text{ex}}$.

It is also easy to establish that the strong limits

$$\Omega_\pm^J = \operatorname*{s-lim}_{t\to\mp\infty} e^{iHt} J_\pm e^{-iH^{\text{ex}}t}, \tag{8.58}$$

$$H^{\text{in}} = \bigoplus_\alpha H^\alpha, \qquad H^{\text{out}} = \bigoplus_\beta H^\beta, \tag{8.59}$$

exist as bounded operators from $\mathscr{H}^{\text{ex}}$ into $\mathscr{H}$, and that in fact

$$\Omega_+^J = \bigoplus_\alpha \Omega_+^{(\alpha)}, \qquad \Omega_-^J = \bigoplus_\beta \Omega_-^{(\beta)}. \tag{8.60}$$

Therefore their adjoints

$$(\Omega_+^J)^\dagger = \sum_\alpha \Omega_+^{(\alpha)*}, \qquad (\Omega_-^J)^\dagger = \sum_\beta \Omega_-^{(\beta)*}, \tag{8.61}$$

also exist as bounded operators, but from the Hilbert space $\mathscr{H}$ of the system into the asymptotic Hilbert spaces $\mathscr{H}^{\text{in}}$ and $\mathscr{H}^{\text{out}}$, respectively, so that

$$S^J = (\Omega_-^J)^\dagger \Omega_+^J \tag{8.62}$$

is a bounded operator from $\mathscr{H}^{\text{in}}$ into $\mathscr{H}^{\text{out}}$.

* In general, if A is a bounded linear transformation from $\mathscr{H}_1$ into $\mathscr{H}_2$ (see Chapter III, Definition 1.5) and $\langle\cdot \mid \cdot\rangle_i$, $i = 1, 2$, denote the respective inner products, then by definition $\langle A^\dagger g \mid f\rangle_1 = \langle g \mid Af\rangle_2$, $f \in \mathscr{H}_1$, $g \in \mathscr{H}_2$. The existence and uniqueness of $A^\dagger$ is proven exactly as in the special case where $\mathscr{H}_1 \equiv \mathscr{H}_2$ (cf. Chapter III, Theorem 2.7).

Let us denote by $E_+^{(\alpha)}$ and $E_-^{(\beta)}$ the projectors onto $\mathsf{M}_+^{(\alpha)}$ and $\mathsf{M}_-^{(\beta)}$, respectively, when these two spaces are viewed as subspaces of $\mathscr{H}^{\text{in}}$ and $\mathscr{H}^{\text{out}}$, respectively. Then by (8.60)–(8.62),

$$E_-^{(\beta)} S^J E_+^{(\alpha)} = E_-^{(\beta)} \Omega_-^{(\beta)*} J_-^{\ \dagger} J_+ \Omega_+^{(\alpha)} E_+^{(\alpha)}, \tag{8.63}$$

so that the above operator is essentially* equivalent to $S_{\beta\alpha}$. In fact, we have for f and g given by (8.56),

$$\langle E_-^{(\beta)} g \mid S^J E_+^{(\alpha)} f \rangle_{\text{out}} = \langle g_\beta \mid S_{\beta\alpha} f_\alpha \rangle, \tag{8.64}$$

so that S^J incorporates information about all interchannel scattering operators $S_{\alpha\beta}$.

In Theorem 8.2 we have seen that in potential scattering $\mathsf{M}_+^{(\alpha)} = \mathsf{M}_-^{(\alpha)} = \mathsf{M}_0^{(\alpha)}$ for all arrangement channels, so that we have a single asymptotic Hilbert space $\mathscr{H}^{\text{ex}}$ and a single identification operator J:

$$\mathscr{H}^{\text{in}} = \mathscr{H}^{\text{out}} = \mathscr{H}^{\text{ex}}, \qquad J_+ = J_- = J. \tag{8.65}$$

Moreover, (8.41) is satisfied as part of asymptotic completeness. As easily seen, this implies that S^J is unitary on the Hilbert space $\mathscr{H}^{\text{ex}}$ appearing in (8.65).

Proceeding under the assumption that (8.65) is true, we can make the transition to stationary two-Hilbert space scattering theory by following exactly the same pattern as in §3 in the one-Hilbert space context. Thus we introduce the Bochner integrals

$$\Omega_{\pm\epsilon}^J = \mp\epsilon \int_0^{\mp\infty} e^{\pm\epsilon t}\, e^{iHt} J e^{-iH^{\text{ex}}t}\, dt, \tag{8.66}$$

and then routinely establish that the following results are true (see Prugovečki [1973a, Theorems 2.1–2.4]):

$$\Psi^\pm = \Omega_\pm{}^J \Psi = \operatorname*{s-lim}_{\epsilon\to+0} \Omega_{\pm\epsilon}^J \Psi, \qquad \Psi \in \mathscr{H}^{\text{ex}}, \tag{8.67}$$

$$\Omega_{\pm\epsilon}^J = \int_{-\infty}^{+\infty} \frac{\pm i\epsilon}{\lambda - H \pm i\epsilon}\, J\, d_\lambda E_\lambda^{H^{\text{ex}}} = \int_{-\infty}^{+\infty} d_\lambda E_\lambda{}^H J \frac{\mp i\epsilon}{\lambda - H^{\text{ex}} \mp i\epsilon} \tag{8.68}$$

$$E^H(B)\Omega_\pm{}^J = \Omega_\pm{}^J E^{H^{\text{ex}}}(B), \qquad B \in \mathscr{B}^1, \tag{8.69}$$

* There is an obvious conceptual distinction between $\mathsf{M}_+^{(\alpha)}$ as a subspace of $\mathscr{H}$, and the space $\mathsf{M}_+^{(\alpha)\prime}$ of all vectors $\oplus_{\alpha'} f_{\alpha'}$ in $\mathscr{H}^{\text{in}}$, which have only the component f_α different from zero. To avoid too cumbersome a notation we shall, however, ignore this distinction.

$$(8.70)\qquad J^\dagger\Psi^\pm = \Psi + \operatorname*{w-lim}_{\epsilon\to+0}\int_{-\infty}^{+\infty}(\lambda - H^{\text{ex}} \pm i\epsilon)^{-1}(J^\dagger H - H^{\text{ex}}J^\dagger)\,d_\lambda E_\lambda{}^H\Psi^\pm,$$

$$(8.71)\qquad \Psi^\pm = J\Psi + \operatorname*{s-lim}_{\epsilon\to+0}\int_{-\infty}^{+\infty}(\lambda - H \pm i\epsilon)^{-1}(HJ - JH^{\text{ex}})\,d_\lambda E_\lambda^{H^{\text{ex}}}\Psi.$$

$$(8.72)\qquad T^J = \frac{S^J - 1^{\text{ex}}}{2\pi i} = \frac{1}{2\pi i}\operatorname*{w-lim}_{\epsilon\to+0}(\Omega^J_{-\epsilon} - \Omega^J_{+\epsilon})^\dagger\Omega_+{}^J$$

$$= \frac{1}{\pi}\operatorname*{w-lim}_{\epsilon\to+0}\int_{-\infty}^{+\infty} d_\lambda E_\lambda^{H^{\text{ex}}}(J^\dagger H - H^{\text{ex}}J^\dagger)\frac{\epsilon}{(H-\lambda)^2+\epsilon^2}\,\Omega_+{}^J$$

$$= \frac{1}{2\pi i}\operatorname*{s-lim}_{\epsilon\to+0}(\Omega_-{}^J)^\dagger(\Omega^J_{+\epsilon} - \Omega^J_{-\epsilon})$$

$$= \frac{1}{\pi}\operatorname*{s-lim}_{\epsilon\to+0}(\Omega_-{}^J)^\dagger\int_{-\infty}^{+\infty}\frac{\epsilon}{(H-\lambda)^2+\epsilon^2}(HJ - JH^{\text{ex}})\,d_\lambda E_\lambda^{H^{\text{ex}}}.$$

Furthermore, the spectral-integral expressions (8.72) for the T^J operator can be related to a $\mathscr{T}^J(\zeta)$ operator analogous to the $\mathscr{T}(\zeta)$ operators introduced in §4.8 for the two-body case. For example, the first set of formulas (8.72) yields (see Chandler and Gibson [1973]):

$$(8.73)\qquad T^J = \operatorname*{w-lim}_{\epsilon\to+0}\operatorname*{w-lim}_{\eta\to+0}\int_{-\infty}^{+\infty} d_\lambda E_\lambda^{H^{\text{ex}}}\int_{-\infty}^{+\infty}\delta_\epsilon(\lambda-\mu)\,\mathscr{T}^J(\mu+i\eta)\,d_\mu E_\mu^{H^{\text{ex}}},$$

$$\mathscr{T}^J(\zeta) = J^\dagger(HJ - JH^{\text{ex}}) + (J^\dagger H - H^{\text{ex}}J^\dagger)(\zeta - H)^{-1}(HJ - JH^{\text{ex}}).$$

We note that if $\mathscr{H}^{\text{ex}} = \mathscr{H}$ and $J = 1$ (so that $H^{\text{ex}} = H_0$), then (8.70)–(8.73) reduce to (3.34), (3.41), (3.52), and (4.114), respectively. Clearly, (8.70) and (8.71) can be justifiably called *two-Hilbert space Lippmann–Schwinger equations*. These equations remain valid even in the presence of long-range forces under the proviso that asymptotically compensating operators Z_α [analogous to Z in (4.121)] are incorporated* into J.

8.7. Multichannel Eigenfunction Expansions and T-Matrices

In its fundamental aspects the approach to the formulation and treatment of eigenfunction expansions is the same in the n-body case as it has been in the two-body case, but the level of complexity increases drastically with increasing values of n. Furthermore, the Fredholm

* For details see Prugovečki [1973a]. A variant of this approach has been extensively studied by Gibson and Chandler [1974] and Chandler and Gibson [1974, 1977, 1978].

method of §6, used extensively in the $n = 2$ case, becomes inapplicable in its original form when $n \geqslant 3$.

To formulate the concept of eigenfunction expansions in the multichannel case we have to choose in accordance with Definition 4.2 for each arrangement channel α complete sets of observables $\{X_1^{(\alpha)}, X_2^{(\alpha)},...\}$ and $\{Y_1^{(\alpha)}, Y_2^{(\alpha)},...\}$ that are best suited for that particular channel. Among the $X^{(\alpha)}$-observables, the standard choices always include the center-of-mass observables $\mathbf{Q}_{c,\alpha}^{(k)}$ for each of the $k = 1,..., n_\alpha$ clusters [whose position variables in the corresponding spectral representation spaces for complete sets of observables, which incorporate $\mathbf{Q}_{c,\alpha}^{(k)}$, we have denoted in (8.48) by $\mathbf{R}_\alpha^{(k)}$]. The corresponding center-of-mass momentum observables $\mathbf{P}_{c,\alpha}^{(k)}$ [whose variables we shall denote by $\mathbf{P}_\alpha^{(k)}$] are always included among the $Y^{(\alpha)}$-observables, with the internal energies for each fragment being the next standard choice. Thus, in accordance with (8.48) the corresponding eigenfunction expansion consists of functions (in physical literature usually called *stationary states*) of the form

$$\Phi_{P_\alpha,\omega_\alpha}^{(\alpha)}(R_\alpha, \rho_\alpha) = \prod_{k=1}^{n_\alpha} \Phi_{\mathbf{P}_\alpha^{(k)}}(\mathbf{R}_\alpha^{(k)})\, \psi_{\omega_\alpha^{(k)}}^{(k)}(\rho_\alpha^{(k)}), \tag{8.74}$$

$$R_\alpha = (\mathbf{R}_\alpha^{(1)},..., \mathbf{R}_\alpha^{(n_\alpha)}), \qquad P_\alpha = (\mathbf{P}_\alpha^{(1)},..., \mathbf{P}_\alpha^{(n_\alpha)}),$$

where $\rho_\alpha^{(k)}$ and $\omega_\alpha^{(k)}$ are variables associated with the rest of the complete set of observables for the kth cluster. For example, in case of the arrangement channel $\{\mathfrak{S}_1, \mathfrak{S}_2 - \mathfrak{S}_3\}$ [with which the equations (8.5)–(8.8) and (8.11)–(8.17) are associated], ρ_1 and ω_1 are redundant for the first cluster which consists only of $\mathfrak{S}_1$, whereas for the second cluster $\mathbf{r}_{23} = \mathbf{r}_2 - \mathbf{r}_3$ play the role of ρ_2 and $E_{(2-3)}$ is incorporated among the variables in ω_2 [to be very specific, if $P_2 - P_3$ is an electron–proton pair, and the spin observables are ignored, then in accordance with the results of §7.7 in Chapter II, we could choose $\omega_2 = (l, m, n)$].

The presence in (8.74) of the plane waves

$$\Phi_{\mathbf{P}_\alpha^{(k)}}(\mathbf{R}_\alpha^{(k)}) = (2\pi)^{-3/2} \exp(i\mathbf{P}_\alpha^{(k)}\mathbf{R}_\alpha^{(k)}), \qquad k = 1,..., n_\alpha, \tag{8.75}$$

tells us that, as in the two-body case, $\Phi_{P_\alpha,\omega_\alpha}^{(\alpha)}$ are not elements of $\mathsf{M}_0^{(\alpha)}$. However, in general, if we make in (8.73) the transition to the configuration representation by expressing (for spinless particles) R_α and ρ_α in terms of $\mathbf{r}_1,..., \mathbf{r}_n$ [cf. (5.10) in Chapter II for a specific example] so that $\Phi_{P_\alpha,\omega_\alpha}^{(\alpha)}$ becomes a function $\hat{\Phi}_{P_\alpha,\omega_\alpha}^{(\alpha)}$ of $\mathbf{r}_1,..., \mathbf{r}_n \in \mathbb{R}^3$, then as a function of P_α and ω_α,

$$(\Phi^{(\alpha)}_{P_\alpha,\omega_\alpha} \mid \psi\rangle = \text{l.i.m.} \int_{\mathbb{R}^{3n}} \hat{\Phi}^{(\alpha)*}_{P_\alpha,\omega_\alpha}(\mathbf{r}_1, \dots, \mathbf{r}_n)\, \psi(\mathbf{r}_1, \dots, \mathbf{r}_n)\, d\mathbf{r}_1 \cdots d\mathbf{r}_n,$$

is in general a well-defined element of

$$L^2\left(\mathbb{R}^{N_\alpha}, \prod_{k=1}^{n_\alpha} d\mathbf{P}^{(k)}_\alpha\, d\nu^{(\alpha)}_k(\omega)\right)$$

for any $\psi \in \mathsf{M}^{(\alpha)} \subset L^2(\mathbb{R}^{3n})$ [where N_α denotes the total number of $Y^{(\alpha)}$-observables and $\nu^{(\alpha)}_k$ is the measure over the spectral representation space for the complete set of observables related to the internal degrees of freedom in the kth cluster]. Using the symbolic notation in (4.81), we can express the fact that, for all channels α, $\Phi^{(\alpha)}_{P_\alpha,\omega_\alpha}$ constitutes an eigenfunction expansion in $\mathsf{M}^{(\alpha)}_0$ by the shorthand statement

$$\sum_\alpha \int \mid \Phi^{(\alpha)}_{P_\alpha,\omega_\alpha})\, d\nu_\alpha(P_\alpha, \omega_\alpha)(\Phi^{(\alpha)}_{P_\alpha,\omega_\alpha} \mid = 1^{\text{ex}}, \tag{8.76}$$

$$d\nu_\alpha(P_\alpha, \omega_\alpha) = \prod_{k=1}^{n_\alpha} d\mathbf{P}^{(k)}_\alpha\, d\nu^{(k)}_\alpha(\omega^{(k)}_\alpha),$$

where 1^{ex} denotes the identity operator in $\mathscr{H}^{\text{ex}}$.

Likewise, the Green's operator $G^{\text{ex}}(\zeta)$ for the asymptotic Hamiltonian H^{ex} [given by (8.59), but with $H^{\text{in}} = H^{\text{out}}$] can be expressed as

$$G^{\text{ex}}(\zeta) = (\zeta - H^{\text{ex}})^{-1} = \sum_\alpha \int \frac{\mid \Phi^{(\alpha)}_{P_\alpha,\omega_\alpha})\, d\nu_\alpha(P_\alpha, \omega_\alpha)(\Phi^{(\alpha)}_{P_\alpha,\omega_\alpha} \mid}{\zeta - \sum_{k=1}^{n_\alpha}[(\mathbf{P}^{(k)2}_\alpha/2M^{(k)}_\alpha) + \lambda^{(k)}_\alpha]}, \tag{8.77}$$

where $\lambda^{(k)}_\alpha$ is the internal energy of the kth cluster in the α-channel, and it is in general a function of $\omega^{(k)}_\alpha$. Hence, the Green "function" for the channel Hamiltonian H_α is in general a complicated entity,

$$G_\alpha(R_\alpha, \rho_\alpha; R_\alpha', \rho_\alpha'; \zeta) = \int \frac{\Phi^{(\alpha)}_{P_\alpha,\omega_\alpha}(R_\alpha, \rho_\alpha)\, d\nu_\alpha(P_\alpha, \omega_\alpha)\, \Phi^{(\alpha)*}_{P_\alpha}(R_\alpha', \rho_\alpha')}{\zeta - \sum_{k=1}^{n_\alpha}[\mathbf{P}^{(k)2}_\alpha/2M^{(k)}_\alpha + \lambda^{(k)}_\alpha(\omega^{(k)}_\alpha)]}, \tag{8.78}$$

which might turn out to be well-defined only in the sense of distributions rather than as an ordinary function. Naturally, the explicit use of the theory of distributions can be avoided once it is realized that in general the integrations in $\mathbf{P}^{(k)}_\alpha$ and $\omega^{(k)}_\alpha$, $k = 1, \dots, n_\alpha$, have to be executed after the integration in R_α, ρ_α, R_α', and ρ_α' have been performed upon taking inner products with suitable wave packets.

These complications make the task of reproducing the derivation of §§4.6 and 4.7 in the multichannel context generally unfeasible without the introduction of new techniques. True, on a purely formalistic level, it would appear that (8.70) unambiguously yields

$$\Phi^{(\alpha)(\pm)}_{P_\alpha,\omega_\alpha} = \Phi^{(\alpha)}_{P_\alpha,\omega_\alpha} + G_\alpha(\lambda_\alpha \pm i0)\, V_\alpha \Phi^{(\alpha)(\pm)}_{P_\alpha,\omega_\alpha} \tag{8.79}$$

$$\lambda_\alpha = \sum_{k=1}^{n_\alpha} \frac{\mathbf{P}_\alpha^{(k)2}}{2M_\alpha^{(k)}} + \lambda_\alpha^{(k)}, \qquad V_\alpha = H - H_\alpha, \tag{8.80}$$

$$(G_\alpha(\zeta)\psi)(R_\alpha, \rho_\alpha) = \text{l.i.m.} \int G_\alpha(R_\alpha, \rho_\alpha; R_\alpha', \rho_\alpha'; \zeta)\, \psi(R_\alpha', \rho_\alpha') \prod_{k=1}^{n_\alpha} d\mathbf{R}_\alpha^{(k)'}\, d\mu_\alpha^{(k)}(\rho_\alpha^{(k)'}). \tag{8.81}$$

However, the fact that $V_\alpha \Phi^{(\alpha)}_{P_\alpha,\omega_\alpha}$ is not square-integrable even in the presence of two-body forces with integrable potentials invalidates the technique used in the two-body case. A further warning that something is seriously amiss with purely formal extrapolations from the two-body case is provided by the observation that due to the presence of different arrangement channels, (8.79) does not have in general unique solutions* even in the case where Theorem 8.2 assures us that $\Omega_\pm^{(\alpha)}$ exist on nontrivial initial domains $\mathsf{M}_\pm^{(\alpha)}$, and are unique.

All the previous observations are applicable to multichannel T matrices, to whose derivation we turn next.

* This can be easily established on a heuristic level by using the second resolvent equation to write for $\beta \neq \alpha$,

$$\begin{aligned}\Phi^{(\beta)(\pm\epsilon)}_{P'_\beta,\omega'_\beta} &= \pm i\epsilon(\lambda^{(\alpha)} - H \pm i\epsilon)^{-1}\, \Phi^{(\beta)}_{P'_\beta,\omega'_\beta} \\ &= \pm i\epsilon(\lambda^{(\alpha)} - H_\alpha \pm i\epsilon)^{-1}\Phi^{(\beta)}_{P'_\beta,\omega'_\beta} + (\lambda^{(\alpha)} - H_\alpha \pm i\epsilon)^{-1} V_\alpha \Phi^{(\beta)(\pm\epsilon)}_{P'_\beta,\omega'_\beta}.\end{aligned}$$

When $\epsilon \to +0$ we obtain

$$\Phi^{(\beta)(\pm)}_{P'_\beta,\omega'_\beta} = G^{(\alpha)}(\lambda^{(\alpha)} \pm i0) V_\alpha \Phi^{(\beta)(\pm)}_{P'_\beta,\omega'_\beta}$$

since $\Phi^{(\beta)}_{P'_\beta,\omega'_\beta}$ is not an eigenfunction of H_α, and therefore the first term on the right-hand side of the previous equation vanishes. Thus, $\Phi^{(\alpha)(\pm)}_{P_\alpha,\omega_\alpha} + a\Phi^{(\beta)(\pm)}_{P'_\beta,\omega'_\beta}$ is another solution of (8.79) for any $a \in \mathbb{C}^1$ and any $\beta \neq \alpha$ (see Lippmann [1956] and Epstein [1957] for further details).

According to (8.72), we have for $f \in \mathsf{M}_0^{(\alpha)}$ and $g \in \mathsf{M}_0^{(\beta)}$,

$$(8.82) \qquad \langle g \mid T^J f \rangle = \lim_{\epsilon \to +0} \langle g \mid T_{\beta\alpha}^{(1)}(\epsilon) f \rangle = \lim_{\epsilon \to +0} \langle g \mid T_{\beta\alpha}^{(2)}(\epsilon) f \rangle,$$

$$(8.83) \quad 2\pi i T_{\beta\alpha}^{(1)}(\epsilon) = (\Omega_{-\epsilon}^{(\beta)} - \Omega_{+\epsilon}^{(\beta)})^* \Omega_+^{(\alpha)}, \qquad 2\pi i T_{\beta\alpha}^{(2)}(\epsilon) = \Omega_-^{(\beta)*}(\Omega_{+\epsilon}^{(\alpha)} - \Omega_{-\epsilon}^{(\alpha)}).$$

By applying the same procedure as in §4.7, we formally obtain for $i = 1, 2$,

$$(8.84) \quad \langle g \mid T_{\beta\alpha}^{(1)}(\epsilon) f \rangle = \int d\nu_\beta(P_\beta', \omega_\beta')\, g^*(P_\beta', \omega_\beta') \int d\nu_\alpha(P_\alpha, \omega_\alpha) \times \delta_\epsilon(\lambda_\beta' - \lambda_\alpha)(\Phi_{P'_\beta, \omega'_\beta}^{(\beta)} \mid V_\beta \mid \Phi_{P_\alpha, \omega_\alpha}^{(\alpha)(+)})\, f(P_\alpha, \omega_\alpha),$$

$$(8.85) \quad \langle g \mid T_{\beta\alpha}^{(2)}(\epsilon) f \rangle = \int d\nu_\beta(P_\beta', \omega_\beta')\, g^*(P_\beta', \omega_\beta') \int d\nu_\alpha(P_\alpha, \omega_\alpha) \times \delta_\epsilon(\lambda_\beta' - \lambda_\alpha)(\Phi_{P'_\beta, \omega'_\beta}^{(\beta)(-)} \mid V_\alpha \mid \Phi_{P_\alpha, \omega_\alpha}^{(\alpha)})\, f(P_\alpha, \omega_\alpha),$$

where in accordance with the notation in (4.81) and (4.89) we have written

$$(8.86) \quad f(P_\alpha, \omega_\alpha) = (\Phi_{P_\alpha, \omega_\alpha}^{(\alpha)} \mid f \rangle = \text{l.i.m.} \int_{\mathbb{R}^{3n}} \hat{\Phi}_{P_\alpha, \omega_\alpha}^{(\alpha)*}(\mathbf{r}_1, ..., \mathbf{r}_n) f(\mathbf{r}_1, ..., \mathbf{r}_n)\, d\mathbf{r}_1 \cdots d\mathbf{r}_n,$$

$$(8.87) \quad (\Phi_{P'_\beta, \omega'_\beta}^{(\beta)} \mid V_\beta \mid \Phi_{P_\alpha, \omega_\alpha}^{(\alpha)(+)}) = \text{l.i.m.} \int \hat{\Phi}_{P'_\beta, \omega'_\beta}^{(\beta)*}(\mathbf{r}_1, ..., \mathbf{r}_n)\, V_\beta(\mathbf{r}_1, ..., \mathbf{r}_n)\, \hat{\Phi}_{P_\alpha, \omega_\alpha}^{(\alpha)(+)}(\mathbf{r}_1, ..., \mathbf{r}_n)\, d\mathbf{r}_1 \cdots d\mathbf{r}_n,$$

with remaining symbols being defined along similar lines. We note that whereas, e.g., $\Phi_{P_\alpha, \omega_\alpha}^{(\alpha)}$ was defined in (8.74) as a function of the center-of-mass position variables $\mathbf{R}_\alpha^{(k)}$ and additional internal motion variables $\rho_\alpha^{(k)}$ for each cluster k in the arrangement channel α, $\hat{\Phi}_{P_\alpha, \omega_\alpha}$ is a function of the position variables $\mathbf{r}_1, ..., \mathbf{r}_n$ of each of the individual particles $\mathfrak{S}_1, ..., \mathfrak{S}_n$ in the system—the transition from one set of variables to the other resulting from equations such as those in §5.2 of Chapter II, relating two such sets of variables. Naturally, if $\alpha = \beta$ then one can use in (8.87) the cluster variables $\mathbf{R}_\alpha^{(k)}$ and $\rho_\alpha^{(k)}$, making sure, however, to also express V_α in term of these variables.

It should be recalled now that, whereas in our treatment of the two-body case we had already eliminated in the initial stages of our considerations the center-of-mass variables, so that we have been working the Hilbert space $\mathscr{H}^{(1)}$ (see §1.7) of the reduced particle, in the present n-body case, $n \geqslant 3$, that reduction has not been performed since, in general, it does not prove to be as advantageous as in the $n = 2$ case. However, if there are no external forces [and therefore V as well as each V_α do not depend on the center-of-mass variables $\mathbf{R}$ of the entire system; cf. Exercises 5.1 and 5.3 in Chapter II], we can always easily dispose of three integrations in (8.87) by introducing $\mathbf{R}$ as one set of integration variables. Indeed, we only have to note that $\Phi^{(\alpha)}_{P_\alpha,\omega_\alpha}$ in (8.74) contains the product of exponential functions (8.75), from which the factor

$$(2\pi)^{-3/2} \exp\left(i\mathbf{R} \sum_{k=1}^{n_\alpha} \mathbf{P}_\alpha^{(k)}\right) \tag{8.88}$$

can be extracted, and the same holds true for $\Phi^{(\beta)}_{P'_\beta,\omega'_\beta}$. Hence, formally speaking, the $\mathbf{R}$-"integration" gives rise to a δ^3-"function," so that we can write

$$(\Phi^{(\beta)}_{P'_\beta,\omega'_\beta} \mid V_\beta \mid \Phi^{(\alpha)(+)}_{P_\alpha,\omega_\alpha}) = \delta^3\left(\sum_{j=1}^{n_\beta} \mathbf{P}_\beta^{(j)'} - \sum_{k=1}^{n_\alpha} P_\alpha^{(k)}\right) \langle P_\beta', \omega_\beta' \mid T^J_{(1)} \mid P_\alpha, \omega_\alpha \rangle, \tag{8.89}$$

$$(\Phi^{(\beta)(-)}_{P'_\beta,\omega'_\beta} \mid V_\alpha \mid \Phi^{(\alpha)}_{P_\alpha,\omega_\alpha}) = \delta^3\left(\sum_{j=1}^{n_\beta} \mathbf{P}_\beta^{(j)'} - \sum_{k=1}^{n_\alpha} \mathbf{P}_\alpha^{(k)}\right) \langle P_\beta', \omega_\beta' \mid T^J_{(2)} \mid P_\alpha, \omega_\alpha \rangle, \tag{8.90}$$

where the two *bona fide* functions of P_α, ω_α, P_β', ω_β' appearing on the right-hand sides of (8.89) and (8.90) are called, respectively the *"post" and "prior" versions of the on-shell T matrix* for the multichannel collision process in question. Naturally, the appearance of the δ^3-"functions" merely signifies that in deriving (8.84) and (8.85), the orders of all the P_α and R_α integrations could not be actually reversed, so that, for example, the l.i.m. in (8.87) had to be interpreted in the context of the last integral in (8.84): in that integral the integration in R_α and P_α in fact have to be restricted to compact sets in order to justify the reversal of integrations in R_α and P_α, and upon letting those sets tend to $\mathbb{R}^{3n_\alpha}$, Fourier theory (e.g., Theorem 4.5 in Chapter III) has to be invoked in relation to the $\mathbf{R}$-variables [the procedure being reminiscent to that leading from (5.9) to (5.10)]. Thus, the use of δ^3-"functions" could be avoided, but in physical literature their explicit appearance is favored due to their providing a graphic illustration of the conservation of the total momentum for the entire system.

In this same spirit of notational expediency, it is customary to combine the first operator relation in (8.72), i.e.,

$$S^J = 1^{\mathrm{ex}} - 2\pi i T^J, \tag{8.91}$$

with (8.82), (8.84), (8.85), (8.89), and (8.90) into a so-called S-matrix formula that generalizes (4.107):

$$\begin{aligned} \langle P_\beta', \omega_\beta' \mid S^J \mid P_\alpha, \omega_\alpha\rangle &= \delta_{\beta\alpha} \prod_{k=1}^{n_\alpha} \delta^3(\mathbf{P}_\alpha^{(k)\prime} - \mathbf{P}_\alpha^{(k)})\, \delta_{\omega_\alpha', \omega_\alpha} \\ &\quad - 2\pi i\, \delta(\lambda_\beta(P_\beta', \omega_\beta') - \lambda_\alpha(P_\alpha, \omega_\alpha)) \\ &\quad \times \delta^3\left(\sum_{j=1}^{n_\beta} \mathbf{P}_\beta^{(j)\prime} - \sum_{k=1}^{n_\alpha} \mathbf{P}_\alpha^{(k)}\right) \langle P_\beta', \omega_\beta' \mid T^J \mid P_\alpha, \omega_\alpha\rangle. \end{aligned} \tag{8.92}$$

Clearly, the first term on the right-hand side represents a symbolic variant of the formula (8.76) for 1^{ex}, and either one of the two versions of the on-shell T matrix in (8.89) and (8.90), respectively, can be used.

**8.8. Multichannel Born Approximations and Faddeev Equations

A "prior" version of the T matrix could have been introduced in the two-body potential scattering as well, but nothing new would have been gained due to the relationships

$$\Phi_{\mathbf{k}}^*(\mathbf{r}) = \Phi_{-\mathbf{k}}(\mathbf{r}), \qquad \Phi_{\mathbf{k}}^{(-)*}(\mathbf{r}) = \Phi_{-\mathbf{k}}^{(+)}(\mathbf{r}), \tag{8.93}$$

and to the presence of a single arrangement channel and therefore of a single interaction channel Hamiltonian, namely V. That this is not so in the multichannel case becomes clear as soon as we consider extrapolating the Born approximation method to the n-body case with $n \geqslant 3$.

Indeed, by repeatedly iterating the Lippmann–Schwinger equation (8.79) we obtain a Born series for retarded stationary states,

$$\Phi_{P_\alpha, \omega_\alpha}^{(\alpha)(+)} = \Phi_{P_\alpha, \omega_\alpha}^{(\alpha)} + \sum_{n=1}^{\infty} (G_\alpha(\lambda_\alpha + i0)V_\alpha)^n \Phi_{P_\alpha, \omega_\alpha}^{(\alpha)} \tag{8.94}$$

and by the same token, also a Born series for advanced stationary states,

$$\Phi^{(\beta)(-)}_{P'_\beta,\omega'_\beta} = \Phi^{(\beta)}_{P'_\beta,\omega'_\beta} + \sum_{n=1}^{\infty} (G_\beta(\lambda_\beta - i0)V_\beta)^n \Phi^{(\beta)}_{P'_\beta,\omega'_\beta} \tag{8.95}$$

Inserting these results in (8.89) and (8.90), respectively, we obtain "post" and "prior" Born series for the corresponding "post" and "prior" versions of the on-shell T matrix. We immediately see that in rearrangement collisions already the first post and prior Born approximations are going to be markedly different even on the energy shell, namely

$$(\Phi^{(\beta)}_{P'_\beta,\omega'_\beta} \mid V_\beta \mid \Phi^{(\alpha)}_{P_\alpha,\omega_\alpha}) \neq (\Phi^{(\beta)}_{P'_\beta,\omega'_\beta} \mid V_\alpha \mid \Phi^{(\alpha)}_{P_\alpha,\omega_\alpha}), \qquad \alpha \neq \beta, \tag{8.96}$$

so that computationally the two versions of the Born series can lead to significantly different approximation schemes.

Despite its wide usage in atomic and molecular physics, there is almost no knowledge of the convergence properties of the general Born series (8.94) and (8.95). The main difficulty stems from the fact that the Fredholm method of §§6.1–6.9 is no longer applicable already in the three-body case.

This can be easily verified, for example, for the channel $\alpha = \{\mathfrak{S}_1, \mathfrak{S}_2, \mathfrak{S}_3\}$ where, assuming two-body forces, we have by (8.2) and (8.80),

$$V_\alpha = V_{\{1,2,3\}} = H - H_0 = V_{12} + V_{23} + V_{13}. \tag{8.97}$$

It is convenient to work in the momentum representation so that assuming that $V_{ij}(\mathbf{r}) \in L_{(2)}(\mathbb{R}^3)$ we obtain

$$(V_{ij}\psi)^\sim(\mathbf{p}_{ij}, \mathbf{P}_{ij}, \mathbf{p}_k) = (2\pi)^{-3/2} \int_{\mathbb{R}^3} \tilde{V}_{ij}(\mathbf{p}_{ij} - \mathbf{p}'_{ij})\, \tilde{\psi}(\mathbf{p}'_{ij}, \mathbf{P}_{ij}, \mathbf{p}_k)\, d\mathbf{p}'_{ij} \tag{8.98}$$

by using center-of-mass coordinates for $\{\mathfrak{S}_i, \mathfrak{S}_j\}$ and noting that

$$\mathbf{r}_i\mathbf{p}_i + \mathbf{r}_j\mathbf{p}_j = \mathbf{R}_{ij}\mathbf{P}_{ij} + \mathbf{r}_{ij}\mathbf{p}_{ij}, \tag{8.99}$$

$$\mathbf{r}_{ij} = \mathbf{r}_i - \mathbf{r}_j, \qquad \mathbf{R}_{ij} = (m_i\mathbf{r}_i + m_j\mathbf{r}_j)/(m_i + m_j), \tag{8.100}$$

$$\mathbf{p}_{ij} = (m_i\mathbf{p}_j - m_j\mathbf{p}_i)/(m_i + m_j), \qquad \mathbf{P}_{ij} = \mathbf{p}_i + \mathbf{p}_j. \tag{8.101}$$

Hence, $G_\alpha(\zeta)V_\alpha = (\zeta - H_0)^{-1}V_{\{1,2,3\}}$ equals the sum in $i < j =$

1, 2, 3 of the following three operators:

$$(8.102)\quad (G_0(\zeta)\, V_{ij}\psi)^\sim(\mathbf{p}_{ij}, \mathbf{P}_{ij}, \mathbf{p}_k)$$
$$= (2\pi)^{-3/2} \int \frac{\tilde{V}_{ij}(\mathbf{p}_{ij} - \mathbf{p}'_{ij})}{\zeta - \left(\dfrac{\mathbf{p}_{ij}^2}{2m_{ij}} + \dfrac{\mathbf{P}_{ij}^2}{2M_{ij}} + \dfrac{\mathbf{p}_k^2}{2m_k}\right)} \tilde{\psi}(\mathbf{p}'_{ij}, \mathbf{P}_{ij}, \mathbf{p}_k)\, d\mathbf{p}'_{ij}.$$

The question is whether upon making the transition in (8.102) to the variables $\mathbf{p}_1$, $\mathbf{p}_2$, $\mathbf{p}_3$ [through the use of (8.101)] and then eliminating the center of mass motion of the three particles by factoring the Hilbert space $\mathscr{H}^{(3)} = L^2(\mathbb{R}^9)$ of the entire system into a tensor product $\mathscr{H}_c^{(1)} \otimes \mathscr{H}^{(2)}$ [representing an extension of (1.26) to the $n = 3$ case], we are going to end up with Hilbert–Schmidt operators. Now, working in the center-of-mass frame amounts to holding

$$(8.103)\qquad \mathbf{P} = \mathbf{p}_1 + \mathbf{p}_2 + \mathbf{p}_3 = \mathbf{P}_{ij} + \mathbf{p}_k$$

fixed, and by Theorem 8.12 in Chapter IV and Exercise 5.3, the resulting operator in $\mathscr{H}^{(2)}$ will be Hilbert–Schmidt if and only if it is an integral operator with square-integrable kernel. We see, however, from (8.102) that this will not be the case except if $\tilde{V}_{ij}(\mathbf{p}) = 0$ almost everywhere.

If $G(\zeta)$ and $G_0(\zeta)$ denote the Green's operators of H and H_0, respectively, in the present three-body case, then by Theorem 3.2,

$$(8.104)\qquad G(\zeta) = G_0(\zeta) + G_0(\zeta)(V_{12} + V_{13} + V_{23})\, G(\zeta),$$

and our earlier conclusion implies that we cannot apply at present the Fredholm method as we had been able to do in the two-body case due to Theorem 5.7. To circumvent this difficulty, Faddeev [1961a] rewrote (8.104) in the form

$$(8.105)\qquad \begin{aligned} G(\zeta) &= G_0(\zeta) + G^{(1)}(\zeta) + G^{(2)}(\zeta) + G^{(3)}(\zeta),\\ G^{(i)}(\zeta) &= G_0(\zeta)[V_{jk} + V_{jk}G_0(\zeta)\,\mathscr{T}^J_{\alpha\alpha}(\zeta)]\, G_0(\zeta), \qquad i \neq j \neq k, \end{aligned}$$

where $\mathscr{T}^J_{\alpha\alpha}(\zeta)$ is the three-body counterpart of (4.101) corresponding to incoming and outgoing arrangement channel $\alpha = \{\mathfrak{S}_1, \mathfrak{S}_2, \mathfrak{S}_3\}$. Indeed, in view of (3.61) and (4.102),

$$(8.106)\qquad \sum_{i=1}^{3} G^{(i)}(\zeta) = G_0(\zeta)[V + VG_0(\zeta)\,\mathscr{T}^J_{\alpha\alpha}(\zeta)]\, G_0(\zeta) = G_0(\zeta)\, VG(\zeta),$$

and we recover (8.104) from (8.105)—the whole procedure is reversible. We note that $\mathscr{T}^J_{\alpha\alpha}(\zeta)$ can be also viewed as a component of the $\mathscr{T}^J(\zeta)$

in (8.73),

(8.107) $$\mathscr{T}^J_{\alpha\alpha}(\zeta) = E^{(\alpha)}_- \mathscr{T}^J(\zeta)\, E^{(\alpha)}_+, \qquad \mathsf{M}^{(\alpha)}_{\pm} = \mathsf{M}^{(\alpha)}_0,$$

for the incoming and outgoing arrangement channel $\alpha = \{\mathfrak{S}_1, \mathfrak{S}_2, \mathfrak{S}_3\}$.

Following Faddeev [1961a, b, 1963], we can proceed by using the two-body problem to solve the three body one. Hence we introduce in $\mathscr{H}^{(3)}$ the $\mathscr{T}$-operators for the arrangement channel $\{\mathfrak{S}_i, \mathfrak{S}_j - \mathfrak{S}_k\}$,

(8.108) $$\mathscr{T}_{jk}(\zeta) = V_{jk} + V_{jk} G_{jk}(\zeta) V_{jk}, \qquad G_{jk}(\zeta) = (\zeta - H_0 - V_{jk})^{-1},$$

These $\mathscr{T}$-operators correspond to the physical process in which $\mathfrak{S}_j$ and $\mathfrak{S}_k$ collide, whereas $\mathfrak{S}_i$ is ever present but never interacts with either $\mathfrak{S}_j$ or $\mathfrak{S}_k$.

If we now write (8.106) in the form

(8.109) $$G^{(i)}(\zeta) = G_0(\zeta)\, \mathscr{T}^{(i)}_{\alpha\alpha}(\zeta)\, G_0(\zeta), \qquad \mathscr{T}^{(i)}_{\alpha\alpha}(\zeta) = V_{jk} + V_{jk} G_0(\zeta)\, \mathscr{T}(\zeta),$$

then in view of (4.103),

(8.110) $$\begin{aligned} \mathscr{T}^J_{\alpha\alpha}(\zeta) &= V_{12} + V_{12} + V_{23} + (V_{12} + V_{13} + V_{23})\, G_0(\zeta)\, \mathscr{T}^J_{\alpha\alpha}(\zeta) \\ &= \mathscr{T}^{(1)}_{\alpha\alpha}(\zeta) + \mathscr{T}^{(2)}_{\alpha\alpha}(\zeta) + \mathscr{T}^{(3)}_{\alpha\alpha}(\zeta). \end{aligned}$$

According to one of the second resolvent equations we have

(8.111) $$G_0(\zeta) = G_{jk}(\zeta) - G_{jk}(\zeta)\, V_{jk} G_0(\zeta),$$

and substituting this result into the defining expression of $\mathscr{T}^{(i)}_{\alpha\alpha}(\zeta)$ we get by (8.108),

(8.112) $$\begin{aligned} \mathscr{T}^{(i)}_{\alpha\alpha}(\zeta) &= V_{jk} + V_{jk} G_{jk}(\zeta)[\mathscr{T}^J_{\alpha\alpha}(\zeta) - V_{jk} G_0(\zeta)\, \mathscr{T}^J_{\alpha\alpha}(\zeta)] \\ &= V_{jk} + V_{jk} G_{jk}(\zeta)[\mathscr{T}^J_{\alpha\alpha}(\zeta) - \mathscr{T}^{(i)}_{\alpha\alpha}(\zeta) + V_{jk}] \\ &= \mathscr{T}_{jk}(\zeta) + V_{jk} G_{jk}(\zeta)[\mathscr{T}^J_{\alpha\alpha}(\zeta) - \mathscr{T}^{(i)}_{\alpha\alpha}(\zeta)]. \end{aligned}$$

Finally, taking into consideration that by (8.108) and (4.102),

(8.113) $$V_{jk} G_{jk}(\zeta) = \mathscr{T}_{jk}(\zeta)\, G_0(\zeta),$$

we arrive at the *Faddeev equations for* $\mathscr{T}^J_{\alpha\alpha}$*-operators*:

(8.114) $$\mathscr{T}^{(i)}_{\alpha\alpha}(\zeta) = \mathscr{T}_{jk}(\zeta) + \mathscr{T}_{jk}(\zeta)\, G_0(\alpha)[\mathscr{T}^{(j)}_{\alpha\alpha}(\zeta) + \mathscr{T}^{(k)}_{\alpha\alpha}(\zeta)],$$

$$i \neq j \neq k = 1, 2, 3.$$

The operators $\mathcal{T}^{(i)}_{\alpha\alpha}(\zeta)$ are of no direct physical significance, but if we can compute them, then by (8.110) we have computed $\mathcal{T}^{(i)}_{\alpha\alpha}(\zeta)$, and in turn by (8.105) and (8.106) we have also computed the full Green's operator. In fact, by (8.113),

$$G_{jk}(\zeta) - G_0(\zeta) = G_0(\zeta)\, V_{jk} G_{jk}(\zeta) = G_0(\zeta)\, \mathcal{T}_{jk}(\zeta)\, G_0(\zeta), \tag{8.115}$$

so that the combination of (8.109) and (8.114) yields Faddeev equations directly for the components $G^{(i)}(\zeta)$ of the full Green's operator (8.105):

$$G^{(i)}(\zeta) = G_{jk}(\zeta) - G_0(\zeta) + G_0(\zeta)\, \mathcal{T}_{jk}(\zeta)[G^{(j)}(\zeta) + G^{(k)}(\zeta)]. \tag{8.116}$$

In turn, proceeding heuristically, by (8.105) we can write a counterpart of the solution-type Lippmann–Schwinger equations for the advanced and retarded eigenfunction expansions in the $\alpha = \{\mathfrak{S}_1, \mathfrak{S}_2, \mathfrak{S}_3\}$ channel,

$$\begin{aligned} \Phi^{(\alpha)(\pm)}_{P_\alpha} &= \pm\lim_{\epsilon\to+0} i\epsilon G(\lambda_\alpha \pm i\epsilon)\, \Phi^{(\alpha)}_{P_\alpha} \\ &= \Phi^{(\alpha)}_{P_\alpha} + \sum_{i=1}^{3} \Phi^{(\alpha)(\pm)}_{(i)P_\alpha}, \end{aligned} \tag{8.117}$$

$$\Phi^{(\alpha)(\pm)}_{(i)P_\alpha} = \pm\lim_{\epsilon\to+0} i\epsilon G^{(i)}(\lambda_\alpha \pm i\epsilon)\, \Phi^{(\alpha)}_{P_\alpha}, \qquad P_\alpha = (\mathbf{p}_1, \mathbf{p}_2, \mathbf{p}_3), \tag{8.118}$$

and thus derive from (8.116) Faddeev equations for these advanced and retarded stationary states:

$$\Phi^{(\alpha)(\pm)}_{(i)P_\alpha} = \Phi^{(\alpha)(\pm)}_{(jk)P_\alpha} - \Phi^{(\alpha)}_{P_\alpha} + G_0(\lambda_\alpha \pm i0)\, \mathcal{T}_{jk}(\lambda_\alpha \pm i0)[\Phi^{(\alpha)(\pm)}_{(j)P_\alpha} + \Phi^{(\alpha)(\pm)}_{(k)P_\alpha}] \tag{8.119}$$

$$\Phi^{(\alpha)(\pm)}_{(jk)P_\alpha} = \pm\lim_{\epsilon\to+0} i\epsilon G_{jk}(\lambda_\alpha \pm i\epsilon)\, \Phi^{(\alpha)}_{P_\alpha}, \qquad \lambda_\alpha = \sum_{i=1}^{3} (\mathbf{p}_i^2/2m_i). \tag{8.120}$$

In view of the developments in §6.2, it is obvious that $\Phi^{(\alpha)(\pm)}_{(jk)P_\alpha}$ are in fact the distorted plane waves for the two-body scattering between $\mathfrak{S}_j$ and $\mathfrak{S}_k$, in which, however, the (free) center-of-mass motion has not been removed, and which in addition incorporates the presence of the (free) particle $\mathfrak{S}_i$:

$$\Phi^{(\alpha)(\pm)}_{(jk)P_\alpha}(\mathbf{r}_1, \mathbf{r}_2, \mathbf{r}_3) = \Phi_{\mathbf{p}_i}(\mathbf{r}_i)\, \Phi_{\mathbf{P}_{jk}}(\mathbf{R}_{jk})\, \Phi^{(\pm)}_{\mathbf{p}_{jk}}(\mathbf{r}_{jk}). \tag{8.121}$$

Faddeev [1965] has shown that for a large class of potentials $V_{ij}(\mathbf{r}_i - \mathbf{r}_j)$, the operators $G_0(\zeta)\mathcal{T}_{jk}(\zeta)$ in (8.114) and (8.116) are completely continuous at positive energies Re $\zeta \geqslant 0$, and that in fact the fifth powers of these operators are compact at all energies, so that the solutions of these equations can proceed by already well-established mathematical methods.

All these statements stay true for any pair of incoming and outgoing arrangement channels α and β, respectively. In fact, in general,

$$\mathscr{T}^J_{\beta\alpha}(\zeta) = E_-^{(\beta)}\mathscr{T}^J(\zeta)\, E_+^{(\alpha)} = V_\alpha + V_\beta(\zeta - K)^{-1}V_\alpha\,,$$

and the derivation of the Faddeev equations for all pairs of incoming and outgoing arrangement channels can proceed simultaneously through the use of the two-Hilbert space formalism of §8.6.

Yakubowskiĭ [1967] has extended the Faddeev technique in its basic aspects to the n-body problem for $n \geqslant 4$, so that in gross outlines, mathematically exact and yet computationally implementable methods are available for the treatment of nonrelativistic scattering of n (point) particles interacting via two-body short-range potentials.

*8.9. Appendix: von Neumann's Mean Ergodic Theorem

The following theorem was first derived by von Neumann [1932] in order to prove the quasi ergodic hypothesis of classical Hamiltonian mechanics, and it was used by us to obtain (8.35).

Theorem 8.3. Suppose A is a self-adjoint operator in the Hilbert space $\mathscr{H}$ and

$$U_t = e^{iAt} = \int_{\mathbb{R}^1} e^{it\lambda}\, dE_\lambda\,. \tag{8.122}$$

Then for any $f, g \in \mathscr{H}$

$$\lim_{t_2 - t_1 \to +\infty} \frac{1}{t_2 - t_1} \int_{t_1}^{t_2} \langle f \mid U_t g\rangle\, dt = \langle f \mid E(\{0\}) g\rangle, \tag{8.123}$$

irrespective of the mode in which $t_2 - t_1$ tends to infinity.

We note that there is a unique bounded linear operator $B(t_1\,, t_2)$ (a Bochner integral) for which the relation

$$\langle f \mid B(t_1\,, t_2) g\rangle = \frac{1}{t_2 - t_1} \int_{t_1}^{t_2} \langle f \mid U_t g\rangle\, dt \tag{8.124}$$

holds for all $f, g \in \mathscr{H}$ (see Exercise 8.7). Hence, (8.123) states that

$$E(\{0\}) = \operatorname*{w-lim}_{t_2-t_1\to+\infty} B(t_1, t_2).$$

In order to prove Theorem 8.3, we write

$$(8.125)\quad \langle f \mid B(t_1, t_2)g\rangle = \langle f \mid B(t_1, t_2)\, E(\{0\})g\rangle + \langle f \mid B(t_1, t_2)\, E(\mathbb{R}^1 - \{0\})g\rangle.$$

From (8.122) we get

$$\langle f \mid U_t E(\{0\})g\rangle = \int_{\mathbb{R}^1} e^{it\lambda}\, d\langle f \mid E_\lambda E(\{0\})g\rangle = \langle f \mid E(\{0\})g\rangle,$$

and consequently, we have

$$(8.126)\qquad \langle f \mid B(t_1, t_2)\, E(\{0\})g\rangle = \langle f \mid E(\{0\})g\rangle.$$

Thus, we see from (8.125) and (8.126) that (8.123) is true if and only if

$$(8.127)\qquad \lim_{t_2-t_1\to+\infty} \langle f \mid B(t_1, t_2)h\rangle = 0, \qquad h = E(\mathbb{R}^1 - \{0\})g.$$

We shall prove that (8.127) holds by showing that

$$\| B(t_1, t_2)h \|^2 = \frac{1}{(t_2 - t_1)^2} \int_{t_1}^{t_2} dt' \int_{t_1}^{t_2} dt'' \langle U_{t'}h \mid U_{t''}h\rangle$$

converges to zero when $t_2 - t_1 \to +\infty$.

Since $U_{t'}^* = U_{-t'}$, we get

$$\| B(t_1, t_2)h \|^2 = \frac{1}{(t_2 - t_1)^2} \int_{t_1}^{t_2} dt' \int_{t_1}^{t_2} dt'' \langle h \mid U_{t''-t'}h\rangle.$$

After introducing in the above integral the new variables $t = t'' - t'$ and $s = t' + t''$, we arrive at the following relations:

$$\begin{aligned}
\| B(t_1, t_2)h \|^2 &= \frac{1}{(t_2 - t_1)^2} \int_{-(t_2-t_1)}^{+(t_2-t_1)} dt \int_{2t_1+|t|}^{2t_2-|t|} ds\, \frac{\langle h \mid U_t h\rangle}{2} \\
&= \frac{1}{(t_2 - t_1)^2} \int_{-(t_2-t_1)}^{+(t_2-t_1)} (t_2 - t_1 - |t|)\langle h \mid U_t h\rangle\, dt \\
&= \frac{1}{(t_2 - t_1)^2} \int_{-(t_2-t_1)}^{+(t_2-t_1)} dt (t_2 - t_1 - |t|) \int_{\mathbb{R}^1} e^{it\lambda}\, d\| E_\lambda h \|^2.
\end{aligned}$$

Since the integral

$$\int_S | t_2 - t_1 - | t | | \, d \| E_\lambda h \|^2 \, dt, \qquad S = \mathbb{R}^1 \times [-(t_2 - t_1), +(t_2 - t_1)]$$

obviously exists, we can apply Fubini's theorem to interchange the order of integration in t and λ. Thus, we obtain, after carrying out the integration in t,

$$\begin{aligned} \| B(t_1, t_2)h \|^2 &= \frac{2}{(t_2 - t_1)^2} \int_{\mathbb{R}^1} \frac{1 - \cos(t_2 - t_1)\lambda}{\lambda^2} \, d \| E_\lambda h \|^2 \\ &= \int_{\mathbb{R}^1} \left(\frac{\sin \frac{1}{2}(t_2 - t_1)\lambda}{\frac{1}{2}(t_2 - t_1)\lambda} \right)^2 d \| E_\lambda h \|^2. \end{aligned}$$

Let us split the domain $\mathbb{R}^1$ of integration in three parts $(-\infty, -\eta]$, $(-\eta, +\eta]$, and $(\eta, +\infty)$, and majorize the integrand by 1 on $(-\eta, +\eta]$ and by $(\frac{1}{2}(t_2 - t_1)\eta)^{-2}$ on the other two intervals. Then we arrive at the estimate

$$\begin{aligned} (8.128) \quad \| B(t_1, t_2)h \|^2 &\leqslant \int_{(-\eta,\eta]} d \| E_\lambda h \|^2 + \frac{4}{(t_2 - t_1)^2 \eta^2} \int_{\mathbb{R}^1} d \| E_\lambda h \|^2 \\ &= \| (E_\eta - E_{-\eta})h \|^2 + \frac{4}{(t_2 - t_1)^2 \eta^2} \| h \|^2. \end{aligned}$$

By choosing η small enough, we can make the first term on the right-hand side of the above inequality smaller than $\epsilon/2$ for any *a priori* given $\epsilon > 0$; this is due to the fact that

$$\lim_{\eta \to +0} \| (E_\eta - E_{-\eta})h \| = \| E(\{0\})h \| = 0.$$

Then, for such a value of η, the second term in (8.60) can be made smaller than $\epsilon/2$ by choosing $| t_2 - t_1 |$ sufficiently large. Hence

$$\lim_{t_2 - t_1 \to +\infty} \| B(t_1, t_2)h \| = 0.$$

Thus, (8.127) holds, and therefore (8.123) is true.

Exercises

8.1. (a) How many arrangement channels can there be in the scattering process of three distinct particles?

(b) How many arrangement channels have actually been observed in the scattering of a proton p, a neutron n, and an electron e, i.e., for which channels α do we have $M_0^{(\alpha)} = \{0\}$?

(c) Give the conventional names of the fragments in all experimentally realizable arrangement channels in the scattering of n, p, and e interacting by means of nuclear and Coulomb forces.

8.2. (a) Count the arrangement channels in four-particle scattering in which there are only two pairs of distinct particles.

(b) How many of these channels are not empty in the scattering of two protrons and two neutrons (as far as our present experimental knowledge extends)?

8.3. Suppose that $f(t)$ is a continuous function for $t \in [0, +\infty)$, and that $\lim_{t\to+\infty} f(t) = a$ exists. Show that

$$\lim_{T\to+\infty} \frac{1}{T} \int_0^T f(t)\, dt = a.$$

8.4. Consider multichannel scattering with a finite number of arrangement channels α. Show that the operator

$$S' = \sum_{\alpha} \Omega_+^{(\alpha)} \Omega_-^{(\alpha)*}$$

commutes with e^{iHt} and with $E^H(B)$.

Remark. This operator was introduced by Jauch [1958b] as a candidate for a scattering operator in multichannel scattering. However, its inadequacy for this role is reflected in the fact that its knowledge is not sufficient to compute transition probabilities for scattering processes in which we have transitions between distinct arrangement channels (see also Exercise 8.6).

8.5. Show that if $\mathsf{R}_+^{(\alpha)} \perp \mathsf{R}_+^{(\beta)}$ and $\mathsf{R}_-^{(\alpha)} \perp \mathsf{R}_-^{(\beta)}$ for $\alpha \neq \beta$, and if in addition (8.41) holds, then

$$S'^* S' = S' S'^* = E_{\mathsf{R}}\,.$$

8.6. Show that if $\Psi_-(t)$ and $\Psi_+(t)$ have the respective strong asymptotic states $\Psi_-^{\text{in}}(t) = e^{-iH_\alpha t}\, \Psi_-^{\text{in}}(0)$ and $\Psi_+^{\text{out}}(t) = e^{-iH_\beta t}\, \Psi_+^{\text{out}}(0)$, then the transition amplitude $\langle \Psi_+(0) \mid \Psi_-(0)\rangle = \langle \Psi_+^{\text{out}}(0) \mid S_{\beta\alpha} \Psi_-^{\text{in}}(0)\rangle$ can be written in the form

$$\begin{aligned}\langle \Psi_+(0) \mid \Psi_-(0)\rangle &= \langle \Psi_+^{\text{out}}(0) \mid \Omega_+^{(\beta)*} S' \Omega_+^{(\alpha)} \Psi_-^{\text{in}}(0)\rangle \\ &= \langle \Psi_+^{\text{out}}(0) \mid \Omega_-^{(\beta)*} S' \Omega_-^{(\alpha)} \Psi_-^{\text{in}}(0)\rangle,\end{aligned}$$

provided that the conditions stipulated in the preceding exercise are satisfied, and that $\mathsf{M}_+^{(\beta)} = \mathsf{M}_-^{(\beta)}$.

8.7. Prove that for any family U_t of unitary operators defined by (8.122) there is a unique bounded operator $B(t_1, t_2)$ which satisfies (8.124) for all $f, g \in \mathscr{H}$.

References for Further Study

Quantum scattering theory is the subject of a multitude of textbooks and monographs, and its basic aspects are routinely treated in just about every single textbook on quantum mechanics. That being so, we shall mention only a few monographs especially well suited as supplementary reading for this chapter.

Newton [1966] treats all the principal aspects of the subject (concentrating, however, exclusively on physical ideas), and also provides an extensive bibliography covering most of the relevant material published by the mid-1960s. Taylor [1972] gives a very lucid exposition that incorporates as well as enhances the physicist's traditional approach to the subject, and he also supplies a mathematically careful presentation of complex-analysis methods in scattering theory. De Alfaro and Regge [1965] treat this last subject in detail, and are doing full justice to its mathematical aspects. Amrein, Jauch, and Sinha [1977] give a mathematically rigorous presentation of most of the basic aspects of quantum scattering theory, whereas Reed and Simon [1979] concentrate on an array of topics of special mathematical interest.

Mathematically closely related to the subject of scattering theory is perturbation theory for linear operators (see Kato [1966], Reed and Simon [1978]), as well as the theory of the structure of the spectra of Schroedinger operators (see Jörgens and Weidmann [1973]).

Titchmarsh [1962] provides a detailed treatment of the theory of eigenfunction expansions for second-order differential operators. A general treatment of the subject is supplied by Berezanskiĭ [1968, 1978], and his second book covers many topics of direct interest in quantum mechanics.

Some of the early papers laying the mathematical foundations of quantum scattering theory are those by Kodaira [1949], Povzner [1953, 1955], Jauch [1958a, b], Jauch and Zinnes [1959], Hack [1959], Zhislin [1960], Ikebe [1960, 1965], Green and Lanford [1960], Faddeev [1961a, b], Jordan [1962a, b], Grossmann and Wu [1962], Scadron *et al.* [1964], Hunziker [1964, 1965], and van Winter [1964, 1965]. Since the mid-1960s, the number of mathematical papers on scattering theory has been escalating at an ever-increasing rate. Hence the reader desiring to familiarize himself with more recent research in this area is well advised to consult published conference proceedings on the subject (such as those edited by La Vita and Marchand [1974], Nuttal [1978], and de Santo, Saenz, and Zachary [1980]), in addition to the references quoted in the appropriate context in §§3.8, 4.9, 4.10, 5.7, 8.6, and 8.8.

Hints and Solutions to Exercises

CHAPTER I

1.3. In order to prove that $\mathscr{C}^0(\mathbb{R}^1)$ is infinite dimensional note that all the polynomials x^n, $n = 1, 2, \ldots$, belong to $\mathscr{C}^0(\mathbb{R}^1)$ and are linearly independent.

1.5. $\mathscr{V}_S$ is the vector subspace of $\mathscr{V}$ which is a subspace of any other vector subspace containing S.

2.2. In order to prove that $\mathscr{C}^0_{(2)}(\mathbb{R}^1)$ is closed under the operation of vector addition in $\mathscr{C}^0(\mathbb{R}^1)$, apply the triangle inequality (Definition 2.2, point 4) on $\mathscr{C}^0([-a, +a])$, and then let $a \to +\infty$ in order to obtain

$$\left(\int_{-\infty}^{+\infty} |f(x) + g(x)|^2\, dx\right)^{1/2}$$
$$\leqslant \left(\int_{-\infty}^{+\infty} |f(x)|^2\, dx\right)^{1/2} + \left(\int_{-\infty}^{+\infty} |g(x)|^2\, dx\right)^{1/2} < +\infty$$

in case that $f(x), g(x) \in \mathscr{C}^0_{(2)}(\mathbb{R}^1)$.

2.3. Apply the Schwarz–Cauchy inequality on $\mathscr{C}^0([-a, +a])$, and then let $a \to \infty$ in order to prove that the integral

$$\int_{-\infty}^{+\infty} f^*(x)\, g(x)\, dx$$

is convergent.

3.1. In checking the transitivity condition make use of the triangle inequality.

3.2. Employ the triangle inequality to obtain

$$d(\xi_1, \xi_2) \leqslant d(\xi_1, \eta_1) + d(\eta_1, \eta_2) + d(\eta_2, \xi_2),$$

and then reverse the roles of ξ_1, ξ_2 and η_1, η_2.

3.3. Apply the triangle inequality first on ξ, ζ, η and then on ξ, η, ζ.

3.5. $|d(\xi, \eta) - d(\xi_n, \eta)| \leqslant d(\xi, \xi_n)$ according to Exercise 3.3.

4.7. If S is countable then $S \times S$ is also countable. If we have a countable number of countable sets S_1, S_2,..., their union $S_1 \cup S_2 \cup \cdots$ is also a countable set.

4.8. Show this for $l^2(n)$ by specializing the methods employed in proving Theorem 4.3.

4.9. In proving that $[S] = \overline{(S)}$, show that the closure $\overline{(S)}$ of a linear space (S) is also a linear space.

4.10. Derive $|\langle f \mid g \rangle - \langle f_n \mid g_n \rangle| \leqslant \|g\| \|f - f_n\| + \|f_n\| \|g - g_n\|$.

5.2. In order to prove that the above integral converges for any $f, g \in \mathscr{C}^1_{(2)}(\mathbb{R}^1)$ use the same hint as in solving Exercise 2.2.

5.5. Note that the distance $d(f, g)$ of any two distinct vectors $f, g \in \mathsf{T}$ is $\|f - g\| = \sqrt{2}$. Let $S = \{h_1, h_2, ...\}$ be a sequence dense in $\mathscr{H}$. We can map T in S by assigning to each $f \in \mathsf{T}$ a vector $h(f) \in S$ such that $d(f, h) = \|f - h\| < 1/\sqrt{2}$. This mapping between T and a subset of S is one to one, since $\|h(f) - h(g)\| > 0$ for $f \neq g$.

CHAPTER II

1.1. Consider R and S to be the subsets of $\mathscr{X} = R \cup S$. If T' denotes the complement of $T \subset \mathscr{X}$ with respect to $\mathscr{X}$, then

$$(R \cup S) - (R \triangle S) = (R \cup S) \cap (R \triangle S)'.$$

By applying Lemma 1.1 we get

$$(R \cup S) \cap (R \triangle S)' = (R \cap (R \triangle S)') \cup (S \cap (R \triangle S)').$$

Further applications of Lemmas 1.1 and 1.2 yield

$$R \cap (R \triangle S)' = R \cap S = S \cap (S \triangle R)'.$$

1.3. Prove first that if $R, S \in \mathscr{R}$, then $R - S$ and $R \triangle S$ are in $\mathscr{R}$. Then use the result of Exercise 1.1 to show that $R \cap S \in \mathscr{R}$.

1.4. Use precisely the same method as in proving Theorem 1.2.

1.5. Prove that

$$I_1 \cup \cdots \cup I_k = \left(I_1 - \bigcup_{m=2}^{k} I_m\right) \cup \left(I_2 - \bigcup_{m=3}^{k} I_m\right) \cup \cdots \cup I_k .$$

1.9. Note that $\bigcap_{k=1}^{\infty} S_k = (\bigcup_{k=1}^{\infty} S_k')'$.

2.4. Consider the case $B_k = [k, +\infty)$, when $\lim_{k\to\infty} B_k = \varnothing$ and $\mu_l(B_k) = +\infty$, $k = 0, 1, \ldots$.

3.1. Note that

$$\{\xi : f(\xi) \geqslant g(\xi)\} = \{\xi : f(\xi) < g(\xi)\}',$$
$$\{\xi : f(\xi) = g(\xi)\} = \{\xi : f(\xi) \geqslant g(\xi)\} - \{\xi : f(\xi) > g(\xi)\}.$$

3.3. Note that

$$\{\xi : \sup_{n=1,2,\ldots} f_n(\xi) > c\} = \bigcup_{n=1}^{\infty} \{\xi : f_n(\xi) > c\},$$

$$\{\xi : \inf_{n=1,2,\ldots} f_n(\xi) < c\} = \bigcup_{n=1}^{\infty} \{\xi : f_n(\xi) < c\},$$

and use the results of Exercise 3.2 and Lemma 3.2.

3.4. We can write

$$f(x) = 2\chi_{\mathfrak{R}_1}(x) + 4\chi_{\mathfrak{J}_1}(x)$$

where $\mathfrak{R}_1(\mathfrak{J}_1)$ is the set of all rational (irrational) numbers inside [0, 1]. Compute the integral from (3.7) by noting that $\mu_l^{(1)}(\mathfrak{R}_1) = 0$ because $\mathfrak{R}_1$ is a countable union of all sets $\{r\}$, $r \in \mathfrak{R}_1$, for which $\mu_l^{(1)}(\{r\}) = 0$, while $\mathfrak{J}_1 = [0, 1] - \mathfrak{R}_1$.

Note that the Riemann integral $\int_0^1 f(x)\,dx$ is not defined.

3.5. Note that if

$$s_1(\xi) = \sum_{i=1}^{n_1} a_i \chi_{R_i}(\xi), \qquad s_2(\xi) = \sum_{j=1}^{n_2} b_j \chi_{S_j}(\xi),$$

we can write $s_1(\xi)$ and $s_2(\xi)$ in the form

$$s_1(\xi) = \sum_{i=1}^{n_1} \sum_{j=1}^{n_2} a_i \chi_{R_i \cap S_j}(\xi),$$

$$s_2(\xi) = \sum_{i=1}^{n_1} \sum_{j=1}^{n_2} b_j \chi_{R_i \cap S_j}(\xi).$$

3.8. Note that for a real function $f(x) = f^+(x) - f^-(x)$ while $|f(x)| = f^+(x) + f^-(x)$.

3.9. Note that $f(\xi) - g(\xi) \geqslant 0$ for $\xi \in R$.

4.1. In proving transitivity, note that for any functions $f(x)$, $g(x)$, $h(x)$,

$$\{x: f(x) \neq h(x)\} \subset \{\{x: f(x) \neq g(x)\} \cup \{x: g(x) \neq h(x)\}\}.$$

4.2. To show that $\|f\| > 0$ for $f \not\equiv 0$, use Lemma 4.1.

4.4. Prove first that C_Ω^r is countable by noting that $\mathfrak{R} \times \cdots \times \mathfrak{R}$ ($2n$ times) is countable since $\mathfrak{R}$ is countable. Then show that $\mathfrak{D}_S$ is countable by using the fact that $\mathfrak{R} \times \mathfrak{R} \times C_\Omega^r$ is countable, and that $S_1 \cup S_2 \cup \cdots$ is denumerable whenever $S_1, S_2, \ldots$ are denumerable.

4.5. Note that $\mu(R \triangle S) = \int_\Omega |\chi_R(x) - \chi_S(x)| \, d\mu(x)$.

4.6. Show that the functions

$$\chi_{\{y\}}(x) = \begin{cases} 0 & \text{for } x \neq y \\ 1 & \text{for } x = y, \end{cases}$$

corresponding to all real-number y, constitute an uncountable orthonormal system in $L^2(\mathbb{R}^1, \mu)$. Note that for this measure μ, the Hilbert space $L^2(I, \mu)$ is nonseparable even when we choose I to be a finite interval.

4.7. Prove first the statement that $C_{\mathbb{R}^n}^r$ is a Boolean algebra of subsets of $\mathbb{R}^n$ by using the same technique as in proving Theorem 1.3.

4.8. It is obvious that $\mathscr{B}_{0,\Omega}^n \subset \mathscr{B}^n$, and consequently $\mathscr{A}_\sigma(\mathscr{B}_{0,\Omega}^n) \subset \mathscr{B}_\Omega{}^n$. Conversely, $\mathscr{I}_\Omega{}^n = \{I \cap \Omega : I \in \mathscr{I}^n\} \subset \mathscr{A}_\sigma(\mathscr{B}_{0,\Omega}^n)$ and therefore $\mathscr{B}_\Omega{}^n = \mathscr{A}_\sigma(\mathscr{I}_\Omega{}^n) \subset \mathscr{A}_\sigma(\mathscr{B}_{0,\Omega}^n)$. In fact, $I \cap \Omega \in \mathscr{A}_\sigma(\mathscr{B}_{0,\Omega}^n)$ for any $I \in \mathscr{I}^n$ since $I \cap \Omega$ can be written as a countable union of intervals contained in Ω. For example, for integer $k > 0$ and finite I divide each one of the edges of I into k equal parts and take the union S_k of all thus obtained subintervals of I which are also contained in Ω; then, in view of the fact that Ω is open, we obviously have $\bigcup_{k=1}^\infty S_k = I \cap \Omega$.

5.1. If we denote by $\mathbf{F}_{ik}$ the force with which the ith particle acts on the kth one, note that

$$\sum_{i=1}^{n} \mathbf{F}_{ik} = m_k \ddot{\mathbf{r}}_k , \qquad \mathbf{F}_{ii} = 0.$$

Employ afterwards the action–reaction principle (Newton's third law), according to which $\mathbf{F}_{ik} = -\mathbf{F}_{ki}$, in order to derive that $\mathbf{F} = \mathbf{0}$.

5.2. Note that since $\partial_t \mid \psi \mid^2 = (\partial_t \psi^*) \psi + \psi^* \partial_t \psi$ is square integrable and continuous, we deal with Riemann integration and by theorems of the calculus we have (justify)

$$\frac{d}{dt} \int_{\mathbb{R}^{3n}} \mid \psi \mid^2 d\mathbf{r}_1 \cdots d\mathbf{r}_n = \int_{\mathbb{R}^{3n}} \partial_t \mid \psi \mid^2 d\mathbf{r}_1 \cdots d\mathbf{r}_n .$$

Choose a finite domain D of $\mathbb{R}^{3n}$ of the form $D_1 \times \cdots \times D_n$ where $D_1 ,..., D_n$ are spheres centered at the origin, and denote with S_k the boundary of $D_k \subset \mathbb{R}^3$. By using (5.5) and Green's theorem, derive

$$\frac{d}{dt} \int_{D} \mid \psi \mid^2 d\mathbf{r}_1 \cdots d\mathbf{r}_n = i\hbar \sum_{k=1}^{n} \frac{1}{2m_k} \int_{S_k} [(\nabla_k \psi^*)\psi - \psi^* \nabla_k \psi] \, d\mathbf{S}_k .$$

The above integrals along the boundaries S_k vanish, when each S_k expands to infinity in all directions, provided $\mid \psi \nabla_k \psi \mid = O(r_k^{-\alpha_k})$, $\alpha_k > 2$.

5.3. We have $\partial V / \partial X = \sum_{k=1}^{n} (\nabla_k V)(\partial \mathbf{r}_k / \partial X)$; prove that $\partial \mathbf{r}_k / \partial X = \mathbf{e}_x$. Since similar results are valid for $\partial V / \partial Y$ and $\partial V / \partial Z$, one can write $\nabla_R V = \sum_{k=1}^{n} \nabla_k V = -\sum_{k=1}^{n} \mathbf{F}_k = -\sum_{i,k=1}^{n} \mathbf{F}_{ik}$ (see the notation in Exercise 5.1). Since $\mathbf{F}_{ik} = -\mathbf{F}_{ki}$ by the action–reaction principle, one gets $\nabla_R V = 0$.

6.2. Show that the mapping $(f_1 ,..., f_n) \to (f_{k_1} ,..., f_{k_n})$, $f_1 \in \mathscr{E}_1 ,...,$ $f_n \in \mathscr{E}_n$, provides an isomorphism between $\mathscr{E}_1 \oplus \cdots \oplus \mathscr{E}_n$ and $\mathscr{E}_{k_1} \oplus \cdots \oplus \mathscr{E}_{k_n}$.

6.3. Verify that the mapping $(f_1 , (f_2 , f_3)) \to (f_1 , f_2 , f_3)$, $f_1 \in \mathscr{E}_1$, $f_2 \in \mathscr{E}_2$, $f_3 \in \mathscr{E}_3$, provides an isomorphism between $\mathscr{E}_1 \oplus (\mathscr{E}_2 \oplus \mathscr{E}_3)$ and $\mathscr{E}_1 \oplus \mathscr{E}_2 \oplus \mathscr{E}_3$.

6.6. $\operatorname{Im} Q(f + g) = 0$ implies that $\operatorname{Im}(f \mid g) = -\operatorname{Im}(g \mid f)$ and $\operatorname{Im} Q(f + ig) = 0$ implies that $\operatorname{Re}(f \mid g) = \operatorname{Re}(g \mid f)$.

6.9. Note that for any $f, f_1 \in \mathscr{H}_1$, $g, g_1 \in \mathscr{H}_2$,

$$\| f \otimes g - f_1 \otimes g_1 \| \leqslant \| (f - f_1) \otimes g \| + \| f_1 \otimes (g - g_1) \|.$$

6.10. If $g \in \mathscr{H}_1$ and $g_1, g_2, \ldots \in \mathscr{V}_1$ converges in norm to g, then $\hat{g}_1, \hat{g}_2, \ldots$ converges in norm to some limit $\hat{g} \in \mathscr{H}_2$; $\hat{g}$ is independent of the chosen sequence converging to g. The mapping $g \to \hat{g}$ of $\mathscr{H}_1$ into $\mathscr{H}_2$ is linear and preserves the inner product. For every $\hat{h} \in \mathscr{H}_2$, there is a sequence $\hat{h}_1, \hat{h}_2, \ldots \in \mathscr{H}_2$ converging to $\hat{h}$, and consequently $\hat{h}$ is the image of the limit h of $h_1, h_2, \ldots \in \mathscr{V}_1$, where h_k is that element of $\mathscr{V}_1$ which is mapped into $\hat{h}_k$.

7.3. Use the ratio test which states that a series $\sum_{k=0}^{\infty} a_k$ converges if $\lim_{k\to\infty} | a_{k+1}/a_k | < 1$.

7.4. In order to show that A is not a Hermitian operator, take $f(r, \theta, \phi)$ everywhere twice continuously differentiable, e.g., $f(r) = e^{-r^2}$, and take $g(r) = (1/r)\, e^{-r^2}$.

7.6. Prove by induction that for $m < n$

$$\begin{aligned} \langle e^{-\rho/2}\rho^m \mid l_n \rangle &= \frac{1}{n!} \int_0^\infty e^{-\rho}\rho^m L_n(\rho)\, d\rho \\ &= (-1)^m \frac{m!}{n!} \int_0^\infty \frac{d^{n-m}}{d\rho^{n-m}} (\rho^n\, e^{-\rho})\, d\rho = 0, \end{aligned}$$

thus showing that $\langle l_k \mid l_n \rangle = 0$ for $k < n$.

Use the above result in the first step of the following derivation:

$$\begin{aligned} \| l_n \|^2 &= \int_0^\infty (l_n(\rho))^2\, d\rho = \left(\frac{1}{n!}\right)^2 \int_0^\infty (-1)^n\, \rho^n \frac{d^n}{d\rho^n} (\rho^n\, e^{-\rho})\, d\rho \\ &= \frac{1}{n!} \int_0^\infty \rho^n\, e^{-\rho}\, d\rho = 1, \end{aligned}$$

which shows that the Laguerre functions are normalized in $L^2([0, +\infty))$].

7.8. The substitution $x = e^{-\rho}$ yields

$$\int_0^\infty f^*(\rho)\, g(\rho)\, d\rho = \int_0^1 \frac{f^*(-\ln x)}{\sqrt{x}} \frac{g(-\ln x)}{\sqrt{x}}\, dx.$$

This shows that the mapping

$$f(\rho) \to f_1(x) = \frac{f(-\ln x)}{\sqrt{x}}$$

defines a unitary transformation of $L^2([0, +\infty))$ onto $L^2([0, 1])$. Since by Lemma 7.4 the family of all polynomials $p(x) = a_0 + a_1 x + \cdots + a_n x^n$ is dense in $L^2([0, 1])$, the family of all functions

$$e^{-\rho/2} p(e^{-\rho}) = e^{-\rho/2}(a_0 + a_1\, e^{-\rho} + \cdots + a_n\, e^{-n\rho})$$

is dense in $L^2([0, +\infty))$. Each $e^{-\rho/2}p(e^{-\rho})$ can be approximated arbitrarily well in the mean by a function of the form

$$b_0 l_0(\rho) + b_1 l_1(\rho) + \cdots + b_i l_i(\rho),$$

since it follows from the last result of Exercise 7.7 (by taking an appropriate s) that each $e^{-\rho/2}\, e^{-k\rho}$ can be approximated arbitrarily well in the mean by a linear combination of Laguerre functions.

7.9. The orthonormality can be deduced from the formula (see Chapter I, Exercise 5.4)

$$h_n(u) = (-1)^n\, e^{u^2/2} \frac{d^n}{du^n} (e^{-u^2})$$

by the methods employed in solving Exercise 7.6. To prove completeness, write for any $f \in L^2(\mathbb{R}^1)$

$$f(u) = f_+(u) + f_-(u), \qquad f_\pm(u) = \tfrac{1}{2}[f(u) \pm f(-u)].$$

Since we have, after the substitution $u = \sqrt{\rho}$,

$$\int_{-\infty}^{+\infty} f_\pm{}^*(u)\, g_\pm(u)\, du = 2\int_0^\infty f_\pm{}^*(u)\, g_\pm(u)\, du$$
$$= \int_0^\infty (\rho^{-1/4} f_\pm(\sqrt{\rho}))^* (\rho^{-1/4} g_\pm(\sqrt{\rho}))\, d\rho,$$

the mapping

$$f_\pm(u) \to \rho^{-1/4} f_\pm(\sqrt{\rho}), \qquad f(u) \in L^2(\mathbb{R}^1),$$

is a unitary transformation of the subspaces $L^2_\pm(\mathbb{R}^1)$ of even and odd functions, respectively, onto $L^2([0, +\infty))$. The completeness of $h_0(u)$, $h_1(u),\ldots$ in $L^2(\mathbb{R}^1)$ can be deduced from the completeness of $l_0(\rho)$, $l_1(\rho),\ldots$ in $L^2([0, +\infty))$.

CHAPTER III

1.2. Prove first that $(Kf)(x)$ is continuous on I for any given $f(x)$, $f \in L^2(I)$, and therefore that Kf is square integrable in I. Show next that if $f_1(x) = f_2(x)$ almost everywhere, then $(Kf_1)(x) = (Kf_2)(x)$ almost everywhere. Then check the linearity of the above mapping.

1.3. Derive, by applying the Schwarz–Cauchy inequality in $L^2(I)$, that

$$\left| \int_I K(x, y) f(y)\, dy \right|^2 \leqslant \left[\int_I |K(x, y)|^2\, dy \right] \left[\int_I |f(y)|^2\, dy \right].$$

By using Fubini's theorem (Chapter II, Theorem 3.13) derive from the above inequality

$$\langle Kf \mid Kf \rangle = \int_I dx \left| \int_I K(x, y) f(y)\, dy \right|^2$$

$$\leqslant \left[\int_I dx \int_I | K(x, y)|^2\, dy \right] \| f \|^2, \qquad \| f \|^2 = \int_I | f(y)|^2\, dy.$$

1.5. Apply B to a sequence of nonnegative functions $f_n(x) \in \mathscr{C}_b^\infty(\mathbb{R}^1)$, $n = 1, 2, \ldots$, with support inside $[-1/n, +1/n]$, for which

$$\| f_n \|^2 = \int_{-\infty}^{+\infty} | f_n(x)|^2\, dx = 1,$$

and show that $\| Bf_n \| \to \infty$ when $n \to \infty$.

2.5. Note that for fixed f, ϕ_f defined by $\phi_f(g) = (f \mid g)$ is a continuous linear functional. Use Riesz' theorem (Theorem 2.3) to establish that for each $f \in \mathscr{H}$ there is a vector $A(f)$ such that $(f \mid g) = \phi_f(g) = \langle A(f) \mid g \rangle$, and check that the mapping $f \to A(f)$ is a linear operator on $\mathscr{H}$. Use the boundedness and Hermiticity of $(f \mid g)$ to prove the boundedness (by taking $g = Af$) and Hermiticity of A.

3.3. If E projects on M, $\dim \mathsf{M} = m$, choose $\{e_1, \ldots, e_m\}$ and $\{e_{m+1}, \ldots, e_n\}$ to be orthonormal systems spanning M and $\mathsf{M}^\perp$, respectively.

4.1. If $\dim \mathscr{H} = n$, introduce an orthonormal basis $\{e_1, \ldots, e_n\}$ in $\mathscr{H}$. Prove that $\{Ve_1, \ldots, Ve_n\}$ is also an orthonormal basis, and, therefore, that $\mathscr{R}_V = \mathscr{H}$.

4.2. Use the procedure employed in proving Theorem 5.6 in Chapter II. The difference between that case and the present case is that Theorem 5.6 applies to $I = \mathbb{R}^1$, while at present $I = [a, b]$, and the norm to be taken is $\| f \| = \int_a^b | f(x)|\, dx$.

4.3. Write the equivalent of (4.8) for the general case, and apply Fubini's theorem (Chapter II, Theorem 3.13) in order to be able to proceed in the same fashion as for $n = 1$. For instance,

$$\int_\alpha^\infty dx_1 \int_{-\alpha}^{\alpha} dx_2, \ldots \int_{-\alpha}^{+\alpha} dx_m f(u + x) \frac{\sin \lambda x_1}{x_1} \cdots \frac{\sin \lambda x_m}{x_m}$$

$$= \int_\alpha^\infty dx_1 \frac{\sin \lambda x_1}{x_1} \int_{-\alpha}^{+\alpha} \cdots \int_{-\alpha}^{+\alpha} dx_2 \cdots dx_m f(u + x) \frac{\sin \lambda x_2}{x_2} \cdots \frac{\sin \lambda x_m}{x_m}$$

approaches zero when $\lambda \to \infty$, due to Lemma 4.1.

4.4. If $f_1, f_2, \ldots \in \mathscr{D}_A$ is a Cauchy sequence, then, due to the isometry of A, the sequence $Af_1, Af_2, \ldots$ is a Cauchy sequence also. Hence, if $g_1, g_2, \ldots \in \mathscr{D}_A$ converges to $g \in \mathscr{D}_B$, then $Ag_1, Ag_2, \ldots$ has a strong limit. The isometry of B can be deduced from the isometry of A and the continuity of B.

5.2. According to Exercise 5.1, the eigenvalue problem for A is equivalent to solving

$$\sum_{k=1}^{n} (A_{ik} - \lambda\delta_{ik})\, x_k = 0$$

in λ and $x_1, \ldots, x_n$. The eigenvalues λ are the roots of the characteristic polynomial of $\| A_{ik} \|$, i.e., they satisfy

$$\det \| A_{ik} - \lambda\delta_{ik} \| = 0,$$

thus being the roots a polynomial of the nth order, which are at most n in number. Explain why the vectors $f = x_1e_1 + \cdots + x_ne_n$, corresponding to all the solutions $x_1, \ldots, x_n$ of the matrix eigenvalue equation, span $\mathscr{H}$.

5.3. Note that $\| A \|$ is a norm in $\mathfrak{A}(\mathscr{H})$ (see Exercise 1.6), and, therefore (Chapter I, Theorem 3.1), every sequence convergent in the norm is a Cauchy sequence.

5.5. Note that

$$\Big| \| A(t) f \| - \| g \| \Big| \leqslant \| A(t) f - g \|$$

for any $g \in \mathscr{H}$.

5.6. Write $AB - A(t)\, B(t) = (A - A(t))B + A(t)(B - B(t))$ and note that

$$\|(AB - A(t)\, B(t)) f \| \leqslant \|(A - A(t))\, Bf \| + \| A(t)(B - B(t)) f \|$$

while $\| A(t)(B - B(t))f \| \leqslant c \,\|(B - B(t))f \|$.

5.7. See Chapter I, Exercise 1.4.

5.8. Use the continuity from above and below of the measures $\mu_f(B) = \langle f \mid E(B) f \rangle$, $f \in \mathscr{H}$, and the relation

$$\Big\| \langle f \mid (E(B) - E(B_n)) f \rangle \Big| = \|(E(B) - E(B_n)) f \|^2,$$

valid whenever $B_n \subset B$ or $B_n \supset B$.

5.9. First derive that $\| A^* \| \leqslant \| A \|$ from

$$\| A^*f \|^2 = \langle f \mid AA^*f \rangle \leqslant \| f \| \, \| AA^*f \| \leqslant \| A \| \, \| f \| \, \| A^*f \|,$$

and then reverse the roles of A and A^* in the above derivation.

5.10. Recall that the operator bound $\| A \|$ has all the properties of a norm. In proving the last of the assumed relations use the result of Exercise 5.9.

5.11. Note that $\| e_{m+1} - e_m \| = \sqrt{2}$, so that Cauchy's criterion is not satisfied. On the other hand, $\lim \langle f \mid e_n \rangle = 0$ for any $f \in \mathscr{H}$, since $\sum_{n=1}^{\infty} |\langle f \mid e_n \rangle|^2 \leqslant \| f \|^2$ by Lemma 4.1 of Chapter I.

6.1. Note that $(f \mid g) = \langle f \mid Ag \rangle$ is a positive-definite bilinear form; apply Theorem 6.4 of Chapter II.

6.2. The problem can be reduced to showing that if $C \geqslant 0$ and $C \leqslant 0$, where $C = C^*$, $\mathscr{D}_C = \mathscr{H}$, then $C = 0$. This follows from the fact that $\langle f \mid Cf \rangle \geqslant 0$ and $\langle f \mid Cf \rangle \leqslant 0$ imply $\langle f \mid Cf \rangle = 0$, $f \in \mathscr{H}$, and from the generalized Schwarz–Cauchy inequality of Exercise 6.1.

6.3. Note that for any $f, g \in \mathscr{H}$

$$\begin{aligned} \langle f \mid Ag \rangle &= \langle f \mid \operatorname*{s-lim}_{n\to\infty} A_n g \rangle = \lim_{n\to\infty} \langle f \mid A_n g \rangle \\ &= \lim_{n\to\infty} \langle A_n f \mid g \rangle = \langle Af \mid g \rangle. \end{aligned}$$

6.5. It is sufficient to consider the case of strictly positive polynomials $p(e^{i\varphi}) > 0$, $\varphi \in [0, 2\pi]$, since the more general case can be reduced to the above case by adding to the polynomial an $\epsilon > 0$, proving the existence of a polynomial $q_\epsilon(x)$ for which $| q_\epsilon(x)|^2 = p(x) + \epsilon$, and letting $\epsilon \to 0$. In case that $p(e^{i\varphi}) = \sum_{k=-n}^{n} c_k e^{ik\varphi} > 0$, the polynomial $P(z) = z^n p(z) = \sum_{k=0}^{2n} c_{k-n} z^k$ has no zeros on the unit circle $| z | = 1$. Moreover, we have

$$P(z) = [(z^*)^{2n} P(1/z^*)]^*$$

which shows that to each root α of $P(z)$ in the interior of the unit circle corresponds a root $1/\alpha^*$ in the exterior of the unit circle. Thus, $P(z)$ has the roots $\alpha_1, \ldots, \alpha_n$, $1/\alpha_1^*, \ldots, 1/\alpha_n^*$ and

$$P(z) = c' \prod_{k=1}^{n} (z - \alpha_k)(z - 1/\alpha_k^*).$$

Therefore

$$p(e^{i\varphi}) = c \prod_{k=1}^{n} (e^{i\varphi} - \alpha_k)(e^{-i\varphi} - \alpha_k^*),$$

where necessarily $c > 0$. Hence, one can take

$$q(e^{i\varphi}) = \sqrt{c} \prod_{k=1}^{n} (e^{i\varphi} - \alpha_k).$$

6.6. For each $\lambda \in I$, denote by $k(\lambda)$ the greatest integer for which

$$u_{k(\lambda)}(\lambda) - v_n(\lambda) < 1/n;$$

such an integer exists since $\lim_{n\to\infty} u_n(\lambda) = \lim_{n\to\infty} v_n(\lambda)$. Denote by $I(\lambda)$ the neighborhood of λ consisting of all points x for which

$$u_{k(\lambda)}(x) - v_n(x) < 1/n;$$

$I(\lambda)$ is nondegenerate since $u_{k(\lambda)}(x)$ and $v(x)$ are continuous. The family of all $I(\lambda)$ constitutes a covering of I. By Borel's covering theorem, any covering [in our case $\{I(\lambda)\}$] of a finite interval I has a finite subset [denote it by $I(\lambda_1),..., I(\lambda_n)$ in the present case] which also covers I. The smallest of the number $k(\lambda_1),..., k(\lambda_n)$ is the k_n we are seeking.

6.7. Start with $f(\xi)$ and $g(\xi)$ positive, and prove the statement by going to the original definition of integration and approximating these functions by simple functions. Proceed then to more complex cases by retracing the definitions of integration in Chapter II, §3.

6.8. Note that $Up(U) = p(U)U$ for any polynomial $p(U)$ in U, and then follow the construction of E_λ in Lemmas 6.1–6.3, using the fact that A s-lim B_k = (s-lim B_k)A.

6.10. Take $g = Af$ in (6.25), consider Exercise 6.7, and note that

$$\langle Af \mid E_\lambda f\rangle = \int_{\mathbb{R}^1} \mu \, d\langle E_\mu f \mid E_\lambda f\rangle = \int_{(-\infty,\lambda]} \mu \, d\langle f \mid E_\mu f\rangle.$$

CHAPTER IV

1.1. Note that for any $f, g \in \mathscr{H}$

$$\langle f \mid A_1 E_\mu^{(2)} g\rangle = \int_{\mathbb{R}^1} \lambda \, d\langle f \mid E_\lambda^{(1)}(E_\mu^{(2)} g)\rangle$$
$$= \int_{\mathbb{R}^1} \lambda \, d\langle E_\mu^{(2)} f \mid E_\lambda^{(1)} g\rangle = \langle E_\mu^{(2)} f \mid A_1 g\rangle.$$

1.2. The commutativity of the spectral functions can be derived from the commutativity of the Cayley transforms

$$V_k = (A_k - i)(A_k + i)^{-1}$$

by noting first that $[p_1(V_1), p_2(V_2)] = 0$ for any polynomials [(6.10) of Chapter III], and then retracing the construction of the spectral function $E_\lambda^{(k)}$ (Chapter III, Lemmas 6.1, 6.3) to obtain $[E_{\lambda_1}^{(1)}, E_{\lambda_2}^{(2)}] = 0$.

1.3. It is straightforward to verify that the statement is true for simple functions $f(x)$. Follow, step by step, the generalization of the concept of integral contained in Definitions 3.4 and 3.5 of Chapter II, to generalize the result to arbitrary μ_ρ-integrable functions $f(x)$.

1.4. Taking in Exercise 1.3 $\rho(x) = |\psi(x)|^2$ one derives

$$\int_{\mathbb{R}^n} x_k^2 \,|\psi(x)|^2\, d^n x = \int_{\mathbb{R}^n} x_k^2 \, d\mu_\rho(x) = \int_{\mathbb{R}^1} \lambda^2 \, d\langle \psi \mid E_\lambda^{(k)} \psi\rangle$$

so that $\mathscr{D}_{Q_k}$ coincides with the domain of the self-adjoint operator with the spectral measure $E^{(k)}(B)$. Furthermore, applying the decomposition (5.13) of Chapter III to the complex measure $\langle \psi_1 \mid E^{(k)}(B)\, \psi_2\rangle$, one reduces it to a linear combination of measures of the form $\langle \psi \mid E^{(k)}(B)\psi\rangle$ to which the result of Exercise 1.3 can be applied to obtain

$$\int_{\mathbb{R}^n} x_k \psi_1^*(x)\, \psi_2(x)\, d^n x = \int_{\mathbb{R}^1} \lambda \, d\langle \psi_1 \mid E_\lambda^{(k)} \psi_2\rangle.$$

1.5. Verify that the statement is true for Borel sets B in $\mathbb{R}^n$ of the form $B = B_1 \times \cdots \times B_n$, and therefore also in the Boolean algebra $\mathscr{B}^n$ generated by such sets. According to Theorem 5.5 of Chapter III, this implies that the two measures coincide on $\mathscr{B}^n$, which is the Boolean σ algebra generated by $\mathscr{B}^n$.

1.6. Verify that $U^{-1}E^A(B)U = E(B)$ is such that $\langle f \mid A_1 g\rangle = \int \lambda \, d\langle f \mid E_\lambda g\rangle$ for any $f \in \mathscr{H}$ and $g \in \mathscr{D}_{A_1}$, and that (cf. p. 225)

$$U^{-1}\mathscr{D}_A = \mathscr{D}_{A_1} = \left\{ g\colon \int_{\mathbb{R}^1} \lambda^2 \, d\| E_\lambda g \|^2 < +\infty \right\}.$$

2.1. Retrace Definitions 3.4 and 3.5 of integrals, as given in §3 of Chapter II.

2.2. Note that $S^{A_1} \times \cdots \times S^{A_n}$ is a closed set in $\mathbb{R}^n$, and consequently it is Borel measurable, so that the integration in (2.3) can be limited to $S^{A_1} \times \cdots \times S^{A_n}$.

2.4. Take A unbounded, so that $\mathscr{D}_A \neq \mathscr{H}$. For $F_1(\lambda) = -F_2(\lambda) = \lambda$ we have $\mathscr{D}_{F_1(A)} = \mathscr{D}_{F_2(A)} = \mathscr{D}_A$, whereas $\mathscr{D}_{(F_1+F_2)(A)} = \mathscr{H}$. For $F_2(\lambda) = 1/F_1(\lambda) = \lambda$ we have $\mathscr{D}_{F_1(A)F_2(A)} \subset \mathscr{D}_{F_2(A)} \neq \mathscr{H}$, whereas $\mathscr{D}_{(F_1 \cdot F_2)(A)} = \mathscr{H}$.

2.5. Use the result of Theorem 2.4 to show that $\langle f \mid F_1 F_2 g\rangle = \langle f \mid F_2 F_1 g\rangle$ for all $f,\ g \in \mathscr{H}$.

2.6. Show first that A is linear. $\mathscr{D}_A$ is dense since the set of vectors f for which $(E_{\lambda_2} - E_{\lambda_1})f = f$ for some real $\lambda_1 < \lambda_2$ is contained in $\mathscr{D}_A$ (because then the above integral converges for all $g \in \mathscr{H}$) and is dense in $\mathscr{H}$. Prove that $A^* \supseteq A$ and then show that $\mathscr{D}_{A^*} \subset \mathscr{D}_A$ by duplicating the method used in proving Theorem 2.6.

2.7. Use the result of Exercise 1.3 to establish that $\mathscr{D}_A \equiv \mathscr{D}_{F(Q_1,\ldots,Q_n)}$ and that $\langle \psi_1 \mid F(Q_1,\ldots,Q_n)\,\psi_2\rangle = \langle \psi_1 \mid A\psi_2\rangle$ for all $\psi_1 \in \mathscr{H}$ and all $\psi_2 \in \mathscr{D}_A$.

3.1. Write $f(x,t) = \text{s-lim}_{\Delta t \to 0}(1/\Delta t)[\psi(x, t+\Delta t) - \psi(x,t)]$. According to the mean-value theorem of differential calculus for $0 \leqslant \theta_1$, $\theta_2 \leqslant 1$

$$\delta(\Delta t) = \int_{\mathbb{R}^{3n}} \left| \frac{\psi(x, t+\Delta t) - \psi(x,t)}{\Delta t} - f(x,t) \right|^2 d^{3n}x$$

$$= \int_{\mathbb{R}^{3n}} | \operatorname{Re} \psi_t(x, t+\theta_1\,\Delta t) + i \operatorname{Im} \psi_t(x, t+\theta_2\,\Delta t) - f(x,t)|^2\, d^{3n}x.$$

If we consider, when letting $\Delta t \to 0$, a sequence $(\Delta t)_1$, $(\Delta t)_2$, $\ldots \to 0$, then by definition $\delta((\Delta t)_n) \to 0$ for $n \to +\infty$. Use Theorem 4.5 of Chapter II to derive

$$f(x,t) = \lim_{n\to\infty} \frac{\psi(x, t+(\Delta t)_n{}') - \psi(x,t)}{(\Delta t)_n{}'}$$

almost everywhere, where $(\Delta t)_1{}'$, $(\Delta t)_2{}'$,... is a subsequence of $(\Delta t)_1$, $(\Delta t)_2$,... . Since by continuity

$$\psi_t(x,t) = \lim_{n\to\infty} [\operatorname{Re}\psi_t(x, t+\theta_1(\Delta t)_n') + i \operatorname{Im} \psi_t(x, t+\theta_2(\Delta t)_n')],$$

the result $\psi_t = f$ follows.

3.2. Apply Parseval's formula in Theorem 4.6 of Chapter I to derive

$$(AB)_{ik} = \langle A^* e_i \mid Be_k\rangle = \sum_{j=1}^{\infty} \langle A^* e_i \mid e_j\rangle\langle e_j \mid Be_k\rangle = \sum_{j=1}^{\infty} A_{ij}B_{jk}\,.$$

3.4. Employ the identity

$$\frac{1}{\Delta t}\,(e^{(i/\hbar)H_0(t+\Delta t)}\;U(t+\Delta t, 0) - e^{(i/\hbar)H_0 t}\;U(t,0))\,\tilde{\Psi}(0)$$

$$= e^{(i/\hbar)H_0(t+\Delta t)}\,\frac{U(t+\Delta t,0) - U(t,0)}{\Delta t}\,\tilde{\Psi}(0)$$

$$+ \frac{1}{\Delta t}\,(e^{(i/\hbar)H_0(t+\Delta t)} - e^{(i/\hbar)H_0 t})\;U(t,0)\,\tilde{\Psi}(0),$$

and consult the methods employed in proving Theorem 3.2 to show that in the limit when $\Delta t \to 0$ the above vector-valued function of t converges strongly to

$$e^{(i/\hbar)H_0 t}\left(-\frac{i}{\hbar} H(t)\, U(t, 0)\, \tilde{\Psi}(0)\right) + \frac{i}{\hbar} H_0\, e^{(i/\hbar)H_0 t}\, U(t, 0)\, \tilde{\Psi}(0)$$

$$= -\frac{i}{\hbar} e^{(i/\hbar)H_0 t}\, V(t)\, U(t, 0)\, \tilde{\Psi}(0)$$

$$= -\frac{i}{\hbar} e^{(i/\hbar)H_0 t}\, V(t)\, e^{-(i/\hbar)H_0 t}\, \tilde{\Psi}(t) = -\frac{i}{\hbar} H_{\mathrm{I}}(t)\, \tilde{\Psi}(t).$$

3.5. Write $R_n(\lambda, t) = e^{i\lambda t} - \sum_{k=0}^{n} (i\lambda t)^k/k!$. According to Theorems 2.6 and 2.7,

$$\| R_n(A, t)f \|^2 = \langle f \mid R_n^*(A, t)\, R_n(A, t)f \rangle = \int_{[-\alpha,+\alpha]} | R_n(\lambda, t)|^2\, d\langle f \mid E_\lambda^A f \rangle.$$

Since it is easily shown by using Taylor's formula that (for fixed t) $| R_n(\lambda, t)| \to 0$ uniformly in $\lambda \in [-\alpha, +\alpha]$ when $n \to \infty$, it follows that $\| R_n(A, t)f \| \to 0$.

3.6. First note that $[(H_0 t)^n\, \psi](\mathbf{r})$ vanishes outside B_0 for all $t \in \mathbb{R}^1$. Hence, if the expansion were correct and the series were convergent in some sense, then we would have $e^{-(i/\hbar)H_0 t}\psi$ vanish outside B_0 at all times t. Since $e^{-(i/\hbar)H_0 t}$ is unitary, this would imply that

$$\int_{B_0} |(e^{-(i/\hbar)H_0 t}\, \psi)(\mathbf{r})|^2\, d\mathbf{r} = \int_{\mathbb{R}^3} |(e^{-(i/\hbar)H_0 t}\, \psi)(\mathbf{r})|^2\, d\mathbf{r} = 1$$

for all $t \in \mathbb{R}^1$.

4.1. For a vector of the form $f_1 \otimes \cdots \otimes f_n$ we have

$$\begin{aligned} &\|(A_1 \otimes \cdots \otimes A_n)(f_1 \otimes \cdots \otimes f_n)\| \\ &= \| A_1 f_1 \|_1 \cdots \| A_n f_n \|_n \\ &\leqslant \| A_1 \| \, \| f_1 \|_1 \cdots \| A_n \| \, \| f_n \|_n \\ &= \| A_1 \| \cdots \| A_n \| \, \| f_1 \otimes \cdots \otimes f_n \|. \end{aligned}$$

Extend the above inequality to $\mathscr{H}_1 \otimes \cdots \otimes \mathscr{H}_n$ by using an orthonormal basis of vectors $e_{i_1}^{(1)} \otimes \cdots \otimes e_{i_n}^{(n)}$ (see Chapter II, Theorem 6.10).

4.2. Use Theorem 6.10 in Chapter II, and take advantage of the manner in which $\mathscr{H}_1^{\otimes S n}$ and $\mathscr{H}_1^{\otimes A n}$ are constructed from the vectors (4.5) and (4.6), respectively.

4.3. Note that

$$\left\langle \sum_{(i_1,\ldots,i_n)} f_{i_1} \otimes \cdots \otimes f_{i_n} \,\middle|\, \sum_{(k_1,\ldots,k_n)} \pi(k_1,\ldots,k_n)\, g_{k_1} \otimes \cdots \otimes g_{k_n} \right\rangle$$

$$= \sum_{(k_1,\ldots,k_n)} \pi(k_1,\ldots,k_n) \left(\sum_{(i_1,\ldots,i_n)} \langle f_{i_1} \mid g_{k_1} \rangle_1 \cdots \langle f_{i_n} \mid g_{k_n} \rangle_n \right) = 0,$$

since the sum by which each $\pi(k_1,\ldots,k_n)$ is multiplied on the right-hand side has the same value for each $(k_1,\ldots,k_n)$.

4.5. Resort to Definition 4.1.

4.9. Show that $U_1 \otimes \cdots \otimes U_n$ is isometric on $\mathscr{H}_1' \otimes_a \cdots \otimes_a \mathscr{H}_n'$, due to the isometry of $U_1,\ldots,U_n$. To prove that the range of $U_1 \otimes \cdots \otimes U_n$ is $\mathscr{H}_2$, note that $\mathscr{H}_1' \otimes_a \cdots \otimes_a \mathscr{H}_n'$ is mapped *onto* $\mathscr{H}_1'' \otimes_a \cdots \otimes_a \mathscr{H}_n''$, and that the algebraic tensor product is dense in the Hilbert tensor product.

5.1. Prove first the statement for $L^2(\mathbb{R}^n, \mu)$ by following the procedure used in proving Theorem 5.6 of Chapter II; the only difference in the present case is that, since the boundary of an interval is not necessarily of μ-measure zero, the first part of the proof should be carried out not only for semiclosed finite intervals of the form (5.28) in Chapter II, but also for any finite closed interval (take all combinations of closed and open end points). Then prove the statement for any $\hat{L}^2(\mathbb{R}^n, \mu)$ by using the definition of symmetric and antisymmetric tensor products.

5.3. Let U_ν be the unitary transformation of $\mathscr{H}_\nu$ onto $L^2(\mathbb{R}^{n_\nu}, \mu_\nu)$ which is such that $A_r'^{(\nu)} = U_\nu A_r^{(\nu)} U_\nu^{-1}$, where $A_r'^{(\nu)}$ is the multiplication operator in Definition 5.2. Then $U = U_1 \otimes \cdots \otimes U_m$ (see Exercise 4.9) is a unitary transformation of $\mathscr{H}_1 \otimes \cdots \otimes \mathscr{H}_m$ onto

$$L^2(\mathbb{R}^{n_1}, \mu_1) \otimes \cdots \otimes L^2(\mathbb{R}^{n_m}, \mu_m),$$

having the property that $\tilde{A}_r'^{(\nu)} = U \tilde{A}_r^{(\nu)} U^{-1}$, where

$$\tilde{A}_r'^{(\nu)}(\psi_1^{(1)} \otimes \cdots \otimes \psi_r^{(\nu)} \otimes \cdots \otimes \psi_{k_m}^{(m)})$$
$$= \psi_1^{(1)} \otimes \cdots \otimes (x_r^{(\nu)} \psi^{(\nu)}{}_r) \otimes \cdots \otimes \psi_{k_m}^{(m)}.$$

According to Theorem 6.9 of Chapter II, there is a unitary transformation $\hat{U}$ of $L^2(\mathbb{R}^{n_1}, \mu_1) \otimes \cdots \otimes L^2(\mathbb{R}^{n_m}, \mu_m)$ onto $L^2(R^{n_1+\cdots+n_m}, \mu_1 \times \cdots \times \mu_m)$ such that $\psi_1^{(1)} \otimes \cdots \otimes \psi_{k_m}^{(m)}$ is mapped into

$$\psi_1^{(1)} \cdots \psi_{k_m}^{(m)} \in L^2(\mathbb{R}^{n_1+\cdots+n_m}, \mu_1 \times \cdots \times \mu_m).$$

The operator $(\hat{U}U)\,\tilde{A}_r^{(\nu)}\,(\hat{U}U)^{-1}$ is the canonical form of $\tilde{A}_r^{(\nu)}$ in $L^2(\mathbb{R}^{n_1+\cdots+n_m}, \mu_1 \times \cdots \times \mu_m)$.

5.4. Apply Theorem 2.7 to answer the second part of the exercise.

5.5. Use the spectral theorem, and in particular (6.24) of Chapter III, to find the domains of $AE(R)$ and $E(R)A$, and afterwards make use of Theorem 2.7 to derive that $\langle g \mid AE(R)f\rangle = \langle g \mid E(R)Af\rangle$ for all $f \in \mathscr{D}_A$ and $g \in \mathscr{H}$.

6.3. Note that it is sufficient to prove the statement for $t_0 = 0$. Using Theorems 2.2 and 2.4, one derives (cf. (3.11))

$$\|(e^{iAt}-1)f\|^2 = \int_{\mathbb{R}^1} |\, e^{i\lambda t} - 1\,|^2\, d\langle f \mid E_\lambda^A f\rangle, \qquad f \in \mathscr{H},$$

where $|\, e^{i\lambda t} - 1\,|$ converges pointwise to zero when $t \to 0$ and is bounded by 2. Hence, Lebesgue's bounded convergence theorem (Lemma 3.1) can be applied.

6.4. Note that since $U^t(U^{t_0})^{-1} = U^{t-t_0}$, it is sufficient to prove the statement for $t_0 = 0$. Use the decomposition (6.1) of Chapter III, to establish that

$$\|(U^t - 1)f\|^2 \leqslant \lim \sum_k |\, e^{it\mu_k} - 1\,|^2\, \|(E_{\lambda_k} - E_{\lambda_{k-1}})f\|^2 \leqslant \epsilon^2 \|f\|^2$$

if t is chosen sufficiently small so that $|\, e^{it\mu} - 1\,| \leqslant \epsilon$ for all $\mu \in [0, 2\pi]$. Hence, $\|\, U^t - 1\,\| \leqslant \epsilon$ for all such t.

6.5. Show that $\hat{E}_\lambda$ commutes with E_n by proving that $U(2\pi)$ commutes with E_n. Then check that all the requirements of Definition 5.5 in Chapter III are satisfied.

6.6. Approximate $f(\lambda)$ within $[-n, n]$, within the limits of accuracy $1/n$, by a trigonometric polynomial of period $2n$. When $n \to \infty$, $p_1(\lambda)$, $p_2(\lambda)$,... converges to $f(\lambda)$.

If $f(\lambda) = \chi_I(\lambda)$, where I is finite, note that $\chi_I(\lambda)$ can be approximated pointwise arbitrarily well by continuous functions of compact support.

6.8. Choose a finite interval I. The function $\chi_I(x)\,\rho(x)$ is square integrable on $\mathbb{R}^n$. Select a monotonically decreasing sequence of continuous functions $h_1(x)$, $h_2(x)$,... converging pointwise to $\chi_I(x)$. Then one obtains $\int |F(x)|\,|\,h_n(\lambda) - \chi_I(x)|\,|\,\rho(\lambda)|\,dx \to 0$, i.e., $\int F(x)\,\chi_I(x)\,\rho(x)\,dx = 0$ for any continuous function $F(x)$ of compact support. Since $\mathscr{C}_b^0(\mathbb{R}^n)$ is dense in $L^2(\mathbb{R}^n)$, one concludes that $\chi_I(x)\,\rho(x) = 0$ almost everywhere.

Consequently, if $I_1, I_2, \ldots$ are finite disjoint intervals and $\bigcup I_n = \mathbb{R}^n$, we have $\rho(x) = \sum_n \chi_{I_n}(x)\,\rho(x) = 0$ almost everywhere.

6.9. Show first that $(e^{iQv}\psi)(x) = e^{ixv}\psi(x)$ by proving that (2.3) is satisfied when $n = 1$, $A_1 = Q$, $(E^Q(B)\,\psi)(x) = \chi_B(x)\,\psi(x)$. Then use the relation $\exp(iPu) = \exp(-iU_F^{-1}QU_F u) = U_F^{-1}\exp(-iQu)\,U_F$, to establish that $(e^{iPu}\psi)(x) = \psi(x + \hbar u)$.

7.1. Observe that if $\langle\cdot \mid \cdot\rangle_i'$ denotes the inner product in $\mathscr{H}_i'$ and $\langle\cdot \mid \cdot\rangle_j''$ denotes the inner product in $\mathscr{H}_j''$, then

$$\left(\sum_{i=1}^{m} \langle f_i' \mid g_i'\rangle_i'\right)\left(\sum_{j=1}^{n} \langle f_j'' \mid g_j''\rangle_j''\right) = \sum_{i=1}^{m}\sum_{j=1}^{n} \langle f_i' \mid g_i'\rangle_i' \langle f_j'' \mid g_j''\rangle_j''.$$

7.2. Apply to

$$\sum_{i,j=1}^{\infty} \langle f_i' \mid g_i'\rangle_i' \langle f_j'' \mid g_j''\rangle_j''$$

the theorem on the absolute convergence of double series [Randolph, 1968, Sec. 3–11, Theorem 3, p. 162], which states that $\sum_{i,j=1}^{\infty} a_{ij}$ converges absolutely if and only if $\sum_{j=1}^{\infty} a_{ij} = b_i$ and $\sum_{i=1}^{\infty} b_i$ converge absolutely, and in that case $\sum_{i,j=1}^{\infty} a_{ij} = \sum_{i=1}^{\infty} b_i$.

7.3. Note that $\sum_{k=1}^{\infty} \langle f_k \mid g_k\rangle_k = \sum_{k=1}^{\infty} \langle U_k f_k \mid U_k g_k\rangle_k$, and show that the range of U is $\mathscr{H}_2$.

7.4. Start by proving the assertions for appropriately symmetrized functions of the form (7.6), and then, using the fact that the set of such functions is dense in the considered subspace, extend the result to the entire subspace.

7.5. Show that the spectral representation space $L^2(\mathbb{R}^4, \mu)$ of $\{Q^{(x)}, Q^{(y)}, Q^{(z)}, S^{\mathrm{n}}\}$ is also a spectral representation space of $\{P^{(x)}, P^{(y)}, P^{(z)}, S^{\mathrm{n}}\}$ by showing that

$$(P^{(x)}\psi)^{\sim}(p_1, p_2, p_3, s) = p_1\tilde{\psi}(p_1, p_2, p_3, s),$$

with similar formulas for $P^{(y)}$ and $P^{(z)}$, where $\tilde{\psi}$ is the Fourier–Plancherel transform of ψ at each fixed $s = -\sigma, \ldots, +\sigma$.

7.6. Obviously $A^{***} \supseteq A^*$. If $f \in \mathscr{D}_{A^{***}}$ and $f^* = A^{***}f$, we must have $\langle f^* \mid g\rangle = \langle f \mid A^{**}g\rangle$ for all $g \in \mathscr{D}_{A^{**}}$, and therefore also for all $g \in \mathscr{D}_A$. Hence, $f \in \mathscr{D}_{A^*}$.

7.7. Compute first

$$\frac{1}{(2\pi\hbar)^{n/2}} \int_{\mathbb{R}^{3n}} \exp[-(i/\hbar)(\mathbf{r}_1\mathbf{p}_1 + \cdots + \mathbf{r}_n\mathbf{p}_n - \tfrac{1}{2}(\mathbf{r}_1^2 + \cdots + \mathbf{r}_n^2))]\, d\mathbf{r}_1 \cdots d\mathbf{r}_n$$
$$= \exp[-(1/2\hbar)(\mathbf{p}_1^2 + \cdots + \mathbf{p}_n^2)].$$

Then observe that the Fourier transform of $P(\mathbf{r}_1, \ldots, \mathbf{r}_n) f_0(\mathbf{r}_1, \ldots, \mathbf{r}_n)$ is $P(-(i/\hbar)\nabla_1, \ldots, -(i/\hbar)\nabla_n)\tilde{f}_0(\mathbf{p}_1, \ldots, \mathbf{p}_n)$ in case that f_0 is continuous and of faster than polynomial decrease at infinity (these conditions are obviously satisfied by $f_0 = \exp[(-1/2\hbar)(\mathbf{r}_1^2 + \cdots + \mathbf{r}_n^2)]$).

7.8. Introduce the new variables $\mathbf{r} = \mathbf{r}_1 - \mathbf{r}_2$, $\boldsymbol{\rho} = \mathbf{r}_1 + \mathbf{r}_2$, and apply Theorem 3.13 of Chapter II to write

$$\int \left| V(\mathbf{r}) f\left(\frac{\boldsymbol{\rho} + \mathbf{r}}{2}, \frac{\boldsymbol{\rho} - \mathbf{r}}{2}\right)\right|^2 d\mathbf{r}\, d\boldsymbol{\rho}$$
$$= \int d\mathbf{r} \mid V(\mathbf{r})|^2 \int \left| f\left(\frac{\boldsymbol{\rho} + \mathbf{r}}{2}, \frac{\boldsymbol{\rho} - \mathbf{r}}{2}\right)\right|^2 d\boldsymbol{\rho}.$$

Then treat the integration in $\mathbf{r}$ separately for $r \leqslant R$ and $r > R$.

7.9. Establish that

$$\sum_{k=0}^{\infty} (-A)^k = \operatorname*{u\text{-}lim}_{n\to\infty} \left[1 + \sum_{k=1}^{n} (-A)^k\right]$$

converges uniformly, and that

$$(1 + A) \sum_{k=0}^{\infty} (-A)^k = \left(\sum_{k=0}^{\infty} (-A)^k\right)(1 + A) = 1.$$

8.1. An infinite orthonormal sequence $\{e_1, e_2, \ldots\}$ of vectors from M is bounded but it has no convergent subsequence since $\| e_m - e_n \| = \sqrt{2}$ for $m \neq n$.

8.2. If $\lambda_0 \in S_c^A$ then $E^A(I) \neq 0$ for every interval containing λ_0. Since $E^A(\{\lambda_0\}) = 0$, this implies that every neighborhood contains a point of the spectrum. In fact, if some closed neighborhood I_0 of λ_0 does not contain any points of the spectrum besides λ_0, then for every $\lambda \in I_0$ we could find a subinterval I_λ of I_0, such that $E^A(I_\lambda) = 0$. Use Borel's covering theorem to establish the existence of a finite number $I_{\lambda_1}, \ldots, I_{\lambda_n}$ of such intervals, with the property that $I_0 = I_{\lambda_1} \cup \cdots \cup I_{\lambda_n}$, and, consequently, $E^A(I_0) \leqslant E^A(I_{\lambda_1}) + \cdots + E^A(I_{\lambda_n}) = 0$.

8.3. If $\lambda_1, \lambda_2, \ldots$ are the eigenvalues of A, it follows from the spectral theorem (see Chapter III, Theorem 6.4) that

$$A = \underset{n\to\infty}{\text{u-lim}} \sum_{k=1}^{n} \lambda_k E^A(\{\lambda_k\}).$$

8.5. Assume that the Borel set B_0 of negative numbers with $\lambda_0 = \sup\{\lambda : \lambda \in B_0\} < 0$ is a subset of the spectrum of A. Then $E^A(B_0) \neq 0$, i.e., $E^A(B_0) f_0 = f_0$ for some $f_0 \neq 0$. Show that

$$\langle f \mid Af \rangle = \int_{B_0} \lambda \, d \| E_\lambda^A f \|^2 \leqslant \lambda_0 \int_{B_0} d \| E_\lambda^A f \|^2 \leqslant \lambda_0 \| f \|^2 < 0.$$

8.6. Build an orthonormal basis in $\mathscr{H}$ by selecting orthonormal bases in $E([0, \infty))\mathscr{H}$ and $E((-\infty, 0))\mathscr{H}$ and taking their union. Note that the unconditional convergence of the series in (8.5) implies its absolute convergence, from which the existence of $\mathrm{Tr}(A^{(+)} + A^{(-)})$ can be deduced.

8.7. Apply the results of Exercise 5.10 in Chapter III to $\{A_n B_n\}$ and then to $\{(A_n B_n) C_n\}$.

CHAPTER V

1.1. In deriving $|\mathbf{r}(t)| - t \mid \mathbf{v}^{\text{ex}} \mid \to \mathbf{r}^{\text{ex}}\mathbf{v}^{\text{ex}}/|\, \mathbf{v}^{\text{ex}} \,|$, note that if one sets $|\mathbf{r}(t)| = \mid \mathbf{r}^{\text{ex}} + \mathbf{v}^{\text{ex}} t \mid + \eta^{\text{ex}}(t)$, then $\eta^{\text{ex}}(t) \to 0$. In fact, if $\mid t \mid$ is large enough,

$$\begin{aligned} \mid \mathbf{r}(t) \mid - t \mid \mathbf{v}^{\text{ex}} \mid &= \eta^{\text{ex}}(t) + (\mid \mathbf{r}^{\text{ex}} \mid^2 + 2t\mathbf{r}^{\text{ex}}\mathbf{v}^{\text{ex}} + t^2 \mid \mathbf{v}^{\text{ex}} \mid^2)^{1/2} - t \mid \mathbf{v}^{\text{ex}} \mid \\ &= \eta^{\text{ex}}(t) + t \mid \mathbf{v}^{\text{ex}} \mid \left[\left(1 + \frac{2}{t} \frac{\mathbf{r}^{\text{ex}}\mathbf{v}^{\text{ex}}}{\mid \mathbf{v}^{\text{ex}} \mid^2} + \frac{\mid \mathbf{r}^{\text{ex}} \mid^2}{t^2 \mid \mathbf{v}^{\text{ex}} \mid^2} \right)^{1/2} - 1 \right] \\ &= \eta^{\text{ex}}(t) + t \mid \mathbf{v}^{\text{ex}} \mid \left(\frac{1}{2} \left(\frac{2}{t} \frac{\mathbf{r}^{\text{ex}}\mathbf{v}^{\text{ex}}}{\mid \mathbf{v}^{\text{ex}} \mid^2} + \frac{1}{t^2} \frac{\mid \mathbf{r}^{\text{ex}} \mid^2}{\mid \mathbf{v}^{\text{ex}} \mid^2} \right) + o\left(\frac{1}{t^2} \right) \right) \\ &\to \frac{\mathbf{r}^{\text{ex}}\mathbf{v}^{\text{ex}}}{\mid \mathbf{v}^{\text{ex}} \mid} \qquad \text{for} \quad t \to \mp\infty. \end{aligned}$$

1.2. Recall that for the given potential the general solution for a $\mathbf{r}(t)$ which is not a bound state is a hyperbola or parabola. If $r(t_0) = r_0$ is the minimum of $r(t) = \mid \mathbf{r}(t) \mid$, then $dr/dt \lesseqgtr 0$ for $t \lesseqgtr t_0$. Hence, $r(t)$ has an inverse in $(-\infty, t_0]$ and in $[t_0, +\infty)$. Since

$$\tfrac{1}{2} m_0 \mathbf{v}^2 + 1/r = \tfrac{1}{2} m_0 (\dot{r}^2 + r^2 \dot{\phi}^2) + 1/r = \text{const} \qquad \text{and} \qquad r^2 \dot{\phi} = \text{const},$$

we have

$$\dot{r} = \left[\frac{2}{m_0}\left(b - \frac{1}{r} - \frac{c^2}{r^2}\right)\right]^{1/2},$$

where b and c are constants. Consequently, setting

$$r = \frac{1}{\sqrt{b}}\left(u + \frac{1}{2\sqrt{b}}\right) \quad \text{and} \quad a = \sqrt{\tfrac{1}{4} + bc^2}\,,$$

we obtain

$$t - t_0 = \int_{r_0}^{r} \frac{dr}{\sqrt{\dfrac{2}{m_0}\left(b - \dfrac{1}{r} - \dfrac{c^2}{r^2}\right)}}$$

$$= \left[\sqrt{\frac{m_0}{2b^2}}\,(u^2 - a^2)^{1/2} + \frac{1}{2}\sqrt{\frac{m_0}{2b^3}}\ln|u + \sqrt{u^2 - a^2}|\right]_{u_0}^{u}.$$

Using the above result and recalling that $r(t) \to +\infty$ when $t \to \mp\infty$, we get $r(t)/t \to (2b/m_0)^{1/2}$, so that $|\mathbf{v}^{\text{ex}}|^2 = 2b/m_0$ according to Exercise 1.1. Show that $r(t) - t|\mathbf{v}^{\text{ex}}| = r(t) - t(2b/m_0)^{1/2}$ diverges, instead of converging to $\mathbf{r}^{\text{ex}}\mathbf{v}^{\text{ex}}/|\mathbf{v}^{\text{ex}}|$, which should happen (according to Exercise 1.1) if asymptotic states exist in the sense of (1.5) and (1.6).

1.3. Use Fubini's theorem to derive from $\langle K^*g \mid f\rangle = \langle g \mid Kf\rangle$ that K^* is an integral operator of the above form.

1.4. It follows from Theorem 1.1 that $\Psi(0) = \Omega_+\Psi^{\text{in}}(0)$ and $\Psi(0) = \Omega_-\Psi^{\text{out}}(0)$. Show that $\Omega_-^*\Psi(0) = \Omega_-^*\Omega_-\Psi^{\text{out}}(0) = \Psi^{\text{out}}(0)$ [see (2.10)]. This implies that $\Psi^{\text{out}}(0) = \Omega_-^*\Omega_+\Psi^{\text{in}}(0) = S\Psi^{\text{in}}(0)$.

1.5. The computation will be the same as in the case treated in the text, except that the integration in **P** cannot be eliminated in (1.38), but rather must be retained to the end of the calculation.

1.7. Note that

$$\begin{aligned}&\langle\Psi(t) \mid A\Psi(t)\rangle - \langle\Psi_0(t) \mid A\Psi_0(t)\rangle\\ &\quad = \langle\Psi(t) - \Psi_0(t) \mid A\Psi(t)\rangle + \langle\Psi_0(t) \mid A(\Psi(t) - \Psi_0(t))\rangle.\end{aligned}$$

1.8. Note that $E^{(\mathbf{P})}(B)$ commutes with $H_0^{(2)}$, and consequently

$$\begin{aligned}\langle\Psi^{\text{ex}}(t) \mid E^{(\mathbf{P})}(B)\,\Psi^{\text{ex}}(t)\rangle &= \langle\Psi^{\text{ex}}(0) \mid \exp(iH_0^{(2)}t)\,E^{(\mathbf{P})}(B)\exp(-iH_0^{(2)}t)\,\Psi^{\text{ex}}(0)\rangle\\ &= \langle\Psi^{\text{ex}}(0) \mid E^{(\mathbf{P})}(B)\,\Psi^{\text{ex}}(0)\rangle\end{aligned}$$

for any $B \in \mathscr{B}^3$; then insert $A = E^{(\mathbf{P})}(B)$ and $\Psi_0(t) = \exp(-iH_0^{(2)}t)\Psi^{\text{ex}}(0)$ in Exercise 1.7.

1.9. Consider the case of $\tilde{\psi}_\epsilon(\mathbf{k}) = h_\epsilon(\mathbf{k})\tilde{\psi}(\mathbf{k})$, where (cf. p. 218)

$$h_\epsilon(\mathbf{k}) = \exp\left(-\epsilon \frac{\mathbf{k}^2}{2m} t\right)$$

Taking inverse Fourier transforms of both sides of the first relation appearing in this exercise leads to

$$(\exp(-iH_0 t)\,\psi_\epsilon)(\mathbf{r}) = (2\pi)^{-3/2} \int_{\mathbb{R}^3} \exp\{i[\mathbf{kr} - (\mathbf{k}^2/2m)t]\}\,\tilde{\psi}_\epsilon(\mathbf{k})\,d\mathbf{k}$$

$$= (2\pi)^{-3} \int_{\mathbb{R}^3} d\mathbf{k} \exp\{i[\mathbf{kr} - (\mathbf{k}^2/2m)t]\}\,h_\epsilon(\mathbf{k}) \int_{\mathbb{R}^3} \exp(-i\mathbf{kr}')\,\psi(\mathbf{r}')\,d\mathbf{r}'.$$

Since the above integrand is integrable on $\mathbb{R}^6$ in $(\mathbf{k}, \mathbf{r})$, Fubini's theorem can be applied:

$$(\exp(-iH_0 t)\,\psi_\epsilon)(\mathbf{r}) = (2\pi)^{-3} \int_{\mathbb{R}^3} d\mathbf{r}' \psi(\mathbf{r}') \int_{\mathbb{R}^3} d\mathbf{k} \exp\{i[\mathbf{k}(\mathbf{r} - \mathbf{r}') - (\mathbf{k}^2/2m)t]\} h_\epsilon(\mathbf{k}).$$

After integrating in the (θ, ϕ) spherical coordinates of $\mathbf{k}$, thus arriving at

$$4\pi \int_\theta^\infty \exp[-(\epsilon + i)(k^2/2m)t] \frac{\sin k \mid \mathbf{r} - \mathbf{r}' \mid}{\mid \mathbf{r} - \mathbf{r}' \mid} k\,dk$$

$$= \frac{2\pi m}{\epsilon + i} \sum_{\alpha=-1}^{+1} \int_0^\infty \exp\left(-\frac{\epsilon + i}{2m} k^2 t + \alpha \mid \mathbf{r} - \mathbf{r}' \mid k\right) dk,$$

and then letting $\epsilon \to +0$, one obtains the desired relation. In fact, the limit on the right-hand side of the preceding relation can be taken under the integral sign by applying Lebesgue's bounded convergence theorem (Chapter IV, Lemma 3.1) to infer that for $\psi \in L^1(\mathbb{R}^3) \cap L^2(\mathbb{R}^3)$

$$\lim_{\epsilon\to+0} \int_{\mathbb{R}^3} |\psi(\mathbf{r}')| \left| \int_{\mathbb{R}^3} \exp\{i[\mathbf{k}(\mathbf{r} - \mathbf{r}') - (\mathbf{k}^2/2m)t]\} \mid 1 - h_\epsilon(\mathbf{k})|\,d\mathbf{k} \right| d\mathbf{r}' = 0.$$

On the left-hand side, the continuity of the operator $e^{-iH_0 t}$ implies that

$$\operatorname*{s-lim}_{\epsilon\to+0} e^{-iH_0 t}\,\psi_\epsilon = e^{-iH_0 t}\,\psi.$$

1.10. Since it is quite obvious that $U_t^{(1)}$ and $U_t^{(2)}$ map $L^2(\mathbb{R}^3)$ *onto* itself, it is only necessary to check that these operators are isometries; for example, setting $\boldsymbol{\rho} = m\mathbf{r}/t$, one gets

$$\| U_t^{(1)}\psi \|^2 = (m/t)^3 \int_{\mathbb{R}^3} |\tilde{\psi}(m\mathbf{r}/t)|^2\,d\mathbf{r} = \int_{\mathbb{R}^3} |\tilde{\psi}(\boldsymbol{\rho})|^2\,d\boldsymbol{\rho} = \| \psi \|^2.$$

Apply $U_t^{(1)} U_t^{(2)}$ to integrable functions $\psi(\mathbf{r})$, $\psi \in L^2(\mathbb{R}^3)$, and use the result of Exercise 1.9 to write that almost everywhere in $\mathbb{R}^3$

$$\begin{aligned}(U_t^{(1)} U_t^{(2)} \psi)(\mathbf{r}) &= (m/it)^{3/2} \exp[im(\mathbf{r}^2/2t)](U_t^{(2)} \psi)^{\sim}(m\mathbf{r}/t) \\ &= (m/it)^{3/2} \exp[im(\mathbf{r}^2/2t)]\, 1/(2\pi)^{3/2} \int_{\mathbb{R}^3} \exp[-i(m\mathbf{r}/t)\mathbf{r}'] \\ &\quad \times \exp[im(\mathbf{r}'^2/2t)]\, \psi(\mathbf{r}')\, d\mathbf{r}' \\ &= (m/2\pi it)^{3/2} \int_{\mathbb{R}^3} \exp[i(m/2t)(\mathbf{r} - \mathbf{r}')^2]\, \psi(\mathbf{r}')\, d\mathbf{r}' \\ &= (\exp(-iH_0 t)\psi)(\mathbf{r}).\end{aligned}$$

Since the family of all such functions is dense in $L^2(\mathbb{R}^3)$, and since both $e^{-iH_0 t}$ and $U_t^{(1)} U_t^{(2)}$ are bounded operators, the above relation can be extended by operator continuity to all of $L^2(\mathbb{R}^3)$.

1.11. Note that due to the unitarity of $U_t^{(1)}$

$$\begin{aligned}\|(\exp(-iH_0 t) - U_t^{(1)})\psi\|^2 &= \| U_t^{(1)}(U_t^{(2)} - 1)\psi\|^2 \\ &= \|(U_t^{(2)} - 1)\psi\|^2 \\ &= \int_{\mathbb{R}^3} |\exp[im(\mathbf{r}^2/2t)] - 1|^2\, |\psi(\mathbf{r})|^2\, d\mathbf{r}.\end{aligned}$$

When $t \to \mp\infty$, the above integral converges to zero by Lebesgue's bounded convergence theorem (Chapter IV, Lemma 3.1).

1.12. By taking in Exercise 1.7 $\Psi_0(t) = e^{-iH_0 t}\psi$, $\Psi(t) = U_t^{(1)}\psi$, and A equal to the projector on $C^{(+)}$, i.e., $(A\psi)(\mathbf{r}) = \chi_{C^{(+)}}(\mathbf{r})\, \psi(\mathbf{r})$, one gets

$$\lim_{t \to \mp\infty} \left[\int_{C^{(+)}} |(e^{-iH_0 t}\psi)(\mathbf{r})|^2\, d\mathbf{r} - \int_{C^{(+)}} |(U_t^{(1)}\psi)(\mathbf{r})|^2\, d\mathbf{r}\right] = 0$$

The substitutions $\mathbf{k} = \mp m\mathbf{r}/t$ yield

$$\int_{C^{(+)}} |(U_t^{(1)}\psi)(\mathbf{r})|^2\, d\mathbf{r} = (m/t)^3 \int_{C^{(+)}} |\tilde{\psi}(m\mathbf{r}/t)|^2\, d\mathbf{r} = \int_{C^{(\mp)}} |\tilde{\psi}(\mathbf{k})|^2\, d\mathbf{k}.$$

2.1. Take a bounded interval $I \subset [-\alpha, +\alpha]$, and approximate $\chi_I(\lambda)$ in the mean on $[-\alpha, +\alpha]$ by a trigonometric polynomial $p_n(\lambda) = \sum_{k=-n}^{n} a_k \exp[(\pi ik/\alpha)\lambda]$. If

$$p_n(A_j) = \sum_{k=-n}^{n} a_k \exp[iA_j(\pi k/\alpha)],$$

one easily derives $p_n(A_1)C = Cp_n(A_2)$. Since by Theorems 2.5–2.7 in Chapter IV $p_n(A_j) \to E^{A_j}(I \cup I_\alpha)$, $I_\alpha = \{\lambda + 2\alpha j : \lambda \in I,\ j = \pm 1, \pm 2, ...\}$; letting $\alpha \to \infty$ so that $E^{A_j}(I_2) \to 0$, one obtains $E^{A_1}(I)C = CE^{A_2}(I)$. This result can be easily extended from finite intervals I to arbitrary Borel sets $B \in \mathscr{B}^1$ (see Chapter III, Theorem 5.6).

2.2. Use the method employed in proving Theorem 4.1 of Chapter III.

2.3. Prove that $(1 - E)f = \mathbf{0}$ by establishing that

$$\langle f \mid f \rangle = \langle Ef \mid Ef \rangle + \langle (1 - E)f \mid (1 - E)f \rangle.$$

2.4. Use (2.16), (2.20), and Theorem 2.7 of Chapter IV.

2.5. Note that the bounded operators $A_N = \int_{[-N,+N]} \lambda\, dE^A$ leave $\mathsf{M}_N = \mathsf{M} \cap E^A([-N, +N])\mathscr{H}$ invariant. Hence, any polynomial $p(A_N)$ leaves M_N invariant and therefore $E^{A_N}(B) = E^A(B)$, $B \subset [-N, +N]$, leaves M_N invariant. Since $\mathsf{M}_N \to \mathsf{M}$ when $N \to \infty$ we get $E^A(B)f \in \mathsf{M}$ for $f \in \mathsf{M}$, and $E^A(B)g \perp \mathsf{M}$ for $g \perp \mathsf{M}$, i.e., $E_{\mathsf{M}}E^A(B) = E_{\mathsf{M}}$. Consequently, $E^A(B)E_{\mathsf{M}} = E_{\mathsf{M}}^* = E_{\mathsf{M}}E^A(B)$.

2.6. Note that E_λ^A commutes with e^{iAt} and therefore

$$\int_{\mathbb{R}^1} \lambda^2\, d\, \| E_\lambda^A e^{iAt} f \|^2 = \int_{\mathbb{R}^1} \lambda^2\, d\, \| e^{iAt}\, E_\lambda^A f \|^2 = \int_{\mathbb{R}^1} \lambda^2\, d\, \| E_\lambda^A f \|^2.$$

3.1. The argument runs along the same lines as the proof of the corresponding statements about $\Omega_{-\epsilon}$ and $\Omega_{+\epsilon}$.

3.2. Note that

$$\|(\Omega_{+\epsilon} - \Omega_{+\epsilon}(s))f\|^2 = \epsilon \int_s^\infty e^{-\epsilon t} \langle (\Omega_\epsilon - \Omega_\epsilon(s))f \mid \Omega(t)f \rangle\, dt$$
$$\leqslant \|(\Omega_\epsilon - \Omega_\epsilon(s))f\|\, \epsilon \int_s^\infty e^{-\epsilon t}\, dt.$$

3.3. Use Definitions 3.1–3.5 of Chapter II.

3.4. Note that the linearity property is shared by every Riemann–Stieltjes sum in the sequences whose limits define these integrals.

3.5. Establish these properties for the corresponding Riemann–Stieltjes sums in (3.18) and (3.20), and take the limit.

3.6. Prove first that if $F(\lambda)$ and $F(\lambda, \lambda_0)$ are nonnegative, then the corresponding Riemann–Stieltjes sums are nonnegative, and therefore the corresponding integrals are also nonnegative. Then set $F = F_1 - F_2$ and apply the linearity property (see Exercise 3.4).

3.7. Prove the statement first for proper Riemann–Stieltjes integrals by establishing it for Riemann–Stieltjes sums and then taking the limit; afterwards, generalize the result to improper Riemann–Stieltjes integrals.

3.8. If $m \leqslant F(\lambda) \leqslant M$ on $[a, b]$, then using the result of Exercise 3.6, one can derive

$$m \int_a^b d_\lambda \sigma(\lambda) \leqslant \int_a^b F(\lambda)\, d_\lambda \sigma(\lambda) \leqslant M \int_a^b d_\lambda \sigma(\lambda).$$

Consider the nontrivial case when $\sigma(a) < \sigma(b)$. If m and M are the minimum and the maximum of $F(\lambda)$ in $[a, b]$, respectively, then $F(\lambda)$ assumes all the values between m and M on $[a, b]$; hence, there is a point λ' at which

$$F(\lambda') = \int_a^b F(\lambda)\, d_\lambda \sigma(\lambda) \Big/ \int_a^b d_\lambda \sigma(\lambda).$$

For the existence part, prove that the limit (3.17) exists and is independent of the chosen sequence of subdivisions $\{\lambda^{(k)}\}$ by showing that if $\{\lambda^{(k,l)}\}$ is a finer subdivision than $\{\lambda^{(k)}\}$ [with $\lambda^{(k,0)} = \lambda^{(k)}$], then

$$\Big| \sum_k F(\lambda'^{(k)})\, \sigma((\lambda^{(k-1)}, \lambda^{(k)}]) - \sum_{k,l} F(\lambda'^{(k,l)})\, \sigma((\hat{\lambda}^{(k,l)}, \lambda^{(k,l)}]) \Big|$$

can be made arbitrarily small for all sufficiently fine subdivisions $\{\lambda^{(k)}\}$ [here $\hat{\lambda}^{(k,l)}$ denotes the point preceding $\lambda^{(k,l)}$ in the subdivision $\{\lambda^{(k,l)}\}$]. To obtain this result, exploit the uniform continuity of $F(\lambda)$ in $[a, b]$, according to which for given $\epsilon > 0$ we have

$$| F(\lambda'^{(k)}) - F(\lambda'^{(k,l)})| < \epsilon$$

for all sufficiently fine subdivisions $\{\lambda^{(k)}\}$. Hence, the above difference is smaller than

$$\sum_{k,l} | F(\lambda'^{(k)}) - F(\lambda'^{(k,l)})|\, \sigma((\hat{\lambda}^{(k,l)}, \lambda^{(k,l)}]) < \epsilon \sigma((a, b]).$$

3.9. Note that

$$\langle f \mid A_R{}^* g\rangle = \langle A_R f \mid g\rangle = \int_R \langle A(\xi) f \mid g\rangle\, d\mu(\xi) = \int_R \langle f \mid A^*(\xi) g\rangle\, d\mu(\xi)$$

for all $f, g \in \mathscr{H}$.

3.10. Note that $\langle g \mid e^{i(A-\zeta)t} f\rangle$ is continuous and therefore Borel measurable in t, and that

$$\int_0^{+\infty} \| e^{i(A-\zeta)t} f \|\, dt \leqslant \| f \| \int_0^{\infty} e^{t \operatorname{Im} \zeta}\, dt < +\infty$$

for Im $\zeta < 0$. Hence, $e^{i(A-\zeta)t}$ is Bochner integrable. Using the spectral theorem, we obtain for any $g \in \mathscr{H}$

$$\begin{aligned}\langle g \mid R_A(\zeta)f\rangle &= \int_{\mathbb{R}^1} \frac{1}{\lambda - \zeta}\, d\langle g \mid E_\lambda^A f\rangle \\ &= -i \int_{\mathbb{R}^1} \left[\int_0^{+\infty} e^{i(\lambda-\zeta)t}\, dt\right] d\langle g \mid E_\lambda^A f\rangle \\ &= -i \int_0^{+\infty} dt \int_{\mathbb{R}^1} e^{i(\lambda-\zeta)t}\, d\langle g \mid E_\lambda^A f\rangle \\ &= -i \int_0^{\infty} \langle g \mid e^{i(A-\zeta)t} f\rangle\, dt,\end{aligned}$$

where the order of integration could be reversed by Fubini's theorem.

4.1. Use Fubini's theorem to carry out the integration in (4.41) successively, first with respect to $\mu^{(k)}$ and then with respect to $\mu^{(\omega)}$. In view of the definition of $\mu^{(\omega)}$ in (4.40), we obtain the desired result by applying Theorem 3.11 of Chapter II with $R_n = \{(l, m)\}$.

4.2. Let B_+ and B_- be the Borel sets in $\mathbb{R}^{2m}$ on which $G^{(1)}(x, x'; \zeta) - G^{(2)}(x, x'; \zeta)$ is positive or negative, respectively. For any Borel sets B_1, $B_2 \in \mathscr{B}^m$ of finite measure we get, using Fubini's theorem,

$$\left\langle \chi_{B_1} \,\middle|\, \frac{1}{\zeta - A}\, \chi_{B_2} \right\rangle = \int_{B_1 \times B_2} G_A^{(i)}(x, x'; \zeta)\, d(\mu \times \mu), \qquad i = 1, 2.$$

Hence, if $B_1 \times B_2 \subset B_\pm$, we infer (see Chapter II, Lemma 4.1) from

$$\int_{B_1 \times B_2} (G_A^{(1)}(x, x'; \zeta) - G_A^{(2)}(x, x'; \zeta))\, d(\mu \times \mu) = 0$$

that $B_1 \times B_2$ is of $\mu \times \mu$ measure zero. Since by the definition in Theorem 2.4, Chapter II, of products of measure, we get from Theorem 1.7 and (2.26) of Chapter II

$$(\mu \times \mu)(B_\pm) = \sup \left\{ \sum_{k=1}^{\infty} (\mu \times \mu)(B_1^{(k)} \times B_2^{(k)}) \colon \sum_{k=1}^{\infty} B_1^{(k)} \times B_2^{(k)} \subset B_\pm \right\},$$

we thus deduce that $(\mu \times \mu)(B_\pm) = 0$.

4.3. Note that

$$\begin{aligned}&| \psi_0^*(\mathbf{r})(G_0(\mathbf{r}, \mathbf{r}'; \lambda + i\eta) - G_0^{(\pm)}(\mathbf{r}, \mathbf{r}'; \lambda))\, V(\mathbf{r}')\, \Phi_{\mathbf{k}}^{(\pm)}(\mathbf{r})\, \tilde{\psi}(\mathbf{k})| \\ &\qquad \leqslant M_B(\mathbf{r})\, |\psi_0(\mathbf{r})| \frac{|V(\mathbf{r}')|}{|\mathbf{r} - \mathbf{r}'|}\, |\tilde{\psi}(\mathbf{k})|, \qquad \operatorname{supp} \tilde{\psi} \subset B,\end{aligned}$$

where the function on the left-hand side of the above inequality converges pointwise to zero when $\eta \to \pm 0$, while the function on the right-hand side is integrable on $\mathbb{R}^9$ in **r**, **r**′ and **k**. The desired result is then obtained by applying Lemma 3.1 of Chapter IV.

4.4. If $\chi_j(y)$ denotes the characteristic function of $\Lambda^{-1}((\lambda_{j-1}, \lambda_j])$ then

$$\sum_{j=1}^{n} \int_{\lambda_{j-1}<\Lambda(y)\leqslant\lambda_j} \check{g}^*(y)(\Phi_y \mid H_1 F(\lambda_j')f\rangle$$

$$= \int_{\mathbb{R}^n} \sum_{j=1}^{n} \check{g}^*(y)\langle H_1{}^x\Phi_y \mid F(\lambda_j')f\rangle\, \chi_j(y)\, d\nu(y).$$

The integrand on the right-hand side of this equality converges pointwise to the integrand on the right-hand side of (4.84). Lebesgue's dominated convergence theorem can be applied since on supp $\check{g}(y)$, $\| H_1{}^x\Phi_y \| \leqslant C_{\check{g}}$, and consequently,

$$| \langle \Phi_y \mid H_1 F(\lambda_j')f\rangle | \leqslant \| H_1{}^x\Phi_y \| \, \| F(\lambda_j')f \| \leqslant C_{\check{g}} C \, \| f \|.$$

4.5. Write (4.103) in terms of matrix elements with $f, g \in \mathscr{D}_{H_0} \subset \mathscr{D}_V$,

$$\langle g \mid \mathscr{T}(\zeta)f\rangle = \langle g \mid Vf\rangle + \langle Vg \mid (\zeta - H_0)^{-1}\, \mathscr{T}(\zeta)f\rangle$$

and express both sides of this equation in the momentum representation, using (4.82):

$$\int_{\mathbb{R}^3} d\mathbf{k}\, \tilde{g}^*(\mathbf{k})(\Phi_{\mathbf{k}} \mid \mathscr{T}(\zeta)f\rangle = \int_{\mathbb{R}^3} d\mathbf{k}\, \tilde{g}^*(\mathbf{k})(\Phi_{\mathbf{k}} \mid Vf\rangle$$

$$+ \int_{\mathbb{R}^3} d\mathbf{k}''\langle Vg \mid \Phi_{\mathbf{k}''})[\zeta - \mathbf{k}''^2/2m)]^{-1}(\Phi_{\mathbf{k}''} \mid \mathscr{T}(\zeta)f\rangle.$$

Since $V(\mathbf{r})\Phi_{\mathbf{k}}(\mathbf{r}) \in L_{(2)}(\mathbb{R}^3)$ if $V(\mathbf{r}) \in L_{(2)}(\mathbb{R}^3)$, we have

$$(\Phi_{\mathbf{k}} \mid Vf\rangle = \langle V\Phi_{\mathbf{k}} \mid f\rangle = \int \langle V\Phi_{\mathbf{k}} \mid \Phi_{\mathbf{k}'})\, d\mathbf{k}'(\Phi_{\mathbf{k}'} \mid f\rangle,$$

with a similar result holding for $\langle Vg \mid \Phi_{\mathbf{k}})$, and also

$$(\Phi_{\mathbf{k}''} \mid \mathscr{T}(\zeta)f\rangle = \langle [1 + VG^*(\zeta)]V\Phi_{\mathbf{k}''} \mid f\rangle$$

$$= \int \langle [1 + VG^*(\zeta)]V\Phi_{\mathbf{k}''} \mid \Phi_{\mathbf{k}'})\, d\mathbf{k}'(\Phi_{\mathbf{k}'} \mid f\rangle.$$

Inserting these results in the preceding equation, and noting that

$$\langle V\Phi_{\mathbf{k}} \mid \Phi_{\mathbf{k}'}\rangle = (\Phi_{\mathbf{k}} \mid V \mid \Phi_{\mathbf{k}'}) = \langle \mathbf{k} \mid V \mid \mathbf{k}'\rangle,$$

$$[1 + VG^*(\zeta)]V\Phi_{\mathbf{k}''} \mid \Phi_{\mathbf{k}'}) = (\Phi_{\mathbf{k}''} \mid V + VG(\zeta)V \mid \Phi_{\mathbf{k}'}) = \langle \mathbf{k}'' \mid \mathscr{T}(\zeta) \mid k'\rangle,$$

we arrive at the result that the left and right hand sides of (4.105) multiplied by $\tilde{g}^*(\mathbf{k})\tilde{f}(\mathbf{k}')$ and integrated over $\mathbb{R}^6$ have to be equal at least for all $\tilde{f}$, $\tilde{g} \in \mathscr{C}_b{}^0(\mathbb{R}^3)$, and therefore (by the same argument as in Exercise 4.2), that (4.105) has to be satisfied almost everywhere in $\mathbb{R}^6$.

4.6. Write (4.113) in terms of matrix elements for f, $g \in L^2(\mathbb{R}^3)$, use (4.96), and apply the argument of Exercise 4.2.

5.1. The function $f(\mathbf{r}, \mathbf{r}') = | \mathbf{r} - \mathbf{r}' |$ is continuous and therefore Borel measurable in $\mathbb{R}^6$ (Chapter II, Theorem 3.1). Since $D_{G_0} = f^{-1}(\{0\})$, it follows that D_{G_0} is a Borel set. To establish that its Lebesgue measure $\mu_l^n(D_{G_0})$ is zero, note that

$$\mu_l^n(D_{G_0}) = \int_{D_{G_0}} d\mathbf{r}\, d\mathbf{r}',$$

and write the above integral as an iterated integral by applying Fubini's theorem.

5.2. Since $K(\alpha, \alpha')$ as well as $f(\alpha)\, g(\alpha')$ are square integrable in $\mathbb{R}^{2n}$ with respect to $\mu \times \mu$, the function $K(\alpha, \alpha')f(\alpha)\, g(\alpha')$ is integrable in $\mathbb{R}^{2n}$. Hence, Fubini's theorem can be applied in the following derivation:

$$\begin{aligned}\langle f \mid Kg\rangle &= \int_{\mathbb{R}^n} d\mu(\alpha) f^*(\alpha) \int_{\mathbb{R}^n} d\mu(\alpha')\, K(\alpha, \alpha')\, g(\alpha') \\ &= \int_{\mathbb{R}^{2n}} f^*(\alpha)\, K(\alpha, \alpha')\, g(\alpha')\, d\mu(\alpha)\, d\mu(\alpha') \\ &= \int_{\mathbb{R}^n} d\mu(\alpha')\, g(\alpha') \int_{\mathbb{R}^n} d\mu(\alpha)(K^*(\alpha, \alpha')f(\alpha))^* = \langle K^*f \mid g\rangle,\end{aligned}$$

which shows that K^* has the kernel $K^*(\alpha', \alpha)$.

For arbitrary $f \in L^2(\mathbb{R}^n, \mu)$, the function

$$\int_{\mathbb{R}^n} | K(\alpha'', \alpha')f(\alpha') |\, d\mu(\alpha') \leqslant \Big[\int_{\mathbb{R}^n} | K(\alpha'', \alpha')|^2\, d\mu(\alpha')\Big]^{1/2} \Big[\int_{\mathbb{R}^n} |f(\alpha')|^2\, d\mu(\alpha)\Big]^{1/2}$$

is μ-square integrable in $\alpha'' \in \mathbb{R}^n$, and consequently

$$\int_{\mathbb{R}^n} d\mu(\alpha'') | K^*(\alpha'', \alpha) | \int_{\mathbb{R}^n} d\mu(\alpha') | K(\alpha'', \alpha') f(\alpha') |$$

exists. Thus, we can apply Tonelli's and Fubini's theorems in succession to infer that

$$\int_{\mathbb{R}^n} d\mu(\alpha'') K^*(\alpha'', \alpha) \int_{\mathbb{R}^n} d\mu(\alpha') K(\alpha'', \alpha') f(\alpha')$$
$$= \int_{\mathbb{R}^n} d\mu(\alpha') f(\alpha') \int_{\mathbb{R}^n} d\mu(\alpha'') K^*(\alpha'', \alpha) K(\alpha'', \alpha').$$

5.3. According to Theorem 8.6 of Chapter IV, K is a Hilbert–Schmidt operator if and only if K^*K is of trace class. In order to establish that K^*K is of trace class, it is sufficient to establish that

$$\operatorname{Tr}(K^*K) = \sum_{k=1}^{\infty} \langle \psi_k \mid K^*K\psi_k \rangle = \sum_{k=1}^{\infty} \| K\psi_k \|^2 < +\infty$$

is some orthonormal basis $\{\psi_1, \psi_2, \ldots\}$.

In view of the fact that the function $K(\alpha, \alpha')$ is an element of $L_{(2)}(\mathbb{R}^{2n}, \mu \times \mu)$ and that $\{\psi_i \otimes \psi_k^*; i, k = 1, 2, \ldots\}$ is an orthonormal basis in $L^2(\mathbb{R}^{2n}, \mu \times \mu)$ (see Chapter II, Theorems 6.9 and 6.10) we obtain

$$\int_{\mathbb{R}^{2n}} | K(\alpha, \alpha')|^2 \, d\mu(\alpha) d\mu(\alpha')$$
$$= \sum_{i,k=1}^{\infty} \left| \int_{\mathbb{R}^{2n}} \psi_i^*(\alpha) \psi_k(\alpha') K(\alpha, \alpha') \, d\mu(\alpha) \, d\mu(\alpha') \right|^2$$
$$= \sum_{k=1}^{\infty} \left(\sum_{i=1}^{\infty} \left| \int_{\mathbb{R}^n} d\mu(\alpha) \psi_i^*(\alpha) \int_{\mathbb{R}^n} d\mu(\alpha') K(\alpha, \alpha') \psi_k(\alpha') \right|^2 \right)$$
$$= \sum_{k=1}^{\infty} \int_{\mathbb{R}^n} d\mu(\alpha) \left| \int_{\mathbb{R}^n} K(\alpha, \alpha') \psi_k(\alpha') \, d\mu(\alpha') \right|^2 = \sum_{k=1}^{\infty} \| K\psi_k \|^2 = \operatorname{Tr}(K^*K).$$

In deriving the above equation we have used in the first step Parseval's relation in $L^2(\mathbb{R}^{2n}, \mu \times \mu)$, in the second step the unconditional convergence of a double series with positive terms, as well as Fubini's theorem, and in the third step Parseval's relation in $L^2(\mathbb{R}^n, \mu)$.

5.4. Apply the Riesz–Fisher theorem (Chapter II, Theorem 4.4) to conclude that $K(\alpha, \alpha')$ is square integrable on $\mathbb{R}^{2m}$.

5.5. The operator BA is completely continuous by Chapter IV, Theorem 8.1. Suppose that $A = A^*$ and that $\{e_1, e_2, \ldots\}$ is an orthonormal basis for which $Ae_k = \lambda_k e_k$. Then

$$\sum_k \| BAe_k \|^2 = \sum_k \lambda_k^2 \| Be_k \|^2 \leqslant \| B \|^2 \sum_k \lambda_k^2 < +\infty,$$

which implies that BA is a Hilbert–Schmidt operator.

In general, if $A \neq A^*$, then we can write $A = A_1 + iA_2$, where $A_1 = A_1^*$ and $A_2 = A_2^*$. Then BA is a Hilbert–Schmidt operator since BA_1 and BA_2 are Hilbert–Schmidt operators by the above, and a family of Hilbert–Schmidt operators is a linear (normed) space (see Theorem 8.10 in Chapter IV).

5.6. Note that for any $f \in \mathscr{H}$

$$\left\| \sum_{k=n+1}^{m} A_k f \right\| \leqslant \sum_{k=n+1}^{m} \| A_k \| \| f \|$$

can be made arbitrarily small for sufficiently large n and all $m > n$. Hence,

$$Af = \operatorname*{s-lim}_{n \to +\infty} \sum_{k=1}^{n} A_k f$$

exists for all $f \in \mathscr{H}$, and obviously defines a linear operator on $\mathscr{H}$. Moreover, since by letting below $n \to +\infty$ we see that

$$\| Af \| \leqslant \left\| \sum_{k=1}^{n} A_k f \right\| + \left\| \sum_{k=n+1}^{\infty} A_k f \right\| \leqslant \sum_{k=1}^{\infty} \| A_k \| \| f \|,$$

we conclude that A is bounded, and the desired estimate is obtained for $\| A \|$. By the same procedure we get

$$\left\| A - \sum_{k=1}^{n} A_k \right\| = \left\| \sum_{k=n+1}^{\infty} A_k \right\| \leqslant \sum_{k=n+1}^{\infty} \| A_k \| \to 0,$$

when $n \to +\infty$.

5.7. Note that

$$\left\| \frac{1}{M} c_k A^k \right\| \leqslant \frac{| c_k |}{M} \| A \|^k \leqslant \| A \|^k$$

and $\sum \| A \|^k < +\infty$.

5.8. Since m_A is a lower bound of S^A,

$$\langle f \mid Af\rangle = \int_{m_A}^{+\infty} \lambda d\| E_\lambda^A f\|^2 \geqslant m_A \int_{m_A}^{+\infty} d\| E_\lambda^A f\|^2 = m_A \| f\|^2.$$

Conversely, if $\langle f \mid Af\rangle \geqslant \gamma \| f\|^2$, then γ must be a lower bound of S^A. In fact, if that were not so, then we would have $\lambda_0 \in S^A$ for some $\lambda_0 < \gamma$. This means that we could find for $\epsilon_0 = \frac{1}{2}(\gamma - \lambda_0)$ a nonzero vector $f_0 \in \mathscr{D}_A$ such that $E^A([\lambda_0 - \epsilon_0 , \lambda_0 + \epsilon_0]) f_0 = f_0$. Consequently, we would have

$$\begin{aligned}\langle f_0 \mid Af_0\rangle &= \int_{\lambda_0-\epsilon_0}^{\lambda_0+\epsilon_0} \lambda d\langle f_0 \mid E_\lambda^A f_0\rangle \\ &\leqslant (\lambda_0 + \epsilon_0)\langle f_0 \mid E^A([\lambda_0 - \epsilon_0 , \lambda_0 + \epsilon_0] f_0\rangle < \gamma \| f_0 \|^2,\end{aligned}$$

which contradicts the initial assumption that $\langle f_0 \mid Af_0\rangle \geqslant \gamma \| f_0 \|^2$.

6.1. Introduce spherical coordinates (r', θ, ϕ), orienting the z axis along the vector $\mathbf{r}$. Then

$$\begin{aligned}&\int_{r' \geqslant R_1} \frac{d\mathbf{r}'}{| \mathbf{r}' |^{2+\epsilon_0} | \mathbf{r} - \mathbf{r}' |} \\ &= 2\pi \int_{R_1}^{\infty} dr' \int_0^{\pi} d\theta \frac{\sin\theta}{(r')^{\epsilon_0}[(r')^2 - 2rr' \cos\theta + r^2]^{1/2}} \\ &= 2\pi \int_{R_1}^{+\infty} \frac{[r^2 + 2rr' + (r')^2]^{1/2} - [r^2 - 2rr' + (r')^2]^{1/2}}{r(r')^{1+\epsilon_0}} dr' \\ &= 4\pi \int_{R_1}^{r} \frac{dr'}{r(r')^{\epsilon_0}} + 4\pi \int_r^{+\infty} \frac{dr'}{(r')^{1+\epsilon_0}} \\ &= \frac{4\pi}{(\epsilon_0 - 1)R_1^{\epsilon_0 - 1}} \frac{1}{r} + \frac{4\pi}{(1 - \epsilon_0)\epsilon_0} \frac{1}{r^{\epsilon_0}} \\ &= \begin{cases} O\left(\dfrac{1}{r^{\epsilon_0}}\right) & \text{if } 0 < \epsilon_0 \leqslant 1 \\ O\left(\dfrac{1}{r}\right) & \text{if } \epsilon_0 > 1. \end{cases}\end{aligned}$$

6.2. Let R_0 be such that $r^{2+\epsilon_0} \mid V(\mathbf{r})| \leqslant C_0$ for all $r \geqslant R_0$. Divide the domain of integration into two parts, so that

$$\int_{\mathbb{R}^3} \frac{|V(\mathbf{r}')|}{|\mathbf{r}-\mathbf{r}'|} d\mathbf{r}' \leqslant \int_{r'\leqslant R_0} \frac{|V(\mathbf{r}')|}{|\mathbf{r}-\mathbf{r}'|} d\mathbf{r}' + C_0 \int_{r'>R_0} \frac{d\mathbf{r}'}{|\mathbf{r}'|^{2+\epsilon_0} |\mathbf{r}'-\mathbf{r}|},$$

and note that

$$\int_{r'\leqslant R_0} \frac{|V(\mathbf{r}')|}{|\mathbf{r}-\mathbf{r}'|} d\mathbf{r}' \leqslant \left\{\int_{r'\leqslant R_0} \frac{d\mathbf{r}'}{|\mathbf{r}-\mathbf{r}'|^2}\right\}^{1/2} \left\{\int_{\mathbb{R}^3} |V(\mathbf{r}')|^2 \, d\mathbf{r}'\right\}^{1/2}.$$

Constants C_1 and C_2 can be chosen which are such that (see Exercise 6.1)

$$\int_{r'\leqslant R_0} \frac{d\mathbf{r}'}{|\mathbf{r}-\mathbf{r}'|^2} \leqslant C_1, \qquad \int_{r'\geqslant R_0} \frac{d\mathbf{r}'}{|\mathbf{r}'|^{2+\epsilon_0} |\mathbf{r}'-\mathbf{r}|} \leqslant C_2$$

for all $\mathbf{r} \in \mathbb{R}^3$. The continuity can be proven as in Exercise 6.9.

6.3. Since $V(\mathbf{r})$ is square integrable for $r_1 \leqslant R_0$, we obtain (by using the Schwarz–Cauchy inequality) for $r > R_0$

$$\int_{r_1\leqslant R_0} \frac{|V(\mathbf{r}_1)|}{|\mathbf{r}-\mathbf{r}_1|^2} d\mathbf{r}_1 \leqslant \left\{\int_{r_1\leqslant R_0} |V(\mathbf{r}_1)|^2 \, d\mathbf{r}\right\}^{1/2} \left\{\int_{r_1\leqslant R_0} \frac{d\mathbf{r}_1}{|\mathbf{r}-\mathbf{r}_1|^4}\right\}^{1/2}$$
$$= \text{const} \frac{1}{|\mathbf{r}-\mathbf{r}_{\text{av}}|^2} = O\left(\frac{1}{r^2}\right);$$

here $\mathbf{r}_{\text{av}}$, $r_{\text{av}} \leqslant R_0$, is a vector (whose existence follows from the mean-value theorem of the integral calculus) for which $|\mathbf{r}-\mathbf{r}_1|^{-4}$ assumes its average when integrated in $\mathbf{r}_1$ over the sphere $\{\mathbf{r}_1 : r_1 \leqslant R_0\}$.

Since $|V(\mathbf{r}_1)| \leqslant C_0 r_1^{-2-\epsilon_0}$ for $r_1 \geqslant R_0$, we easily carry out the following computation by orienting the z_1 axis in the direction of $\mathbf{r}$:

$$\int_{r_1\geqslant R_0} \frac{|V(\mathbf{r}_1)|}{|\mathbf{r}-\mathbf{r}_1|^2} d\mathbf{r}_1 \leqslant 2\pi C_0 \int_{R_0}^{+\infty} dr_1 \int_0^{\pi} \frac{\sin\theta}{r_1^{\epsilon_0}(r^2 - 2rr_1\cos\theta + r_1^2)} d\theta$$
$$= \frac{\pi C_0}{r} \int_{R_0}^{+\infty} \frac{1}{r_1^{1+\epsilon_0}} \ln\left(\frac{1+r_1/r}{1-r_1/r}\right)^2 dr_1 = \frac{\pi C_0}{r^{1+\epsilon_0}} \int_{R_0/r}^{+\infty} \frac{1}{u^{1+\epsilon_0}} \ln\left(\frac{1+u}{1-u}\right)^2 du.$$

It is immediately seen that the above integral is $O(1/r^{1+\epsilon_0})$ when $0 < \epsilon_0 < 1$. For $\epsilon_0 \geqslant 1$, the integral on the right-hand side of the above equalities does not converge when the lower limit R_0/r is replaced by zero. However, then

$$\frac{1}{r^{1+\epsilon_0}} \int_{R_0/r}^{+\infty} \frac{1}{u^{1+\epsilon_0}} \ln\left(\frac{1+u}{1-u}\right)^2 du$$

$$\leqslant \frac{1}{r^{1+\epsilon_0}} \left(\frac{r}{R_0}\right)^{\epsilon_0} \int_0^{+\infty} \frac{1}{u} \ln\left(\frac{1+u}{1-u}\right)^2 du = O\left(\frac{1}{r}\right).$$

To prove the local integrability in $\mathbf{r}$ of the function under consideration, note that the function $|V(\mathbf{r}_1)| \, |\mathbf{r}-\mathbf{r}_1|^{-2}$ is square integrable on the set $\{(\mathbf{r}, \mathbf{r}_1): r \leqslant R < +\infty, \mathbf{r}_1 \in \mathbb{R}^3\}$; this follows (by Tonelli's theorem, Chapter II, Theorem 3.14) from the existence of the iterated integral

$$\int_{\mathbb{R}^3} d\mathbf{r}_1 |V(\mathbf{r}_1)| \int_{r \leqslant R} \frac{d\mathbf{r}}{|\mathbf{r}-\mathbf{r}_1|^2}.$$

Hence, the desired result follows by Fubini's theorem.

6.4. By using the Schwarz–Cauchy inequality we obtain

$$\int_{\mathbb{R}^3} \frac{|V(\mathbf{r}_2)| \, d\mathbf{r}_2}{|\mathbf{r}-\mathbf{r}_2| \, |\mathbf{r}_2-\mathbf{r}_1|} \leqslant \left\{\int_{\mathbb{R}^3} \frac{|V(\mathbf{r}_2)| \, d\mathbf{r}_2}{|\mathbf{r}-\mathbf{r}_2|^2}\right\}^{1/2} \left\{\int_{\mathbb{R}^3} \frac{V(\mathbf{r}_2) \, d\mathbf{r}_2}{|\mathbf{r}_2-\mathbf{r}_1|^2}\right\}^{1/2}.$$

Thus, the present case is reduced to the case considered in Exercise 6.3.

6.5. Combine the results of Exercises 6.2 and 6.3; note also that $V(\mathbf{r})$ is square integrable.

6.6. Since $V(\mathbf{r})$ is locally square integrable it is measurable on $\mathbb{R}^3$ (see Chapter II, Theorem 4.1). Since $|\mathbf{r}-\mathbf{r}'|^{-2} V(\mathbf{r})$ satisfies $|\mathbf{r}-\mathbf{r}'|^{-2} |V(\mathbf{r})| \leqslant C_0 |\mathbf{r}-\mathbf{r}'|^{-2} r^{-2-\epsilon_0}$ for $r \geqslant R_0$, it is integrable on the set $\{\mathbf{r}: r \geqslant R_0\}$ due to the integrability of $|\mathbf{r}-\mathbf{r}'|^{-2} r^{-2-\epsilon_0}$ on that set (see Chapter II, Theorem 3.9). By Fubini's theorem the function

$$\int_{\mathbb{R}^3} |V(\mathbf{r})| \, |\mathbf{r}-\mathbf{r}'|^{-2} \, d\mathbf{r}$$

is locally integrable in $\mathbf{r}'$, and majorized by const $|\mathbf{r}'|^{-1-\epsilon_0}$ for $|\mathbf{r}'| \geqslant R$ (see Exercise 6.3). Hence, it is integrable on the set $\{\mathbf{r}' : r' \geqslant R_0\}$.

The integrability of $|V(\mathbf{r})| \, |\mathbf{r}-\mathbf{r}'|^{-2} \, |V(\mathbf{r}')|$ on $\mathbb{R}^6$ now follows by Tonelli's theorem from the existence of the iterated integral.

6.7. According to (6.29)

$$| v_{\mathbf{k}}(\mathbf{r})| \leqslant F_V(\mathbf{r}; \mathbf{k}^2/2m).$$

Use the result of Exercise 6.5 and the continuity of $\check{d}(k^2/2m)$ to prove that $F_V(\mathbf{r}; \mathbf{k}^2/2m) \leqslant$ const for all $r \geqslant R_0$ and all $\mathbf{k} \in D_0$.

6.8. By the mean-value theorem of differential calculus

$$\frac{\Delta I}{\Delta u} = \frac{1}{\Delta u} \int_{\mathbb{R}^3} [h(\mathbf{r}, u + \Delta u) - h(\mathbf{r}, u)] f(\mathbf{r})\, d\mathbf{r}$$
$$= \int_{\mathbb{R}^3} h_u(\mathbf{r}, u + \theta\, \Delta u) f(\mathbf{r})\, d\mathbf{r}, \qquad 0 \leqslant \theta \leqslant 1.$$

For given $\epsilon > 0$, choose $R_1 > R_0$ so that

$$\left| \int_{r \geqslant R_1} h_u(\mathbf{r}, u) f(\mathbf{r})\, d\mathbf{r} \right| \leqslant \int_{r \geqslant R_1} \rho(\mathbf{r})\, d\mathbf{r} < \epsilon/2.$$

Since $h_u(\mathbf{r}, u)$ is uniformly continuous over the region $\{(\mathbf{r}, u + \Delta u): r \leqslant R_1, | \Delta u | \leqslant \text{const}\}$, we can choose Δu so small that

$$\int_{r \leqslant R_1} | h_u(\mathbf{r}, u + \theta\, \Delta u) - h_u(\mathbf{r}, u)|\, |f(\mathbf{r})|\, d\mathbf{r} < \epsilon/2.$$

Hence, we have

$$\left| \frac{\Delta I}{\Delta u} - \int_{\mathbb{R}^3} h_u(\mathbf{r}, u) f(\mathbf{r})\, d\mathbf{r} \right| < \epsilon$$

for sufficiently small $| \Delta u |$.

6.9. For given $\epsilon > 0$, choose a so small that

$$\left| \int_{|\mathbf{r}'-\mathbf{r}| \leqslant a} \frac{\exp(ik\, | \mathbf{r}_1 - \mathbf{r}' |)}{| \mathbf{r}_1 - \mathbf{r}' |} f(\mathbf{r}')\, d\mathbf{r}' \right| \leqslant \int_{|\mathbf{r}'-\mathbf{r}| \leqslant a} \frac{|f(\mathbf{r}')|}{| \mathbf{r}_1 - \mathbf{r}' |}\, d\mathbf{r}'$$
$$\leqslant \left(\int_{|\mathbf{r}'-\mathbf{r}| \leqslant a} | \mathbf{r}_1 - \mathbf{r}' |^{-2}\, d\mathbf{r}' \right)^{1/2} \left(\int_{|\mathbf{r}'-\mathbf{r}| \leqslant a} |f(\mathbf{r}')|^2\, d\mathbf{r}' \right)^{1/2} < \frac{\epsilon}{6}.$$

for all $\mathbf{r}_1$ from some closed neighborhood $\mathscr{N}$ of $\mathbf{r}$. Then select $R_0 > 0$ so large that for all those $\mathbf{r}_1$,

$$\left| \int_{r' \geqslant R_0} \frac{\exp(ik| \mathbf{r}_1 - \mathbf{r}' |)}{| \mathbf{r}_1 - \mathbf{r}' |} f(\mathbf{r}')\, d\mathbf{r}' \right| \leqslant \text{const} \int_{r' \geqslant R_0} \frac{d\mathbf{r}'}{| \mathbf{r}' |^{3+\epsilon_0}} < \frac{\epsilon}{6}.$$

Since $| \mathbf{r}_1 - \mathbf{r}' |^{-1} \exp(ik\, | \mathbf{r}_1 - \mathbf{r}' |)$ is uniformly continuous for $\mathbf{r}_1 \in \mathscr{N}$

and $\mathbf{r}' \in S_0 = \{\mathbf{r}': |\mathbf{r}' - \mathbf{r}| \geqslant a, r' \leqslant R_0\}$, we can obtain

$$\int_{S_0} \left| \frac{\exp(ik|\mathbf{r} - \mathbf{r}'|)}{|\mathbf{r} - \mathbf{r}'|} - \frac{\exp(ik|\mathbf{r}_1 - \mathbf{r}'|)}{|\mathbf{r}_1 - \mathbf{r}'|} \right| |f(\mathbf{r}')| \, d\mathbf{r}' < \frac{\epsilon}{3}$$

for all sufficiently small values of $|\mathbf{r}_1 - \mathbf{r}|$. Combining these three results, we easily obtain $|g(\mathbf{r}_1) - g(\mathbf{r})| < \epsilon$ for such values of $|\mathbf{r}_1 - \mathbf{r}|$.

6.10. Since $\mathscr{S}_V^3$ is a closed set of measure zero, it is possible to construct a sequence $h_1(\mathbf{r}), h_2(\mathbf{r}),\ldots$ of infinitly many times differentiable functions with supports disjoint from $\mathscr{S}_V^3$, which is such that $\int |h_n(\mathbf{r}) - 1|^2 \, d\mathbf{r} \to 0$. An example of such a sequence would consist of the functions $h_n(\mathbf{r}) = 0$ for $\mathbf{r} \in \mathscr{S}_V^3$, $h_n(\mathbf{r}) = 1$ for $\mathbf{r} \notin \mathscr{S}_V^3$ at a distance $d(\mathbf{r}) > 1/n$ from $\mathscr{S}_V^3$, and $h_n(\mathbf{r}) = \exp(-n^2 d^2(\mathbf{r})[n^2 d^2(\mathbf{r}) - 1]^{-1})$ for $\mathbf{r} \notin \mathscr{S}_V^3$ and $d(\mathbf{r}) \leqslant 1/n$. Since $f_n(\mathbf{r}) = f(\mathbf{r}) \, h_n(\mathbf{r}) \in \mathscr{C}_V^\infty$ for any $f(\mathbf{r}) \in \mathscr{C}_b^\infty(\mathbb{R}^3)$, and $\mathscr{C}_b^\infty(\mathbb{R}^3)$ is dense in $L^2(\mathbb{R}^3)$, so is $\mathscr{C}_V^3$.

7.1. Setting

$$F(u) = \int_0^u f(t) \, dt, \qquad u \geqslant 0,$$

and integrating by parts, we obtain

$$\begin{aligned}
\int_0^{+\infty} e^{-\epsilon t} f(t) \, dt &= \lim_{\alpha \to +\infty} \int_0^{+\alpha} e^{-\epsilon t} f(t) \, dt \\
&= \lim_{\alpha \to +\infty} \left\{ e^{-\epsilon t} F(t) \Big|_{t=0}^{t=\alpha} + \epsilon \int_0^{+\alpha} e^{-\epsilon t} F(t) \, dt \right\} \\
&= \epsilon \int_0^{+\alpha} e^{-\epsilon t} F(t) \, dt.
\end{aligned}$$

This establishes the existence of I_ϵ for $\epsilon > 0$. We note that the above limit for $\alpha \to +\infty$ exists since $\lim_{t \to +\infty} F(t) = I_0$ exists [and consequently $\lim_{\alpha \to +\infty} e^{-\epsilon \alpha} F(\alpha) = 0$], and since from the continuity of $F(t)$ it follows that $|F(t)| \leqslant C_0$ for $0 \leqslant t < +\infty$, so that $\int_0^{+\infty} e^{-\epsilon t} F(t) \, dt$ exists.

Using a method already employed in proving Theorem 3.1, we write

$$\begin{aligned}
|I_\epsilon - I_0| &= \left| \epsilon \int_0^{+\infty} e^{-\epsilon t} F(t) \, dt - \epsilon \int_0^{+\infty} e^{-\epsilon t} I_0 \, dt \right| \\
&\leqslant \left| \int_0^{+\infty} e^{-u} \left(F\left(\frac{u}{\epsilon}\right) - I_0 \right) du \right| \leqslant \int_0^{+\infty} e^{-u} \left| F\left(\frac{u}{\epsilon}\right) - I_0 \right| du.
\end{aligned}$$

For any given $\eta > 0$, select $\alpha > 0$ so that $\alpha < (\eta/2)(C_0 + |I_0|)$,

and consequently

$$\int_0^{\alpha} e^{-u} \left| F\left(\frac{u}{\epsilon}\right) - I_0 \right| du \leqslant |\alpha| (C_0 + |I|) < \frac{\eta}{2}.$$

For such a fixed $\alpha > 0$, choose $\epsilon > 0$ so small that $| F(u/\epsilon) - I_0 | < \eta/2$ for all $u \geqslant \alpha$. Hence

$$\int_{\alpha}^{+\infty} e^{-u} \left| F\left(\frac{u}{\epsilon}\right) - I_0 \right| du \leqslant \frac{\eta}{2} \int_0^{+\infty} e^{-u}\, du = \frac{\eta}{2},$$

and consequently

$$| I_{\epsilon} - I_0 | \leqslant \int_0^{\alpha} + \int_{\alpha}^{+\infty} < \frac{\eta}{2} + \frac{\eta}{2} = \eta.$$

We note that the proof would be much more straightforward if $|f(t)|$ were integrable on $[0, +\infty)$. However, in the present case we deal with an improper Riemann integral and not with a Lebesgue integral, and the convergence of the improper Riemann integral of $f(t)$ does not imply the convergence of the improper Riemann integral of $|f(t)|$.

7.2. Note that since $\tilde{\psi}(\mathbf{k})$ is integrable,

$$(2\pi)^{3/2} |(\exp(-iH_0 t)\psi)(\mathbf{r})|$$

$$= \left| \int_{\mathbb{R}^3} \exp[-i(t\mathbf{k}^2/2m - \mathbf{kr})]\, \tilde{\psi}(\mathbf{k})\, d\mathbf{k} \right| \leqslant \int_{\mathbb{R}^3} | \tilde{\psi}(\mathbf{k})|\, d\mathbf{k}.$$

7.3. Using (6.58) and (6.38) we get (see also Excercise 6.5)

$$\int_{\Omega_S} d\omega' \int_{\Omega_S} d\omega\, | T^{(1)}(k; \boldsymbol{\omega}, \boldsymbol{\omega}')|^2$$

$$\leqslant \text{const} \int_{\Omega_S} d\omega' \int_{\Omega_S} d\omega \left\{ \int_{\mathbb{R}^3} | V(\mathbf{r}')(1 + F_V(\mathbf{r}'; k^2/2m)|\, d\mathbf{r}' \right\}^2$$

$$= 16\pi^2 \,\text{const} \left\{ \int_{\mathbb{R}^3} | V(\mathbf{r}')|\, (1 + F_V(\mathbf{r}'; k^2/2m)\, d\mathbf{r}' \right\}^2.$$

Hence, the square integrability of $T^{(1)}(k; \boldsymbol{\omega}, \boldsymbol{\omega}')$ follows by Tonelli's and Fubini's theorems.

7.4. Using (7.63) we get

$$\begin{aligned}
&\langle T(k) f_1 \mid T(k) f_2\rangle \\
&\quad = (1/4\pi^2)\langle (S(k)-1) f_1 \mid (S(k)-1) f_2\rangle \\
&\quad = (1/4\pi^2) \int_0^{2\pi} (e^{i\lambda}-1)\, d\langle (S(k)-1) f_1 \mid E_\lambda(k) f_2\rangle \\
&\quad = (1/4\pi^2) \int_0^{2\pi} (e^{i\lambda}-1)\, d_\lambda\langle (S(k)-1) f_1 \mid (E_{\lambda_2'}(k) - E_{\lambda_2}(k))\, E_\lambda(k) f_2\rangle.
\end{aligned}$$

On the other hand,

$$\begin{aligned}
&\langle (E_{\lambda_2'}(k) - E_{\lambda_2}(k))\, E_{\lambda'}(k) f_2 \mid (S(k)-1) f_1\rangle \\
&\qquad = \int_0^{2\pi} (e^{i\lambda}-1)\, d_\lambda\langle (E_{\lambda_2'}(k) - E_{\lambda_2}(k))\, E_{\lambda'}(k) f_2 \mid E_\lambda(k) f_1\rangle = 0
\end{aligned}$$

since $[E_{\lambda_2'}(k) - E_{\lambda_2}(k)]\, E_\lambda(k) f_1 = \mathbf{0}$.

7.5. The Legendre polynomials $P_l(u)$ are the solutions of a Sturm–Liouville problem (see Chapter II, Theorem 7.5). Hence, the expansion

$$e^{i\alpha u} = \frac{1}{2} \sum_{l=0}^{\infty} (2l+1)\, P_l(u) \int_{-1}^{+1} P_l(u)\, e^{i\alpha u}\, du$$

is a Sturm–Liouville expansion, and as such it converges uniformly (see Titchmarsh [1962, Theorem 1.9]). To calculate the coefficients in this series, expand $e^{i\alpha u}$ in a power series and show first that

$$\int_{-1}^{+1} u^m\, P_l(u)\, du$$

is zero unless $l + m$ is even. For the case of even $l + m$, the above integral can be computed by using standard relations obeyed by Legendre polynomials (see Butkov [1968, Chapter 9, in particular Problem 17]).

7.6. Note that as in the case of (7.2)

$$(e^{-iH_0 t}\phi_{\mathbf{r}_0})(\mathbf{r}) = (2\pi)^{-3/2} \int_{\mathbb{R}^3} \exp[-(\beta^2 + it)(\mathbf{k}^2/2m) + i\mathbf{k}(\mathbf{r} - \mathbf{r}_0)]\, d\mathbf{k}$$

and carry out the integration in $\mathbf{k} \in \mathbb{R}^3$.

7.7. Note that the required result has been established for $\psi_0^{\mathrm{f}}(\mathbf{r}) = \phi_{\mathbf{r}_0}(\mathbf{r})$ in Exercise 7.6. Now, according to Lemma 7.1, the set $\{\phi_{\mathbf{r}_0} : \mathbf{r}_0 \in \mathbb{R}^3\}$ is dense in $L^2(\mathbb{R}^3)$. Since $\| E^{\mathrm{Q}}(B)\, e^{-iH_0 t} \| \leqslant 1$, we can apply Lemma 1.1 to reach the desired conclusion for arbitrary $\psi_0 \in L^2(\mathbb{R}^3)$.

7.8. Since $\| \psi_t - \psi_t^{\mathrm{ex}} \| \to 0$ when $t \to \mp\infty$ and

$$\| E^{\mathrm{Q}}(B)\psi_t \| \leqslant \| \psi_t - \psi_t^{\mathrm{ex}} \| + \| E^{\mathrm{Q}}(B)\psi_t^{\mathrm{ex}} \|,$$

the desired result follows from Exercise 7.7.

8.1. (a) Four channels: (n, p, e), $(n\text{–}p, e)$, $(n\text{–}e, p)$, and $(p\text{–}e, n)$, plus the bound states $(n\text{–}p\text{–}e)$.

(b) $(n\text{–}e)$ has not been observed so that $\mathsf{M}_0^{(p, n-e)} = \{\mathbf{0}\}$ as far as present experimental knowledge goes.

(c) $(p\text{–}e)$ = hydrogen atom, $(p\text{–}n)$ = deuteron, $(p\text{–}n\text{–}e)$ = deuterium.

8.2. (a) Eight channels: (p, p, n, n), $(p, p, n\text{–}n)$, $(n, n, p\text{–}p)$, $(n, p, n\text{–}p)$, $(n\text{–}p, n\text{–}p)$, $(p\text{–}p, n\text{–}n)$, $(n, n\text{–}p\text{–}p)$ and $(p, p\text{–}n\text{–}n)$ plus the bound states $(n\text{–}n\text{–}p\text{–}p)$.

(b) (p, p, n, n), all four particles are free; $(n, p, n\text{–}p)$ = neutron, protron, and deuteron; $(n\text{–}p, n\text{–}p)$ = two deuterons; $(p, p\text{–}n\text{–}n)$ = protron and triton $(p\text{–}n\text{–}n)$; $(n, p\text{–}p\text{–}n)$ = neutron and the ionized helium isotope $(^3\mathrm{He})^{2+}$ of $^3\mathrm{He}$; $(n - n - p - p)$ = alpha particle $(^4\mathrm{He})^{2+}$.

8.3. Since $f(t)$ is continuous and has a limit at $+\infty$, it is bounded; i.e., $|f(t)| \leqslant M$ for all $0 \leqslant t < +\infty$. For given $\epsilon < 0$ choose $N(\epsilon)$ so that $|f(t) - a| < \epsilon/2$ for all $t \geqslant N(\epsilon)$.

Then

$$\left| \frac{1}{T} \int_0^T f(t)\, dt - a \right| \leqslant \frac{1}{T} \left[\int_0^{N(\epsilon)} |f(t) - a|\, dt + \int_{N(\epsilon)}^T |f(t) - a|\, dt \right]$$
$$\leqslant 2M \frac{N(\epsilon)}{T} + \frac{\epsilon}{2}.$$

Hence, for all $T \geqslant T(\epsilon) = 4M[N(\epsilon)/\epsilon]$, we have

$$\left| \left(\frac{1}{T}\right) \int_0^T f(t)\, dt - a \right| < \epsilon.$$

8.4. Use the relations (8.23) and (8.24).

8.5. Follow the line of argument employed in proving (8.42) and (8.43).

8.6. Note that

$$\Omega_+^{(\beta)*} S' \Omega_+^{(\alpha)} = \sum_\gamma (\Omega_+^{(\beta)*} \Omega_+^{(\gamma)})(\Omega_-^{(\gamma)*} \Omega_+^{(\alpha)}) = \Omega_-^{(\beta)*} \Omega_+^{(\alpha)}$$

since $\Omega_+^{(\beta)*} \Omega_+^{(\beta)} = E_{\mathsf{M}_\pm^{(\beta)}}$, while $\Omega_+^{(\beta)*} \Omega_+^{(\gamma)} = 0$ for $\beta \neq \gamma$ due to the fact that the initial domain $\mathsf{R}_+^{(\beta)}$ of $\Omega_+^{(\beta)*}$ is orthogonal to the final domain $\mathsf{R}_+^{(\gamma)}$ of $\Omega_+^{(\gamma)}$. A similar argument leads to the conclusion that

$$\Omega_-^{(\beta)*} S' \Omega_-^{(\alpha)} = \Omega_-^{(\beta)*} \Omega_+^{(\alpha)}.$$

8.7. Prove that

$$(f \mid g) = \int_{t_1}^{t_2} \langle f \mid U_t g \rangle \, dt$$

is a bounded bilinear form, and apply the result of Exercise 2.5 in Chapter III.

References

Akhiezer, N. I., and Glazman I. M. (1961). "Theory of Linear Operators in the Hilbert Space," translated by M. Nestell. Ungar, New York.

Ali, S. T., and Prugovečki, E. (1977a). Systems of imprimitivity and representations of quantum mechanics on fuzzy phase spaces. *J. Math. Phys.* **18**, 219–228.

Ali, S. T., and Prugovečki, E. (1977b). Classical and quantum statistical mechanics in a common Liouville space. *Physica* **89A**, 501–521.

Ali, S. T., and Prugovečki, E. (1980). *In* "Mathematical Methods and Applications of Scattering Theory" (J. A. De Santo, A. W. Saenz, and W. W. Zachary, eds.), pp. 197–203. Springer, New York.

Alsholm, P. (1977). Wave operators for long-range scattering. *J. Math. Anal. Appl.* **59**, 550–572.

Alt, E. O., Sandhas W., and Ziegelman, H. (1978). Coulomb effects in three-body reactions with two charged particles. *Phys. Rev.* **C17**, 1981–2005.

Amrein, W. O. (1974). *In* "Scattering Theory in Mathematical Physics" (J. A. LaVita and J. P. Marchand, eds.), pp. 97–140. Reidel, Dordrecht, Holland.

Amrein, W. O., Martin, Ph. A., and Misra, B. (1970). On the asymptotic condition in scattering theory. *Helv. Phys. Acta* **43**, 313–344.

Amrein, W. O., Georgescu, V., and Jauch, J. M. (1971). Stationary state scattering theory. *Helv. Phys. Acta* **44**, 407–434.

Amrein, W. O., Jauch, J. M., and Sinha, K. B. (1977). "Scattering Theory in Quantum Mechanics." Benjamin, Reading, Mass.

Antoine, J.-P. (1969). Dirac formalism and symmetry problems in quantum mechanics I. General Dirac formalism, *J. Math. Phys.* **10**, 53–69.

Ascoli, R., Gabbi, P., and Palleschi, G. (1978), On the representation of linear operators in L^2 spaces by means of "generalized matrices." *J. Math. Phys.* **19**, 1023–1027.

Balescu, R. (1975). "Equilibrium and Nonequilibrium Statistical Mechanics." Wiley, New York.

Ballentine, L. E. (1970), The statistical interpretation of quantum mechanics. *Rev. Mod. Phys.* **42**, 358–381.

Barut, A. O., and Rączka, R. (1977). "Theory of Group Representations and Applications." PWN-Polish Scientific Publ., Warsaw.

Bateman, H. (1953). "Higher Transcendental Functions." McGraw-Hill, New York.

Belinfante, J. G. (1964). Existence of scattering solutions for the Schrödinger equation. *J. Math. Phys.* **5**, 1070–1074.

Berezanskiĭ, Yu. M. (1968). "Expansions in Eigenfunctions of Self-Adjoint Operators," translated from Russian by R. Bolstein, J. M. Danskin, J. Rovnyak, and L. Shulman. Amer. Math. Soc., Providence, Rhode Island.

Berezanskiĭ, Yu. M. (1978). "Self-Adjoint Operators in Spaces of Functions of Infinitely Many Variables" (in Russian). Naukova Dumka, Kiev.

Birkhoff, G., and MacLane, S. (1953). "A Survey of Modern Algebra. Macmillan, New York.

Birman, M. S., and Kreĭn, M. G. (1962). On the theory of wave operators and scattering operators. *Dokl. Akad. Nauk. SSSR* **144**, 475–478.

Birman, M. S., and Solomyak, M. Z. (1967). *In* "Topics in Mathematical Physics," Vol. 1, pp. 25–54. Consultants Bureau, New York.

Birman, M. S., and Solomyak, M. Z. (1968). *In* "Topics in Mathematical Physics," Vol. 2, pp. 19–46. Consultants Bureau, New York.

Bogolubov, N. N., Logunov, A. A., and Todorov, I. T. (1975). "Introduction to Axiomatic Quantum Field Theory" (transl. by S. A. Fulling and L. G. Popova). Benjamin, Reading, Mass.

Born, M. (1926a). Zur Quantenmechanik der Stossvorgänge, *Z. Physik* **37**, 863–867.

Born, M. (1926b). Quantenmechanik der Stossvorgänge, *Z. Physik* **38**, 803–827.

Buslaev, V. S., and Matveev, V. B. (1970). Wave operators for the Schroedinger equation with slowly a decreasing potential. *Theor. Math. Phys.* **2**, 266–274.

Butkov, E. (1968). "Mathematical Physics." Addison-Wesley, Reading, Massachusetts.

Cattapan, G., Pisent, G., and Vanzani, V. (1975). Exact soluble multichannel scattering problems with Coulomb interactions. *Z. Physik* **A274**, 139–144.

Chandler, C., and Gibson, A. G. (1973). Transition from time-dependent to time-independent multichannel quantum scattering theory. *J. Math. Phys.* **14**, 1328–1335.

Chandler, C., and Gibson, A. G. (1974). Time-dependent multichannel Coulomb scattering theory. *J. Math. Phys.* **15**, 291–294.

Chandler, C., and Gibson, A. G. (1977). *N*-body quantum scattering theory in two Hilbert spaces I. Basic equations. *J. Math. Phys.* **18**, 2336–2347.

Chandler, C., and Gibson, A. G. (1978). *N*-body quantum scattering theory in two Hilbert spaces II. Some asymptotic limits. *J. Math. Phys.* **19**, 1610–1616.

Coombe, D. A., Sanctuary, B. C., and Snider, R. F. (1975). Definitions and properties of generalized collision cross-sections. *J. Chem. Phys.* **63**, 3015–3030.

Corbett, J. V. (1970). Galilean symmetry, measurement, and scattering as an isomorphism between two subalgebras of observables. *Phys. Rev.* **D1**, 3331–3344.

Courant, R., and Hilbert, D. (1953). "Methods of Mathematical Physics." Vol. I. Wiley (Interscience), New York.

Daletskiĭ, Y. L., and Krein, S. G. (1956). The integration and differentiation of functions of Hermitian operators and applications to perturbation theory. *Voronezh. Gos. Univ. Trudy Sem. Funk. Anal.*, No. 1, 81–105.

de Alfaro, V., and Regge, T. (1965). "Potential Scattering." North-Holland Publ., Amsterdam.

de Dormale, B. M., and Gautrin, H.-F. (1975). Spectral representation and decomposition of self-adjoint operators. *J. Math. Phys.* **16**, 2328–2332.

Dennery, P., and Krzywicki, A. (1967). "Mathematics for Physicists." Harper, New York.

De Santo, J. A., Saenz, A. W., and Zachary, W. W., eds. (1980). "Mathematical Methods and Applications of Scattering Theory." Springer, Berlin.

d'Espagnat, B. (1976). "Conceptual Foundations of Quantum Mechanics," 2nd ed., Benjamin, Reading, Mass.

Dietrich, C. F. (1973). "Uncertainty, Calibration and Probability." Wiley, New York.

Dirac, P. A. M. (1930). "The Principles of Quantum Mechanics." Oxford Univ. Press (Clarendon), London and New York.
Dixmier, J. (1952). "Les Algèbres d'Opérateurs dans l'Espace Hilbertien." Gauthier-Villars, Paris.
Dollard, J. D. (1964). Asymptotic convergence and the Coulomb interaction. *J. Math. Phys.* **5**, 729–738.
Dollard, J. D. (1969). Scattering into cones I: Potential scattering. *Comm. Math. Phys.* **12**, 193–203.
Dunford, N., and Schwartz, J. T. (1957). "Linear Operators I: General Theory." Wiley (Interscience), New York.
Ekstein, H. (1956). Theory of time-dependent scattering for multichannel processes. *Phys. Rev.* **101**, 880–890.
Emch, G. G. (1972). "Algebraic Methods in Statistical Mechanics and Quantum Field Theory." Wiley (Interscience), New York.
Epstein, S. T. (1957). Theory of rearrangement collisions. *Phys. Rev.* **106**, 598.
Eu, B. C. (1975). Kinetic equations for reacting chemical system: Exchange reaction. *J. Chem. Phys.* **63**, 303–315.
Faddeev, L. D. (1961a), Scattering theory for a three-particle system, *Soviet Phys. JETP* **12**, 1014–1019.
Faddeev, L. D. (1961b), The resolvent of the Schroedinger operator for a system of three particles interacting in pairs, *Soviet Phys. Dokl.* **6**, 384–386.
Faddeev, L. D. (1963). The construction of the resolvent of the Schrödinger operator for a three-particle system, and the scattering problem. *Soviet Phys. Dokl.* **7**, 600–602.
Faddeev, L. D. (1965). "Mathematical Aspects of the Three-Body Problem in Quantum Scattering Theory." Israel Program for Scientific Translations, Jerusalem.
Fano, U. (1963). Pressure broadening as a prototype of relaxation. *Phys. Rev.* **131**, 259–268.
Fine, T. L. (1973). "Theories of Probability: An Examination of Foundations." Academic, New York.
Friedrichs, K. (1939). On differential operators in Hilbert space. *Amer. J. Math.* **61**, 523–544.
Gärding, L., and Wightmann, A. S. (1954a). Representations of anticommutation relations. *Proc. Nat. Acad. Sci. U. S.* **40**, 617–621.
Gärding, L., and Wightmann, A. S. (1954b). Representations of commutation relations. *Proc. Nat. Acad. Sci. U. S.* **40**, 622–626.
Gelfand, I. M., and Vilenkin, N. Ya (1968). "Generalized Functions," Vol. IV (transl. by A. Feinstein). Academic, New York and London.
Gersten, A. (1976). A modified Lippmann–Schwinger equation for Coulomb-like interactions. *Nucl. Phys.* **B103**, 465–476.
Gibson, A. G., and Chandler, C. (1974). Time-independent multichannel scattering theory for charged particles. *J. Math. Phys.* **15**, 1366–1377.
Gottfried, K. (1966). "Quantum Mechanics." Benjamin, New York.
Green, T. A., and Lanford, III, O. E. (1960). Rigorous derivation of the phase-shift formula for the Hilbert space scattering operator of a single particle. *J. Math Phys.* **1**, 39–48.
Grossman, A., and Wu, T. T. (1962). Schroedinger scattering amplitude III. *J. Math. Phys.* **3**, 684–689.
Hack, M. N. (1959). Wave operators in multichannel scattering. *Nuovo Cimento* **13**, 231–236.

Hagedorn, G. (1978). Asymptotic completeness for a class of four particle Schrödinger operators. *Bull. Amer. Math. Soc.* **84**, 155–156.

Halmos, P. R. (1950). "Measure Theory." Van Nostrand, Princeton, New Jersey.

Heisenberg, W. (1925). Über quanten theoretische Umdeutung kinematischer und mechanischer Beziehungen. *Z. Physik* **33**, 879–893.

Hörmander, L. (1969). "Linear Partial Differential Operators." Springer, New York.

Hörmander, L. (1976), The existence of wave operators in scattering theory. *Math. Z.* **146**, 69–91.

Howland, J. S. (1967). Banach space techniques in the perturbation theory of self-adjoint operators with continuous spectrum. *J. Math. Anal. Appl.* **20**, 22–47.

Hunziker, W. (1964). Proof of a conjecture of S. Weinberg. *Phys. Rev. B* **135**, 800–803.

Hunziker, W. (1965). Cluster properties of multiparticle systems. *J. Math. Phys.* **6**, 6–10.

Hunziker, W. (1966). On the spectra of Schrödinger multiparticle Hamiltonians. *Helv. Phys. Acta* **39**, 451–462.

Ikebe, T. (1960). Eigenfunction expansions associated with the Schroedinger operators and their applications to scattering theory. *Arch. Rational Mech. Anal.* **5**, 1–34.

Ikebe, T. (1965). On the phase shift formula for the scattering operator. *Pacific J. Math.* **15**, 511–523.

Ikebe, T. (1975), Spectral representations for Schrödinger operators with long-range potentials. *J. Func. Anal.* **20**, 158–177.

Ince, E. L. (1956). "Ordinary Differential Equations." Dover, New York.

Iorio, R. J., Jr. (1978). On the discrete spectrum of the *N*-body quantum mechanical Hamiltonian I. *Commun. Math. Phys.* **62**, 201–212.

Jammer, M. (1974). "The Philosophy of Quantum Mechanics." Wiley, New York.

Jauch, J. M. (1958a). Theory of the scattering operator I. *Helv. Phys. Acta* **31**, 127–158.

Jauch, J. M. (1958b). Theory of the scattering operator II: Multichannel scattering. *Helv. Phys. Acta* **31**, 662–684.

Jauch, J. M. (1964). The problem of measurement in quantum mechanics. *Helv. Phys. Acta* **37**, 293–316.

Jauch, J. M. (1968). "Foundations of Quantum Mechanics." Addison-Wesley, Reading, Massachusetts.

Jauch, J. M., and Misra, B. (1965). The spectral representation. *Helv. Phys. Acta* **38**, 30–52.

Jauch, J. M., and Zinnes, I. I. (1959). The asymptotic condition for simple scattering systems. *Nuovo Cimento* **11**, 553–567.

Jauch, J. M., Misra, B., and Gibson, A. G. (1968). On the asymptotic condition of scattering theory. *Helv. Phys. Acta* **41**, 513–527.

Joachain, C. J. (1975). "Quantum Collision Theory." North-Holland, Amsterdam.

Jordan, T. F. (1962a). The quantum mechanical scattering problem I. *J. Math. Phys.* **3**, 414–428.

Jordan, T. F. (1962b). The quantum mechanical scattering problem II: Multichannel scattering. *J. Math. Phys.* **3**, 429–439.

Jörgens, K. (1964). Wesentliche Selbstadjungierthert singulärer elliptischer Differential-operatoren zweiter Ordnung in $C_0^\infty(G)$. *Math. Scand.* **15**, 5–17.

Jörgens, K., and Weidmann, J. (1973). "Spectral Properties of Hamiltonian Operators." Springer, Berlin.

Kato, T. (1951). Fundamental properties of Hamiltonian operators of the Schrödinger type. *Trans. Amer. Math. Soc.* **70**, 196–211.

Kato, T. (1959). Growth properties of solutions of the reduced wave equation with a variable coefficient. *Comm. Pure Appl. Math.* **12**, 403–425.
Kato, T. (1966). "Perturbation Theory for Linear Operators." Springer, New York.
Kato, T. (1967). Some mathematical problems in quantum mechanics. *Progr. Theoret. Phys. Suppl.* **40**, 3–19.
Kato, T., and Kuroda, S. T. (1959). A remark on the unitarity property of the scattering operator. *Nuovo Cimento* **14**, 1102–1107.
Kirczenov, G., and Marro, J. (1974). "Transport Phenomena." Springer, Berlin.
Kitada, H. (1977). Scattering theory for Schrödinger operators with long-range potentials. *J. Math. Soc. Japan* **29**, 665–691.
Klein, J. R., and Zinnes, I. I. (1973). On nonrelativistic field theory: Interpolating fields and long-range forces. *J. Math. Phys.* **14**, 1205–1212.
Kodaira, K. (1949). The eigenvalue problem for ordinary differential equations of the second order and Heisenberg's theory of *S*-matrices. *Amer. J. Math.* **71**, 921–945.
Kolmogorov, A. N., and Fomin, S. V. (1961). "Measure, Lebesgue Integrals, and Hilbert Space," translated by N. A. Brunswich and A. Jeffrey. Academic Press, New York.
Kuroda, S. T. (1959a). On the existence and unitary property of the scattering operator. *Nuovo Cimento* **12**, 431–454.
Kuroda, S. T. (1959b). Perturbation of continuous spectra by unbounded operators, I. *J. Math. Soc. Japan* **11**, 247–262.
Kuroda, S. T. (1959c). Perturbation of continuous spectra by unbounded operators, II. *J. Math. Soc. Japan* **12**, 243–257.
Kuroda, S. T. (1967). An abstract stationary approach to perturbation of continuous spectra and scattering theory. *J. Analyse Math.* **20**, 57–177.
Landau, L. D., and Lifshitz, E. M. (1958). "Quantum Mechanics," translated from Russian by J. B. Sykes and J. S. Bell. Addison-Wesley, Reading, Massachusetts.
Lapicki, G., and Losonsky, W. (1979). Coulomb deflection in ion-atom collisions. *Phys. Rev.* **A20**, 481–490.
Lavine, R. (1973). Absolute continuity of positive spectrum for Schrödinger operators with long-range potentials. *J. Func. Anal.* **12**, 30–54.
LaVita, J. A., and Marchand, J. P., eds. (1974). "Scattering Theory in Mathematical Physics." Reidel, Dordrecht, Holland.
Lax, P. D., and Phillips, R. S. (1967). "Scattering Theory." Academic Press, New York.
Leaf, B. (1975). *S*-operator (matrix) formalism for the Liouville and von Neumann equations and the approach to equilibrium. *Physica* **81A**, 163–189.
Lévy-Leblond, J.-M. (1971). *In* "Group Theory and Its Applications," (E. M. Loebl, ed.), Vol. II, pp. 222–299. Academic, New York.
Limić, N. (1963). On the existence of the scattering operator. *Nuovo Cimento* **28**, 1066–1090.
Lippmann, B. A. (1956). Rearrangement collisions. *Phys. Rev.* **102**, 264–268.
Loebl, E. M., ed. (1971). "Group Theory and Its Applications," Vol. II. Academic, New York.
Manoukian, E. B., and Prugovečki, E. (1971). On the existence of wave operators in time-dependent potential scattering for long-range potentials. *Can. J. Phys.* **49**, 102–107.
Margenau, H. (1958). Philosophical problems concerning the meaning of measurement in quantum mechanics. *Philos. Sci.* **25**, 23–33.
Marlow, A. R., ed. (1978). "Mathematical Foundations of Quantum Theory," Academic, New York and London.

Massey, H. S. W. (1956). Theory of atomic collisions. *In* "Handbuch der Physik" (S. Flügge, ed.), Vol. 36. Springer, Berlin.
Masson, D., and Prugovečki, E. (1976). An effective-potential approach to stationary scattering theory for long-range potentials. *J. Math. Phys.* **17**, 297–302.
Matveev, V. B., and Skriganov, M. M. (1972). Scattering problem for radial Schrödinger equation with a slowly decreasing potential. *Theor. Math. Phys.* **10**, 156–164.
Messiah, A. (1962). "Quantum Mechanics," translated by J. Potter. Wiley, New York.
Miles, J. R. N., and Dahler, J. S. (1970). Classical scattering theory. Elastic collisions. *J. Chem. Phys.* **52**, 616–621.
Miller, W., Jr. (1972). "Symmetry Groups and Their Applications." Academic, New York and London.
Munroe, M. E. (1953). "Introduction to Measure and Integration." Addison-Wesley, Reading, Massachusetts.
Naimark, M. A. (1959). "Normed Rings," translated by L. F. Boron, P. Noordhoff. N. V. Groningen, The Netherlands.
Narnhofer, H. (1975). Continuity of the *S*-matrix. *Nuovo Cimento* **30B**, 254–266.
Nelson, E. (1959). Analytic vectors. *Ann. Math.* **70**, 572–615.
Nering, E. D. (1963). "Linear Algebra and Matrix Theory." Wiley, New York.
Newton, R. G. (1966). "Scattering Theory of Waves and Particles." McGraw-Hill, New York.
Newton, R. G. (1979). The density matrix of scattered particles. *Found. Phys.* **9**, 929–935.
Nuttal, J., ed. (1978). "Atomic Scattering Theory: Mathematical and Computational Aspects." Univ. Western Ontario Press, London, Ontario.
Pap, A. (1962). "An Introduction to the Philosophy of Science." The Free Press of Glencoe, New York.
Park, J. L., and Margenau, H. (1968). Simultaneous measurability in quantum mechanics. *Int. J. Theor. Phys.* **1**, 211–283.
Pearson, D. (1975). An example in potential scattering illustrating the breakdown of asymptotic completeness. *Commun. Math. Phys.* **40**, 125–146.
Povzner, A. Ya. (1953). On the expansions of arbitrary function in terms of eigenfunctions of the operator $-\Delta u + cu$ (in Russian). *Mat. Sb.* (*N.S.*) **32**(74), 109–156.
Povzner, A. Ya. (1955). Expansions in functions which are solutions of the scattering problem. *Dokl. Akad. Nauk SSR* **104**, 360–363.
Price, W. C., and Chissick, S. S. (1977). "The Uncertainty Principle and Foundations of Quantum Mechanics." Wiley, New York, London.
Prigogine, I. (1962). "Non-Equilibrium Statistical Mechanics." Wiley (Interscience), New York.
Prugovečki, E. (1966). An axiomatic approach to the formalism of quantum mechanics. *J. Math. Phys.* **7**, 1054–1096.
Prugovečki, E. (1967). On a theory of measurement of incompatible observables in quantum mechanics. *Can. J. Phys.* **45**, 2173–2219.
Prugovečki, E. (1969a). Complete sets of observables. *Can. J. Phys.* **47**, 1083–1093.
Prugovečki, E. (1969b). Rigorous derivation of generalized Lippmann-Schwinger equations from time-dependent scattering theory. *Nuovo Cimento B* **63**, 569–592.
Prugovečki, E. (1971a). On time dependent scattering theory for long-range interactions. *Nuovo Cimento* **4B**, 105–123.
Prugovečki, E. (1971b). Integral representations of the wave and transition operators in nonrelativistic scattering theory. *Nuovo Cimento* **4B**, 124–134.
Prugovečki, E. (1972a). Riemann–Stieltjes integration of operator-valued functions with respect to vector-valued functions in Banach spaces. *SIAM J. Math. Anal.* **3**, 183–205.

Prugovečki, E. (1972b). Scattering theory in Fock space. *J. Math. Phys.* **13**, 969–976.
Prugovečki, E. (1973a). Multichannel stationary scattering theory in two-Hilbert space formulation. *J. Math. Phys.* **14**, 957–962; erratum, *ibid.* **16**, 442.
Prugovečki, E. (1973b). The bra and ket formalism in extended Hilbert space. *J. Math. Phys.* **14**, 1410–1422.
Prugovečki, E. (1976). Quantum two-particle scattering in fuzzy phase space. *J. Math. Phys.* **17**, 1673–1681.
Prugovečki, E. (1978a). Liouville dynamics for optimal stochastic phase-space representations of quantum mechanics. *Ann. Phys. (N.Y.)* **110**, 102–121.
Prugovečki, E. (1978b). A unified treatment of dynamics and scattering in classical and quantum statistical mechanics. *Physica* **91A**, 202–228.
Prugovečki, E. (1978c). A quantum-mechanical Boltzmann equation for one-particle Γ_s-distribution functions. *Physica* **91A**, 229–248.
Prugovečki, E. (1978d). Consistent formulation of relativistic dynamics for massive spin-zero particles in external fields. *Phys. Rev.* **D18**, 3655–3675.
Prugovečki, E. (1979). Stochastic phase spaces and master Liouville spaces in statistical mechanics. *Found. Phys.* **9**, 575–587.
Prugovečki, E. (1981a). Quantum spacetime operationally based on propagators for extended test particles. *Hadronic J.* **4**, 1018–1104.
Prugovečki, E. (1981b). Quantum action principle and functional integration over paths in stochastic phase space. *Nuovo Cimento* **61A**, 85–118.
Prugovečki, E. (1981c). A self-consistent approach to quantum field theory for extended particles. *Found. Phys.* **11**, 355–382.
Prugovečki, E., and Tip, A. (1974). Scattering theory in a time-dependent external field. *J. Phys. A: Math. Nucl. Gen.* **7**, 572–596.
Prugovečki, E., and Tip, A. (1975). Semi-groups of rank-preserving transformers on minimal norm ideals in $\mathscr{B}(\mathscr{H})$. *Comp. Math.* **30**, 113–136.
Prugovečki, E., and Zorbas, J. (1973a). Modified Lippmann–Schwinger equations for two-body scattering theory with long-range forces. *J. Math. Phys.* **14**, 1398–1409; erratum, *ibid.* **15**, 268.
Prugovečki, E., and Zorbas, J. (1973b). Many-body modified Lippmann–Schwinger equations for Coulomb-like potentials. *Nucl. Phys.* **A213**, 541–569.
Putman, C. R. (1967). "Commutation Properties of Hilbert Space Operators and Related Topics." Springer, New York.
Randolph. J. F. (1968). "Basic Real and Abstract Analysis." Academic Press, New York.
Reed, M., and Simon, B. (1972). "Methods of Modern Mathematical Physics I: Functional Analysis." Academic, New York.
Reed, M., and Simon, B. (1975). "Methods of Modern Mathematical Physics II: Fourier Analysis, Self-Adjointness." Academic, New York.
Reed, M., and Simon, B. (1978). "Methods of Modern Mathematical Physics IV: Analysis of Operators." Academic, New York.
Reed, M., and Simon, B. (1979). "Methods of Modern Mathematical Physics III: Scattering Theory." Academic, New York.
Rejto, P. A. (1976). On a theorem of Titchmarsh–Kodaira–Weidmann concerning absolute continuous operators. *Ind. Math. J.* **25**, 629–658.
Rellich, F. (1939). Störungs theorie der Spektralzerlegung II. *Math. Ann.* **116**, 555–570.
Resibois, P. (1959). Theorie formelle du scattering clasique. *Physica* **25**, 725–732.
Riesz, F., and Sz. Nagy, B. (1955). "Functional Analysis," translated by L. F. Boron. Ungar, New York.
Rollnik, H. (1956). Streumaxima und gebundene Zustande. *Z. Phys.* **145**, 639–653.

Rosenberg, L. (1979). Coulomb scattering in a laser field. *Phys. Rev.* **A20**, 457–464.

Rys, F. (1965). Theory of scattering of identical particles. *Helv. Phys. Acta* **38**, 457–468.

Sáenz, A. W., and Zachary, W. W. (1976). Wave operators for long-range hard-core potentials. *J. Math. Phys.* **17**, 954–957.

Saito, Y. (1977). Eigenfunction expansions for the Schrödinger operator with long-range potentials, $Q(y) = O(|y|^{-\epsilon})$, $\epsilon > 0$. *Osaka J. Math.* **14**, 37–53.

Scadron, M., Weinberg, S., and Wright, J. (1964). Functional analysis and scattering theory. *Phys. Rev.* **135**, B202–B207.

Schatten, R. (1960). "Norm Ideals of Completely Continuous Operators," Springer, Berlin.

Schechter, M. (1971). "Principles of Functional Analysis." Academic, New York and London.

Schechter, M. (1975). Scattering theory for second order elliptic operators. *Ann. Mat. Pura Appl.* **105**, 313–331.

Schechter, M. (1978). Another look at scattering theory. *Int. J. Theor. Phys.* **17**, 33–41.

Schminke, U.-W. (1972). Essential self-adjointness of a Schrödinger operator with strongly singular potentials. *Math. Z.* **124**, 47–50.

Schroeck, F. E., Jr. (1978). *In* "Mathematical Foundations of Quantum Theory" (A. R. Marlow, ed.), pp. 299–327. Academic, New York and London.

Schroedinger, E. (1926a). Quantisierung als Eigenwert Problem. *Ann. Physik* **79**, 361–376, 489–527.

Schroedinger, E. (1926b). Über das Verhältnis der Heisenberg–Born–Jordanschen Quantenmechanik zu der meinen von Erwin Schrödinger. *Ann. Physik* **79**, 734–756.

Schwartz, J. (1960). Some non-self-adjoint operators. *Commun. Pure Appl. Math.* **13**, 609–639.

Segal, I. E. (1947). Postulates for general quantum mechanics. *Ann. Math.* **48**, 930–948.

Semon, M. D., and Taylor, J. R. (1975). Scattering by potentials with Coulomb tails. *Nuovo Cimento* **26A**, 48–58.

Semon, M. D., and Taylor, J. R. (1976). Cross sections for screened potentials. *J. Math. Phys.* **17**, 1366–1370.

She, C. Y., and Hefner, H. (1966). Simultaneous measurement of noncommuting observables. *Phys. Rev.* **152**, 1103–1110.

Sigal, I. (1978a). On quantum mechanics of many-body systems with dilation analytic potentials. *Bull. Amer. Math. Soc.* **84**, 152–154.

Sigal, I. M. (1978b). Mathematical foundations of quantum scattering theory for multiparticle systems. *Mem. Am. Math. Soc.* **16**, No. 209.

Smithies, F. (1965). "Integral Equations." Cambridge Univ. Press, London and New York.

Snider, R. F., and Sanctuary, B. C. (1971). Generalized Boltzmann equation for molecules with internal states. *J. Chem. Phys.* **55**, 1555–1566.

Srinivas, M. D., and Wolf, E. (1975). Some nonclassical features of phase-space representations of quantum mechanics, *Phys. Rev.* **D11**, 1477–1485.

Stone, M. H. (1964). "Linear Transformations in Hilbert Space." Amer. Math. Soc., Providence, Rhode Island.

St. Pierre, A. G. (1973). Classical phase space description of rotationally inelastic collisions. *J. Chem. Phys.* **59**, 5364–5372.

Stummel, F. (1956). Singuläre elliptische differential operatoren in Hilbertschen Räumen. *Math. Ann.* **132**, 150–176.

Taylor, J. R. (1972). "Scattering Theory: The Quantum Theory of Nonrelativistic Collisions." Wiley, New York.

Thomas, L. E. (1975). Asymptotic completeness in two- and three-particle quantum mechanical scattering. *Ann. Phys. (N.Y.)* **90**, 127–165.

Tip, A. (1971). Transport equations for dilute gases with internal degrees of freedom. *Physica* **52**, 493–522.

Titchmarsh, E. C. (1962). "Eigenfunction Expansions." Oxford Univ. Press (Clarendon), London and New York.

Tixaire, A. G. (1959). Scattering integral equations in Hilbert space. *Helv. Phys. Acta* **32**, 412–422.

Treves, F. (1967). "Topological Vector Spaces, Distributions and Kernels." Academic, New York and London.

Trotter, H. (1959). On the product of semigroups of operators. *Proc. Amer. Math. Soc.* **10**, 545–551.

van Winter, C. (1964). Theory of finite systems of particles I. The Green function. *Mat.-Fys. Skr. Danske Vid. Selsk.* **2**, No. 8.

van Winter, C. (1965). Theory of finite systems of particles II. Scattering theory. *Mat.-Fys. Skr. Danske Vid. Selsk.* **2**, No. 10.

van Winter, C., and Brascamp, H. J. (1968). The N-body problem with spin-orbit or Coulomb interactions. *Comm. Math. Phys.* **11**, 19–55.

Varadarajan, V. S. (1968). "Geometry of Quantum Mechanics." Van Nostrand, Princeton, New Jersey.

von Neumann, J. (1931). Die eindentigheit der Schrödingerschen Operatoren. *Math. Ann.* **104**, 570–578.

von Neumann, J. (1932). Proof of the quasi-ergodic hypothesis. *Proc. Nat. Acad. Sci. U.S.* **18**, 70–82.

von Neumann, J. (1936). On an algebraic generalization of the quantum mechanical formalism I. *Mat. Sb.* **1**, 415–484.

von Neumann, J. (1955). "Mathematical Foundations of Quantum Mechanics," translated by R. T. Beyer. Princeton Univ. Press, Princeton, New Jersey.

Wick, G. C., Wigner, E. P., and Wightman, A. S. (1952). Intrinsic parity of elementary particles. *Phys. Rev.* **88**, 101–105.

Wienholtz, E. (1958). Halbbeschränkte partielle Differentialoperatoren zweiter Ordnung vom elliptischen Typus. *Math. Ann.* **135**, 50–80.

Wilcox, C. E., ed. (1966). "Perturbation Theory and Its Applications to Quantum Mechanics." Wiley, New York.

Yakubovskiĭ (1967). On the integral equations in the theory of N-particle scattering. *Soviet J. Nucl. Phys.* **5**, 937–942.

Yosida, K. (1974). "Functional Analysis." 4th edition. Springer, Berlin.

Zachary, W. W. (1972). Nonrelativistic time-dependent scattering theory and von Neumann algebras I. Single channel scattering. *J. Math. Phys.* **13**, 609–615.

Zachary, W, W, (1976). Wave operators for multichannel scattering by long-range potentials. *J. Math. Phys.* **17**, 1056–1063.

Zhislin, G. M. (1960). An investigation of the spectrum of the Schrödinger operator for many-particle systems (in Russian). *Tr. Moskov. Mat. Obšč.* **9**, 81–120.

Zinnes, I. I. (1959). Two theorems on scattering. *Nuovo Cimento Suppl.* **12**, 87–99.

Zorbas, J. (1974). A space cut-off approach to scattering involving Coulomb-like potentials. *J. Phys. A: Math. Nucl. Gen.* **7**, 1557–1567.

Zorbas, J. (1976a). Stationary scattering for N-body systems involving Coulomb potentials. *J. Math. Phys.* **17**, 498–502.

Zorbas, J. (1976b). The Gell–Mann–Goldberger formula for long-range potential scattering. *Rept. Math. Phys.* **9**, 309–320.

Zorbas, J. (1977). Renormalized off-energy-shell Coulomb scattering. *J. Math. Phys.* **18**, 1112–1120.

Zwanzig, R. (1963). Method for finding the density expansion of transport coefficients of gases. *Phys. Rev.* **129**, 486–494.

Index

D

E

F

G

H

I

K

L

T

A CATALOG OF SELECTED

DOVER BOOKS

IN SCIENCE AND MATHEMATICS

Mathematics

FUNCTIONAL ANALYSIS (Second Corrected Edition), George Bachman and Lawrence Narici. Excellent treatment of subject geared toward students with background in linear algebra, advanced calculus, physics and engineering. Text covers introduction to inner-product spaces, normed, metric spaces, and topological spaces; complete orthonormal sets, the Hahn-Banach Theorem and its consequences, and many other related subjects. 1966 ed. 544pp. 6⅛ x 9¼. 0-486-40251-7

ASYMPTOTIC EXPANSIONS OF INTEGRALS, Norman Bleistein & Richard A. Handelsman. Best introduction to important field with applications in a variety of scientific disciplines. New preface. Problems. Diagrams. Tables. Bibliography. Index. 448pp. 5⅜ x 8½. 0-486-65082-0

VECTOR AND TENSOR ANALYSIS WITH APPLICATIONS, A. I. Borisenko and I. E. Tarapov. Concise introduction. Worked-out problems, solutions, exercises. 257pp. 5⅝ x 8¼. 0-486-63833-2

AN INTRODUCTION TO ORDINARY DIFFERENTIAL EQUATIONS, Earl A. Coddington. A thorough and systematic first course in elementary differential equations for undergraduates in mathematics and science, with many exercises and problems (with answers). Index. 304pp. 5⅜ x 8½. 0-486-65942-9

FOURIER SERIES AND ORTHOGONAL FUNCTIONS, Harry F. Davis. An incisive text combining theory and practical example to introduce Fourier series, orthogonal functions and applications of the Fourier method to boundary-value problems. 570 exercises. Answers and notes. 416pp. 5⅜ x 8½. 0-486-65973-9

COMPUTABILITY AND UNSOLVABILITY, Martin Davis. Classic graduate-level introduction to theory of computability, usually referred to as theory of recurrent functions. New preface and appendix. 288pp. 5⅜ x 8½. 0-486-61471-9

ASYMPTOTIC METHODS IN ANALYSIS, N. G. de Bruijn. An inexpensive, comprehensive guide to asymptotic methods–the pioneering work that teaches by explaining worked examples in detail. Index. 224pp. 5⅜ x 8½ 0-486-64221-6

APPLIED COMPLEX VARIABLES, John W. Dettman. Step-by-step coverage of fundamentals of analytic function theory–plus lucid exposition of five important applications: Potential Theory; Ordinary Differential Equations; Fourier Transforms; Laplace Transforms; Asymptotic Expansions. 66 figures. Exercises at chapter ends. 512pp. 5⅜ x 8½. 0-486-64670-X

INTRODUCTION TO LINEAR ALGEBRA AND DIFFERENTIAL EQUATIONS, John W. Dettman. Excellent text covers complex numbers, determinants, orthonormal bases, Laplace transforms, much more. Exercises with solutions. Undergraduate level. 416pp. 5⅜ x 8½. 0-486-65191-6

RIEMANN'S ZETA FUNCTION, H. M. Edwards. Superb, high-level study of landmark 1859 publication entitled "On the Number of Primes Less Than a Given Magnitude" traces developments in mathematical theory that it inspired. xiv+315pp. 5⅜ x 8½. 0-486-41740-9

CALCULUS OF VARIATIONS WITH APPLICATIONS, George M. Ewing. Applications-oriented introduction to variational theory develops insight and promotes understanding of specialized books, research papers. Suitable for advanced undergraduate/graduate students as primary, supplementary text. 352pp. 5⅜ x 8½. 0-486-64856-7

COMPLEX VARIABLES, Francis J. Flanigan. Unusual approach, delaying complex algebra till harmonic functions have been analyzed from real variable viewpoint. Includes problems with answers. 364pp. 5⅜ x 8½. 0-486-61388-7

AN INTRODUCTION TO THE CALCULUS OF VARIATIONS, Charles Fox. Graduate-level text covers variations of an integral, isoperimetrical problems, least action, special relativity, approximations, more. References. 279pp. 5⅜ x 8½. 0-486-65499-0

COUNTEREXAMPLES IN ANALYSIS, Bernard R. Gelbaum and John M. H. Olmsted. These counterexamples deal mostly with the part of analysis known as "real variables." The first half covers the real number system, and the second half encompasses higher dimensions. 1962 edition. xxiv+198pp. 5⅜ x 8½. 0-486-42875-3

CATASTROPHE THEORY FOR SCIENTISTS AND ENGINEERS, Robert Gilmore. Advanced-level treatment describes mathematics of theory grounded in the work of Poincaré, R. Thom, other mathematicians. Also important applications to problems in mathematics, physics, chemistry and engineering. 1981 edition. References. 28 tables. 397 black-and-white illustrations. xvii + 666pp. 6⅛ x 9¼. 0-486-67539-4

INTRODUCTION TO DIFFERENCE EQUATIONS, Samuel Goldberg. Exceptionally clear exposition of important discipline with applications to sociology, psychology, economics. Many illustrative examples; over 250 problems. 260pp. 5⅜ x 8½. 0-486-65084-7

NUMERICAL METHODS FOR SCIENTISTS AND ENGINEERS, Richard Hamming. Classic text stresses frequency approach in coverage of algorithms, polynomial approximation, Fourier approximation, exponential approximation, other topics. Revised and enlarged 2nd edition. 721pp. 5⅜ x 8½. 0-486-65241-6

INTRODUCTION TO NUMERICAL ANALYSIS (2nd Edition), F. B. Hildebrand. Classic, fundamental treatment covers computation, approximation, interpolation, numerical differentiation and integration, other topics. 150 new problems. 669pp. 5⅜ x 8½. 0-486-65363-3

THREE PEARLS OF NUMBER THEORY, A. Y. Khinchin. Three compelling puzzles require proof of a basic law governing the world of numbers. Challenges concern van der Waerden's theorem, the Landau-Schnirelmann hypothesis and Mann's theorem, and a solution to Waring's problem. Solutions included. 64pp. 5⅜ x 8½. 0-486-40026-3

THE PHILOSOPHY OF MATHEMATICS: AN INTRODUCTORY ESSAY, Stephan Körner. Surveys the views of Plato, Aristotle, Leibniz & Kant concerning propositions and theories of applied and pure mathematics. Introduction. Two appendices. Index. 198pp. 5⅜ x 8½. 0-486-25048-2

Math–Decision Theory, Statistics, Probability

ELEMENTARY DECISION THEORY, Herman Chernoff and Lincoln E. Moses. Clear introduction to statistics and statistical theory covers data processing, probability and random variables, testing hypotheses, much more. Exercises. 364pp. 5⅜ x 8½. 0-486-65218-1

STATISTICS MANUAL, Edwin L. Crow et al. Comprehensive, practical collection of classical and modern methods prepared by U.S. Naval Ordnance Test Station. Stress on use. Basics of statistics assumed. 288pp. 5⅜ x 8½. 0-486-60599-X

SOME THEORY OF SAMPLING, William Edwards Deming. Analysis of the problems, theory and design of sampling techniques for social scientists, industrial managers and others who find statistics important at work. 61 tables. 90 figures. xvii +602pp. 5⅜ x 8½. 0-486-64684-X

LINEAR PROGRAMMING AND ECONOMIC ANALYSIS, Robert Dorfman, Paul A. Samuelson and Robert M. Solow. First comprehensive treatment of linear programming in standard economic analysis. Game theory, modern welfare economics, Leontief input-output, more. 525pp. 5⅜ x 8½. 0-486-65491-5

PROBABILITY: AN INTRODUCTION, Samuel Goldberg. Excellent basic text covers set theory, probability theory for finite sample spaces, binomial theorem, much more. 360 problems. Bibliographies. 322pp. 5⅜ x 8½. 0-486-65252-1

GAMES AND DECISIONS: INTRODUCTION AND CRITICAL SURVEY, R. Duncan Luce and Howard Raiffa. Superb nontechnical introduction to game theory, primarily applied to social sciences. Utility theory, zero-sum games, n-person games, decision-making, much more. Bibliography. 509pp. 5⅜ x 8½. 0-486-65943-7

INTRODUCTION TO THE THEORY OF GAMES, J. C. C. McKinsey. This comprehensive overview of the mathematical theory of games illustrates applications to situations involving conflicts of interest, including economic, social, political, and military contexts. Appropriate for advanced undergraduate and graduate courses; advanced calculus a prerequisite. 1952 ed. x+372pp. 5⅜ x 8½. 0-486-42811-7

FIFTY CHALLENGING PROBLEMS IN PROBABILITY WITH SOLUTIONS, Frederick Mosteller. Remarkable puzzlers, graded in difficulty, illustrate elementary and advanced aspects of probability. Detailed solutions. 88pp. 5⅜ x 8½. 65355-2

PROBABILITY THEORY: A CONCISE COURSE, Y. A. Rozanov. Highly readable, self-contained introduction covers combination of events, dependent events, Bernoulli trials, etc. 148pp. 5⅜ x 8¼. 0-486-63544-9

STATISTICAL METHOD FROM THE VIEWPOINT OF QUALITY CONTROL, Walter A. Shewhart. Important text explains regulation of variables, uses of statistical control to achieve quality control in industry, agriculture, other areas. 192pp. 5⅜ x 8½. 0-486-65232-7

Physics

OPTICAL RESONANCE AND TWO-LEVEL ATOMS, L. Allen and J. H. Eberly. Clear, comprehensive introduction to basic principles behind all quantum optical resonance phenomena. 53 illustrations. Preface. Index. 256pp. 5⅜ x 8½. 0-486-65533-4

QUANTUM THEORY, David Bohm. This advanced undergraduate-level text presents the quantum theory in terms of qualitative and imaginative concepts, followed by specific applications worked out in mathematical detail. Preface. Index. 655pp. 5⅜ x 8½. 0-486-65969-0

ATOMIC PHYSICS (8th EDITION), Max Born. Nobel laureate's lucid treatment of kinetic theory of gases, elementary particles, nuclear atom, wave-corpuscles, atomic structure and spectral lines, much more. Over 40 appendices, bibliography. 495pp. 5⅜ x 8½. 0-486-65984-4

A SOPHISTICATE'S PRIMER OF RELATIVITY, P. W. Bridgman. Geared toward readers already acquainted with special relativity, this book transcends the view of theory as a working tool to answer natural questions: What is a frame of reference? What is a "law of nature"? What is the role of the "observer"? Extensive treatment, written in terms accessible to those without a scientific background. 1983 ed. xlviii+172pp. 5⅜ x 8½. 0-486-42549-5

AN INTRODUCTION TO HAMILTONIAN OPTICS, H. A. Buchdahl. Detailed account of the Hamiltonian treatment of aberration theory in geometrical optics. Many classes of optical systems defined in terms of the symmetries they possess. Problems with detailed solutions. 1970 edition. xv + 360pp. 5⅜ x 8½. 0-486-67597-1

PRIMER OF QUANTUM MECHANICS, Marvin Chester. Introductory text examines the classical quantum bead on a track: its state and representations; operator eigenvalues; harmonic oscillator and bound bead in a symmetric force field; and bead in a spherical shell. Other topics include spin, matrices, and the structure of quantum mechanics; the simplest atom; indistinguishable particles; and stationary-state perturbation theory. 1992 ed. xiv+314pp. 6⅛ x 9¼. 0-486-42878-8

LECTURES ON QUANTUM MECHANICS, Paul A. M. Dirac. Four concise, brilliant lectures on mathematical methods in quantum mechanics from Nobel Prize-winning quantum pioneer build on idea of visualizing quantum theory through the use of classical mechanics. 96pp. 5⅜ x 8½. 0-486-41713-1

THIRTY YEARS THAT SHOOK PHYSICS: THE STORY OF QUANTUM THEORY, George Gamow. Lucid, accessible introduction to influential theory of energy and matter. Careful explanations of Dirac's anti-particles, Bohr's model of the atom, much more. 12 plates. Numerous drawings. 240pp. 5⅜ x 8½. 0-486-24895-X

ELECTRONIC STRUCTURE AND THE PROPERTIES OF SOLIDS: THE PHYSICS OF THE CHEMICAL BOND, Walter A. Harrison. Innovative text offers basic understanding of the electronic structure of covalent and ionic solids, simple metals, transition metals and their compounds. Problems. 1980 edition. 582pp. 6⅛ x 9¼. 0-486-66021-4

A TREATISE ON ELECTRICITY AND MAGNETISM, James Clerk Maxwell. Important foundation work of modern physics. Brings to final form Maxwell's theory of electromagnetism and rigorously derives his general equations of field theory. 1,084pp. 5⅜ x 8½. Two-vol. set. Vol. I: 0-486-60636-8 Vol. II: 0-486-60637-6

QUANTUM MECHANICS: PRINCIPLES AND FORMALISM, Roy McWeeny. Graduate student-oriented volume develops subject as fundamental discipline, opening with review of origins of Schrödinger's equations and vector spaces. Focusing on main principles of quantum mechanics and their immediate consequences, it concludes with final generalizations covering alternative "languages" or representations. 1972 ed. 15 figures. xi+155pp. 5⅜ x 8½. 0-486-42829-X

INTRODUCTION TO QUANTUM MECHANICS With Applications to Chemistry, Linus Pauling & E. Bright Wilson, Jr. Classic undergraduate text by Nobel Prize winner applies quantum mechanics to chemical and physical problems. Numerous tables and figures enhance the text. Chapter bibliographies. Appendices. Index. 468pp. 5⅜ x 8½. 0-486-64871-0

METHODS OF THERMODYNAMICS, Howard Reiss. Outstanding text focuses on physical technique of thermodynamics, typical problem areas of understanding, and significance and use of thermodynamic potential. 1965 edition. 238pp. 5⅜ x 8½. 0-486-69445-3

THE ELECTROMAGNETIC FIELD, Albert Shadowitz. Comprehensive undergraduate text covers basics of electric and magnetic fields, builds up to electromagnetic theory. Also related topics, including relativity. Over 900 problems. 768pp. 5⅜ x 8¼. 0-486-65660-8

GREAT EXPERIMENTS IN PHYSICS: FIRSTHAND ACCOUNTS FROM GALILEO TO EINSTEIN, Morris H. Shamos (ed.). 25 crucial discoveries: Newton's laws of motion, Chadwick's study of the neutron, Hertz on electromagnetic waves, more. Original accounts clearly annotated. 370pp. 5⅜ x 8½. 0-486-25346-5

EINSTEIN'S LEGACY, Julian Schwinger. A Nobel Laureate relates fascinating story of Einstein and development of relativity theory in well-illustrated, nontechnical volume. Subjects include meaning of time, paradoxes of space travel, gravity and its effect on light, non-Euclidean geometry and curving of space-time, impact of radio astronomy and space-age discoveries, and more. 189 b/w illustrations. xiv+250pp. 8⅜ x 9¼. 0-486-41974-6

STATISTICAL PHYSICS, Gregory H. Wannier. Classic text combines thermodynamics, statistical mechanics and kinetic theory in one unified presentation of thermal physics. Problems with solutions. Bibliography. 532pp. 5⅜ x 8½. 0-486-65401-X